D1333195

Molecular Endocrinology

THIRD EDITION

Molecular Endocrinology

THIRD EDITION

Franklyn F. Bolander, Jr.

Department of Biological Sciences
University of South Carolina
Columbia, South Carolina

ELSEVIER
ACADEMIC
PRESS

AMSTERDAM • BOSTON • HEIDELBERG • LONDON
NEW YORK • OXFORD • PARIS • SAN DIEGO
SAN FRANCISCO • SINGAPORE • SYDNEY • TOKYO

Elsevier Academic Press
525 B Street, Suite 1900, San Diego, California 92101-4495, USA
84 Theobald's Road, London WC1X 8RR, UK

This book is printed on acid-free paper. ∞

Library of Congress Cataloging-in-Publication Data
Application submitted.

British Library Cataloguing in Publication Data
A catalogue record for this book is available from the British Library

ISBN: 0-12-111232-2

1605075744

For all information on all Academic Press publications
visit our Web site at www.academicpress.com

Printed in the United States of America
03   04   05   06   07   08   9   8   7   6   5   4   3   2   1

Contents

Part 4
Gene Regulation by Hormones   385

Part 5
Special Topics   493

Preface to the Third Edition

The first edition of *Molecular Endocrinology* was published 15 years ago. It was originally written out of desperation: the author was teaching a course of the same name and could not find a textbook that covered this field in the depth and breadth that he required. About a decade later, several texts on signaling began to appear; although their coverage of signaling pathways was adequate, they still failed to embrace the full range of hormone action and did not closely relate signaling to hormone effects and interactions, which are the core of endocrine control. This deficiency, coupled with tremendous advances in the field of molecular endocrinology, provided the impetus for writing the third edition.

Molecular Endocrinology is first and foremost an endocrinology text. Indeed, it begins with an introductory unit on basic endocrinology for those readers who have not yet been exposed to the discipline. Such an introduction is critical for understanding how hormones act at the molecular level and why their signaling pathways synergize or antagonize each other in the manner they do. This section is followed by units on receptors, second messengers, transcription, and a final section on a few special topics. However, the emphasis is always on hormone action and integration. For example, in the latter half of Chapter 12, there is a comprehensive discussion of several hormones that have signaling pathways linked to their cellular actions. In addition, glycogen metabolism and smooth muscle contraction are discussed with respect to how the second messengers of the regulating hormones interact. In Chapter 13, this discussion is carried into the nucleus, where it is applied to transcription. Through these discussions the molecular aspects of signaling can be related to the gross effects of hormones in organisms.

This edition contains several new sections. As noted previously, in Chapter 12 there is a discussion of step-by-step pathways for several important hormones. Such a molecular analysis of hormone action would have been impossible just a few years ago; its presentation in Chapter 12 attests to how far molecular endocrinology has advanced. There is also a new chapter: Chapter 4 is a brief synopsis of signaling and prefaces Part 2. Its creation arose from the fact that the processes of endocrinology are so integrated that it is impossible to present one topic without encountering several others. For example, receptors are presented first because they are the first cellular structure hormones encountered. However, receptors are regulated by phosphorylation, and the relevant kinases are not discussed until later. If the kinases were covered first, then the readers would have no appreciation of the receptors and signaling pathways that activated them. In other words, every subject appears to be a prerequisite for discussing any other subject. Chapter 4 is designed to provide the reader with key concepts and components of receptors and second messengers prior to their detailed discussions in subsequent chapters.

The text has also been extensively updated. Since the second edition, the three-dimensional structure of many important receptors and transcription factors has been published, and the mechanisms of action of these receptors and factors have been elucidated. In addition, there have been major advances in the understanding of the modular nature of signaling and the importance of compartmentalization. Finally, the new discipline of proteomics combined with the sequence of the genomes from several species has added flesh to the chapter on the evolution of the endocrine system and comparative endocrinology.

There is also a major stylistic change: detailed references to specific facts have been replaced by recent review articles. This change freed up considerable space for the updated material and was made possible by the abundance of reviews currently available in the literature. However, if readers would like to have the specific reference for any statement of fact in this text, they are welcome to contact the author.

PART *1*

Introduction and General Endocrinology

CHAPTER **1**

Introduction

Definitions

Endocrinology is the study of hormones; but what are hormones? The question is far more difficult to answer today than it was a few decades ago. The classic definition is that hormones are chemical substances produced by specialized tissues and secreted into blood, where they are carried to target organs. However, this definition was constructed when most of the available knowledge of endocrinology was restricted to vertebrate systems. As the field of endocrinology has expanded, new hormones and new systems that previously would not have been included under this definition have been discovered. It is useful to describe these discrepancies so that a more functional definition can be developed.

1. *Specialized tissues for hormone synthesis.* Discrete endocrine glands exist only in arthropods, mollusks, and vertebrates, even though chemical substances that have hormonal activity have been identified throughout the animal, plant, and fungal kingdoms. Even in vertebrates, there exists a class of hormones, the *parahormones*, designed to act locally. Because parahormones are made wherever they are needed, they tend to have a nearly ubiquitous distribution. Finally, many vertebrate growth factors are synthesized in multiple locations.

2. *Blood for hormone distribution.* First, blood is unique to vertebrates. The addition of hemolymph to the definition would permit arthropod hormones to be included in the definition, but those of plants and lower animals would still be omitted. Second, even in vertebrates, the parahormones diffuse through the extracellular fluid to reach their local targets. Other hormones are released by neurons and also have local effects. Finally, the classic definition would exclude *ectohormones*, hormones that traverse air or water to act between or among individuals. These hormones are particularly well developed in certain insect species and include *pheromones* (sexual attractants), *gamones* (inducers of sexual development), and *allomones* and *kairomones* (interspecies attractants).

3. *A separate target organ.* Some parahormones, once secreted, not only diffuse to surrounding cells but also stimulate the cells originally synthesizing them. This positive feedback is referred to as *autocrine* function, and it results in the synthesizing cell becoming its own target organ. Furthermore, bacteria make several regulatory molecules for internal use. These signal molecules, called *alarmones*, are usually modified nucleotides and are produced in response to a particular stress such as starvation or a vitamin deficiency.

Because of these limitations, a broader definition is used in this book: a *hormone* is a chemical, nonnutrient, intercellular messenger that is effective at micromolar concentrations or less. In other words, hormones are chemical substances that carry information between two or more cells. This definition includes all of the preceding examples except the alarmones. This exclusion is clearly the bias of mine, but the essence of endocrinology is the chemical coordination of bodily functions, and alarmones are used exclusively with single cells. However, other bacterial hormones that signal sporulation, competence (ability to take up exogenous DNA), conjugation, and other activities that are coordinated among individual bacteria are included. The restriction of hormones to chemical substances seems initially to be a logical one, even though species such as fireflies can use light to induce behavioral patterns in others. However, because the visual pigment rhodopsin and the G protein-coupled receptors (GPCRs) are homologous, one could also argue that, under certain circumstances, light is a hormone. Finally, metabolic pathways can be induced or repressed by substrate levels; indeed, substrate flow is an important regulator in many systems. Therefore nutrients are also excluded in the hormone definition. The inclusion of the concentration clause is used to eliminate other miscellaneous inducers; the one thing that sets hormones apart from other chemical regulators is their effectiveness at extremely low concentrations, usually in the nanomolar range or below. Plant hormones are unusual in that a few are required in larger amounts; it is for that reason that the micromolar limit is used.

The importance of endocrine regulation is apparent from the examination of the genomes of those organisms for which a complete sequence is available. For example, the genome of the nematode *Caenorhabditis elegans* contains about 20,000 genes. The single most abundant group, at 3.5% of the total, is the group of genes for GPCRs, which are receptors for hormones and other small molecules. The second most abundant group, at 2.6%, is the group of protein kinases, which are integral components of many signaling pathways. Finally, the third most abundant group, at 1.4%, is a transcription factor class that includes the nuclear receptors for steroids and other hydrophobic hormones. In metazoans, cellular communication and coordination are essential for successful development and survival, and their significance is reflected in the proportion of the genome allotted for endocrine functions.

The study of hormone action at the cellular and molecular level is called *molecular endocrinology*, which is the subject of this treatise. In particular, this book concentrates on the molecular mechanisms of hormone action and interaction. However, the topic of hormonal synergism and antagonism at the molecular level is better understood against a background knowledge of hormone action in the whole organism; for example, the progesterone inhibition of prolactin receptors and second messengers in the mammary gland is just an isolated fact unless one knows the general function of these hormones in the reproductive cycle. Therefore the function of this unit is to provide the reader, in general, and the novice, in particular, with sufficient background information to appreciate the molecular interrelationships that are discussed in later units. It is obvious that a complete presentation of general endocrinology cannot be accomplished in only three chapters: the coverage is specifically oriented and just sufficient to prepare the reader for the remainder of the book. However, it is hoped that the reader will become interested enough to consult any of the excellent and far more comprehensive texts listed in the General References section at the end of this chapter.

The rest of this chapter is concerned with identifying the basic characteristics of hormones and their regulation, and it concludes with an illustrative example, the hormonal control of calcium metabolism. Then, in Chapter 2, the other classical endocrine systems are examined. Finally, Chapter 3 briefly covers non-classical and nonvertebrate hormones, such as growth factors, parahormones, and the hormones of plants and insects.

Hormone-Target Relationships

As noted previously, the classic endocrine system involves a hormone being made in one part of the body and reaching its target in another part of the body through the bloodstream (Fig. 1-1, *A*). However, there are many other types of interactions that can occur. In a paracrine system, the hormone remains in the tissue, where it reaches nearby cells by diffusion (Fig. 1-1, *B*). The *juxtacrine* system represents another mechanism for limiting the diffusion of hormones. In this case, the hormone is synthesized as a membrane-bound precursor. Although this precursor is usually cleaved to yield a soluble peptide, it may also remain attached to the plasma membrane, where it retains its biological activity. Therefore its effects are limited to the length of its tether (Fig. 1-1, *C*). Hormones that may act in this fashion include the epidermal growth factor, transforming growth factor α, tumor necrosis factor α, colony-stimulating factor 1, and the Kit ligand (see Chapter 3). In some cases, the intracellular domain of the hormone anchor is coupled to second messengers so that receptor engagement generates signals in both cells (Fig. 1-1, *D*). The ephrins are examples of this *bilateral* or *reverse signaling*. Finally, juxtacrine signaling may also include hormones whose diffusion is limited by the fact that they are tightly bound to the extracellular matrix.

The hormone may even influence the cell that originally secreted it; this is often part of either a negative or positive feedback loop (Fig. 1-1, *E*). This is called an *autocrine* system. The *intracrine* system was originally defined as one where hormone synthesis and receptor binding occurs intracellularly; a possible example would be the nuclear receptors for various lipid intermediates. For example, the liver X receptor binds cholesterol-like sterols, whose synthesis it regulates. Normally, this would not be considered a hormone system by this text, but there are three variations of the definition that would qualify: the endogenous generation of hormones from precursors synthesized elsewhere by (1) executing the final synthetic steps, (2) hydrolyzing hormones inactivated by conjugation, or (3) cleaving the hormone from its protein precursor (Fig. 1-1, *F*). The thyroid

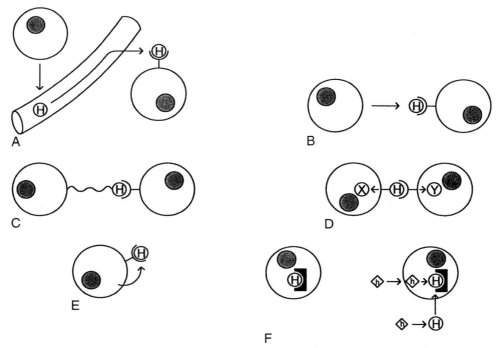

Fig. 1-1. Hormone-target relationships. (A) Classical endocrine; (B) paracrine; (C) juxtacrine; (D) juxtacrine with bilateral signaling; (E) autocrine; (F) intracrine; (G) transsignaling; (H) cryptocrine; and (I) neurocrine. *H*, Hormone; *encircled H*, active hormone; *h within a diamond*, prohormone; *X and Y*, second messengers.

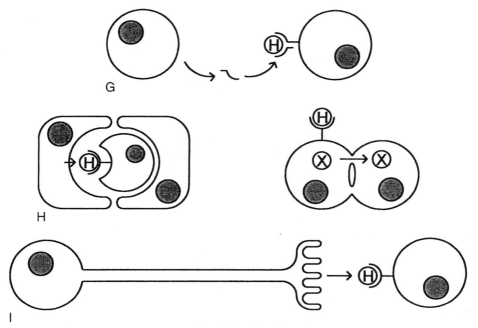

Fig. 1-1 *(continued)*

hormones represent an example of the first variation: the thyroid gland secretes predominantly thyroxine; this compound is converted to the active form, triiodothyronine, by peripheral enzymes. The sex steroids represent another example; 40% of all androgens in males and 75% to 100% of estrogens in postmenopausal females are generated in target tissues from adrenal precursors. This generation is accomplished by 5α-reductase and aromatase, respectively. Steroids can also be produced locally by steroid sulfatases: many steroids are inactivated by sulfation, and some peripheral tissues, like the breast, reactivate these steroids by deconjugating them. Finally, certain peptide hormones, such as the hepatocyte growth factor and the transforming growth factor β, are secreted as inactivate precursors that are cleaved locally to generate the activate hormone. These enzymes allow the tissue to adjust the hormone levels to local conditions.

Transsignaling is a process by which one cell supplies the high specificity, high affinity receptor subunit to another cell, which has the core receptor subunits. For example, many cytokine receptors have multiple subunits, some of which may be cleaved to produce soluble proteins. The receptor for interleukin 6 (IL-6) consists of two membrane-bound receptors, a 130 kDa glycoprotein (gp130) and IL-6R. IL-6 and gp130 are constitutively synthesized, but the affinity between IL-6 and gp130 is too low to generate a signal. During inflammation, defense cells invade the tissue and shed IL-6R, which then combines with gp130 to form the active receptor (Fig. 1-1, *G*). Essentially, the soluble receptor is analogous to a parahormone. A variation of this process is seen in monocytes, which constitutively produce IL-15. IL-15 is not secreted but remains tightly bound to its high affinity subunit, IL-15Rα. This pair is presented to target cells that only possess the other two components of the receptor complex, IL-2Rβ and γ_c. In this example, the IL-15Rα subunit acts like a juxtacrine hormone. A similar phenomenon is observed in the IL-12 family, except that the ligand and constitutively bound receptor α subunit are soluble.

All of the systems discussed thus far are open; that is, there are no diffusion barriers, and selectivity of target cells is determined by the presence or absence of receptors to that hormone. In contrast, a *cryptocrine* system involves the secretion of a hormone into a closed environment. This system obviously requires a very special intimacy between cells, such as that between Sertoli's cells and spermatids or thymic nurse cells and T lymphocytes (Fig. 1-1, *H*). Another example of this phenomenon is the transfer of second messengers, such as cyclic nucleotides or inositol trisphosphates, through gap junctions between adjacent cells. Finally, the *neurocrine* system is the secretion of chemical messengers by neurons (Fig. 1-1, *I*). However, some authorities consider the synapse to be a restricted environment and neurotransmission to be a variation of cryptocrine signaling.

Chemical Nature

Structurally, hormones are extremely diverse (Fig. 1-2). The most abundant and most versatile of these are the peptide and protein hormones, which range in size from a simple tripeptide (thyrotropin-releasing hormone) to 198 amino acids (prolactin). Some protein hormones, such as human chorionic gonadotropin, are even larger because of multiple subunits and glycosylation. In addition to full proteins, individual amino acids have been modified to yield hormones; the most common amino acid precursors are tyrosine (the catecholamines and thyroid hormones), histidine (histamine), and tryptophan (serotonin and indoleacetic acid).

The lipids are another rich source of hormones. The steroids form an entire group by themselves. Fatty acid derivatives include the prostaglandins and related compounds; some insect pheromones are also synthesized from fatty acids. Finally, the structure of platelet activating factor is similar to that of phosphatidylcholine.

The nucleotides would seem to be an unusual source, but they too are well represented: some pheromones, the cytokinins (plant hormones), 1-methyladenine (a starfish hormone), and cyclic AMP (cAMP) (in slime molds). In addition, several purine derivatives act as parahormones in mammals.

Oligosaccharide hormones were first characterized in plants, where they are produced from the breakdown of the plant cell wall in response to certain plant infections. These elicitors then trigger a defense response within the plant cell. Carbohydrate hormones have now been postulated to occur in animals: the aggregation factor of sponges is a glycan, one of the mediators of insulin action appears to be an oligosaccharide (see Chapter 11), and β-glucan and pectin can act as secretagogues (secretion stimulators) in vertebrates. In the latter case, the physiological relevance is still uncertain.

Finally, even gases are represented. Ethylene ripens fruit in plants, and nitric oxide is a potent vasodilator in mammals. Carbon monoxide and hydrogen sulfide have also been proposed as physiological gaseous hormones; both are vasodilators.

Although the structural diversity of hormones is great, there is one property that is particularly important: water solubility (Table 1-1). Hydrophobic hormones are difficult to store because they pass through membranes so easily; as a result, they are synthesized as they are needed. The thyroid hormones are an exception and are discussed further in Chapter 2. Hydrophobic hormones do not

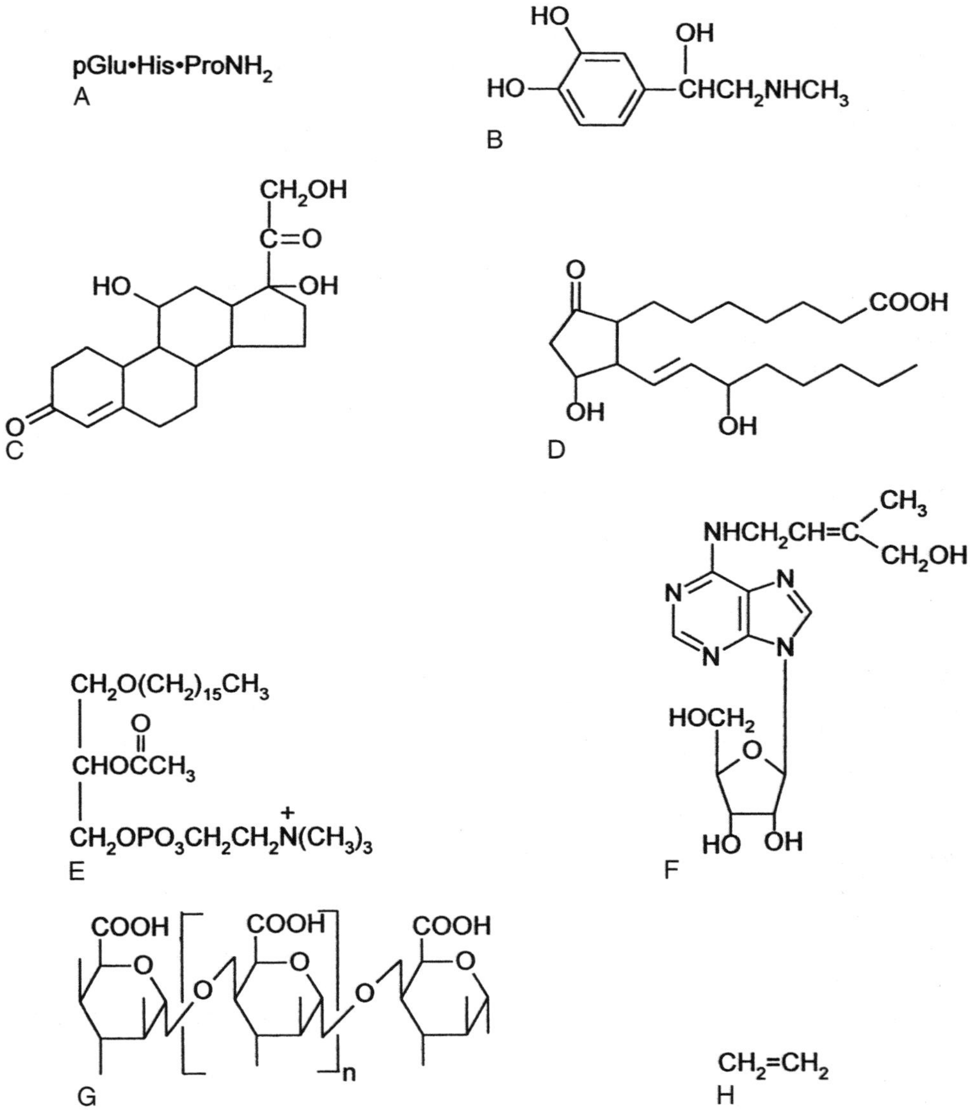

pGlu•His•ProNH$_2$

A

Fig. 1-2. Structural diversity of hormones. (A) Thyrotropin-releasing hormone; (B) epinephrine; (C) cortisol; (D) prostaglandin; (E) platelet-activating factor; (F) zeatin (a cytokinin); (G) α-1,4-oligoga-lacturonide (an elicitor); (H) ethylene.

Table 1-1
A Comparison of Hydrophobic and Hydrophilic Hormones

Characteristic	Hydrophobic	Hydrophilic
Examples	Steroids and thyronines	Peptides and catecholamines
Storage after synthesis	Minimal except for thyronines	Yes
Binding proteins	Always	Sometimes, especially smaller peptides
Half-life	Long (hours or days)	Short (minutes)
Receptors	Cytoplasmic or nuclear	Plasma membrane
Mechanism of action	Direct	Indirect (second messenger)

dissolve readily in water; therefore they require serum transport proteins with hydrophobic pockets. Because they are partially hidden in these pockets, they are protected, and their half-lives are long. Finally, their hydrophobicity allows them to cross the plasma membrane, bind to cytoplasmic or nuclear receptors, and elicit direct cellular effects. Again, the thyroid hormones are an exception: in spite of the hydrophobic rings, the thyroid hormones are still zwitterions and must be transported by either the L-type amino acid transporter or the organic anion transporters. However, this process is virtually automatic and does not represent any real obstacle to the cellular penetration by these hormones.

Hydrophilic hormones, however, can be contained within membrane vesicles, so they can be stored. Although a few of the smaller peptides are known to bind to serum proteins, most of the water-soluble hormones are transported free in the serum, but as a result they are rapidly eliminated from the circulation. Because they cannot cross the plasmalemma, they must interact with their receptors at the cell surface and generate a second signal to affect cellular processes; that is, their mechanism of action is indirect.

Biological Activity

What are the functions of hormones? Hormones coordinate nearly all of the biological activities within an organism; these activities are primarily metabolism, growth, and reproduction (Table 1-2). Metabolism is the sum of all processes that handle or alter materials within living organisms and can be divided into (1) mineral and water metabolism and (2) energy metabolism. Hormones involved in the former regulate the absorption, storage, and secretion of electrolytes and water; their function is to maintain a constant ionic environment inside the body. Hormones involved in energy metabolism regulate the flow of organic substrates through chemical pathways to maintain appropriate adenosine 5'-triphosphate (ATP) levels within the cell. Insulin is a hormone of energy storage because it shunts substrates into macromolecular reservoirs: glucose into glycogen, amino acids into protein, and fatty acids into triglycerides. Most of the other hormones regulating energy metabolism are involved in energy expenditure; that is, they break down these reservoirs and shunt the liberated substrates into chemical pathways that generate ATP.

Growth is the enlargement of a cell, tissue, or organism by the net accumulation of material, an increase in cell number, or both. It is a very complex process requiring the coordination of both mitosis and metabolism; the latter supplies the necessary materials and energy for the former. As such, it should not be surprising that some of these hormones, such as growth hormone, are involved in both growth and metabolism. In addition to generalized growth, hormones can selectively affect certain tissues, such as epidermal or neural tissues.

Reproduction is the process by which an organism generates and (sometimes) nourishes a new member of the species. It too is a very complex process: sex steroids and gonadotropins promote gametogenesis; relaxin and oxytocin stimulate lactation and suckling.

In addition to these broad functions, many more specialized functions are served by hormones. Both the tropic hormones and the releasing and inhibiting factors participate in a regulatory hierarchy and merely stimulate or block the synthesis and secretion of hormones from other glands. The parahormones are

Table 1-2
Some Major Vertebrate Hormones and Their Characteristics

Hormones	Structure	Mechanism[a]	Source	Target	Action
Calcium Metabolism					
Parathormone (PTH)	Peptide	3',5'-cyclic AMP, calcium	Parathyroid gland	Bone, kidney	Bone resorption; renal calcium resorption
PTH-related protein	Peptide	cAMP	Mammary gland and uterus	Mammary gland, placenta	Local calcium transport
Calcitonin (CT)	Peptide	cAMP, calcium	CT cells (C cells) (thyroid gland)	Bone	Inhibits bone resorption
1,25-Dihydroxycholecalciferol (1,25-DHCC)	Sterol derivative	DNA binding	Skin, liver, kidney	Intestine, kidney	Intestinal calcium absorption and renal resorption
Tropic Hormones					
Adrenocorticotropic hormone (ACTH)	Protein	cAMP	Adenohypophysis	Adrenal cortex	Stimulates glucocorticoid synthesis and secretion
Luteinizing hormone (LH)/human chorionic gonadotropin (hCG)	Protein	cAMP	Adenohypophysis/placenta	Gonads	Stimulates progesterone (♀) and testosterone (♂) synthesis and secretion
Follicle-stimulating hormone (FSH)	Protein	cAMP	Adenohypophysis	Gonads	Stimulates estrogen synthesis and secretion (♀) and gamete development
Thyroid-stimulating hormone (TSH)	Protein	cAMP	Adenohypophysis	Thyroid	Stimulates thyroid hormone synthesis and secretion
Melanocyte-stimulating hormone (MSH)	Peptide	cAMP	Hypophysis	Melanocyte	Skin darkening
Sodium and Water Metabolism/Blood Pressure					
Antidiuretic hormone (ADH)/vasopressin (VP)	Peptide	cAMP, calcium	Neurohypophysis	Kidney, liver	Renal water resorption and glycogenolysis
Angiotensin II (AT)	Peptide	Calcium	Liver	Adrenal cortex, vasculature	Stimulates aldosterone synthesis and secretion; vasoconstriction

continues

continued

Hormones	Structure	Mechanism[a]	Source	Target	Action
Aldosterone	Steroid	DNA binding	Adrenal cortex	Kidney	Renal sodium and water resorption
Atrial natriuretic peptide (ANP)	Peptide	cGMP	Atria	Kidney, vasculature	Sodium diuresis; AT antagonist
Endothelins (ETs)	Peptide	Calcium	Endothelium	Vasculature	Vasoconstriction and smooth muscle growth
Nitric oxide (NO)	NO	cGMP	Endothelium and other tissues	Vasculature	Vasodilates and decreases platelet adhesion
Reproductive Hormones					
Estrogens	Steroids	DNA binding	Ovary, placenta	Reproductive tract etc.	Sexual characteristics
Androgens	Steroids	DNA binding	Testis, adrenal cortex	Reproductive tract etc.	Sexual characteristics; Spermatogenesis
Progesterone	Steroid	DNA binding	Ovary, placenta	Reproductive tract etc.	Pregnancy maintenance
Relaxin	Peptide	cAMP	Ovary	Pubic symphysis	Parturition
Prolactin (PRL)	Peptide	Tyrosine kinase	Adenohypophysis	Mammary gland	Lactation
Oxytocin (OT)	Peptide	Calcium	Neurohypophysis	Mammary gland, uterus	Milk ejection, parturition
Inhibin/activin	Peptide	Serine kinase	Gonads	Hypothalamo-pituitary axis; gametes	Inhibition of FSH (inhibin); Gametogenesis
Energy Metabolism					
Growth hormone (GH)/ somatotropin	Peptide	Tyrosine kinase	Adenohypophysis	Multiple	Lipolysis, glucose sparing, general body growth
Glucocorticoids	Steroids	DNA binding	Adrenal cortex		Gluconeogenesis
Triiodothyronine (T₃)	Tyrosine derivative	DNA binding	Thyroid gland	Multiple	Lipolysis, glycogenolysis
Glucagon	Peptide	cAMP	α Cells (pancreas)	Liver	Glycogenolysis, gluconeogenesis
Epinephrine (E)	Tyrosine derivative	cAMP, calcium	Adrenal medulla	Fat, muscle	Lipolysis, glycogenolysis
Insulin	Peptide	Tyrosine kinase	β Cells (pancreas)	Multiple	Lipid, protein, and glycogen Synthesis
Neurotransmitters					
Norepinephrine (NE)	Tyrosine derivative	cAMP, calcium	Central and sympathetic nervous system	Multiple	Fight-or-flight response

			Hypothalamus, etc.	Adenohypophysis, etc.	
Dopamine	Tyrosine derivative	cAMP, calcium	Central and parasympathetic nervous system	Muscles etc.	Inhibition of PRL secretion
Acetylcholine (ACh)	Amino alcohol	Sodium, calcium	Central and parasympathetic nervous system	Muscles etc.	Maintenance of involuntary activity and muscle contraction
Glutamate (Glu)	Amino acid	Calcium	Central nervous system	Central nervous system	Stimulatory transmitter
γ-Aminobutyric acid (GABA)	Glutamate derivative	Chloride	Central nervous system	Central nervous system	Inhibitory transmitter
Glycine (Gly)	Amino acid	Chloride	Spinal cord	Spinal cord	Inhibitory transmitter
5-Hydroxytryptamine (serotonin)	Amino acid derivative	Cations	Central nervous system	Central nervous system	Inhibitory transmitter; affects mood and sleep

Release and inhibiting factors: see Table 2–1
Local gastrointestinal hormones: see Table 2–5
Growth and growth-inhibiting factors: see Table 3–1
Immunoinflammatory hormones: see Table 3–3
Parahormones: see Table 3–4

[a]The mediators listed are the primary or best characterized ones; the list is not meant to be all-inclusive. The given mediators may either be positively or negatively affected, depending on the particular hormone.

made and act locally; for example, most of the eicosanoids are involved in inflammation, blood clotting, or smooth muscle contraction. Finally, the gastrointestinal tract contains many hormones that act regionally to facilitate the digestion and absorption of ingested material.

There is one other group of hormones listed in Table 1-2—the neurotransmitters. At first, this group may seem somewhat out of place because the endocrine and nervous systems have classically been considered distinct entities. However, neurotransmitters satisfy the definition of a hormone developed in the Definitions section. In addition, many molecules can function as either hormones or as neurotransmitters; for example, the catecholamines and the gastrointestinal hormones. Consequently, the two systems have become partially fused into a neuroendocrine system, which may be considered appropriate subject matter for an endocrinology text.

Control

In the following sections, many different hormones are discussed. Although their actions may vary considerably, their regulation will conform to a limited number of mechanisms. The simplest mechanism is negative feedback: rising levels of a hormone shut off its production so that a constant or desired concentration can be maintained. Because many glands in the body are under hierarchical control, this negative feedback can occur at several levels (Fig. 1-3, A). For example, the hypothalamus produces releasing factors that stimulate the secretion of hormones from the anterior pituitary; most of these hormones will, in turn, stimulate other glands in the body (see Chapter 2, Hypothalamus and Pituitary Gland). The hormones from these peripheral glands may have a feedback effect primarily on the anterior pituitary gland or on the hypothalamus.

Positive feedback, in which rising hormone levels stimulate further hormone production, is less common because it can produce a vicious cycle. However, if there is a clear termination point in the cycle, such positive feedback can greatly augment the initial stimulus without going out of control. An example is the control of oxytocin, a hormone secreted by the posterior pituitary gland (Fig. 1-3, B). As parturition nears, uterine contractions begin and stimulate the release of oxytocin, a potent inducer of smooth muscle contraction. As a result, the uterus contracts harder and further stimulates oxytocin secretion. When the fetus is finally expelled, the cycle is broken.

Finally, the type of feedback may be dependent on other physiological parameters. For example, estrogen normally has a negative feedback effect on the hypothalamic-pituitary axis (Fig. 1-3, C). However, at midcycle, the axis suddenly becomes stimulated by estrogen, a positive feedback is established, and estrogen levels rise until they trigger a surge of luteinizing hormone, which leads to ovulation.

Hormonal Control of Calcium Metabolism

The hormonal regulation of calcium homeostasis represents an ideal introduction to endocrinology because it is a relatively simple, closed system primarily involving three hormones (parathormone [PTH], calcitonin, and the active form of

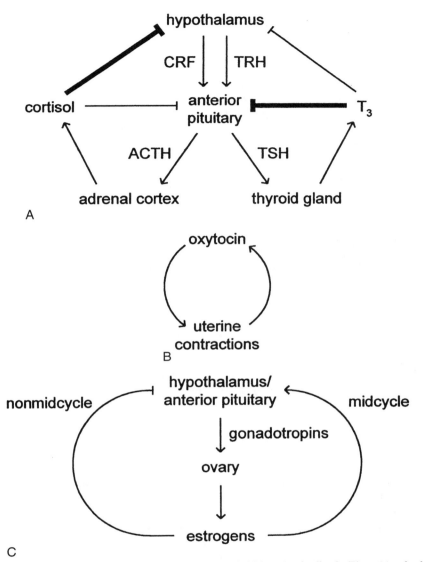

Fig. 1-3. Feedback regulation of hormone secretion. (A) Negative feedback; (B) positive feedback; and (C) cycle-dependent feedback. Arrows depict stimulation; flat heads represent inhibition. *ACTH,* Adrenocorticotropic hormone; *CRF,* corticotropin-releasing factor; T_3, triiodothyronine; *TRH,* thyrotropin-releasing hormone; *TSH,* thyroid-stimulating hormone (thyrotropin).

vitamin D) and three organs (bone, kidney, and the gastrointestinal tract). The function of these hormones is to maintain a constant calcium concentration in the blood because the calcium ion is critically involved in the activity of excitable tissues, such as nerves and muscles. Elevated calcium levels depress electrical activity, whereas low calcium levels enhance such activity. For example, patients with hypercalcemia exhibit muscular weakness, bradycardia (slow heartbeat), lethargy, and confusion, and, in severe cases, they may become comatose. Patients with hypocalcemia, however, may demonstrate muscular spasms (tetany), irritability, psychosis, and even seizures. Another problem requiring tight regulation of calcium levels is calcium's poor solubility; if the concentration of calcium in biological fluids becomes too high, it can precipitate out of solution anywhere. This phenomenon is called *ectopic calcification*. Therefore it is readily understandable why the body controls calcium levels so rigidly.

The concentration of calcium in the blood is kept between 2.2 and 2.55 mM and exists in three forms: bound, complexed, and free. About 30% of the calcium is bound to protein, primarily serum albumin; another 10% is complexed with various chelators, such as citrate; and the remaining 60% is free or ionized. Only the latter form is important in biological activities, and therefore free calcium is the fraction that is tightly regulated. For example, in certain liver diseases, serum albumin levels decline; with less protein to bind this cation, total calcium also falls. However, the patient shows no evidence of hypocalcemia because the free calcium is still normal.

Although it is the blood level of calcium that is regulated, blood and the extracellular fluid constitute the smallest calcium reservoir in the body (0.1%). Some calcium is stored intracellularly (1%), but most resides in bone and teeth (99%). Therefore an understanding of bone composition, structure, and metabolism is essential to any discussion of calcium metabolism.

Bone

Bone is composed of osteoid, calcium salts, and cells. Osteoid is the organic matrix into which the calcium salts are deposited. The osteoid gives the bone resilience, whereas the minerals produce rigidity. These are best seen in cadaveric long bones treated with either alkali, which digests the osteoid, or acid, which removes the calcium; the alkali-treated bone is rigid but crumbles easily, whereas the acid-treated bone is so flexible that it can be tied into a knot.

The primary component (90%) of the osteoid is collagen, a fibrous protein composed of three chains. Each mature chain contains about 1000 amino acids, and all three chains tightly twist around each other to form a long rod, 1.5×300 nm (Fig. 1-4). This triple helix is possible because the individual chains possess a large number of small amino acids, which eliminate steric interference; approximately one-third is glycine and another third is proline or hydroxyproline. Collagen is initially synthesized as a precursor, procollagen, by the bone-forming cells, or osteoblasts. After secretion, the ends of the procollagen are proteolytically removed to form tropocollagen, which then polymerizes by lining up in a quarter-staggered array with other tropocollagen rods. Initially, the polymerization is noncovalent, but later, covalent cross-links are established. The remainder of the osteoid is composed of miscellaneous glycoproteins, mucopolysaccharides, and osteocalcin. Osteocalcin, a 49-amino acid globular peptide, contains an unusual amino acid, γ-carboxyglutamic acid, which is synthesized from glutamic acid by means of a vitamin K-dependent reaction. This amino acid is an excellent calcium chelator. It is thought that osteocalcin may sequester calcium to allow for a more gradual and more controlled precipitation of bone mineral.

The major calcium salt is hydroxyapatite, which has the following formula: $3Ca_3(PO_4)_2 \cdot Ca(OH)_2$. The salt is deposited in the gaps between the collagen molecules and within the fibrils (Fig. 1-4); the crystals have their long axes aligned with those of the collagen.

Both the synthesis of the osteoid and its calcification are performed by the third component, the cells. The osteoblast arises from the osteoprogenitor cell, a fibroblastlike cell located on the bone surface. In addition to synthesizing colla-

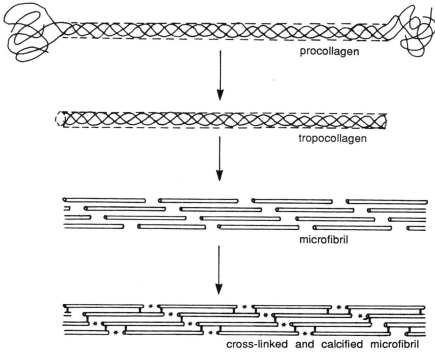

Fig. 1-4. Formation of collagen. The asterisks indicate calcification sites.

gen, this cell also secretes calcium into the extracellular fluid until the salt concentration exceeds its solubility. This supersaturation is possible because various pyrophosphates are also secreted; these compounds act as crystal growth inhibitors by absorbing onto and stabilizing the hydroxyapatite crystal embryos. According to one hypothesis, calcium precipitation is initiated when these pyrophosphates are hydrolyzed by alkaline phosphatase, an abundant enzyme in osteoblasts. As calcification proceeds, some osteoblasts become entrapped and are relegated to the maintenance of the bone in their immediate vicinity; these are the osteocytes. Finally, because bone acts, in part, as a calcium reserve for the body, there must be a mechanism to reclaim these salts during calcium deprivation or loss. This bony dissolution is accomplished by large, multinucleated cells with ruffled borders. These *osteoclasts*, as they are called, are rich in lysosomal and mitochondrial enzymes.

Hormones

Parathormone (PTH) is an 84-amino acid peptide, although only the first 34 amino acids are required for full biological activity. The parathyroid glands, which synthesize and secrete PTH, are derived from the third and fourth pharyngeal pouches and migrate caudally until they become embedded in the posterior wall of the thyroid gland (Fig. 1-5), one each in the superior and inferior pole of each lateral lobe. A single gland measures only $6 \times 4 \times 2$ mm and histologically contains two cell types. The chief cells synthesize PTH, whereas the oxyphil cells, which do not appear until after puberty, have no known function. A homologous peptide, parathyroid hormone–related protein (PTHrP), was first

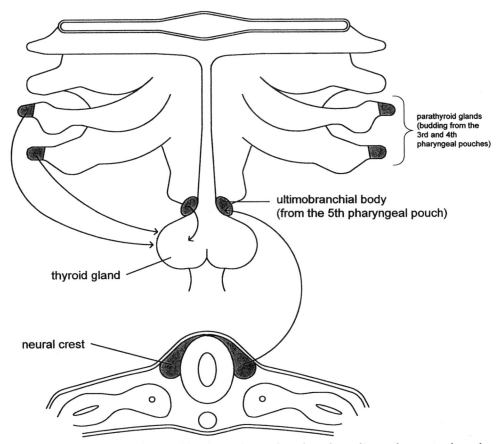

Fig. 1-5. Embryonic pharynx. The skin and mesoderm have been dissected away to show the outside of the pharyngeal lining with its pouches and the thyroid diverticulum. Future glandular tissue is stippled. Adapted and reproduced by permission from McClintic, J.R. (1983). *Human Anatomy*. The C.V. Mosby Co., St. Louis, Missouri.

identified in tumors associated with hypercalcemia, but has now been found in many normal tissues, such as the mammary gland, bladder, and uterus. It acts as a parahormone for local calcium transport and also functions in smooth muscle relaxation.

Calcitonin (CT) is a 32-amino acid peptide with an amino terminal disulfide loop and an amidated carboxy terminus. The cells that synthesize CT originate from the neural crest and initially migrate to the ultimobranchial body of the fifth pharyngeal pouch. From there they invade the thyroid gland, which is descending the neck after having invaginated from the floor of the primitive mouth (Fig. 1-5). Histologically, the cells are located between, or partially embedded in, the thyroid follicles; for this reason, they are called *parafollicular cells*. Because they secrete CT, they are also known as *C cells*. Like PTH, CT also has several other family members: calcitonin gene-related peptide (CGRP), adrenomedullin (ADM), and amylin. CGRP is a neurotransmitter that produces vasodilation and increased heart rate and atrial contractility. ADM induces smooth muscle relaxation in blood vessels and bronchioli. Amylin (also called *islet amyloid polypeptide* [IAPP]) has anti-insulin effects, inhibits bone resorption, and acts as a growth factor for renal epithelial cells.

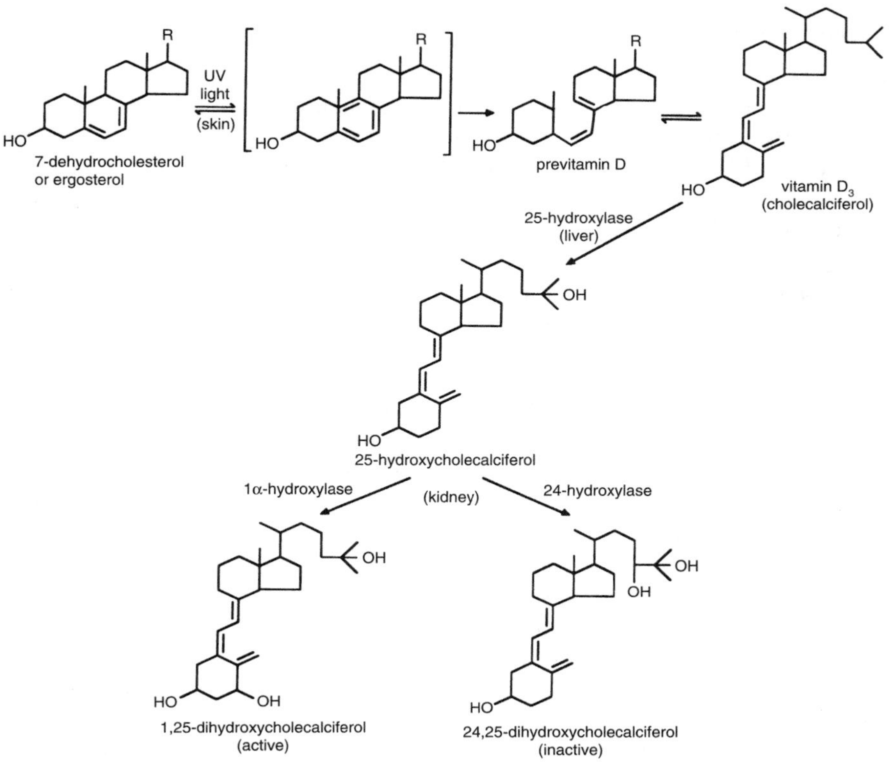

Fig. 1-6. Synthetic pathway for vitamin D and its active metabolites.

The active form of vitamin D is 1,25-dihydroxycholecalciferol (1,25-DHCC), and it is synthesized by a fascinating pathway involving three different tissues and a nonenzymatic step (Fig. 1-6). The precursor is either 7-dehydrocholesterol from animals or ergosterol from plants; they differ only in their side chains. Although 7-dehydrocholesterol can be synthesized in some human tissues, the amounts may be inadequate to meet the needs of the body, especially during rapid growth. Consequently, either 7-dehydrocholesterol or ergosterol is required in the diet—thus, their designation as vitamins. After absorption from the digestive tract, the sterol travels to the skin, where ultraviolet light aromatizes the B ring, causing it to rupture. After rearrangement of the double bonds, vitamin D goes to the liver, where it is 25-hydroxylated. Finally, the compound is carried to the kidneys, where it is hydroxylated on either the 1 or 24 position. The form active in elevating serum calcium is 1,25-DHCC, whereas 24,25-DHCC was merely thought to be an inactive metabolite. However, it is now known that 24,25-DHCC actively opposes the effects of 1,25-DHCC. First, 24,25-DHCC stimulates the metabolism of 1,25-DHCC, thereby lowering the serum levels of the latter. Second, in chondrocytes, 24,25-DHCC inhibits a transduction signal activated by 1,25-DHCC. Therefore the choice between the 1 and 24 positions has dramatically different effects, and it is highly regulated (see later). It is worth noting that the increased incidence of rickets in children during the Industrial Revolution was a result of the children spending all day working in factories. Inadequate exposure to the sun would clearly impair the synthesis of 1,25-DHCC at the first step. Without 1,25-DHCC, the body absorbs calcium very poorly and rapidly growing bones become soft because of inadequate calcification.

Hormonal Regulation

PTH elevates calcium levels by dissolving the salts in bone and preventing their renal excretion. In bone, PTH activates the osteoclasts and induces their lysosomal enzymes. In the kidney, PTH stimulates calcium resorption while promoting phosphate and bicarbonate excretion. The hormone acts, in part, through cyclic AMP (cAMP) and protein kinase C (PKC), second messengers that are discussed in Chapters 9 and 10.

CT was once believed to stimulate bone synthesis, but it only blocks the action of PTH; that is, it is simply a PTH antagonist. CT also stimulates cAMP and PKC. This paradox of an agonist and antagonist both acting through the same second messengers was initially explained with the hypothesis that the two hormones affected different cell types. However, it is now known that both hormones act on the same cell, the osteoclast. It is likely that these and possibly other second messengers are differentially activated by hormone concentration, temporal exposure, or other factors (see Chapter 9 for a more detailed discussion). Another unresolved problem is the apparent insignificance of CT in calcium metabolism: total thyroidectomy without CT replacement does not result in any abnormality of calcium regulation. For this reason, some authorities consider CT to be an evolutionary vestige; but if this is true, the gene for this hormone should have been inactivated or lost through random mutation because enough time has elapsed for these changes to occur. The answer may lie in its gene, which actually codes for two peptides: CT and CGRP. In the parafollicular cells, the mRNA is processed and translated to give CT, but in certain parts of the nervous system, such as the trigeminal ganglion, the mRNA is processed and translated to yield CGRP. This peptide induces analgesia and has multiple effects on the cardiovascular system. These, as well as other yet undiscovered activities, may be sufficiently important to the organism to cause selection for the entire gene. Seemingly useless DNA that is maintained in the genome because of its close association with vital DNA has been termed *selfish DNA,* and CT may be an example of this phenomenon.

At least two target organs of 1,25-DHCC are related to calcium homeostasis: the gastrointestinal tract and bone. In the digestive tract, 1,25-DHCC induces a calcium-binding peptide, which is required for calcium absorption. This peptide is a member of the calmodulin family (see Chapter 10). In bone, pharmacological doses of 1,25-DHCC mimic the effects of PTH; in physiological concentrations, it synergizes with PTH. Previously, no effects of 1,25-DHCC could be documented in the kidney. However, in vitamin D–deficient animals, this sterol stimulates calcium resorption in the distal renal tubule. Although this resembles the action of PTH, they use different mechanisms: PTH activates the (Na^+, Ca^{2+}) exchanger, whereas 1,25-DHCC stimulates resorption that is ATP-dependent.

Interestingly, 1,25-DHCC is also involved with epithelial differentiation; it is especially important in hair follicle development. This is strikingly apparent in animals carrying mutations in the gene for the 1,25-DHCC receptor: the animals are afflicted with both rickets and hair loss. In addition, this sterol can induce differentiation in mammary epithelium. This activity has led to the evaluation of 1,25-DHCC as a potential drug treatment in breast cancer; *in vitro* it inhibits the growth of breast cancer cells by inducing their differentiation.

These hormones also affect phosphate metabolism. Calcitonin facilitates the transport of phosphate into cells, and 1,25-DHCC promotes the absorption of phosphate from the intestines. Finally, PTH induces the excretion of phosphate in the kidneys. This effect was originally thought to facilitate the solubilization of calcium from bone, but it is now known that PTH can still resorb bone without increasing phosphate excretion. Another possible explanation for this effect is that it enables PTH to elevate calcium concentrations without risking ectopic calcification, which would be more likely to occur if both calcium and phosphate levels were elevated.

Integration

These actions can now be integrated into a general scheme (Fig. 1-7). Low serum calcium levels stimulate PTH and inhibit CT secretion; these two hormones are always reciprocally regulated because their effects antagonize one another (Fig. 1-8, *A*). PTH releases calcium from bone and reduces its excretion from the kidney. Both hypocalcemia and PTH induce the 1α-hydroxylase; hypophosphatemia, which frequently accompanies low serum calcium levels, further augments this induction while inhibiting the 24-hydroxylase. The final result is a

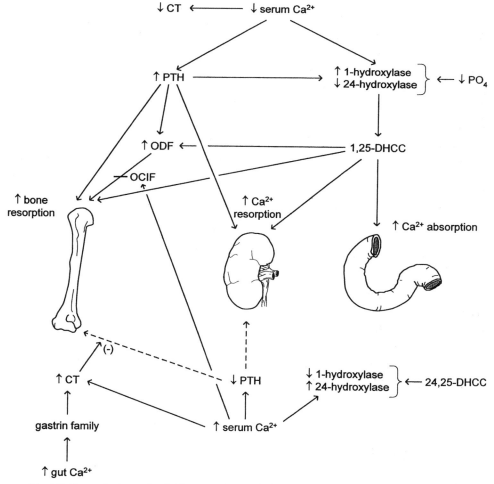

Fig. 1-7. General scheme for the hormonal control of calcium metabolism. *CT,* Calcitonin; *DHCC,* dihydroxycholecalciferol; *OCIF,* osteoclastogenesis inhibitory factor; *ODF,* osteoclast differentiation factor; *PO₄,* phosphate; *PTH,* parathormone.

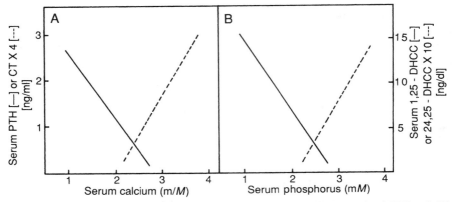

Fig. 1-8. Relationship between serum calcium concentrations and the levels of PTH and CT (A) and between serum phosphate concentrations and the levels of 1,25-DHCC and 24,25-DHCC (B).

shift in vitamin D from the inactive 24,25-DHCC form to the active 1,25-DHCC form (Fig. 1-8, *B*). The latter stimulates calcium uptake from the digestive tract and resorption from the kidney, while synergizing with PTH in releasing calcium from bone. Therefore hypocalcemia is corrected by recruiting calcium from the bone and digestive tract, while restricting its loss through the kidney. If calcium levels are high, the reverse occurs. PTH secretion is inhibited and CT is stimulated to antagonize what little PTH may still be circulating. Furthermore, the 24-hydroxylase is induced, while the 1α-hydroxylase is inhibited, resulting in a shift to 24,25-DHCC. This metabolite induces the degradation of 1,25-DHCC and blocks the activity of any residual 1,25-DHCC.

Although PTH, CT, and 1,25-DHCC are the primary hormones involved in calcium metabolism, there are several other hormones that play supporting roles. For example, gastrin functions in anticipatory signaling by the gastrointestinal tract; one of the functions of the digestive tract is to identify the contents of a meal and relay this information to the body to prepare it metabolically for these substrates (see Chapter 2, Gastrointestinal Hormones). In this case, calcium in the digestive tract stimulates the release of members of the gastrin family of hormones; these, in turn, stimulate the secretion of CT. After all, if calcium is about to be absorbed from the intestines, it does not have to be resorbed from bone.

Osteoclast differentiation factor (ODF), also called *osteoprotegerin ligand (OPG-L)*, is a member of the tumor necrosis factor (TNF) family. It stimulates osteoclast differentiation, fusion, activation, and survival; the resulting osteoclasts will then dissolve the bone to release calcium. ODF is elevated by PTH and 1,25-DHCC. It can also be elevated by cortisol and is responsible for the osteoporosis than develops during long-term glucocorticoid therapy. Osteoprotegerin (OPG), also called *osteoclastogenesis inhibitory factor (OCIF)*, is homologous to the binding domain of the ODF receptor. However, it is not coupled to any biological activity; as such, it is a decoy receptor that sequesters and inactivates ODF. Its elevation is inhibited by PTH, 1,25-DHCC, and cortisol, but it is stimulated by calcium, transforming growth factor-β (TGF-β), and estradiol. This induction of OPG is the basis for the treatment of osteoporosis with estrogens.

Stanniocalcin is a glycoprotein produced in many tissues where it acts in a paracrine manner; its levels are especially high in the kidney, skeleton, and gonads. It is stimulated by elevated calcium levels; it promotes the resorption of renal phosphate to chelate the calcium and inhibits any further calcium uptake from the gut. Fibroblast growth factor 23 (FGF23) is another parahormone. It reduces serum phosphate by increasing renal excretion, reduces 1α-hydroxylase and serum 1,25-DHCC, and reduces bone mineralization. Finally, prolactin, which is primarily involved with lactogenesis, also stimulates calcium uptake in the duodenum during lactation; this activity ensures adequate calcium supplies for milk production.

Summary

A hormone is a chemical, nonnutrient, intercellular messenger that is effective at micromolar concentrations or less. Hormones can have virtually any chemical structure; however, functionally they can be divided into hydrophobic and hydrophilic groups. The former diffuses through the plasma membrane, binds intracellular receptors, and induces transcription. The latter binds membrane receptors and acts through second messengers. Although hormones can affect any biological activity, most of their effects are related to metabolism, growth, and reproduction. Finally, hormone levels can be regulated by positive or negative feedback; the former controls events that have a clear end point so that an uncontrolled vicious cycle does not occur.

Calcium metabolism is primarily regulated by three hormones. In response to hypocalcemia, PTH elevates calcium by dissolving bone and resorbing calcium from the urine. 1,25-DHCC synergizes with PTH on bone and enhances calcium absorption from the gastrointestinal tract. During hypercalcemia, CT blocks the function of PTH.

References

General References

DeGroot, L.J., and Jameson, J.L. (eds.) (2001). *Endocrinology*. 4th ed. Saunders, Philadelphia, Pennsylvania.

Griffin, J.E., and Ojeda, S.R. (eds.) (2000). *Textbook of Endocrine Physiology*. 4th ed. Oxford University Press, Oxford, United Kingdom.

Hadley, M.E. (2000). *Endocrinology*. 5th ed. Prentice Hall, Englewood Cliffs, New Jersey.

Henry, H., and Norman, A.W. (2003). *Encyclopedia of Hormones*. Academic Press, San Diego, California.

Norris, D.O. (1997). *Vertebrate Endocrinology*. 3rd ed. Academic Press, San Diego, California.

Larsen, P.R., Kronenberg, H.M., Melmed, S., and Polonsky, K.S. (2003). *Textbook of Endocrinology*. 10th ed. Saunders, Philadelphia, Pennsylvania.

Hormone-Target Relationships

Labrie, F., Luu-The, V., Labrie, C., and Simard, J. (2001). DHEA and its transformation into androgens and estrogens in peripheral target tissues: Intracrinology. *Front. Neuroendocrinol.* 22, 185-212.

Nobel, S., Abrahmsen, L., and Oppermann, U. (2001). Metabolic conversion as a pre-receptor control mechanism for lipophilic hormones. *Eur. J. Biochem.* 268, 4113-4125.

Seckl, J.R., and Walker, B.R. (2001). Minireview: 11β-hydroxysteroid dehydrogenase type 1-a tissue-specific amplifier of glucocorticoid action. *Endocrinology (Baltimore)*. 142, 1371-1376.

Hormonal Control of Calcium

Epstein, F.H. (2000). The physiology of parathyroid hormone-related protein. *N. Engl. J. Med.* 342, 177-185.

Hofbauer, L.C. (1999). Osteoprotegerin ligand and osteoprotegerin: Novel implications for osteoclast biology and bone metabolism. *Eur. J. Endocrinol.* 141, 195-210.

Wimalawansa, S.J. (1997). Amylin, calcitonin gene-related peptide, calcitonin, and adreno-medullin: A peptide superfamily. *Crit. Rev. Neurobiol.* 11, 167-239.

Wysolmerski, J.J., and Stewart, A.F. (1998). The physiology of parathyroid hormone-related protein: An emerging role as a developmental factor. *Annu. Rev. Physiol.* 60, 431-460.

CHAPTER **2**

Classical Endocrinology

CHAPTER OUTLINE

In Chapter 1, the basic characteristics of hormones and their regulation were reviewed and illustrated with examples. In this chapter, the other classical hormones are discussed (see Table 1-2). The chapter begins with the most centralized endocrine system, the hypothalamic-pituitary axis, whose output controls the adrenal glands, the thyroid gland, and the gonads. After this axis and its dependent glands are discussed, the hormones involved with energy metabolism are examined. These hormones are closely associated with the gastrointestinal tract and include insulin and glucagon, among others.

Hypothalamus and Pituitary Gland

The pituitary gland, or hypophysis, is really two glands fused together; each gland has a different embryonic origin, secretes a different class of hormones, and is regulated differently. The posterior pituitary, or neurohypophysis, is an outgrowth of the floor of the third ventricle and is still connected to the ventricular floor through the infundibulum (Fig. 2-1). In most species, the anterior pituitary, or adenohypophysis, arises as an ectodermal invagination (Rathke's pouch) from the primitive mouth, the stomodeum. However, there are exceptions: the anterior pituitary arises from the endoderm in the hagfish and from the ectoderm of the face in the lamprey. The intermediate lobe of the pituitary gland is really just a subdivision of the adenohypophysis; in birds and in some mammals, it is completely absent.

Posterior Pituitary

The neurohypophysis secretes two nonapeptides, vasopressin and oxytocin, each of which contains a carboxy-terminal amide and a disulfide loop between residues one and six. However, the two peptides are actually synthesized in the peptidergic neurons of the supraoptic and paraventricular nuclei of the hypothalamus. The sequence of these peptides is encoded within a larger protein that also

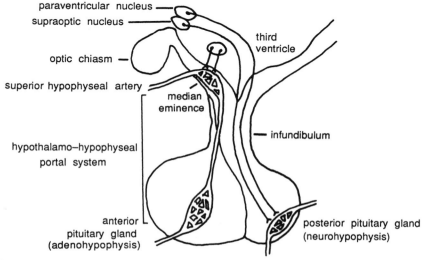

Fig. 2-1. Anatomy of the hypothalamus and hypophysis, showing the hypothalamo-hypophysial portal system to the adenohypophysis and the neural pathways to the neurohypophysis.

contains the sequences of other biologically active peptides; such a molecule is known as a *polyprotein*. The polyprotein precursor for the neurohypophysial hormones is cleaved into three pieces; the nonapeptide is the amino terminus, a neurophysin occupies the central region, and a 40-amino-acid glycoprotein forms the carboxy terminus. The neurophysin is a 10-kDa protein that binds the peptide hormone and protects it from rapid degradation. Larger proteins can form enough internal bonds to create a stable, tight globular structure with no exposed amino or carboxy termini. Such a structure is relatively resistant to proteolysis. The neurohypophysial hormones overcome the handicap of their small size by binding to a larger carrier protein. Further protection is afforded by the primary structure of the hormones themselves: each has its amino acids and carboxy termini blocked, as previously noted. The third product of the polyprotein, the carboxy-terminal glycoprotein, has no known function but may be involved in the processing of the precursor. After synthesis and cleavage of the polyprotein, the hormone and its neurophysin are packaged into vesicles, travel down the axons through the infundibulum, and are stored in the nerve endings in the posterior pituitary until they are released.

Vasopressin (VP), or antidiuretic hormone (ADH), is involved in water conservation. The most important stimulus for its secretion is an elevated blood osmolarity. Secretion can also be elicited by a 10% to 25% decrease in blood volume or by stress and nausea. The major effects of this hormone include the stimulation of water resorption in the kidneys and glycogenolysis in the liver. The former would obviously dilute the blood concentration and the latter may be part of the fight-or-flight response to stress (see the section on Adrenal Glands). ADH may also cause vasoconstriction, but this effect requires pharmacological concentrations of the hormone. Clinically, ADH deficiency results in diabetes insipidus, which is the inability to concentrate urine. As a result, patients can excrete as much as 15 liters of urine daily and must consume equal amounts of liquids to prevent dehydration.

The other peptide hormone is oxytocin, which stimulates smooth muscle contraction; it functions in both parturition and suckling. Uterine contractions at the time of parturition stimulate oxytocin release through a positive feedback loop that is broken when the fetus is finally expelled. Suckling triggers another neural reflex leading to oxytocin secretion, which stimulates the contraction of the myoepithelial cells around the alveoli and ducts of the mammary gland. This contraction forces milk toward the nipple, resulting in the milk "letdown," and facilitates suckling. This is another example of positive feedback; the loop is interrupted when the infant is sated and stops nursing. Maternal deficiency of oxytocin does not impair delivery, but it is likely that fetal oxytocin crosses into the maternal circulation.

Hypothalamus

The hypothalamus also controls the anterior pituitary, although there are no neural pathways connecting the two structures. Instead, the control is exerted by hormones, which are carried from the hypothalamus to the adenohypophysis through a special circulatory system, the hypothalamo-hypophyseal portal system (Fig. 2-1). The superior hypophyseal artery supplies both the pituitary stalk and the median eminence; the latter forms part of the floor of the third ventricle. The primary capillary plexus is drained by the hypophyseal portal vessel, which opens into a second capillary bed in the anterior pituitary. The neurons in the hypothalamic-hypophysiotropic nuclei send their axons to the median eminence, where they secrete releasing and inhibiting factors into the primary plexus. These factors are then delivered directly to the anterior pituitary, where they regulate the secretion of hormones synthesized in the adenohypophysis.

The structures for many of these factors are known (Table 2-1). Most are small peptides that are synthesized as a larger precursor. In the case of the thyrotropin-releasing hormone (TRH), the precursor contains five copies of the sequence, Gln-His-Pro-Gly, flanked by pairs of basic amino acids. Couplets of basic residues indicate cleavage sites, a carboxy-terminal glycine is a signal for amidation, and the amino-terminal glutamine forms an intra-amino acid peptide bond between the α-amino group and the γ-carboxy group to produce pyroglutamic acid (pGlu). Amino-terminal pGlu's and amidated carboxy termini protect the free ends of these peptides from degradation by exopeptidases. This is important for the releasing and inhibiting factors because their small size precludes any significant secondary or tertiary structure into which loose ends could be tucked. In the case of the gonadotropin-releasing hormone (GnRH) precursor, two factors may be produced: the GnRH forms the amino terminus, whereas a GnRH-associated peptide forms the carboxy terminus. This latter peptide has prolactin (PRL)–inhibiting activity, but its physiological significance has not been determined.

Table 2-1
Releasing and Inhibiting Factors Synthesized by the Hypothalamus

Hormone	Action	Structure	Precursor
Thyrotropin-releasing hormone (TRH)	↑ TSH, PRL	Blocked tripeptide (pGlu-His-Pro-NH$_2$)	29-kDa precursor containing five copies of TRH
Prolactin (PRL)-releasing peptide (PrRP)	↑ PRL	20- or 31-Amino-acid peptide with amidated carboxy terminus	Amino terminus of an 87-amino-acid precursor
Dopamine	↓ PRL, ↑ growth hormone (GH)	Catechol	
Gonadotropin-releasing hormone (GnRH, LHRH)	↑ Luteinizing hormone (LH), follicle-stimulating hormone (FSH)	Blocked decapeptide (pGlu . . . Gly-NH$_2$)	Amino terminus of 90-amino-acid precursor; carboxy-terminal 56 amino acids may be a PRL-inhibiting factor
Growth hormone-releasing factor (GHRF)	↑ GH	44-Amino-acid peptide with amidated carboxy terminus	Residues 32-75 from a 107-amino-acid precursor
Ghrelin	↑ GH, ↓ insulin and SRIF	26-Amino-acid peptide with an *O*-n-octanoyl	Amino terminus of a 117-amino-acid precursor
Somatotropin (GH) release-inhibiting factor (SRIF, somatostatin)	↓ GH	14-Amino-acid peptide with disulfide bond between residues 3 and 14	Carboxy terminus of a 92-amino-acid precursor
Corticotropin-releasing factor (CRF)	↑ Adrenocorticotropic hormone (ACTH)	41-Amino-acid peptide with amidated carboxy terminus	Carboxy terminus of a 190-amino-acid precursor
Pituitary adenylate cyclase-activating polypeptide (PACAP)	↑ LH, PRL, GH, and ACTH	27- or 38-Amino acid peptide with amidated carboxy terminus	Residues 132-169 from a 176-amino-acid precursor

The pituitary adenylate cyclase-activating polypeptide (PACAP) is unique in its wide distribution and broad specificity. This peptide is a member of the vaso-active intestinal peptide family (see section on Adipokines) and can stimulate the release of several pituitary hormones from the adenohypophysis, as well as epinephrine from the adrenal medulla and insulin from the pancreas.

Several releasing factors have similar activity; for example, either TRH or the prolactin-releasing peptide (PrRP) stimulates PRL secretion, and either ghrelin or the growth hormone-releasing factor (GHRF) triggers the release of growth hormone (GH). Other factors have multiple activities: GnRH releases both luteinizing hormone (LH) and follicle-stimulating hormone (FSH), and PACAP stimulates the secretion of LH, PRL, GH, and the adrenocorticotropic hormone (ACTH). In some cases, the pattern of secretion provides signaling specificity; for example, rapid pulses ($>1/h$) of GnRH favor LH release.

Anterior Pituitary

The cells of the anterior pituitary can be histologically classified by the stains they take up. The chromophobes do not stain at all; at least one group, the folliculo-stellate cells, produces a variety of parahormones for local regulation. The acidophils stain with acid dyes and synthesize members of the growth hormone–prolactin family. These cells may be specialized: somatotrophs secrete only GH, whereas mammotrophs or lactotrophs secrete PRL. However, a few cells may secrete both hormones. In female mice, the somatotrophs and mammo-trophs are nearly equal in abundance, but in male mice the somatotrophs out-number the mammotrophs 6:1; this ratio may explain the larger size of the male in most species. The basophils stain with basic dyes and secrete the tropic hor-mones; a tropic hormone is one that regulates another gland. The thyrotrophs synthesize thyroid-stimulating hormone (TSH), gonadotrophs make LH and FSH, and the corticotroph-lipotroph cells produce ACTH and related peptides.

Growth Hormone–Prolactin Family

Growth hormone (GH) and PRL are homologous hormones, which arose from gene duplication. Each hormone contains nearly 200 amino acids and has a large, central disulfide loop and a small, carboxy-terminal one. In addition, PRL has another small loop at the amino terminus. GH has both direct and indirect actions. Many of its metabolic effects are direct and include the stimulation of lipolysis, amino acid uptake, and protein synthesis; the latter two effects are especially pronounced in muscle. GH also induces peripheral resistance to insu-lin such that glucose cannot be used and blood glucose levels rise; this is known as the *diabetogenic effect*. GH also stimulates linear growth of the skeleton; origin-ally, this action was thought to be indirect—GH stimulated the secretion of insulin-like growth factor-I (IGF-I), previously called *somatomedin C*. The IGF-I, in turn, stimulated both chondrocyte mitosis and sulfate incorporation into the cartilage matrix; the growth of long bones occurs at the cartilaginous growth plates. Although the role of IGF-I in chondrocyte proliferation is clearly estab-lished, it is now known that GH plays an equally important and direct role in chondrocyte differentiation. Finally, IGF-I promotes the uptake of amino acids and glucose into muscle. Many other hormones also stimulate IGF-I secretion, but in these cases IGF-I acts as a parahormone. For example, parathormone stimu-lates the production of IGF-I in bone, and ACTH does the same in adrenal cortex. In summary, GH has two major actions: (1) direct metabolic effects that facilitate muscle growth and glucose sparing and (2) skeletal growth effects that are partially direct (differentiative) and partially mediated by IGF-I (proliferative).

PRL is an ancient hormone with multiple functions within the vertebrate lineage; these functions are discussed more fully in Chapter 18. In mammals, the three major activities of PRL are lactogenic, gonadotropic, and immunological functions. In all mammals, PRL is essential for the development of the mammary glands in preparation for milk production. However, PRL has also been implicated in the growth and development of nonmammary tissues; for example, PRL knockout mice have delayed ossification of the calvaria, and PRL in the amniotic fluid and milk of rats stimulates the growth and differentiation of the gastrointestinal tract. In excess, PRL can cause prostate hyperplasia.

In rodents, PRL is also a gonadotropic agent and is essential for the maintenance of pregnancy. Except for its role in lactation, PRL is not required for human reproduction; nevertheless, the human ovary still has abundant receptors for PRL, and pathologically elevated levels of this hormone can cause amenorrhea. Hyperprolactinemia can also produce impotence in men, but at physiological concentrations, PRL is not thought to play a role in male reproduction.

The newest activities of PRL to receive attention are its immunological and hematological effects. PRL is synthesized and secreted by several parts of the immune system where it may act as a parahormone: it is mitogenic for some immune cells, it ameliorates some forms of anemia, and its receptor is homologous to those of the cytokine family of immune and hematopoietic hormones (see Chapter 3). However, its effects are most obvious only after stress, which explains why PRL receptor knockout mice exhibit normal basal immune functions. PRL can also stimulate adrenal steroidogenesis, which may affect immunity (see later).

The PRL family has diversified greatly, and not all forms are present in all species. Some forms are synthesized in the placenta, but many are parahormones. However, all these hormones reflect the activity profiles of GH and PRL. For example, proliferin is angiogenic, but proliferin-related protein is antiangiogenic. Bovine PRL-related protein is immunosuppressive, rodent PRL-like proteins B and E stimulate erythropoiesis, and rodent PRL-like proteins E and F stimulate thrombopoiesis. The GH gene has also undergone several duplications in a species-specific manner; a variant GH is involved with fetal growth and the primate placental lactogen may affect fetal growth, maternal metabolism, and mammary growth during pregnancy.

PRL is the only adenohypophyseal hormone that is under tonic inhibitory control; that is, the mammotrophs are programmed to secrete PRL unless otherwise inhibited. This is dramatically demonstrated when the hypothalamo-hypophyseal portal system is disrupted. In the absence of any releasing or inhibiting factors, serum levels of all the adenohypophyseal hormones, except PRL, fall; however, PRL serum levels rise. PRL secretion is also stimulated by suckling by means of a neural reflex; this is essential for the continuation of lactation. Thus, if a woman does not wish to nurse her infant, after delivery she is given a dopamine agonist, which suppresses PRL secretion and prevents lactation (Table 2-1). The secretion of both GH and PRL occurs as short pulses, which are most abundant during sleep. For GH, this pattern is necessary for its biological activity. For example, clinically, there are certain patients with short stature but normal basal and stimulated GH serum levels; however, the spiking pattern is absent. These patients will grow if exogenous GH is administered intermittently. However, other biological activities may require chronically elevated levels. For example, in experimental animals, the continuous administration of GH induces PRL receptors but inhibits a major urinary protein, whereas a pulsatile administration induces the protein but inhibits the receptors.

Glycoprotein Hormones

The hormones from the basophilic cells form another family: they are all dimers sharing a common 89-amino acid α-subunit. The β-subunit, containing 112–115 amino acids, is different and provides specificity to the biological actions of each hormone, but it is inactive alone. All these hormones are glycosylated to generate a final molecular mass of 32 kDa, and all stimulate their target organs through 3′,5′ cyclic AMP (cAMP). Thyrotropin stimulates the synthesis and release of thyroxine (T_4) from the thyroid gland; FSH promotes gametogenesis in both sexes and estradiol secretion in the female, and LH stimulates progesterone production in the corpus luteum of the ovary and testosterone production in the Leydig cells of the testis.

Recently a new glycoprotein has been identified in the anterior pituitary: thyrostimulin is dimer composed of a unique α- and β-subunit. It is most closely related to TSH and can stimulate TSH receptors; however, it is not as effective as TSH and cannot compensate for TSH deficiency. As such, its function is still unknown.

Proopiomelanocortin

The final tropic hormone, ACTH, is synthesized as part of a 31-kDa polyprotein containing 265 amino acids. This polyprotein is called proopiomelanocortin (POMC) and contains three melanocyte-stimulating hormones (MSH-α, -β, and -γ), three endorphins (α-, β-, and γ-endorphin), and ACTH (Fig. 2-2). Many of these sequences are overlapping, so that only a certain subset of hormones can be produced from a single precursor. Therefore posttranslational processing regulates the expression of these hormones.

MSH is generated in the cells of the intermediate lobe of the pituitary gland in fish, reptiles, and amphibians. Human beings do not have this lobe, but these cells are still present and scattered in the remaining anterior and posterior lobes. In lower-order vertebrates, MSH causes the pigment granules in melanocytes to disperse so that the skin will darken. Because genetic factors are much more important in the skin color of mammals and birds, MSH appears to be less important, although it still transiently increases pigment synthesis in the higher-order vertebrates. MSH is also anti-inflammatory and anorexic, and it can induce penile erection.

ACTH stimulates the synthesis and secretion of adrenal steroids. Primarily it promotes the uptake of cholesterol and its conversion to pregnenolone. This is the first, and rate-limiting, step in the pathway (see section on Adrenal Glands). Like the other tropic hormones, ACTH uses cAMP as a second messenger (see Chapter 9). The endorphins are discussed in Chapter 3.

Fig. 2-2. Proopiomelanocortin and the individual hormones into which it can be cleaved.

Adrenal Glands

Anatomy

Like the pituitary gland, the adrenal gland is two glands in one: the outer layer, the adrenal cortex, synthesizes steroids, and the center, the adrenal medulla, synthesizes catecholamines, primarily epinephrine. The medulla originates from the neural crest and is, in fact, a sympathetic ganglion modified to secrete its "neurotransmitter" into the circulatory system rather than into a synaptic cleft. The cortex develops from the mesothelium of the abdominal cavity and envelops the medulla during embryogenesis. Eventually, the pyramidal gland comes to rest at the superior pole of each kidney.

Each part of the adrenal gland has its own blood supply. The cortex is divided into three layers: the zona glomerulosa contains ball-like clusters of cells and lies just underneath the adrenal capsule, the zona fasciculata contains cells arranged in columns and is the middle layer, and the zona reticularis contains cells in a netlike pattern and is the innermost layer (Fig. 2-3). The capsular arteries penetrate the capsule and form the capsular plexus in the zona glomerulosa. From there the straight capillaries carry the blood between the columns of the zona fasciculata to the reticular plexus of the zona reticularis. Finally, the blood is drained into the alar and emissary veins. The medulla is supplied by the medullary artery, which gives rise to the medullary capillaries; the medulla is drained by the central vein. There is one important connection between these two circulatory systems: the reticular plexus also drains into the medullary capillaries, exposing the medulla to high concentrations of corticosteroids. These steroids are involved in regulating epinephrine synthesis (see the section on catecholamine synthesis under Adrenal Medulla).

Zonae Fasciculata and Reticularis

The major product of these layers are the glucocorticoids (cortisol in human beings and corticosterone in rodents) and the androgens (Fig. 2-4). The pathway is regulated by ACTH, which stimulates P450scc (*P450* cytochrome responsible for *s*ide-*c*hain *c*leavage) in the first step. This is accomplished in two ways: first, ACTH induces the enzyme itself through its second messengers, cAMP and calcium. Second, ACTH increases the translation of the steroidogenic acute regulatory protein (StAR), a cholesterol shuttle that transfers this lipid from the outer to the inner mitochondrial membrane. This transport is the rate-limiting step in steroid synthesis. ACTH can also induce several other enzymes in this pathway. Acutely, ACTH can activate StAR and several enzymes through cAMP-directed phosphorylation of these proteins. Initially, it was believed that each of the reactions shown in Fig. 2-4 was catalyzed by a separate enzyme; however, it is now known that the steroidogenic enzymes all belong to a family of P450 cytochromes that have multiple activities. For example, a desmolase cleaves carbon-carbon bonds when each carbon contains an oxygen. Originally, the hydroxylation of adjacent carbon and the subsequent cleavage of the intervening bond were thought to be catalyzed by separate enzymes, but single proteins can possess both hydroxylase and desmolase activities. ACTH is only one leg of a regulatory loop; ACTH elevates serum cortisol, which inhibits the hypothalamus from secreting corticotropin-releasing factor (CRF). Without CRF, the adenohypophysis will not release any more ACTH. Cortisol also has a feedback effect on the

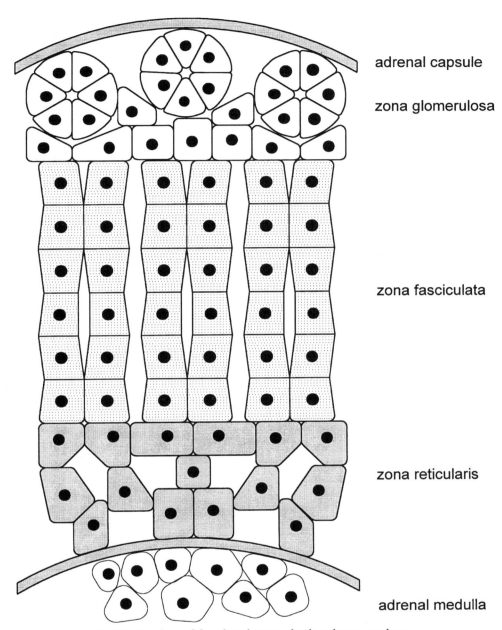

adrenal capsule

zona glomerulosa

zona fasciculata

zona reticularis

adrenal medulla

Fig. 2-3. Histology of the adrenal cortex; the three layers are shown.

pituitary gland. Superimposed on this control is a circadian rhythm such that ACTH is secreted during sleep; thus cortisol levels are highest in the morning. Finally, stress also induces the adenohypophysis to secrete ACTH; both the circadian rhythm and the stress effect are mediated through the hypothalamus and CRF.

Glucocorticoids have one major function: gluconeogenesis (Fig. 2-5). These steroids inhibit protein synthesis and, at higher concentrations, may even stimulate protein degradation; the resulting amino acids will serve as precursors for glucose. The steroids then induce enzymes to remove and detoxify the amino acids' nitrogen so that only the carbon skeletons remain. Glucocorticoids can also

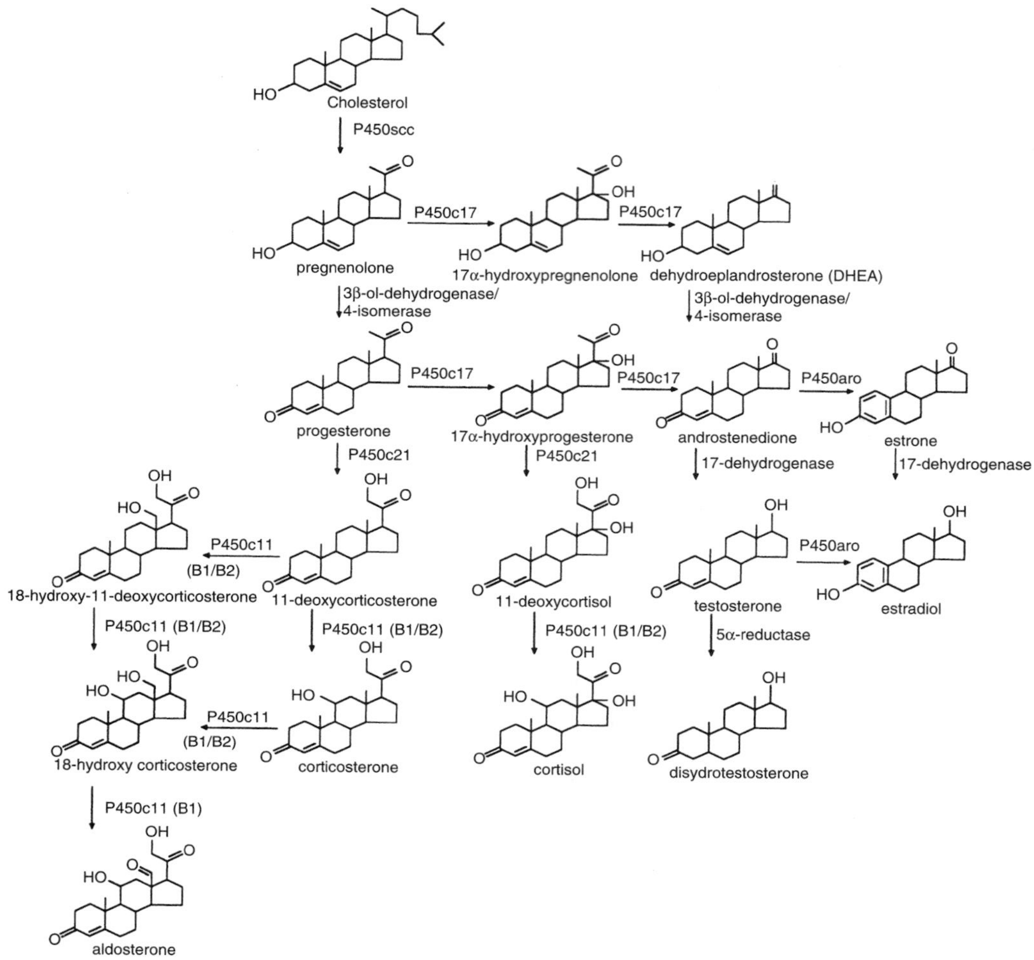

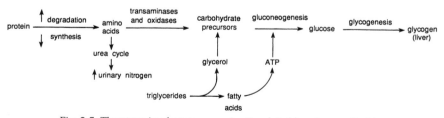

Fig. 2-4. Steroid biosynthetic pathway. This is a composite scheme; not all of these steps are in each layer.

Fig. 2-5. The steps in gluconeogenesis stimulated by glucocorticoids.

promote lipolysis, which provides another carbon skeleton, glycerol, for gluconeogenesis. Finally, these steroids induce enzymes critical in reversing glycolysis (see section on Gastrointestinal Hormones). Therefore these steroids are involved with every facet of gluconeogenesis. Glucocorticoids are also potent anti-inflammatory agents; they can stabilize lysosomes, inhibit leukocyte migration, lower antibody levels, and destroy lymphocytes. These actions make glucocorticoids very useful in the treatment of inflammatory diseases and certain leukemias and lymphomas. However, these effects only manifest at pharmacological concentrations and probably have no physiological relevance. Indeed, at low concentrations, glucocorticoids have been shown to actually stimulate lymphocyte differentiation, such as antibody production.

The sex steroids are only by-products of the glucocorticoid pathway. Dehydroepiandrosterone is the major androgen secreted by the adrenal gland. It is of no importance in males, whose primary sources of androgens are the testes. In women, however, the adrenal glands are a major source of androgens, which are important in the development of sexual hair. Only insignificant amounts of estrogens are synthesized. Even in males, the major sources of estrogens are not the adrenal glands but the testicular androgens, which are aromatized in peripheral tissues.

Because steroids are so hydrophobic, they must have serum carrier proteins. Both the steroid- and thyroid hormone–binding proteins are of two types: (1) proteins specifically designed for a particular group of steroids and (2) general carriers with hydrophobic pockets. The former usually have a single, high-affinity binding site per molecule, whereas the latter have many lower-affinity sites. The most important general carrier is albumin: although its affinity is only $10^3 - 10^6/M$ toward most steroids, its serum concentration is so high that it still binds a significant portion of these hydrophobic hormones (Table 2-2). In contrast, more specific carriers have affinities in the range of 10^7 to $10^{10}/M$. In addition to solving the solubility problem, these proteins also protect the hormone from degradation. Indeed, the half-life of these molecules is inversely related to the proportion of free hormone: 36% of aldosterone is free and its half-life is only 20 to 30 min, whereas only 0.02% of T_4 is free and its half-life is 6 days. Another contributing factor to the half-lives of hydrophobic hormones is their membrane permeability, which allows them to be resorbed passively from the urine. Nonetheless, the free hormone is the most important fraction because it is biologically active. For example, during pregnancy, estrogens induce many of the carrier

Table 2-2
Steroid-Binding Proteins in Serum

Steroid-Binding Protein	Serum Concentration (mg/dl)	Bound or Free (%)					
		Cortisol	Aldosterone	Testosterone	Estrogen	T_3	T_4
Corticosteroid-binding globulin (CBG)	3.0-3.5	70	17	—	—	—	—
Sex hormone-binding globulin (SHBG)	0.09-0.28	—	—	30	38	—	—
Thyronine-binding globulin (TBG)	1-2	—	—	—	—	60	78
Thyronine-binding prealbumin (TBPA)	25-33	—	—	—	—	30	3
Albumin	3500-4500	22	47	68	60	10	19
Free		8	36	2	2	0.2	0.02

T_3, Triiodothyronine; T_4, thyroxine.

proteins so that *total* hormone levels rise but pregnant women have no signs or symptoms of hormone excess because the *free* fraction remains the same.

Binding proteins may also influence the biological activity of their ligands in more specific and highly regulated ways; one of the best examples is the binding protein for a peptide, IGF. The IGF binding protein (IGFBP) can either enhance or inhibit the activity of IGF (see Chapter 3). In addition, the sex hormone binding globulin (SHBG) can mediate some of the nongenomic effects of sex steroids (see Chapter 6).

Zona Glomerulosa

This layer makes only aldosterone. This restriction was originally thought to be due to both the absence of P450c17, so it cannot synthesize the sex steroids or many glucocorticoids, and the exclusive presence of an 18-hydroxylase, which is essential for the synthesis of aldosterone. However, with the cloning of the enzymes of steroid synthesis, it became apparent that the 11β-hydroxylase, 18-hydroxylase, and 18-ol-dehydrogenase activities all resided in a single enzyme, renamed P450c11. Because the zonae fasciculata and reticularis have 11β-hydroxylase activities, they must also have 18-ol-dehydrogenase activity because both catalytic activities occur within the same enzyme. So why is aldosterone made exclusively in the zona glomerulosa? The answer lies in the fact that there are actually two isoforms of this enzyme, CYP11B1 and CYP11B2. Although they are 93% identical, only the CYP11B2 has 18-ol-dehydrogenase activity, and this isoform is unique to the zona glomerulosa. The function of this mineralocorticoid is to stimulate sodium resorption from the distal renal tubule; water is passively resorbed with the sodium. The major stimuli for aldosterone secretion are hypovolemia (low blood volume), hyponatremia (low sodium levels), and low renal perfusion pressure. However, none of these stimuli directly affect the zona glomerulosa; instead, they are channeled through the kidney. Why are the kidneys involved? The reason is simple: they have the most to lose if blood pressure falls too much. The kidneys remove wastes from the body, and to accomplish this they receive 25% of the cardiac output. If blood pressure falls excessively, they will not be able to perform their function. In addition, the kidneys are metabolically very active, and a severe drop in blood pressure, even for a short period of time, can lead to tubular necrosis.

Therefore the kidney has a special sensor in the juxtaglomerular complex, which is located where the distal convoluted tubule meets the afferent arteriole. The juxtaglomerular cells in the arteriolar part of the complex can measure the renal perfusion pressure, whereas the macula densa in the tubular part of the complex detects osmolarity (Fig. 2-6). The complex does not monitor blood volume directly, but the sympathetic nervous system does and stimulates this complex when the volume gets too low. The secretory product of the juxtaglomerular cells is a proteolytic enzyme, renin. Its substrate is angiotensinogen, an α_2-globulin made in the liver and secreted into the blood. Angiotensinogen is actually a prohormone; prohormones are inactive precursors and must be modified in some way to yield the active hormone. In this case, angiotensinogen is initially cleaved in the blood by renin to form angiotensin I, an inert peptide. Then, in the lungs, a converting enzyme removes two more amino acids to form the active angiotensin II. This octapeptide is the most potent vasopressor known, and the resulting vasoconstriction helps to restore the renal perfusion pressure. Angiotensin II also stimulates aldosterone synthesis by inducing the desmolase complex, and both angiotensin II and aldosterone inhibit further renin secretion via a short feedback loop. As previously noted, aldosterone promotes sodium and water resorption, which corrects the hyponatremia and hypovolemia.

There is a counterregulatory mechanism involving another hormone, atrial natriuretic hormone or peptide (ANP). It is synthesized as a 126-amino-acid precursor in the atria of the heart. The active hormone is cleaved from the carboxy terminus; various fragments are produced, but the most likely form *in vivo* has about 30 amino acids. It is released when the atria become distended, as by hypervolemia (volume overload), and its function is to eliminate this excess volume. To produce this effect, it dilates the afferent arterioles and constricts the efferent arterioles of the nephron, resulting in an increased glomerular capillary pressure. This higher pressure increases filtration and induces a sodium and water diuresis (Fig. 2-6). Its other effects are related to opposing the renin-angiotensin-aldosterone system; it inhibits both renin and aldosterone secretion and it antagonizes the vasoconstriction of angiotensin II. Its actions are mediated by cyclic GMP (cGMP) (see Chapter 7). Two other isoforms of ANP are produced in the central nervous system; their roles are less clear than those of the atrial hormone.

Nitric oxide (NO) is another hormone that opposes the renin-angiotensin-aldosterone system. NO is a parahormone that is synthesized in the endothelium from arginine by progressive oxidation; and it induces relaxation in the adjacent smooth muscle layer of the blood vessels by elevating cGMP, like ANP. It can also

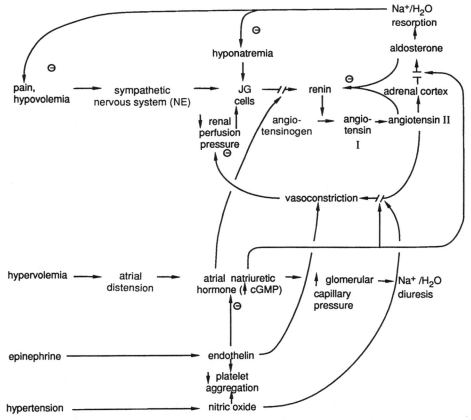

Fig. 2-6. The control of water-sodium balance and blood pressure by the renin-angiotensin-aldosterone system, atrial natriuretic hormone, endothelin, and nitric oxide. Parallel lines indicate those steps blocked by the indicated hormone. *JG*, Juxtaglomerular; *NE*, norepinephrine.

decrease platelet aggregation and adhesion, and its presence in macrophages and neural tissues suggests that other functions will be discovered, as well. In the endothelium, NO synthesis is stimulated by hypertension and several hormones, and the vasodilation induced by NO would facilitate the reduction of blood pressure.

Like NO, endothelins are parahormones made in the endothelium. These 21-amino-acid peptides are cleaved from a larger precursor in response to several hormones, such as epinephrine. This catechol is released during stress (see later) and elevates blood pressure, in part, through endothelins, which induce vasoconstriction and smooth muscle growth, suppress ANF serum levels, and inhibit the production of cGMP by ANF. As such, they complement the renin-angiotensin-aldosterone system. Like NO, they also inhibit platelet aggregation.

There are several other hormones that play a role in blood pressure and sodium homeostasis. As noted in Chapter 1, adrenomedullin is a member of the calcitonin (CT) family. It is found in adrenal glands, lungs, kidneys, and heart, and it induces vasodilation through cAMP. Uroguanylin links the intestines with the kidneys; it is secreted from the intestines after sodium ingestion, travels to the kidneys, and stimulates sodium excretion.

Adrenal Medulla

Autonomic Nervous System

The adrenal medulla developed from the autonomic nervous system (ANS) and remains functionally linked to it; therefore one cannot discuss one without the other. The ANS is concerned with regulating unconscious activities; it is divided into two opposing components. The parasympathetic nervous system is concerned with the maintenance of bodily activities under basal conditions; for example, it slows the heart rate and stimulates visceral functions, such as the secretion of digestive enzymes and increased gut motility. The sympathetic nervous system is concerned with energy expenditure during times of stress; for that reason it is frequently referred to as the *fight-or-flight system*. For example, it mobilizes substrates for conversion into energy; it dilates bronchioles to increase oxygen uptake; and it increases the perfusion of liver, brain, skeletal muscle, and heart to facilitate the delivery of these substrates and oxygen.

Most organ systems in the body are innervated by both components of the ANS; the net effect on the system is determined by a balance between the two. Each component has a two-neuron pathway from the central nervous system to the target organ. In the sympathetic nervous system, the first neuron synapses with the second in a ganglion; the second then travels to its target organ, where it secretes norepinephrine. The adrenal medullae are actually derived from a pair of these sympathetic ganglia, except that the second neurons empty their secretions into the blood instead of leaving the ganglia for some peripheral tissue. Another difference is that the medullae secrete epinephrine rather than norepinephrine.

Catecholamine Synthesis

Catecholamine synthesis in the second neuron begins with tyrosine hydroxylase, which catalyzes the conversion of tyrosine to dihydroxyphenylalanine (DOPA) (Fig. 2-7). ACTH is required to maintain basal levels of this enzyme; this effect is achieved through cAMP-dependent phosphorylation. After DOPA is decarboxylated, the product, dopamine, is transported into the chromaffin granule, where dopamine is further hydroxylated to form norepinephrine, the distal

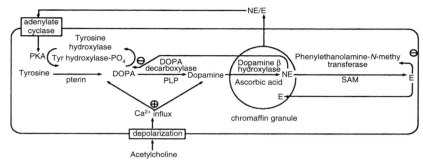

Fig. 2-7. Catecholamine biosynthetic pathway in the adrenal medulla. *DOPA*, Dihydroxyphenylalanine; *E*, epinephrine; *NE*, norepinephrine; *PKA*, cAMP-dependent protein kinase; *PLP*, pyridoxal phosphate; *SAM*, S-adenosylmethionine.

neurotransmitter in the sympathetic nervous system. As norepinephrine accumulates, it acts through a feedback mechanism to inhibit the tyrosine hydroxylase. In the adrenal medulla, norepinephrine is converted to epinephrine in a methylation step catalyzed by phenylethanol-N-methyl transferase; S-adenosylmethionine (SAM) is the methyl donor. This reaction takes place in the cytoplasm, but the epinephrine reenters the granule after its methylation. The methyltransferase is induced by glucocorticoids and inhibited by epinephrine. The former originate in the adrenal cortex, which drains, in part, through the medulla (see the section on anatomy in Adrenal Glands); the latter represents product inhibition.

Synthesis and secretion is stimulated by the first neuron, whose neurotransmitter is acetylcholine (ACh). Acetylcholine stimulates both the tyrosine hydroxylase and the dopamine β-hydroxylase by means of a calcium-dependent mechanism. Once secreted, the catecholamines not only enter the blood but also have a feedback effect on the secretory cell, elevating cAMP levels and activating tyrosine hydroxylase. In summary, intracellular catecholamines represent an accumulated product, which turns synthesis off, but extracellular catecholamines represent a depleted reserve and stimulate further synthesis.

Catecholamine Actions

Because norepinephrine has very specific effects, it is the molecule of choice for neurotransmission. Its major activity is to increase total peripheral resistance in the circulatory system. Epinephrine has more general effects, and therefore it is distributed in the blood. Its major actions can be divided into metabolic, cardiovascular, and smooth muscle effects. The metabolic activity involves the mobilization of substrates for energy expenditure: it stimulates lipolysis in adipocytes and glycogenolysis in liver and muscle. It actually plays an ancillary role in liver glycogenolysis, which is more strongly affected by glucagon. Furthermore, because muscle lacks glucose-6-phosphatase, the glucose liberated from glycogen cannot leave the cell; instead, the glucose is broken down to lactate, which is released into blood and resynthesized into glucose in the liver. In the cardiovascular system, epinephrine stimulates the vasoconstriction of the subcutaneous, splanchnic, renal, and mucosa beds, although the skeletal muscle bed vasodilates; this shunts blood from the viscera to the skeletal muscles, which get circulatory priority in the fight-or-flight response. Epinephrine also increases the heart rate and cardiac output and thus facilitates the delivery of nutrients to the skeletal muscles.

Finally, epinephrine relaxes the smooth muscle of the uterus, bladder, and bronchioles, although it constricts the sphincters. Dilated airways improve oxygenation of the blood. As for the rationale behind the sphincter and bladder effects, it is presumed that when one's life is threatened, one cannot be bothered by the more earthy bodily functions.

The Thyroid Gland

The thyroid gland originates as an invagination from the floor of the mouth and migrates down the neck. It consists of two lateral lobes connected by an isthmus; occasionally, there is also a central pyramidal lobe. The thyroid is yet another gland containing two different secretory components; during its migration, it incorporates the C cells, which originate from the neural crest after passing through the ultimobranchial body (see Fig. 1-5). The C cells synthesize CT and are discussed in the section on Hormonal Control of Calcium in Chapter 1. Histologically, the thyroid gland contains follicles: epithelial cells surrounding a colloid material. The C cells are located between or attached to the follicle (Fig. 2-8).

Synthesis of Thyroid Hormones

T_3 and T_4 are basically iodinated tyrosines with an extra phenyl ring. However, their synthesis presents a problem: both T_3 and T_4 are very hydrophobic, which makes their storage difficult. The gland could simply synthesize the hormones on demand; this is how the adrenal glands and gonads solve the same problem with the hydrophobic steroids. Unfortunately, this solution is not satisfactory for the thyroid hormones because one of the essential ingredients, iodide, is not always readily available. Ideally, the gland should build up a reserve of T_3 and T_4 when iodide is abundant; then it could draw on this store when iodide is scarce. The solution is to synthesize T_3 and T_4 from tyrosines, which are already incorporated into a protein; the residues may be hydrophobic, but the entire protein is hydrophilic and can be easily compartmentalized. When thyroid hormones are needed,

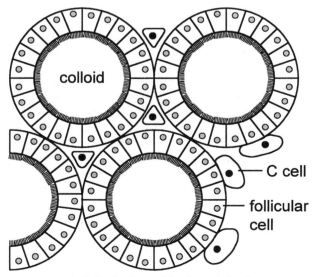

Fig. 2-8. Histology of the thyroid gland.

some of the protein is digested to liberate the T_3 and T_4. In this way, as much as a 6-month supply of thyroid hormones can be maintained.

The first step in the biosynthetic pathway is the procurement of the iodide (Fig. 2-9). This is accomplished by an iodide pump that can concentrate iodide against a 25– to 40–fold gradient. Another preliminary step is the synthesis of the iodination enzyme and substrate. The substrate is thyroglobulin, a huge protein composed of two identical 330-kDa subunits. It is a glycoprotein with a large number of cysteines. The enzyme is thyroid peroxidase, and it is packaged within the same vesicles as the thyroglobulin. However, it is not active until it is released into the colloid at the apical surface of the follicular cell. There the enzyme oxidizes I^- to I^+. *In vitro*, the oxidized iodide spontaneously incorporates itself into the tyrosine ring; however, *in vivo* there is evidence that this organification is also facilitated by the peroxidase. Finally, the iodinated ring from one tyrosine is coupled to that of another (Fig. 2-10). *In vitro*, the coupling can also occur spontaneously, but the existence of a coupling enzyme has been postulated for the *in vivo* reaction. Because the sole purpose of thyroglobulin is to be a precursor for thyroid hormones, one might suppose it to be rich in tyrosines; however, this is not the case. One molecule has 140 tyrosines; about 25 of them are iodinated, and only 8 iodinated tyrosines are coupled to form four T_3 and/or T_4. Four to ten times more T_4 than T_3 is made.

When thyroid hormones need to be secreted, the iodinated thyroglobulin is taken up by the follicular cell. The endocytotic vesicle fuses with lysosomes and the protein is degraded. T_3 and T_4 are released and diffuse into the blood. The uncoupled tyrosines, both monoiodotyrosines and diiodotyrosines, are deiodinated and the iodide recycled. There is one final activation step: the conversion of T_4 to T_3. Several data support the importance of this step:

1. In all systems studied thus far, T_3 is more potent than T_4.

2. Nuclear receptors for the thyroid hormones prefer T_3 to T_4 (see Chapter 6).

3. T_3 deficiency produces hypothyroidism even in the presence of normal T_4 levels.

4. Iopanoic, a specific 5'-deiodinase inhibitor, blocks all the biological actions of T_4.

Therefore it is generally assumed that T_3 is responsible for most, if not all, of the biological activities of the thyroid hormones and that T_4 is only a prohormone. This conversion from T_4 to T_3 takes place in the peripheral tissues, which contain the iodothyronine deiodinases. The type 1 deiodinase is located in the plasma membrane, has the lowest affinity for T_4, and regulates T_3 blood levels. The type 2 enzyme is found in the endoplasmic reticulum, has the highest affinity for T_4, and regulates nuclear T_3 levels. For example, hypothyroidism induces this enzyme so as to maintain T_3 levels, whereas hyperthyroidism has the opposite effect. The type 3 deiodinase is the major deactivating enzyme.

The activity of the follicular cell is controlled by TSH from the pituitary gland. This tropic hormone stimulates synthesis by promoting iodide uptake and inducing the genes for thyroid peroxidase and thyroglobulin. It stimulates secretion by increasing the number, height, and activity of the microvilli, which internalize the iodinated thyroglobulin. Finally, TSH stimulates the pentose phosphate pathway to provide the energy for all the aforementioned processes.

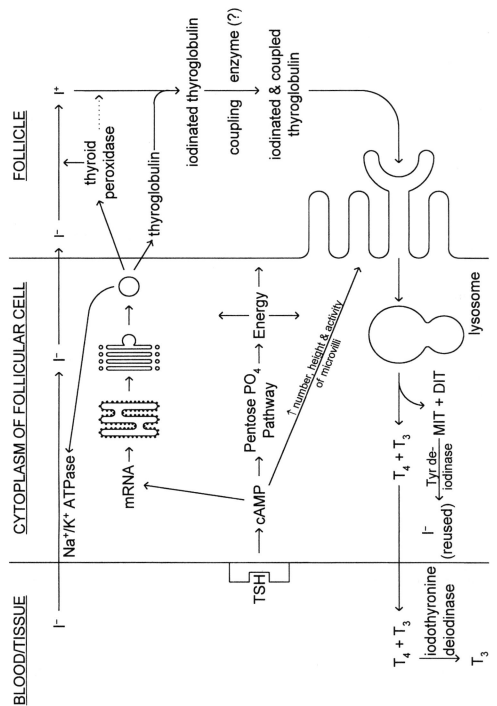

Fig. 2-9. Biosynthetic pathway of the thyroid hormones. *MIT*, Monoiodotyrosine; *DIT*, diiodothyronine.

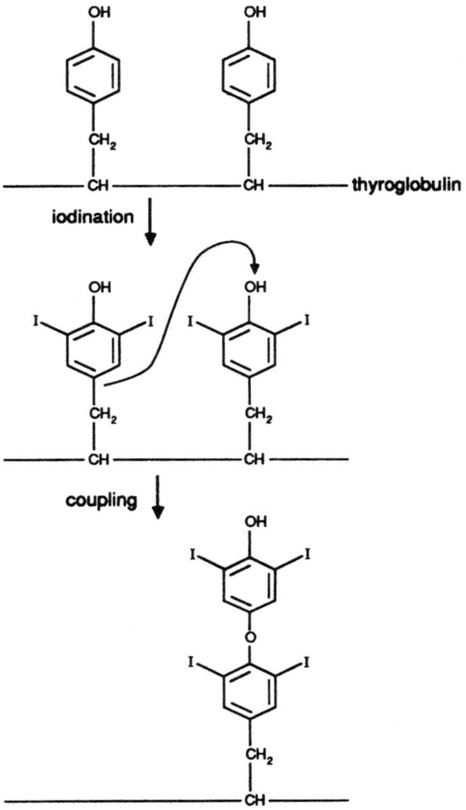

Fig. 2-10. Coupling reaction in thyroid hormone synthesis.

Actions of Thyroid Hormones

Thyroid hormones elicit a bewildering, and often conflicting, array of actions in vertebrates. One of the first and most prominent effects is the stimulation of calorigenesis, as evidenced by increased oxygen consumption. The muscular, gastrointestinal, and renal systems appear to be the most responsive, whereas the nervous, reproductive, and immunological systems are relatively unresponsive. The molecular basis for this effect appears to be a series of futile cycles activated by T_3. For example, in carbohydrate metabolism, T_3 potentiates the effect of insulin on glucose uptake and glycogen synthesis; but it also potentiates the effects of the catecholamines on glycogenolysis. The latter predominates, but blood glucose levels do not rise as much as one would expect because much of the glucose is oxidized. T_3 also stimulates both lipid synthesis and lipolysis; again, degradation predominates. These effects on lipolysis are both direct and indirect; the latter actions are a result of the potentiation of the effects of catecholamines, GH, glucocorticoids, and glucagon. In fact, T_3 potentiates the effects of so many hormones that it has frequently been called a *permissive hormone*. At least part of this potentiation is due to the T_3-induction of the receptors for these hormones; more receptors would increase the responsiveness of the tissue toward these hormones. In an action also related to lipid metabolism, T_3 stimulates hydroxy-methylglutaryl coenzyme A reductase, the rate-limiting enzyme in cholesterol synthesis; however, cholesterol levels actually fall because its elimination through bile acids exceeds its enhanced synthesis. It may be that T_3 does not exert any specific metabolic effects; rather, it may simply adjust the overall metabolic rate in the tissues. In addition to its metabolic and calorigenic effects, the thyroid hormones also exert many important developmental and growth effects. At physiological concentrations, T_3 promotes protein synthesis, linear growth, and skeletal maturation. In amphibians it stimulates molting and metamorphosis. However, even in growth and development the specific role played by T_3 is unclear. For example, T_3 is known to exert a "permissive" effect on other growth-related hormones. Furthermore, some of its effects on metamorphosis can be explained by assuming that T_3 is merely triggering a preprogrammed event. For example, a T_3 pellet implanted in frog skin will cause the epidermis to thicken everywhere except the opercular window, where the forelimb will emerge; here the epidermis thins. This result has been interpreted to mean that each cell in the tadpole is genetically programmed to execute certain activities during metamorphosis. Although T_3 triggered this program, the response of the cell was determined by its genetic program.

Reproduction

Androgens

In males, androgens are primarily synthesized in the Leydig cells of the testes and secreted as testosterone. In females, the adrenal cortex is the major source and secretes dehydroepiandrosterone (DHEA); the ovary makes a minor contribution.

Actions

Androgens act in one of three molecular forms: dihydrotestosterone (DHT), testosterone (T), and estradiol (E_2). The peripheral tissues possess a 5α-reductase that converts T to DHT. This is the active androgen in all adult tissues, except muscle; that is, T is usually a prohormone. Nonetheless, T does have some direct effects in several embryonic structures and in adult muscles. Finally, T can be converted to E_2 by an aromatase in peripheral tissues; this is the form in which T acts on certain parts of the brain.

Although the sex chromosomes determine whether the gonad will become a testis or an ovary, the development of the other genital structures results from the interaction between the genetic elements and hormones. Genetically, all embryos are programmed to develop a female phenotype. As a consequence, female embryos do not require the presence of estrogen or any other sex steroid to generate a female genital pattern; indeed, castration of female embryos does not interfere with the development of this pattern. However, in male embryos, androgens will stimulate the growth of the genital tubercle into a penis, induce the fusion of the labioscrotal swellings to form the scrotum, and promote the descent of the testes into this sac. If the testes in a male embryo are inactivated, the genital tubercle forms a clitoris, the swellings never fuse and become the labia majora, and the gonads remain in the abdomen; that is, a female phenotype prevails. In male embryos, androgen secretion by the Leydig cells is initiated by a placental gonadotropin, human chorionic gonadotropin (hCG), but hCG is only elevated during the first trimester. The embryonic pituitary gland must take over or the genitalia will be incompletely masculinized; that is, for example, the penis will be small and the testes undescended. The Leydig cells secrete testosterone, which stimulates the external genitalia, including the penis and scrotum. DHT is responsible for the development of the internal genitalia, including the epididymis, vas deferens, seminal vesicles, and prostate gland. In nonprimates, androgens may have two other important targets: the mammary gland and the hypothalamus. In the mammary gland, DHT destroys all or part of the epithelium and thwarts nipple formation. In the hypothalamus, T, after conversion to E_2, changes the secretory pattern of the gonadotropins from the cyclic female pattern to the tonic male pattern. In females, serum E_2 is sequestered in α-fetoprotein, so the cyclic pattern persists. If female animals are given enough estrogen to exceed the binding capacity of this protein, the hypothalamus will become masculinized; that is, gonadotropins will be secreted tonically. In human fetuses, serum gonadotropin and T levels return to basal levels by 28 weeks of gestation, and the testes will remain quiescent until puberty.

There are several other hormones that are also important in the embryonic and fetal development of the male reproductive tract. The default female phenotype includes the müllerian ducts, which will form the oviducts and uterus. In males, these structures must be destroyed. The Sertoli cells synthesize antimüllerian hormone (AMH), also called *müllerian inhibiting substance* (MIS), a peptide that belongs to the transforming growth factor β (TGFβ) family and that induces the regression of müllerian ducts. The insulin-like 3 protein (INSL3), also called the *relaxin-like factor* (RLF) or the *Leydig insulin-like protein* (LEYI-L), is a member of the insulin family and is synthesized in the Leydig cells. Synthesis is inhibited by estrogens. INSL3 stimulates the growth and differentiation of the gubernaculum, which leads to testicular descent into the scrotum. Inactivating mutations of INSL3 are the cause of some forms of cryptorchidism (undescended testes).

In adolescence, androgens have important developmental, psychological, and sexual effects. Androgens stimulate both linear growth and skeletal maturity; that is, although the longitudinal growth of bones is accelerated, the growth plate shrinks and eventually becomes obliterated when the epiphyses and diaphyses fuse. This fusion marks the permanent termination of linear growth. Estrogens synergize with androgens: men unable to synthesize estrogens have a prolonged period of skeletal growth resulting in a eunuchoid habitus. The anabolic actions of androgens also facilitate general growth; for example, androgens increase protein synthesis, which is most marked in muscles. Behaviorally, androgens induce aggression and libido. Finally, androgens promote the development of both primary and secondary male sexual characteristics. Secondary sexual characteristics are those not required for reproduction per se, although they may facilitate the attraction of mating partners. They include sexual hair and a deep voice. Primary male sexual characteristics are those essential for successful intromission and insemination. They include the penis, testes, and accessory sexual structures.

Control

The testes are composed of seminiferous tubules and Leydig cells. The tubules are responsible for spermatogenesis and contain both the spermatogonia and the Sertoli cells. The latter nurture the sperm as they progress through their developmental stages. Spermatogenesis requires both FSH and very high levels of T. The T is supplied by the surrounding Leydig cells and is concentrated in the tubules by a special androgen-binding protein induced by FSH. LH stimulates T synthesis by inducing P450scc in the Leydig cells. As in the case with ACTH, this is accompanied by the stimulation of cholesterol transport into the mitochondria (see section on Zonae Fasciculata and Reticularis). It should be noted that all steroid-synthesizing tissues are regulated at the initial P450scc step; LH in the testis and ovary, as well as ACTH in the zonae fasciculata and reticularis, work through cAMP, whereas the effects of angiotensin II in the zona glomerulosa are mediated by calcium. The differences in the type of steroids secreted are determined by which other enzymes are present in the tissues.

The regulation of T secretion is relatively simple (Fig. 2-11); low steroid levels release the hypothalamus and pituitary from feedback inhibition. The resulting GnRH secretion stimulates LH release, which, in turn, activates steroid synthesis in the Leydig cells. When steroid levels have been restored, feedback inhibition is reestablished. Control of spermatogenesis is more problematic; the product of the pathway is not a hormone that can exert feedback effects to the brain but is a cell that remains in the tubule. It appears that the Sertoli cells monitor spermatogenesis and, whenever appropriate, release a hormone, inhibin, which specifically inhibits FSH secretion. Inhibin, like AMH, is a member of the TGFβ family; it has a dual function based on its subunit composition. As a heterodimer containing a

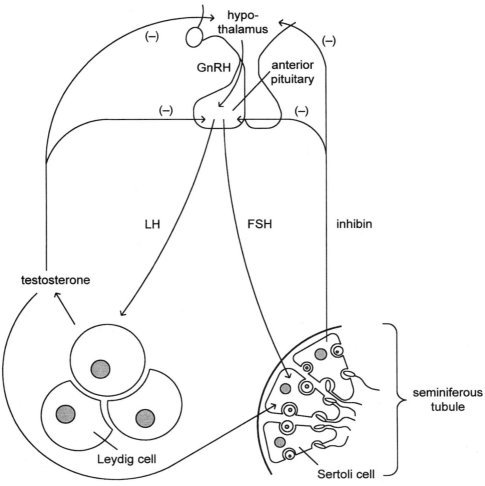

Fig. 2-11. Hormonal regulation of steroidogenesis and spermatogenesis in the testis. Adapted and reprinted by permission from Smith, E.L., Hill, R.L., Lehman, I.R., Lefkowitz, R.J., Handler, P., and White, A. (1983). *Principles of Biochemistry: Mammalian Biochemistry*. 7th ed. McGraw-Hill, New York, p. 500. Copyright © 1983 McGraw-Hill Book Company.

disulfide-linked 18-kDa α-subunit and a 14-kDa β-subunit, it inhibits FSH secretion, but homodimers of the β-subunit act as parahormones on the development of spermatogonia. Because the β-subunit homodimer also selectively stimulated FSH secretion *in vitro*, it was called *activin;* however, there is some controversy over whether this activity is physiological.

Estrogens and Progestins

The major sources of estrogens in women are the ovaries; in men, estrogens are formed from the peripheral conversion of androgens. The adrenal glands do not synthesize estrogens in any significant amounts. As they do in male embryos, hCG and the embryonic pituitary gonadotropins stimulate sex steroid production from the gonads. However, there is no known role for estrogens in embryogenesis because the basic phenotype is already female. For example, females with metabolic defects that prevent estrogen synthesis are morphologically normal at birth.

During puberty, estrogens have actions similar to the androgens. They stimulate linear growth and skeletal maturation, but they are much less active in inducing muscle protein synthesis. They cause behavioral changes and promote the development of both primary and secondary female sexual characteristics. Primary female sexual characteristics include the internal and external genitalia; secondary sexual characteristics include breast development and the female pattern of fat deposition.

Oogenesis

In male human beings, spermatogenesis requires 10 weeks to progress from spermatogonia to spermatozoa. However, the entire testis is not synchronized, although short stretches of the seminiferous tubule are; therefore, because the development of the male gametes is distributed throughout the gonad, there is a continuous supply of spermatozoa. Furthermore, because of attrition during storage and, eventually, during their trek in the female reproductive tract, the spermatozoa are made in large numbers.

Female mammals, however, exemplify a different reproductive strategy. Because a female will carry the fetuses to term within her body, they have an excellent chance for survival, and therefore females need to produce only a few mature ova. In the case of a female human being, she usually produces only one per reproductive cycle. Furthermore, because the female retains the fetuses, the development of her reproductive tract must be coordinated with oogenesis, intromission, and implantation. All of this requires a synchronized series of hormonal and anatomical events called the *estrous* or *menstrual cycle,* depending on whether the mammal sheds blood during the cycle.

In human beings, oogenesis begins *in utero*. The oogonia divide to produce between 6 and 7 million cells by 5 months of gestation. Then all of them become primary oocytes and enter prophase of meiosis I, at which stage they will remain until puberty. In males, spermatogonia are always reproducing themselves and thus maintain a reserve population; this allows spermatogenesis to continue well into old age. However, because there is no such reserve of oogonia, no further oocytes can be produced. In fact, almost immediately, oocytes begin to undergo atresia so that only 2 million are present at birth and only 300,000 remain by the time puberty approaches. Further atresia takes place during each menstrual cycle (see later), so women generally deplete their supply of oocytes when they are about 50 years of age.

In human beings, the menstrual cycle begins during puberty. It consists of four phases: the follicular phase, ovulation, the luteal phase, and the menstrual phase. The follicular phase is also known as the *estrogenic* or *proliferative phase* and lasts 10 to 14 days. It begins when FSH recruits 10 to 20 primordial follicles to develop into primary follicles; this development involves proliferation of the granulosa cells surrounding the primary oocyte (Fig. 2-12). The follicles also become invested with stroma: the outer layer, or theca externa, is merely a connective tissue capsule, but the inner layer, or theca interna, is steroidogenic. LH, aided by inhibin, stimulates P450scc in the theca interna, but because these cells lack aromatase activity, the major products are androstenedione and T. However, this androgen production is partially checked by activin. These steroids are then transported to the adjacent granulosa cells, which convert them to E$_2$ by a FSH-induced aromatase.

The E$_2$ further stimulates LH secretion but inhibits FSH secretion. FSH activity is also inhibited by bone morphogenetic protein 15 (BMP15), also called *growth differentiation factor 9B*. BMP15 is another member of the TGFβ family; its name originated from the fact that one of the first members of this group was involved with bone development. However, the BMPs are important in many diverse developmental processes. BMP15 is a parahormone secreted by the oocyte to stimulate follicular growth. In addition, BMP15 blocks the actions of

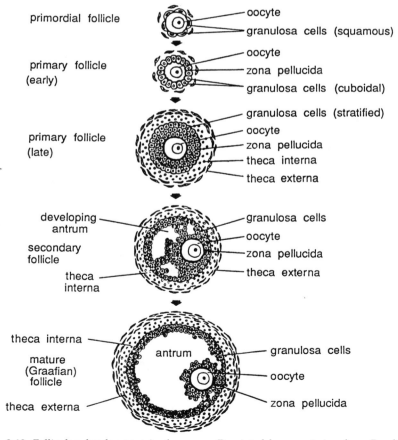

Fig. 2-12. Follicular development in the ovary. Reprinted by permission from Bruck-Kan, R. (1979). *Introduction to Human Anatomy.* Harper & Row, New York, p. 512.

FSH by reducing its receptor. The FSH activity declines and the follicles compete with one another for the available hormone. At this stage the follicles begin to develop a fluid-filled cavity, the antrum, and are now called *secondary follicles*. As FSH concentration becomes a limiting factor, the less successful follicles degenerate and, in human beings, only one usually survives. When the antrum is complete, the entire structure is called a *mature* or *graafian follicle*. In the meantime, the primary oocyte completes meiosis I to form the secondary oocyte and the first polar body, which usually degenerates. The secondary oocyte then enters meiosis II but becomes arrested in metaphase; it will remain in this state until fertilization. The rising levels of E_2 (Fig. 2-13) thicken the vagina and stimulate mucus production in preparation for intromission by the male copulatory organ. E_2 also stimulates hypertrophy (increase in cell size) and hyperplasia (increase in cell number) of the myometrium and endometrium, which are the muscle and inner lining of the uterus, respectively. Finally, E_2 promotes myometrial contractility, which may facilitate the transfer of the sperm from the vagina to the oviducts.

The positive feedback by E_2 results in a dramatic rise in LH levels at midcycle (Fig. 2-13). The LH alters the integrity of the follicle, which ruptures and releases the oocyte and its surrounding granulosa cells. This is ovulation.

The luteal phase is also known as the *progestational* or *secretory phase* and lasts another 10 to 14 days. After ovulation, the follicle collapses and a new protein, follistatin, becomes elevated. This molecule binds activin, thereby neutralizing its inhibition of the androgen production needed for sex steroid synthesis. LH, now unhampered by activin, stimulates the collapsed follicle to produce large amounts of progesterone. In fact, it is so rich in cholesterol and steroids that it has a yellow color, which gives the structure its name—corpus luteum (yellow body). The elevation of serum progesterone is augmented by PRL, which suppresses the expression of 20α-hydroxysteroid dehydrogenase, a major progesterone-metabolizing enzyme. Progesterone stimulates the proliferation of the uterine glands, the secretion of glycogen, and the development of the spiral arterioles, all of which prepare the uterus for implantation and nourishment of

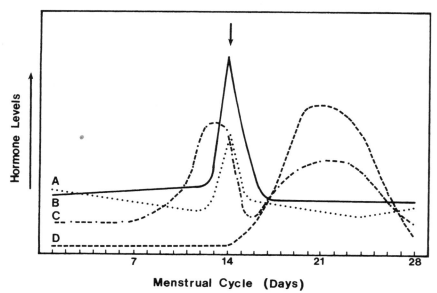

Fig. 2-13. Hormone levels during the menstrual cycle. The hormones depicted are (A) follicle-stimulating hormone (FSH), (B) luteinizing hormone (LH), (C) estradiol, and (D) progesterone.

the embryo. It also suppresses uterine contractions, which might otherwise endanger implantation. Finally, it inhibits both the synthesis and secretion of LH but only the secretion of FSH; that is, although the levels of both gonadotropins decline, the pituitary content of FSH increases because synthesis continues. However, the pituitary LH content is totally depleted. Because LH is required for progesterone synthesis by the corpus luteum, the synthesis of this steroid likewise declines.

The superficial layer of the endometrium, the stratum functionalis, is totally dependent on progesterone for its maintenance. When progesterone levels fall, the stratum functionalis atrophies and is shed along with the unfertilized ovum. This is the menstrual phase. The decline of progesterone levels also releases the pituitary gland from inhibition. FSH is immediately secreted because intracellular stores are filled and it recruits more primordial follicles for the next cycle. LH secretion is delayed because it must first be synthesized. If LH were to be secreted too quickly, the LH-induced E_2 production would inhibit FSH secretion and follicular development.

Pregnancy

If fertilization takes place, the secondary oocyte will complete meiosis II and the female and male pronuclei will fuse to form a zygote. Fertilization occurs in the upper third of the oviducts and cleavage begins promptly; by the time the embryo reaches the uterus, it is a blastocyst. The blastocyst is divided into two components: the inner cell mass will become the embryo proper, and the trophoblast will become the fetal placenta. The human trophoblast invades the endometrium during implantation and secretes hCG (Fig. 2-14), which is a placental homologue of LH. Therefore when the pituitary LH level falls, LH is replaced by a placental gonadotropin, which continues to stimulate the corpus luteum to produce progesterone. This prevents menstruation and maintains the pregnancy. Eventually, the fetoplacental unit will take over all steroid synthesis. In fact, if the mother undergoes an ovariectomy after 30 days of gestation or a hypophysectomy after 12 weeks, the pregnancy still continues normally.

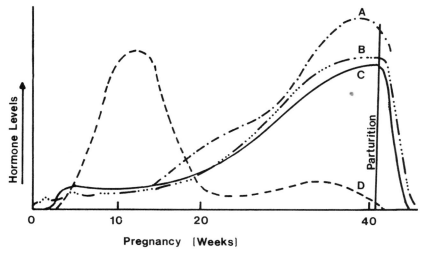

Fig. 2-14. Hormone levels during pregnancy. The hormones depicted are (A) human placental lactogen, (B) estrogen, (C) progesterone, and (D) human chorionic gonadotropin (hCG).

The major sources of steroids during pregnancy are the placenta, the maternal adrenal cortex, and the fetal adrenal cortex and liver. The fetal adrenal cortex is located between the adult cortex and the medulla; it makes up a large portion of the fetal adrenal gland but disappears by 1 year postpartum. The interrelationships among these sources are depicted in Fig. 2-15.

Although the placenta is the immediate source of the sex steroids during pregnancy, it lacks certain enzymes and relies heavily on other tissues for appropriate precursors. For example, it lacks hydroxymethylglutaryl coenzyme A reductase, the rate-limiting enzyme in cholesterol synthesis. The placental cholesterol comes from the mother, who also supplies at least 50% of the cholesterol for the fetal adrenal. From cholesterol, the placenta can synthesize progesterone, but it cannot go beyond this step because it lacks 17α-hydroxylase. This second enzyme deficiency is bypassed if both the fetal and maternal adrenal glands supply the placenta with DHEA. The placenta, which has very high aromatase activity, converts DHEA to estrone and E_2. The fetal adrenal gland also sulfates the DHEA; the resulting steroid is hydroxylated in the fetal liver and aromatized in the placenta to form estriol.

The function of progesterone is to prevent abortion by maintaining the endometrium. However, the functions of the other hormones, especially at the high concentrations found in the serum, are unknown. Indeed, human beings are among the very few species that have high estrogen levels during pregnancy. The placenta also secretes large amounts of human placental lactogen (hPL), a member of the GH-PRL family, but its role is likewise unclear because patients with hPL gene deletions and no serum hPL levels can still have normal pregnancies.

Near parturition, the corpus luteum secretes relaxin, a member of the insulin family. Relaxin induces the dissolution of collagen to relax the pubic symphysis and dilate the cervix; these changes will facilitate the expulsion of the fetus at birth. In some species, it also inhibits the myometrium until parturition arrives. In the male, relaxin is synthesized in the prostate gland, but no function has been associated with it.

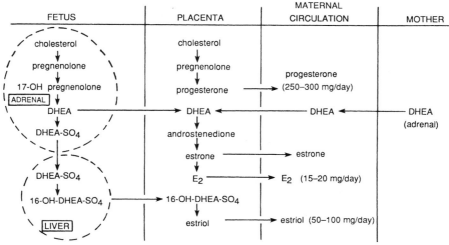

Fig. 2-15. Interrelationships among the placenta, fetus, and mother with respect to steroidogenesis during pregnancy. *DHEA*, Dehydroepiandrosterone; E_2, estradiol.

Gastrointestinal Hormones

It is the function of the gastrointestinal (GI) tract to digest and absorb ingested nutrients. This act requires the close coordination of the entire GI system, and much of this coordination is provided by hormones. Furthermore, because eating is intermittent, the body must be prepared to handle periodic surges of nutrients from the GI tract, and again, hormones are the major regulators. Finally, it is only appropriate that some of these hormones also control metabolism during times of fasting. This section is divided into three parts: the first deals with the major hormones of the pancreas and the metabolic shifts that they induce throughout the body, the second part concerns the hormones produced by adipose tissue, and the third part focuses on the GI tract itself and examines the local endocrine control of digestion.

Pancreas

The pancreas is both an endocrine and an exocrine gland. This chapter is concerned only with the endocrine function, which resides in nests of cells between the acini. These islets of Langerhans are histologically composed of several cell types, each occupying a specific position in the islet and each secreting a different hormone (Table 2-3).

Insulin

Although insulin consists of two peptides linked by disulfide bonds, it is originally synthesized as an 81-amino-acid precursor, proinsulin. The entire molecule is required to approximate the appropriate cysteines, and the disulfide bridges form while the prohormone is in the rough endoplasmic reticulum. Once this linkage occurs, the intervening piece, or C peptide, must be removed because proinsulin is only about 10% as active as insulin. Cleavage is accomplished by a trypsin-like protease attracted to two pairs of basic amino acids (Fig. 2-16). Cleavage begins when the molecule is in the Golgi apparatus and continues in the secretory granules, although it is never complete; about 6% of insulin is secreted as proinsulin. Finally, zinc is transported into the granules and triggers the crystallization of insulin as dimers and hexamers.

For the control of insulin secretion to be understood, it is necessary to discuss briefly its actions. Insulin is an anabolic hormone; that is, it is involved with energy storage. Primarily, this action is manifested as an increase in glucose and amino acid transport into cells and as the conversion of these precursors into storage forms, such as glycogen, protein, and triacylglycerides. Therefore elevated blood levels of glucose, fatty acids, or amino acids will stimulate insulin

Table 2-3
Histochemical Characteristics of Some Pancreatic Hormones

Hormone	Location	Structure
Glucagon	A (α) cells: outermost rim of islets	29-Amino-acid, linear polypeptide
Insulin	B (β) cells: core of islet	Two chains (21 and 30 amino acids) connected by disulfide bonds
SRIF	D (δ) cells: inner rim of islet	14-Amino-acid, cyclic peptide (via disulfide bond)
Pancreatic polypeptide	F cells	36-Amino acid, linear polypeptide

SRIF, Somatotropin (GH) release-inhibiting factor.

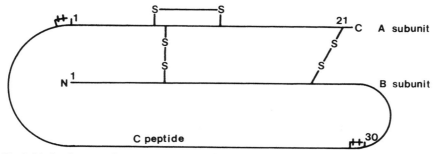

Fig. 2-16. A schematic diagram of the proinsulin molecule showing the A, B, and C peptides, as well as the pairs of basic residues that are attacked by proteases during the conversion of proinsulin to insulin.

release. However, this would result in the secretion of insulin after the substrate levels had already risen in the blood. For these blood concentrations to be regulated more smoothly, several anticipatory signals have been developed. For example, GI tract motility and secretion are stimulated by the parasympathetic nervous system (see the section on the adrenal medulla under Adrenal Glands), whose activity means that the GI tract is actively digesting food and that nutrients will soon be absorbed into the blood stream. As a result, Ach, which is the distal neurotransmitter of the parasympathetic nervous system, stimulates insulin secretion. Finally, the presence of carbohydrate in the intestines results in the release of gastric inhibitory peptide, which is also an insulin secretagogue.

Mechanistically, the actions of insulin are complex. Some of the effects merely involve blocking or reversing the actions of other hormones. For example, several of the cAMP-dependent hormones stimulate glycogenolysis and lipolysis by phosphorylating critical enzymes in these pathways (see also Chapter 12). However, insulin promotes the storage of glucose and fatty acids and inhibits the breakdown of glycogen and triacylglycerides. It does so, in part, by activating specific phosphatases that remove the phosphate from these enzymes; that is, it reverses the phosphorylation induced by the cAMP-dependent protein kinase.

Insulin can also affect metabolism by altering substrate flow. For example, insulin stimulates the uptake of glucose within cells and the resulting high levels of glucose-6-phosphate allosterically activate glycogen synthase (Fig. 2-17). In another example, insulin stimulates glycolysis; but eventually the cell becomes sated with ATP, which shuts down the tricarboxylic acid cycle by allosterically inhibiting several important enzymes, including isocitrate dehydrogenase. As a result, citrate begins to accumulate and will allosterically activate the committed step in fatty acid synthesis.

Insulin can also have more direct effects; for example, the immediate stimulation of metabolite transport and certain enzymes is mediated by second messengers (see Chapter 12). Finally, the delayed activation of other enzymes occurs through gene induction.

Glucagon

Glucagon is considerably simpler to discuss. It is a 29-amino-acid, linear peptide, whose major action is to elevate blood glucose levels; therefore it is catabolic and antagonizes the actions of insulin. More specifically, it stimulates liver glycogenolysis and inhibits glycolysis; it is aided by epinephrine, which

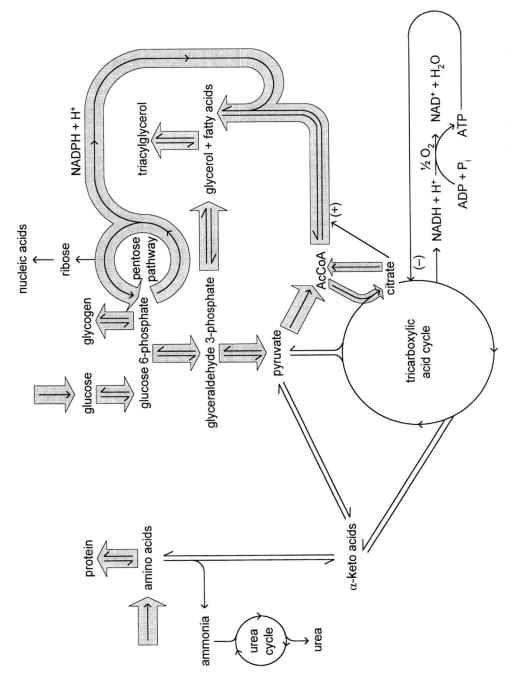

Fig. 2-17. A schematic representation of the biochemical actions of insulin. The arrow overlays indicate substrate flow during insulin stimulation.

triggers glycogenolysis in muscle and lipolysis in adipose tissue. Both hormones act through a cAMP-dependent protein kinase, which phosphorylates critical enzymes in the above pathways (see also Chapter 12). Cortisol also plays a complementary role by stimulating gluconeogenesis in the liver and providing this organ with amino acid metabolites. At low concentrations, cortisol inhibits protein synthesis; at higher concentrations, it promotes protein breakdown. The resulting amino acids are deaminated and the ammonia is detoxified in the urea cycle. Except for its effects on protein synthesis, all of these processes are accomplished by enzyme induction via steroid receptors. The effects of all these hormones are listed in Table 2-4 and schematically depicted in Fig. 2-18. It should be noted that Table 2-4 only summarizes the major actions of each hormone so as to emphasize the complementary nature of these hormones. In truth, each hormone has a wide variety of overlapping activities; for example, epinephrine can stimulate hepatic glycogenolysis, although not to the same extent as glucagon. Similarly, glucocorticoids can induce lipolysis and glucagon can promote hepatic proteolysis.

Table 2-4
Biochemical Effects of Some Metabolic Hormones

Metabolic pathway	cAMP	Cortisol	Insulin
Glycolysis	↓ Fructo-2,6-phosphate kinase		↑ Glucokinase
			↑ Fructo-2,6-phosphate kinase
	↑ Pyruvate kinase		↓ Pyruvate kinase
Gluconeogenesis		↑ Glucose-6-phosphate phosphatase	↓ Glucose-6-phosphate phosphatase
		↑ Fructose-1,6-phosphate phosphatase	↑ Glucose transport
		↑ Pyruvate carboxylase	
		↑ PEP carboxykinase	
Glycogen	↓ Glycogen synthase		↑ Glycogen synthase
	↑ Phosphorylase kinase, etc.		↑ Phosphoprotein phosphatase, etc.
Pentose cycle			↑ Glucose-6-phosphate dehydrogenase
			↑ 6-Phosphogluconate dehydrogenase
Fatty acids	↑ Triacylglycerol lipase		↑ ATP-citrate lyase
	↓ AcCoA carboxylase		↑ Phosphoprotein phosphatase
Protein synthesis		↑ Protein breakdown	↑ eIF-2B
			↑ eIF-4E
		↑ Amino acid catabolic enzymes[a]	↓ 4E-BP
			↑ S6
Urea cycle		↑ Argininosuccinate lyase	
		↑ Arginase	
		↑ Transaminases	

[a] For example, alanine and tyrosine aminotransferases and tryptophan oxygenase. *AcCoA*, Acetyl coenzyme A; *ATP*, adenosine 5′-triphosphate; *cAMP*, 3′,5′-cyclic AMP; *PEP*, phosphoenolpyruvate.

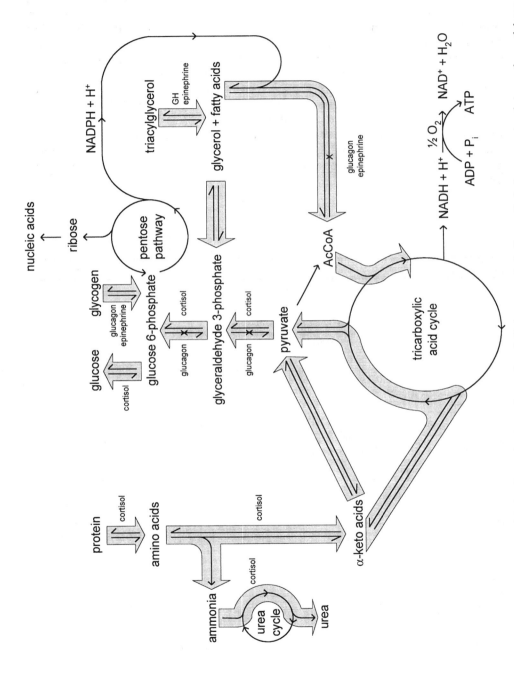

Fig. 2-18. A schematic representation of the biochemical actions of several catabolic hormones, including cortisol, glucagon, epinephrine, and growth hormone (GH). The arrow overlays indicate substrate flow during stimulation by these hormones.

Adipokines

At first, adipose tissue would seem to be an unlikely source of hormones. However, because this tissue represents the single largest reservoir of potential energy in the body, it is only appropriate that it should play an active role in the endocrine regulation of metabolism. The peptide hormones originating from this tissue are called *adipokines* and include resistin, tumor necrosis factor α (TNFα), leptin, adiponectin, and acylation stimulating protein (ASP).

Resistin and TNFα are elevated in obesity, and both induce insulin resistance. It was once thought that resistin represented negative feedback by insulin, but two of the major mediators of insulin action, the peroxisome proliferator-activated nuclear receptor γ (PPARγ) and protein kinase B, actually inhibit the expression of resistin.

On the other hand, leptin and adiponectin increase insulin sensitivity. Leptin is secreted by adipose tissue to signal the hypothalamus concerning the nutritional status and energy balance of the organism. For example, leptin levels are elevated in obesity and are low during starvation. Its major actions include the inhibition of food intake, the promotion of energy expenditure, and the increase in insulin sensitivity. It may play an ancillary role in reproductive and immune functions: both are impaired during starvation and improved after leptin administration. Adiponectin is homologous to complement factor C1q; it induces the genes for enzymes involved in fatty acid oxidation and energy expenditure, and it increases insulin sensitivity. Adiponectin levels are decreased in obesity, and secretion is stimulated by PPARγ agonists and inhibited by catecholamines, cortisol, and TNFα.

ASP is derived from complement C3, and it functions to clear serum lipid by stimulating triacylglyceride synthesis and inhibiting the lipase. Its secretion is stimulated by insulin and chylomicrons.

Local Gastrointestinal Hormones

Several hormones are secreted by the GI tract in response to nutrients or acidity. Their major function is the coordination of digestive secretions, GI tract motility, and visceral blood flow (Table 2-5).

For example, amino acids stimulate gastrin release, which in turn elevates gastric acid secretion to facilitate further proteolysis. If hydrogen ion concentrations become too high, gastrin secretion is inhibited. In another example, cholecystokinin (CCK), formerly called *pancreozymin*, is concerned with the digestion of triacylglycerides; therefore its secretion is stimulated by fatty acids. CCK, in turn, stimulates the secretion of pancreatic enzymes, including the pancreatic lipase. However, the effectiveness of the lipase alone is limited because triacylglycerides are not soluble in water. However, CCK also stimulates contraction of the gallbladder, which contains bile. Bile salts are amphipathic and thus are excellent detergents; they emulsify the fat, which then becomes more susceptible to the lipase. The action of CCK is facilitated by neurotensin (NT), which is also elevated in response to fat ingestion and which stimulates pancreatic secretions. One final example of the regulation of digestion is the case of secretin. When the stomach empties its acidic contents into the duodenum, the acid stimulates the release of secretin, which promotes bicarbonate secretion in bile and pancreatic fluids. The bicarbonate is used to neutralize the acid.

Table 2-5
Physiochemical and Physiological Characterization of Several Gastrointestinal Hormones

Hormone	Size (amino acids)	Location	Stimulators	Actions
Gastrin	17	Antrum (G cells)	Amino acids (H^+, inhibitor)	Stimulates HCl secretion; stimulates mucosal and pancreatic growth
Gastrin-releasing peptide (GRP)	27	Stomach and small intestine (K cells)	Vagus	Stimulates gastrin secretion and growth of gastrointestinal tract
Cholecystokinin (CCK)	33	Duodenum and jejunum (CCK cells)	Fatty acids	Stimulates gallbladder contraction and pancreatic enzyme secretion; inhibits sphincter of Oddi and gastric emptying
Neurotensin (NT)	14	Ileum (N cells)	Fatty acids	Decreases gastric emptying; increases insulin and pancreatic secretions; vasodilation
Secretin	27	Duodenum (S cells)	Acid	Stimulates pancreatic and bile secretion of bicarbonate
Gastric inhibitory peptide (GIP)	43	Duodenum and jejunum (I cells)	Carbohydrate	Stimulates insulin secretion
Glucagon-like peptide (GLP)	37	Intestine (L cells)	Mixed meal	Stimulates insulin and inhibits glucagon secretion
Vasoactive intestinal peptide (VIP)	28	Nerve endings in jejunum and colon	Nerve stimulation	Visceral vasodilator; inhibits the smooth muscle of GI and genitourinary tract; general hormone releaser (gut, pancreas, and pituitary gland)
PP-fold family	36	Gut (P cells)	Vagus (acute); amino acids	Vasoconstriction; inhibition of pancreatic and intestinal secretions
Substance P	11	Gut (ECL cells)	Mixed meal	Vasodilator and sialogogue; stimulates GI motility
Motilin	22	Duodenum (ECL cells)	Unknown	Stimulates GI motility in the fasting state
Somatostatin (SRIF)	14	Pancreas and intestine (D cells)	Unknown	General hormone release inhibitor (gut, pancreas, and pituitary gland)

ECL, Enterochromaffin-like; *HCl*, Hydrochloric acid; *GI*, gastrointestinal; *PP*, pancreatic polypeptide.

Digestion involves the synthesis and secretion of enzymes, muscular activity, and active transport. To support this high metabolic rate, vasoactive intestinal peptide (VIP), NT, and substance P dilate the mesenteric circulatory system. This increased blood flow also helps to dilute the absorbed nutrients, which are present in the hepatic portal system at high concentrations. These two hormones, along with motilin, also regulate GI motility. On the other hand, as digestion proceeds toward completion, the GI system must be shutdown. The members of the PP–fold family are potent vasoconstrictors and inhibitors of pancreatic and intestinal secretion.

Finally, there is coordination among the hormones themselves. The role of gastric inhibitory peptide (GIP) in alerting the pancreas that carbohydrate is in the gut has already been mentioned. Gastrin-releasing peptide (GRP), a member of the bombesin peptide family, is another anticipatory signal. The sight or odor of food activates the parasympathetic nervous system, which then triggers the release of GRP via the vagus nerve. Somatostatin, or somatotropin release-inhibiting factor (SRIF), which was first discovered as an inhibitor of GH release, also inhibits the release of many other hormones. It is synthesized in the D cells of the islets, where it affects the secretion of insulin and glucagon. On the other hand, VIP is a general stimulator of hormone release.

Many of these hormones are members of one of four families. Every hormone in the gastrin family has the same carboxy-terminal pentapeptide (-Gly-Trp-Met-Asp-Phe-NH_2). All the activity of gastrin is found in this pentapeptide, whereas the actions of CCK can be fully mimicked by its carboxy-terminal octapeptide. Cerulein, a decapeptide found in the skin and GI tract of amphibians, is also a member of this family. The tachykinin family includes substance P and neurokinins A and B. They only have about a dozen amino acids, and all end with a carboxy-terminal sequence of -Phe-X-Gly-Leu-Met-NH_2, where X represents any amino acid. In contrast, the members of the PP-fold and secretin families require the entire molecule for activity. The PP-fold family includes the pancreatic polypeptide (PP), peptide YY (PYY), and neuropeptide Y (NPY); the family name arises from the three-dimensional structure of these hormones. The secretin family includes secretin, glucagon, GLP, VIP, GIP, and GHRF.

References

General References

DeGroot, L.J., and Jameson, J.L. (eds.) (2001). *Endocrinology*. 4th ed. Saunders, Philadelphia, Pennsylvania.

Griffin, J.E., and Ojeda, S.R. (eds.) (2000). *Textbook of Endocrine Physiology*. 4th ed. Oxford University Press, Oxford.

Hadley, M.E. (2000). *Endocrinology*. 5th ed. Prentice Hall, Englewood Cliffs, New Jersey.

Henry, H., and Norman, A.W. (2003). *Encyclopedia of Hormones*. Academic Press, San Diego, California.

Larsen, P.R., Kronenberg, H.M., Melmed, S., and Polonsky, K.S. (2003). *Textbook of Endocrinology*. 10th ed. Saunders, Philadelphia, Pennsylvania.

Norris, D.O. (1997). *Vertebrate Endocrinology*. 3rd ed. Academic Press, San Diego, California.

Hypothalamus and Pituitary Gland

Freeman, M.E., Kanyicska, B., Lerant, A., and Nagy, G. (2000). Prolactin structure, function, and regulation of secretion. *Physiol. Rev.* 80, 1523-1631.

Gantz, I., and Fong, T.M. (2003). The melanocortin system. *Am. J. Physiol.* 284, E468-E474.

Goffin, V., Shiverick, K.T., Kelly, P.A., and Martial, J.A. (1996). Sequence-function relationships within the expanding family of prolactin, growth hormone, placental lactogen, and related proteins in mammals. *Endocr. Rev.* 17, 385-410.

Soares, M.J., Müller, H., Orwig, K.E., Peters, T.J., and Dai, G. (1998). The uteroplacental prolactin family and pregnancy. *Biol. Reprod.* 58, 273-284.

Yu-Lee, L. (1997). Molecular actions of prolactin in the immune system. *Proc. Soc. Exp. Biol. Med.* 215, 35-52.

Adrenal Glands

Beltowski, J. (2001). Guanylin and related peptides. *J. Physiol. Pharmacol.* 52, 351-375.

Forte, L.R. (1999). Guanylin regulatory peptides: Structures, biological activities mediated by cyclic GMP and pathobiology. *Regul. Pept.* 81, 25-39.

Hinson, J.P., Kapas, S., and Smith, D.M. (2000). Adrenomedullin, a multifunctional regulatory peptide. *Endocr. Rev.* 21, 138-167.

Stocco, D.M. (2001). StAR protein and the regulation of steroid hormone biosynthesis. *Annu. Rev. Physiol.* 63, 193-213.

Thyroid Gland

Bianco, A.C., Salvatore, D., Gereben, B., Berry, M.J., and Larsen, P.R. (2002). Biochemistry, cellular and molecular biology, and physiological roles of the iodothyronine selenodeiodinases. *Endocr. Rev.* 23, 38-89.

Reproduction

Nef, S., and Parada, L.F. (2000). Hormones in male sexual development. *Genes Dev.* 14, 3075-3086.

Riggs, B.L., Khosla, S., and Melton, L.J. (2002). Sex steroids and the construction and conservation of the adult skeleton. *Endocr. Rev.* 23, 279-302.

Welt, C., Sidis, Y., Keutmann, H., and Schneyer, A. (2002). Activins, inhibins, and follistatins: From endocrinology to signaling. A paradigm for the new millennium. *Exp. Biol. Med.* 227, 724-752.

Gastrointestinal Hormones

Ahima, R.S., and Flier, J.S. (2000). Leptin. *Annu. Rev. Physiol.* 62, 413-437.

Cianflone, K., Xia, Z., and Chen, L.Y. (2003). Critical review of acylation-stimulating protein physiology in humans and rodents. *Biochem. Biophys. Acta* 1609, 127-143.

Greeley, G.H. (ed.) (1999). *Gastrointestinal Endocrinology.* Humana Press, Totowa, New Jersey.

Havel, P.J. (2002). Control of energy homeostasis and insulin action by adipocyte hormones: Leptin, acylation stimulating protein, and adiponectin. *Curr. Opin. Lipid.* 13, 51-59.

Jiang, G., and Zhang, B.B. (2003). Glucagon and regulation of glucose metabolism. *Am. J. Physiol.* 284, E671-E678.

Prins, J.B. (2002). Adipose tissue as an endocrine organ. *Best Pract. Res. Clin. Endocrinol. Metab.* 16, 639-651.

Stefan, N., and Stumvoll, M. (2002). Adiponectin-Its role in metabolism and beyond. *Horm. Metab. Res.* 34, 469-474.

Nonclassical Endocrinology

CHAPTER OUTLINE

In the preceding chapters, chemicals closely fitting the classical definition of a hormone were described. In this chapter, the "renegades" are discussed. These hormones are frequently synthesized at multiple sites and may act locally. However, their most bewildering characteristics are their extensively overlapping biological activities, their large repertoire of effects, and their occasional contradictory activities. All of these properties are probably related to the fact that most of these hormones are growth factors: all hormones stimulating mitosis are going to activate a number of common pathways; therefore some overlap in activity is inevitable. Furthermore, growth is a complex process requiring coordination with metabolism, which will supply the materials and energy for growth, and with reproduction, which will be its chief competitor for these limited nutrients. As such, these hormones will have a wide range of effects.

Finally, many of these hormones can exert opposite activities, such as growth promotion and inhibition. These phenomena can be explained if one assumes that the action of a hormone will be influenced by the history and environment of its target. All growth is not the same; for example, the growth manifested during embryogenesis is very different than that seen during simple enlargement or that required for the repair of tissue damage. In addition, it must be remembered that hormones usually have no intrinsic activity; only a few are known to be enzymes and none are transport proteins. Their effect is actually the cellular response to their presence. This response is not stereotyped; the cell integrates all of the information it has received and then initiates action appropriate to the circumstances.

Invertebrate and plant hormones are also discussed here because many of them are somewhat atypical with respect to the classical definition of a hormone. In addition, their coverage will help prepare the reader for the chapter on the evolution of the endocrine system (Chapter 18). This discussion does not cover every growth factor or nonvertebrate hormone; rather, it is designed to present a representative sample of the various groups and to concentrate on those hormones that are discussed later in this text.

Growth Factors

Cell Cycle

Actively dividing cells undergo a repetitive sequence of events known as the *cell cycle*. This cycle is divided into four stages (Fig. 3-1): M, G_1, S, and G_2. M and S are easy to recognize. *M* is mitosis, when the cell is visibly undergoing cell division. The mitotic stages, prophase through telophase, can be followed by light microscopy. *S* is when DNA synthesis is occurring and has no visible correlate; however, it can be detected with radiolabeled thymidine incorporation. G_1 and G_2 are the gaps between M and S and between S and M, respectively. Unfortunately, these gaps are still "black boxes." Although the actions of various growth factors can be mapped within these stages, it is not always clear exactly what processes they are affecting.

Most growth factors act in G_1 to speed the advance of an already actively dividing cell toward S, but they often have no effect on nondividing cells. These factors are known as *hormones of progression*. Other growth factors can make nondividing cells sensitive to the hormones of progression, although they themselves cannot actually initiate cell division. These latter factors are called *hormones of competency,* and their requirement has led to the speculation that nondividing

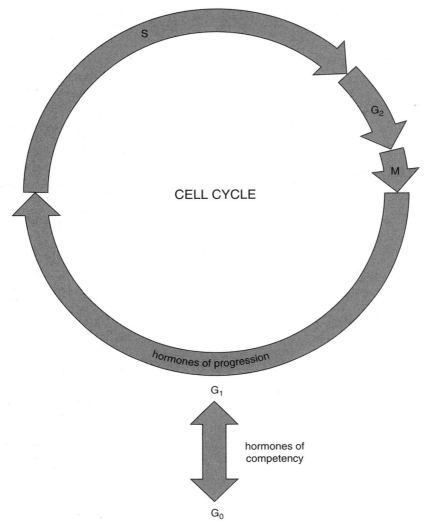

Fig. 3-1. The cell cycle.

cells are not simply stalled in G_1 but actually lie outside of the cell cycle in a stage known as G_0. They must first be brought back into the cycle by the hormones of competency before they can be stimulated by the hormones of progression.

General Growth Factors

Epidermal Growth Factor Family

The epidermal growth factor (EGF) family consists of EGF, transforming growth factor α (TGFα), epiregulin, amphiregulin, heparin-binding EGF-like growth factor (HB-EGF), Cripto-1, betacellulin, schwannoma-derived growth factor (SDGF) (also called acetylcholine receptor inducing activity [ARIA]), neuregulins, heregulin (also called Neu differentiation factor [NDF]), glial-derived growth factor (GGF-II), and sensory and motor neuron-derived growth factor (SMDF) (Table 3-1). Both EGF and TGFα are synthesized as membrane-bound

Table 3-1
Some Growth and Growth-Inhibiting Factors in Vertebrates

Hormone	Source	Action
EGF Family: Single Peptide Chain		
EGF	Salivary glands	Epithelial proliferation; anchorage independence; neovascularization
TGFα	Many	As above
NDF	Many	Mammary epithelial proliferation
Epiregulin	Many	Fibroblast proliferation
ARIA	Schwann cells	Induce ACh receptor
TGFβ Family: Identical or Homologous Dimers		
TGFβs	Ubiquitous	Inhibits proliferation of epithelial and immune cells; induces mesoderm
Müllerian inhibiting substance	Mammalian testes	Regression of müllerian ducts in males
Inhibins/activins	Gonads and hypothalamus	Inhibits FSH secretion; spermatogenesis
Bone morphogenetic protein	Many	Osteogenesis; bone repair; embryonic axis establishment; nervous system gradients
PDGF Family: Identical or Homologous Dimers		
PDGF	Platelets	Proliferation of connective tissue; tissue repair
VEGF	Neural tissues and vascular smooth muscle	Vascular endothelial mitogen; increases vascular permeability
Insulin Family: Intact or Cleaved Peptide Chain		
IGF-I	Many	GH-dependent cartilage growth; paracrine growth factor
IGF-II	Many	Fetal growth
Relaxin	Ovary	Inhibits uterine contractions, elongates interpubic ligament, and softens cervix
INSL3	Testis	Testicular descent
FGF Family: Single Peptide Chain		
FGF	Many	Development of the nervous, musculoskeletal, integumentary, cardiovascular, and respiratory systems
KGF (FGF-7)	Stroma of epithelia	Epithelial cell mitogen
NGF Family: Peptides		
NGF	Brain, heart, and spleen	Neuronal survival, outgrowth, and differentiation; especially, dorsal root and paravertebral ganglia
BDNF	Brain and heart	As above, especially, dorsal root and nodose ganglia
NT-3	Brain, heart, kidney, liver, and thymus	As above, including all three ganglia
Hematopoietic Growth Factors: Glycoproteins (Not Homologous)		
Erythropoietin	Kidney	Proliferation of RBC progenitor cell
CSFs	Endothelium, T cells, fibroblasts and macrophages	Proliferation of WBC progenitor cells

continues

continued

Hormone	Source	Action
IL-3	Helper T cell	Proliferation of multi-lineage colonies
SCF	Many	Proliferation of hematopoietic (especially mast cells), germ cells and melanoblasts
Thrombopoietin	Liver	Proliferation of platelet progenitor
Miscellaneous Factors		
Retinoic acid	Diet (stored in liver)	Epithelial differentiation

ARIA, Acetylcholine receptor inducing activity (also called Schwannoma-derived growth factor [SDGF]); *BDNF*, brain-derived neurotrophic factor; *CSFs*, colony-stimulating factors; *EGF*, epidermal growth factor; *FGF*, fibroblast growth factor; *IGF*, insulin-like growth factor; *IL*, interleukin; *INSL3*, insulin-like 3 protein; *KGF*, keratinocyte growth factor; *NDF*, Neu differentiation factor; *NGF*, nerve growth factor; *NT-3*, neurotrophin-3; *PDGF*, platelet-derived growth factor; *SCF*, stem cell factor (also called Kit ligand [KL], mast cell growth factor [MGF], or steel factor [SLF]); *TGF*, transforming growth factor; *VEGF*, vascular endothelial growth factor.

precursors that are cleaved to yield 6-kDa soluble hormones. However, both precursors are active, and the EGF precursor appears to persist in some tissues, such as the mammary gland, where this form is required for directional growth of the epithelium. Therefore EGF may have a juxtacrine, as well as an endocrine, mode of action. Both EGF and TGFα bind to the same receptor and have very similar activity. Less is known about the other members of this family; many appear to be tissue-restricted growth factors. For example, NDF stimulates the proliferation of mammary epithelium; epiregulin stimulates fibroblast proliferation; and SDGF, GGF-II, and SMDF induce neuromuscular development. Both EGF and TGFα are hormones of progression that act early in G$_1$ to stimulate epithelial cell proliferation. They can also stimulate anchorage-independent growth and neovascularization, which is the formation of new blood vessels to supply the expanding tissue.

Transforming Growth Factor β Family

The transforming growth factor β (TGFβ) family is part of the cysteine knot superfamily, which is characterized by a conserved three-dimensional structure that includes interlocking disulfide bonds. TGFβ is a 25-kDa homodimer that forms the carboxy terminus of a larger precursor. After cleavage, the homodimer remains noncovalently attached to the amino-terminal halves as an inactive tetramer. It is eventually liberated by the protease, plasmin, which is elevated during injury; this association has led to the speculation that TGFβ may play a role in wound repair.

TGFβ and the other members of this family are interesting for their ability to inhibit, as well as stimulate, growth. In addition, many members have important functions in embryogenesis. TGFβ itself inhibits the proliferation of most epithelia and immune cells by arresting them in G$_1$. However, it has two other effects that can reverse this inhibition: first, it can stimulate the synthesis and secretion of growth factors, which along with TGFβ can induce mesoderm and blood vessel formation and stimulate the proliferation of keratinocytes and intestinal epithelium. In part, this switch can be regulated by TGFβ concentration; low levels favor growth, whereas high TGFβ concentrations inhibit growth. Second, it can also promote the production and deposition of extracellular matrix, which can further promote mitogenesis.

Antimüllerian hormone (AMH), also called *müllerian inhibiting substance* (MIS), is a 70-72-kDa homodimer, whose carboxy terminus exhibits homology to TGFβ. In males, it is produced by the testes during embryogenesis where it thwarts the development of the müllerian duct derivatives, including the uterus and oviducts. In persistent müllerian duct syndrome, this hormone is absent, and male pseudohermaphrodism results. Such males appear to be normally virilized with the sole exception of possessing uteri and oviducts that are usually discovered during an incidental operation.

The inhibins and activins were covered in Chapter 2 in the section on Reproduction, where their role in regulating follicle-stimulating hormone (FSH) secretion was discussed. However, these hormones are also active in mesoderm induction, as well as in the embryonic formation of axial and dorsal structures, such as the notochord, neural tube, and segmented somites.

Myostatin restricts muscle growth during embryogenesis. Certain cattle breeds have mutations in this hormone and exhibit marked hyperplasia of their musculature (see Chapter 17).

There are more than 30 bone morphogenetic proteins (BMPs) in this family. Although initially named for their activity in bone formation, they have effects in many tissues and are important in development. For example, in embryogenesis BMPs establish the ventral-dorsal axis and determine cell lineage and positional gradients in the central and peripheral nervous systems. In reproduction, BMP8A is required for maintenance of spermatogenesis, and BMP15 is involved with oocyte maturation (see Chapter 2). The cartilage-derived morphogenetic proteins are most closely related to BMP5 and BMP6 and are restricted to cartilage.

Platelet-Derived Growth Factor Family

The platelet-derived growth factor (PDGF) family is another member of the cysteine knot superfamily. Like inhibin/activin, PDGF is composed of two homologous subunits that can form either homodimers or heterodimers. The A subunit is 12 kDa, whereas the B subunit is 18 kDa. The B subunit has a carboxy-terminal extension that electrostatically interacts with the plasma membrane, where it remains bound. As such, it resembles the EGF precursor. This membrane-binding region is also present in the A subunit, but it is proteolytically removed during processing; as a result, PDGF A homodimers are secreted and not retained on the cell surface. The ratio of these isoforms depends on the species and the tissue being examined; for example, in human platelets, the AB isoform represents 70% of PDGF, whereas the BB isoform accounts for the remainder. The PDGF receptor also has two isoforms that form homodimers and heterodimers and that show relative specificity for different PDGF isoforms (see Chapter 7). PDGF is involved with chemotaxis and the proliferation of connective tissue, especially in wound repair. PDGF is a hormone of competency.

Vascular endothelial growth factor (VEGF) is distantly related to PDGF and is a dimer of apparently identical 23-kDa subunits. It is produced by a number of neural cell types, as well as by vascular smooth muscle. It stimulates mitogenesis in vascular endothelium and increases vascular permeability. The latter action probably allows new matrix precursors to enter tissues to support angiogenesis. Its site of action in the cell cycle has not yet been investigated.

Fibroblast Growth Factor Family

The fibroblast growth factor (FGF) family has more than 20 members and is closely associated with body form. Indeed, duplications in the genes for FGF and its receptor correspond to major evolutionary changes in body structure, such as the appearance of limbs. Various FGFs have been implicated in the development of the nervous, musculoskeletal, integumentary, cardiovascular, and respiratory systems, as well as in the formation of the limbs, inner ear, and eye lens. Like PDGF, it is a hormone of competency. Some members of this family are bound to the cell surface, presumably by their cationic carboxy termini, in a manner similar to PDGF.

Insulin Family

Insulin is a metabolic hormone that was covered in Chapter 2 in the section on Gastrointestinal Hormones. However, this family has several other members that are better known for their growth-promoting activity. Insulin-like growth factors-I and -II (IGF-I and IGF-II) are both insulin homologues whose C peptide has been reduced in length and retained. They are both found in the serum associated with six known IGF binding proteins (IGFBPs), which serve several functions (Table 3-2). First, because of the small size of IGF, binding to IGFBPs can prolong the half-life of IGF. Second, IGFBPs can act as a reservoir for IGF, especially when bound to the extracellular matrix. Third, they can control IGF activity; for example, they can inhibit IGF by sequestering it, potentiate IGF by concentrating it for the IGF receptor, and control its release, usually by the phosphorylation or proteolysis of the IGFBP. Finally, IGFBPs can have activity independent of IGF. For example, membrane receptors that have been identified for IGFBP-1 and IGFBP-3 may mediate some of these activities. In addition, both IGFBP-3 and IGFBP-5 have nuclear localization signals and may have functions in the nucleus.

It was originally thought that the liver was the predominant source of IGF-I; however, the major effects of IGF-I are mediated by local concentrations. IGF-I is produced in many tissues under the control of tropic hormones to promote local growth and/or differentiation, for example, in leukocytes and the mammary gland by growth hormone; in bone by parathormone and sex steroids; in adrenocortical cells by adrenocorticotropin hormone (ACTH); in testis by LH and FSH; in granulosa cells by LH, FSH, and estradiol; in uteri by estradiol and progesterone; and in fibroblasts by EGF, FGF, and PDGF. However, the liver is the major source of circulating IGF-I, which is required for its effects on carbohydrate and lipid metabolism and for feedback inhibition. IGF-II appears to be important for fetal growth. Both factors are hormones of progression and act late in G_1.

There are several other members of the insulin family; many are involved with reproduction. Relaxin was discussed in Chapter 2 in the Reproduction section; however, it also affects the cardiovascular system. It is synthesized in the heart and is stimulated by increased ventricular filling pressure. It reduces this pressure by inducing vasodilation through the inhibition of endothelin and angiotensin and by promoting diuresis through the stimulation of the atrial natriuretic peptide (ANP). It also impairs clot formation by increasing the plasminogen activator. The insulin-like 3 protein was also discussed in Chapter 2 and promotes testicular descent.

Nerve Growth Factor Family

The nerve growth factor (NGF) family also belongs to the cysteine knot superfamily and consists of NGF, brain-derived neurotrophic factor (BDNF),

Table 3-2

Functions and Regulation of Some Insulin-like Growth Factor-Binding Proteins

IGFBP	IGF-I function	IGF-I-independent function	Regulation
IGFBP-1	Minor serum carrier; may shuttle IGF-I between serum and tissue	Cell migration via integrin binding	Phosphorylation increases IGF-I binding and sequestration Dephosphorylation or proteolysis releases IGF-I
IGFBP-2	Significant serum carrier; inhibits IGF-I		GH repressible
IGFBP-3	Major serum carrier; inhibits IGF-I (if bound) or potentiates (if released)	Inhibits growth in breast cancer cells and stimulates apoptosis in prostate cancer cells	IGF-I and GH inducible Proteolysis releases IGF-I CK2 phosphorylation increases affinity for ALS[a]; increases IGFBP-3 half-life and inhibits IGF-I by sequestration Binding to cell surface receptor decreases affinity for IGF-I
IGFBP-4	Minimal serum carrier; inhibits IGF-I		IGF-II induces IGFBP-5 protease to release IGF
IGFBP-5	Minimal serum carrier	Stimulates osteoblast proliferation; may act as nuclear shuttle	Binding to extracellular matrix releases IGF-I Proteolysis releases IGF-I

[a] The acid-labile subunit (ALS) binds to IGF-I and IGFBP3 to form a ternary complex. CK2, Casein kinase 2; GH, growth hormone; IGF, insulin-like growth factor; IGFBP, insulin-like growth factor binding protein.

and neurotrophin-3 (NT-3). All three peptides are synthesized as a precursor that is cleaved to generate an approximately 120-amino-acid hormone; the sequences of the mature peptides are about 50% identical to each other. Their biological activities also significantly overlap; all three hormones are involved with the survival, outgrowth, and differentiation of neurons. Their major differences are found in their target tissues, and their sites of synthesis. NGF acts on the dorsal root and paravertebral sympathetic ganglia and is made in many tissues, especially the brain, heart, and spleen. BDNF affects the dorsal root and nodose ganglia and has a more restricted distribution; BDNF is found in high levels in the brain and heart, with lesser concentrations in the lung and muscle. NT-3 stimulates all three ganglia and, like NGF, is synthesized in most tissues. However, the levels in these tissues are much higher, often equaling or exceeding those found in the brain. It was originally believed that biological activity was restricted to these mature hormones; however, both pro-NGF and pro-BDNF can activate programmed cell death, or apoptosis. As such, the processing of these precursors is a critical switch that determines whether the neuron will live or die.

A fourth neurotrophin, ciliary neurotrophic factor (CNTF), is functionally, but not structurally, related to the NGF family. As the name implies, it stimulates ciliary ganglion growth; it can also promote cholinergic and glial differentiation. Unlike the NGF family, it is not cleaved from a precursor. With its 200 amino acids, it is larger than NGF and it has a very restricted distribution; it is only found in abundance in the sciatic nerve, whereas the spinal cord has low levels.

Hematopoietic Growth Factors

All blood cells are generated from a single stem cell in the bone marrow, the hematocytoblast (Fig. 3-2). This, in turn, gives rise to five blast cells: the proerythroblast develops into the erythrocyte, or red blood cell, which is important in transporting oxygen. The myeloblast produces cells with secretory granules, the granulocytes; they include the neutrophil, eosinophil, and the basophil. The neutrophil phagocytizes bacteria, and the eosinophil is involved in allergic reactions and parasitic infections. The basophil, called a *mast cell* when it is found in tissues, contains histamine and heparin and is involved with allergic responses and blood clotting. The monoblast gives rise to monocytes, which are elevated during viral and tuberculous infections. Monocytes can also develop into macrophages, which remove debris from injured or infected tissues. The lymphoblast produces both the B lymphocytes, which produce antibodies, and the T lymphocytes, which are involved with cell-mediated immunity (see Immunoinflammatory Hormones later). The megakaryoblast will eventually give rise to a cell that fragments. These cellular pieces are called *thrombocytes* or *platelets* and form a critical part of the blood clotting system.

There are several unusual characteristics of the hematopoietic growth factors. First, they are not stored but are rapidly synthesized when they are needed. In addition, they are produced in many different cell types. Second, the actions of many cytokines are redundant and pleiotropic. Their specificity is determined, in part, by the fact that their synthesis and release is spatially and temporally restricted; the expression of their receptors is also limited. Finally, their activities are influenced by other factors, such as the growth state of the target cells, neighboring cells, the presence of other cytokines, their relative concentrations, and their temporal sequence. These characteristics make the analysis of cytokine action very difficult.

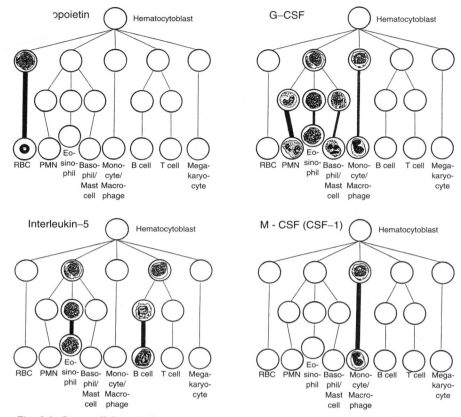

Fig. 3-2. Stem cell lines in hematopoiesis and the hormones that regulate them. Thick lines represent major effects; medium lines represent effects only observed at higher concentrations.

Erythropoietin and Thrombopoietin

Erythropoietin (EPO) is a 39-kDa glycoprotein secreted by the juxtaglomerular cells of the renal arterioles in response to low oxygen tension (hypoxia). It is also produced in astrocytes in response to hypoxia; astrocytes are nonneural support cells in the central nervous system. The kidney and brain are highly metabolic organs; as such, they are susceptible to damage from hypoxia. Under these circumstances, the juxtaglomerular cells and astrocytes release EPO, which stimulates erythrocyte progenitor cell proliferation, increases the hemoglobin content per red blood cell, and promotes the early release of erythrocytes from the marrow. It is the absence of this hormone that produces the severe anemia seen in end-stage kidney disease.

Thrombopoietin most closely resembles EPO, although it has an additional, glycosylated, carboxy-terminal extension. It is synthesized in the liver and stimulates megakaryocytopoiesis, which generates platelets.

Colony-Stimulating Factors

The colony-stimulating factors (CSFs) stimulate the proliferation of one or more leukocytic lines and are produced by a wide variety of cell types, including endothelium, fibroblasts, T lymphocytes, and macrophages, especially when these cells originate in the lungs, kidneys, spleen, or salivary glands.

The granulocyte–macrophage CSF (GM-CSF) is a glycoprotein with 118 to 127 amino acids having a molecular weight of 16 to 41 kDa, depending on

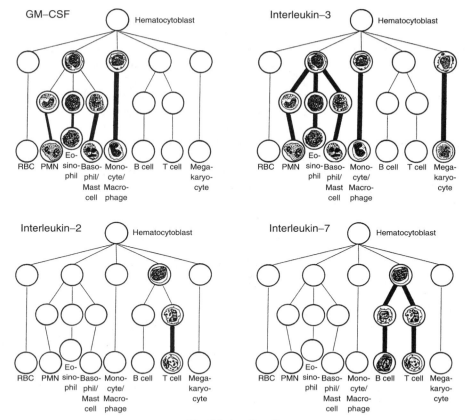

Fig. 3-2. *(continued).*

the species and the degree of glycosylation. The carbohydrate portion of the molecule is not necessary for biological activity, a fact that also applies to all of the other CSFs. It has the most wide-ranging effects: at low concentrations it selectively stimulates the proliferation of the monocyte-macrophage cell line, whereas at higher concentrations it affects all of the granulocytes, especially the basophils and mast cells.

The macrophage CSF (M-CSF or CSF-1) is more selective and only stimulates the monoblast cell line and adipocytes. It is a homodimer of 47 to 70 kDa, depending on glycosylation, and is initially synthesized as a membrane-bound precursor. Both the membrane-bound and the soluble forms enhance the survival, proliferation, and differentiation of monocytes and macrophages.

The effect of the granulocyte CSF (G-CSF) depends on the concentration of this 25- to 30-kDa glycoprotein. At low concentrations, it selectively stimulates the proliferation of the neutrophil progenitor cell; however, at higher concentrations, it can affect all of the granulocytes, as well as the monoblast lineage.

The last CSF was discovered by several groups working independently; unfortunately, each group gave it a different name and, as yet, no consensus has arisen concerning the nomenclature. The following are the synonyms with an explanation of each: Kit ligand (KL) (its receptor is the Kit gene product), mast cell growth factor (MGF), stem cell growth factor (SCF), and steel factor (SLF) (it complements the *steel* mutation in mice). This factor stimulates hematopoiesis, especially the mast cell line; but unlike other CSFs, it also promotes mitosis in other stem cells, such as germ cells and melanoblasts.

Interleukins

The interleukins are primarily concerned with the proliferation and differentiation of T and B lymphocytes and are therefore discussed later (see section on Immunoinflammatory Hormones). However, several of these hormones are important for other cell types. In fact, interleukin-3 (IL-3 or multi-CSF) does not have any effect at all on lymphocytes, although it stimulates all of the other white blood cells and even the megakaryoblast cell line. IL-5 does affect the B lymphocyte, but it also promotes the production of eosinophils. Growth hormone, prolactin, and leptin are also members of the cytokine family; they were covered in Chapter 2.

Structurally, these hormones fall into one of four groups. Class I cytokines are four α-helical bundles having an up-up-down-down arrangement and long loops between the first and last pair of helices, called *A-B* and *C-D*, respectively. Class I is further divided into the long- and short-chain cytokines. The former have approximately 25-amino-acid helices and include the IL-6 and IL-12 families, as well as growth hormone and prolactin. The latter have only 15-amino-acid helices and include the IL-2 family, IL-3, IL-5, and GM-CSF. Class II cytokines have the same general structure except that the type I hormones have a fifth helix in the C-D loop; the helices are 13-24 amino acids long, and helices A, B, C, and E form the classic four α-helical bundle. Type I cytokines include interferon α and β. Type II cytokines are dimers of six interdigitated helices; the helices are 6–18 amino acids long, and helices A, C, D, and F form the classic four α-helical bundle. Type II cytokines include interferon γ and the IL-10 family.

The IL-17 family is a cytokine by function but not structure. They have the same three-dimensional structure of the cysteine knot superfamily but without the signature cysteines. Their receptors do not resemble the classic cytokine receptors (see Chapter 7).

Immunoinflammatory Hormones

To understand the immunoinflammatory hormones (Table 3-3), one must first understand the immunoinflammatory reaction. In general, this reaction is two-fold: first, there is the production of antibodies, which are proteins having a high affinity for a foreign substance, the antigen. Because many antibodies are secreted into the blood, this type of response is known as *humoral immunity*. In another type of defense response, there is the generation of cytotoxicity, which is called *cell-mediated immunity*.

Humoral Immunity

Antibodies are modular proteins; each of the two heavy chains consists of four repeating domains, and each of the two light chains have two such repeating units (Fig. 3-3). The amino termini line up, and the first domain of all four chains forms two antigen-binding pockets, the immunoglobulin folds. The carboxy termini of the heavy chains determine antibody localization or trigger other immune response, such as complement fixation. There are five major classes of heavy chains, each with its own unique spectrum of immune activity.

There are only two genes for the light chains, and one for the heavy chains. The origin of the different antibody classes and antigen specificity lies in the genomic organization of these genes (Fig. 3-4). Each light chain gene contains one exon coding for the second domain, known as C_L for constant region. However, there are several hundred exons coding for the first domain, or V_L for variable

Table 3-3
Some Immunoinflammatory Hormones in Vertebrates

Hormone	Source	Action
IL-1	Macrophages	Stimulates helper T cells to produce IL-2; inflammatory
IL-2	Helper T cells	Stimulates T cell proliferation
IL-3	Helper T cells	Hematopoeisis; mast cell growth factor
IL-4	Helper T cells	Activates B cells; immunoglobulin class switching
IL-5	Helper T cells	Stimulates B cell proliferation; eosinophil growth and differentiation
IL-6	Helper T cells, fibroblasts, epithelium, and macrophages	Stimulates B cell maturation to antibody secreting cells; inflammatory
IL-7	Bone marrow stromal cells	B cell precursor and T cell proliferation
IL-8	Fibroblasts, epithelium, hepatocytes, and macrophages	Neutrophil chemotaxis
IL-9	T cells	Synergize with IL-2, IL-3, and IL-4
IL-10	Th2 and B cells; macrophages; skin	Antagonize IFNγ
IL-11	Fibroblasts, endothelium, and some blood lines	Synergize with IL-6; induce acute phase proteins; stimulate neurons and osteoclast differentiation; inhibit fat differentiation
IL-12	B cells and monocytes	Stimulate NK cells; immunity against intracellular organisms
IL-13	T cells	Induce antibody switching in B cells; decrease inflammatory response in macrophages
IL-14	T cells	Stimulate B cell proliferation
IL-15	Monocytes, macrophages, glia, and skeletal muscle	Stimulate proliferation of T, B, and NK cells; myocyte differentiation
IL-16	T cells	Induce migration of T cells, eosinophils, and monocytes
IL-17 (A-F)	T cells and monocytes	Synergize with TNF in bone resorption; stimulate cytokine production; inhibit angiogenesis
IL-18	Macrophages, osteoblasts, and epithelium	Induce IFNγ in T cells and enhance NK cytotoxicity
IL-19	B cells and monocytes	Unknown; possibly anti-inflammatory (IL-10 family)
IL-20	Skin	Keratinocyte proliferation (IL-10 family)
IL-21	T and mast cells	Increase acute phase proteins in liver
IL-22[a]	T cells	Small inhibition of IL-4 production by T cells; induce acute phase reactive proteins (IL-10 family)
IL-23	Dendritic cells	Stimulate NK and memory T cells (IL-12 family)
IL-24	T cells	Wound healing; differentiation (IL-10 family)
IL-25[b]	Th2 cells	Allergic inflammatory response (IL-17 family)
IL-26	T cells	Unknown; possibly autocrine growth factor (IL-10 family)
IL-27	Bone marrow stroma	Lymphocyte proliferation (IL-12 family)
IL-28[c]	Monocytes	Antiviral activity (IL-10 family)
IL-29[c]	Monocytes	Antiviral activity (IL-10 family)

continues

continued

Hormone	Source	Action
TNFs	Immune cells	B and T cell activation, growth, and differentiation; inflammation; cytotoxicity; proliferation of vascular epithelium and fibroblasts; catabolism
CSFs	Immune cells and fibroblasts	Hematopoiesis; inflammation
IFNs	Helper T cells	B and T cell activation, growth and differentiation; inflammation; cytotoxicity; inhibition of protein synthesis

[a] Also called IL-TIF (T cell-derived inducible factor).
[b] Also called IL-17E.
[c] IL-28A, IL-28B, and IL-29 are also called IFNλ2, IFNλ3, and IFNλ1, respectively.
CSF, Colony-stimulating factors; *IFN,* interferon; *IL,* interleukin; *NK,* natural killer cells; *TNF,* tumor necrosis factor.

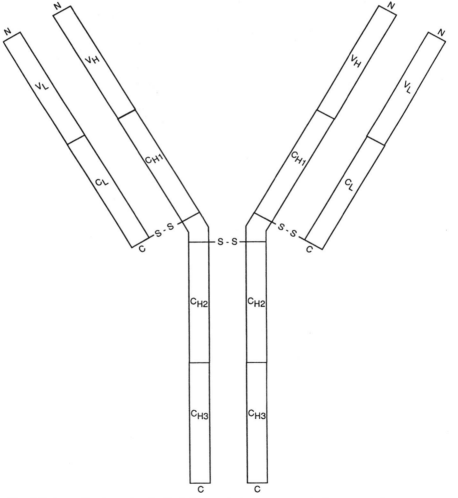

Fig. 3-3. An antibody molecule. Each heavy chain is composed of one variable (V_H) and three constant domains (C_H); each light chain has only one of each type of domain (V_L, C_L). N and C represent the amino termini and carboxy termini, respectively.

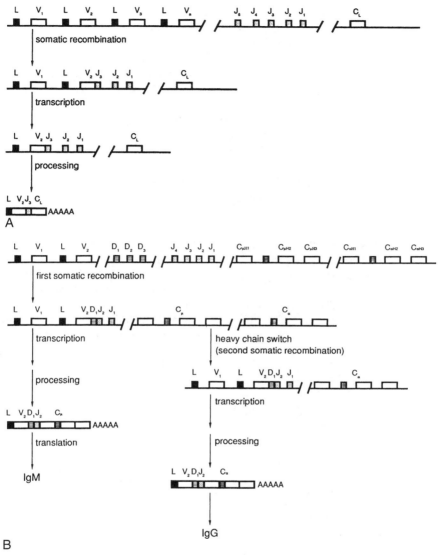

Fig. 3-4. Somatic recombination: (A) light chain recombination; (B) heavy chain recombination (see the text for description). *C*, Constant exon; *D*, diversity exon; *hnRNA*, heterologous nuclear RNA (mRNA precursor); *Ig*, immunoglobulin; *J*, junctional exon; *L*, leader sequence; *V*, variable exon.

region. There are also several small exons that code for a few amino acids located at the junction between V_L and C_L; these are the J (junctional) exons. The heavy chain gene has a similar organization except that there is an additional group of small exons like the J exons; these are the D (diversity) exons and they are located between the V_L and J exons. In addition, the heavy chain gene has three constant domains. Each one has its own exon ($C_{H1}, C_{H2},$ and C_{H3}) with a small hinge exon between C_{H1} and C_{H2}. The first set of constant exons codes for the immunoglobulin M (IgM) heavy chain; this is followed by sets for the other antibody classes and their isoforms.

During embryogenesis, the B lymphocytes (B cells) differentiate. The heavy and light chain genes in each cell randomly couple one V_L and J exon to the C_L exon and one V_H, J, and D exon to the first C_H set. All intervening exons are spliced out and lost; this process is called *somatic recombination.* The resulting antibody will be of the IgM class and have an antigen specificity determined by the random association of multiple exons. Each B lymphocyte will produce a unique IgM and display it on its plasma membrane.

If the B lymphocyte ever encounters an antigen matching the specificity of its antibody, the antigen will bind to the IgM, be internalized, be partially degraded, and finally be displayed on its cell surface bound to Class II major histocompatibility complexes (MHCs) (Fig. 3-5). The Class II MHCs are merely two light chains with hydrophobic carboxy termini that anchor the chains in the plasma membrane. At the same time, this entire process is being mimicked by antigen-presenting cells, including macrophages, dendritic cells of lymphoid organs, and Langerhans cells of the skin. This antigen processing is necessary because the T cell receptor (see later) only recognizes unfolded peptides.

Another group of lymphocytes, the helper T cells, also have receptors that resemble the Class II MHCs and are called *T cell receptors.* The helper T cell having a receptor with appropriate specificity will bind to the partially degraded antigen on the macrophage cell surface; this interaction is achieved with special adhesion proteins, such as CD4. CD4 is particularly noteworthy because it is used by the human immunodeficiency virus (HIV) to gain entry to T cells, which the virus eventually destroys. As a result of its interaction with the T cell, the macrophage secretes IL-1, which mediates several inflammatory responses, such as fever, and which also stimulates the helper T cell to produce IL-2. IL-2 acts in an autocrine manner to stimulate the proliferation of helper T cells; this is called *clonal selection* because it amplifies only those cells producing a helper T cell receptor that has a specificity for the antigen in question.

The helper T cell can now bind to the same antigen on the B cell; this interaction leads to the secretion of IL-4, IL-5, and IL-6 by the helper T cell. IL-4

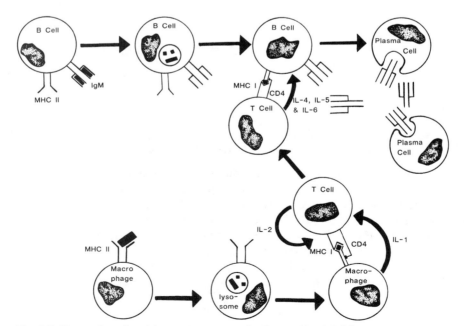

Fig. 3-5. Humoral mediated immunity; see text for description. Solid figures represent intact or processed antigens. *IL,* Interleukins; *MHC I and II,* class I and II major histocompatibility complexes.

stimulates a second somatic recombination event where the V_H-D-J exon complex is now randomly spliced to a constant domain set. In this way, the antigen specificity of the original IgM can be transferred to an IgG or any other antibody class; this phenomenon is known as *heavy chain switching*. IL-5 stimulates B cell proliferation, which results in the clonal selection of B cells making antibodies that have specificity for this particular antigen. Finally, IL-6 matures the B cell into an antibody-producing cell.

Cell-Mediated Immunity

Cell-mediated immunity is often used for viral and fungal infections. The initial response is similar to that for humoral immunity: antigen-presenting cells phagocytize the antigen, partially degrade it, and display it on the cell surface through Class II MHC. However, the macrophage interacts with a different T cell, the cytotoxic T cell. Nonetheless, the response is the same: the macrophage secretes IL-1, which stimulates the cytotoxic T cell to produce IL-2. The latter acts in an autocrine manner to stimulate T cell proliferation (clonal selection).

Meanwhile, cells infected with viruses or fungi are leaking foreign antigens to the cell surface, where they are bound to Class I MHC. Class I MHC is a single light chain with a carboxy-terminal hydrophobic extension that anchors it in the plasma membrane. The immunoglobulin fold is intact because this light chain has an extra variable domain at the amino terminus. The cytotoxic T cell binds the antigen on the infected cell and injects perforins into the target cell membrane where they polymerize into pores; this phenomenon is similar to the formation of C9 pores in complement fixation. The cellular contents begin to leak out and the infected cell dies.

Miscellaneous Immunoinflammatory Hormones

Tumor Necrosis Factor Family. This family consists of tumor necrosis factor α (TNFα) and TNFβ. TNFα has 157 amino acids and is not glycosylated in human beings, although this modification does occur in the murine hormone. It is synthesized as a membrane-bound precursor. TNFβ is larger because of a 17-amino acid amino-terminal extension and the presence of glycosylation; it is not synthesized as a membrane-bound precursor. Both hormones are trimers, forming a pyramid with threefold symmetry. They are secreted from macrophages, natural killer cells, and some T lymphocytes.

The most remarkable biological effect of the TNFs is their ability to induce regression and, sometimes, total destruction of certain tumors. This is accomplished by redirecting electrons in mitochondria toward the formation of oxygen radicals that are primarily responsible for the damage. In healthy cells, it induces protective proteins, such as the manganese superoxide dismutase, but malignant cells are often deficient in this detoxifying enzyme and are sensitive to oxygen radicals. Finally, TNF exerts some of its effects through other hormones. For example, it activates phospholipase A_2, which liberates arachidonic acid for the formation of eicosanoids (see later), and it stimulates the release of IL-6, which is a major inducer of acute phase proteins. Acute phase proteins are associated with the acute phase response that accompanies profound systemic injury or infection. They include protease inhibitors; complement components; fibrinogen; C-reactive protein, which promotes phagocytosis; and haptoglobulin and hemopexin, both of which bind hemoglobin released during hemolysis to prevent kidney damage and recycle the iron.

A functionally related, although structurally distinct, peptide is the leukemia inhibitory factor (LIF). This is a 179-amino-acid, 58-kDa glycoprotein with a unique amino acid sequence. Its biological activity is similar to that of IL-6: it stimulates the proliferation and differentiation of various blood cells and it induces the acute phase response. However, it differs from IL-6 in its inhibition of adipogenesis, sometimes leading to wasting (cachexia). In this respect, it resembles TNF. Oncostatin M is another member of the LIF family. It stimulates the growth of fibroblasts and HIV-infected cells, inhibits proliferation of normal endothelial and cancer cells, and increases IL-6 production by endothelium. It is produced by activated T lymphocytes and monocytes.

Interferon Family.   The interferon (IFN) family consists of the type I IFNs, IFNα, and IFNβ, and type II IFN, IFNγ. The type I IFNs have 166 amino acids and share about 30% homology with each other; their genes have no introns. However, IFNα is not *N*-glycosylated, whereas IFNβ is, leading to a slightly higher molecular weight for the latter. The type II IFN, IFNγ, is a homodimer, whose glycosylated subunits have 143 amino acids each.

The IFNs are best known for their antiviral activity. One major effect of these hormones is to halt protein synthesis via two routes. First of all, they induce a eukaryotic (translation) initiation factor (eIF)–2 kinase that phosphorylates and inactivates the translation initiation factor, eIF-2. Second, they induce an oligo-2′,5′-adenylate synthetase that, when activated by double-stranded RNA, synthesizes oligo-2′,5′-adenylates. These short polynucleotides activate an endonuclease that degrades mRNA. Furthermore, in macrophages, they are potent inducers of MHC antigens and trigger a respiratory burst that generates superoxides and peroxides for phagocytosis. The latter effect is reminiscent of the activity of TNF; indeed, IFN and TNF are highly synergistic in killing tumor cells.

Parahormones

Parahormones are hormones that have their effects within the tissues that synthesize them; that is, they are local hormones. There are simply too many parahormones for all to be described in a brief synopsis. However, three groups will become important in future sections: the eicosanoids, the opiate peptides, and the purine derivatives.

Eicosanoids

The eicosanoids are derivatives of arachidonic acid and include the prostaglandins, thromboxanes (both from the cyclooxygenase pathway), leukotrienes (products of the 5-lipoxygenase enzyme), lipoxins (from the 12-lipoxygenase pathway), and the epoxy derivatives (from the epoxygenase pathway). According to standardized nomenclature, the first two letters signify the particular group to which the compound belongs. For example, PG designates a prostaglandin; TX, a thromboxane; and LX, a lipoxin. The third letter denotes a series within each group. Each compound within a particular series has an identical head (ring) group, including substitutions and double bonds (Fig. 3-6). The numerical subscript indicates the number of double bonds in the side chains.

The major synthetic steps in the formation of eicosanoids are shown in Fig. 3-7. The prostaglandins are made everywhere, but the synthesis of the thromboxanes, leukotrienes, and lipoxins has a more restricted distribution. All three groups are made in platelets, neutrophils, and the lung; thromboxanes are also synthesized

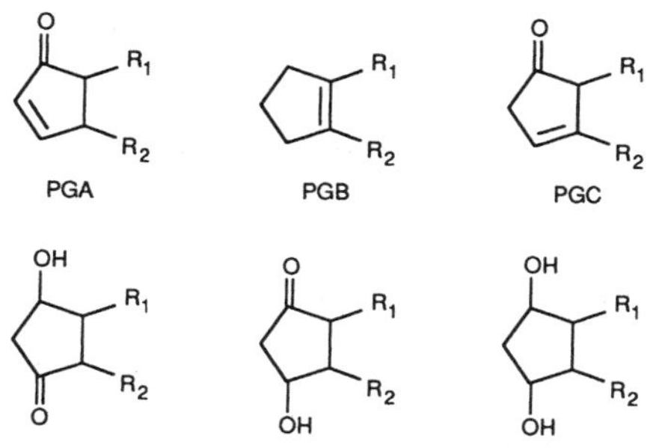

Fig. 3-6. Head (ring) groups for the different series of prostaglandins.

in the brain and the leukotrienes and lipoxins, in mast cells. Eicosatetraynoic acid (ETYA) is an arachidonic acid analog containing triple bonds in place of the double bonds, and it inhibits any enzyme with this fatty acid as substrate. Although ETYA is not used clinically, there are several other inhibitors that are useful drugs. Certainly, the most common one is aspirin, which inhibits the cyclooxygenase involved in prostaglandin synthesis. Because many prostaglandins mediate an inflammatory response (see later), aspirin acts as an anti-inflammatory agent. Another drug is dipyridamole, a coronary vasodilator. This compound inhibits the synthesis of the thromboxanes, which are vasoconstrictors. The lipoxygenase pathway can be selectively inhibited by nordihydroguaiaretic acid (NDGA).

Eicosanoids perform three major actions: mediation of inflammation, prevention of blood loss, or contraction of smooth muscle. Prostaglandin A (PGA), the leukotrienes, and 11,12-epoxyeicosatrienoic acid (11,12-EET) (also called *endothelium-derived hyperpolarizing factor* [EDHF]) belong to the first group and elicit all of the classic signs of inflammation. They dilate blood vessels to produce erythema, increase their permeability such that they leak and produce edema, provoke pain and fever, stimulate lysosome release, and are chemotactic. The lipoxins appear to counteract the inflammatory activity of these other eicosanoids. As such, they may play a role in maintaining tight control over this process. Thromboxanes fall into the second group; they constrict blood vessels and promote platelet aggregation, thereby facilitating blood clotting. PGF is a member of the last group; it is a potent stimulator of virtually all smooth muscle, including that in the vasculature, bronchioles, and gastrointestinal (GI) and reproductive tracts. The function of this latter group depends on the particular tissue and circumstances. For example, the prostaglandins in the serum stimulate uterine contractility, which is thought to aid in the transport of sperm up the female reproductive tract.

Opiate Peptides

While studying the mechanism of action of narcotics, pharmacologists discovered that animals possessed receptors for these drugs. Why would animals have a specific receptor for a plant alkaloid, unless these receptors were originally designed for an endogenous, narcotic-like compound? Such a compound was sought, and [Met^5]enkephalin and [Leu^5]enkephalin were eventually isolated. Two other groups of opiate peptides have now also been purified and characterized. All three groups have the same amino terminal pentapeptide: H_2N-Tyr-Gly-Gly-Phe-Met- or H_2N-Tyr-Gly-Gly-Phe-Leu-. Indeed, this is the part of the molecule that is recognized by all of the different opiate receptors; the

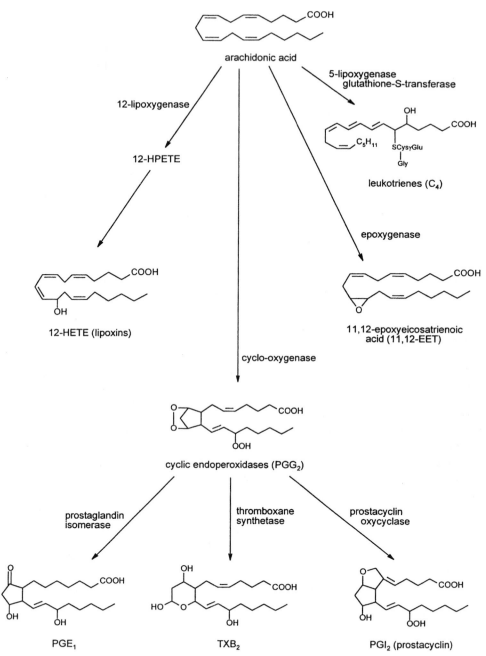

Fig. 3-7. Biosynthetic pathway for the eicosanoids. *HETE*, 5-Hydroxy-6,8,11,14-eicosa-tetraenoic acid; *HPETE*, 5-hydroxy-peroxy-6,8,11,14-eicosatetraenoic acid; *TXB₂*, tromboxane B₂.

carboxy-terminal extension merely enhances the binding of each class of opiate peptides to a particular receptor (see later).

What do a plant alkaloid and a pentapeptide have in common? The three-dimensional conformation of the narcotics has been determined, and much information on structure–function relationships has been obtained: all potent narcotics have a phenol ring, an amine, and, usually, a phenyl ring. These groups are also present in the opiate peptides (Fig. 3-8). Presumably, the three-dimensional

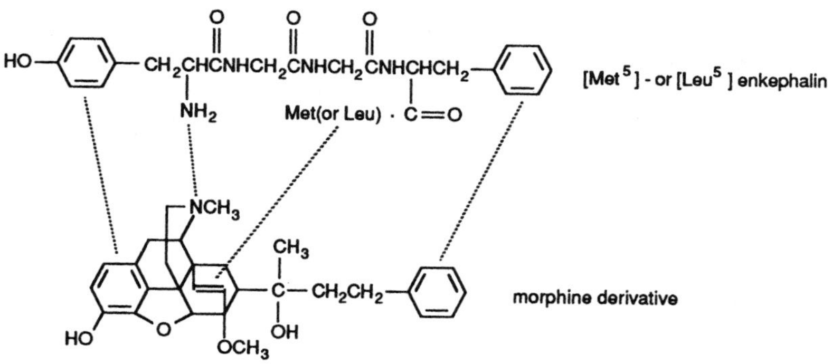

Fig. 3-8. Structural comparison of enkephalin and a morphine derivative.

structure of these peptides is such that these groups occupy the same relative positions as they do in the narcotics. However, small peptides rarely form stable three-dimensional structures, and although some structures for opiate peptides have been proposed, there is still considerable controversy over which, if any, are correct.

The opiate peptides can be divided into three groups based on their origin. The endorphins are incorporated in the polyprotein proopiomelanocortin (POMC), which can give rise to ACTH, three different melanocyte-stimulating hormones (MSHs), or three different endorphins, depending on how the protein is processed (see Chapter 2). There is only one copy of the endorphin sequence within POMC; the three forms are generated by proteases, which give rise to carboxy termini of different lengths. The endorphins have between 16 and 27 amino acids, depending on how much of the carboxy terminus is removed. The enkephalins were the first opiate peptides isolated and are only 5 to 7 amino acids long; they are derived from another precursor, proenkephalin. Finally, the dynorphins and neoendorphins are 10 to 17 amino acids long and are encoded within prodynorphin. Both proenkephalin and prodynorphin contain multiple copies of their respective opiate peptide, and all three precursors are evolutionarily related. Interestingly, there is another protein that is related to the opiate precursors: pronociceptin. However, this precursor gives rise to nociceptin, which mediates pain rather than analgesia. The receptor for nociceptin is also related to those for the opiate peptides.

Like many parahormones, opiate peptides are made in multiple locations; the highest concentrations are found in the pituitary gland, adrenal medulla, and peptidergic neurons. There are at least three different receptors for these peptides. The δ receptor mediates analgesia, the μ receptor mediates euphoria, and the κ receptor mediates sedation, dysphoria, and anorexia. In general, the δ receptor has a higher affinity for [Leu^5]enkephalin; the μ receptor, for β-endorphin and [Met^5]enkephalin; and the κ receptor, for dynorphin. However, there is considerable overlap in receptor binding activities. All three receptors belong to the G protein-coupled receptor family (see Chapter 7).

Purines

The metabolic role of adenine compounds is well-known and the function of 3′5′-cyclic AMP (cAMP) as a second messenger is commonly appreciated. However, some of these nucleotides also act as parahormones. Adenosine is released during times of stress, such as hypoxia, and acts on the P_1 purinergic receptors. It appears to exert a protective effect on highly metabolic tissues by reducing their energy expenditures; for example, it slows the heart, relaxes smooth muscle, and raises the threshold of certain neurons (Table 3-4). For this reason, it has often been called a "retaliatory metabolite." Adenosine 5'-triphosphate (ATP) also appears to be released during stress, but its effects are not easily classified. ATP binds to P_2 purinergic receptors, where it usually induces vasodilation and platelet aggregation and affects glandular secretion and the immune system. It is secreted by certain neurons, platelets, and the adrenal medulla.

Miscellaneous Factors

Retinoic Acid

Retinoic acid, like 1,25-dihydroxycholecalciferol (DHCC), cannot be made by the human body; its precursor, β-carotene, must be consumed in the diet. For that reason it is also known as *vitamin A*. It is perhaps best recognized as a component of rhodopsin, a retinal pigment necessary for light detection; however, it is also important in epithelial cell differentiation. *In vivo*, vitamin A deficiency causes secretory epithelium to become highly keratinized, whereas vitamin A replacement restores mucous and ciliated epithelium. *In vitro*, retinoic acid can inhibit the proliferation of some tumor cell lines by inducing terminal differentiation.

Because derivatives of vitamin A are hydrophobic, they must be carried by specialized transport proteins, like those for steroids and thyroid hormones. β-Carotene is absorbed by the intestinal mucosa, oxidized to retinol, esterified, and packaged into chylomicrons. In the liver, the retinol is liberated and repackaged into a 21-kDa globulin, called *retinol-binding protein*. Each protein has one binding site for retinol and complexes with one prealbumin molecule, a T_3-binding protein. Epithelial cells have high-affinity binding sites for the retinol binding protein; this cell-protein interaction releases the retinol, which is transported inside the cell, where it is bound to cellular retinol-binding protein (CRBP). It can also be oxidized to the retinoic acid and be bound to another

Table 3-4
Purinergic Receptors, Agonists, Mediators, and Actions

Receptor class	Subclass	Endogenous agonists	Second messengers	Actions
P_1	A_1	Adenosine	Decreases cAMP; calcium; arachidonic acid	Bradycardia; anticonvulsant; negative inotropism (atria); antiadrenergic
	A_2	Adenosine	cAMP; calcium	Vasodilation; anti-inflammatory
	A_3	Adenosine	Decreases cAMP; calcium	Mast cell secretion; preconditioning
P_2	P_{2T}	ADP	Decreases cAMP; calcium	Platelet aggregation
	P_{2X}	ATP	Cation channels	Vasoconstriction
	P_{2Y}[a]	ATP, UTP, UDP-glucose	cAMP; calcium; nitric oxide; prostaglandins	Smooth muscle relaxant; exocrine-endocrine secretion
	P_{2Z}	ATP^{4-}	Ion channels	Mast cell degranulation and increased membrane permeability; inhibits lymphocyte cytotoxicity

[a] P_{2D} and P_{2U} are very similar to P_{2Y}.

ADP, Adenosine 5'-diphosphate; ATP, adenosine 5'-triphosphate; cAMP, 3'5'-cyclic AMP; UDP, uridine diphosphate.

carrier, cellular retinoic acid-binding protein (CRABP). The function of these cytoplasmic proteins is not clear because retinoic acid exerts its biological activity by binding to nuclear receptors. Although these proteins may help transport the retinoic acid to the nucleus, steroids, which have an identical mechanism of action, do not need such help. Perhaps the proteins maintain a cytoplasmic reservoir for retinoids or aid in some nongenomic effects of these hormones.

Local Inflammatory Peptides

The *N*-formylpeptides are unusual in eukaryotic systems; they are thought to arise from the breakdown of both bacterial proteins during infection and mitochondrial proteins after tissue damage. In either case, infection or injury represents a potentially dangerous situation that requires a prompt defensive response. The formyltripeptide formylmethionylleucylphenylalanine (FMLP) primarily affects neutrophils and macrophages; its effects include the induction of chemotaxis, phagocytosis and degranulation, as well as the generation of superoxides, which are used to kill ingested organisms.

The chemokines have a similar spectrum of activity. There are four chemokine families: the α or CXC chemokines, the β or CC chemokines, the C chemokines, and the fractalkines or CX_3C chemokines. The letter designations refer to the arrangement of conserved cysteines within each group. The CXC chemokines include IL-8 and GRO (also called *melanocyte growth-stimulatory activity* [MGSA]); they specifically activate neutrophils. The CC chemokines include RANTES (*r*egulated upon *a*ctivation, *n*ormal *T* cell *e*xpressed and *s*ecreted), MIP (*m*acrophage *i*nflammatory *p*rotein), and MCP (*m*onocyte *c*hemoattractant *p*rotein); they specifically activate monocytes. The C chemokines include lymphotactin; they specifically activate T lymphocytes and natural killer cells. Finally, fractalkine is membrane bound via a mucin-like stalk. The membrane-bound form mediates adhesion between monocytes and T cells, whereas the soluble form mediates chemotaxis for monocytes and T cells.

Bradykinin is one of several small inflammatory peptides, the kinins, released from precursors, called kininogens. Bradykinin is generated by the action of a serine protease, kallikrein, which circulates in the blood as an inactive precursor, prekallikrein. As such, bradykinin is produced at the end of a cascade where the blood coagulation factor XIIa converts prekallikrein to kallikrein, which then digests kininogen to release bradykinin. Bradykinin is a local vasodilator and increases capillary permeability. It also produces pain and prostaglandins, which mediate some of the actions of bradykinin. Finally, it can affect smooth muscle contractility by inducing contractions or relaxation, depending on the source of the smooth muscle.

Thrombin is a most unlikely growth factor. It is a serine protease that is most commonly associated with the blood coagulation cascade, where it catalyzes the conversion of fibrinogen to fibrin during clot formation. However, it is a very potent activator of platelets, which it induces to aggregate, secrete serotonin, and produce prostaglandins. It is also mitogenic for fibroblasts and monocytes–macrophages, increases endothelial permeability, and is chemotactic for monocytes–macrophages. Finally, it can induce a slow sustained contraction in smooth muscle, although it can also promote relaxation in previously contracted smooth muscle. Most, if not all, of these actions are associated with the binding of thrombin to a thrombin receptor in target cells. Receptor activation involves thrombin cleavage of the amino terminus of the receptor (see Chapter 7).

Another unusual hormone is platelet-activating factor (PAF). It is a phospholipid with a *sn*-1 ether, a *sn*-2 acetate, and a choline head group (see Fig. 1-2, *E*). Many of its actions are involved with mediating inflammation and allergic reactions. In addition to activating platelets, neutrophils, monocytes, and macrophages, it has major effects on the circulatory system, including increased vascular permeability and decreased cardiac output and blood pressure. It can also stimulate hepatic glycogenolysis and induce the contraction of uterine and bronchiolar smooth muscle.

Miscellaneous Growth Factors

Hepatocyte growth factor (HGF), formerly called *scatter factor*, is also related to a serine protease involved with blood coagulation. HGF is synthesized from a 728-amino-acid precursor having a 38% identity with plasminogen, a protease that dissolves blood clots. The precursor is cleaved to separate the four kringle domains in the original amino terminus from the pseudocatalytic domain in the carboxy terminus. The subunits remain together to form the heavy and light chains of the mature hormone. Regulation of HGF production occurs at the proteolytic step. HGF induces cell dissociation and motility; influences epithelial morphology, especially the formation of branching tubules; and is involved with tissue repair and organ regeneration.

Another enzyme that was thought to have hormonal activity is platelet-derived endothelial cell growth factor (PD-ECGF), a 45-kDa protein with thymidine phosphorylase activity. However, it is now known that this enzyme actually synthesizes the real factor, 2-deoxy-D-ribose. *In vitro,* this sugar stimulates chemotaxis and thymidine incorporation in endothelial cells; *in vivo,* it induces angiogenesis.

Pleiotrophin, also called *heparin-binding growth-associated molecule* (HB-GAM) or *heparin-binding neurite-promoting factor* (HBNF), is an 18-kDa protein that promotes neurite outgrowth and increases plasminogen activity in endothelial cells. This family also includes midkine, which is involved with angiogenesis, neurogenesis, and mesoderm-epithelial interactions.

Invertebrate Hormones

Insects

Many of the known hormones involved in insect endocrinology are given in Table 3-5. One of the first things one notices is the complexity of the insect endocrine system. A second observation is the conserved nature of many hormones: homologues of the insulin, calcitonin, and several neuropeptide hormone families are represented in insects. Only a few insect hormones are mentioned later in this text; many of these hormones are related to either ecdysis (molting) or reproduction.

Ecdysis

Ecdysis and metamorphosis are regulated by two major hormones: ecdysone and juvenile hormone. The neurosecretory cells and the corpus cardiacum, a transformed ganglion closely associated with the heart, produce a peptide, called the prothoracicotropic hormone (PTTH), which stimulates the prothoracic gland to produce ecdysone, a sterol hormone (Fig. 3-9).

Table 3-5
The Major Insect Hormones

Hormone	Structure	Source	Target	Effect[a]	Mechanism
Energy Metabolism					
Hyperglycemic hormone (HTH)	Peptide (AKH/RPCH family)	Corpora cardiaca	Fat body	Glycogenolysis	IP_3
AKH	Peptide (AKH/RPCH family)	Corpora cardiaca	Fat body	Lipolysis	cAMP, IP_3
ILP, IRP	Peptides (insulin family)	Corpora cardiaca	Fat body	Glycogenolysis, lipolysis, hypotrehalosemia, growth of imaginal disks	
			Corpora allata	JH secretion	PPI, PKB
			Prothoracic glands	Ecdysone secretion	PPI, PKB
Water Metabolism					
CRF-DH	Peptide	Abdominal ganglion	Malpighian tubules	Diuresis via Na^+ transport	cAMP
DH31 (also called *calcitonin-like peptide*)	Peptide	Corpora cardiaca	Malpighian tubules	Diuresis via Cl^- transport	Calcium
5-HT	Trp derivative	Widespread	Malpighian tubules	Diuresis after feeding	cAMP, calcium
ADF	Peptide	Brain	Malpighian tubules	Water resorption	cGMP
ADH	Peptide	Corpora cardiaca	Rectum	Water resorption	cAMP
ITP	Peptide (CHH family)	Corpora cardiaca	Ileum	Water resorption	cAMP
Muscle Activity					
Proctolin	Peptide	Ganglia	Foregut, hindgut, oviduct, antennal heart, hyperneural muscle	Contraction	
Cardioacceleratory hormone	Peptide (AKH/RPCH family)	Abdominal ganglion to perivisceral organ	Heart and hindgut	Contraction	

continues

continued

Hormone	Structure	Source	Target	Effect[a]	Mechanism
Corazonin	Peptide	Corpora cardiaca	Heart, antennal heart, hyperneural muscle	Contraction	
Sulfakinins	-(D/E)Y(SO3H) GHMRF-amide (CCK family)	Corpora cardiaca	Heart and hindgut	Contraction; α-amylase release	Calcium
Kinins	-FXSWGamide (X = H,N,S,Y)	Corpora cardiaca	Hindgut Malpighian tubules	Contraction diuresis via anion transport	Calcium
Pyrokinin/myotropins	-(F/Y)XPRLamide (PBAN family)	Corpora cardiaca	Heart, hindgut, oviduct, antennal heart, hyperneural muscle	Contraction	
Tachykinin-related peptides	-GFXGXRamide (tachykinin family)	Viscera (especially midgut)	Gut	Contraction and AKH release	
Periviscerokinin	GXSGLI.... RX	Perivisceral organ	Foregut, heart, antennal heart, hyperneural muscle	Contraction	
Cardioactive peptide (CAP)	Peptide (CCAP family)	Subesophageal ganglion	Heart, nervous system	Contraction, adult ecdysis	
Accessory gland myotropin	Peptide (allatotropin family)	Accessory sex glands, brain, and ganglia	Smooth muscle	Contraction	
Locustamyoinhibitory peptide	Peptide	Ventral nerve cord	Cardiac and smooth muscle	Relaxation	
Myosuppressins	Peptide (CCK family)	Thoracic gland	Muscles	Relaxation of cardiac and smooth muscle; skeletal muscle contraction	
Octopamine	Amine derivative	Neurotransmitter	Heart	Decrease rate	cAMP, calcium
PTTH	Peptide	Neurosecretory cells of brain to corpora cardiaca	Prothoracic glands	Ecdysone secretion	cAMP, PPI

Molting

Hormone	Type	Source	Target	Action	Second messenger
Prothoracicostatic hormone (ecdysiostatin)	Peptide	Vitellogenic oocytes	Prothoracic glands	Inhibits ecdysone secretion and follicular development	
Ecdysone (precursor) → 20-hydroxyecdysone (active)	Sterol	Prothoracic glands to fat body and epidermis	Epidermis	Molting	Nuclear receptors
		Follicle cells of ovary to fat body and epidermis	Fat body	Vitellogenin synthesis	
		Cellular sheath of testis to fat body and epidermis	Testes	Spermatogenesis	
Bursicon	Peptide	Abdominal ganglion (major); Subesophageal ganglion	Epidermis	Cuticle tanning	
MRCH	Peptide	Subesophageal ganglion	Epidermis	Cuticular melanization	
EH	Peptide	Tritocerebral neurosecretory cells	Epitracheal glands	ETH secretion	cGMP
JH	Sesquiterpenoid	Corpora allata	Epidermis; Fat body; Oocyte; Male and female accessory glands; Brain	Inhibits developmental switches; Inhibit metamorphosis; Vitellogenin synthesis; Vitellogenesis; Secretion; Larval diapause (by inhibiting PTTH)	Nuclear receptors and IP_3
ETH	Peptide	Epitracheal glands	Ventral nerve	Ecdysis behavior and eclosion	
PETH	Peptide	Epitracheal glands	Dorsal nerve	Ecdysis behavior and eclosion	

continues

continued

Hormone	Structure	Source	Target	Effect[a]	Mechanism
Allatotropin	Peptide	Medial neurosecretory cells of brain	Corpora allata	JH secretion	IP_3
Allatostatin	Peptide	Lateral neurosecretory cells of brain	Corpora allata	Inhibits JH secretion (transient)	
Allatinhibin	Unknown peptide	Brain	Corpora allata	Inhibits JH (prolonged)	
GBP	Peptide	Brain and fat body		Blocks pupation	
Reproduction					
PBAN	Peptide	Subesophageal ganglion	Terminal abdominal ganglion → pheromone glands	Pheromone synthesis	
Neuroparsin	Peptide	Corpora cardiaca	Corpora allata	Inhibits JH	
			Rectum	Water resorption	Calcium
EDNH (also called OEH)	Peptide	Brain	Ovary	Ecdysone secretion	
Ovary-maturing parsin	Peptide	Brain	Oocyte	Oocyte maturation	
TMOF	Protein	Vitellogenic oocytes	Fat body	Inhibits vitellogenesis	
DH	Peptide (PBAN family)	Subesophageal ganglion	Oocyte	Diapause	cAMP
Defense					
ENF peptides (GBP, PSP, etc.)	Peptide	Hemolymph	Plasmatocyte	Reduces feeding and weight gain, delays pupation, and stimulates plasmatocytes	

ADF, Antidiuretic factor; *ADH*, antidiuretic hormone; *AKH*, Adipokinetic hormone; *cAMP*, 3'5'-cyclic AMP; *CAP*, cardioactive peptide; *CCAP*, crustacean cardioactive peptide; *CCK*, cholecystokinin; *cGMP*, 3'5'-cyclic GMP; *CRF-DH*; corticotropin (ACTH)–releasing factor diuretic hormone; *DH*, diuretic hormone; *DH31*, diuretic hormone 31; *EDNH*, egg development neurosecretory hormone; *EH*, eclosion hormone; *ENF*, defense peptides beginning with the sequence (E)-asparagine (N)-phenylalanine (F); *GBP*, growth-blocking peptide; *5-HT*, 5-hydroxytryptamine; *HTH*, hypertrehalosemic hormone; *ILP*, insulin-like peptide; *IP₃*, inositol 1,4,5-trisphosphate; *IRP*, insulin-like related peptide; *ITP*, ion-transport peptide; *JH*, juvenile hormone; *MRCH*, melanization and reddish coloration hormone; *OEH*, ovarian ecdysteroidogenic hormone; *PBAN*, pheromone biosynthesis activating neuropeptide; *PETH*, proecdysis-triggering hormone; *PKB*, protein kinase B; *PPI*, polyphosphoinositide; *PSP*, plasmocyte-spreading peptide; *PTTH*, prothoracicotropic hormone; *RPCH*, red pigment concentrating hormone; *TMOF*, trypsin-modulating oostatic factor.

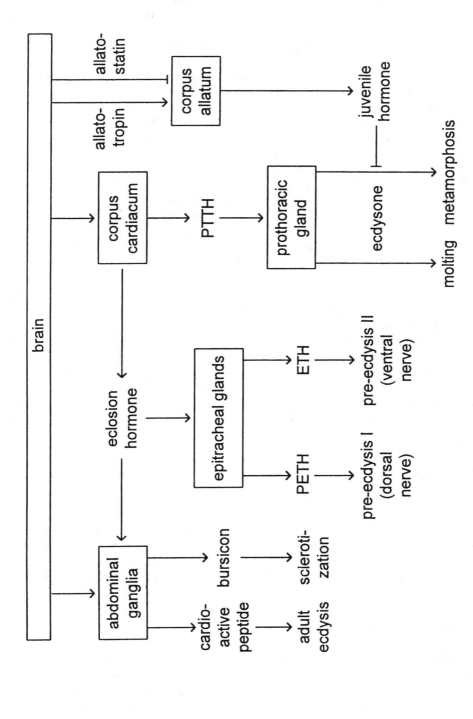

Fig. 3-9. Insect ecdysis; see text for description. Arrows indicate stimulation, and flat heads represent inhibition. *ETH*, Ecdysis-triggering hormone; *PETH*, preecdysis triggering hormone (PETH); *PTTH*, prothoracicotropic hormone.

Ecdysone alone will produce both molting and metamorphosis. However, this is undesirable during early development; insect larvae need several molts to grow before they metamorphose. For metamorphosis to be delayed until the last larval molt, a metamorphosis-inhibiting hormone must be produced at each ecdysis; this is juvenile hormone. It is secreted by an epithelial gland, the corpus allatum, in response to two tropic hormones from the brain: allatotropin stimulates juvenile hormone secretion, whereas allatostatin inhibits it. Only when ecdysone is produced and juvenile hormone is suppressed does metamorphosis accompany a molt.

There are also several ancillary hormones. Bursicon is produced by the brain and is responsible for tanning the new cuticle. Eclosion hormone (EH) is secreted by the corpus cardiacum and induces either molting behavior or behavior leading to the emergence of the insect from its pupa (eclosion). Molting behavior is accomplished by stimulating the epitracheal glands to produce preecdysis triggering hormone (PETH) and ecdysis-triggering hormone (ETH), which stimulate the dorsal and ventral nerves, respectively. These nerves stimulate the peristalsis required for molting. Like molting, eclosion is mediated by other hormones induced by EH: EH stimulates the abdominal ganglia to produce the cardioactive peptide, which triggers a rotatory movement characteristic of adult ecdysis.

Reproduction

Both ecdysone and juvenile hormone are also involved with reproduction. The following is a general discussion of insect oogenesis (Fig. 3-10); details in any given species may differ. Various tropic hormones from the brain stimulate the prothoracic gland to produce ecdysone and the corpus allatum to secrete juvenile hormone; other tropic hormones trigger oocyte development. Juvenile hormone stimulates progression of the oocyte to the vitellogenic stage and, along with ecdysone, promotes vitellogenin production by the fat body. The vitellogenic oocytes also provide negative feedback to both the fat body through trypsin-modulating oostatic factor (TMOF) and the prothoracic gland through ecdysiostatin.

Crustaceans and Mollusks

Many of the hormones in crustaceans are similar to those in insects (Table 3-6). Those in mollusks appear to be quite different; it is not within the scope of this text to discuss many of them, although one is very interesting from an evolutionary point of view. Insulin-like substance is a molluscan peptide that is not only homologous to mammalian insulin but also has much the same function: it is made in the gut and associated structures and it stimulates glycogen synthesis.

Plant Hormones

Although growth factors have often been relegated to the periphery of classical endocrinology, plant hormones have sometimes had a worse fate with some authorities even denying their existence. The controversy is both conceptual and technical. Conceptually, animals and plants are organized differently: animals are unitary creatures requiring a high degree of coordination to function. Coordination requires central planning, the generation of precise instructions, a

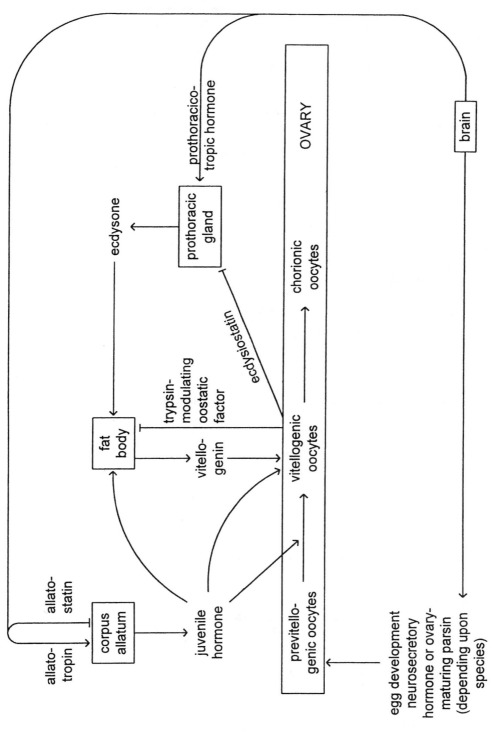

Fig. 3-10. Insect reproduction; see text for description. Arrows indicate stimulation, and flat heads represent inhibition.

Table 3-6
The Major Hormones in Crustaceans

Hormones	Structure	Source	Target	Action
Molting				
Molt inhibiting hormone (MIH)	CHH peptide family	X-organ-sinus gland	Y gland	Inhibits ecdysone synthesis
Ecdysterone	Sterol	Y gland	Epidermis	Ecdysis
Reproduction				
Gonad-inhibiting hormone (GIH)	Peptide	X-organ-sinus gland	Fat body and oocyte	Inhibits oocyte growth and vitellogenesis
Androgenic hormone (AH)	Insulin family	Androgenic gland		Male secondary sexual development
Vitellogenin-stimulating ovarian hormone (VSOH)	Unknown	Ovary	Fat body, etc.	Vitellogenesis and female secondary sexual characteristics
Permanent ovarian hormone (POH)	Unknown	Ovary		Induction of oostegites
Methyl farnesoate (MF)	Sesquiterpene	Mandibular organs	Fat body and oocyte	Inhibits vitellogenesis and oocyte growth
Mandibular organ-inhibiting hormone (MOIH)	CHH peptide family	X-organ-sinus gland	Mandibular organs	Inhibits methyl farnesoate synthesis
Pigmentation				
Red pigment-concentration hormone (RPCH)	AKH/RPCH peptide family	X-organ-sinus gland and postcommis-sural organs	Chromatophores	Concentration of red pigment granules
Black pigment-concentration hormone (BPCH)	Peptide	As above	Chromatophores	Concentration of black pigment granules
Pigment dispersing hormone (PDH)	AKH/RPCH peptide family	As above	Chromatophores	Pigment granule dispersion and retinal light adaptation
Enkephalins	Peptide	X-organ-sinus gland	Chromatophores	Concentrates red and black pigment granules
Metabolism				
Crustacean hyperglycemic hormone (CHH)	CHH peptide family	X-organ-sinus gland	Hepatopancreas and abdominal muscles	Glycogenolysis and secretion
Myotropism				
Proctolin	Peptide	Pericardial organ and neurons	Heart	Cardioexcitatory
Crustacean cardioactive peptide (CCAP)	Peptide	Pericardial organ and abdominal ganglion	Heart and hindgut	Stimulatory
FLRFamide	Peptide	Pericardial organ	Heart and muscle	Stimulatory
Orcokinin	Peptide	Abdominal nerve cord	Hindgut	Stimulatory

way to disseminate those instructions, and feedback. In animals, part of this coordination is provided by chemical signals, the hormones.

On the other hand, plants are modular organisms; for example, growth of various parts of the plant is determined as much by local environment as by any genetic plan. In addition to this apparent lack of central coordination, it was thought that plant growth substances had very imprecise biological activities, were not to be transported, and were not subject to feedback regulation. Furthermore, hormone concentrations did not correlate with biological activity.

Some of these concerns are similar to those raised for vertebrate growth factors and can be dismissed if one takes a broader view of endocrinology as regulation at a local, as well as organismal level. Other concerns arise from technical problems involved with plant research; for example, plant hormones can now be measured accurately, revealing several systems where hormone concentrations do correlate with biological activities. These techniques have also shown that many plant hormones are transported to other plant structures (Fig. 3-11). Therefore the use of the term *plant hormone* is justified.

Growth Hormones

Auxins and cytokinins are major growth promoters and morphogens (Table 3-7, Fig. 3-12). Auxin, or indoleacetic acid, is synthesized in young leaves and in developing seeds from the amino acid tryptophan. In this portion of the plant, it is responsible for apical dominance and opening stomata, pores through which gaseous exchange takes place and water vapor is lost. It also travels downs the phloem to reach the roots. Growth is achieved by activating polysaccharide hydrolases that increase the extensibility of the cell wall, which allows for cell enlargement.

Cytokinins stimulate growth by inducing cell division. These adenine derivatives are synthesized in the root tips and travel up the xylem to the shoots. The cytokinin and auxin set up a gradient along the plant and their ratios determine form (Fig. 3-13); a predominance of auxin favors root formation, whereas a high ratio of cytokinin to auxin leads to shoot development. Low concentrations of both hormones will produce flowers.

Gibberellins stimulate another kind of growth: they promote internodal elongation and germination. They are products of the isoprenoid, or polyprenyl, pathway, which also produces steroids, retinoic acid, and juvenile hormone. They are synthesized in young shoots and developing seeds.

Peptide hormones were once thought to be nonexistent in plants because they lacked a rapid transport system for distributing them. However, several growth factors have now been shown to be peptides. The phytosulfokines stimulate cell proliferation, POLARIS stimulates cell expansion, and ENOD40 induces nodulation in roots in response to nitrogen-fixing bacteria. ENOD40 may also regulate the nutrient supply to these bacteria. Some peptide hormones inhibit growth: CLAVATA3 inhibits meristem growth, whereas RALF inhibits root growth.

Stress Hormones

Abscisic acid is a derivative of carotene, another product of the isoprenoid pathway. It is synthesized in mature leaves in response to water stress; it reduces water loss by closing the stomata. It also induces dormancy and accelerates

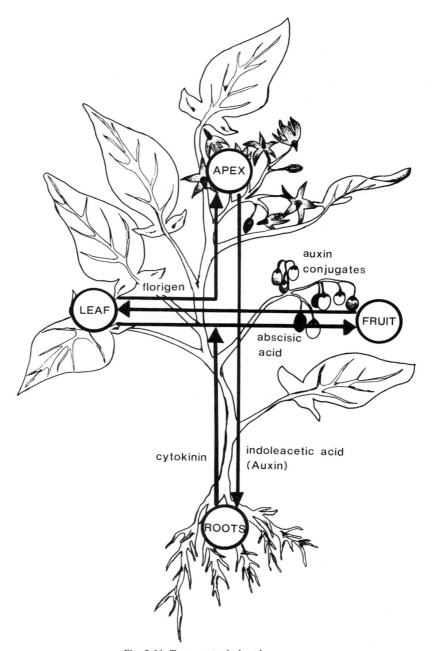

Fig. 3-11. Transport of plant hormones.

senescence. Interestingly, this hormone has been found in sponges where it mediates the response to temperature stress.

Ethylene is one of the few gaseous hormones in nature. It can be synthesized from the amino acid, methionine, by most tissues in response to stress (Fig. 3-14). The 1-aminocyclopropane-1-carboxylic acid synthase is the rate-limiting step and is highly regulated. The pathway uses S-adenosylmethionine, an intermediate in polyamine metabolism. Because methionine is in limited supply in plants, the ethylene and polyamine pathways compete against one another for this

Table 3-7
The Major Plant Hormones

Hormones	Structure	Action
Growth		
Auxin	Indoleacetic acid	Stimulates shoot growth, root initiation, and xylem/phloem differentiation; responsible for apical dominance and phototropism and gravitropism; opens stomata
Cytokinins	Substituted purines	Cell division, bud induction, and leaf morphology; embryonic growth
Gibberellins	Diterpenes	Germination, internodal elongation, and flowering
Brassinosteroids	Sterol derivatives	Elongation, curvature, and splitting of the internode, especially in seedlings; light-regulated development
Phytosulfokines	Peptide	Cell proliferation
CLAVATA3	Peptide	Growth inhibition of meristem
POLARIS	Peptide	Cell expansion
Dormancy/Senescence		
Abscisic acid	Terpenoid	Dormancy; water conservation by stomatal closure; embryonic growth; accelerates senescence
Ethylene	Alkene	Fruit ripening; abscission and dehiscence; wound responses; thigmotropism; induces macrocyst formation in certain fungi
Defense		
Traumatic acid	Dicarboxylic acid	Abscission and adventitious bud formation, especially in response to wounding
Jasmonic acid	Cyclopentanone derivative	Inhibits germination and promotes senescence, tuberization, and male gamete development; inhibits chloroplast genes; induces defense response
12-Oxo-phytodienoic acid	Jasmonic acid precursor	Tendril coiling
Salicylic acid	2-Hydroxybenzoic acid	Thermogenic in *Arum* lilies; mediates pathogen resistance; induces flowering
Systemin	Peptide	Induces defense genes
Oxylipins	Oxylipins	Cell death; induces defense genes
Elicitors		
β-Glucan and pectin fragments	Oligosaccharides	Elicitors from microorganisms; induce defense responses; morphogenesis, such as flower formation in tobacco (pectin); IAA antagonists
Bruchin	Fatty acid derivative	Elicitor from insects; induces callus formation
Volicitin	Fatty acid derivative	Elicitor from insects; induces release of attractant for parasitic wasp of feeding insect larvae
Other		
ENOD40	Peptide	Nodule formation; reduced apical dominance
RALF	Peptide	Rapid alkalinization; inhibits root growth

IAA, Indoleacetic acid.

compound; this parallels their antagonistic biological activities (see Chapter 7). Ethylene is responsible for the abscission and dehiscence of flowers, leaves, and fruit. It also induces fruit ripening and local wound responses.

Elicitor is a general term for a group of chemicals that are produced during infection or injury. They include β-glucan and chitin from fungal cell walls; pectin

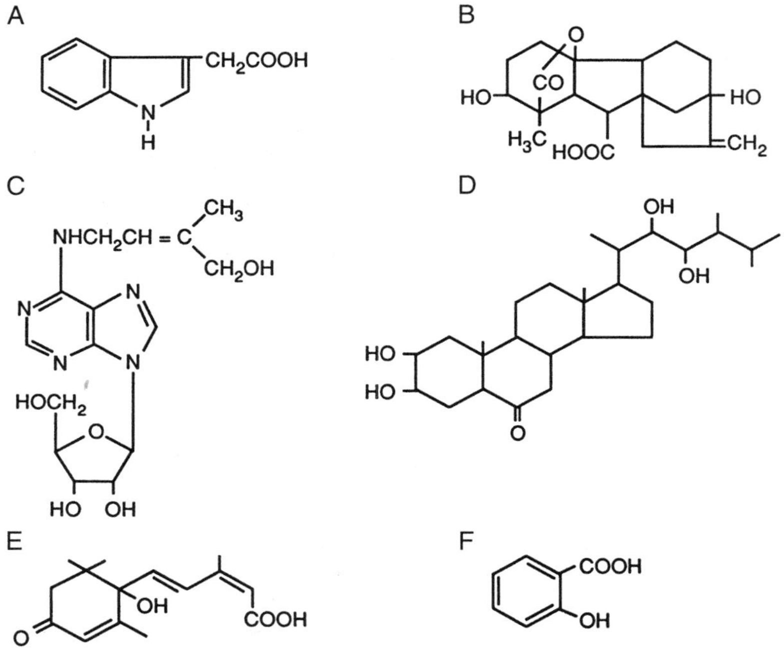

A

B

C

D

E

F

Fig. 3-12. Chemical structure of some plant hormones.

fragments from plant cell walls; N-acyl-homoserine lactone from bacteria; and the fatty acid derivatives, bruchin and volicitin, from insects. The most active oligosaccharides have between 5 and 13 sugar monomers and are not transported away from their site of production. Their effects include the induction of (1) β-glucanases and chitinases that degrade fungal cell walls, (2) protease inhibitors to protect the plant, (3) lignification (wood formation) to wall off the damage, (4) ethylene, and (5) several isoprenoid pathway enzymes responsible for the synthesis of phytoalexins, compounds that are toxic to fungi and bacteria.

Salicylic acid is probably best known as a thermogenic compound in some species of the genus *Arum*. These plants are pollinated by carrion-feeding insects that are attracted to the heat and pungent odor produced by these plants. Salicylic acid can also induce resistance to infections. For example, in the tobacco plant, salicylic acid induces the hypersensitive response, whereby cells adjacent to a viral infection die so that the infection is locally restricted. Salicylic acid is induced by infection and not just wounding. Finally, salicylic acid can undergo interconversion to volatile methyl salicylate to create an ectohormone; in this form, it may also feedback to distant parts of the same plant.

Jasmonic acid, another gaseous hormone, inhibits germination, promotes senescence, and repartitions nitrogen in plant tissues, although some have attributed its inhibitory effects to toxic doses. However, it is far more interesting as a likely example of convergent chemical evolution (Fig. 3-15). Jasmonic acid is essentially a "prostaglandin" synthesized from linolenic acid; angiosperms do not have arachidonic acid, from which authentic prostaglandins are produced. Elicitors can elevate jasmonic acid, which in turn can induce another stress hormone, systemin.

Systemin, an 18-amino-acid peptide, was the first peptide hormone discovered in plants. Despite its size, it has been shown to migrate throughout the

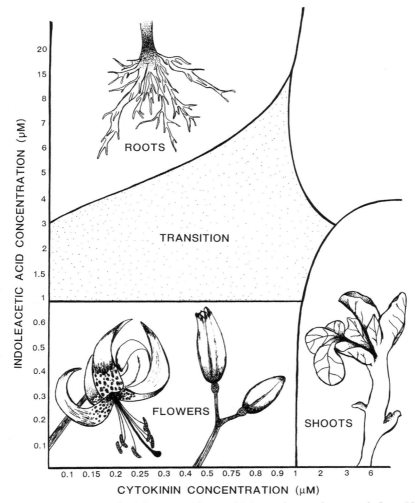

Fig. 3-13. Effect of ratios of indoleacetic acid to cytokinin on plant explant morphology. Note that the scale is arbitrary. Adapted and reprinted by permission from Palme, K., Hesse, T., Moore, I., Campos, N., Feldwisch, J., Garbers, C., Hesse, F., and Schell, J. (1991). Hormonal modulation of plant growth: the role of auxin perception. *Mech. Dev.* 33, 97–106. Copyright © 1989 the American Society of Plant Physiologists.

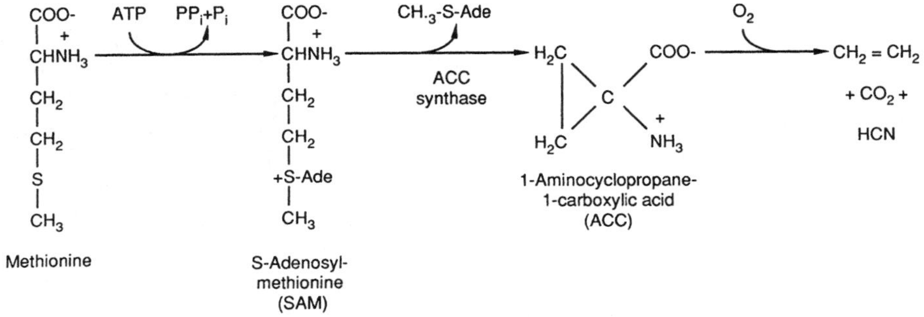

Fig. 3-14. Ethylene synthesis.

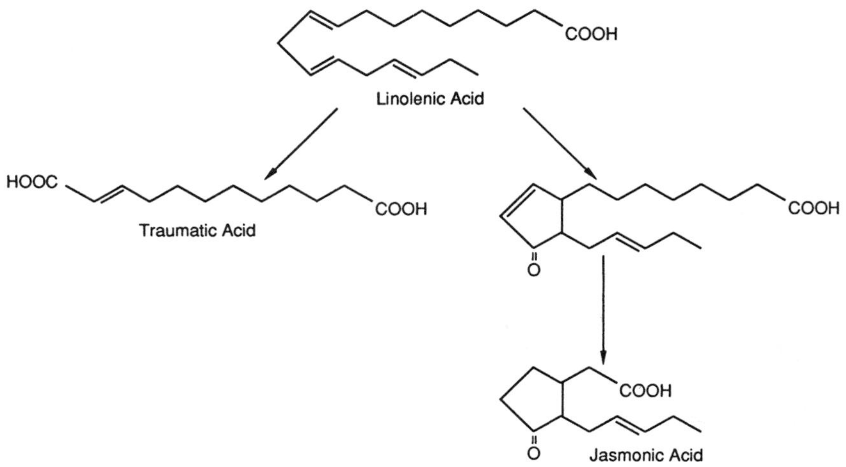

Fig. 3-15. Synthesis of traumatic and jasmonic acid.

plant, where it induces protease inhibitors involved with defense mechanisms. Ethylene, jasmonic acid, and systemin are associated with both infection and wounding, unlike salicylic acid, which is only associated with infection.

Miscellaneous Hormones

Brassinosteroids are yet another product of the extremely versatile isoprenoid pathway. Their biological activity is similar to the cytokinins; they stimulate the elongation, curvature, and splitting of the internodes. However, their greatest effect is in seedlings; indeed, although made in most tissues, the brassinosteroids are found in the highest concentrations in pollen and seeds.

Traumatic acid comes from the same pathway as jasmonic acid, but it is not cyclized. As the name suggests, traumatic acid induces abscission and adventitious bud formation, especially in response to wounding.

Summary and Prospective

Not all hormones fit the classic definition. Growth factors in particular tend to be produced locally. In addition, many growth factors have a broad range of activities, which frequently overlap with other factors. Finally, these factors are important in many developmental programs, and their actions can change depending on the stage of the program. The immunoinflammatory hormones have undergone rapid diversification because they are involved with defending the body against ever-adapting pathogens. Finally, the examination of nonvertebrates provides perspective for the development of the endocrine system. In particular, the occurrence of insulin-like molecules, neuropeptides, and neurotransmitters in many disparate phyla demonstrates the continuity of life.

In the following parts, the actions and relationships described in the first part are carried to the cellular and molecular level. The topics are discussed in the same order that the biochemical information flows: the first molecule that a hormone encounters during its interaction with a cell is its receptor (Part 2), and, if that receptor is membrane bound, the information must be transferred to some other mediator, which can act directly on cellular processes (Part 3). These mediators, as well as the soluble receptors, can also act on the genome to induce transcription and affect other posttranscriptional processes (Part 4). Finally, several special topics are presented in the last unit (Part 5).

References

General Growth Factors

Benito, M., Valverde, A.M., and Lorenzo, M. (1996). IGF-I: A mitogen also involved in differentiation processes in mammalian cells. *Int. J. Biochem. Cell Biol.* 28, 499-510.

Firth, S.M., and Baxter, R.C. (2002). Cellular actions of the insulin-like growth factor binding proteins. *Endocr. Rev.* 23, 824-854.

Galzie, Z., Kinsella, A.R., and Smith, J.A. (1998). Fibroblast growth factors and their receptors. *Biochem. Cell Biol.* 75, 669-685.

Harris, R.C., Chung, E., and Coffey, R.J. (2003). EGF receptor ligands. *Exp. Cell Res.* 284, 2-13.

Heldin, C.H., and Westermark, B. (1999). Mechanism of action and in vivo role of platelet-derived growth factor. *Physiol. Rev.* 79, 1283-1316.

Levi-Montalcini, R., Skaper, S.D., Dal Toso, R., Petrelli, L., and Leon, A. (1996). Nerve growth factor: From neurotrophin to neurokine. *Trends Neurosci.* 19, 514-520.

Matsumoto, K., and Nakamura, T. (1996). Emerging multipotent aspects of hepatocyte growth factor. *J. Biochem. (Tokyo)* 119, 591-600.

Mohan, S., and Baylink, D.J. (2002). IGF-binding proteins are multifunctional and act via IGF-dependent and -independent mechanisms. *J. Endocrinol.* 175, 19-31.

Neufeld, G., Cohen, T., Gengrinovitch, S., and Poltorak, Z. (1999). Vascular endothelial growth factor (VEGF) and its receptors. *FASEB J.* 13, 9-22.

Hematopoietic Growth Factors

Haddad, J.J. (2002). Cytokines and related receptor-mediated signaling pathways. *Biochem. Biophys. Res. Commun.* 297, 700-713.

Nicola, N.A., and Hilton, D.J. (1999). General classes and functions of four-helix bundle cytokines. *Adv. Protein Chem.* 52, 1-65.

Immunoinflammatory Hormones

Aggarwal, S., and Gurney, A.L. (2002). IL-17: prototype member of an emerging cytokine family. *J. Leukocyte Biol.* 71, 1-8.

Fickenscher, H., Hör, S., Küpers, H., Knappe, A., Wittmann, S., and Sticht, H. (2002). The interleukin-10 family of cytokines. *Trends Immunol.* 23, 89-96.

Moseley, T.A., Haudenschild, D.R., Rose, L., and Reddi, A.H. (2003). Interleukin-17 family and IL-17 receptors. *Cytokine Growth Factor Rev.* 14, 155-174.

Orlinick, J.R., and Chao, M.V. (1998). TNF-related ligands and their receptors. *Cell. Signal.* 10, 543-551.

Parahormones

Di Virgilio, F., Chiozzi, P., Ferrari, D., Falzoni, S., Sanz, J.M., Morelli, A., Torboli, M., Bolognesi, G., and Baricordi, O.R. (2001). Nucleotide receptors: an emerging family of regulatory molecules in blood cells. *Blood* 97, 587-600.

Miscellaneous Factors

Muramatsu, T. (2002). Midkine and pleiotrophin: Two related proteins involved in development, survival, inflammation and tumorigenesis. *J. Biochem. (Tokyo)* 132, 359-371.

Rollins, B.J. (1997). Chemokines. *Blood* 90, 909-928.

Invertebrate Hormones

Coast, G.M., Orchard, I.,Phillips, J.E., and Schooley, D.A. (2002). Insect diuretic and antidiuretic hormones. *Adv. Insect Physiol.* 29, 279-409.

Mesce, K.A., and Fahrbach, S.E. (2002). Integration of endocrine signals that regulate insect ecdysis. *Front. Neuroendocrinol.* 23, 179-199.

Predel, R., Nachman, R.J., and Gäde, G. (2001). Myostimulatory neuropeptides in cockroaches: structures, distribution, pharmacological activities, and mimetic analogs. *J. Insect Physiol.* 47, 311-324.

Plant Hormones

Hooykaas, P.J.J., Hall, M.A., and Libbenga, K.R. (eds.) (1999). *Biochemistry and Molecular Biology of Plant Hormones*. Elsevier, Amsterdam.

Lindsey, K., Casson, S., and Chilley, P. (2002). Peptides: new signalling molecules in plants. *Trends Plant Sci.* 7, 78-83.

Matsubayashi, Y., Yang, H., and Sakagami, Y. (2001). Peptide signals and their receptors in higher plants. *Trends Plant Sci.* 6, 573-577.

Porta, H., and Rocha-Sosa, M. (2002). Plant lipoxygenases. Physiological and molecular features. *Plant Physiol.* 130, 15-21.

Ryan, C.A., Pearce, G., Scheer, J., and Moura, D.S. (2002). Polypeptide hormones. *Plant Cell* 14, S251-S264.

PART 2

Receptors

CHAPTER 4

Overview of Receptors and Transducers

CHAPTER OUTLINE

Hormones play a crucial role in the control of virtually every process in an organism. As such, endocrine pathways are highly regulated. Both feedback control and cross talk, where parallel signaling pathways influence each other, use second messengers to affect early events. This produces a pedagogical dilemma; in what order does one present a pathway where any step may affect any other? This book presents these topics in chronological order to emphasize the continuity of the signaling pathways: hormones → receptors → transducers → transcription. However, this detailed presentation is preceded by a brief overview of all major pathways. Such a synopsis will enable the reader to understand interrelationships, even when certain components have yet to be covered in depth. For example, when receptor regulation by kinases is covered (Part 2), the novice will be able to understand the discussion, even though kinases are not presented in detail until later (Part 3).

Receptors and Transducers

There are five major groups of receptors. Nuclear receptors are the simplest; basically, they are ligand-regulated transcription factors. Because they are intracellular, their ligands are the lipid-soluble hormones, which can easily traverse the plasma membrane. The remaining four receptor groups are integral membrane proteins: the enzyme-linked receptors, the cytokine receptors, the G protein-coupled receptors, and the ligand-gated ion channels (Fig. 4-1).

Like the nuclear receptors, the enzyme-linked receptors are also relatively simple in structure: they have an amino-terminal extracellular domain, a single transmembrane α-helix, and a carboxy-terminal intracellular domain that contains an intrinsic enzymatic activity, such as a tyrosine kinase, a serine-threonine kinase, or a guanylate cyclase. In the case of the receptor tyrosine kinase (RTK), the hormone binds to the amino terminus and induces aggregation. This brings the catalytic domains, which have low basal activity, close enough to cross-phosphorylate each other's activation loop, resulting in full kinase activity. The serine-threonine kinase receptors are activated in a similar manner. However, activation of the membrane-bound guanylate cyclases, which synthesize $3'5'$-cyclic GMP (cGMP), is less clear but appears to involve ligand-induced neutralization of an autoinhibitory domain in the cytoplasmic region.

The major substrate for the RTK is itself; the phosphorylated tyrosines (pYs) attract effectors, which have special binding modules that recognize pYs. For this brief introduction, only three are mentioned: phospholipase Cγ (PLCγ), phophatidylinositol-3 kinase (PI3K), and docking proteins.

PLCγ hydrolyzes membrane phosphoinositides to yield the soluble head group inositol 1,4,5-trisphosphate (IP_3) and diacylglycerol (DG) (Fig. 4-2). The former opens calcium channels, which elevate cytoplasmic calcium concentrations. Although calcium can affect proteins directly, it more frequently acts through calmodulin (CaM), a calcium-binding protein. The CaM-dependent protein kinase, type II (CamKII), is a classic example of how CaM works. The CaMKII has an autoinhibitory domain that acts as a pseudosubstrate for both the adenosine 5'-triphosphate (ATP) and the catalytic sites. The CaM binding site is adjacent to the autoinhibitory domain. When calcium levels rise, calcium binds CaM and exposes its protein-binding site; CaM then binds to CaMKII and prevents the autoinhibitory domain from occupying the ATP and catalytic sites, thereby acti-

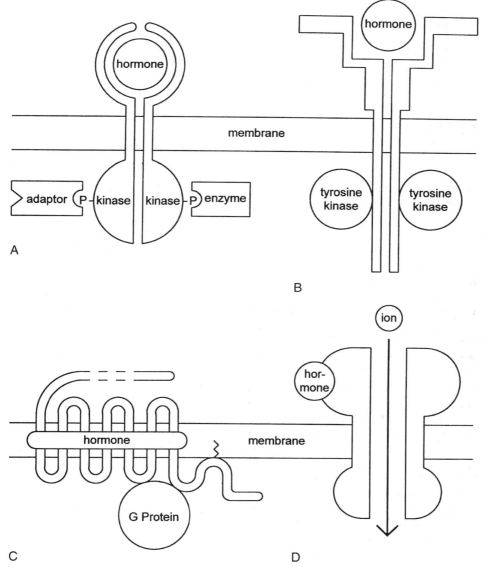

Fig. 4-1. The four major groups of integral membrane receptors: (A) enzyme-linked receptors, (B) cytokine receptors, (C) G protein-coupled receptors, and (D) ligand-gated ion channels.

vating the kinase. Many enzymes, channels, and other proteins have these CaM binding-autoinhibitory domains. The DG is also a second messenger; its best-known target is protein kinase C (PKC), also called *calcium-activated, phospholipid-dependent protein kinase*. However, this latter term is not completely correct because only some PKC isozymes require calcium for activation.

Other lipases are also involved in signaling. Phospholipase A_2 (PLA$_2$) removes the second fatty acid from phospholipids. This is usually arachidonic acid, which is both a second messenger and a precursor for eicosanoids. Phospholipase D (PLD) cleaves phospholipids between the phosphate and the head group to produce phosphatidic acid, which is both a transducer and a precursor for DG. Finally, sphingomyelinase (SMase) hydrolyzes sphingomyelin into

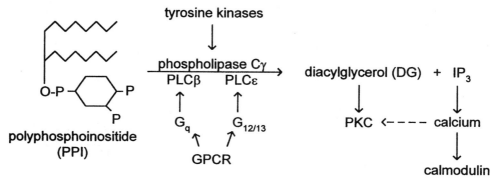

Fig. 4-2. Phosphoinositide signaling pathway. *GPCR*, G protein-coupled receptor; *IP₃*, inositol 1,4,5-triphosphate; *PKC*, protein kinase C (calcium-activated, phospholipid-dependent protein kinase C); *PLC*, phospholipase.

ceramide and phosphocholine; the former is an important trigger for programmed cell death, or apoptosis. PI3K phosphorylates phosphoinositides at the 3′ position (PI(3)P$_n$). Polyphosphoinositides (PPIs) with a phosphate at this position are not substrates for any known PLC; rather, the intact phospholipid acts as a second messenger. As with pY, there are binding modules that recognize PI(3)P$_n$ and cause the proteins containing them to localize to the plasma membrane. The binding may also allosterically activate the proteins. Three examples are given (Fig. 4-3). First, PI(3)P$_n$ can activate a protein kinase cascade; PI(3)P$_n$ initially binds both PIP$_2$-dependent kinase (PDK) and protein kinase B (PKB) and localizes them to the membrane. The former then phosphorylates and activates the latter. Second, PI(3)P$_n$ can stimulate several small guanosine 5′-triphosphate (GTP)–binding proteins, or simply G proteins. G proteins are molecular switches; they are inactive when bound to guanosine diphosphate (GDP) but active when bound to GTP. GNP exchange facilitators (GEFs), also called *GDP dissociation stimulators* (GDSs), affect the exchange of GDP for GTP, thereby stimulating the G proteins. The G proteins are weak GTPases; that is, eventually the activated G proteins will hydrolyze the GTP to GDP and return to their inactive state. As such, G proteins are switches with intrinsic timers that turn themselves off. PI(3)P$_n$ can bind and stimulate GEFs for the G proteins Rho and Rac, both of which are involved with remodeling the cytoskeleton. A major output of these G proteins is kinases; Rho binds and activates a Rho-associated coiled coil protein kinase (ROCK), whereas Rac stimulates a novel PKC-like protein kinase (PKN) and p21-activated protein kinase (Pak).

Finally, RTKs can stimulate another small G protein, Ras, via a different pathway (Fig. 4-4). First, a few terms need to be defined. An *adaptor* is a small protein that consists of only a few binding modules and bridges two other proteins. A *docking protein* is a larger, more versatile version of an adaptor; it contains many binding modules and acts as a working platform for many transducers. RTKs phosphorylate a docking protein, which then attracts an adaptor that possesses a binding module recognizing pY. Its other binding module recognizes a polyproline helix in RasGEF. Once Ras is activated, it initiates a protein kinase cascade culminating in the stimulation of the extracellular-signal regulated kinase (ERK), which is a member of the mitogen-activated protein kinase (MAPK) family. Many RTKs are used by growth factors, and Ras plays an important role in cell proliferation.

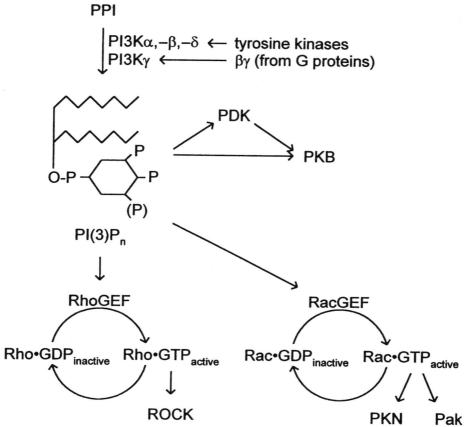

Fig. 4-3. Phosphatidylinositol 3-phosphate signaling pathway. *GDP*, Guanosine diphosphate; *GEF*, GNP exchange facilitator; *GTP*, guanosine 5′-triphosphate; *Pak*, p21-activated protein kinase; *PDK*, PIP₂-dependent kinase; *PI3K*, phosphatidylinositol-3 kinase; *PI(3)Pₙ*, phosphoinositides at the 3′ position; *PKB*, protein kinase B; *PKN*, protein kinase N (novel PKC-like protein kinase); *PPI*, polyphosphoinositide; *ROCK*, Rho-associated coiled coil protein kinase.

The cytokine receptors are basically a variation of the RTKs (see Fig. 4-1, *B*); rather than the tyrosine kinase being an intrinsic part of the receptor, a separate soluble tyrosine kinase (STK) associates with the intracellular domain. Otherwise, the mechanism of activation and the transducers used are very similar. These receptors are primarily involved in the growth and differentiation of blood and immune cells.

The third group of membrane receptors crosses the plasmalemma seven times via α-helices; as such, they have been called the *seven transmembrane segment, heptahelical,* or *serpentine receptors* (see Fig. 4-1, *C*). However, because their best-studied effectors are G protein trimers, they are most commonly referred to as the *G protein-coupled receptors* (GPCRs). G protein trimers have three subunits: the α-subunit is the actual GTP-binding protein and operates in the same manner the as the small G proteins; the βγ dimer acts to anchor the α-subunit to the plasma membrane and can also have some separate effector functions. There are four major G protein trimers: G_s and G_i stimulate and inhibit the adenylate cyclase to increase and decrease 3′5′-cyclic AMP (cAMP) levels, respectively (Fig. 4-5, *A*). This second messenger regulates a cAMP-dependent protein kinase (PKA). This kinase is highly homologous to a cGMP-dependent protein

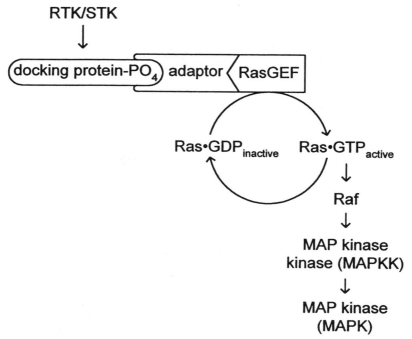

Fig. 4-4. Ras signaling pathway. *GDP*, Guanosine diphosphate; *GEF*, GNP exchange facilitator; *GTP*, guanosine 5'-triphosphate; *MAPK(K)*, mitogen-activated protein kinase (kinase); *PO₄*, phosphate; *STK*, soluble tyrosine kinase; *RTK*, receptor tyrosine kinase.

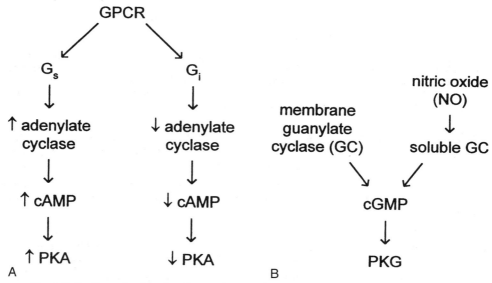

Fig. 4-5. Cyclic nucleotide signaling pathways. (A) cAMP pathways and (B) cGMP pathways. *GPCR*, G protein-coupled receptor; *PKA*, protein kinase A (cAMP-dependent protein kinase); *PKG*, protein kinase G (cGMP-dependent protein kinase).

kinase (PKG). The cGMP can be elevated either by the membrane-bound guanylate cyclase described previously or by a soluble guanylate cyclase that is stimulated by nitric oxide (Fig. 4-5, *B*). G_q and $G_{12/13}$ activate PLCβ and PLCε, respectively (see Fig. 4-2); these PLCs act in the same manner as PLCγ. Finally, the βγ dimer can activate an isoform of PI3K (see Fig. 4-3).

The ligand-gated ion channels are the last group of membrane-bound receptors. Basically, these receptors regulate calcium levels directly or indirectly. The calcium concentration in the cytosol is very low compared with the extracellular fluid and certain cellular organelles. In the simplest scenario, the receptors are calcium channels gated by hormones or second messengers. However, the channels may transport other ions and affect calcium levels indirectly through, for example, changes in cell polarization. The effects of calcium on PKC and CaM were described under RTK (see Fig. 4-2).

Finally, there are several important kinases that are not closely associated with any single pathway. Casein kinase 2 (CK2) is most commonly associated with polyamines, although it is not clear that polyamines are the physiological activators of this kinase. Glycogen synthase kinase 3 (GSK3) was first identified as a kinase involved with glycogen metabolism; however, it is now understood that GSK3 has a broader role in transduction. In general, GSK3 tends to inhibit its substrates, while it is inhibited when other kinases phosphorylate it. ERK has already been mentioned as a member of the MAPK family; other members include p38 and the stress-activated protein kinase (SAPK), also called the *Jun amino-terminal kinase* (JNK). Like ERK, these enzymes are regulated by a kinase cascade activated by various hormones. Ribosomal S6 protein kinase I (S6KI) enhances translation and is activated by nearly all growth factors through the MAPK and PDK pathways. Lastly, cell cycle-dependent kinases (Cdks) regulate mitosis but can also interact with signaling pathways.

General Principles

There are several general principles of transduction that become readily apparent after even a cursory examination of the main signaling pathways. First, protein phosphorylation is a major mechanism for cellular regulation. Every pathway discussed in this chapter has terminated in one or more protein kinases. Indeed, Chapter 12 is entirely devoted to this topic and contains an excellent summary table (Table 12-1) to which the reader may want to refer as he or she progresses through Parts 2 and 3.

The second general principle is that transduction is modular: each signaling step is located on a separate component and these components can be assembled in almost endless combinations. This phenomenon increases the versatility of the pathways, but it can also blur distinctions among pathways by converting direct, linear paths into confusing nets. This topic is more thoroughly covered in Chapter 7.

Third, many signaling molecules have modular components that are auto-inhibitory. Such an arrangement ensures low basal activity. Activation occurs when this domain is neutralized by cleavage, phosphorylation or dephosphorylation, or the binding of proteins or small molecules. For example, PKA has a pseudosubstrate site that occupies and inhibits the kinase domain; cAMP binding displaces this site. In PKC, phospholipids can displace the pseudosubstrate site; however, in this kinase, the site can also be removed by cleavage. CaM binding to the autoinhibitory domain of CaMKII represents neutralization by protein-protein interactions and was described previously. Another example of this mechanism involves STKs, which are inhibited by an intramolecular bond between a pY binding domain and a pY in their carboxy termini. Dephosphorylation of the carboxy termini or binding to a pY on another protein relieves the inhibition.

Finally, location is everything. The cytoplasm was once envisioned as a homogenous soup. There is now a much greater appreciation of the compartmentalization that occurs within the cell. RTKs attract transducers to the plasma membrane through their pYs and through the generation of $PI(3)P_n$. Furthermore, docking proteins can act as assembly platforms for various transducers.

Summary

Lipophilic hormones have no problem traversing the plasma membrane. Their receptors are soluble cellular proteins that function as ligand-regulated transcription factors. Hydrophilic hormones cannot cross the plasmalemma and must interact with the cell at its surface. Because their receptors are membrane bound, a second messenger must be generated to carry the signal deeper into the cell. These receptors recruit many local molecules for this job: membrane phospholipids, calcium, and membrane enzymes such as adenylate cyclase. Finally, most of these transduction pathways involve protein kinases because phosphorylation is a major way to regulate protein function.

References

Bolander, F.F. (1996). Molecular endocrinology. *In* Meyers RA (ed). *Encyclopedia of Molecular Biology and Molecular Medicine*, vol 2, pp. 206-216. VCH Publishers, New York, New York.

Conn, P.M., and Means, A.R. (eds.) (2000). *Principles of Molecular Regulation*, Humana Press, Totowa, New Jersey.

Gomperts, B.D., Kramer, I.M., and Tatham, P.E.R. (2002). *Signal Transduction*, Academic Press, San Diego, California.

Helmreich, E.J.M. (2001). *The Biochemistry of Cell Signalling*, Oxford University Press, Oxford, U.K.

Krauss, G. (2001). *Biochemistry of Signal Transduction and Regulation*, 2nd ed., Wiley-VHC, Weinheim, Germany.

Pufall, M.A. and Graves, B.J. (2002). Autoinhibitory domains: Modular effectors of cellular regulation. *Annu. Rev. Cell Dev. Biol.* 18, 421-462.

Weintraub, B.D. (ed.) (1995). *Molecular Endocrinology: Basic Concepts and Clinical Correlations*, Raven Press, New York, New York.

CHAPTER **5**

Kinetics

The first step in the action of a hormone is its interaction with a specific binding protein in or on the target cell; such a protein is called a *receptor*. This chapter describes the basic characteristics of these receptors and how to determine their number and affinity. The structure, function, and metabolism of the better-known receptors are discussed in the ensuing chapters; nuclear receptors are covered in Chapter 6 and membrane receptors in Chapter 7. Finally, mechanisms for the regulation of receptor activity are presented in Chapter 8.

What are some of the major characteristics of receptors? Because hormone concentrations are very low, the receptor should have a *high affinity*. A *high specificity* ensures that closely related hormones will still preferentially bind to their own receptors and remain functionally distinct. Closely related hormones may still cross-bind, but the affinity for the unintended ligand is usually low enough so as not to present any problems under physiological conditions. The receptor should be *saturable*; that is, there should be a finite number of them. This characteristic distinguishes receptor binding from nonspecific binding. The effects of hormones frequently decay rapidly following hormonal removal; this temporal pattern is a result of the *reversibility* of the hormone-receptor binding. The receptor for a particular hormone should have a *tissue distribution* appropriate to the actions of that hormone; that is, it should be present in the target organs of that hormone and absent from the tissues unresponsive to the hormone. Finally, receptor binding should be correlated with some *biological effect*.

If a binding protein possesses these characteristics, it is probably a hormone receptor. However, there are always special circumstances that can make any individual case problematic. For example, the aldosterone receptor binds both aldosterone and cortisol with equal affinity, but the target cells for this mineralocorticoid only respond to aldosterone. On the basis of this loose binding specificity, one might assume that this protein was not the true aldosterone receptor. However, the specificity lies not with the receptor but with the metabolizing enzymes in the target cells. These tissues are rich in 11β-hydroxysteroid dehydrogenase, which inactivates cortisol; therefore the nuclear receptor is normally never exposed to glucocorticoids and does not need to distinguish between them and mineralocorticoids. In another example, many cytokine receptors exist as heterooligomers. High affinity requires the intact complex; individual subunits only display low to moderate affinity. If these proteins disassembled during purification, one might question their physiological relevance as receptors.

Reversibility is another criterion that may not apply to all receptors. Thrombin is not the actual ligand for the thrombin receptor; rather, thrombin, a protease, cleaves the amino terminus of the receptor to reveal an internal sequence, which functions as the ligand. Because the ligand and the binding site are part of the same protein, they cannot be separated. However, these receptors are located on platelets and are activated during blood clotting; because coagulation is a terminal event for platelets, reversibility is not relevant.

Historically, receptors were postulated by Langley as early as 1878. In one experiment, he noted that curare could block the effects of nerve stimulation in muscle contraction but did not interfere with direct stimulation. He reasoned that curare could not be acting directly on the "chief substance" (that is, those factors involved with contraction) but must act on something between the nerve stimulation and the chief substance; he called it the "receptive substance." He then universalized this mechanism for all hormones. In 1948, Ahlquist discovered that various adrenergic agonists exhibited two different orders of potency, depending on the tissue tested, and he proposed the existence of two types of adrenergic receptors, α and β. However, the actual measurement of receptors did not come until 1962, when Jensen demonstrated intracellular receptors for estradiol. It was also in 1962 that Hunter and Greenwood developed an easy and reliable way of radioiodinating peptide hormones. Their intention was to use these hormones in radioimmunoassays; but in 1969 Lefkowitz used ^{125}I-labeled adrenocorticotropin hormone (ACTH) to demonstrate the presence of ACTH receptors in the adrenal gland. Since then, the scientific literature has become replete with the measurement of receptors.

Receptor Assays

Receptors can be measured with three basic techniques: (1) nucleic acid hybridization, (2) immunoassay, and (3) kinetic analysis. Each assay has its advantages and disadvantages. The first two techniques are easy and accurate; however, the hybridization assays are actually determining receptor mRNA, which may not always correlate with the levels of receptor protein. The various immunoassays do measure receptor protein, but receptors can exist in inactive forms that cannot bind hormone (*cryptic receptors*), trigger a transduction system (*desensitized receptors*), or bind DNA in the case of nuclear receptors (*untransformed receptors*) (see Chapters 6 and 7). Monoclonal antibodies directed against phosphotyrosines (pYs) of receptor tyrosine kinases (RTKs) can recognize active receptors because only activated RTKs are autophosphorylated. However, antibodies to other receptors may not be able to distinguish among these various forms.

In kinetic analysis the binding of a radiolabeled ligand is used to detect the receptor. It has the advantage of not requiring that the receptor be purified or its gene cloned. It has the disadvantage of relying on several assumptions that some biological systems may not be able to satisfy (see section on Kinetics). The two most common types of kinetic measurement are (1) percentage of specific binding and (2) receptor number and affinity. Percentage of specific binding is the easiest technique and merely consists of adding labeled hormone to a receptor sample and determining how much specifically binds. Unfortunately, molecules may nonspecifically adhere to almost anything, including other proteins and even the walls of the reaction vessel; total binding must be corrected for this phenomenon. Nonspecific binding can be determined in the presence of an excess of unlabeled hormone because the latter will displace the labeled hormone from its receptors but not from nonspecific binding sites. Such a determination reveals that nonspecific binding is linear with respect to hormone concentration; it is also unsaturable (Fig. 5-1). Specific binding is then calculated by subtracting nonspecific binding from total binding, and the percentage of specific binding is the ratio of specific binding to total labeled hormone added.

The percentage of specific binding is a reflection of both receptor number and affinity. For example, assume that the percentage of total labeled hormone specifically bound to a tissue increases after some experimental treatment. The enhanced binding could be a result of (1) more receptors, (2) a higher affinity of the same number of receptors, or (3) an increase in both receptor number and affinity. Therefore, although the percentage of specific binding is easy to determine, the data are somewhat ambiguous.

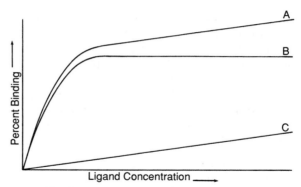

Fig. 5-1. Percentage of binding as a function of ligand concentration. Total binding (A) is a combination of both specific (B) and nonspecific binding (C).

Kinetics

Scatchard Analysis

The determination of receptor number and affinity involves analyses based on the laws of mass action; the hormone (H) and the receptor (R) bind in a reversible reaction:

$$H + R \rightleftharpoons HR$$

The rate of formation equals $k_f[H][R]$ and the rate of dissociation equals $k_r[HR]$. At equilibrium, the rate of formation equals the rate of dissociation:

$$k_f[H][R] = k_r[HR] \tag{1}$$

or

$$k_f/k_r = K_a = 1/K_d = [HR]/[H][R] \tag{2}$$

In equation (2), the two rate constants are combined into a single one: the association constant (K_a) or its reciprocal, the dissociation constant (K_d). The total number of receptors (n) is equal to the number of free receptors ([R]) plus the number bound to the hormone ([HR]). By solving for [R], one obtains the following relationship:

$$[R] = n - [HR] \tag{3}$$

This is then substituted into equation (2):

$$K_a = [HR]/[H](n - [HR]) \tag{4}$$

This substitution is necessary because there is no way to measure free receptor. In contrast, the hormone can be radioactively labeled, and free label represents free hormone (F = [H]), whereas bound label represents bound hormone (B = [HR]). For operational convenience, these new abbreviations are used in equation (5), and both sides are multiplied by ($n - B$):

$$K_a = B/F(n - B) \tag{5}$$

or

$$K_a n - K_a B = B/F \tag{6}$$

This is the equation for a straight line where B/F is the ordinate, and B is the abscissa (Fig. 5-2, *A*). The *x*-intercept (B/F = 0) is the total number of receptors (*n*), whereas the slope is the negative value of the association constant $(-K_a)$. This is the Scatchard plot. However, some authorities recommend rearranging equation (6) and performing a direct linear plot (Fig. 5-2, B):

$$n/B - K_d/F = 1 \tag{7}$$

In this technique, the B and F from each sample are plotted as separate points along the axes: B on the ordinate (0,B) and F on the abscissa $(-F, 0)$. These two points are connected to give a straight line. There will be one such line for each sample, and all these lines will intersect at a single point. The reflection of this intersecting point onto the ordinate (B) is the total number of receptors (n); the reflection onto the abscissa $(-F)$ is the dissociation constant (K_d). Table 5-1 contains actual data from the scientific literature (1); these are the data plotted in Fig. 5-2, B.

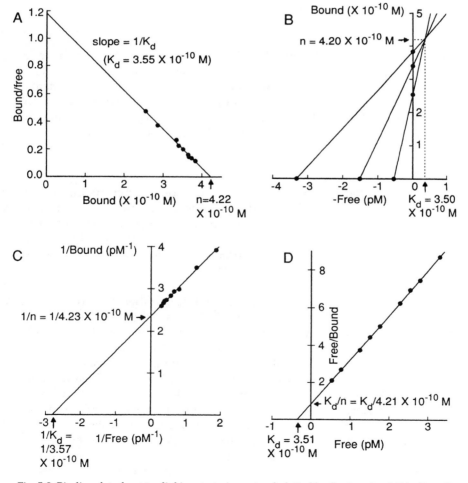

Fig. 5-2. Binding data for estradiol in rat uterine cytosol plotted by the Scatchard (A), direct linear (B), Lineweaver-Burk (C) or Hanes method (D). The original data are given in Table 4-1; for clarity, only three data points are plotted in (B): the first, the fourth, and the last. Plots (A) and (B) were adapted and reprinted by permission from Ref. 1. Copyright © 1976 Academic Press, Inc.

Table 5-1

Equilibrium Concentrations of Receptor-Bound (B) and Free (F) 17β-Estradiol in Samples of Rat Uterine Cytosol

Bound (fM)	1/Bound (pM^{-1})	Free (fM)	1/Free (pM^{-1})	B/F	F/B
255	3.92	538	1.86	0.473	2.11
286	3.5	771	1.3	0.371	2.7
334	2.99	1252	0.8	0.267	3.75
340	2.94	1510	0.66	0.225	4.44
352	2.84	1763	0.57	0.199	5.03
365	2.74	2279	0.44	0.16	6.25
367	2.72	2541	0.39	0.144	6.94
374	2.67	2798	0.36	0.134	7.46
384	2.6	3312	0.3	0.115	8.7

Others have advocated the Lineweaver-Burk plot (equation [8]; Fig. 5-2, C), the Hanes plot (equation [9]; Fig. 5-2, D), or computer analysis of nontransformed, nonlinear data:

$$1/B = K_d/nF + 1/n \tag{8}$$
$$F/B = K_d/n + F/n \tag{9}$$

As can be seen in Fig. 5-2, excellent data yield identical results by any method, but when data are less than perfect, some claim that the Scatchard plot accentuates the error. In truth, no method is perfect. In a recent comparison of these methods, it was shown that the Lineweaver-Burk plot gave the best accuracy but the poorest reproducibility, whereas the Eadie-Hofstee plot, which is very similar to the Scatchard plot, gave the best reproducibility but the poorest accuracy.

Assumptions

Recently, the Scatchard analysis has come under some criticism, but most of this criticism is misplaced. The Scatchard plot is only the result of a mathematical derivation; the problem is that several assumptions must be made and most real systems do not satisfy all of them. Therefore for the limitations of the Scatchard analysis to be appreciated, these assumptions and other cautions are discussed in detail.

  1. The labeled hormone is biologically identical to the native hormone. Most peptide hormones are iodinated with Na[^{125}I]; this rather large, electronegative atom can alter both the physical and biological properties of the hormone. For example, ^{125}I-labeled calcitonin has 9.3 times the affinity and 25 times the biological activity as unmodified calcitonin. In addition, this isotope is a powerful gamma emitter that will progressively damage the hormone. Even tritiated hormones cannot be assumed to be indistinguishable from unmodified hormones; heavily tritiated steroids can have a significantly higher molecular weight than the endogenous hormone, and the tritium could affect hydrogen bonding with the receptor. Furthermore, tritiated compounds older than 2 months have shown signs of degradation. Ideally, the labeled hormones should be tested in a sensitive biological assay to ensure that the biological activity is preserved; then they should be periodically checked for damage and repurified as needed.

  2. The labeled hormone is homogenous. Although iodination conditions can be adjusted to give an "average" of one iodide atom per hormone molecule, the procedure actually generates a mixture of uniodinated, monoiodinated, and

Table 5-2
The Kinetic and Biological Properties of Insulin Iodinated by Chloramine T[a]

Subunit and residue iodinated	Yield (%)	Receptor data		Biological activity[b]		Conclusion
		$K_a(M)$	n (nM)	Antilipolysis (%)	Glucose oxidation (%)	
A14	50	2.0×10^9	>0.65	100	102	Same as native insulin
A19	30	1.2×10^9	0.43	75	—	Less active
B16	10	0.9×10^9	2	94	93	Slightly less active
B26	10	1.4×10^9	1.1	129	119	More active

[a]From Kienhuis, C.B.M., Heuvel, J.J.T.M., Ross, H.A., Swinkels, L.M.J.W., Foekens, J.A., and Benraad, T.J. (1991). *Clin. Chem.* 37, 1749-1755; Laduron, P.M. (1984). *Biochem. Pharmacol.* 33, 833-839.
[b]Versus unmodified insulin.

diiodinated species. Even monoiodinated hormones may have the iodide located on different tyrosines, and each species may have different properties. For example, insulin has four tyrosines, and iodination with chloramine T will yield a mixture of products, each with a different tyrosine labeled. As can be seen in Table 5-2, these species have slightly different biological activities, and each generates slightly different kinetic data. Similar results have also been shown for epidermal growth factor (EGF) and glucagon.

3. The receptor is homogenous. Many receptors exist in multiple forms, such as the α- and β-adrenergic receptors (see also Chapter 7). Fortunately, in most tissues, one subtype predominates; for new systems, this possibility must always be eliminated. If both receptors are present in a sample and if their numbers and/or affinities are sufficiently different, the Scatchard analysis will yield a curvilinear plot (Fig. 5-3). Under these circumstances, there are

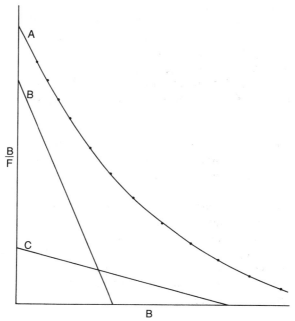

Fig. 5-3. A curvilinear Scatchard plot (A) generated by a mixture of receptors in the sample: a high-affinity, low-capacity group (B) and a low-affinity, high-capacity group (C).

statistical methods that can dissect out each receptor component from the single, concave curve.

4. *The receptor acts independently.* Many of the cytokines have receptors with multiple subunits; alone, each subunit has a low-to-intermediate affinity, whereas only the intact complex has high affinity (see Chapter 7). In other cases, ligand binding will induce the receptor to form a homodimer that has a higher affinity than the monomer. In still other examples, receptors can exhibit negative cooperativity, which would result in a nonlinear Scatchard plot (see Chapter 8) similar to that produced by multiple receptor types (Fig. 5-3). Finally, receptors can bind with accessory proteins that can then affect ligand affinity; for example, the β-adrenergic receptor interacts with a G protein that can increase the affinity of that receptor for its ligand. There are computer programs and mathematical models that can be used to distinguish among these different types of aggregating systems.

5. *The receptor is unoccupied.* If some of the receptor is already occupied by endogenous hormone and if the affinity is sufficiently high to prevent dissociation during sample preparation, then these occupied receptors will not be able to bind the labeled hormone. Such receptors are said to be "masked," and their existence results in an underestimation of n.

6. *The reaction is at equilibrium.* This assumption entails several other assumptions: (1) both the hormone and receptor are stable, (2) the reaction is reversible, and (3) the equilibrium is not perturbed when F and B are separated. The destruction or loss of either the hormone or the receptor would decrease the concentration of the "reactants" and shift the equilibrium toward dissociation. Many peptide hormones, especially insulin and glucagon, are susceptible to proteolysis; certain tissues, such as liver and fat, are rich in such proteases. Therefore protease inhibitors are frequently added to the assay tubes, and the labeled hormone in the supernatant should always be checked for degradation. For steroids, a related problem exists: tissues that have estradiol receptors also have an enzyme that can convert estradiol into estrone. This conversion can be prevented by adding an excess of dihydrotestosterone, which is a competitive inhibitor. Steroid receptors are also very labile and must be stabilized by molybdenum salts or sulfhydryl-protecting reagents, depending on the system. Some investigators attempt to circumvent this problem by using nonmetabolizable ligands, but the affinity of these synthetic compounds may differ from the natural ligand.

The measurement of membrane receptors in living cells is fraught with another problem: hormone binding to membrane receptors triggers the internalization and recycling of these receptors, and the hormone is destroyed (see Chapter 7). If the rate constants allow, incubations can be terminated before internalization begins. However, if equilibration requires a long time, the incubations can be performed at low temperatures, when membrane recycling does not occur. If membrane receptors are measured in broken cell preparations, the plasma membrane receptors will become contaminated with the intracellular receptors. In some systems, these latter receptors may represent 95% of the total receptors in the cell; they can be separated from the plasma membrane receptors by cell fractionation techniques.

Not even the assumption of reversibility is safe; about 5% to 10% of insulin becomes bound to its receptor by disulfide bonds. Whether this has any physiological importance or whether it is just a result of random disulfide interchange is unknown, but it obviously affects the equilibrium. EGF can also covalently bind to its receptor. In this case, the binding is an artifact of the chloramine T iodination. Chloramine T is a potent oxidizing agent (see later) that can activate several amino acids; this activation can then result in the formation of covalent bonds. In testicular membranes, the dissociation of labeled follicle-stimulating hormone (FSH) is enhanced by inhibitors of transglutaminase, suggesting that some FSH molecules become cross-linked to the receptor by isopeptide bonds. Lastly, this phenomenon has also been reported for steroids; 16α-hydroxyestrone, an estradiol metabolite, binds to a lysine in the estrogen receptor by means of Schiff base with subsequent Heyns rearrangement.

Finally, for F and B to be determined, they must be separated. For membrane receptors, the cells or membranes (B) can be centrifuged, leaving the free hormone (F) in the supernatant. For steroids, the receptors can be adsorbed onto one of several matrices. In either case, once the separation begins, a new equilibrium may become established; therefore separations are usually performed rapidly and at reduced temperatures. There are two techniques that can overcome the separation problem. In equilibrium dialysis, separation is virtually instantaneous. This technique uses a cell with two chambers separated by a dialysis membrane whose pore size will allow the free passage of hormone but not receptor. The receptor is placed in one chamber while the labeled hormone becomes freely distributed in both chambers. After equilibrium is reached, the solution in the receptor-free chamber is quickly evacuated and counted. The value of F is twice the measured counts (assuming the two chambers are of equal size), and that of B is the total number of counts added to the cell less the value of F. Unfortunately, this is an expensive technique when a large number of samples are being assayed. Furthermore, this method only works well when the receptor concentration is high relative to the ligand.

The second technique, laser scanning imaging, does not require separation at all. Cells are placed in the wells of microtiter plates and a fluorescent-tagged ligand is added. After equilibrium has been reached, the wells are scanned: ligand bound to receptors produces blips because the receptors are concentrated on cells. However, free ligand is uniformly distributed to generate a flat background; that is, free and bound hormone can be detected and measured without the need to physically separate them. Furthermore, this technique is rapid, sensitive, and capable of handling large numbers of samples.

7. There is no specific nonreceptor binding. The classic way to determine specific binding is to incubate the receptor and labeled hormone with and without an excess of unlabeled hormone; specific binding is equated with receptor binding (see previous section). Unfortunately, there are other, nonreceptor proteins that can exhibit high affinity and specific binding for hormones. Degrading enzymes are examples: tissues containing receptors for catecholamines or acetylcholine are also rich in catechol-*O*-methyltransferase or acetylcholinesterase, respectively. Partly for this reason, the natural hormone cannot be used in binding studies (see assumption 6); instead, labeled agonists are used. Although these substitutes are not degraded by these enzymes, they can still bind to them and do so with high affinity and specificity. Therefore this enzyme binding would show up experimentally as specific binding and compli-cate the interpretation of the results. Indeed, a putative auxin "receptor" isolated from henbane was recently shown to be glutathionine S-transferase, an enzyme that conjugates auxin. Other sources of specific, nonreceptor binding are the serum-binding globulins for steroids and thyroid hormones. When these receptors are measured in fresh tissue, care must be taken to remove as much blood as possible from the sample. Clearance receptors, which bind hormones for the purpose of internalization and degradation, represent another group of high affinity binding proteins that are not coupled to a biological effect. Finally, some receptors have truncated variants that do not couple to transducers; rather, they act as receptor inhibitors by binding and sequestering the hormone.

Iodination

There are several iodinated steroid and catecholamine derivatives available com-mercially, but peptide hormones are still usually labeled by individual investi-gators. Only the most commonly used methods are discussed. Central to most of these techniques is the oxidation of a radioisotope of iodide (I^-) to I^+, which then spontaneously incorporates into tyrosines (see Chapter 2). There are three popu-lar oxidizing agents: chloramine T, iodogen, and peroxide (Fig. 5-4). Chloramine T (N-monochloro-p-toluenesulfonamide) dissociates in water to form hypochlor-ous acid, the actual oxidizing agent; the reaction must be terminated by the addition of a reducing agent, sodium metabisulfite. This method is easy to perform, but the strong oxidizing and reducing agents often damage proteins. Iodogen (1,3,4, 6-tetrachloro-3a,6a-diphenylglycoluril) is very hydrophobic; it is usually dissolved in an organic solvent and coated onto the reaction vessel as the solvent evaporates. In essence, this is a solid-phase iodination, which can be terminated by simply decanting the supernatant. It is reported to be less dam-aging to proteins. Finally, peroxide and lactoperoxidase can be used. This is also a gentle technique because the enzyme allows for the gradual use of the peroxide. In truth, hormones vary greatly in their susceptibility to damage by oxidation, and the best iodination method is usually determined by trial and error.

If a hormone is very sensitive to oxidation, such that none of these techniques works, or if it lacks a tyrosine, or if the tyrosine is critically involved with receptor binding, several indirect iodination methods are available. One technique uses a preiodinated compound, which is basically a deaminated tyrosine whose carb-oxylic group is activated by succinylimide; this compound is the Bolton-Hunter reagent (N-succinylimidyl 3-[4-hydroxyphenyl]propionate). It is available in both the monoiodinated and diiodinated forms (Fig. 5-4). This compound will react with any free amino group such as the amino terminus or the ε-amino group of lysine. The hormone is never exposed to an oxidizing agent; however, the reagent is relatively expensive. If amino groups are unavailable, cysteines can be used. SDPE is a modified benzamide that possesses a sulfhydryl group and can form a disulfide bond with a cysteine in the peptide (Fig. 5-4). Finally, chelating groups can be reacted with hormones enabling them to complex with either radioactive (^{111}In) or fluorescent metals (Eu^{3+}). For example, the latter can bind diethylene triaminepentaacetic acid (DTPA), whose anhydride readily reacts with peptides.

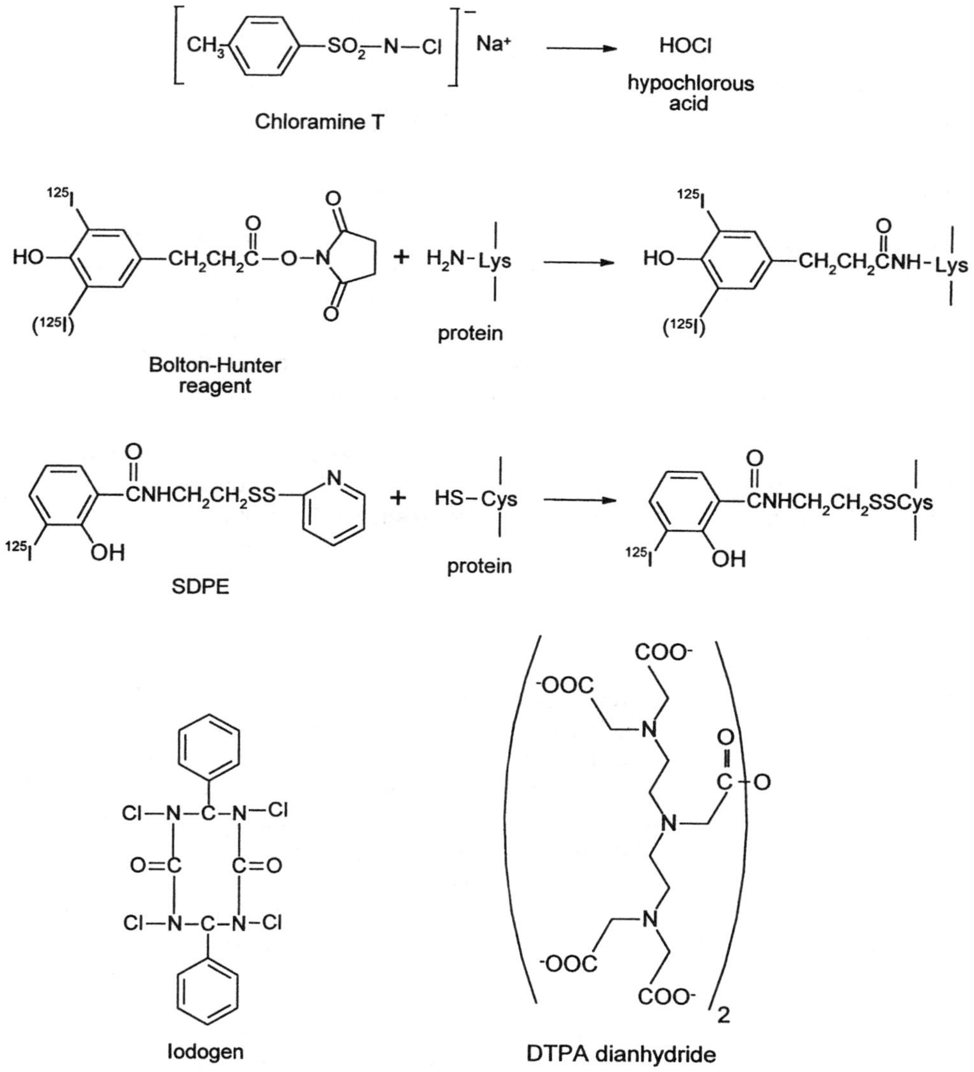

Fig. 5-4. Chemical structure and reactions of several iodinating reagents. DTPA, Diethylene triaminepentaacetic acid; SDPE, 1-(2'-dithiopyridyl)-2-(3'-iodo-2'-hydroxybenzamide)ethane.

Receptor Preparations

Membrane Receptors

Membrane receptors can be assayed in many different sources. Tissue fragments, freshly isolated cells, and established cell lines may all internalize the hormone-receptor complex (see assumption 6 in numbered list). Explants have the additional problems of having high nonspecific binding and of presenting the labeled hormone with a diffusion barrier. These problems can be circumvented with freshly isolated cells, but these cells are obtained by treatment with collagenases contaminated with proteases. The latter enzymes have been shown to damage membrane receptors. Established cell lines lack all of these problems, but data derived from them may not accurately reflect the situation in normal tissues.

These receptors can also be assayed in membrane fragments, but if there is a sizable intracellular pool of receptors, they will have to be removed by fractionation. The sensitivity of a cell is determined by its surface receptors, so the presence of contaminating, internally located receptors will give misleading results.

Nuclear Receptors

Classically, nuclear receptors were assayed in either the cytosol or the nucleus. Untransformed receptors resided in the cytoplasm; after ligand binding and activation, the hormone-receptor complex migrated to the nucleus. However, it is now known that these receptors continuously cycle between the cytosol and nucleus, although the individual transit times may differ significantly from one receptor to another. For some steroids, the unoccupied receptors had a weak affinity for chromatin, and homogenization leached them into the buffer-diluted cytosol. After steroid binding, the affinity increased and the complex remained in the nucleus even during homogenization; therefore the hormone-receptor complex appeared to translocate from the cytoplasm to the nucleus. This phenomenon is discussed further in Chapter 6.

Summary

Receptors can be measured with several techniques; each has its advantages and disadvantages. Hybridization assays for receptor mRNA and immunoassays for the protein are easy and accurate, but the former may not correlate with the actual number of receptors and the latter cannot give kinetic data and may not be able to distinguish among the different possible states that a receptor may occupy.

Receptors can also be determined by kinetics. Using the laws of mass action, one can derive equations that describe the interaction between a hormone and its receptor and from which one can calculate receptor number and affinity. However, there are several assumptions implicit in these derivations:

1. The labeled hormone is both homogeneous and biologically identical to the native hormone.

2. The receptor is homogeneous and acts independently.

3. The reaction is at equilibrium; that is, the hormone and receptor are stable, the reaction is reversible, and the equilibrium is not perturbed during the separation of F and B.

4. There is no specific nonreceptor binding.

The accuracy of the data derived from these equations is dependent on how well any given system satisfies these assumptions.

Membrane receptor determination in living cells may be complicated by receptor internalization, diffusion barriers, or receptor damage during the cell isolation procedure. Furthermore, receptor measurement in cell fragments is complicated by a heterogeneous membrane preparation, which should be fractionated before assay.

Iodination involves the oxidation of radioactive iodide and its incorporation into protein. The protein can be directly iodinated by chloramine T, iodogen, or peroxide and lactoperoxidase. The first is a harsher method than the other two. Proteins can also be indirectly iodinated by coupling them to a preiodinated species that reacts with either amino or sulfhydryl groups, or they can be coupled to chelating groups that can complex with either radioactive or fluorescent metals.

References

General References

Hall, H. (1992). Saturation analysis in receptor binding assays: An evaluation of six different calculation techniques. *Pharmacol. Toxicol.* 71, 45-51.

Kienhuis, C.B.M., Heuvel, J.J.T.M., Ross, H.A., Swinkels, L.M.J.W., Foekens, J.A., and Benraad, T.J. (1991). Six methods for direct radioiodination of mouse epidermal growth factor compared: Effect of nonequivalence in binding behavior between labeled and unlabeled ligand. *Clin. Chem.* 37, 1749-1755.

Laduron, P.M. (1984). Criteria for receptor sites in binding studies. *Biochem. Pharmacol.* 33, 833-839.

Schumacher, C., and von Tscharner, V. (1994). Practical instructions for radioactively labeled ligand receptor binding studies. *Anal. Biochem.* 222, 262-269.

Wofsy, C., and Goldstein, B. (1992). Interpretation of Scatchard plots for aggregating receptor systems. *Math. Biosci.* 112, 115-154.

Cited References

1. Keefer, L.M., De Meyts, P., and Frank, B.H. (1982). Receptor binding properties of the four isomers of 125I-monoiodoinsulin. *Program Abstr. 64th Annu. Meet. Endocr. Soc.*, p. 333.

2. Peavy, D.E., Abram, J.D., Frank, B.H., and Duckworth, W.C. (1984). Receptor binding and biological activity of specifically labeled [^{125}I]- and [^{127}I]monoiodoinsulin isomers in isolated rat adipocytes. *Endocrinology (Baltimore)* 114, 1818-1824.

3. Woosley, J.T., and Muldoon, T.G. (1976). Use of the direct linear plot to estimate binding constants for protein-ligand interactions. *Biochem. Biophys. Res. Commun.* 71, 155-160.

CHAPTER *6*

Nuclear Receptors

CHAPTER OUTLINE

As noted in Chapter 1, all hormones can be classified according to their water solubility. Hydrophobic hormones have direct access to the cellular interior, where they bind soluble proteins. These receptors are actually ligand-regulated transcription factors. As such, they are covered in two chapters. In this chapter, their structure, classification, and metabolism are examined, and in Chapter 13, their transcriptional activity is discussed. These proteins include the receptors for a large number of small, hydrophobic molecules, including triiodothyronine, vitamins A and D, fatty acids, and the sterols and their derivatives.

Structure

The nuclear receptors (NRs) are modular in construction, having four to five distinct domains: (1) an amino-terminal A/B region, (2) a DNA-binding C region, (3) a hinge D region, (4) a ligand-binding E region, and occasionally (5) an F region extending beyond the E region (Fig. 6-1). This linear arrangement of functions is reflected in the structural organization; the receptor can be proteolytically cleaved into three pieces, the A/B, C and E regions, suggesting separately folded domains. This hypothesis was initially confirmed by electron microscopic study results that showed that the glucocorticoid receptor is composed of two globular domains, presumably the A/B and E regions, separated by a tether, the C region. This structural independence was further reflected in the functional interaction between the domains; the various regions can be rearranged in recombinant molecules, and the receptor will still retain its function. Finally, the C and E regions have had their three-dimensional structures determined; each region forms a separate module that folds independently.

A/B Region

The amino terminus is the most variable part of the NR, and it does not appear to have any stable structure. It is believed that the amino terminus assumes its final conformation only upon interacting with other parts of the receptor, with the DNA, and with transcriptional coactivators. The major function of the A/B region is in transcription activation. In all NRs, there are at least two transcription activating domains (TADs): one in the A/B domain (TAD1) and one in the E domain (TAD2).* The two are quite different based on their dependence on cell type, promoter, hormone response element, receptor concentration, and hormones. For example, TAF1 from the estrogen receptor (ER) is active in yeast but not HeLa cells, whereas TAF2 works in HeLa cells but not yeast; both function in fibroblasts. Part of this difference may be related to the fact that TAF1 can act at simple promoters but TAF2 requires complex promoters. The number and type of DNA sequences that these receptors bind are also important; such sequences are generally called *hormone response elements* (HREs). Specific HREs are designated by replacing the H by the first letter of the particular hormone; for example, GRE is the glucocorticoid HRE and ERE is the estrogen HRE. TAF1 in ER requires multiple copies of the ERE for optimal activity. TADs can also exhibit gene selectivity; for example, TAF1 is required for S2 gene transcription but not for

---

* Although this text uses the designation TAD for most nuclear receptors, for historical reasons, it uses τ1 and τ2 for glucocorticoid receptors and TAF1 and TAF2 for estrogen receptors.

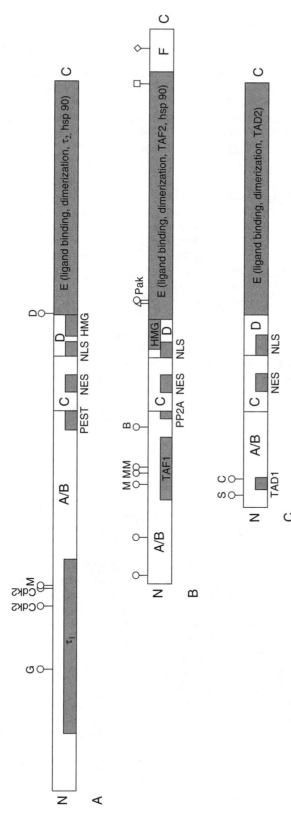

Fig. 6-1. Schematic representation of the (A) glucocorticoid, (B) estrogen, and (C) triiodothyronine receptors, showing the various functional domains. The receptors contain *777* (A), *595* (B), and *456* amino acids (C). *Circles* represent serine or threonine phosphorylation sites. *A*, PKA; *B*, PKB; *C*, PKC; *Cdk2*, cell cycle-dependent kinase 2; *D*, DNA-dependent protein kinase; *G*, glycogen synthase 3 kinase; *M*, MAPK; *S*, casein kinase 2); *squares* are tyrosine phosphorylation sites; *diamonds* represent *N*-acetylglucosylated amino acids; *triangles* are acetylated residues. *HMG*, High mobility group nuclear protein; *hsp*, heat shock protein; *NES*, nuclear exit signal; *NLS*, nuclear localization signal; *PEST*, proline, glutamic acid, serine, threonine-rich domain; *PP2A*, protein phosphatase 2A; *TAD*, *TAF*, τ, transcription activation domains.

vitellogenin gene activation, and TAD1 in the retinoic acid receptor β (RARβ) is required for transcribing the cellular retinal binding protein II (CRBP II) gene but not the cellular retinoic acid binding protein II (CRABP II) gene. Finally, TAD activity can be influenced by receptor concentration: TAF1 of the ER is estradiol (E_2)–independent and requires high ER concentrations, whereas TAF2 is E_2-dependent and acts at lower ER levels.

The androgen receptors (ARs) and mineralocorticoid receptors (MRs) actually contain two TADs in their long amino termini; TAD1a is ligand-dependent, like TAD2, whereas TAD1b is ligand-independent.

Several other functions have frequently been attributed to the A/B domain, although two of them are strongly related to the transcriptional activity of this region: receptor synergism and gene selectivity. Receptor synergism refers to the phenomenon where transcription activation by multiple HREs is greater than the sum of the effects of the individual HREs. In the glucocorticoid receptor (GR), ER, and progesterone receptor (PR), this synergism has been mapped to TAD1. The mechanism of transcription activation is thought to involve the binding of TAD1 to transcription factors through coactivators. This could easily explain the synergism: the more receptors bound near the gene, the more likely that the necessary factors will be attracted and stably retained in the vicinity. Gene selectivity may reflect the fact that many induced genes have requirements for specific factors; therefore the constellation of transcription factors bound by TAD1 could activate a different set of genes than those bound by TAD2.

In the GR, the amino terminus also contains a PEST domain. This name arises from the one-letter designation of the four amino acids predominant in this domain: proline (P), glutamic acid (E), serine (S), and threonine (T). These regions control the stability of proteins. Finally, the extreme amino terminus of the PR A contains a transcription repressor that is lacking in the amino-terminal–truncated PR B isoform.

C Region

The C region, or DNA-binding domain (DBD), consists of two zinc-binding sites (Fig. 6-2, *A*). Each zinc-binding site is coordinated by two pairs of cysteines located 10 to 15 amino acids apart. These structures were christened "zinc fingers" because it was thought that the peptide between the cysteine pairs would loop out like a finger and wrap around the DNA. When the actual structure of the DBD was solved, it was discovered that the structure is dominated by two α-helices; each one begins in the carboxy-terminal knuckle of the fingers and extends past the fingers. The amino-terminal helix lies in the major groove and makes sequence-specific contacts with the DNA bases, whereas the carboxy-terminal helix lies on top at right angles (Figs. 6-2, *B* and 6-3). As such, the structure resembles the helix-turn-helix transcription factors of prokaryotes. The P box is located at the amino terminus of the first helix and contains the amino acids that actually recognize the HRE.

In several members of the thyroid receptor (TR) family, there are two additional domains carboxy-terminal to the second finger: the T box participates in dimerization (see later), and the A box (also called the *H box*) binds in the minor groove to the AT-rich region upstream of (5′ to) the HRE. In the nuclear magnetic resonance (NMR) structure this A box forms an α-helix, whose amino terminus abuts the first helix and zinc site. The basic residues in this helix make contact with the deoxyribose-phosphate backbone of the DNA and enhance DNA

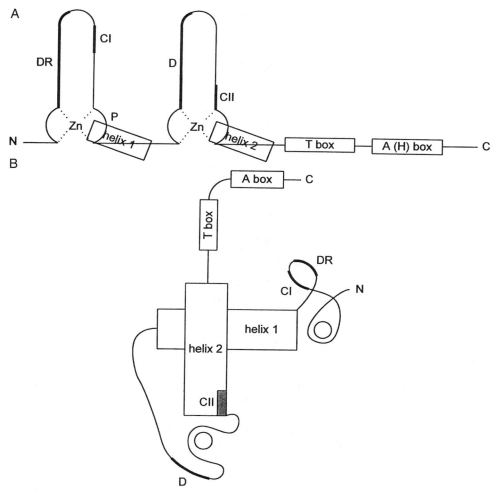

Fig. 6-2. A linear (A) and three-dimensional representation (B) of the DNA binding domain and adjacent regions of nuclear receptors.

binding. The A box may also improve DNA binding by stabilizing the DBD because some of its residues help form the hydrophobic core holding the first two helices together. Although this region is unstructured in GR and ER, functional studies indicate that an A box also exists in these steroid receptors. These NRs have DNA affinities in the range of 10^{-10} to $10^{-11}\,M$.

The DBD also possesses several minor dimerization domains: CI and the direct repeat (DR) box are located on the first finger; the D box and CII reside on the second finger; and the A box is carboxy-terminal to the second finger (Fig. 6-2). The simplest mode of dimerization is represented by the ER and the GR families, which bind palindromic HREs in a head-to-head fashion. In this orientation, the amino-terminal half of each second finger lines up in an antiparallel manner to form hydrogen and ionic bonds (Fig. 6-3). This bonding is often too weak to stabilize the dimer in solution, but it may participate in cooperative binding on HREs.

The members of the TR family can bind DR, suggesting that they dimerize head-to-tail. In this case, the D boxes would not juxtapose one another. In addition, many of these receptors bind the same DR; in part, their specificity

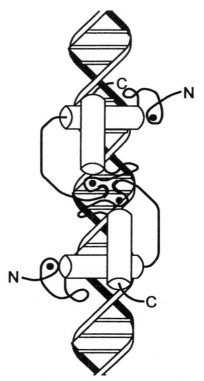

Fig. 6-3. A nuclear receptor dimer binding a palindromic HRE in DNA. The heavy dots represent zinc ions and the thickened segments are the D boxes.

actually resides in the spacing between the DRs rather than in the nucleotide sequence itself. These DRs are identified by a numerical suffix that represents the number of nucleotide base pairs (bp) between the repeats; for example, DR5 has 5 bp between direct repeats. This variable spacing introduces another problem: because DNA is a double helix, the monomers will not only vary in their distance from each other but also in their lateral translation about the spiral. These problems are solved by the use of multiple dimerization domains (Fig. 6-4). Indeed, these domains can actually dictate the DR spacing required by any given dimer, thereby imparting DNA binding specificity onto the receptors.

Although the major TADs have been associated with the A/B and E domains, several transcription coactivators and repressors have been shown to bind the DBD. They include, for example, elements of the basic RNA transcriptional machinery, other transcription factors, and histone acetyltransferases. Finally, a nuclear exit signal (NES) is located between the two fingers and shifts the receptor into the cytoplasm when it is unoccupied.

D Region

This region is also known as *the hinge* because it is located between the DBD and the ligand-binding domain (LBD). Its major function is in nuclear localization. This task is accomplished by a specific sequence, the nuclear localization signal (NLS), which is recognized by the nuclear transport system: $(+)-(+)-X_{10}-[(+)_{3-4}, X_{2-1}]$, where $(+)$ represents a basic amino acid and X, any amino acid. This sequence is often preceded by a casein kinase (CK2) site, whose

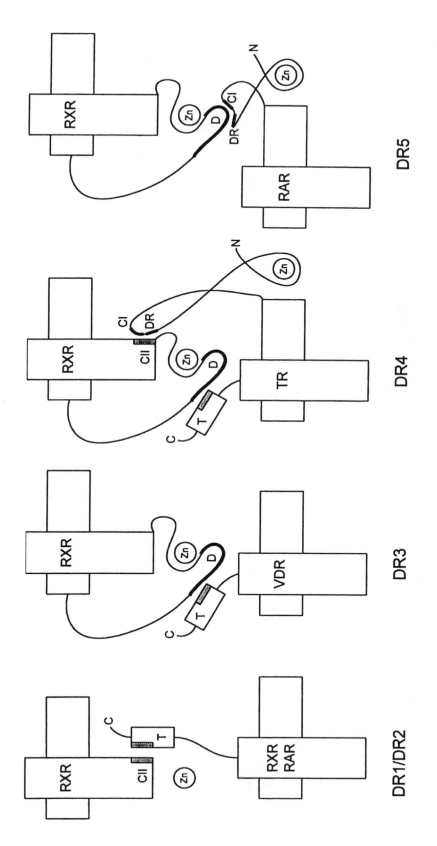

Fig. 6-4. Nuclear receptor dimerization on HREs that are direct repeats (DRs) separated by various distances. *RAR*, Retinoic acid receptor; *RXR*, retinoic acid X receptor; *TR*, thyroid receptor; *VDR*, vitamin D receptor.

phosphorylation accelerates nuclear translocation. There is still some controversy over whether the NLS is constitutively active or activated on ligand binding; this question is covered in the section on Metabolism later.

Several other functions have also been associated with the hinge. First, because the A and T boxes technically reside in the D region, the hinge participates in DNA binding and dimerization as previously described. Second, some ligand-binding determinants extend into the D region from the LBD. Finally, a few transcriptional coactivators and repressors have been shown to bind the hinge region. The most notable one for the steroid receptors is the high mobility group B protein (HMGB), which increases the DNA binding affinity of these NRs.

E Region

The E region is also known as the LBD because hormone binding is one of the four major functions residing in this region. The other three functions are transcriptional activation, dimerization, and heat shock protein (hsp) binding. The LBD is a three-tiered, α-helical sandwich (Fig. 6-5); each layer of helices is approximately perpendicular to the adjacent layers. In the aporeceptor, helix 11/12 (so named because it is kinked and could be considered two helices) lies outside the plane of the third layer and exposes the hydrophobic core. This is the door through which the hydrophobic ligand enters. When helix 11/12 closes behind the ligand, it creates two important surfaces. First, the third layer of helices becomes flat and acts as a dimerization interface. Second, it completes a groove at the edge of the LBD; this groove represents TAD2. The size and character of this groove is such that it attracts α-helices having the sequence,

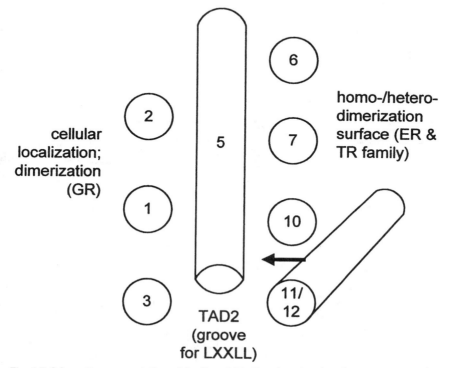

Fig. 6-5. Schematic representation of the ligand-binding domain of nuclear receptors. Helices 4, 8 and 9 are aligned behind helix 5 and are therefore not visible.

LXXLL (called either a *nuclear receptor interaction domain* [NID] or *an NR box*). NIDs are present in almost all transcriptional coactivators and repressors that bind NRs. In the aporeceptor, the repressors preferentially bind TAD2 and maintain basal activities at very low levels. When helix 11/12 completes the groove, additional structural limitations are created that lead to a switch in preference for coactivators. Both the AR and PR have NID-like sequences in their TAD1; these sequences intramolecularly bind TAD2 and compete with some coactivators. Since not all genes require the same coactivators for expression, this competition results in the use of different coactivators, which leads to additional specificity and selectivity in transcriptional responses. The amino terminus of the ER also occupies this groove; however, it is displaced after ligand binding. In this case, its purpose is to inhibit ER until E_2 binds. This groove is not the only site in the LBD where coactivators can bind; for example, because Ski interacting protein (SKIP) lacks an NID, it binds to helix 10 of the vitamin D receptor (VDR).

Recent data suggest that the picture just presented is overly simplistic. Rather, NRs appear to be in a dynamic equilibrium between the active and inactive conformations. Ligands and coactivators stabilize the active conformation, corepressors stabilize the inactive conformation, and various covalent modifications may stabilize either structure depending on the site and the particular NR. The final conformation is determined by the integration of these factors. In fact, this mechanism is entirely logical when it is realized that ligand binding was a late, acquired property of NRs (see Chapter 18). Before ligands, these transcription factors were regulated in other ways; ligand binding has not replaced these other factors but has simply been added to them.

The major dimerization site in the ER and TR families is also found in the E region, and it is strong enough to stabilize receptor dimers in solution. The site is located on the third helical layer and is created when helix 11/12 moves into the plane (Fig. 6-5). The same site is used whether the receptors bind palindromes (head-to-head) or DRs (head-to-tail), suggesting that the linker between the DBD and the LBD is very flexible. In the TR and VDR, there is a second dimerization domain in helices 9 and 10. These helices intertwine with a similar structure in the LBD of the retinoic acid X receptor (RXR) to form a coiled coil. Because this interaction does not involve the flat plane of the third layer, it is ligand-independent. The GR differs from the other two families in that the dimer interface is on the opposite side of the LBD, where several short β-strands between the helices from each dimer form a β-sheet.

The surface of the first helical layer is involved with the binding of hsps, as well as the nuclear matrix and cytoskeleton. In addition to these well-known functions, several other activities are thought to reside in this region. These include a second NLS in the GR and transcription repressor activities located in the extreme ends of the TR E domain.

F Region

The F domain is simply defined as the region carboxy-terminal to the ligand-binding site. In many NRs, the entire carboxy terminus is involved with hormone binding; therefore there is no F region. Even in those receptors that have this domain, very few functions have been associated with it. However, in the ER, it is known to bind a coactivator.

Families

Glucocorticoid

There are 10 recognized NR families that contain members that bind ligands (Table 6-1). However, on the basis of physical and functional characteristics, they can be placed into three major groups (Table 6-2). The most recently evolved and most closely related family is the glucocorticoid family, which includes the receptors for glucocorticoid, aldosterone, progesterone, and androgen. All four members have an identical P box, which recognizes the HRE. Indeed, all four bind to the same HRE, which occurs as an inverted repeat or palindrome separated by 3 bp. Probably because of their recent evolutionary appearance, they have not yet diversified and only a few isoreceptors are known. *Isoreceptors* are structurally and functionally distinct receptors for the same hormone; they may arise from separate genes or from the same gene through alternate processing. For example, in chickens and human beings, there are two progesterone receptors that are identical except that the short form (PR A) lacks the amino-terminal 127–164 amino acids, depending on the species. It is still not clear whether PR A is a result of the use of an alternate promoter for transcription or the use of an alternate start site for translation. There is no difference between the two forms with respect to ligand binding and tissue distribution; in fact, the two forms can heterodimerize. However, the long form, PR B, is more active at a minimum promoter and in the presence of multiple progesterone response elements (PREs). Recently, a third PR (PR C) and a second GR (GRβ) and AR (AR-A) have been identified, but their significance is unknown. Some of these isoreceptors are species specific; for example, mice do not have GRβ.

The mineralocorticoid receptor (MR) is unusual in that it binds cortisol as well as aldosterone. Because there is much more cortisol than aldosterone in the serum, the MR should be primarily regulated by cortisol. However, aldosterone target tissues contain an enzyme that degrades cortisol, preventing it from ever reaching the MR. In these tissues, the MR is truly an aldosterone receptor. In other tissues that possess an MR but lack this enzyme, the MR becomes a second GR. Furthermore, because the MR binds cortisol with an affinity 10 times that of the classical GR, the MR can function as a supersensitive GR; for this reason, the MR is sometimes referred to as *GR I*, whereas the classical GR becomes GR II. These two GRs allow the body to respond to cortisol over a 100-fold concentration range.

There is a single AR for both testosterone and dihydrotestosterone (DHT), although its affinity for DHT is 10 times that for testosterone. As such, the AR is primarily a DHT receptor and functions as a testosterone receptor only in those tissues with high testosterone concentrations. The AR is unique among steroid receptors in that it can bind DRs in addition to palindromes. Binding to DRs requires the second finger and a region in the hinge. As such, it probably resembles the DR1/DR2 configuration in Fig. 6-4.

Thyroid

The members of the thyroid hormone receptor family are probably the oldest NRs and this is reflected by their extreme heterogeneity. They are distinguished from the glucocorticoid family by having a short A/B region and a unique P box. They also have a high affinity for DNA and can often bind the HRE in the absence of a

Table 6-1
Nuclear Receptors and their Ligands and Function

Family[a]	Receptor	Ligand	Function
1A	TR	T_3	Thyroid hormone activity
1B	RAR	All *trans* RA	Retinoid activity
1C	PPARα	Polyunsaturated fatty acids (PUFA), eicosanoids, LTB_4, oleylethanolamide	Fatty acid metabolism; LTB_4 negative feedback and inflammation inhibition
	PPARβ	Unknown	Placental implantation and decidualization
	PPARγ	PGJ_2, PUFA, lysophosphatidic acid	Adipogenesis
	PPARδ	Prostacyclin	Blastocyst implantation; inhibits PPARα and PPARγ
1F	ROR	Cholesterol	Cholesterol homeostasis
1H	LXR	Oxysterols	Cholesterol metabolism
	FXR	Bile acids, lanosterol	Bile acid and cholesterol metabolism
	EcR	Ecdysone	Ecdysis, etc.
1I	VDR	1,25-DHCC	Intestinal calcium absorption, etc.
	PXR	Pregnanes, lithocholic acid[b]	Steroid and xenobiotic metabolism
	SXR	Corticosterone[b]	Steroid and xenobiotic metabolism
	BXR	Alkyl esters of benzoic acids	Steroid and xenobiotic metabolism
	CAR	Androstenol, androstenal[b,c]	Steroid and xenobiotic metabolism
2A	HNF4α	Fatty acids	Lipid and carbohydrate metabolism
2B	RXR	9-*cis*-RA, docosahexaenoic acid	Common dimerization partner; embryogenesis
3A	ERα	Estradiol	Estrogenic activity
	ERβ	E_2, phytoestrogens, 5α-androstane-3β, 17β-diol	Estrogenic activity
3C	AR	DHT	Androgenic activity
	GR	Cortisol	Glucocorticoid activity
	MR	Aldosterone, cortisol	Renal sodium resorption; glucocorticoid activity
	PR	Progesterone	Progestional activity

[a] Families not having any receptors with known ligands have been omitted.
[b] Including several other exogenous xenobiotic agents.
[c] Inverse agonists; that is, they inhibit CAR, although only at unphysiological concentrations.

ligand. Finally, this group does not form stable complexes with the hsps, although they may transiently associate during synthesis. Despite having the same P box, the HREs bound by this group are very diverse both in sequence and arrangement; indeed, a single receptor may be capable of binding to several different HREs. Like all of the other NRs, these receptors dimerize, but there is considerable flexibility in this interaction, which allows them to bind HREs that occur as either direct or inverted repeats (Fig. 6-4). Some receptors can even bind to a singlet or $\frac{1}{2}$ HRE.

Most members of this family have multiple isoforms, usually as a result of separate genes or alternate mRNA splicing. In some cases, these forms can have dramatically different tissue distributions, activities, or regulation. For example, TRα1 is primarily found in skeletal muscles and brown fat and is downregulated

Table 6-2
Characteristics of the Major Nuclear Receptor Families

Characteristic	Glucocorticoid	Estrogen	Thyroid
Members	GR II, MR (=GR I), PR, and AR	ER and ERRs	TR, VDR, RAR, RXR, EcR, PPAR, FXR, JH, PXR, SXR, CAR, and LXR
Amino terminus	Long	Short-medium	Short
Dimers	Homodimers[a]	Homodimers[a]	Homodimers and heterodimers
Isoreceptors	Uncommon	Rare (ERα and ERβ)	Many
P box sequence	CGSCKV	CEGCK(A/S)	CEGCKG[b]
Hormone response element	Inverted repeat (TGTTCT) with 3-bp spacer	Inverted repeat (TGACCT) with 3-bp spacer	Solitary or repeat (inverted or direct) with variable spacer (TGACC but variable)
Ligand-dependence of DNA binding	Yes	Yes	Many bind DNA without ligand
Hsp 90	Yes	Yes, but poorly	No

[a] Heterodimers have been reported but are uncommon.
[b] With a few exceptions.

by triiodothyronine (T_3). TRα2 is an alternately spliced form of TRα1; TRα2 lacks part of the LBD and cannot bind T_3. As a result, it cannot activate transcription and will actually repress gene induction by the intact form. It is found in the brain. TRβ1 is ubiquitous and its levels are generally unaffected by T_3. Finally, TRβ2 is predominantly found in the pituitary gland.

This group is also known for its promiscuous dimerization; not only can the isoreceptors heterodimerize, but also totally different classes can interact. The RXR occupies a central position in this phenomenon because it can dimerize with every other known member in this group. Some authorities think that RXR is an obligatory partner; that is, all active members in this family are heterodimers with RXR. However, it is not clear how absolute this requirement is. In most cases, the homodimer has a low activity that is significantly elevated if RXR is added to the reaction. Until now, this low activity had been attributed to endogenous RXR, but it can be reproduced in the yeast system, where there is no endogenous RXR. Furthermore, RXR heterodimers and non-RXR homodimers have different HRE preferences; for example, only VDR homodimers bind perfect repeats. These data suggest that the RXR is an important modulator of receptor activity in this family but that it is not absolutely essential. When RXR is present, it usually binds to the first (5′) repeat; 9-*cis* retinoic acid is not required as a coagonist, but its effect is often additive to that of the ligand for the other member of the receptor pair.

Two members of this group have interesting ligand-activity relationships. The hepatic nuclear factor 4α (HNF4α) is involved with lipid and carbohydrate metabolism and can bind a variety of fatty acids. Myristoyl and palmitoyl coenzyme A (CoA) activate HNF4α, whereas stearoyl and some omega polyunsaturated acyl CoAs inhibit it. The constitutive androstane receptor (CAR) does not require a ligand for activity; however, either androstenol or androstenal can inhibit this constitutive activity. Such ligands are called inverse agonists. Although many receptors in this family can act as repressors in the absence of ligands, these are the only two examples where ligand binding can decrease activity.

There are two other NRs whose family membership is unclear: the ecdysone receptor (EcR) in insects and the peroxisome proliferator activated receptor (PPAR). Both have P boxes identical to the thyroid hormone receptors, but the EcR has a long A/B domain and prefers its HRE as an inverted repeat separated by a single bp. Despite this unique HRE, EcR must heterodimerize with a RXR homologue to induce transcription. The PPAR mediates the induction of cytochrome P450 IV and enzymes involved with the β-oxidation of long-chain fatty acids. Its ligand appears to include saturated and unsaturated fatty acids, especially arachidonic acid, which is a known hormone transducer (see Chapter 10). Therefore PPAR may represent either an intracrine or a second messenger receptor. Its HRE is completely unrelated to those for the other members of this family, but it does require heterodimerization with the RXR for DNA binding.

Estrogen

The estrogen receptor appears to occupy a position intermediate between the glucocorticoid and thyroid hormone families. Although it has a unique P box, its sequence is closer to that for TR and it has the same HRE except that the ERE is arranged as an inverted repeat with a 3-bp spacer. In fact, TR can bind to the ERE. Furthermore, the ER can bind DNA in a ligand-independent manner like TR. However, like GR it binds hsps, although poorly; it does not bind RXR; and its HRE is arranged similar to the GRE. This group also includes several estrogen receptor-related (ERR) sequences; it is not known if the ERRs bind ligands.

Orphans

Orphan receptors are proteins homologous to the NRs, but for whom no ligand has been identified. The ERRs and the chicken ovalbumin upstream promoter (COUP) transcription factor are examples. The latter is homologous to the seven-up (svp) transcription factor in the genus *Drosophila*; for this reason, the factor is sometimes designated COUP/svp. Because ligand binding is believed to be an evolutionarily acquired trait (see Chapter 18), these receptors may represent the more primitive condition where regulation occurs developmentally or by posttranslational modification. Alternatively, they may be intracrine receptors whose ligands are cellular metabolites.

Metabolism

Heat Shock Proteins

As the members of the GR and ER families are synthesized, they are incorporated into large complexes containing several hsps. Therefore before any discussion of activation and recycling of NRs, it would be worthwhile to first examine these proteins. Heat shock proteins are so named because they are induced during heat shock or other stress. Because such conditions often denature proteins, it should not be surprising to discover that most of the hsps are concerned with renaturing proteins: they may help to refold them, rearrange their disulfide bonds, or disaggregate them. As such, they are also called *protein chaperones*. These proteins are simply designated by their molecular weight in kilodaltons; the most important for NRs are hsp 90 and hsp 70. Both are ATPases and undergo cyclic binding to, and dissociation from, their substrates in coordination with adenosine 5'-triph-osphate (ATP) binding and hydrolysis. They may be required by some NRs because the hydrophobic core of the LBD must be exposed to allow the ligand access to its binding site, but the exposure of this hydrophobic surface predis-poses to aggregation. As such, these receptors are intrinsically unstable and require chaperones until the hormone binds and the core is isolated from the hydrophilic exterior.

Most NRs are believed to cycle between the nucleus and the cytosol. This equilibrium is not identical for all receptors; the unoccupied GR is almost exclu-sively located in the cytoplasm, whereas many members of the thyroid hormone receptor family are predominantly nuclear. The cycle described below is based on the GR, for which there is the most information (Fig. 6-6). The cycle begins when hsp 70·ATP binds GR as it begins to exit the ribosome. Heat shock protein 40, an hsp 70 cochaperone also called *YDJ1*, positions and locks hsp 70 in place. Heat shock protein 40 also stimulates the ATPase in the presence of GR to begin opening the ligand-binding pocket. Another cochaperone, p48, also called *Hip* (hsp 70 interacting protein), then stabilizes the ADP form of hsp 70 and promotes the interaction with GR. Finally, p60, also called *Hop* (hsp 70/hsp 90 organizing protein), binds and recruits the hsp 90 dimer and stabilizes the ADP form. Heat shock protein 90·ADP further opens the binding pocket.

p23 replaces p60 and catalyzes the exchange of ADP by ATP on hsp 70 and hsp 90. At this point, hsp 70 is supposed to leave, but there is some evidence that it may remain with the PR, and possibility with the GR and ER, too. Indeed, it may even be required to dissociate the ligand for receptor recycling. By compet-ing with coactivator binding, p23 has also been implicated in receptor recycling, thereby promoting disassembly of the DNA-bound receptor complex. Because immunophilins and p60 compete for the same site on the receptor, the exit of p60 allows the immunophilins to bind the receptor complex. Organ transplantation is

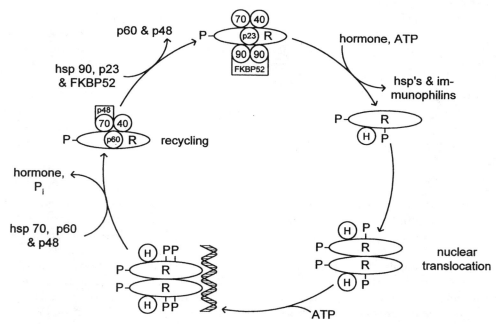

Fig. 6-6. Activation and recycling of a nuclear receptor. *40, 70, 90, p23, p48,* and *p60,* heat shock proteins; *FKBP52,* an immunophilin; *H,* hormone; *P,* phosphorylation site; *R,* receptor.

an important medical procedure but can be complicated by graft rejection. Several drugs have been empirically found to suppress the immune system and graft rejection. In an attempt to understand how these drugs worked, scientists determined with what cellular proteins the drugs interacted: these molecules became known as the *immunophilins*. There are three classes: Ess1, cyclophilin A, and FK506 binding protein (FKBP). They have several activities. First, they can facilitate nuclear transport: for example, FKBP52 recruits dynein to move GR along microtubules to the nucleus. Members from all three classes can also interact with histone deacetylases, which are involved with transcriptional control (see Chapter 13). Finally, they are *cis-trans* isomerases; this activity is required for FKBP52 to enhance the transcriptional activity of GR in the presence of limiting hormone. This potentiation is a result of an increased affinity toward glucocorticoids.

Activation and Recycling

The first step in activation is ligand binding, which has several effects. First, it triggers a change in conformation; the shift in helix 11/12 causes the receptor to become more compact and protease resistant. Second, it results in the dissociation of many of the hsps, their cochaperones, and the immunophilins. Third, it induces dimer formation, which is a direct result of the shift in helix 11/12 at the dimer-binding site. Fourth, if the complex is predominantly cytoplasmic, the dimer is translocated into the nucleus. This translocation is thought to result from exposure of the NLS after the hsps dissociate. Finally, ligand binding results in receptor phosphorylation (see later). The change in the LBD conformation and the dissociation of hsps may allow kinases greater access to phosphorylation sites. The temporal order of these effects is unknown.

Until recently, ligand delivery to the receptors had not been seriously considered. Because these hormones were hydrophobic, they were thought to diffuse freely throughout the cell. Although this assumption may be correct for some hormones, other ligands appear to need special transport proteins to carry them into the nucleus. For example, retinoic acid and T_3 are conveyed to the nucleus through CRABP-II and the thyroid hormone binding protein, respectively.

Once in the nucleus, the dimer binds an HRE via its DBD and binds coactivators or corepressors via the TADs; the latter leads to altered transcriptional activity. The shift in helix 11/12 after hormone binding is responsible for activating TAD2. However, there is still controversy as to whether the ligand is directly involved in DNA binding or whether the only function of the ligand is to remove the hsps and translocate the receptor into the nucleus. The fact that the isolated DBD is constitutively active might indicate that the ligand is unnecessary for this phenomenon. Unfortunately, these studies involve transfecting cells with the DBD fragment. This process can result in very high receptor concentrations, which have been shown to drive transcription when more physiological levels will not. Indeed, in the presence of normal receptor levels, ligands can increase the specificity of DNA binding and activate transcription.

When the ligand is withdrawn, the receptor complexes reappear faster than can be accounted for by *de novo* synthesis, demonstrating that the receptors must be recycled. This process is still an enigma, but growing evidence suggests that the receptor must be dephosphorylated. For example, if GR dephosphorylation is blocked, the GR will not recycle. Dephosphorylation may be required for exiting the nucleus and/or reassembly of the receptor complex.

Phosphorylation

NRs are phosphoproteins. Although many of the sites have been mapped and the relevant kinases identified, the scientific literature on the function of this phosphorylation remains very confusing. Some of the problems are methodological; for example, many studies actually measure phosphate turnover rather than stoichiometry. In addition, homologous sites in different receptors may have different functions, and adjacent sites in the same receptor may have opposite effects. For example, mitogen-activated protein kinase (MAPK) phosphorylation of GR in TAD1 stimulates transcription, but cell-cycle dependent kinase (Cdk) phosphorylation in the same region inhibits transcription. Phosphorylation may also have indirect effects. For example, phosphorylation often occurs in a defined sequence, where only the final modification has an effect on function. The first modification simply serves to facilitate or inhibit the subsequent modifications. This phenomenon is known as *hierarchical phosphorylation*. If investigators are unaware that a phosphorylation site is merely the initial one in such a sequence, they may conclude that this modification has no function. Finally, some phosphorylation sites are species, cell, and/or promoter specific.

There are three distinct phases of phosphorylation: basal phosphorylation is constitutive and occurs during or soon after receptor synthesis; hormone-induced phosphorylation occurs after ligand binding; and in the third round, phosphorylation by the DNA-dependent kinase (DNA-PK) occurs after DNA binding. The following are functions that have been attributed to NR phosphorylation; they are not necessarily mutually exclusive.

First, phosphorylation can affect receptor localization. As just noted, phosphorylation near the NLS can enhance nuclear transport, and dephosphorylation is required for the nuclear exiting and recycling of GR. Second, it can affect stability. GR has a PEST domain whose phosphorylation triggers receptor degradation. These serines and threonines can also be O-glycosylated with a single N-acetylglucosamine. This will block phosphorylation and increase the half-life of the receptor. In other contexts, phosphorylation can stabilize NRs; for example, MAPK phosphorylation of RXR prolongs its half-life.

One of the most studied effects of phosphorylation is on transcription: the phosphorylation of TAD1 by MAPK or related kinases recruits coactivators that enhance transcription. Interestingly, this form of activation can occur in the absence of ligands. For example, many growth factors synergize with steroid hormones. The basis for this interaction is the growth factor stimulation of MAPK, which then phosphorylates and activates an NR. Occasionally, receptor tyrosine kinases (RTKs) can directly modify some NRs; for example, the EGF receptor can directly phosphorylate the ER. Actually, this should not seem surprising; it is thought that NRs evolved from general transcription factors by the acquisition of an allosteric site for hydrophobic hormones. Before this step, these factors had to be regulated in some other way, such as by phosphorylation or stress (see Chapter 18), and these effects may be vestiges of this regulation.

Finally, phosphorylation may affect ligand binding, DNA binding, or dimerization. Any of these parameters may be either positively or negatively regulated, depending on the receptor and the site of modification.

Phosphorylation is not the only posttranslational modification that NRs can undergo. As previously noted, O-glycosylation with N-acetylglucosamine blocks phosphorylation in PEST domains and prolongs receptor half-life. In addition, both AR and ER can be acetylated on lysines. In the AR, acetylation is required for the ligand-dependent displacement of corepressors, which is a prerequisite for transcription. However, acetylation of ER is assumed to be inhibitory because mutants incapable of being acetylated exhibit increased E_2 sensitivity. This negative effect may be related to the fact that the modification site resides within a calmodulin (CaM)–binding domain and that CaM both stabilizes the ER and is required for its transcriptional activity. If acetylation blocks CaM binding, the ER would be inactivated and/or degraded. Finally, GR, PR, and AR can be sumoylated. Sumoylation is the process by which the carboxy terminus of a peptide called *Sumo* is coupled to the ε-amino group of lysines in a target protein. Among other effects (see Chapter 8), it can affect the transcriptional activity of NRs; for example, sumoylation of GR enhances transcription and facilitates synergism among multiple GRs. Conversely, transcription activation by PR is inhibited, but sumoylation is required for PR-induced transcription repression. Reports on the modified AR are conflicting.

Nongenomic Actions

Criteria

The best characterized binding proteins for hydrophobic hormones are transcription factors. As a result, the activity of these hormones became synonymous with transcription. However, there is now considerable evidence that in some instances ligands for NRs may act more directly on cellular metabolism without the need for gene induction. Because there is still some controversy about some of these effects, it is worth reviewing the evidence that can be used to support the existence of a nongenomic effect.

1. The effect does not require RNA or protein synthesis. This criterion is sufficient but not necessary. For example, a steroid may activate a transcription factor, but it may do so through second messengers rather than through an NR.

2. Levels of second messengers increase within seconds or minutes of hormone exposure. For example, rapid alterations in phospholipase C (PLC), inositol 1,4,5-trisphosphate (IP$_3$), diacylglycerol (DG), calcium, protein kinase C (PKC), soluble tyrosine kinase (STK), 3'5'-cyclic AMP (cAMP), 3'5'-cyclic GMP (cGMP), nitric oxide synthase (NOS), and phospholipase A$_2$ have all been documented in response to various NR ligands.

3. The effect exhibits a steroid specificity different from that of the classic receptor. This criterion would only be valid if the nongenomic effect were being mediated by a protein other than the classic NR; some nongenomic actions may be mediated by the classic receptor independent of direct transcription.

4. The hormone effector is immunologically distinct from the classic receptor. The same caveat listed for criterion (3) is applicable here.

5. The effect can be seen, for example, in cells lacking the classic receptor, in knockout (KO) mice, and in enucleated cells. Again, the same caveat listed for criterion (3) is applicable here. This problem can be circumvented with knock-in experiments using an NR with a discreet mutation that

selectively abolishes DNA binding; however, this solution assumes that the nongenomic effects are not mediated by the DBD. In addition, in KO mice there may be yet unidentified isoreceptors that partially compensate for the loss of the major isoform.

6. The hormone is only active on plasma membranes; this criterion is appropriate only when the effect is mediated by a plasma membrane receptor. A membrane-restricted effect could be demonstrated by showing that (1) the hormone is still active when it is bound to impermeable polymers, (2) it is inactive when injected intracellularly, or (3) it is active in cells transfected with NRs genetically engineered to contain a membrane localization signal.

7. Finally, if the effect is not mediated by the classic NR, the ultimate proof would be the purification and sequencing of the effector protein. Unresponsive cells could then be transfected with this gene to determine if they now displayed the nongenomic activity.

Some authorities have proposed using the terms *nuclear-initiated steroid signaling* (NISS) and *membrane-initiated steroid signaling* (MISS) instead of genomic and nongenomic activity. However, as described later, not all nongenomic action originates at the membrane, and the terms would technically exclude nonsteroids. Therefore this text will continue to use the terms *genomic* and *nongenomic*.

Mechanisms

Nonreceptor

There are several mechanisms by which a hydrophobic hormone may exert a nongenomic action. First, it may directly bind enzymes and alter their function without any intermediate receptor. The thyroid hormones provide the best examples of this mechanism: they directly stabilize the inactive monomeric form of pyruvate kinase, activate 3α-hydroxysteroid dehydrogenase, and abolish the ATP inhibition of cytochrome c oxidase. Both glucocorticoids and 1,25-dihydroxycholecalciferol (1,25-DHCC) can bind and translocate PKC. In addition to enzymes, several of these hormones have been reported to affect the cytoskeleton; for example, an E_2 metabolite binds microtubules and inhibits nucleation and propagation. Thyroxine (T_4) also binds microtubules and inhibits polymerization.

Receptor

Membrane receptors.   A hydrophobic hormone can also act by means of three types of membrane receptors: (1) its own classic receptor, (2) a nonclassic receptor, or (3) the receptor of another ligand. In the latter case, the hormone becomes an allosteric modulator for a receptor that primarily mediates the actions of another ligand.

The most economical situation would be to use the classic receptor to mediate the nongenomic actions because this protein already exists. However, because these proteins are designed to migrate to the nucleus, they have to be altered. For example, the sperm membrane PR has a novel amino terminus that may be a transmembrane domain, and the mitochondrial TR and RXR are truncated at the amino terminus. In contrast, the membrane ER from the uterus is generated from the same gene as the classic ER and has the same size. It is suspected that both the membrane and nuclear ERs in this tissue are identical and simply distribute between the two locations based on their relative affinity for either DNA or caveolin, a membrane protein. This conclusion is supported by the observation that E_2 binding causes the ER to dissociate from the plasmalemma.

However, there is also evidence that some hydrophobic hormones act through nonclassic receptors. The membrane VDR has an agonist profile distinctly different from the classic receptor; and several putative membrane receptors for progesterone have been cloned. The vascular smooth muscle membrane PR has a sequence that is unique but that topologically resembles the cytokine receptors, whereas the fish oocyte membrane PR has similarity to a G protein-coupled receptor (GPCR).

Finally, hydrophobic hormones can allosterically modulate receptors for other ligands. The best-known examples are the ligand-gated ion channels (Table 6-3). In addition, E_2 can bind the ANP receptor and enhance ANP stimulation of its guanylate cyclase; and progesterone binds and inhibits the oxytocin receptor, a GPCR.

In all of these examples, the hormone directly binds the receptor. There is one example of the hormone acting through an intermediary. Because of their hydrophobicity, steroids are transported in blood bound to the sex steroid binding globulin (SHBG), which has its own membrane receptor in reproductive tissues. E_2 and DHT bind SHBG in opposite orientations and induce different protein conformations: E_2 acts as an agonist allowing SHBG to elevate cAMP, whereas DHT acts as an antagonist by causing SHBG to dissociate from its receptor.

Intracellular receptors. Some nongenomic activities are mediated by intracellular receptors, which, like membrane receptors, may be either classic or nonclassic receptors. For example, there is a scaffolding protein with both NIDs and polyproline domains. The former domain binds TAF2 of E_2-bound ER and the latter attracts STKs with polyproline-binding modules. The STK then phosphorylates docking proteins, thereby initiating a cascade that ultimately activates MAPK (see Chapter 4). PR displays a more direct pathway that eliminates the

Table 6-3
Summary of the Direct Effects of Steroids on Ion Channels

Channel	Steroid	Effect
nAChR	Progesterone	Inhibition
	17β-Estradiol	Stimulation
GABA$_A$	3α, 5α and 3α, 5β derivatives of progesterone, deoxycorticosterone, and testosterone	Stimulation
	Pregnenolone sulfate and DHEAS	Inhibition
Glycine	Progesterone and pregnenolone sulfate	Inhibition
5-HT$_3$	17α-Estradiol, 17β-estradiol, progesterone, testosterone, and 3α, 5α-tetrahydroprogesterone	Inhibition
NMDA	DHEA and pregnenolone sulfate	Stimulation
	17β-Estradiol	Inhibition
AMPA	Pregnenolone sulfate	Inhibition
Kainate	17β-Estradiol and progesterone	Stimulation
	Pregnenolone sulfate	Inhibition

AMPA, α-Amino-3-hydroxy-5-methyl-4-isoxazole propionic acid; *DHEA*, dihydroepiandosterone; *DHEAS*, dihydroepiandosterone sulfate; *GABA$_A$*, γ-aminobutryic acid receptor, type A; *5-HT$_3$*, 5-hydroxytryptamine (serotonin) receptor, type 3; *nAChR*, nicotinic acetylcholine receptor; *NMDA*, N-methyl-D-aspartate.

need for a scaffold: the A/B domain of PR has a polyproline helix that becomes exposed after progesterone binding and that attracts STKs with polyproline binding modules. Again, STK activation eventually leads to MAPK stimulation. Finally, GR can bind and activate the first kinase in the MAPK cascade through an adaptor that recognizes phosphoserines in the GR.

NRs can also activate several other nongenomic pathways; for example, some NRs can directly bind and stimulate both phosphatidylinositol-3-kinase (PI3K) and p21-activated protein kinase (Pak). The VDR can bind and activate the serine-threonine phosphatases, PP1 and PP2A. It has even been reported that ER can bind and activate G_i.

Less work has been done on nonclassic intracellular receptors, but there is an ER in the uterus that differs significantly from the nuclear ER: it does not bind DNA, and classic E_2 antagonists are agonists on this receptor. It appears to be involved in nuclear export.

Summary

All known nuclear receptors consist of a single protein, which is modular in construction. The amino terminus contains one or two transcriptional activation domains; the midsection binds DNA, and the hinge is responsible for nuclear localization. The carboxy terminus is the most complex: it contains the LBD, the hsp-binding sites, the major dimerization domain, and another transcriptional activation motif.

The nuclear receptors are grouped into several families. The most recent to emerge is the glucocorticoid family, whose members are highly homologous, bind hsp 90, and recognize the same palindromic DNA sequence. In contrast, the thyroid family is probably the oldest because of the great diversity among its members. These receptors have many isoforms, exhibit promiscuous dimerization, do not bind hsp 90, and recognize a variety of DNA sequences, including direct repeats. The E_2 receptor seems to straddle these two families.

The unoccupied receptor exists within a complex and cannot bind DNA. The composition may vary, but classically it contains a single receptor, an hsp 90 dimer, one hsp 70, and various cochaperones and immunophilins. After hormone binding, the components dissociate, the receptor dimerizes, undergoes phosphorylation, and binds DNA, where it activates transcription. When the hormone is withdrawn, the receptor must be dephosphorylated before it can be recycled.

Steroids and related hormones may also have nongenomic actions mediated by hormone binding to membrane receptors, to enyzmes, or to allosteric sites on ion channels. In addition, the classic NRs can directly interact with elements of the cytoplasmic signaling pathways to generate second messengers.

References

Structure

Bourguet, W., Germain, P., and Gronemeyer, H. (2000). Nuclear receptor ligand-binding domains: Three-dimensional structures, molecular interactions and pharmacological implications. *Trends Pharmacol. Sci.* 21, 381-388.

Renaud, J.P., and Moras, D. (2000). Structural studies on nuclear receptors. *Cell. Mol. Life Sci.* 57, 1748-1769.

Families

Goodwin, B., and Kliewer, S.A. (2002). Nuclear receptors: I. Nuclear receptors and bile acid homeostasis. *Am. J. Physiol.* 282, G926-G931.

Lu, T.T., Repa, J.J., and Mangelsdorf, D.J. (2001). Orphan nuclear receptors as eLiXiRs and FiXeRs of sterol metabolism. *J. Biol. Chem.* 276, 37735-37738.

Noy, N. (2000). Retinoid-binding proteins: Mediators of retinoid action. *Biochem. J.* 348, 481-495.

Rosen, E.D., and Spiegelman, B.M. (2001). PPARγ: A nuclear regulator of metabolism, differentiation, and cell growth. *J. Biol. Chem.* 276, 37731-37734.

Tzameli, I., and Moore, D.D. (2001). Role reversal: New insights from new ligands for the xenobiotic receptor CAR. *Trends Endocrinol. Metab.* 12, 7-10.

Xie, W., and Evans, R.M. (2001). Orphan nuclear receptors: The exotics of xenobiotics. *J. Biol. Chem.* 276, 37739-37742.

Metabolism

Flint, A.P.F., Sheldrick, E.L., and Fisher, P.A. (2002). Ligand-independent activation of steroid receptors. *Domest. Anim. Endocrinol.* 23,13-24.

Freeman, B.C., and Yamamoto, K.R. (2001). Continuous recycling: A mechanism for modulatory signal transduction. *Trends Biochem. Sci.* 26, 285-290.

Hennemann, G., Docter, R., Friesema, E.C.H., de Jong, M., Krenning, E.P., and Visser, T.J. (2001). Plasma membrane transport of thyroid hormones and its role in thyroid hormone metabolism and bioavailability. *Endocr. Rev.* 22, 451-476.

Kato, S., Masuhiro, Y., Watanabe, M., Kobayashi, Y., Takeyama, K., Endoh, H., and Yanagisawa, J. (2000). Molecular mechanisms of a cross-talk between oestrogen and growth factor signalling pathways. *Genes Cells* 5, 593-601.

Kimmins, S., and MacRae, T.H. (2000). Maturation of steroid receptors: An example of functional cooperation among molecular chaperones and their associated proteins. *Cell Stress Chaperones* 5, 76-86.

Lannigan, D.A. (2002). Estrogen receptor phosphorylation. *Steroids* 68, 1-9.

Nollen, E.A.A., and Morimoto, R.I. (2002). Chaperoning signaling pathways: Molecular chaperones as stress-sensing 'heat shock' proteins. *J. Cell Sci.* 115, 2809-2816.

Pratt, W.B., and Toft, D.O. (2003). Regulation of signaling protein function and trafficking by the hsp90/hsp70-based chaperone machinery. *Exp. Biol. Med.* 228, 111-133.

Richter, K., and Buchner, J. (2001). Hsp90: Chaperoning signal transduction. *J. Cell. Physiol.* 188, 281-290.

Rochette-Egly, C. (2003). Nuclear receptors: Integration of multiple signalling pathways through phosphorylation. *Cell. Signal.* 15, 355-366.

Young, J.C., Moarefi, I., and Hartl, F.U. (2001). Hsp90: A specialized but essential protein-folding tool. *J. Cell Biol.* 154, 267-273.

Nongenomic

Borski, R.J. (2000). Nongenomic membrane actions of glucocorticoids in vertebrates. *Trends Endocrinol. Metab.* 11, 427-436.

Bramley, T. (2003). Non-genomic progesterone receptors in the mammalian ovary: Some unresolved issues. *Reproduction* 125, 3-15.

Harvey, B.J., Alzamora, R., Healy, V., Renard, C., and Doolan, C.M. (2002). Rapid responses to steroid hormones: From frog skin to human colon. *Biochim. Biophys. Acta* 1566, 116-128.

Heinlein, C.A., and Chang, C. (2002). The roles of androgen receptors and androgen-binding proteins in nongenomic androgen actions. *Mol. Endocrinol.* 16, 2181-2187.

Levin, E.R. (2003). Bidirectional signaling between the estrogen receptor and the epidermal growth factor receptor. *Mol. Endocrinol.* 17, 309-317.

Lösel, R., and Wehling, M. (2003). Nongenomic actions of steroid hormones. *Nature Rev. Mol. Cell Biol.* 4, 46-56.

Mellon, S.H., and Griffin, L.D. (2002). Neurosteroids: Biochemistry and clinical significance. *Trends Endocrinol. Metab.* 13, 35-43.

Norman, A.W., Henry, H.L., Bishop, J.E., Song, X.D., Bula, C., and Okamura, W.H. (2001). Different shapes of the steroid hormone $1\alpha, 25(OH)_2$-vitamin D_3 act as agonists for two different receptors in the vitamin D endocrine system to mediate genomic and rapid responses. *Steroids* 66, 147-158.

Pietras, R.J., Nemere, I., and Szego, C.M. (2001). Steroid hormone receptors in target cell membranes. *Endocrine* 14, 417-427.

Segars, J.H., and Driggers, P.H. (2002). Estrogen action and cytoplasmic signaling cascades. Part I: Membrane-associated signaling complexes. *Trends Endocrinol. Metab.* 13, 349-354.

Simoncini, T. (2003). Nongenomic actions of sex steroid hormones. *Eur. J. Endocrinol.* 148, 281-292.

Wrutniak-Cabello, C., Casas, F., and Cabello, G. (2001). Thyroid hormone action in mitochondria. *J. Mol. Endocrinol.* 26, 67-77.

CHAPTER **7**

Membrane Receptors

CHAPTER OUTLINE

This chapter covers the receptors for the hydrophilic hormones. Because these hormones cannot pass through the plasma membrane, their receptors are integral membrane proteins whose extracellular domain contains the hormone-binding activity. These receptors must also generate an output such that the hormone signal can be transmitted into the interior of the cell. This chapter is concerned with the structure, synthesis, and metabolism of these receptors; several other related topics are also covered.

Because membranes are lipid bilayers, all membrane receptors must have one or more hydrophobic domains that can traverse the plasmalemma. These domains most commonly assume the form of an α-helix. The existence of transmembrane regions and other structural domains is often based solely on cDNA sequence analysis. Although the cloning of receptors is currently commonplace, x-ray crystallographic studies are tedious, and such studies of transmembrane domains in their lipid environment are extremely difficult. Therefore the nature of the transmembrane domains is often inferred from the sequence data. Unfortunately, not all of the final characteristics of proteins can be determined from this approach; for example, one can only make intelligent guesses as to where the signal sequence ends, where cleavage may take place, which asparagines (Asn, N) might be glycosylated and which serines (Ser, S), threonines (Thr, T), and/or tyrosines (Tyr, Y) might be phosphorylated. Even the assignment of the transmembrane region can be difficult. Four such domains were originally postulated for the nicotinic acetylcholine receptor (nAChR); later the sequence data were reinterpreted as showing five transmembrane regions; then more complete data suggested that the original hypothesis of four helices was correct, but subsequent x-ray crystallographic data suggested that there was not enough room for four helices and that some transmembrane elements may have a β-structure. However, the most recent model has once again returned to the four-helical model.

Signaling Modules

The coding regions of most eukaryotic genes (exons) are interrupted by long stretches of noncoding sequences (introns). In many proteins, these exons appear to encode functionally independent domains; the wide spacing between them allows for their easy transfer to other proteins through nonhomologous recombination. In this way, proteins are built from smaller domains. Nowhere is this more apparent than in signaling molecules. Indeed, some modules are so important and widespread that they will be covered before the intact receptors are discussed.

The extracellular domain of receptors is designed to bind ligands and nothing appears to be suited better for this function than β-sheets (Table 7-1). The binding activity usually resides in the loops, which can vary their lengths, charge, and other characteristics to accommodate any ligand. The sheets stabilize the loops. The immunoglobulin (Ig), fibronectin type III (FNIII), and cytokine domains are thought to be evolutionarily related. The leucine-rich repeat (LRR) is also a β-sheet but differs from the others in its organization. It is composed of repeating α-helix/β-strand units; the β-strands meet to form a sheet, whereas the α-helices are pushed to one side. In many globular proteins, the sheet would then roll up into a β-barrel to form the hydrophobic core lined with α-helices on the outer surface. However, in the LRR, the circle is incomplete, creating a C. The inner β-sheet forms a binding pocket for the ligand.

Table 7-1
Ligand-Binding Motifs Found in Hormone Receptors

Motif	Structure	Example
EGF-like	6 β-Turns but only 2 β-strands are stabilized by disulfide bonds and calcium	EGFR
Ig	7-Stranded β-sandwich	PDGFR family; Trk
FNIII	7-Stranded β-sandwich	Cytokine receptors; Eph/Elk/Eck
Cytokine	FNIII domain with 2 disulfide bonds and a WS box; FNIII modules are nonlinear	Cytokine receptors
TNF	5-Stranded β-sandwiches end-to-end	TNFR; DR5
LRR	24-Amino acid αβ-unit with conserved leucine and aliphatic amino acids arranged in a parallel β-barrel	Glycoprotein receptors; Trk
Sema	7-Blade β -propeller	Semaphorin, HGF and MSP receptors

EGF, Epidermal growth factor; *EGFR*, EGF receptor; *Eck*, epithelial cell kinase; *Elk*, ephrin-like kinase; *Eph*, ephrin; *FNIII*, fibronectin III; *HGF*, hepatocyte growth factor, *Ig*, immunoglobulin; *LRR*, leucine-rich repeat; *MSP*, macrophage stimulating protein; *PDGFR*, platelet-derived growth factor receptor; *TNF*, tumor necrosis factor; *TNFR*, TNF receptor; *Trk*, tropomyosin-related kinase.

The intracellular domain of receptors, as well as the soluble transducers that associate with them, are involved with interactions with proteins, amino acids modified after translation, and membrane phospholipids (Table 7-2). β-Sheets also dominate the modules identified with these functions. As noted in Chapter 4, phosphorylation is a major regulatory mechanism and many motifs recognize phosphorylated amino acids. Phosphotyrosine (pY) can be recognized either by a phosphotyrosine-binding domain (PTB) or SH2. The latter was first discovered in the Src* tyrosine kinase; as such, they are called *Src homology* (SH) *domains*. There are actually three conserved domains in this soluble tyrosine kinase (STK): the SH1 is the catalytic site that is shared by all STKs, SH2 is the pY-binding site, and SH3 binds proline-rich sequences. The polyphosphoinositides (PPIs) are also phosphorylated rings and have been reported to bind PTB and SH2, but they appear to be minor ligands under most physiological conditions.

Phosphoserine (pS) and phosphothreonine (pT) are bound by several modules. Protein 14-3-3 is actually an entire peptide whose antiparallel α-helices form a tube slit lengthwise. Dimers of 14-3-3 create an omega in cross section; each trough can bind a pS from separate proteins. There are several isoforms with different substrate and dimerization properties. As homodimers, they sequester phosphoproteins in the cytoplasm; as heterodimers, they can facilitate signaling by bringing different elements of a transduction pathway together.

Because receptor tyrosine kinases (RTKs) are integral membrane proteins, all signaling pathways start at the plasmalemma; as a result, many of these transducers bind membrane lipids. Furthermore, one of the first signals involves alterations in PPI (see Chapter 10); therefore these phospholipids became natural targets for many of these binding domains. In fact, many directly bind PPI; however, C2 (protein kinase C [PKC] conserved region 2) uses calcium as an intermediary between membrane phospholipids and itself.

Binding motifs for proline-rich domains and polyproline helices are also abundant. There are several properties of these structures that make them attractive. First, proline is a hydrophobic amino acid that creates a hydrophobic strip along the helix. This favors protein-protein interactions without requiring

---

* By convention, the names of genes are designated by lower-case letters and sometimes italicized, whereas the protein products of those genes are capitalized.

Table 7-2

Motifs Found in Postreceptor Transducers

Motif	Specificity[a]	Structure	Example
Phosphotyrosine Binding			
SH2	pY and carboxy-terminal amino acids; DDXXY; PIP$_2$ (can displace pY)	100 Amino acids; central β-sheet with an α-helix on each side	Adaptors (e.g., Grb2, Nck)
PTB	NPX(pY); PIP > PIP$_2$	100–150 Amino acids; 2 perpindicular β-sheets with an α-helix at one end	Docking proteins (e.g., IRS, Shc)
Phosphoserine/Threonine Binding			
14-3-3	RSX(pS/pT)XP; RXXX(pS/pT)XP	9 Antiparallel α-helices form a C; dimer forms an omega	14-3-3
MH2	(pS)X(pS)	130–140 Amino acids; 11 β-stranded sandwich	Smad
FHA	(HΦ)X(HΦ)(pT)QX(D/HΦ)	130–140 Amino acids; 11 β-stranded sandwich	FoxO3a
WW (type IV)	(pS/pT)P	38 Amino acid β-sheet	Pin1 (prolyl isomerase)
Membrane Binding			
PH	PIP$_2$; pS; WD-40	100 Amino acids; 2 perpendicular β-sheets with an α-helix at one end	RasGAP, RasGEF, PKB, PLCγ
FYVE	Singly phosphorylated PI(3)P	70 Amino acids; 2 antiparallel β-sheets with an α-helix and 2 zinc clusters	Sara (TGFβR adaptor), an AKAP
PX	PI(3)P	130 Amino acids; 3 β-sheets; on one side is a polyproline helix between 2 pairs of α helices	PLD, PI3K, CISK
ENTH	PI(4,5)P$_2$	9 α-Helices in a solenoid	AP180
FERM	PI(4,5)P$_2$; protein-protein interactions	300 Amino acids; 3-domain unit	ERM (cytoskeleton), Jak
C1	Diacylglycerol	50 Amino acids; 5 short β-strands, 1 short α-helix and 2 zincs	PKC, Raf, DG kinase
C2	Calcium (intermediary between protein and phospholipids)	130 Amino acids; β-sandwich	PKC, PLCs, RasGAP, PLA$_2$
Polyproline Binding			
SH3	L-handed polyproline helix, type II (PXXP); self (dimerization); α-helix	60 Amino acids; 2 perpendicular β-sheets	Adaptors (e.g., Grb2, Nck)
WW (types I-III)	Proline-rich domains (XPPXY)	38 Amino acid β-sheet	Phosphoinositide phosphatases
RING	Proline-rich domains [PX(L/V)XP(A/S)XP]	50 Amino acids with 2 zincs in a cross-brace	Adaptor (e.g., TRAF)

continues

continued

Motif	Specificity[a]	Structure	Example
Calcium Binding			
EF hand	Calcium	30 Amino acids; helix-loop-helix	CaM, PLCs
Protein-Protein			
PDZ	Protein-protein; carboxy-terminal 4-5 amino acids with HΦ at carboxy terminus	80-90 Amino acids; β-sandwich with α-helix on 2 edges	PTP (FAP), adaptor (GRIP)
DD, DED, and CARD	Protein-protein interactions	90 Amino acids; 6 α-helices (2 perpendicular layers of 3 α-helices, DD)	TNFR and associated adaptors
Ankyrin	Protein-protein interactions	Antiparallel β-sheets backed by 2 rows of α-helices[b]	IκB
Armadillo	Protein-protein interactions	α-Helices in a triangular stack to form a superhelix	β-catenin
HEAT	Protein-protein interactions	37-43 Amino acid, antiparallel, kinked α-helices; 3-22 pairs stacked at 15°	A subunit of PP2A, PI3K, TOR
WD-40	Protein-protein interactions; WD-40 (dimerization)	β-Propeller; wedges form disk	Gβ, PICK
PAS	Protein-protein; protein-signaling molecule	5-Stranded, antiparallel β-sheet flanked on one side by several α-helices	Aromatic hydrocarbon receptor
Modified Amino Acid Binding			
Bromo domain	Acetylated lysine	110 Amino acid, 4 α-helical bundle	Histone acetyltransferase
Chromo domain	Methylated arginine	3-Stranded β-sheet perpendicular to a carboxy-terminal α-helix	HP1

[a] HΦ, Hydrophobic amino acid; X, any amino acid. For other abbreviations, see the Abbreviations table at the end of the book.
[b] Similar to LRR (Table 7-1) except that the repeating unit consists of two α-helices and a β-turn instead of one α-helix and β-turn; the units are arranged in two rows.

complementary surfaces. Furthermore, these van der Waals forces are weak, which permits rapid reversibility and interchange. Finally, phosphorylation can disrupt protein interactions by introducing a charge into this hydrophobic strip. Second, polyproline helices are relatively rigid so that they generate extended, exposed structures; this further facilitates protein interactions. Third, proline is a secondary amine that is poorly hydrated; this leaves the carbonyls free to form hydrogen bonds with protein partners.

There are many other domains designed to assist protein-protein interactions; most of them are structurally and functionally heterogeneous. The exceptions are the death domain (DD), death effector domain (DED), and caspase activity-recruiting domain (CARD). They all have very similar structures and are found on signaling molecules that are involved with programmed cell death, or apoptosis.

Finally, there are domains that recognize acetylated lysines and methylated arginines. These modifications are important in the nucleus (see Chapter 14).

These modules occur on receptors, effectors, intermediaries that connect receptors to effectors, and accessory proteins. The major intermediaries and accessory proteins include adaptors, docking proteins (or dockers), scaffolds, anchoring proteins, and coactivators.

Adaptors are small proteins that usually consist of only one or two types of modules; they basically act as a bridge between two other signaling components. For example, Grb2 (growth factor receptor-bound protein 2) has two SH2 domains that bind pYs on an RTK and one SH3 domain that binds a polyproline helix in a RasGEF. Docking proteins are larger and have multiple modules and phosphorylation sites. IRS (insulin receptor substrate) can be phosphorylated by RTKs; these pYs then become docking sites for transducers possessing SH2 domains, including Grb2, PI3K (phosphatidylinositol-3 kinase), and an SH2-containing protein tyrosine phosphatase (SHP). It also has a pleckstrin homology (PH) and a PTB domain.

Scaffolds are dockers that are specialized for binding components in a single signaling pathway. The best-known scaffolds bind multiple kinases in the mitogen-activated protein kinase (MAPK) cascade. The placement of successive transduction elements in physical proximity has several advantages. First, it overcomes the slow diffusion that occurs when multiple signaling components have to interact. Second, it spatially restricts signaling. Third, it increases specificity. For example, there are actually several MAPK families, and the specificity of some of the kinases is such that cross-activation of pathways could occur. However, this possibility is minimized by keeping all of the components of a single family together on a scaffold. On the other hand, other scaffolds may allow the regulated binding of elements from other pathways to permit signal integration. Finally, scaffolds can compensate for low activity of transducers by creating high local concentrations. One disadvantage of scaffolds is that they decrease amplification.

Some scaffolds are further specialized. A *superscaffold* is a scaffold that binds other scaffolds, creating a massive signaling complex. For example, the Shank family possesses an ankyrin, an SH3, and PDZ (PSD-95/DLG/ZO-1) domains; a polyproline helix; and a sterile α-motif (SAM). Many of these domains bind other scaffolds that can then bind various effectors. *Transducisome, signalosome,* and *signalplex* are loose terms referring to preassembled signaling complexes from the molecular (such as scaffolds) to subcellular level (such as the specialized membrane domains called caveolae). Some authors restrict these terms to transcription factor complexes, although the labels *enhanceosome* or *transcriptosome* are more appropriate for this latter function.

Anchoring proteins are proteins that localize effectors to specific cellular compartments. A kinase (PKA) anchoring protein 85 (AKAP85) fixes PKA to the Golgi complex. However, as more is learned about these anchoring proteins, it is being discovered that many of them are in reality docking and scaffolding proteins that also contain a localization module. Finally, coactivators connect specific transcription factors, which recognize hormone response elements (HREs), to general transcription factors, which actually initiate transcription. Many of these coactivators also have intrinsic activities, such as acetyltransferase activity.

Enzyme-Linked Receptors

Tyrosine Kinase Receptors

Receptors can be classified according to their output and structural relationships (see Fig. 4-1): (1) those that have intrinsic enzymatic activity, (2) those that interact with STKs, (3) those that interact with G proteins, (4) those that form ion channels, and (5) those that do not functionally or structurally fit into any of these categories.

All of the enzyme-linked receptors have the same basic structural organization (Fig. 7-1): an extracellular, amino terminus that binds the hormone, a single transmembrane domain, and an intracellular, carboxy terminus that possesses the catalytic site. Cellular tyrosine kinases can be both membrane bound and soluble; the latter is discussed in the section about cytokine receptors.

Three major functions reside in the extracellular domain: ligand binding, oligomerization, and glycosylation. Ligand binding and subsequent activation require oligomerization. In the simplest case, the only purpose of the ligand is to induce *de novo* dimerization, and this results in cross-phosphorylation of the activation loops in the intracellular kinase domains. However, some receptors exist as preassembled complexes and ligand binding tightens the subunits; the insulin and erythropoietin (EPO) receptors are examples of this mechanism. Finally, ligand binding may be required to properly orient the subunits, such as that occurring with the epidermal growth factor receptor (EGFR).

Dimerization is achieved in one of three ways. First, the ligand itself may be a dimer with each one binding one receptor; for example, platelet-derived growth factor (PDGF) is a dimeric ligand. Second, the ligand may be monomeric with multiple binding sites for the receptor; for example, growth hormone (GH) can bind two receptors simultaneously. Finally, a monomeric ligand may recruit proteoglycans to help assemble the receptor complex; this mechanism is discussed later in the section on the fibroblast growth factor (FGF) receptor. Although the extracellular domain plays a major role in oligomerization, the transmembrane and intracellular domains can also contribute in some receptors.

Because the amino terminus is extracellular, it is glycosylated. This posttranslational modification is important in receptor folding, transportation to the plasma membrane, and processing, such as cleavage. In general, it does not participate in specific binding. The transmembrane domain also serves three functions: it acts as a membrane anchor; it can facilitate dimerization and internalization through its role in lateral mobility; and it may help orient the kinase for activation.

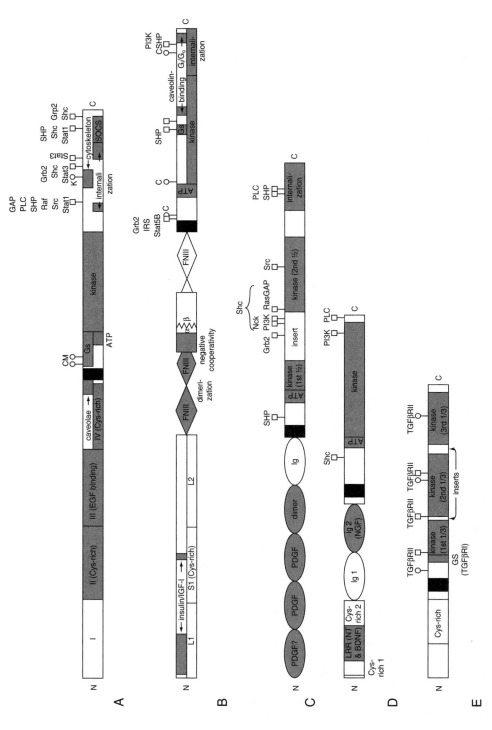

Fig. 7-1. Enzyme-linked receptors: (A) EGFR, (B) insulin/IGF-IR, (C) PDGF receptor (Trk), and (E) TGFβRI. These receptors have a single transmembrane region (black), with the amino terminus (N) extracellular and the carboxy terminus (C) cytoplasmic. The kinase domain of Trk is split, but the gap is so short that it is not visible at this scale. Immunoglobulin-like loops are represented by ovals, functional domains are shaded and labeled, and phosphorylation sites are depicted as either circles (pS or pT) or squares (pY). The responsible kinases are as follows: A, PKA; C, PKC; K, CaMKII; M, MAPK. The receptors contain 1186 (A), 1343 (B), 1067 (C), 790 (D), and 542 amino acids (E). *GAP*, GTPase activating protein; *Grb2*, an adaptor; *IRS*, insulin receptor substrate (a docking protein); *Nck*, an adaptor; *PI3K*, phosphatidylinositol-3 kinase; *PLC*, phospholipase C; *Raf*, a S-T kinase; *Shc*, a docking protein; *SHP*, SH2-containing protein tyrosine phosphatase; *SOCS*, suppressor of cytokine signaling; *Src*, a STK; *Stat*, a transcription factor.

The intracellular domain is the effector domain. It contains the tyrosine kinase, the autophosphorylation sites that both activate the kinase and attract signaling molecules having SH2 modules, and other effector-binding sites. In addition to its own kinase, RTK often bind and activate other kinases. Most STKs have SH2 modules and bind pY on RTKs; subsequent STK activation would allow tyrosine phosphorylation to become more widespread. RTKs can also bind S-T kinases; for example, the EGFR and platelet-derived growth factor receptor (PDGFR) bind p21-activated protein kinase (Pak) through the adaptors Grb2 and Nck. Finally, the intracellular domain possesses an internalization signal that triggers receptor endocytosis and either degradation or recycling after ligand binding.

There are six RTK families, although the classification is subject to some variation (Table 7-3). For example, some authorities group the nerve growth factor (NGF) and collagen receptors with the insulin receptor family because of the similarity in their kinase domains. Others split the FGF receptors from the PDGF receptor family. Finally, the recent discovery that the extracellular domains of the EGF and insulin families resemble each other would argue for a common grouping, although they are often still discussed separately for historical reasons. The classification in Table 7-3 is primarily based on the similarity in the extracellular domains and historical considerations.

RTKs can autophosphorylate themselves, and these pYs then become binding sites for proteins with SH2 domains. The binding is relatively specific between a particular pY and a given SH2 domain because the amino acids surrounding the pY are also important for SH2 recognition (Table 7-4). In addition, SH2 domains usually occur in pairs so that there are two sites that must match for efficient binding. Essentially, it is the substrate that has a binding site for the enzyme and not vice versa. There are several ways that this binding can affect the activity of the SH2-containing protein. First, binding the pY of a RTK brings the protein within reach of the RTK catalytic site, which could phosphorylate it. For example, phospholipase γ (PLCγ) is activated in part by being phosphorylated on tyrosine. Second, mere binding to pY may allosterically activate the protein; PLCγ, PI3K, and SHP are all allosterically activated by pY binding. Finally, because the RTK is an integral membrane protein, any protein attracted to it would be localized to the plasmalemma. Both PLCγ and PI3K are enzymes that act on phospholipids; by binding pYs, they are brought into proximity to their substrates. Essentially, they are activated by the high local concentration of phospholipids. The biological activity of any given RTK will depend on what substrates it attracts and activates.

The Epidermal Growth Factor Receptor Family

The EGFR family is characterized by two clusters of cysteines in the extracellular domain, no internal cleavage, and an intact kinase domain (see Fig. 7-1, *A*). There are four members of this family: human EGFR 1 (HER1 [human EGF receptor homologue]), HER2 (also called *Neu*), HER3, and HER4. These receptors form homodimers and heterodimers to bind an increasingly large number of hormones, including EGF, transforming growth factor α (TGFα), heparin-binding EGF-like growth factor (HB-EGF), amphiregulin (AR), betacellulin (BTC), Neu differentiation factor (NDF), and the neuregulins (NRGs). For example, EGF and TGFα bind HER1 homodimers, whereas most of the neuregulins prefer HER2-HER3 dimers. HER2 and HER3 are unusual. First, no high affinity, endogenous agonists have been found for HER2 homodimers. Second, HER3 has acquired deleterious mutations in its catalytic domain. As a result it lacks any kinase

Table 7-3
Receptor Tyrosine Kinase Families

Family	Ligands	Extracellular motifs	Kinase insert	Cleavage	Other
EGF	EGF, TGFα, HB-EGF, AR, BTC, NDF, NRG	(β-Helix)-(Cys-rich)-(β-helix)-(Cys-rich)	No	No	HER2 is common subunit; HER3 is kinase-deficient and acts as a binding subunit and adaptor
Insulin	Insulin, IGF-I, HGF, MSP, pleiotrophin	(β-Helix)-(Cys-rich)-(β-helix)	No	Yes	Disulfide-linked ($\alpha_2\beta_2$); IGF-IR may exist in IGF-I-insulin receptor hybrids
PDGF	PDGF, FGF, CSF-1, SCF, KGF, VEGF, Flt3 ligand, GDNF, NRTN, ARTN, PSPN, angiopoietin-1	2-7 Ig domains; ligand binds to Ig 1-3; Ig 4 involved with dimerization	Yes	No	Frequent isoforms; often use proteoglycans; GDNF group binds to nonhomologous receptor subunit
NGF	NGF, BDNF and NT (Trk family); agrin	(Cys-rich)-(3 Leu)-(Cys-rich)-(2 Ig); 4 Ig (MuSK)	Small	No	Common LNGFR subunit; Ig loops bind ligand
Eph	EphrinA, ephrinB	(β-taco)-(Cys-rich)-(2 FNIII)	No	No	Ligands frequently membrane-bound or membrane-associated; bidirectional signaling; slow kinetics
DDR	Collagen	Discoidin (8-stranded β-sandwich)	Small	Yes	Kinase related to NGF-insulin family; slow kinetics

Table 7-4
Sequence Specificity for SH2 Domains from Various Proteins[a]

Sequence	Binding proteins
Y-E-E-I	Src, Lck, Fyn, and Fgr[b]
Y-(-)-Hπ-P	Crk[c], Nck[c], and Abl[b]
Y-HΦ-X-M	Regulatory subunit of PI3K
Y-HΦ-X-HΦ	PLCγ and SHP
Y-X-X-L	Lck and ZAP-70[b]
P-X-Y-(V/I)-N-(V/I)	Grb2[c]

[a] HΦ, Hydrophobic amino acids; Hπ, hydrophilic amino acids; X, any amino acid.
[b] Soluble tyrosine kinases.
[c] Adaptors.

activity and must dimerize with one of the other intact receptors. It can still affect ligand specificity and output; the latter is possible because HER3 can be phosphorylated by its kinase-active partner and behaves like a scaffold.

The major function of the extracellular domain is ligand binding. The three-dimensional structure of these domains resembles the conformation determined for the insulin receptor family (see later). Domains I and III are β-helices, and domains II and IV are chains of short interlocking β-strands (Fig. 7-2). A β-helix is really a parallel β-sheet that is rolled up into a cylinder. It should not be confused with a β-barrel where the β-strands run parallel to the axis of the cylinder; in a β-helix, the strands run around the perimeter of the cylinder. Domains I-III form a C into which the hormone fits; the primary contacts are between the ligand and the β-helices. The major difference between the EGFR and the insulin-like growth factor-I receptor (IGF-IR) is a loop that is inserted into domain II and that extends away from the first β-helix. This loop makes contact with domain IV, which bends backwards to reach it; this pulls the intervening, second β-helix out of position for ligand binding. It is assumed that this interaction maintains basal activity at a low level; upon ligand binding the domain II-IV linkage is broken. This change in conformation has two effects: first, it allows the two β-helices to come closer to each other to form the binding pocket. Second, it exposes the domain II loop, which anchors the dimerization interface between the two domain IIs. HER2 differs slightly from this prototypical structure; domains II and IV do not make contact so that the ligand-binding site constitutively assumes the "bound" conformation. This may explain why no hormone that specifically binds HER2 homodimers has been identified; basically, there is no need for a ligand to relieve an autoinhibitory interaction. As a result, HER2 exists in a primed state ready to

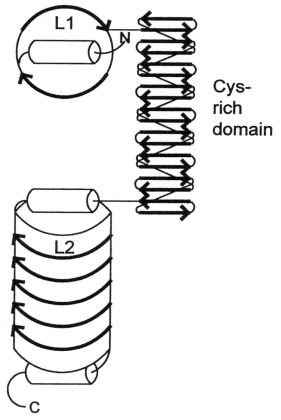

Fig. 7-2. A schematic depiction of the ligand-binding site of the insulin and insulin-like growth factor I (IGF-I) receptors. Basically, it consists of two β-helices (L1 and L2) connected by seven repeats of a three β-strand module.

dimerize with any of the other isoreceptors. Finally, domain IV has another function: the extreme carboxy-terminal end of this domain is responsible for targeting the EGFR to caveolae, which are membrane compartments that are specialized for signal transduction (see later).

The transmembrane domain appears to serve as a membrane anchor and to properly orient the kinase domains for intermolecular phosphorylation. In addition, a natural point mutation in this region in HER2 favors dimerization and results in the constitutive activation of the kinase. Because dimerization is associated with EGFR activation, it is possible that, by favoring oligomerization in the absence of ligand, the native transmembrane domain helps to maintain a low basal activity.

The intracellular region adjacent to the membrane, the juxtamembrane domain, has several critical phosphorylation sites. Phosphorylation by protein kinase C (PKC) inhibits the kinase activity, whereas the MAPK sites are required for receptor internalization. This region is also involved with mitogenesis, perhaps by affecting substrate specificity. Finally, both the phosphatidylinositol and phosphatidylinositol 4-phosphate kinases, as well as calmodulin and G_s, bind in this domain; calmodulin inhibits tyrosine kinase activity, but the binding of the kinases does not affect either their own intrinsic activity or that of the EGFR. Nonetheless, because these phospholipid kinases are being brought to the membrane where their substrates are located, one would expect their catalytic efficiencies to be enhanced. Although G proteins are more closely associated with G protein-coupled receptors (GPCRs), there is considerable promiscuity among receptor classes and signaling pathways. The EGFR phosphorylates G_s at two sites and activates it.

The adenosine 5′triphosphte (ATP)–binding and kinase domains are carboxy-terminal to the juxtamembrane domain. In the remaining carboxy terminus are several very important phosphorylation sites. S-1046 and S-1047 are phosphorylated by CaMKII (calmodulin [CaM]–dependent protein kinase II); this modification reduces the tyrosine kinase activity by half and is, in part, responsible for agonist-induced desensitization. Y-992 is the binding site for many molecules: PLCγ; Raf; the GTPase-activating protein (GAP); SHP; the STK, Src; and the transcription factor, Stat1. PLCγ initiates the calcium-signaling pathway (see Chapters 4 and 10) and Raf is in the Ras-MAPK cascade. GAP is a protein that enhances the GTPase activity of Ras; because GDP· Ras is inactivated, GAP was originally viewed solely as a negative regulator and the EGFR was assumed to be sequestering GAP to prolong Ras activity. However, it is now known that GAP is a more versatile protein and can actually mediate some of the actions of Ras (see Chapter 9); as such, the EGFR may be recruiting GAP for Ras signaling. SHP is another bifunctional molecule; as a tyrosine phosphatase, it was assumed to be responsible for returning the system to baseline after ligand dissociation. However, it can also serve as an adaptor to couple the RTK to other molecules; for example, it links the PDGFR to Grb2 and the insulin receptor to insulin receptor substrate (IRS). Finally, Stat is a transcription factor that is more closely associated with another family of receptors, the cytokine receptors. The EGFR does not directly tyrosine phosphorylate Stat; rather, this is accomplished by a bound STK, like Src.

Y-1068 binds Stat3 and an adaptor known as Grb2, which is also involved in the Ras pathway by recruiting a RasGEF. A cluster of tyrosines in the vicinity of position 1100 is associated with internalization; and the suppressor of cytokine signaling (SOCS) binds at the extreme carboxy terminus. SOCS inhibits Stat activation and is discussed in Chapter 8. Finally, the EGFR can interact with the cytoskeleton through its carboxy terminus. This relationship is probably bidirectional: the cytoskeleton can localize the EGFR to specific parts of the plasma membrane, but the EGFR can also act on the cytoskeleton to remodel it (see Chapter 11).

Not all these phosphorylation sites are on all EGFR isoforms; this allows each HER dimer to have specific activities. For example, MAPK is the major output for both HER1-HER1 and HER1-HER3, but the latter will also activate PI3K to a limited degree. On the other hand, both MAPK and PI3K are major outputs of HER2-HER3, and this dimer is the most potent subunit combination for stimulating mitogenesis.

Finally, it should be noted that the determination of the function of particular tyrosines within a RTK can be difficult; the scientific literature contains numerous conflicting reports that mutation of a given tyrosine to phenylalanine does or does not impair a particular activity. It is now known that RTKs have latent phosphorylation sites that can become modified, if the primary site is mutated. Because these new sites can often partially replace the original ones, the investigator may erroneously conclude that the original site is unimportant.

Insulin Receptor Family

Like the EGFR, members of the insulin receptor family have clustered cysteines in the ligand-binding region and an intact kinase domain (see Fig. 7-1, *B*); however, most of these receptors are cleaved in the extracellular domain to give rise to a membrane-bound subunit (β) attached by disulfide bonds to an exclusively extracellular subunit (α). Another difference is that although the EGFR forms dimers only on ligand binding, the insulin and IGF-I receptors form a stable "dimer" between two αβ-pairs; this tetramer is linked by disulfide bonds. It is not known if the other RTKs in this group also form covalently linked tetramers. In addition to "homodimers," an insulin receptor αβ-pair can form a "heterodimer" with an IGF-IR αβ; indeed, 80% of IGF-IRs are found in IGF-IR/IR hybrids, whereas only 15% of insulin receptors are found here. These hybrids only bind IGF-I.

The members of this family include the receptors for insulin; IGF-I; hepatocyte growth factor (HGF, also called *scatter factor*), whose receptor is called *Met*; macrophage stimulating protein (MSP), whose receptor is called *Ron*; pleiotrophin; and midkine. The latter two hormones share a common receptor subunit: the anaplastic lymphoma kinase (ALK). These receptors all have kinase domains highly homologous to the insulin receptor. However, it is now known that the extracellular domains of some of these RTKs are not related to the insulin receptor; for example, the amino termini of the HGF and MSP receptors are actually homologous to that of the plexin receptors (see later). Nonetheless, they are still included in this group because of their other shared characteristics.

The three-dimensional structure of the extracellular domain of IGF-IR has been determined, and modeling studies have strongly suggested that the domain in the insulin receptor is very similar (see Fig. 7-2). The ends of the ligand-binding domain (LBD) (L1 and L2) are β-helices capped by short α-helices on either end. The two β-helices are perpendicular to each other and connected by the cysteine-rich region (S1). S1 consists of seven modules, each of which has three interlocking β-strands. S1 begins along L1 and fixes its position; however, L2 is completely mobile. Although the ligand was not cocrystallized with the extracellular domain, it is presumed to fit in the notch between L1 and L2, as has been shown for the EGFR (see previous discussion).

Between the LBD of the insulin receptor and the membrane are three FNIII domains (Fig. 7-3). The disulfide bonds are located between the FNIII, which float like pontoons on the outer leaflet of the plasmalemma and keep the kinases in the αβ-dimers apart; this prevents activation in the absence of a ligand. The α-subunits form an open funnel in the unoccupied state. The α-subunits come together to bind a single hormone molecule; this movement brings the kinases together and results in the phosphorylation of the activation loops. However, the movement also displaces the FNIII and puts a strain on the disulfide bonds, such that the receptor springs back to its inactive state after the ligand dissociates.

The only known function of the transmembrane region of the insulin receptor is membrane anchorage. The juxtamembrane domain has a pY that binds Grb2, Stat, and IRS. As with the EGFR, Grb2 initiates the Ras-MAPK cascade, and Stat migrates to the nucleus to stimulate gene expression. However, the IGF-IR uses the STK, Jak, to phosphorylate Stat; on the other hand, the insulin receptor uses both Jak and direct phosphorylation to activate Stat. IRS is a scaffold that mediates many of the metabolic functions of insulin. The ATP-kinase domain that follows also contains several important phosphorylation sites. Y-1162 and Y-1163 are required for immediate internalization, insulin-stimulated kinase activity, glucose transport, and diacylglycerol production, whereas Y-1146 participates in internalization and growth, but not metabolic activity. In addition, there are two other binding domains in this region: one domain localizes the receptor to caveolae by binding caveolin, the signature protein in these structures. The other domain binds G_s. In fact, the insulin receptor can bind, phosphorylate, and activate three types of G proteins: G_s binds to the kinase domain, $G_{i/o}$ associates with the carboxy terminus, and G_q binds to an unknown site in the insulin receptor.

The carboxy terminus may affect substrate specificity. Deletion of the last 43 amino acids eliminates metabolic activity and the activation of protein phosphatase 1 (PP1), but it markedly enhances MAPK activation and mitogenesis. This region may form part of the substrate-binding pocket that selects metabolic messengers over mitogenic ones; its deletion would allow mitogenic substrates access to the catalytic site. The carboxy terminus is also important in internalization. The extreme carboxy terminus (residues 978-1300) is required for microaggregation on the plasma membrane, whereas the adjacent region (residues 944-965) is involved with the migration to coated pits and internalization. The extreme carboxy terminus of IGF-IR binds PDZ-RhoGEF. Rho is a small G protein involved with cytoskeletal remodeling (see Chapter 9). The carboxy terminus also contains a pY that binds PI3K and SHP. Finally, there are several PKC phosphorylation sites on the receptor; the one in the carboxy terminus inhibits receptor function, whereas the one in the kinase domain is involved with internalization.

Platelet-Derived Growth Factor Receptor Family

There are several characteristics that set the PDGF receptor family apart from the other RTKs (see Fig. 7-1, C): (1) the extracellular domain consists of a series of immunoglobulin-like loops, (2) the protein is uncleaved, and (3) the kinase domain is split. The known members include receptors for PDGF, FGF, colony-stimulating factor-1 (CSF-1) (also called *macrophage colony-stimulating factor* or M-CSF), stem cell factor (SCF, whose receptor is called *Kit*), keratinocyte growth factor (KGF), vascular endothelial growth factor (VEGF) (also called *vascular*

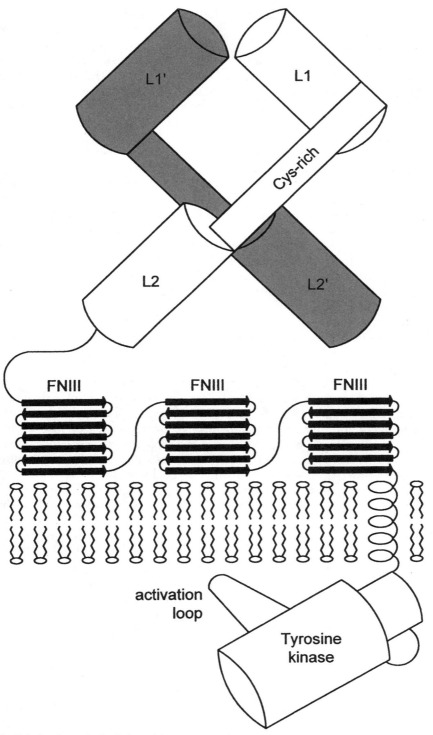

Fig. 7-3. A schematic depiction of the entire insulin receptor; for clarity, only the ligand-binding domain (shaded gray) of the second αβ-dimer is shown.

permeability factor or VPF), angiopoietin-1, the Flt3 ligand, neurturin, artemin (also called *enovin*), persephin, and the glial cell line-derived neurotropic factor (GDNF) (whose receptor is called Ret). Many of these receptors have multiple forms. The PDGF receptor (PDGFR) has two major isoforms, the A (or α) and B (or β) receptors, which are products of distinct genes. The FGF isoreceptors (FGFRs) are the result of both variable mRNA splicing and separate genes. For example, the FGF2R has two exons that code for the carboxy half of Ig3: one exon produces a receptor that binds only FGF, whereas the other one generates a receptor that binds KGF.

The extracellular region consists entirely of immunoglobulin-like domains: the number varies from two in some FGFR variants to seven in the VEGF receptor. Mutation studies in all PDGFR members examined so far indicate that Ig2 and Ig3 are required for binding; the data on Ig1 are conflicting. The VEGFR was the first to be crystallized, and its three-dimensional structure readily explains the results from the mutation studies (Fig. 7-4, *A*). Ig2 and Ig3 form a notch into which VEGF binds. Ig1 has no apparent contact with VEGF, but it may fold down and protect the hydrophobic surface of Ig2 in the unoccupied state; that is, it could indirectly participate in ligand binding. However, there are some FGFR variants that lack Ig1 altogether without affecting binding affinity. Ig4 is involved with dimerization. Mutation studies have suggested that subsequent Ig loops inhibit RTKs; they may play a role in reducing basal activity.

The FGFR requires the proteoglycan heparin sulfate as a coreceptor (Fig. 7-4, *B*). Actually, heparin sulfate has been shown to bind to several hormones, including PDGF, VEGF, EGF, HGF, pleiotrophin, TGFβ, and some chemokines. Three functions have been attributed to proteoglycans. First, by binding these hormones, proteoglycans can sequester and inhibit their activity until needed; essentially, they act like a storage depot. Second, because there are regulated enzymes that can degrade proteoglycans to release the hormones, proteoglycan binding is a way to control the distribution and concentration of the hormones. Finally, the proteoglycan can act as a coreceptor by facilitating the formation, stabilization, and/or transduction of the hormone-receptor complex. The latter is possible because some of these proteoglycans have intracellular domains that interact with scaffolds and other signaling molecules. In the FGFR, heparin serves as a coreceptor.

The Ret family also requires a coreceptor—GDNF family receptor α (GFRα). This family of coreceptors provides binding specificity. For example, GDNF binds GFRα1 and Ret, neurturin binds GFRα2 and Ret, enovin or artemin binds GFRα3 and Ret, and persephin binds GFRα4 and Ret. The extracellular GFRα has four cadherin-like domains and a six-stranded β-sandwich. The former is a motif found in cellular adhesion proteins and the latter resembles an Ig loop. Another unusual aspect of this family is that the ligands are distant relatives of the TGFβ family, but use a RTK rather than the S-T kinases used by the TGFβ group (see later).

As with the other members of this family, FGF binds Ig2 and Ig3 called D2 and D3 in the FGFR. In particular, each FGF monomer binds the carboxy-terminal loops of D2 and the amino-terminal loops of D3 of its own FGFR. Each FGF also has a second site by which it binds the other FGFR, but because of its small size, it is a weak dimerization surface. However, the D2 domains of the FGFR dimer splay apart creating a positively charged trough along which the negatively charged heparin lies (Fig. 7-4, *C*). This proteoglycan also makes contact with each FGF and these interactions stabilize the complex (Fig. 7-4, *B*). D3 domains do not participate in dimerization; although they appear apposed in Fig. 7-4, *C*,

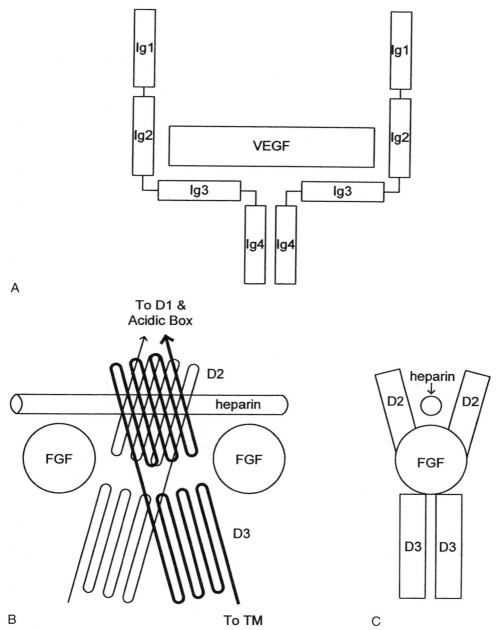

Fig. 7-4. A schematic depiction of the ligand-binding site in the VEGFR (A) and FGFR (B,C). *TM*, Transmembrane domain.

this is an illusion due to the foreshortening of the lateral view. Because of the angle of the D2-D3 domains in the two FGFRs, the D3 do not interact (Fig. 7-4, *B*). There is an acidic box between D1 and D2 that may block heparin binding in absence of FGF and prevent premature activation.

Unlike the members of the EGF and insulin RTKs, mutations in the transmembrane region of PDGFR render the receptor inactive, suggesting that this domain may participate in dimer formation. The ATP-kinase domain that follows is split by intervening sequences of unknown function. These sequences frequently contain pYs that bind to substrates containing SH2 motifs, but there is no obvious reason why these amino acids could not have been located elsewhere. The SH2-binding sites in PDGFR have been thoroughly mapped, and the pYs bound by SHP, PI3K, GAP, Src, PLCγ, and various adaptors have been determined (see Fig. 7-1). The PDGFR can also bind and phosphorylate Stat, but the exact binding site has not been determined. Similar mapping has been done in many of the other RTKs in this family. Surprisingly, the sites for any given SH2 protein do not always occur in the same region of these different RTKs. As with all the other RTKs, the carboxy terminus is involved with internalization.

Receptor activation occurs on homodimerization; the only exception is the angiopoietic receptor, which requires tetramerization. As with the EGFR family, heterodimerization can also occur within several PDGFR family members. For example, there are four isoforms of PDGF (A, B, C, and D) that can form five possible hormones: AA, AB, BB, CC, and DD. There are also two isoreceptors, A and B, that can form three possible receptors: AA, AB, and BB. PDGF-AA and PDGF-CC will induce only PDGFR-AA formation; PDGF-DD will induce only PDGFR-BB formation; PDGF-AB will trigger the dimerization of both PDGFR-AA and PDGFR-AB; and PDGF-BB can bind to all three receptor types.

Nerve Growth Factor Receptor Family

The NGF receptor family is difficult to classify: the kinase domain is most closely related to the insulin receptor family, but the kinase domain is split by a short insert and the extracellular region is composed, in part, of immunoglobulin-like loops, like the PDGFR family. It has been proposed that these RTKs are actually chimeras between an insulin receptor-like ancestor and an immunoglobulin-like molecule. Another characteristic of this family is a very short carboxy-terminal tail following the kinase domain. All of the members of this family are involved with growth and differentiation of the nervous system and include the receptors for NGF, brain-derived neurotrophic factor (BDNF), and several neurotropins (NTs). The receptors for these hormones are designated tropomyosin-related kinase A (TrkA), TrkB, and TrkC; NGF binds TrkA and BDNF binds TrkB. The various NTs bind either TrkB or TrkC.

The amino-terminal half of the extracellular domain consists of a LRR flanked by cysteine clusters; however, these cysteine clusters are not as large or dense as those in the insulin receptor. The carboxy-terminal half contains two Ig loops. Crystallization studies have shown that NGF binding occurs through Ig2 (Fig. 7-5). All of these hormones begin with a short, amino-terminal helix followed by four long β-strands that are tied together at the waist by a "cysteine knot." Ig2 touches NGF at two sites: the carboxy-terminal loops of Ig2 bind the body of NGF and the side of the Ig2 β-sheet binds the amino-terminal α-helix. This latter interaction provides some specificity in ligand binding: NGF has a hydrophobic α-helix that fits into the hydrophobic pocket of the TrkA Ig2. However, NT has a hydrophilic 3_{10} helix that prevents it from binding TrkA. Additional specificity is provided by a coreceptor, the low affinity NGF receptor (low-affinity nerve growth factor receptor [LNGFR] or p75). In the presence of the LNGFR, NGF also appears to bind the cysteine domain 1 of TrkA, and BDNF and NT bind both the cysteine domain 1 and the LRR of TrkB. LNGFR is actually a member of the cytokine receptor family and is discussed further later.

As noted previously the kinase domain has an insert of only 14 to 15 amino acids (not shown in Fig. 7-1 because of its very short length). In some isoforms, this insert is variable and determines, in part, the substrate specificity; for

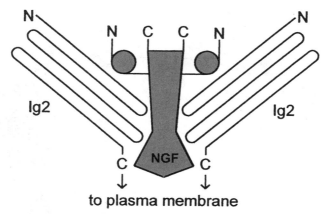

to plasma membrane

Fig. 7-5. A schematic depiction of the NGF-TrkA interaction. NGF is shaded, and the Ig2s are from separate TrkA receptors. *Ig2*, Immunoglobulin 2; *NGF*, nerve growth factor; *TrkA*, tropomyosin-related kinase A.

example, Trk isoreceptors with longer inserts will not phosphorylate PLCγ. As with the other RTKs, the NGF receptor dimerizes.

Miscellaneous Receptor Tyrosine Kinases

There are several other RTK families. The first is the ephrin (Eph) family, which is divided into two classes: the EphA group includes Eph, epithelial cell kinase (Eck), Hek, and Sek and the EphB group includes Eph-like kinase (Elk), Nuk, and Htk. The EphA group binds ephrin-A ligands (also called *ligands for Eph-related kinases* or *LERK1*), and EphB group binds ephrin-B ligands (also called *LERK2*). The ligands are membrane bound; ephrin-A is covalently coupled to a membrane lipid and ephrin-B is an integral membrane protein. EphB6 is kinase-dead like HER3; as such, it only acts as a binding subunit and scaffold. Many of these RTKs are involved with morphogenesis. The protracted nature of morphogenesis probably explains the slow binding kinetics of this family.

The extracellular domain consists of a twisted β-sandwich called a *β-taco* followed by a cysteine-rich region and two FNIII domains. The ligand is a modified Greek key β-barrel from which a long and a short loop extend. Each extended loop interacts with separate RTKs through the loops of the β-taco to form a tetrameric ring (Fig. 7-6).

The intracellular domain contains the tyrosine kinase. In the EphAs, the kinase is followed by a sterile alpha motif (SAM). This is a five α-helical module that mediates dimerization. In the EphBs, the kinase is followed by a PDZ domain, which is involved with protein-protein interactions.

One of the amazing characteristics of this family is the bidirectional signaling; that is, the ligands and receptors are located on separate cells and their engagement triggers signals in both cells. Signaling by Eph is straightforward: dimerization activates the tyrosine kinase leading to autophosphorylation. The intracellular domains of the LERK2s are also phosphorylated on tyrosine, but they do not have kinase domains. They do associate with other RTKs, such as PDGFR and FGFR. Presumably, aggregation of ephrin leads to aggregation of the attendant RTKs, which become activated and phosphorylate the intracellular domain of the transmembrane ligand or other substrates. This family is involved with neurogenesis and is coupled to cytoskeletal remodeling. For example, pYs bind Grb4, an adaptor that couples the LERK2 to Ras and Rac exchange factors

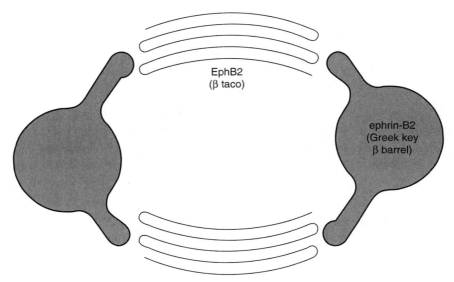

Fig. 7-6. A schematic depiction of the ephrin-B2-EphB2 interaction. The reader is looking down onto the surface of the plasma membrane. The ligand, ephrin-B2 (shaded) is a β-barrel that is seen on end. Extended loops interact with loops from a twisted β-sandwich (β-taco) at the amino terminus of the receptor, EphB2.

and to Pak, a Rac-activated kinase. Another way LERK2 signals is through its carboxy terminus, which binds PDZ modules; for example, PAR-3, a scaffold protein, connects LERK2 with Cdc42 and Rac, which are small G proteins associated with the cytoskeleton. These interactions would place the small G proteins, their activators, and at least one effector at the cytoplasmic domain of LERK2. The mechanism of LERK1 transduction is still unknown.

Axl is closely related to the Eph family. It includes Axl and Tyro 3; the latter binds gas6, which is a mitogen for smooth muscle. The extracellular domain contains two Ig loops followed by two FNIII domains.

The last RTK family is the discoidin domain receptor (DDR). The extracellular domain is an eight-stranded β-sandwich first identified in the discoidin protein. DDR binds collagen; in other proteins, the discoidin domain binds sugars or membrane phospholipids. Collagen is glycosylated, but these sugar residues are not critical for binding to DDR. Like receptors in the Eph family, DDR exhibits slow binding kinetics. The kinase domain most closely resembles that in the insulin receptor and Trk, raising the question of whether these RTKs should be grouped together.

Serine-Threonine Kinase Receptors

Transforming Growth Factor β Receptor Family

The serine kinase receptors are very similar to the RTKs: they have an extracellular amino terminus, a single transmembrane helix, and a carboxy-terminal kinase domain. However, in these receptors, the kinase phosphorylates serines and threonines, rather than tyrosines. The TGFβ, activin, antimüllerian hormone, bone morphogenic protein (BMP), and nodal receptors belong to this family.

The TGFβ receptor is actually a complex of three different binding components; the types I and II receptors are both homologous serine kinases. The amino termini bind the hormone and consist of six β-strands that pair up to form three fingers. The type I receptor cannot bind TGFβ by itself but does increase the affinity of the type II receptor toward its ligand. Some type I receptors, also called *activin-like receptors* (ALK), are shared among hormones, whereas the type II receptor is unique for each hormone and provides binding specificity. The type II receptor constitutively autophosphorylates and activates itself; the kinase domain has two short inserts of unknown function. Several hormone-receptor complexes have had their three-dimensional structures elucidated and it appears that each hormone induces a unique orientation of the receptor subunits; for example, activin binding brings the two type II receptors very close together, whereas TGFβ and BMP binding produce a complex where the extracellular domains of the type II receptors are not even touching. It is believed that these different orientations may contribute to the specificity of action of this hormone family.

Hormone binding either induces tetramerization or strengthens the interactions in a preexisting tetramer containing two of each receptor type. The type II receptor will then phosphorylate ALK in the juxtamembrane region called the *GS box* because of its high content of glycine and serine (see Fig. 7-1). This modification activates the ALK kinase and displaces an immunophilin, FKBP12, that normally inhibits ALK. The displacement of FKBP12 then allows a transcription factor, Smad, to bind. The recruitment of Smad to ALK is facilitated by various adaptors, like Sara. Sara is also an adaptor for PP1, which inactivates the receptor by dephosphorylation. There are three classes of Smad. The receptor Smads (R-Smad) bind individual ALKs and provide signaling specificity; for example, TGFβ activates Smad2 and Smad3, whereas BMP activates Smad1, Smad5, and Smad8. Activation occurs by means of ALK phosphorylation, which induces trimerization between two R-Smads and a common Smad (Smad4 or Co-Smad). Co-Smad is not phosphorylated and forms part of all Smad trimers regardless of what other R-Smad isoforms are present. Because Sara only binds the monomer, the trimer dissociates and migrates to the nucleus to initiate transcription. Although it has its own DNA binding domain and TAD, Smad most commonly acts as a coactivator with other transcription factors (see Chapter 13). Smad6 and Smad7 (anti-Smads) inhibit the transduction of these receptors in three ways: first, they bind phosphorylated ALK and block R-Smad activation. Second, they trigger receptor degradation via the ubiquitin pathway (see Chapter 8). Finally, they can bind DNA and act as corepressors. Both ALK and the type II receptor have other substrates; these are discussed in Chapter 12, after the reader has gained more knowledge of second messengers.

The type III receptor is a coreceptor that increases hormone binding affinity by presenting the ligand to the receptor. The TGFβR uses a proteoglycan, called *betaglycan*; Nodal uses Cripto, a member of the EGF family that does not bind EGFRs. Activin does not appear to require a coreceptor.

Plant Serine-Threonine Kinases

Plants do not have RTKs; indeed, they do not have STKs either. They do have MAPKs that must be activated by phosphorylation on both a threonine and a tyrosine, but the activating kinase is a dual function kinase that can phosphorylate both sites. Plants do not have any dedicated tyrosine kinases; however, they do have many receptors with intracellular S-T kinase domains. The structure of the extracellular domains varies, but one of the largest groups has LRRs. This group includes the brassinosteroid, phytosulfokine, systemin, and the CLAVATA3 receptors. In addition, these receptors can bind products of pathogens or symbionts leading to defense or nodulation responses, respectively. The former receptors include FLS2, which binds bacterial flagellin, and CHRK1, which binds chitin from pathogenic fungi; the latter includes the nodulation receptor kinase that recognizes the Nod factor from nitrogen-fixing bacteria.

The output from these receptors is still poorly understood. The brassinosteroid receptor is a heterodimer between two LRR receptors: BRI1 and BAK1. Ligand binding induces or stabilizes the heterodimer and triggers cross-phosphorylation of the activation loop. In this respect, it resembles the TGFβR above; however, downstream substrates have been less well characterized. The brassinosteroid receptor can inhibit the plant homologue of GSK3, but the molecular mechanism for this effect is unknown. The receptor for CLAVATA3 is a heterotetramer of CLAVATA1 and CLAVATA2. It can bind and phosphorylate protein phosphatase type 2C (PP2C), which is thought to dephosphate the receptor in a negative feedback loop. It can also bind a Rho-like G protein.

Histidine Kinase Receptors

Histidine kinase receptors represent one of the oldest transduction systems known. It is the major chemoreceptor system in bacteria and has persisted in fungi and plants, where it is used to detect osmolarity and some hormones, respectively. There is no evidence that it exists in animals.

The ethylene receptor has a short amino terminus that is involved with dimerization by a disulfide bond. Although dimer formation is required for cross-phosphorylation, this interaction need not be covalent. There are three transmembrane α-helices; the second one has a cysteine that binds ethylene through a copper ion. The intracellular domain consists of an amino-terminal histidine kinase and a carboxy-terminal regulator. In most bacteria and fungi, the regulator is a separate protein, but in plants it may either form the carboxy terminus of the receptor or be separate. As with bacterial systems, the ethylene receptor is constitutively active and intermolecularly autophosphorylates itself on a histidine in the kinase domain. This phosphate is then transferred to an aspartic acid in the regulator, which in turn inhibits the Raf-MAPK cascade. The mechanism for this activity is unknown, but it has been suggested that the three-dimensional structure of the regulator resembles Ras, which is known to regulate Raf in animals. Ethylene binding inhibits the histidine kinase, thereby removing the inhibition of Raf. Interestingly, ethylene receptors are divided into two subfamilies; subfamily II lacks kinase activity. These latter receptors may act like HER2; that is, they may simply serve as regulators for members of subfamily I. They may also act as adaptors for other substrates.

The cytokinin receptor represents a more complex histidine kinase system, where there are multiple phosphate transfers. It has two transmembrane α-helices, and the cytokinin binds the large extracellular loop between the two. Intracellularly, it has one or two attached regulators, depending on the isoform. In this case, cytokinin binding actually stimulates kinase activity. The phosphate is transferred from the histidine in the kinase to the aspartic acid in the regulator, just as occurs in the ethylene receptor (Fig. 7-7). However, the regulator then transfers the phosphate to a histidine in the phosphotransmitter, which shuttles to the nucleus and finally transfers the phosphate to the regulator in the amino terminus of several transcription factors. It is believed that the regulator is an autoinhibitory domain that represses transcription and that phosphorylation of the regulator relieves this inhibition.

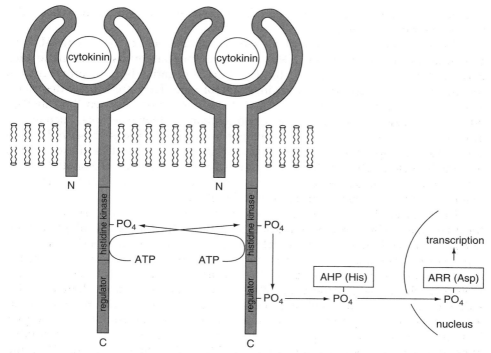

Fig. 7-7. The pathway for cytokinin signaling. *AHP*, *Arabidopsis* histidine phosphtransfer protein; *ARR*, *Arabidopsis* response regulator.

Guanylate Cyclases

Like all of the enzyme-linked receptors, the membrane-bound guanylate cyclases (GCs) traverse the membrane once; the amino termini are extracellular and the cyclase domains are intracellular. In addition to the membrane-bound enzyme, there is a homologous soluble GC that acts as a cytoplasmic receptor for nitric oxide (NO) and carbon monoxide (CO).

Membrane-Bound Guanylate Cyclases

There are three major groups of membrane-bound GCs: (1) guanylate cyclases A and B (GC-A and GC-B), (2) GC-C, and (3) Ret-GC and GC-D. GC-A and GC-B are very closely related and bind different forms of the atrial natriuretic peptide: GC-A binds atrial natriuretic peptide (ANP) from the heart, and GC-B binds C-type natriuretic peptide (CNP) from the brain. GC-C is similar in structural organization to GC-A and GC-B except for a 60-65 amino acid carboxy-terminal extension; however, its amino acid sequence is considerably more divergent. GC-C was initially identified as the receptor for a bacterial endotoxin that triggers fluid secretion through chloride channels. Since then, an endogenous ligand, guanylin, was discovered for GC-C. The third group of GCs is involved with sensory transduction; for example, Ret-GC is a retinal guanylate cyclase that is regulated by a calcium-sensitive peptide, and the GC-D family serves as odor receptors in certain neurons in mammals and as chemosensory receptors in nematodes. Finally, GC-S is homologous to GC-A and GC-B in the extracellular region, but its cyto-plasmic region is extremely short and lacks a cyclase domain. It binds another isoform of ANP, called *BNP*; however, this receptor does not transduce a signal.

Rather, GC-S is a clearance receptor; that is, it merely serves to internalize the hormone for degradation.

The extracellular domain of GC-A is homologous to the bacterial type I periplasmic binding protein. This is a protein that binds small molecules between the inner and outer membranes of gram-negative bacteria and transfers the molecules to transporters in the inner membrane. They are usually described as clamlike, and it has been shown that the ligand binds between the "shells," which then close. These homologues also appear in other receptors, such as the glutamate receptor, where the mechanism for ligand binding is preserved. However, crystallographic studies show that ANP does not bind in this pocket; rather, it binds in a cleft between dimers. This arrangement may be necessitated by the size of ANP, which is considerably larger that the amino acids that are the usual ligands for the periplasmic proteins and the glutamate receptors.

The intracellular domain consists of two regions: the carboxy-terminal one is the guanylate cyclase and the amino-terminal one is the kinase homology domain (KHD). The KHD has no kinase activity, but it can still bind ATP. The KHD represses cyclase activity, and this inhibition is relieved by ATP. As such, the cyclase activities of GC-A and GC-B have an absolute requirement for ATP. However, the cyclase activity of GC-C is not absolutely dependent on ATP, although this nucleotide will still enhance cyclase activity. The receptor exists as a preformed, phosphorylated oligomer; current data suggest that the complex is a tetramer and that phosphorylation is required for activity. Ligand binding alters the receptor conformation, which facilitates ATP binding to the KHD and neutralization of its inhibitory function. In addition, the change in conformation initiates ligand-induced desensitization by decreasing ligand affinity and exposing the phosphorylation sites to phosphatases.

Soluble Guanylate Cyclase

The soluble guanylate cyclase (sGC) has no membrane-spanning region; it consists of a carboxy-terminal cyclase domain and an amino-terminal heme-binding domain. The latter acts as an autoinhibitory domain for the cyclase. The enzyme is activated when either NO or CO binds to the heme moiety and displaces it from the enzyme. The sGC is an obligate dimer between two homologous, but nonidentical, subunits.

NO is produced from arginine by NO synthase (NOS), an NADPH-dependent monoxygenase. In fact, there are three major NOS families: neural NOS (nNOS or NOS1), induced NOS (iNOS or NOS2), and endothelial NOS (eNOS or NOS3). NOS is totally dependent on calcium and CaM so that any hormone activating this pathway may stimulate NO production (Fig. 7-8). The synthase is also regulated by phosphorylation, which can either stimulate or inhibit catalytic activity. For example, protein kinase B (PKB), a product of the PI3K pathway, and protein kinase A (PKA) modify and neutralize the autoinhibitory carboxy terminus, thereby stimulating the NOS. Conversely, PKC, CaMKII, and MAPK are responsible for inhibitory phosphorylation. PKC labels the CaM binding site and blocks direct calcium-CaM activation. On the other hand, CaM, in addition to directly activating NOS, stimulates PP2B, which dephosphorylates these inhibitory sites. Lastly, NOS can be regulated by a direct interaction with GPCRs (see later).

NOS is also regulated by both arginine and calcium availability. Arginine is limiting in cells, especially if there is sustained NO production; elevation of arginine levels can be achieved by transport, synthesis, decreased catabolism,

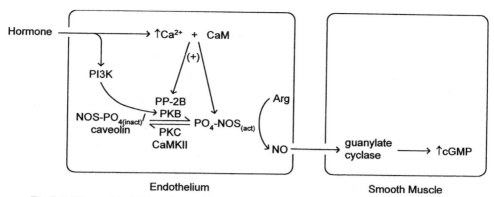

Fig. 7-8. Nitric oxide (NO) pathway. Hormones elevate cytosolic calcium levels in the endothelial cell. The calcium activates calmodulin (CaM), which in turn stimulates the NO synthase (NOS). NOS activity is also regulated by phosphorylation and cellular localization.

and the inhibition of competing, arginine-consuming pathways. NOS is physically coupled to arginine transporters in caveolae so that imported arginine can be channeled directly to the enzyme. This communication is bidirectional in that NO can nitrotyrosinate and stimulate the transporter. On the other hand, arginine transport can be inhibited by 3'5'-cyclic GMP (cGMP) in a negative feedback loop. Arginine synthesis is another control point that hormones can influence; for example, many cytokines that trigger NO production also induce arginine synthetase and argininosuccinate lyase, while inhibiting arginase. Furthermore, many of these enzymes colocalize with NOS in caveolae. Finally, there is one other aspect of arginine metabolism that affects NOS activity: asymmetric dimethylarginine is an endogenous NOS inhibitor that naturally arises from protein degradation. Retinoic acid and interleukin-1β can induce dimethylarginine dimethylaminohydrolase II, which catabolizes this inhibitory molecule and increases NO production. Conversely, NO *S*-nitrosylates a cysteine in the enzyme's active site. This inhibition would allow dimethylarginine to accumulate and inhibit NOS; this would represent another aspect of negative feedback.

There are two other pathways that can compete with NOS for arginine: ornithine synthesis for polyamine production (see Chapter 11), and proline synthesis for collagen production. Macrophages can switch between these outputs in response to hormones and other factors. For example, lipopolysaccharides will stimulate NOS while inhibiting ornithine synthesis; however, TGFβ, interleukin-4, interleukin-10, and macrophage-stimulating protein have the opposite effect. During wound repair, macrophages will produce NO for the first 1 to 2 days to sterilize the wound. Thereafter, NO declines and polyamines and proline increase; the former is required for mitosis, which replaces dead cells, and the latter is required for collagen synthesis, which is involved with stromal repair. Calcium availability is determined by channels and pumps. Extracellular calcium concentrations are greater than cytoplasmic levels, so that opening calcium channels in the plasma membrane will elevate calcium in the vicinity of the channel (see Chapter 10). Conversely, plasma membrane calcium pumps will lower cytoplasmic calcium levels. NOS is associated with both. Indeed, in neurons NOS1 forms a supercomplex that includes the glutamate receptor (a calcium channel), the cationic amino acid transporter 1 (an arginine transporter), sGC, and a plasmalemma calcium pump. The first two components would stimulate NOS and the last would inhibit it, whereas the sGC represents a target of the NO.

In other cells, a similar collection of transducers occurs in caveolae, where NOS3 binds and is inhibited by caveolin. CaM displaces caveolin and activates NOS at a site that can both support and respond to NO production. In addition to regulation, localization plays an important role in determining the effect of the resulting NO. As noted previously, in neurons the NOS is associated with the sGC, ensuring that the NO will result in elevated cGMP levels. In the heart, NOS can have opposite effects depending on its location within the cell; NOS3 is located in caveolae with β-adrenergic receptors (βARs), whereas NOS1 is found in the sarcoplasmic reticulum associated with the ryanodine receptor, a calcium channel. NO from the former modifies and inhibits the βAR; because the βAR stimulates the heart, this modification depresses cardiac activity. On the other hand, NO from NOS1 modifies and activates the ryanodine receptor, which releases calcium and stimulates the heart. In skeletal muscle, NOS1 localizes to the sarcolemma, where it is activated by calcium influx secondary to muscle contraction. The resulting NO diffuses to nearby blood vessels and produces vasodilation to support the increased metabolic requirements of active muscles.

The other activator of sGC, CO, is produced from the degradation of heme by heme oxygenase in the liver (type 1 isozyme) or brain (type 2); the former enzyme is responsible for the bulk degradation of heme, whereas the latter generates CO for sGC activation. There is little known about the regulation of heme oxygenase 2 except that it is stimulated by either PKC or casein kinase 2 (CK2) phosphorylation.

Nitric Oxide Output

NO can have four effects: elevation of cGMP, S-nitrosylation of cysteine, tyrosine nitration, and binding to heme and nonheme iron. Actually, the first and last effects are related because NO activates the sGC by binding and displacing the heme in the amino-terminal autoinhibitory domain. However, cGMP is such an important second messenger that it is considered in greater detail later (see Chapter 9). The major target of cGMP is PKG; however, it can also act through the 3'5'-cyclic AMP (cAMP) pathway through two mechanisms. First, high, but still physiological, cGMP concentrations can cross-activate PKA. Second, cGMP can inhibit cAMP phosphodiesterase 3, which normally hydrolyzes cAMP; its inhibition allows cAMP levels to accumulate.

NO can covalently modify two amino acids within proteins. First, NO can react with the sulfur of cysteines; this reaction requires higher NO concentrations and is slower than that required to generate cGMP. Although this is a nonenzymatic reaction, it is not random: cysteines located in hydrophobic environments or between acidic and basic groups appear to be particularly susceptible to S-nitrosylation. This modification activates Ras and several G protein trimers, the ryanodine and cyclic nucleotide-gated channels, and metalloproteinases. On the other hand, it inhibits several adenylate cyclases, protein tyrosine phosphatases (PTPs), the rate-limiting enzyme in polyamine synthesis, and caspases, which are proteases involved with apoptosis. In addition to modulating the activity of proteins, NO can use proteins as a NO reservoir. NO is extremely labile, but in the form of S-nitrosylcysteine it is stable and the reaction is readily reversible. For example, transglutaminase can react with eight NO under basal conditions; elevated calcium alters the conformation of this enzyme and the NO is released.

Tyrosine nitration is more stable and, possibly, irreversible. One key effect of this modification is that it precludes tyrosine phosphorylation, a major signaling event. This effect is not always inhibitory; Src, a STK, is kept dormant by an intramolecular link between a pY and a SH2 domain. By reacting with this tyrosine, it cannot be phosphorylated, the link with SH2 is broken, and the catalytic site is activated. This modification can also stimulate the MAPK pathway.

Finally, NO can bind heme and nonheme iron. The following three examples will provide a spectrum of effects that this modification can have. First, it activates the sGC by displacing the autoinhibitory, heme domain. Second, it inhibits NOS1 as a form of negative feedback. Third, it can totally change the function of a protein: it converts aconitase, an enzyme in the tricarboxylic acid cycle, into an RNA-binding protein that affects the translation of iron metabolizing proteins (see Chapter 15).

NO can also be associated with cellular damage, but this effect usually occurs only with prolonged or very high NO production. This circumstance leads to the accumulation of more toxic NO metabolites, such as NO_2, N_2O_3, and $OONO^-$.

Cytokine Receptors

Fibronectin is an extracellular matrix protein that is important in cell adhesion and migration. Structurally it is composed of three repeating modules, each of which has specific binding properties. The type III repeat (FNIII) consists of seven β-strands organized into a sandwich, which is very similar to an Ig domain; indeed, these two motifs may have a common ancestor. The FNIII forms a binding surface for cells or heparin. In cytokine receptors, two FNIII domains are modified to form a ligand-binding pocket, known as the *cytokine domain*. Most of these receptors are involved with the immune response and are under heavy selective pressure to defend the organism against rapidly adapting pathogens. As such, they have an accelerated rate of evolution and exhibit almost no sequence homology at all. However, they do cross the membrane once, have modified FNIII motifs in the extracellular domain, and have a similar genomic organization. Finally, alternate processing is frequent; the most common variations are receptors lacking large portions of their intracellular, and sometimes transmembrane, regions. The latter results in the production of soluble receptors, which may have important regulatory functions (see Chapter 8).

Soluble Tyrosine Kinases

In the simplest terms, cytokine receptors are RTKs, whose kinase domains have been severed from the intracellular domain to form a separate subunit. Otherwise, the function of RTKs and cytokine receptors is very similar: a ligand induces receptor aggregation, and the STKs cross-phosphorylate and activate each other. Tyrosine phosphorylation of the receptor and/or STK then initiates the assembly of a signaling complex. As such, STKs are central to the action of these receptors and are discussed first.

There are eight STK families (Table 7-5). Most STKs are involved with signal transduction from the receptors for various hormones, antigens, and substratum. The cytokine receptors primarily use members of the Jak family. STKs can be regulated by several mechanisms: they are maintained in an inactive state by an interaction between an amino-terminal SH2 and a carboxy-terminal pY. This linkage can be broken in one of two ways: the tyrosine can be dephosphorylated

Table 7-5
Structural and Functional Characteristics of Soluble Tyrosine Kinases

Family	Members	Size (kDa)	Structure[a]			Possible function
			Kinase	SH2	SH3	
Src	Blk, Fgr, Fyn, Hck, Lck, Lyn, Src, Yes, Yrk	53-64	1	1	1	Transduction by antigens, RTKs, and cytokines receptors
Jak[b]	Jak1-3, Tyk	130	2	0	0	Cytokine receptor transduction
Syk	Syk, Zap	70-72	1	2	0	Antigen transduction
Fak	Fak, Pyk2[c]	125	1	0	0	Transduction by integrin and GPCRs
Abl	Abl, Arg	150	1	1	1	Cytoskeletal-nuclear transduction
Itk[d]	Bpk, Btk, Itk, Tec, Etk	62-77	1	1	1	Lymphocyte differentiation
Csk	Csk, Chk/Hyl	50	1	1	1	Phosphorylation and inhibition of other STKs
Fes	Fer, Fes	92-98	1	1	0	Dissolve adherens junctional complexes

[a] Number of kinase, SH2, and SH3 domains in each family.
[b] The amino-terminal kinase is inactive. These STKs also contain an amino-terminal FERM domain.
[c] Pyk2 has two Pro-rich, SH3 binding sites and binds Grb2.
[d] These STKs also contain a PH domain.

by a PTP, or the SH2 domain can bind a pY on another molecule, such as an autophosphorylated RTK. As noted previously, aggregation can also stimulate STKs by inducing cross-phosphorylation of the activation loops. Calcium can either stimulate (Src and Pyk2) or inhibit (Yes), depending on the particular STK. Finally, S-T phosphorylation can also affect kinase activity; for example, PKC inhibits Lck.

Class 1 Family

The class 1 family has the classic cytokine motif: two FNIII domains with two small disulfide loops inserted into the amino terminus of the first FNIII, and a tryptophan-serine box (WS) near the membrane at the carboxy terminus of the second FNIII. The WS box is a sequence of W-S-X-W-S, where W is tryptophan and X represents any amino acid. In contrast to the general lack of overall sequence homology, the WS box is highly conserved; the only known exceptions are sequences of L-S-X-W-S in the α-subunit of the interleukin-3 receptor (IL-3Rα) and Y-G-X-F-S in the growth hormone receptor (GHR).

These receptor complexes are composed of as many as three different subunits. The α- and β-subunits are both members of the cytokine receptor family. The α-subunit enhances the ligand binding specificity and affinity of the β-subunit. It may not transduce a signal; indeed, some α-subunits are soluble or tethered to the plasmalemma by membrane lipids. They may not even be present in some receptor complexes, such as those for growth hormone (GH) or erythropoietin (EPO). The β-subunit occurs as either a homodimer or heterodimer and is required for transduction. Many β-subunits are shared among several different complexes; these "common" subunits are often designated by a "c" subscript (Table 7-6). Complexes may be preassembled, in which case the ligand induces the proper

Table 7-6
Subunit Organization of Some Cytokine Receptors

Receptor subfamily	Subunit composition	Cytokine
β_c	β_c + IL-3Rα	IL-3
	β_c + IL-5Rα	IL-5
	β_c + GM-CSFRα	GM-CSF
γ_c	γ_c + IL-2Rβ + IL-2Rα^a	IL-2
	γ_c + IL-4Rα	IL-4
	γ_c + IL-7Rα	IL-7
	γ_c + IL-9Rα	IL-9
	γ_c + IL-4Rα + IL-13Rα^b	IL-13
	γ_c + IL-2Rβ + IL-15Rα^a	IL-15
	γ_c + IL-21Rα	IL-21
	TSLPR^c + IL-7Rα	TSLP
IL-6Rβ (gp130)	gp130 + IL-6Rα	IL-6
	gp130 + IL-11Rα	IL-11
	gp130 + LIFRβ (gp190)^d	LIF
	gp130 + LIFRβ+ CNTFRα	CNTF
	gp130 + OSMRβ^e	OSM
	gp130 + LIFRβ+ CT-1Rα	CT-1
	IL-12Rβ1^d + IL-12Rβ2 + p40^f	IL-12 (NKSF)
	IL-12Rβ1^d + IL-23Rβ+ p40^f	IL-23
	IL-27Rβ^d + EBI3^f	IL-27
	Leptin-R^d (homodimer)	Leptin
Homodimers	EPOR	EPO
	G-CSFR	G-CSF
	GHR	GH
	c-Mpl	TPO
	PRLR	PRL

^{a} Noncytokine receptors but homologous to each other.
^{b} γ_c May be an optional subunit for this receptor.
^{c} Homologous to γ_c.
^{d} Homologous to gp130.
^{e} Some species also have LIFRβ in the complex.
^{f} p40 and EBI3 are actually considered part of the ligand dimer, although they are homologous to other Rα subunits.

orientation of the subunits rather than oligomerization. In addition to the standard α- and β-subunits, receptor complexes for IL-2 and IL-15 contain a third, accessory receptor. They are not members of the cytokine receptor family; instead, these receptors are characterized by a β-sheet motif known as a *Sushi domain*.

The reader is warned that the concept of α- and β-subunits arose only after a considerable number of complexes had been analyzed. As such, some of the earlier labels do not conform to these definitions. Furthermore, because their original designations are frequently kept for historical reasons, confusion can occur. For example, the accessory receptor for IL-2 was the first to be purified and was given the name IL-2Rα. The real "α" subunit, as previously defined, was the next to be isolated, but it was named IL-2Rβ. The "β" subunit was the last to be discovered and became IL-2Rγ. Another interesting discrepancy is represented by the IL-12 family; the receptor α subunit was probably a soluble subunit that became constitutively associated with the ligand rather than the β subunit. As such, when the ligand was first isolated, the α subunit was considered a ligand subunit rather than a receptor subunit.

The three-dimensional structure of the extracellular domain of several cytokine receptors has been determined; the two modified FNIII motifs are oriented at right angles with their ends forming half of a binding pocket (Fig. 7-9). The GHR is a homodimer, and both receptors participate in creating the ligand-binding site. GH, which consists of four α-helices in a bundle, sits in the notch between these dimers and forms contacts with residues along the sides of its helices (Fig. 7-9, A). The major point of interaction is with the sides of this pocket, although this may vary depending on the particular hormone-receptor complex. For example, IL-3 binds primarily to the floor of the pocket (Fig. 7-9, B).

Almost all of the cytokines, whose three-dimensional structures are known, are α-helical bundles having either long or short helices. These cytokines bind their receptors in a very similar manner: ligand helices A and C bind to one receptor site, whereas helix D and, to a lesser extend, helix A bind to another receptor site. Figure 7-9, C, shows a monomeric ligand binding to both sites and inducing or stabilizing the receptor complex. However, other stoichiometries

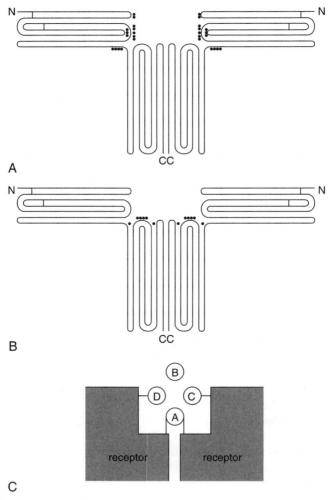

Fig. 7-9. The ligand-binding sites in class 1 cytokine receptors. Critical contact points between the receptors and their ligands are depicted by dots for growth hormone (GH) (A) and interleukin-3 (IL-3) (B). The generic interaction between the individual helices of the cytokines at two sites on their receptors is also depicted (C).

are possible (Table 7-7). For example, the IL-6R complex contains two ligands, two α-subunits, and two β-subunits (Fig. 7-10). In this case, the notch is formed by an amino-terminal Ig domain (D1) and the first half of the cytokine domain (D2). Because the two β-subunits (gp130) bind to opposite ends of IL-6, they do not

Table 7-7
Stoichiometry of Some Cytokine-Cytokine Receptor Complexes

Stoichiometry	Examples
1 Ligand + 2 identical Rβ	GH, PRL, EPO
1 Ligand + 2 different Rβ	LIF[a]
2 Ligands + 2 identical Rβ	Leptin, G-CSF
2 Ligands + 2 identical Rα + 2 identical Rβ	IL-6
2 Ligands + 2 identical Rα + 2 different Rβ	CNTF

[a]gp[190] is really a β-like subunit.

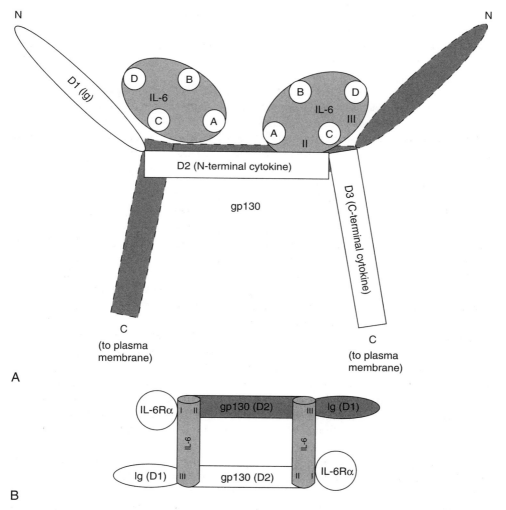

Fig. 7-10. A schematic depiction of the interaction between interleukin 6 (IL-6) and its receptor complex: (A) side view and (B) overhead view. The second gp130 subunit is shown in heavy shade behind the first gp130. The lightly shaded IL-6 contains four α-helices, which are designated by upper case letters; receptor-binding sites are denoted by Roman numerals.

directly interact. Viewed from above, this arrangement forms an open square with the long helices of IL-6 constituting one pair of parallel sides and the D2 of the β-subunits making up the other pair (Fig. 7-10, *B*). The α-subunits interact with a third site (designated I) on IL-6.

Although these receptors are defined by the cytokine domain, other domains can also be found in the extracellular region; they include duplicated cytokine domains, Ig loops, and FNIII motifs (Fig. 7-11, *A*). The Ig domains have been shown to be important in ligand binding (gp130 and gp190), protein transport and cleavage (gp130), and dimer stabilization (G-CSFR). The FNIII domains are involved with receptor spacing and orientation (gp130) and in dimerization (gp130, gp190, and G-CSFR).

The purpose of the WS box is to stabilize the β-barrel of the cytokine domain. As such, it is important in those functions that are dependent on this structure. For example, it is involved with dimerization because the second halves of the cytokine motifs can interact in some receptors. This interaction, in turn, helps orient the cytokine domains. The WS box is also important in the ligand binding of those cytokines that bind to the floor of the notch because the WS box stabilizes these loops. Finally, it is required for proper receptor processing; a mutated WS box will result in an unstable structure that will not be glycosylated and routed to the plasma membrane.

As noted at the beginning of the chapter, many of these extracellular regions are synthesized as soluble products or are generated by proteolysis or alternate RNA splicing. These soluble receptors have several functions. The GHR extracellular domain acts as a GH serum–binding protein that protects the hormone from degradation and delivers it to its target tissues. In excess, the external region can sequester and inhibit the cytokine. Finally, IL-6Rα participates in forming the IL-6 receptor complex. Its solubility allows it to be secreted by macrophages and lymphocytes to regulate the local sensitivity to IL-6.

Although there is no homology among the intracellular regions of these receptors, there are several boxes that possess a conserved amino acid content and are functionally vital for transduction. These boxes are located within 100-150 amino acids of the transmembrane region. Box 1 is rich in hydrophobic amino acids and proline and is required for the activation of Jak (Fig. 7-11, *B*). Box 2 is rich in hydrophobic and negatively charged amino acids; the proximal portion of box 2 contributes to Jak binding. Structure-function studies have shown that these boxes are inactive if they are moved toward the carboxy terminus. Because Jak contains a FERM (4.1-ezrin-radixin-mosein) domain that can bind PIP_2 (phosphatidylinositol 4,5-bisphosphate), it is believed that members of this STK family must bind both boxes and PIP_2 to be active. This phospholipid binding would explain why the boxes must be located close to the membrane. In some receptors there is a third box, which is rich in serines and threonines and binds Stat. The distal region of the cytoplasmic domain is responsible for the specificity of the output. This is the region that is phosphorylated by Jak and attracts SH2-containing transducers.

The output most closely identified with the cytokine receptors is Stat, which is activated by Jak. Jak is constitutively bound to most class 1 receptors. Ligand binding induces receptor aggregation, tightens a preassembled receptor complex, or properly orients the subunits within the complex. Jaks on adjacent subunits will cross-phosphorylate and activate each other; they can then phosphorylate the receptor. Stat can couple to the receptor is several ways: Stat may use its SH2 to bind to the pY on either the receptor or Jak, or it may

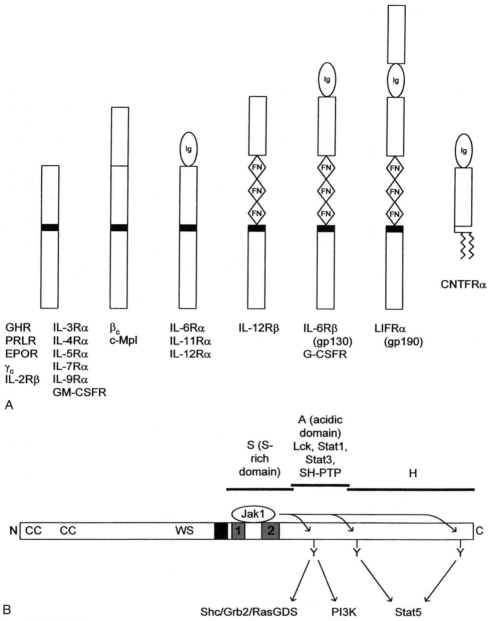

Fig. 7-11. Class 1 cytokine receptors. The different possible variations are shown in the upper panel (A). The rectangle above the black transmembrane helix represents the cytokine domain; diamonds, fibronectin type III regions (FN); and ovals, immunoglobulin-like loops (Ig). CNTFRα is anchored in the membrane by the covalent attachment to a phospholipid. The IL-2Rγ is shown in more detail in the lower panel (B). The conserved cysteines (small C), WSXWS sequence (WS), Jak-binding regions (shaded) and phosphorylation sites (Y) are depicted.

bind to the unphosphorylated receptor. Upon Jak activation, Stat is phosphorylated. Because the SH2 domain of Stat prefers its own pY to that on the receptor complex, Stat dissociates from the receptor, dimerizes by means of this SH2-pY interaction, migrates to the nucleus, and stimulates transcription.

Class 2 Family

Group 2 Subfamily

The class 2 cytokine receptors have FNIII-like domains but these domains are not modified to form cytokine motifs. These receptors are subdivided into two groups; because the group 2 subfamily more closely resembles the class 1 receptors, they will be considered first. The group 2 receptors have 2 FNIII domains that are linked end-to-end rather than at acute angles; however, ligand binding still occurs primarily at the loops and hinges. The intracellular domain has box 1, binds Jak, and activates Stat. The spectrum of subunits and stoichiometry parallels that found in the class 1 receptors: one interferon γ (IFNγ) molecule binds a simple homodimer; but in the IL-10 family, a ligand dimer binds a tetrameric complex composed of two α-receptors and two β-receptors. Other members of this subfamily include receptors for IFNα and IFNβ.

Group 1 Subfamily

The group 1 receptors have a simplified β-sandwich containing only five β-strands; however, like the group 2 receptors, these motifs are aligned end-to-end. The group 1 receptors are involved with programmed cell death, or apoptosis. As a result, their intracellular region is dominated by a death domain (DD); this is a protein-protein interaction domain that binds transducers mediating apoptosis. For example, the DD through adaptors binds procaspases, which are proteases that function in cell death. These zymogens have low basal activity. Ligand binding and receptor aggregation increases their local concentration; they can then cross-cleave each other, converting the caspase precursors to fully active proteases.

There are two well-studied members of this receptor group: tumor necrosis factor (TNF) and LNGFR. The TNFR has four repeats in its extracellular domain; the second and third bind TNF. TNF is a trimer that can exist as either a membrane bound or soluble form; the latter is proteolytically generated from the former. Ligand binding induces TNFR trimerization. Genetic engineering has shown that a TNFR dimer exhibits the same spectrum of activities as the trimer, but the magnitude of the effects is reduced. TNFR aggregation triggers the dissociation of the suppressor of DD (SODD), a protein that maintains low basal activity by inhibiting the DD. After SODD leaves, the DD can begin recruiting adaptors.

There are two TNFRs: TNFR1 (also called p55) and TNFR2 (also called p70). TNFR1 mediates the activity of both the membrane bound and soluble forms of TNF; TNFR2 concentrates the ligand for TNFR1 and helps mediate the activity of the membrane-bound TNF. Details of TNFR transduction are considered in Chapter 12, after the reader has gained more knowledge of second messengers.

LNGFR also has four FNIII-like repeats in its extracellular domain and a DD in its cytoplasmic domain. It was first discovered as a nonspecific binding protein for the NGF family, and considerable controversy erupted as to what its physiological role was. In fact, LNGFR has many roles. First, it alters the affinity and selectivity of the Trk family; for example, when coupled to TrkA, it increases NGF and decreases NT-3 binding; and when coupled to TrkB, it decreases NT-4 and NT-5 binding. Second, it acts as a high affinity receptor for the NGF and BDNF precursors. Third, it generates its own activity via the DD. LNGFR is constitutively active, inhibiting neurite outgrowth and inducing apoptosis. The NGF and BDNF precursors enhance these activities, whereas the mature growth factors inhibit LNGFR and ensure nerve cell survival.

In addition to the NGF family, LNGFR acts as a coreceptor for a structurally heterogeneous group of factors that inhibit axon regeneration. The premier factor, Nogo, lends its name to the primary binding protein, the Nogo receptor, which consists of eight LRRs in the extracellular domain. However, it lacks any transmembrane or intracellular domains; rather, it is covalently attached to a membrane phospholipid that anchors it to the plasma membrane. Because of the absence of a cytoplasmic domain, the Nogo receptor depends on the LNGFR for signal transduction.

Other members of the group 1 receptors include the lymphotoxin-β receptor, Fas, and the receptor activator of nuclear factor-κB (RANK), which is the receptor for osteoclast differentiation factor (ODF), also called *RANK ligand* (RANKL).

G Protein-Coupled Receptors

General Characteristics

The next family of receptors is one of the oldest and probably one of the largest groups; in human beings the number of these receptors is rapidly approaching 1000. There are several names that have been given to this group; because one of the receptors' distinguishing characteristics is the fact that they pass through the membrane seven times (Fig. 7-12), they have been labeled the seven-

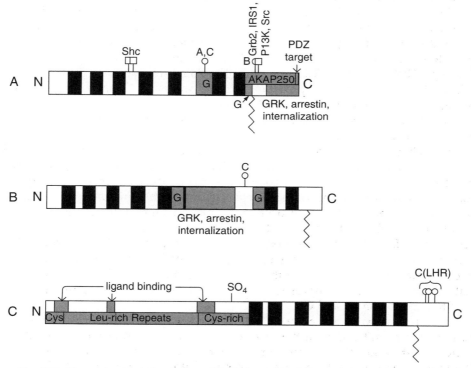

Fig. 7-12. G protein-coupled receptors: (A) β-adrenergic, (B) muscarinic acetylcholine, and (C) glycoprotein receptors. These receptors have seven transmembrane helices (black) and are often palmitoylated (the zigzag line). Other markings are defined in the legend to Fig. 7-1.

transmembrane-segment (7TMS) or heptahelical receptors. Because they assume a snakelike appearance as they weave through the membrane, they are also known as the *serpentine receptors*. Finally, they have no intrinsic enzymatic activity; instead they bind to and activate G protein trimers. Therefore they have been referred to as the G protein-coupled receptors (GPCRs). Despite the fact that G proteins are not the only output for these receptors, GPCR has become the standard designation for this group. In addition to acting as hormone receptors, GPCRs are also involved in sensory perception; they include receptors for light, odor, pain, and taste (sweet, bitter, and amino acids).

Structure

As with RTKs, glycosylation of the extracellular amino terminus is required for receptor processing but not ligand binding or signaling. However, sulfation in this region occurs in some GCPRs and is required for high affinity ligand binding; these sulfated GPCRs include the glycoprotein hormone receptors, gastrin receptor, complement 5a receptor, and most chemokine receptors.

The intracellular domain consists of three loops (designated IC1, IC2, and IC3) and the carboxy terminus. The major determinants for G protein coupling reside in the amino- and carboxy-terminal ends of IC3 and the membrane proximal part of the carboxyl terminus with lesser contributions from IC1 and IC2. There is little homology in these regions, and it is not yet possible to predict from the sequence what receptor will couple to what G protein; apparently, the overall three-dimensional structure is more important than a particular amino acid sequence. Indeed, all of these segments can form amphipathic helices; and these isolated helices can, at very high concentrations, activate G proteins.

The intracellular domain also contains the signals for desensitization and internalization. These topics are covered more completely in Chapter 8. Briefly, there is an S-T-rich region in either the carboxy terminus (β-adrenergic receptor or βAR) or IC3 (muscarinic receptor, mAChR, or simply M), which is phosphorylated by a GPCR kinase (GRK). The phosphorylated residues then attract arrestin, which desensitizes the receptor by blocking access to G proteins and which initiates receptor internalization and recycling.

Many GPCRs are palmitoylated on a cysteine in the proximal region of the carboxyl terminus. This posttranslational modification appears to serve different functions in different receptors. First, it inhibits GRK phosphorylation in the βAR, α_2AR, M2, and LH receptor (LHR); protects the angiotensin receptor 1 from proteolysis; facilitates G protein coupling in the endothelin receptor; is required for thyroid-stimulating hormone receptor (TSHR), vasopressin receptor 2, and histamine receptor 2 (H_2) transport to the plasma membrane; expedites downregulation of H_2; and stabilizes the inactive conformation of the bradykinin receptor 2 (B_2). It is also suspected that it transports GPCRs to specialized membrane domains called *rafts* (see later). One GPCR, the prostacyclin receptor, is prenylated; this is another posttranslational modification where a hydrophobic tail is attached to the receptor (see Chapter 9). Prenylation is required for this GCPR to couple to G_s and G_q.

Palmitoylation is not a static modification. Ligand binding stimulates the turnover of palmitic acid on the βAR; this allows the GRK to phosphorylate the βAR so that desensitization and internalization can occur. NO can *S*-nitrosylate the membrane proximal cysteine in βAR and block the attachment of palmitic acid. Finally, phosphorylation of an adjacent tyrosine will also inhibit palmitoylation in B_1.

The ligand-binding site often depends on the size of the hormone. Several of the transmembrane helices have conserved prolines, which kink the helices to create a binding pocket for small ligands (Fig. 7-13). The GPCRs exist in two states; the inactive state is stabilized by bonds between the transmembrane helices. Ligand binding disrupts these bonds causing the GPCR to convert to the active state. For large ligands, additional specificity is created by interactions with the amino terminus and external loops. For example, the glycoprotein hormone receptors have long amino termini containing LRRs that

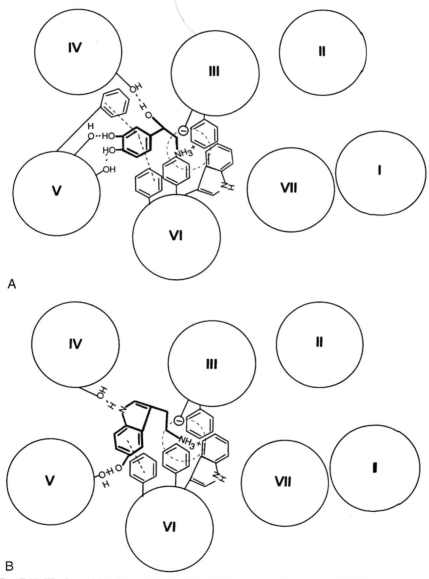

Fig. 7-13. The ligand-binding site in GPCRs: (A) β-adrenergic receptor, (B) 5-hydroxytryptamine (serotonin) receptor, (C) histamine receptor, and (D) muscarinic acetylcholine receptor. The large circles containing Roman numerals are transmembrane helices; the dashed circle represents the hydrophobic box that encloses the ammonium ion; straight dashed lines depict van der Waals interactions.

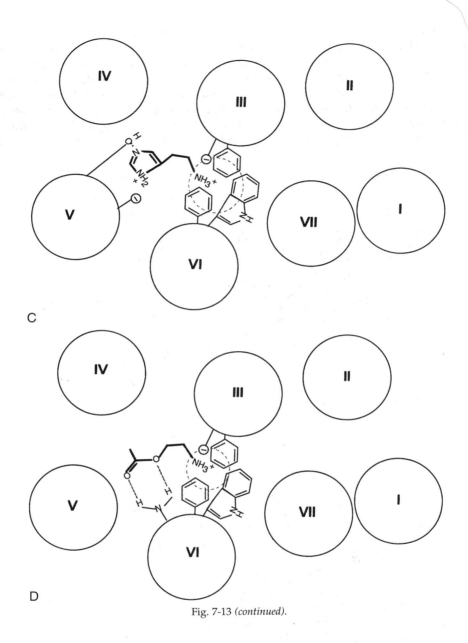

Fig. 7-13 (continued).

wrap around the hormones. Activation by these ligands may occur in any of several ways; first, the extracellular domain may bind one part of the ligand and guide another part into the transmembrane binding pocket. For example, the tachykinins are a family of hormones with a common carboxy terminus; it is believed that this region binds to the transmembrane helices and activates the receptor. The amino terminus and first extracellular loop of the receptor also appear to be involved in recognizing a common motif of the tachykinins, whereas the second and third extracellular loops bind hormone-specific regions that distinguish the different tachykinins. Second, ligand binding may cause the amino terminus to move into the binding pocket and activate the GPCR. For example, thrombin is both a protease in blood clotting and a hormone. When it binds the thrombin receptor, it cleaves off the extreme carboxy terminus to reveal an internal sequence that acts as the ligand. Finally, ligand binding to the amino terminus and external loops can alter the conformation of these structures and the transmembrane helices connected to them. This would disrupt the inter-helical bonds and activate the GPCR without any peptide insertion into the transmembrane region.

The transmembrane helices are predicted to extend into the cytosol and form a binding pocket for G proteins. G proteins bind guanine nucleotides in a cleft between two globular domains. The helical movement induced by ligand binding is thought to pry open the nucleotide-binding site and expose it to the cytoplasm. Because the concentration of GTP in the cytosol is much greater than GDP, GTP displaces GDP and activates the G protein. This is not an all-or-none effect: many drugs can stabilize a wide variety of intermediary GCPR conformations that can affect the magnitude and nature of the response. In addition, some GPCRs are subject to allosteric regulation; for example, progesterone can bind and inhibit the oxytocin (OT) receptor.

Finally, some GPCRs have been reported to exhibit ligand-independent, constitutive activity; these examples include the receptors for calcitonin (CT), histamine, thyrotropin-releasing hormone (TRH), gastrin, angiotensin, bradyki-nin, and glutamate. In some cases, such as presynaptic H_2, the constitutive activity is actually inhibition and helps to maintain a low basal state. However, some of these reports may be artifacts of high receptor concentrations normally found in transfection studies. Such results may also arise from the absence of an endogenous antagonist. For example, the anchoring protein, Homer, binds to and inhibits the constitutive activity of the metabotropic glutamate receptor (mGluR). The absence of Homer in a native cell transfected with mGluR would result in constitutive activity. Finally, constitutive activity may be the result of other interactions; for example, TSHR can be activated by either TSH or fibronectin in the substratum. TSHR on cells cultured on substratum containing fibronectin would show activity even in the absence of TSH, thereby appearing to be consti-tutively active.

Dimerization

Originally, GPCRs were thought to act as monomers because there was nothing in their mechanism of action that would require an oligomer. However, evidence for dimerization came from so many different approaches, that the dimerization of some GPCRs is now accepted as fact. A brief summary of this evidence follows:

1. Cross-linked ligands are more active than monomers.

2. Cross-linked receptors are constitutively active.

3. Receptors mutated at different sites complement each other. This would only be possible if the intact parts of each receptor could interact to create a normal hybrid.

4. Some inactive receptors are dominant negative inhibitors. The only way that a mutant receptor can inhibit a normal one would be to sequester it in an inactive oligomer.

5. Inhibition of dimerization decreases receptor activity.

6. Transfection of high receptor numbers leads to constitutive activity. High concentrations would increase the chance of receptor-receptor collisions and dimerization.

7. Some GPCRs are obligate heterodimers. The GABA$_B$R requires both isoreceptors 1 and 2 for function.

8. In kinetic studies, some agonists measure twice the receptor number than others. This result could arise if some GPCR dimers bind to a single ligand; that is, the ligand-to-receptor ratio is 1:2. Agonists that are unable to induce dimerization would show a ratio of 1:1, allowing the receptor to bind twice the labeled hormone.

9. Some kinetic studies show positive cooperativity that cannot be attributed to G protein interactions. Cooperativity requires protein-protein interactions.

10. Bioluminescence and fluorescence resonance energy transfer studies indicate that GPCRs are in proximity to each other. In these studies, bioluminescent or fluorescent compounds are attached to GPCRs, and the transfer of energy between these molecules on adjacent GPCRs can be used to calculate the distance between these molecules.

11. Finally, GPCRs have been physically linked. For example, Frizzled, the Wnt receptor, has been crystallized as a dimer, the angiotensin receptor 1 (AT$_1$) and B$_2$ have been coimmunoprecipitated, and covalently linked GPCRs have been isolated.

There are several mechanisms that have been shown or postulated for GPCR dimerization. Both the calcium receptor (CaR) and the mGluR have been isolated as disulfide-linked dimers. Other GPCRs appear to interdigitate their transmembrane helices; complementary studies suggest that helices 1 through 5 from one GPCR interact with helices 6 and 7 from another. The GABA$_B$ dimerizes through the α-helices in their carboxy termini; these helices intertwine to form a coiled coil. Dimerization through interactions in IC loops or through lateral packing has also been postulated. Finally, βAR mutants incapable of being glycosylated dimerize poorly, suggesting the involvement of these structures in aggregation. There is no clear association between oligomerization and ligand binding.

Not all GPCRs have been shown to be dimers. Of those that have, four types of oligomers have been identified. The first are simple homodimers. The second are heterodimers between homologous subunits; the heterodimer between GABA$_B$1 and GABA$_B$2 is an example. The third are oligomers containing nonhomologous subunits; for example, the CGRP receptor is a trimer of calcitonin receptor-like receptor (CRLR), receptor activity modifying protein 1 (RAMP1), and receptor component protein (RCP). Only CRLR is a GPCR. The PTH and the vasopressin receptors have also been shown to be associated with members of the RAMP family. In another example of nonhomologous heterodimerization, the Wnt receptor is an oligomer of Frizzled, which is a GPCR, and the low-density lipoprotein receptor related protein 6 (LRP6). Finally, the dimer may represent a cleaved monomer: both TSHR and the LNB-TM7 are cleaved in the extracellular domain to generate a dimer where the amino terminus is noncovalently attached to the transmembrane and cytoplasmic domains.

There are at least four functions that have been attributed to dimerization. First, one of the subunits may act as a chaperone: $GABA_B2$ and RAMP are required for the maturation and membrane targeting of the $GABA_B$ and CLRL complexes, respectively. Second, dimer partners may affect internalization; for example, when the β_2AR is paired with the dopamine receptor (D), agonist-mediated endocytosis is increased, but when β_2AR is paired with the κ opiate receptor, endocytosis is decreased. Third, dimerization may affect the affinity or specificity of ligand binding; for example, the CaR homodimer has a higher affinity for calcium than the monomer, and the D_2-somatostatin 5 receptor (SSTR5) heterodimer binds dopamine or SRIF with a higher affinity than either GPCR alone. The binding specificity can also be modulated: for example, CRLR binds CGRP when coupled to RAMP1 but binds adrenomedullin when coupled to RAMP2, and T1R3 is the taste receptor for sweetness when paired with T1R2 but recognizes amino acids when paired with T1R1. Finally, dimerization can affect receptor activity. For example, neither the CCR2 nor the CCR5 activate G_q, but the heterodimer does; the β_2AR strongly stimulates MAPK, but the stimulation is much weaker when β_2AR is coupled to the κ opiate receptor. RAMP selectively augments phosphoinositide (PI) hydrolysis induced by the vasopressin receptor without affecting cAMP production.

Non-G Protein Output

GPCRs are named for their ability to activate G proteins. However, it is now known that these receptors are far more versatile in their signaling capabilities. As noted previously, several RTKs can also activate G proteins; therefore it should not be surprising that GPCRs can activate STKs. Both Jak and Src can directly bind GPCRs: Jak binds the sequence YIPP in the carboxy terminus of AT_2, and the Src SH3 domain binds the proline-rich region in the IC3 of the β_3AR. Both are presumably activated upon dimerization and cross-phosphorylation, although this mechanism has not been proven. In addition to kinases, GPCRs can bind phosphatases. AT_2 binding of SHP requires the presence of α_s; ligand binding induces the dissociation and activation of this tyrosine phosphatase. Various adaptors and scaffolds can also link GPCRs to many effectors. Grb2 binds the proline-rich region in IC3 to couple GPCR to the MAPK and PLCγ pathways; the βγ subunits of the G protein bind the sequence NPXXY in the last transmembrane helix to link GPCRs to PLD through Rho; and AKAP250 binds the carboxy terminus of β_2AR (see Fig. 7-12, *A*) and forms a complex with PKA, PKC, GRK, β-arrestin, and clathrin. Many adaptors connect GPCRs to the cytoskeleton and to other receptors.

GPCRs can also couple to CaM, NO, and Ras pathways. For example, CaM binds to IC3 of the opiate receptor and is released on agonist binding. On the other hand, CaM binding to the mGluR is calcium dependent: calcium-CaM displaces the βγ dimer from the mGluR, allowing βγ to inhibit calcium channels. NOS3 binds, and is inhibited by, the membrane proximal region of the carboxy terminus of some GPCRs; agonist binding induces GPCR phosphorylation, which releases NOS3. NOS2 becomes active when its PDZ domain binds the carboxy terminus of the 5-hydroxytryptamine (serotonin) 2B receptor ($5\text{-}HT_{2B}$). Finally, the carboxy terminus of β_1AR binds the PDZ domain of CNrasGEF. The latter is an exchange factor for Ras and possesses binding sites for cAMP, which activate the GEF. β_1AR stimulates CNrasGEF by activating adenylate cyclase through G_s to produce cAMP and by localizing the GEF near the cAMP production site.

Phosphorylation of the carboxy terminus of $\beta_1 AR$ by GRK induces the dissociation of CNrasGEF and the termination of signaling.

In addition, GPCRs can interact with channels. The carboxy terminus of D5 binds the second intracellular domain of the $\gamma 2$ subunit of $GABA_A$, a chloride channel (see later); the interaction is mutually inhibitory. The carboxy terminus of D1 binds and inhibits NMDA, a glutamate-activated channel. These associations can also be indirect: the carboxy terminus of the $\beta_1 AR$ binds NMDA via the scaffold, PSD95.

Finally, GPCRs can directly activate translation and transcription factors. As previously noted, AT stimulates Jak, which can then phosphorylate Stat. This mechanism is very similar to that used by the cytokine receptors. Activating transcription factor 4 (ATF4), also called the *cAMP response element binding protein* 2 (CREB2), is attached to the carboxy terminus of $GABA_B 1$ by means of a coiled coil. This is also the dimerization domain with $GABA_B 2$, and binding of ATF4 and $GABA_B 2$ is mutually exclusive. Ligand binding to $GABA_B 1$ induces dimer formation with $GABA_B 2$. ATF4 is displaced and becomes available for MAPK phosphorylation. As with Stat, phosphorylated ATF4 migrates to the nucleus and stimulates transcription. Protein synthesis can also be stimulated: IC3 of M4 binds eEF1A2, which transports charged transfer RNAs (tRNAs) to the ribosome. M4 facilitates the displacement of GDP by GTP, thereby hastening the recycling of this translation factor.

This is just a brief sampling of the non-G protein outputs that these receptors can have. The reader is referred to references at the end of this chapter for extensive reviews on this topic.

Classification

There are several classification schemes that have been proposed for the GPCRs. The one used here divides these receptors into six families. The first family is further divided into three subfamilies. Family 1a includes the adrenergic, H, D, 5-HT, M, adenosine (A), tyramine, opiates, adrenocorticotropic hormone (ACTH), and cannabinoid receptors. All the ligands in this subfamily bind in the transmembrane pocket. These receptors have an S-T-rich internalization domain either in the carboxy terminus (see Fig. 7-12, *A*) or in IC3 (see Fig. 7-12, *B*). Subfamily 1b binds many peptides, interleukins, and the platelet-activating factor (PAF). These ligands bind the amino terminus and extracellular loops. This group contains two notable receptors: the thrombin and GHRH receptors. Thrombin is a protease that removes the amino-terminal 16 to 17 amino acids of its receptor; the new amino terminus acts like an agonist that then binds to the receptor. The binding specificity for thrombin resides exclusively in the cleavage site, and once cleavage has occurred, thrombin is no longer needed. This is beautifully illustrated by changing the sequence at this site to one recognized by enterokinase; this mutated receptor is now activated by enterokinase instead of thrombin. This system has a unique feature: the agonist is tethered to its receptor and provides continuous activation after cleavage; that is, receptor activation is irreversible. The system can be turned off by either metabolizing or inactivating the receptor. The second unusual receptor is the mammalian GHRH receptor, which has a very short carboxy terminus and a markedly decreased rate of internalization. The receptor in fish and chickens has a longer carboxy terminus and a faster rate of internalization. Family 1c includes the receptors for glycoprotein hormones (see Fig. 7-12, *C*) and the relaxin family; as in family 1b, the ligands bind the amino terminus and extracellular loops.

Family 2 is also divided into three groups; all members bind their ligands in the amino terminus and extracellular loops. The secretin family includes secretin, glucagon, PTH, and CT receptors. Some of these GPCRs possess interesting characteristics: first, the CT group uses an accessory receptor, RAMP, which was described previously in the section on Dimerization. Second, some GPCRs have multiple ligands. For example, the CT receptor can also bind calcium, and the cysteinyl leukotriene receptor 1 binds leukotriene C_4 (LTC$_4$), leukotriene D_4(LTD$_4$), or uridine diphosphate (UDP).

The second group in family 2 possesses a very long amino terminus with a variety of binding motifs usually involved in cell-cell or cell-substratum interactions, such as EGF-like repeats, lectins, and Ig domains. These are the LNB-TM7 receptors and some of them are involved with immune and inflammatory responses. The third group is the Methuselah family, whose amino terminus contains three β-domains. A cleft between the first and second domains appears to form a ligand-binding site.

Family 3 includes the mGluR, GABA$_B$, CaR, and some taste and pheromone receptors. The amino termini of the first three families are homologous to bacterial periplasmic proteins (see section on Guanylate Cyclases). The ligands bind entirely within the cleft between the globular halves of this domain. The CaR also binds certain amino acids, especially aromatic ones. This phenomenon explains why protein restriction elevates PTH and why a high-protein diet increases the urinary calcium excretion. The taste receptors represent a group of GPCR dimers whose composition dictates the binding specificity: T2R homodimers detect bitter; T1R2-T1R3 heterodimer, sweet; and T1R1-T1R3, amino acids.

Family 4 includes other pheromones and Family 6 contains the cAMP receptor in slime molds. Family 5 consists of GPCRs involved in development: Smoothened and Frizzled. Both receptors have very unusual mechanisms of action. Smoothened (Smo) is a constitutively active GPCR that is inhibited by Patched (Ptc), which is the receptor for Sonic hedgehog (SHH). The target of Smo is a complex containing an S-T kinase (Fused, or Fu), a transcription factor (Cubitus interruptus or CI), a kinesin-like molecule (Costal2 or Cos-2) and a suppressor of Fused [Su(fu)]. Under basal conditions, the complex is bound to microtubules by Cos-2 and phosphorylation of CI triggers its cleavage, which converts it from a transcription activator to a repressor. When stimulated, SHH binds Ptc and prevents its inhibition of Smo. Smo, which directly binds the Cos-2-Fu-CI complex, inhibits the phosphorylation and cleavage of CI and stimulates Fu to phosphorylate Cos-2; the latter either releases CI to migrate to the nucleus or transports CI to the nucleus along the microtubules. Once in the nucleus, CI activates transcription.

Frizzled is the receptor for Wnt. Under basal conditions, GSK3 phosphorylates axin and APC (adenomatous polyposis of the colon); the former stabilizes the binding between axin and APC and the latter attracts β-catenin. β-Catenin serves two functions: it connects adhesion molecules to actin and acts as a transcription coactivator. When bound to axin-APC, it is phosphorylated by CK2 and GSK3 and is targeted for degradation. When Wnt binds Frizzled, GSK3 is inhibited; β-catenin escapes destruction and migrates to the nucleus, where it becomes a coactivator in transcription (see Chapter 13).

Ion Channel Receptors

The next group of receptors considered in this chapter are the ion channels. Structurally, they are a very heterogeneous group of membrane proteins. Functionally, some are ligand-gated by hormones or neurotransmitters, and others are regulated by second messengers. Some groups do not bind ligands at all, but for completeness all ion channels are covered together.

Cys-Loop Superfamily

Acetylcholine has two types of receptors; the muscarinic acetylcholine receptor is a GPCR and was previously discussed. The nicotinic acetylcholine receptor (nAChR) is a ligand-gated sodium channel in muscle and a calcium channel in the nervous system. Like acetylcholine, 5-hydroxytryptamine and γ-aminobutyric acid have receptors that are both GPCRs and ion channels; the 5-HT$_3$ receptor is a divalent cation channel, and GABA$_A$ and GABA$_C$ are chloride channels. Other members of this family are also chloride channels, which mediate the effects of inhibitory neurotransmitters and include the receptors for glycine, glutamic acid (*Caenorhabditis elegans*), and histamine (*Drosophila*). Note that the vertebrate histamine receptor is a GPCR and that the vertebrate glutamate receptor belongs to a different ion channel superfamily.

These receptors have five homologous subunits: in the muscle nAChR, the subunit structure is $\alpha_2\beta\gamma\delta$ (Fig. 7-14, *A*). However, this can vary depending on the particular isoreceptor or the developmental stage of the organism; for example, the neuronal nAChR has a simpler composition, $\alpha_2\beta_3$, and in the adult, the γ subunit of the muscle nAChR is replaced by the ε subunit. Some functional specialization has occurred among these subunits; the γ/ε subunit stabilizes the closed state, thereby desensitizing the nAChR, and the δ subunit is involved with channel closing and voltage-gating. Both α-subunits in the nAChR represent the acetylcholine binding sites. Finally, there is a 43-kDa peripheral protein, rapsyn, that mediates receptor clustering, such as that occurring at synapses.

The extracellular domain is a twisted, 10-stranded β-sandwich whose strands run perpendicular to the plasma membrane. The sandwich is oriented such that there is an inner and outer sheet. When ACh binds, one of the loops moves toward the binding pocket causing the inner sheet to rotate 15 to 16° clockwise and the outer sheet to tilt 11°. This movement is transmitted to the transmembrane helices described later.

The Cys-loop superfamily of channels has four transmembrane α-helices, and both termini are extracellular. The second helix (M2) from each of five separate subunits forms the channel wall (Fig. 7-14, *B*). Each second helix is thought to be flanked by the first and third helices; however, the x-ray crystallographic structure has not been solved to a resolution sufficient to follow the peptide backbone. As such, these transmembrane regions have been identified from hydropathy plots of the amino acid sequence, mutation studies, and modeling. The four-helix model is the generally accepted model, although other models have also been proposed.

Structural and functional analyses have identified several rings within the ion channel (Fig. 7-14, *C*). There is an outer anion ring in the nAChR beginning on the extracellular side. This ring provides a hydrophilic environment for ions. This is followed by two hydrophobic rings: first a ring of valine, and then one of leucine. Because hydrophobic amino acids would impede the movement of ions, these

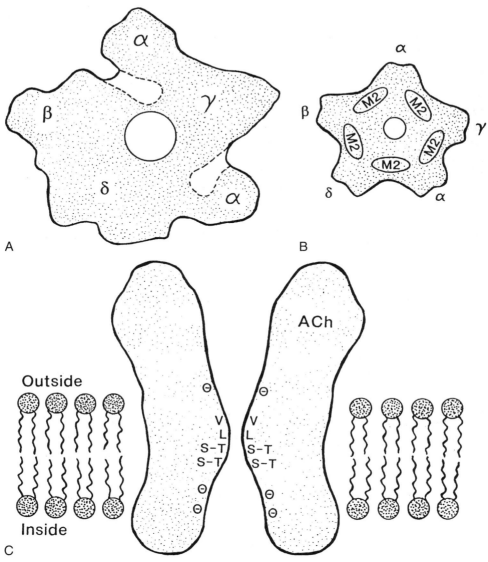

Fig. 7-14. The nicotinic acetylcholine receptor. (A) A transverse cross section through the nAchR above the plasma membrane as viewed from overhead. The low-density areas in the α-subunits (dashed lines) are thought to represent ligand-binding pockets. (B) A transverse cross section at the level of the plasmalemma showing the second transmembrane helix (M2) that lines the channel. (C) A longitudinal section showing the rings forming the ion channel.

rings may block ion flow in the closed state; that is, they may represent the gating mechanism. If these amino acids are replaced by hydrophilic ones, the channel becomes leaky, suggesting that these rings do indeed act as a gate. This region is followed by two S-T rings, although only one of these rings is conserved in the chloride channels. They serve several functions. First, because this region and the leucine ring above it represent the narrowest part of the pore, it acts as a gate. This constriction is a result of an M2 kink that points toward the center of the pore. Second, the arrangement of the hydroxy groups determines cation selectivity. Finally, the ring facilitates ion transport. The intermediate anion ring is located below the S-T rings; it selects monovalent versus divalent ions and may also select for charge. It is negatively charged in the nAChR and positively charged in the chloride channels. The M1-M2 loop is also involved in charge selectivity. The inner anion ring is on the cytoplasmic side of the channel and appears merely to provide a hydrophilic environment at the pore opening.

Ach binding causes M2 to rotate so that the kink and the leucine side chains move to the side. This motion enlarges the pore size like the diaphragm of a camera. It is not known how the ligand-binding and M2 domains are connected, but both the inner β-sheet of the extracellular domain and M2 rotate in the same direction. As such, the simplest explanation would be that the two domains are directly and physically linked.

The nAChR is regulated not just by ligand binding but also by phosphorylation (Table 7-8). Both PKA and PKC can phosphorylate the large intracellular loops in the γ and δ subunits; this is associated with desensitization and the prevention of subunit assembly. In the latter case, dephosphorylation must take place before the receptor can be assembled after synthesis. Tyrosine phosphorylation of this same loop in the β, γ, and δ subunits is associated with synapse formation; the pY also acts as a docking site for Grb2.

P (H5) Superfamily

Voltage-gated Family

The voltage-gated ion channels consists of four domains. Each domain contains six transmembrane α-helices and a loop (P or H5) between helices 5 and 6 (Fig. 7-15, *A*). Originally, P was believed to be a β-loop; β-loops from each of the

Table 7-8
Effect of *In Vitro* Phosphorylation on Some Ion Channels

Ion channel	Kinase	Phosphorylation site[a]	Effect
nAChR	RTK (MuSK[b]), STK (Fyn, Fyk)	IC M3-M4 loop	Receptor clustering; binding site for Grb2
	PKA, PKC	IC M3-M4 loop	8-Fold faster desensitization
GABA$_A$	PKA	IC M3-M4 loop	Decrease current; increase assembly
	PKC	IC M3-M4 loop	Inhibition
	STK	IC M3-M4 loop	Enhance GABA current
Glycine	STK (Src)	IC M3-M4 loop	Increase current, glycine potency, and desensitization
	PKC	IC M3-M4 loop	Increase desensitization
AMPA	PKA	IC M3-M4 loop and carboxy terminus	Potentiate current; insertion into the synapse
	CaMKII, PKC	IC M3-M4 loop and carboxy terminus	Receptor localization
	PKC	Carboxy terminus	GRIP[c] dissociation
NMDA	PKC	Carboxy terminus	Dispersion of receptor clusters
	Cdk5	Carboxy terminus	Long-term potentiation
	STK	Unknown	Increase open probability

[a] IC designates an intracellular loop between two transmembrane helices (M). In GluRs, M3 and M4 actually represent the second and third helices because the helices were named before it was determined that ''M2'' was the P loop.

[b] Muscle-specific RTK.

[c] A PDZ protein that localizes AMPA to the synapse.

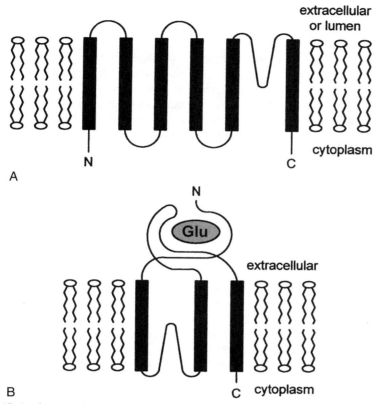

Fig. 7-15. A schematic depiction of the P (H5) superfamily of ion channels: (A) the voltage-gated family and (B) the glutamate family.

four domains would then create an eight-stranded β-barrel to form the wall of the channel. However, KcsA, a bacterial potassium channel believed to be structurally related to the P superfamily, has recently had its three-dimensional conformation determined, and the penetrating edge of the loop is an α-helix, whereas the exiting edge is an uncoiled strand. Whatever the true conformation of the loop, it does not extend all the way through the membrane and acts as selectivity filter. The fourth helix is positively charged and acts as the voltage sensor; the intracellular loop between the fourth and fifth helices blocks the channel. On depolarization, the positively charged fourth helix is attracted to the negatively charged outer surface. The amino acid side chain interactions prevent a straightforward outward movement; instead, the helix rotates 60° as it elevates 5Å. This disrupts the adjacent intracellular loop and the channel becomes unblocked.

The four domains just described may exist either on separate proteins or on a single, large one. The former condition is believed to be the ancestral prototype and persists today in the potassium channel. In time, gene duplication with fusion occurred to produce the voltage-gated sodium and calcium channels. This structural difference affects the mechanism of fast inactivation. The channel can be closed by simply reversing the movement of the fourth helix, but this is relatively slow. In the potassium channel, the extreme amino terminus is rich in basic and hydrophobic amino acids. When the channel opens, negative charges are exposed and attract the amino terminus; the hydrophobic residues then plug the pore. However, the sodium channel only has a single amino terminus because the four domains are fused. In this channel, a cluster of hydrophobic amino acids between the third and fourth domains flip shut over intracellular opening.

Cyclic Nucleotide-Gated Channels.   In photoreceptors, darkness generates an electrical current, whereas light turns the current off. Light accomplishes this by activating a cGMP phosphodiesterase that lowers cGMP levels; cGMP is necessary to keep open a cation channel that carries this dark current. Therefore cGMP hydrolysis inactivates the channel and terminates the dark current. The cyclic nucleotide-gated (CNG) channel diverged early from the potassium channel branch; as such, it consists of four separate subunits, each with six transmembrane helices and a P region. In addition, there is a cGMP-binding domain located in the carboxy terminus and a CaM-binding site in the amino terminus. CaM decreases the affinity of the CNG channels for cyclic nucleotides. CNG channels are also found in other sensory tissue, including olfactory epithelium and taste buds. In the latter, they act as receptors for acidity. The role of CNG channels in nonsensory organs is less well-defined.

Ryanodine receptor family.   The ion channels discussed so far are located on the plasma membrane; the members of the ryanodine receptor family are located on calcium-containing intracellular organelles, such as the sarcoplasmic reticulum of muscle. They are difficult to classify because they have no homology to any of the other channels; even the number of their transmembrane segments is uncertain. Some characteristics, like the spacing of the helices and the tetrameric structure, resemble the voltage-gated channels; for this reason, this text depicts them as having six transmembrane helices.

In muscle, the action potential at the plasma membrane and T-tubules results in the release of calcium from the sarcoplasmic reticulum; this calcium, in turn, triggers contraction. It is the ryanodine receptor that couples these two membranes. The hydrophobic helices are embedded in the sarcoplasmic reticulum, where they form a calcium channel; the extremely long amino terminus acts as a physical bridge to the plasma membrane, where it contacts a typical voltage-gated calcium channel. Apparently, the depolarization brought about by the action potential activates the voltage-gated calcium channel, which then opens the ryanodine receptor, presumably by transmitting a signal through the amino terminus of this receptor. Recently, this receptor has been found in nonmuscle tissue. Furthermore, the ability of cyclic ADP-ribose, an NAD^+ metabolite, to open these channels has suggested that they may be regulated by second messengers (see Chapter 10).

Another member of this family is clearly activated by a second messenger: the inositol 1,4,5-trisphosphate receptor (IP_3R). IP_3 is one of several mediators generated by the hydrolytic activity of PLC. It diffuses from the plasma membrane, where it is liberated, to the endoplasmic reticulum, where it activates its receptor, a calcium channel. The IP_3R is a noncovalently linked tetramer that forms a 25-nM square. Each subunit binds one IP_3 at the extreme amino terminus and binding is cooperative. If the IP_3R is like the ryanodine receptor, the channel is straight as it traverses the membrane, but then splits into four channels that bend 90° and exit the amino-terminal domain laterally. As such, it resembles a rotating sprinkler head, where the water comes up the stem and then is ejected laterally through the spinning arms. This analogy applies only to the three-dimensional structure; there is no evidence that the IP_3R rotates. This lateral dispersion may facilitate the rapid diffusion of the calcium ions into the cytoplasm. In addition to ligand gating, the IP$_3$R is allosterically regulated by ATP and calcium; it is also phosphorylated by several kinases. These controls are discussed in Chapter 10.

Transient receptor potential channels.   Once the internal calcium supplies have been depleted, they must be recharged. Because these channels somehow sense the calcium status of the cell and only open when stores are low, they have been referred to as the *calcium capacitative channel*. It is now believed that this function resides in the transient receptor potential C (TrpC) subfamily of the Trp channels. There are two other subfamilies. TrpM includes a cold sensor and a magnesium transporter, and TrpV includes a renal transporter for calcium, as well as receptors for heat, osmolarity, *N*-arachidonoyl-dopamine, and odorants.

Calcium-activated potassium channels.   Calcium-activated potassium channels are divided into two groups: small conductance (or SK) and large conductance (or BK) channels. The former constitutively bind CaM in the carboxy terminus. Elevated calcium binds the amino-terminal CaM domain to expose the hydrophobic cleft, which then binds an adjacent subunit. This distortion opens the pore. The BK channel directly binds calcium via two carboxy-terminal, bilobed structures. When calcium binds in the cleft between the two lobes, they tilt and expand the diameter of the pore by pulling on the number six helices, to which they are connected. PKG phosphorylation and *S*-nitrosylation also increase the activity of these channels.

Glutamate Receptor Family

As with many other hormones, glutamic acid has receptors in both the GPCR family (the metabotropic glutamate receptor or mGluR) and the ion channel superfamily (the ionotropic glutamate receptor, iGluR, or simply GluR). The GluR family has the same basic structure as the voltage-gated family, but the number of transmembrane helices has been reduced: there are only three such helices, and P is located between the first two (Fig. 7-15, *B*). The amino terminus and the extracellular loop between the last two helices each form separate lobes of the ligand-binding site. This bilobed structure is homologous to the bacterial periplasmic proteins, just described for GC-A and family 3 of the GPCRs. When glutamate binds, the cleft between the lobes narrows and tension is exerted on the transmembrane helices. If modeling based on the bacterial potassium channel, KcsA, is valid, this tension causes the helices to rotate counterclockwise and enlarge the pore size. Most GluRs are calcium channels.

GluRs are divided into four groups; the first three are named for the pharmacological agonist specific for each one. The AMPA receptor specifically binds α-amino-3-hydroxy-5-methyl-4-isoxazole propionic acid, is located on postsynaptic membranes, and is responsible for moment-to-moment signaling. It is composed of the subunits GluR1-4; complexes containing GluR2 transmit sodium instead of calcium. The kainate receptor modulates synaptic activity, is located at several sites in the presynaptic and postsynaptic membranes, and is composed of the subunits GluR5-7 and KA1-2. The NMDA receptor has a duplication in its extracellular domain and is composed of the subunits NMDAR1 and NR2. NMDAR1 specifically binds *N*-methyl-D-aspartate, and NR2 binds glycine, although some studies suggest that D-serine, not glycine, is the real agonist. The NMDA receptor is believed to be a tetramer of two NMDAR1 and two NR2 subunits; as such, it requires both glutamate and glycine/serine as coagonists.

However, glutamate is the actual signaling molecule because the extracellular concentration of glycine/serine is believed to be saturating under physiological conditions. The NMDA receptor is involved with long-term potentiation and depression. Activity is affected by several factors: polyamines increase glycine binding and arachidonic acid potentiates receptor activity. Much less is known about the last GluR, delta. Only two subunits, delta 1 and 2, are known, and this GluR appears to be involved with long-term depression in the Purkinje cells of the cerebellum.

In addition to ion fluxes, these receptors have other outputs: they can activate STK and the MAPK pathway; the AMPA receptor can activate G_i; and the NMDA receptor is coupled to NOS through the scaffold, PSD-95. GluRs can also be regulated by phosphorylation (Table 7-8).

Inward Rectifying Potassium Channel

Rectification refers to the unidirectional flow of ions; as such, the G protein-activated inward rectifying potassium channel (GIRK) only allows potassium to enter the cell. It is truncated even more than the GluRs; it possesses only two transmembrane helices with an intervening P loop. It is activated by the G protein $\beta\gamma$ subunits, which directly bind the channel. This activation requires PIP_2. GIRK can also be inhibited by polyamines and by α_i, independently of $\beta\gamma$.

Epithelial Sodium Channels

Like the GIRK, epithelial sodium channels are tetramers having two transmembrane helices; however, there is no intervening P loop. These channels include the P_{2X} purinergic receptor, SCaMPER, FMRFamide-gated sodium channels, and mechanotransduction receptors. There is still controversy over the former two: first, some authorities claim that the P_{2X} receptor does have a P loop and belongs in the P superfamily. Second, SCaMPER (sphingolipid calcium release mediating protein of the endoplasmic reticulum) was proposed to be a calcium channel activated by metabolites of sphingolipids, which like PPI metabolites can act as second messengers (see Chapter 10). However, at least one report claims that it has only one transmembrane helix and that it is not localized to any known calcium pool.

Miscellaneous Membrane Receptors

Sorting Receptors

The most important sorting receptor that binds a hormone is the IGF-II receptor (IGF-IIR). The extracellular domain consists of 15 repeating units containing 8 cysteines each and binds IGF-II, mannose-6-phosphate, and any protein containing this sugar. Indeed, this protein was first characterized as a cation-independent, mannose-6-phosphate receptor responsible for transporting mannose-6-phosphate containing proteins to the lysosome. IGF-II and the sugar bind to different sites; the latter interacts at two separate regions comprising repeats 1-3 and 7-10. Although these two ligands bind separately, mannose-6-phosphate does synergize with IGF-II in generating second messengers. The extreme carboxy terminus of the intracellular domain binds to an inhibitory G protein adjacent to a region phosphorylated by casein kinase 2; this phosphorylation is required for the receptor to be transported to the prelysosome. The intracellular domain is also palmitoylated.

Because IGF-II can act through the IGF-IR and because the IGF-IIR serves other functions, the relevance of the IGF-IIR for mediating the biological activities of IGF-II has been questioned. First, the IGF-IIR constitutively recycles, suggesting a housekeeping rather than signaling function. Second, the IGF-II/mannose-6-phosphate receptor from chicken and *Xenopus* does not bind IGF-II, although IGF-II is biologically active in these species. Third, during rat embryogenesis, IGF-II is abundant, but only the IGF-IR is present; therefore IGF-II must act through the IGF-IR. Fourth, mutant IGF-IIs that prefer either the IGF-IR or IGF-IIR have been constructed, and biological activity always correlates with IGF-IR binding. Finally, overexpression of IGF-IR enhances IGF-II activity, whereas mutated IGF-IRs that are dominant inhibitors block IGF-II effects. In contrast, overexpression of IGF-IIR actually inhibits the response to IGF-II, suggesting that the IGF-IIR acts like a sink to trap IGF-II and clear it from the tissue.

There are also arguments in support of a role for the IGF-IIR in IGF-II action. First, stimulating antibodies specific for the IGF-IIR induce thymidine uptake and calcium fluxes. Second, the IGF-IIR appears to bind an inhibitory G protein; such coupling is normally associated with signal transduction. Finally, an agonist specific to the IGF-IIR mimics the effect of IGF-II on cell motility, and this effect is not blocked by antibodies to the IGF-IR. These conflicting views may be reconciled if the IGF-IIR acted indirectly; for example, the IGF-IIR may concentrate IGF-II for the IGF-IR. Alternatively, IGF-IIR may act as a signal integrator; in addition to IGF-II, the receptor also binds and clears LIF, M-CSF, and proliferin. Competition for IGF-IIR would determine which hormone was cleared and which one exerted its biological activity. Finally, the IGF-IIR can activate the TGFβ precursor by transporting it to the lysosome for cleavage and release of TGFβ.

Interleukin-1 Receptor

Toll-like receptors (TLRs) are involved with innate immunity. Antibodies represent acquired immunity, where the defense molecule is created and tailored to a specific antigen; unfortunately, this process requires several weeks. However, mammals are also born with proteins that bind common molecules found on pathogens and that provide immediate protection; this protection is called *innate immunity*. For example, TLR2 and TLR6 bind the peptidoglycan in bacterial cell walls, TLR3 binds double-stranded RNA in viruses, TLR4 binds lipopolysaccharide and lipoteichoic acid in bacterial membranes, TLR5 binds flagellin in bacterial flagella, and TLR9 binds unmethylated bacterial CpG DNA. These receptors can also bind endogenous molecules that represent necrosis; for example, TLR2 and TLR4 can bind several heat shock proteins, which are induced and released during stress. Recently in evolution, the defensins and the IL-1 family of cytokines acquired an affinity for some of these receptors. The defensins are small antimicrobial peptides secreted by mucosa and skin; β-defensin 2 has been shown to bind and activate TLR4. The IL-1 family, which includes IL-18, binds to TLRs that have evolved into a distinct receptor group.

These receptors have three extracellular Ig domains that form a C and wrap around the ligand. Unlike most other ligand-Ig interactions, IL-1 binds to the sides of the β-sheets and not to the loops. The receptor complex exists as a heterodimer of homologous subunits. There is a single transmembrane α-helix.

The intracellular domain binds adaptors that will initiate a protein kinase cascade leading to stress-related transcription factors.

Semaphorin Receptors

Semaphorins are involved with morphogenesis of the nervous system and with T and B lymphocyte development. There are interesting similarities between the semaphorins and the LERKs: both are involved with the development of organ systems, have membrane-bound and soluble forms, display bidirectional signaling, and play important roles in cytoskeletal remodeling. Membrane-bound semaphorins bind plexins, whereas soluble semaphorins bind a complex between plexin and neuropilin. Neuropilin is also used as an accessory receptor with the VEGFR. Plexins have an extracellular sema domain, a single transmembrane domain, and an intracellular carboxy terminus. Semaphorins and certain members of the insulin receptor family, like the HGF and MSP receptors, also possess sema domains. The three-dimensional structure of the sema domain is a β-propeller that most closely resembles the β-subunit of G proteins. Plexin B1 inhibits Rac by acting as a RacGAP; however, it stimulates Rho by binding and activating PDZ-RhoGEF. Plexin A1 is a RndGAP. These effects induce actin depolymerization and collapse of the growth cone in neurons. In addition, plexin constitutively binds Fyn and the HGF receptor; ligand binding stimulates these tyrosine kinases, presumably by inducing receptor aggregation and kinase cross-phosphorylation.

Scavenger Receptor-Like Receptors

Scavenger receptors form loose, heterogeneous groups; only two will be mentioned here. The SRCR family contains scavenger receptor cysteine-rich domains in the extracellular region. The SRCR domain is a six-stranded, β-sheet that is partially curved around an α-helix; the ligand is thought to bind the bulging side. There is a single transmembrane helix. In vertebrates, these receptors are primarily found on macrophages and function in the immune system. However, in invertebrates, they mediate hormone signaling; for example, in sea urchins this family includes the receptor for speract, a pheromone. The cytoplasmic domain of the speract receptor contains a guanylate cyclase. In the sponge, this family includes the receptor for the aggregation factor.

Receptor for advanced glycation endproducts (RAGE) is another scavenger family; it is best known as a receptor for proteins glyoxidized during hyperglycemia in diabetes. RAGE mediates the inflammatory activity of these compounds. In addition, it is the receptor for HMGB1. Although HMGs are important chromatin proteins (see Chapter 14), HMGB1 is also secreted by macrophages and monocytes and released by necrotic cells. The extracellular region contains Ig domains and the short intracellular tail mediates an inflammatory response by an unknown mechanism.

Receptor Metabolism

Synthesis

Membrane receptors are proteins, and they are synthesized in the same manner as all other proteins. However, there are some variations that can have a major impact on the ultimate function of the receptor. Protein synthesis actually begins with transcription of the gene followed by processing of the resulting mRNA. Exon 11 in the insulin receptor mRNA codes for 12 amino acids located near the carboxy terminus of the α-subunit. In some molecules, this exon is spliced out, resulting in a shorter receptor with a higher affinity for insulin and a more rapid rate of internalization. This may be physiologically important; the longer, less sensitive form occurs in the liver, which receives blood directly from the pancreas through the hepatic portal system and which is exposed to high concentrations of insulin. The lower affinity receptor may protect the liver from overstimulation. In the FGF2R, there are two alternate exons that code for the carboxy-half of the third Ig domain: one exon produces a receptor that binds only FGF, whereas the other one generates a receptor that binds KGF.

Translation occurs on membrane-bound ribosomes: membrane insertion is initiated by the signal sequence and is stopped by a pair of basic residues at the end of the hydrophobic helix. As with all membrane proteins, the extracellular domain is glycosylation. The function of this carbohydrate has been investigated by several techniques: (1) stripping the mature protein of sugars by glycosidases, and blocking the addition of carbohydrate by either (2) a specific inhibitor like tunicamycin, or (3) mutation of the asparagine acceptor site. In general, stripping the mature receptor has no effect on ligand binding or signal transduction; however, preventing the addition of sugars during translation does impair both functions. In the latter case, the receptors appear to be misfolded and are retained in the endoplasmic reticulum. Glycosylation also appears to be important in the dimerization of βARs and in suppressing basal activity of some RTKs. Therefore glycosylation seems to be necessary for protein processing, such as folding, oligomerization, and membrane targeting, but it is not for specific ligand binding or signal transduction. Many receptors also undergo another posttranslational modification: palmitoylation. Palmitic acid is added in the Golgi complex; however, its high turnover rate suggests that mechanisms to remove and reattach this fatty acid also exist in the cytosol. Its role in the desensitization of GPCRs was previously discussed.

Finally, multisubunit receptors must be assembled. The insulin receptor represents the simplest case because the α- and β-subunits are part of a single polyprotein. Therefore the subunits are necessarily synthesized stoichiometrically and are physically attached until cleavage. Pulse-chase experiments show that this cleavage does not take place until the receptor is at least core glycosylated (Fig. 7-16). The nAChR represents a more complex example; the assembly of subunits is initially prevented by phosphorylation. After dephosphorylation, αγ (or αε) and αδ dimers first form; this interaction is mediated by the amino termini. Then each dimer binds a β-subunit to create the final pentamer. Other methods have suggested alternate pathways: in one, an αβγ trimer forms first; the δ and ε subunits are then inserted sequentially. Regardless of the assembly route, the pentamer must be complete before it is transported to the plasma membrane. This is accomplished by a retention signal located in the first transmembrane helix. This sequence is exposed in monomers, which are retained in the endoplasmic reticulum; however, the sequence is masked in the assembled pentamer, which can then migrate to the cell surface. Finally, several of the subunits are phosphorylated on tyrosine; this appears to attract rapsyn, a protein that is responsible for escorting the receptors to the plasma membrane and ordering them at synapses.

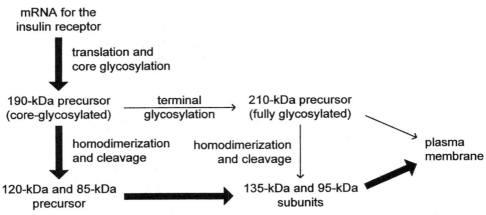

Fig. 7-16. Synthetic pathway for the insulin receptor. The major pathway is shown with heavy arrows.

Coated Pits, Caveolae, and Lipid Rafts

Within the plasma membrane are several microdomains that can concentrate receptors. Coated pits are plasma membrane invaginations 50 to 150 nM in diameter and surrounded by a basket or cage of polygonal units. The basic unit of this cage is a triskelion containing three 180-kDa clathrin molecules and three 33- to 36-kDa light chains, or clathrin-associated proteins; the former form the legs, whereas the latter act as glue. The legs are 445 Å long with a bend 190 Å from the vertex; the legs originally align with each other to form hexagons. Subsequent rearrangement to pentagons correlates with the formation of the membrane bud. The tips of the legs turn in to form struts that bind adaptors in the plasma membrane. These adaptors recognize the receptors. Dynamin, a G protein, forms a ring around the collar and is responsible for pinching off the bud during endocytosis. After the pits bud off they quickly lose their clathrin coat. The major function of coated pits is the metabolism and down-regulation of activated receptors. Transduction can also occur, but the signaling is very simple.

In contrast, the other two domains are specialized for transduction: caveolae, also called *smooth pits*, and rafts, also called *detergent-insoluble, ganglioside-enriched complexes* (DIGs). The two domains are very similar, and their relationship is currently a matter of great debate. There are three possibilities. First, caveolae and rafts are completely distinct. Second, caveolae and DIGs are identical; DIGs were isolated by biochemists, and caveolae were described by electron microscopists. As such, the term *DIG* reflects their biochemical composition, whereas the term *caveolae* describe their shape. Third, DIGs are precursors to caveolae; DIGs are formed first and become caveolae when the protein caveolin is added. Considering their close similarity, the first hypothesis is unlikely, and the two are considered together.

DIGS are characterized by their rigid structure and a cohesiveness so strong that they resist detergent solubilization. These properties arise from their lipid composition.

1. DIGs are rich in sphingolipids, whose fatty acids are usually saturated. Saturated fatty acids pack together more tightly.

2. Unlike phospholipids, most sphingolipids lack a phosphate between the head group and lipid backbone. The negatively charged phosphates would otherwise repel each other and resist close packing.

3. The amide bond to the second fatty acid and the carbohydrate head groups of sphingolipids allow for a wealth of hydrogen bonding between lipid molecules.

4. Gaps may occur between sphingolipids with large carbohydrate head groups; these spaces are filled with cholesterol, whose cyclopentaphenan-threne ring imparts additional rigidity and whose hydroxyl head group is very small. Caveolin and NAP-22 are resident proteins that strongly bind cholesterol.

5. DIGs are also rich in plasmenylethanolamine. In plasmalogens, the first position is occupied by a long chain alcohol in an ether linkage rather than the standard fatty acid in an ester bond. Ethers are much more hydrophobic than esters.

The lipid composition also reflects its signaling function; 50% of all PPI and all agonist hydrolyzable PPI are located in DIGs. It contains 50% to 70% of all ceramide, which is used in apoptosis signaling (see Chapter 10). In addition, plasmenylethanolamine is rich in arachidonic acid, which is both a second messenger and a precursor for eicosanoids.

The unique characteristics of DIGs attract proteins with complementary properties, including palmitoylation, longer transmembrane helices, and glycosylation. Palmitic acid is saturated and fits well into DIGs. Because saturated fatty acids are straighter, DIGs are slightly thicker than the general plasmalemma and it favors proteins with longer transmembrane helices. Finally, glycosylated proteins can hydrogen bind to the sugar residues in resident gangliosides. There is evidence that different kinds of DIGs exist based on their differential solubility in various detergents and their protein composition. In T cells, receptor activation leads to the fusion of DIGs, which would allow different sets of transducers to interact. This is important because the small size of DIGs ($\sim 2100\,nm^2$) would limit the number of proteins present to about 20 per DIG.

Many of the resident proteins are involved with signaling, including STKs, cytokine receptors, Stat, numerous adaptors and dockers, small and trimeric G proteins, MAPK, S6KI, PLC, PLD, SMase, a cyclic nucleotide phosphodiesterase (PDE), and the sodium-hydrogen exchanger (NHE). Some transducers are maintained in an inactive state by binding to caveolin until they are stimulated; these proteins include RTKs, STKs, TGFβR, PKA, PKC, GRK, PI3K, NOS, and adenylate cyclase. Indeed, isolated caveolae will activate MAPK in response to PDGF, demonstrating that all of the intervening components of this cascade are present in these structures.

DIGs and caveolae have several functions. First, they are transduction complexes that bring signaling components together to facilitate their interactions. Second, they can be involved in receptor processing; however, internalization is 2 to 4 times slower than occurs with coated pits and receptor recycling is simpler because there are no intermediates, such as lysosomes, through which receptors pass. Third, caveolae are associated with simple transport phenomena, like potocytosis and transcytosis. Potocytosis is the internalization of small molecules without endocytosis; briefly, caveolae act as concentration chambers for solutes that are then targeted by membrane transporters. Transcytosis is the transport of molecules from one side of a cell to another. An example of the latter is the capillary endothelium, which transports molecules between the blood and extracellular fluid. Fourth, they function in cholesterol homeostasis. Caveolin has a very high affinity for cholesterol and serves to deliver newly synthesized cholesterol to the plasma membrane. Finally, DIGs are required for front-to-back polarity in migrating cells.

Although coated pits and caveolae are structurally distinct, they can be functionally linked. Receptor internalization frequently requires signal activation (see Chapter 8). As a result, some receptors must first localize to caveolae to initiate transduction and then migrate to coated pits, whose internalization is triggered by second messengers generated in the caveolae.

Internalization and Processing

Internalization is a process by which signaling can be terminated and the receptors either destroyed or recycled. There are several routes for internalization and there appears to be some flexibility in their use. For example, some receptors (β_2AR and A_1) are prelocalized to DIGs and, after activation, leave for the coated pits to be internalized. Other receptors are diffusely distributed in the plasma membrane and, after activation, localize to DIGs, where they may (B_2) or may not be internalized (B_1). In fact, any given receptor may shift routes depending upon circumstances. ET_A is normally internalized in caveolae; however, when cholesterol depletion disrupts these structures, ET_A shifts to coated pits. Conversely, βAR is normally internalized in coated pits; but in presence of a dominant inhibitor of dynamin, it can use an alternate route.

Migration to coated pits or DIGs appears to be the result of simple diffusion associated with a change in receptor conformation and/or post-translational modification that favors recruitment to pits or DIGs. For example, many GPCRs and RTKs have a conserved $NPX_{1-2}Y$ sequence that binds adaptors in coated pits, and EGFR kinase activity is required for its localization to coated pits. After receptor stimulation, the endocytotic machinery is activated; this process involves both PI3K and PLD. The products of PI3K are associated with vesicular trafficking; the PIP_n may alter the membrane properties and/or attract proteins required for endocytosis.

Once internalized, there are several processing routes. In one route, the endocytotic vesicle, called an *endosome* or *receptosome*, becomes acidified through an ATP-dependent mechanism. At about the same time, slender tubules begin to emerge from the surface of the vesicle; this entire complex has been called *CURL* (compartment for uncoupling receptors and ligands). This acidification disrupts ligand-receptor binding and the unoccupied receptors migrate into the tubules, whereas the free ligand remains in the vesicle (Fig. 7-17). Actually, the entire process of segregation may be passive; the geometry of these structures is such that 90% of the surface area is in the tubules, although 90% of the volume is in the vesicle. Because receptors are membrane bound and ligands are soluble, their distribution could be entirely explained by simple diffusion. However, some evidence suggests that this separation also involves more active processes. The tubules then detach and return to the plasma membrane, where the receptors are recycled. Finally, the vesicle fuses with lysosomes and the ligand is degraded. This pathway is known as the short loop.

Some hormone-receptor complexes are treated in a similar manner except that the liberated receptors first go to the Golgi apparatus before returning to the plasma membrane; this is called the *long loop*. The reason for this detour is not known; temporary storage, repair, and/or purification from other membrane components have all been postulated. In downregulation, hormone receptors

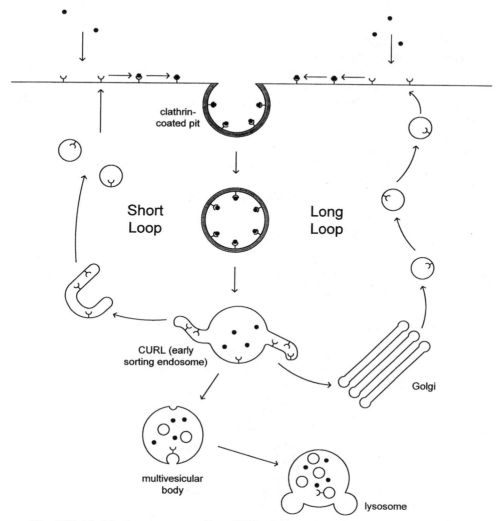

Fig. 7-17. Model of receptor recycling. *CURL,* Compartment for uncoupling receptors and ligands.

are not recycled at all; for example, the EGF receptor is totally degraded in many tissues. This destruction requires autophosphorylation, which attracts a ubiquitin ligase (see Chapter 8). Under normal circumstances, ubiquitinated proteins are targeted to a cytoplasmic, tubular complex of proteases, called the *proteasome.* However, ubiquitination of membrane receptors targets them to multivesicular bodies. These structures are endosomes whose membranes have undergone further endocytosis into the center of the vesicle; this process isolates the receptors from the cytoplasm to completely terminate signaling. These vesicles are subsequently fused to lysosomes and the receptors are destroyed. However, this phenomenon shows tissue specificity: insulin receptors in the liver are recycled, but those in adipocytes are not. The purpose of downregulation is to induce a refractory period in the target tissue after hormone stimulation; the occupied receptors are destroyed and the tissue becomes less sensitive to additional hormone until new receptors can be synthesized (see also Chapter 8).

Relevance

Termination of Signaling

There are numerous functions that internalization performs. Perhaps the foremost is the termination of signaling. Because receptors have a high affinity for their ligands, signaling could be protracted if there were not a mechanism to strip receptors of their hormones. Internalization and acidification of the endosome provides such a mechanism. Receptors can also be downregulated.

Detection of Gradients

Rapid internalization and recycling turn off weak stimulation; this allows a system to detect differences in the concentration of agonists. For example, internalization of the α chemokine receptor, CXCR2, is required for chemotaxis.

Prolongation of Signaling

Many hormones are internalized with some of their transducing components and are known to continue signaling for as long as 2 hours after internalization. In addition, desensitized GPCRs are dephosphorylated when they are internalized; this resensitization allows the GPCRs to recycle and continue signaling.

Cross Talk

Some GPCRs activate MAPK through RTKs; that is, the GPCR activates the RTK, which then stimulates MAPK. This phenomenon requires internalization; presumably colocalization of the GPCR and RTK in the same endosome facilitates cross-activation.

Generation of Second Messengers in Selected Areas

The best example of this function is NGF action in the nerve cell. Neurons have very long cellular appendages, and NGF binding to its receptor at the nerve terminals will not be able to affect processes in the cell body that may be decimeters away. NGF and Trk are internalized in the termini, are transported up the axon, and generate their second messengers in the nerve cell body. Another variation of this is seen in the cytokine receptors. Stats do not have a NLS but cytokines do. The cytokine-receptor-Stat complex localizes to the perinuclear region after internalization; if internalization is blocked, so is transcription.

Access to Sequestered Second Messengers

As just noted, some hormones activate sphingomyelinase, which generates sphingolipid second messengers. The neutral SMase is located in DIGs and is freely accessible to receptors, but the acid SMase is located in lysosomes (see Chapter 10). Stimulation of the latter enzyme by IL-1R3 and TNFR requires internalization to gain access to the lysosome. The reverse is also possible; receptor internalization decreases PLC and PI3K activity because their substrate, PIP_n, remains in the plasma membrane and is no longer accessible to these enzymes.

On-Site Storage of Hormones

Astrocytes take up BDNF through a truncated Trk. Because this Trk has no functional kinase, there is no immediate signal, and because BDNF is not degraded, it remains bioactive and can be released later to regulate BDNF levels.

Miscellaneous Topics

Spare Receptors

Very early in the study of membrane receptors an apparent discrepancy between hormone activity and receptor binding was noted: maximal activity occurred when only a small fraction of the receptors were occupied. For example, human chorionic gonadotropin (hCG) maximally stimulated testosterone synthesis in Leydig cells when only 0.3% of the receptors were bound; insulin maximally stimulated glucose oxidation in adipocytes with a receptor occupancy level of 2%. Similar results have been reported for glucagon and the catecholamines. The reverse phenomenon has also been observed: 80% of the insulin receptors on mammary tumor cells can be destroyed by trypsin, whereas insulin-induced glucose transport decreases by only 13%. Initially, two hypotheses were formulated to explain this discrepancy. One postulated that specific hormone binding was composed of two fractions—a small pool of true receptors coupled to a transduction system and a large pool of receptors with no known function. Clearance receptors, whose sole function would be to internalize and degrade the hormone, would be examples of the latter. The other, more popular hypothesis, claimed that most receptors were simply spare receptors.

Further work has suggested two more hypotheses that seem more logical. According to the first, spare receptors are an artifact produced when only one biological activity is monitored; that is, different responses may require different degrees of receptor occupancy. As just noted, many actions of insulin occur when receptor binding is low; these include glucose transport and oxidation, inhibition of lipolysis, and alterations in protein phosphorylation. Conversely, the stimulation of amino acid transport and RNA synthesis requires nearly maximal binding, and other activities occur at intermediate occupancy levels; these latter effects include the stimulation of glycogen synthase, acetyl coenzyme A carboxylase, pyruvate dehydrogenase, tyrosine aminotransferase, and protein synthesis. However, this phenomenon has been difficult to quantitate. Now that many of transducers of insulin are known, the EC_{50} (effective concentration necessary for 50% of the maximal effect) for these signaling molecules can be measured: tyrosine kinase corresponds exactly to insulin binding, PI3K ($EC_{50} = 4.5\,nM$ insulin), S6KII ($7\,nM$), MAPK ($10\,nM$), and IRS (20 to $30\,nM$).

The other hypothesis claims that these spare receptors modulate the hormone sensitivity of the cell. The kinetics of hormone-receptor binding follow the laws of mass action (see Chapter 5):

$$H + R \rightleftharpoons HR \rightarrow \text{biological response}$$

The critical concentration is that for HR, and this, in turn, is determined by the concentrations of both the hormone and the receptor. With an increase in the concentration of R, the reaction can be driven to the right. In other words, the higher R is, the less H is required to reach the same concentration of HR; that is, the cell is more sensitive to H. One could also increase hormone sensitivity by increasing the affinity, but there are two drawbacks to this option. First, increasing the affinity would decrease hormone-receptor dissociation, resulting in a slower response time. Second, a higher affinity would saturate the receptor at lower hormone concentrations and the system would no longer be responsive to these high ligand levels. As such, "spare receptors" are an elegant way to increase hormone sensitivity while preserving a fast response time and sensitivity to a wide range of hormone concentrations.

Isoreceptors

Isoreceptors are structurally and functionally distinct receptors for the same hormone; such receptors frequently have different tissue distributions, second messengers, and functions (Table 7-9). Because all the receptors in a single class recognize the same natural hormone, other ligands must be used to distinguish among them. In this respect, pharmacologists have been very helpful; they have provided researchers with numerous agonists, some of which may be highly specific for one particular isoreceptor. For example, nicotine preferentially binds to the acetylcholine receptor on skeletal muscle, whereas muscarine is more specific for those receptors in the parasympathetic nervous system; these isoreceptors are now named for their respective agonists. Although isoreceptors are very common, some hormones such as insulin have only one receptor except for possible slight variations in glycosylation.

Why would one hormone have multiple receptors? Isoreceptors allow a hormone to have widely diverse, even opposite, effects in different tissues. The adrenergic receptors provide an excellent example (Table 7-9). Catecholamines, like many hormones, have numerous actions that are related to a single, overall function; in this case, the flight-or-fight response (see Chapter 2). Smooth muscle *contracts* in the bladder sphincter to prevent voiding and in the mesenteric arterioles to shift blood to the skeletal muscles; however, smooth muscle must *relax* in the airways to increase oxygenation and in the intestines because digestion is being deferred. Furthermore, catecholamines not only affect smooth muscle, but also metabolism and cardiovascular activity. These diverse actions would be difficult to elicit with a single receptor and second messenger.

Isoreceptors could also extend the dose-response of a system; for example, fat cells contain β_1, β_2, and β_3 adrenergic receptors, all of which mediate lipolysis. However, their hormone responsiveness differs: the $\beta_1 AR$ is most sensitive, whereas the $\beta_3 AR$ is least sensitive. These isoreceptors allow the fat cell to respond to a wide range of catecholamine concentrations.

Cryptic Receptors

The concept of cryptic receptors is derived from numerous observations where the number of receptors suddenly increases without an obvious explanation. Presumably, these receptors already existed in some latent form incapable of hormone binding; that is, they were cryptic receptors. The inducing factor is envisioned as merely stimulating the conversion of a latent receptor to an overt one. There are several possible mechanisms. First, the receptor could undergo a change in conformation triggered by phosphorylation, sulfhydryl modification, or cleavage. The GnRH receptor can be unmasked by phosphorylation, the neuropeptide Y2 receptor can be exposed by sulfhydryl modification, and cleavage of the insulin receptor may be an example of the latter. In addition, both of the gonadotropin receptors can be unmasked by partial deglycosylation. Perturbations of the environment of receptors can also increase the total number of binding sites in the absence of protein synthesis; detergents, phospholipases, phospholipid methylation, and cholesterol sequestering reagents have all been reported to induce this phenomenon. The latter disrupt DIGs. The ionic

Table 7-9
Examples of Isoreceptors for Selected Hormones

Hormone	Isoreceptors	Mediator	Location	Function
Acetylcholine	Nicotinic (muscle)	Sodium	Skeletal muscle	Contraction
	Nicotinic (neural)	Sodium	Nervous system	Neurotransmission
	M1	Calcium	Parasympathetic nervous system	Neurotransmission
	M2	Decrease cAMP	Parasympathetic nervous system	Neurotransmission
Vasopressin	V1	Calcium	Liver; arterioles	Glycogenolysis; contraction
	V2	cAMP	Kidney	Water resorption
Histamine	H1	Calcium	Bronchioles; arterioles; gut	Contraction
	H2	cAMP	Stomach	Acid secretion
	H3	Lower calcium	Brain; arterioles; bronchioles	Presynaptic inhibition; relaxation
Epinephrine and norepinephrine	$\alpha 1$	Calcium	Liver; smooth muscle of blood vessels, gut, and urinary system	Glycogenolysis; contraction
	α_2	Decrease cAMP	Fat; platelets; smooth muscle of gut	Lipolysis; platelet aggregation; relaxation
	β_1	cAMP	Fat; heart; smooth muscle of gut	Lipolysis; tachycardia; relaxation
	β_2	cAMP	Skeletal muscle; smooth muscle of blood vessels, bronchioles, and urinary system	Glycogenolysis; relaxation
Dopamine	D1	cAMP	Central nervous system; retina; parathyroid gland; vascular smooth muscle	PTH secretion; relaxation
	D2	Decrease cAMP	Sympathetic and central nervous system; pituitary gland	Inhibit PRL and MSH secretion
Opiate peptides	μ	Potassium, calcium, and cAMP	Hypothalamus	Euphoria; decrease gut motility
	δ	Potassium, calcium, and cAMP	Limbic system; basal ganglia	Analgesia; hyperthermia
	κ	Potassium, calcium, and cAMP	Substantia nigra; neurohypophysis	Sedation; dysphoria; anorexia

environment is equally important, and basic compounds can expose EGFRs. Finally, energy depletion is another circumstance that can lead to the appearance of additional receptors. The mechanism for the latter effect is unknown. These membrane proteins may represent a ready reservoir of receptors for rapid cellular responses.

Summary

There are four types of membrane receptors: (1) enzyme-linked, (2) cytokine, (3) GPCRs, and (4) ion channels. All of the RTKs cross the plasmalemma once and have their amino termini extracellular, where the hormone binds. Binding motifs include β-helices (the insulin and EGF receptor families), Ig domains (PDGF and NGF families), and modified β-sheets (Eph and DDR families). The carboxy terminus is intracellular and contains the tyrosine kinase domain and critical tyrosine phosphorylation sites that are responsible for substrate binding. The carboxy terminus is also required for receptor internalization. Other enzyme-linked receptors are the S-T kinases, the histidine kinases, and the guanylate cyclases. The S-T kinases are identical in structural organization to the RTKs; they only differ in substrate specificity. The histidine kinases are not found in animals and signal by transferring a phosphate from a histidine in the kinase domain to an aspartate in the regulator domain. The cyclases have two catalytic sites intracellularly. The carboxy-terminal one is the cyclase, whereas the amino-terminal kinase–like domain has evolved into a regulatory domain. A soluble form of the cyclase is activated by NO and CO through a heme-containing amino terminus.

The cytokine receptors have modified FNIII repeats in the extracellular domain, a single transmembrane helix, and an intracellular domain that lacks sequence homology but is rich in serines, threonines, prolines, and acidic residues. Their biological effects are mediated through the activation of soluble tyrosine kinases. Essentially, they are RTKs whose catalytic site is located on a separate subunit.

The GPCRs cross the plasma membrane seven times, and interact with GTP-binding proteins. This is the largest and probably the oldest receptor family. The ligands primarily bind to the transmembrane helices, although the amino terminus and extracellular loops can also bind when the ligand is large. Transduction occurs through the juxtamembrane portion of the carboxy terminus and the cytoplasmic loop between helices 5 and 6.

There are three types of ion channel receptors. The Cys-loop superfamily consists of five homologous subunits containing four transmembrane helices each. The second helix of each subunit forms the channel wall. Examples include the receptors for ACh, GABA, glycine, and 5-HT. The P (H5) superfamily has four subunits or one protein with four fused domains; each subunit or domain contains from two to six transmembrane helices. The channel wall is formed from a loop between two of the helices. There are three families: the voltage-gated family includes the CNG channels, the ryanodine receptor, and IP_3R; the glutamate family; and GIRK, which is regulated by G proteins. The epithelial sodium channel is the last superfamily. It has two transmembrane helices, but the structure of the pore is unknown.

Many receptors are localized to specialized membrane structures called *caveolae* and *DIGs*. These domains facilitate signaling by bringing receptors and many different types of transducers together. Eventually, activated receptors migrate into coated pits and are internalized. Acidification of these vesicles disrupts hormone-receptor binding; the hormone remains within the central cavity, whereas the receptors migrate into tubules that bud off the vesicle. The tubules break off, are routed through the Golgi apparatus, and then are recycled to the plasma membrane. The vesicles containing the hormone fuse with lysosomes, and the ligand is degraded. In addition to terminating the hormone signaling, internalization can have other important functions, such as the detection of gradients, cross talk, the generation of second messengers in selected cellular compartments, and hormone storage.

Although only a small number of receptors are occupied, even during maximal stimulation, the spare receptors are still important in determining the hormonal sensitivity of a cell. Furthermore, not all receptors for a given hormone are identical; many can be divided into several subgroups. These isoreceptors permit a hormone to have different actions in different tissues. Finally, some receptors are incapable of binding their ligands until some additional event takes place; these cryptic receptors may represent a silent reservoir of binding proteins.

References

Protein Domains

Aitken, A., Baxter, H., Dubois, T., Clokie, S., Mackie, S., Mitchell, K., Peden, A., and Zemlickova, E. (2002). Specificity of 14-3-3 isoform dimer interactions and phosphorylation. *Biochem. Soc. Trans.* 30, 351-360.

Borden, K.L.B. (2000). RING domains: Master builders of molecular scaffolds? *J. Mol. Biol.* 295, 1103-1112.

Cho, W. (2001). Membrane targeting by C1 and C2 domains. *J. Biol. Chem.* 276, 32407-32410.

Cullen, P.J., Cozier, G.E., Banting, G., and Mellor, H. (2001). Modular phosphoinositide-binding domains—their role in signalling and membrane trafficking. *Curr. Biol.* 11, R882-R893.

Downward, J. (2001). The ins and outs of signalling. *Nature (London)* 411, 759-762.

Ellson, C.D., Andrews, S., Stephens, L.R., and Hawkins, P.T. (2002). The PX domain: A new phosphoinositide-binding module. *J. Cell Sci.* 115, 1099-1105.

Harris, B.Z., and Lim, W.A. (2001). Mechanism and role of PDZ domains in signaling complex assembly. *J. Cell Sci.* 114, 3219-3231.

Hung, A.Y., and Sheng, M. (2002). PDZ domains: Structural modules for protein complex assembly. *J. Biol. Chem.* 277, 5699-5702.

Hurley, J.H., and Meyer, T. (2001). Subcellular targeting by membrane lipids. *Curr. Opin. Cell Biol.* 13, 146-152.

Itoh, T., and Takenawa, T. (2002). Phosphoinositide-binding domains: Functional units for temporal and spatial regulation of intracellular signalling. *Cell. Signal.* 14, 733-743.

Kay, B.K., Williamson, M.P., and Sudol, M. (2000). The importance of being proline: The interaction of proline-rich motifs in signaling proteins with their cognate domains. *FASEB J.* 14, 231-241.

Lemmon, M.A. (2003). Phosphoinositide recognition domains. *Traffic* 4, 201-213.

Maffucci, T., and Falasca, M. (2001). Specificity in pleckstrin homology (PH) domain membrane targeting: A role for a phosphoinositide-protein co-operative mechanism. *FEBS Lett.* 506, 173-179.

Muslin, A.J., and Xing, H. (2000). 14-3-3 proteins: Regulation of subcellular localization by molecular interference. *Cell. Signal.* 12, 703-709.

Pawson, T., and Nash, P. (2000). Protein-protein interactions define specificity in signal transduction. *Genes Dev.* 14, 1027-1047.

Sato, T.K., Overduin, M., and Emr, S.D. (2001). Location, location, location: Membrane targeting directed by PX domains. *Science* 294, 1881-1885.

Stenmark, H., and Aasland, R. (1999). FYVE-finger proteins—effectors of an inositol lipid. *J. Cell Sci.* 112, 4175-4183.

Tzivion, G., and Avruch, J. (2002). 14-3-3 proteins: Active cofactors in cellular regulation by serine/threonine phosphorylation. *J. Biol. Chem.* 277, 3061-3064.

Weber, C.H., and Vincenz, C. (2001). The death domain superfamily: A tale of two interfaces? *Trends Biochem. Sci.* 26, 475-481.

Xu, Y., Seet, L.F., Hanson, B., and Hong, W. (2001). The Phox homology (PX) domain, a new player in phosphoinositide signalling. *Biochem. J.* 360, 513-530.

Yaffe, M.B., and Smerdon, S.J. (2001). Phosphoserine/threonine binding domains: You can't pSERious? *Structure* 9, R33-R38.

Tyrosine Kinase Receptors

Adams, T.E., Epa, V.C., Garret, T.P.J., and Ward, C.W. (2000). Structure and function of the type 1 insulin-like growth factor receptor. *Cell. Mol. Life Sci.* 57, 1050-1093.

Bass, M.D., and Humphries, M.J. (2002) Cytoplasmic interactions of syndecan-4 orchestrate adhesion receptor and growth factor receptor signalling. *Biochem. J.* 368, 1-15.

Chao, M.V. (2003). Neurotrophins and their receptors: A convergence point for many signalling pathways. *Nature Rev. Neurosci.* 4, 299-309.

Cowan, C.A., and Henkemeyer, M. (2002). Ephrins in reverse, park and drive. *Trends Cell Biol.* 12, 339-346.

Jorissen, R.N., Walker, F., Pouliot, N., Garrett, T.P.J., Ward, C.W., and Burgess, A.W. (2003). Epidermal growth factor receptor: mechanisms of activation and signalling. *Exp. Cell Res.* 284, 31-53.

Kullander, K., and Klein, R. (2002). Mechanisms and functions of EPH and ephrin signalling. *Nature Rev. Mol. Cell Biol.* 3, 475-486.

Madhani, H.D. (2001). Accounting for specificity in receptor tyrosine kinase signaling. *Cell (Cambridge, Mass.)* 106, 9-11.

Olayioye, M.A., Neve, R.M., Lane, H.A., and Hynes, N.E. (2000). The ErbB signaling network: Receptor heterodimerization in development and cancer. *EMBO J.* 19, 3159-3167.

Park, P.W., Reizes, O., and Bernfield, M. (2000). Cell surface heparin sulfate proteoglycans: Selective regulators of ligand-receptor encounters. *J. Biol. Chem.* 275, 29923-29926.

Patapoutian, A., and Reichardt, L.F. (2001). Trk receptors: Mediators of neurotrophin action. *Curr. Opin. Neurobiol.* 11, 272-280.

Schlessinger, J. (2000). Cell signaling by receptor tyrosine kinases. *Cell (Cambridge, Mass.)* 103, 211-225.

TGFβ Receptor Family

Massagué, J. (2000). How cells read TGF-β signals. *Nature Rev. Mol. Cell Biol.* 1, 169-178.

Miyazono, K., Kusanagi, K., and Inoue, H. (2001). Divergence and convergence of TGF-β/BMP signaling. *J. Cell. Physiol.* 187, 265-276.

Moustakas, A., Souchelnytskyi, S., and Heldin, C.H. (2001). Smad regulation in TGF-β signal transduction. *J. Cell Sci.* 114, 4359-4369.

Piek, E., Heldin, C.H., and ten Dijke, P. (1999). Specificity, diversity, and regulation in TGF-β superfamily signaling. *FASEB J.* 13, 2105-2124.

Zimmerman, C.M., and Padgett, R.W. (2000). Transforming growth factor β signaling mediators and modulators. *Gene* 249, 17-30.

Plant Serine-Threonine Kinases

Becraft, P.W. (2002). Receptor kinase signaling in plant development. *Annu. Rev. Cell Dev. Biol.* 18, 163-192.

Clouse, S.D. (2002). Brassinosteroid signal transduction: Clarifying the pathway from ligand perception to gene expression. *Mol. Cell* 10, 973-982.

DeYoung, B.J., and Clark, S.E. (2001). Signaling through the CLAVATA1 receptor complex. *Plant Mol. Biol.* 46, 505-513.

Møller, S.G., and Chua, N.H. (1999). Interactions and intersections of plant signaling pathways. *J. Mol. Biol.* 293, 219-234.

Müssig, C., and Altmann, T. (2001). Brassinosteroid signaling in plants. *Trends Endocrinol. Metab.* 12, 398-402.

Torii, K.U. (2000). Receptor kinase activation and signal transduction in plants: An emerging picture. *Curr. Opin. Plant Biol.* 3, 361-367.

Histidine Kinase Receptors

Haberer, G., and Kieber, J.J. (2002). Cytokinins. New insights into a classic phytohormone. *Plant Physiol.* 128, 354-362.

Lohrmann, J., and Harter, K. (2002). Plant two-component signaling systems and the role of response regulators. *Plant Physiol.* 128, 363-369.

Schaller, G.E. (2000). Histidine kinases and the role of two-component systems in plants. *Adv. Bot. Res.* 32, 109-148.

Stepanova, A.N., and Ecker, J.R. (2000). Ethylene signaling: From mutants to molecules. *Curr. Opin. Plant Biol.* 3, 353-360.

Guanylate Cyclases

Barañano, D.E., and Snyder, S.H. (2001). Neural roles for heme oxygenase: Contrasts to nitric oxide synthase. *Proc. Natl. Acad. Sci. U.S.A.* 98, 10996-11002.

Bredt, D.S. (2002). Nitric oxide signaling specificity—the heart of the problem. *J. Cell Sci.* 116, 9-15.

Lucas, K.A., Pitari, G.M., Kazerounian, S., Ruiz-Stewart, I., Park, J., Schulz, S., Chepenik, K.P., and Waldman, S.A. (2000). Guanylyl cyclases and signaling by cyclic GMP. *Pharmacol. Rev.* 52, 375-413.

Mayer, B. (ed.) (2000). *Nitric Oxide.* Springer-Verlag, Heidelberg, Germany.

Mori, M., and Gotoh, T. (2000). Regulation of nitric oxide production by arginine metabolic enzymes. *Biochem. Biophys. Res. Commun.* 275, 715-719.

Potter, L.R., and Hunter, T. (2001). Guanylyl cyclase-linked natriuretic peptide receptors: Structure and regulation. *J. Biol. Chem.* 276, 6057-6060.

Silberbach, M., and Roberts, C.T. (2001). Natriuretic peptide signalling: Molecular and cellular pathways to growth regulation. *Cell. Signal.* 13, 221-231.

Wedel, B.J., and Garbers, D.L. (2001). The guanylyl cyclase family at Y2K. *Annu. Rev. Physiol.* 63, 215-233.

Cytokine Receptors

Baud, V., and Karin, M. (2001). Signal transduction by tumor necrosis factor and its relatives. *Trends Cell Biol.* 11, 372-377.

Bodmer, J.L., Schneider, P., and Tschopp, J. (2002). The molecular architecture of the TNF superfamily. *Trends Biochem. Sci.* 27, 19-26.

Bravo, J., and Heath, J.K. (2000). Receptor recognition by gp130 cytokines. *EMBO J.* 19, 2399-2411.

Denecker, G., Vercammen, D., Declercq, W., and Vandenabeele, P. (2001). Apoptotic and necrotic cell death induced by death domain receptors. *Cell. Mol. Life Sci.* 58, 356-370.

Gaffen, S.L. (2001). Signaling domains of the interleukin 2 receptor. *Cytokine* 14, 63-77.

Imada, K., and Leonard, W.J. (2000). The Jak-STAT pathway. *Mol. Immunol.* 37, 1-11.

Ishihara, K., and Hirano, T. (2002). Molecular basis of the cell specificity of cytokine action. *Biochim. Biophys. Acta* 1592, 281-296.

MacEwan, D.J. (2002). TNF receptor subtype signaling: Differences and cellular consequences. *Cell. Signal.* 14, 477-492.

Piwien-Pilipuk, G., Huo, J.S., and Schwartz, J. (2002). Growth hormone signal transduction. *J. Pediatr. Endocrinol. Metab.* 15, 771-786.

Schindler, C., and Brutsaert, S. (1999). Interferons as a paradigm for cytokine signal transduction. *Cell. Mol. Life Sci.* 55, 1509-1522.

Yeh, T.C., and Pellegrini, S. (1999). The Janus kinase family of protein tyrosine kinases and their role in signaling. *Cell. Mol. Life Sci.* 55, 1523-1534.

G Protein-Coupled Receptors

Angers, S., Salahpour, A., and Bouvier, M. (2002). Dimerization: An emerging concept for G protein-coupled receptor ontogeny and function. *Annu. Rev. Pharmacol. Toxicol.* 42, 409-435.

Brady, A.E., and Limbird, L.E. (2002). G protein-coupled receptor interacting proteins: Emerging roles in localization and signal transduction. *Cell. Signal.* 14, 297-309.

Gershengorn, M.C., and Osman, R. (2001). Insights into G protein-coupled receptor function using molecular models. *Endocrinology (Baltimore)* 142, 2-10.

Gether, U. (2000). Uncovering molecular mechanisms involved in activation of G protein-coupled receptors. *Endocr. Rev.* 21, 90-113.

Hamm, H.E. (2001). How activated receptors couple to G proteins. *Proc. Natl. Acad. Sci. U.S.A.* 98, 4819-4821.

Heuss, C., and Gerber, U. (2000). G-protein-independent signaling by G-protein-coupled receptors. *Trends Neurosci.* 23, 469-475.

Milligan, G. (2001). Oligomerisation of G-protein-coupled receptors. *J. Cell Sci.* 114, 1265-1271.

Milligan, G., and White, J.H. (2001). Protein-protein interactions at G-protein-coupled receptors. *Trends Pharmacol. Sci.* 22, 513-518.

Morris, A.J., and Malbon, C.C. (1999). Physiological regulation of G protein-linked signaling. *Physiol. Rev.* 79, 1373-1430.

Pierce, K.L., Premont, R.T., and Lefkowitz, R.J. (2002). Seven-transmembrane receptors. *Nature Rev. Mol. Cell Biol.* 3, 639-650.

Qanbar, R., and Bouvier, M. (2003). Role of palmitoylation/depalmitoylation in G-protein-coupled receptor function. *Pharmacol. Ther.* 97, 1-33.

Rios, C.D., Jordan, B.A., Gomes, I., and Devi, L.A. (2002). G-protein-coupled receptor dimerization: Modulation of receptor function. *Pharmacol. Ther.* 92, 71-87.

Ion Channel Receptors

Dascal, N. (2001). Ion-channel regulation by G proteins. *Trends Endocrinol. Metab.* 12, 391-398.

Davis, M.J., Wu, X., Nurkiewicz, T.R., Kawasaki, J., Gui, P., Hill, M.A., and Wilson, E. (2001). Regulation of ion channels by protein tyrosine phosphorylation. *Am. J. Physiol.* 281, H1835-H1862.

Flynn, G.E., Johnson, J.P., and Zagotta, W.N. (2001). Cyclic nucleotide-gated channels: Shedding light on the opening of a channel pore. *Nature Rev. Neurosci.* 2, 643-652.

Hucho, F., and Weise, C. (2001). Ligand-gated ion channels. *Angew. Chem. Int. Ed.* 40, 3100-3116.

Montell, C., Birnbaumer, L., and Flockerzi, V. (2002). The TRP channels, a remarkably functional family. *Cell (Cambridge, Mass.)* 108, 595-598.

Sheng, M., and Pak, D.T.S. (2000). Ligand-gated ion channel interactions with cytoskeletal and signaling proteins. *Annu. Rev. Physiol.* 62, 755-778.

Tomita, S., Nicoll, R.A., and Bredt, D.S. (2001). PDZ protein interactions regulating glutamate receptor function and plasticity. *J. Cell Biol.* 153, F19-F23.

Miscellaneous Membrane Receptors

Akira, S. (2003). Mammalian Toll-like receptors. *Curr. Opin. Immunol.* 15, 5-11.

Daun, J.M., and Fenton, M.J. (2000). Interleukin-1/Toll receptor family members: Receptor structure and signal transduction pathways. *J. Interferon Cytokine Res.* 20, 843-855.

Imler, J.L., and Hoffman, J.A. (2001). Toll receptors in innate immunity. *Trends Cell Biol.* 11, 304-311.

Martin, M.U., and Wesche, H. (2002). Summary and comparison of the signaling mechanisms of the Toll/interleukin-1 receptor family. *Biochim. Biophys. Acta* 1592, 265-280.

Receptor Metabolism

Alonso, M.A., and Millán, J. (2001). The role of lipid rafts in signalling and membrane trafficking in T lymphocytes. *J. Cell Sci.* 114, 3957-3965.

Brown, D.A., and London, E. (2000). Structure and function of sphingolipid- and cholesterol-rich membrane rafts. *J. Biol. Chem.* 275, 17221-17224.

Ceresa, B.P., and Schmid, S.L. (2000). Regulation of signal transduction by endocytosis. *Curr. Opin. Cell Biol.* 12, 204-210.

Clague, M.J., and Urbé, S. (2001). The interface of receptor trafficking and signalling. *J. Cell Sci.* 114, 3075-3081.

Ginty, D.D., and Segal, R.A. (2002). Retrograde neurotrophin signaling: Trk-ing along the axon. *Curr. Opin. Neurobiol.* 12, 268-274.

Goligorsky, M.S., Li, H., Brodsky, S., and Chen, J. (2002). Relationship between caveolae and eNOS: Everything in proximity and the proximity of everything. *Am. J. Physiol.* 283, F1-F10.

Grimes, M.L., and Miettinen, H.M. (2003). Receptor tyrosine kinase and G-protein coupled receptor signaling and sorting within endosomes. *J. Neurochem.* 84, 905-918.

Harder, T., and Scheiffele, P. (2000). Regulation of raft architecture. *Protoplasma* 212, 1-7.

McPherson, P.S., Kay, B.K., and Hussain, N.K. (2001). Signaling on the endocytic pathway. *Traffic* 2, 375-384.

Masserini, M., and Ravasi, D. (2001). Role of sphingolipids in the biogenesis of membrane domains. *Biochim. Biophys. Acta* 1532, 149-161.

Massimino, M.L., Griffoni, C., Spisni, E., Toni, M., and Tomasi, V. (2002). Involvement of caveolae and caveolae-like domains in signalling cell survival and angiogenesis. *Cell. Signal.* 14, 93-98.

Prieschl, E.E., and Baumruker, T. (2000). Sphingolipids: second messengers, mediators and raft constituents in signaling. *Immunol. Today* 21, 555-560.

Razani, B., Woodman, S.E., and Lisanti, M.P. (2002). Caveolae: From cell biology to animal physiology. *Pharmacol. Rev.* 54, 431-467.

Smart, E.J., Graf, G.A., McNiven, M.A., Sessa, W.C., Engelman, J.A., Scherer, P.E., Okamoto, T., and Lisanti, M.P. (1999). Caveolins, liquid-ordered domains, and signal transduction. *Mol. Cell. Biol.* 19, 7289-7304.

Sorkin, A., and von Zastrow, M. (2002). Signal transduction and endocytosis: Close encounters of many kinds. *Nature Rev. Mol. Cell Biol.* 3, 600-614.

Wiley, H.S., and Burke, P.M. (2001). Regulation of receptor tyrosine kinase signaling by endocytic trafficking. *Traffic* 2, 12-18.

Zajchowski, L.D., and Robbins, S.M. (2002). Lipid rafts and little caves: Compartmentalized signalling in membrane microdomains. *Eur. J. Biochem.* 269, 737-752.

CHAPTER **8**

Receptor Regulation

Because hormone-receptor binding obeys the laws of mass action, the number of receptors determines, in part, the sensitivity of the cell to the hormone (see Chapter 7, Spare Receptors). However, sensitivity can also be influenced by other receptor characteristics, such as receptor affinity or location. These potential molecular mechanisms are discussed in this chapter using specific examples of both homologous and heterologous regulation. Homologous regulation is the control of a receptor by its own ligand; conversely, heterologous regulation is the control of a receptor by a hormone that does not bind to that receptor. Other terms that should be clarified before continuing are downregulation and desensitization. *Downregulation* refers to an actual decrease in receptor number, whereas *desensitization* describes a decreased responsiveness of a system to a hormone. The latter may occur at the receptor or at a postreceptor site.

Receptor Number

Synthesis and Stability

The concentration of any molecule will be a result of the balance between synthesis and degradation; synthesis, in turn, is a product of both transcription and translation. This chapter is not concerned with the mechanisms of synthesis, except to warn the reader that increased receptor mRNA is not necessarily reflected in the amount of translated receptor.

The insulin receptor has been one of the most thoroughly studied systems, and several investigators have reported on the mechanisms involved in the regulation of this receptor by glucocorticoids. This is an example of heterologous control. In most of these experiments, cells are exposed to amino acids containing heavy isotopes (e.g., ^{2}H, ^{13}C, and ^{15}N). These heavy amino acids will be incorporated only into the newly synthesized receptors, which can then be separated from the preexisting receptors by density gradient centrifugation. This procedure allows one to determine both the synthetic rate and turnover of insulin receptors. In hepatocytes, dexamethasone, a synthetic glucocorticoid, increases the insulin receptor number fivefold (Table 8-1). This effect is due exclusively to an increase in the rate of synthesis, which is also elevated fivefold, whereas the half-life is unchanged.

In adipocytes, the mechanism is different. The stimulatory effect of dexamethasone is a result of increasing the receptor half-life instead of its synthesis. The reciprocal effect is observed with insulin, which lowers its own receptor abundance by shortening its receptor half-life. Neither the affinity of the receptor nor its cellular distribution is altered during the experiment. Tissue-specific differences are quite common and should always be considered when reading the literature. In fact, not only can mechanisms be different, but a hormone may also have opposite effects in different tissues. For example, estradiol elevates the progesterone receptor (PR) in the stroma and muscularis of the uterus and oviduct, while it inhibits it in the epithelium. In another example, growth hormone (GH) can lower its own receptor in adipocytes, while increasing it in muscle.

The same kinds of regulation are seen with steroid receptors. Estradiol stimulates the PR in breast cancer cells 10-fold by elevating receptor synthesis. This system also illustrates the fact that receptor number may change through multiple mechanisms operating simultaneously; R5020, a progesterone agonist,

Table 8-1
Regulation of Receptor Synthesis and Stability[a]

Receptor	Tissue	Hormone treatment	Receptor number (%)	Half-life (hr)	Synthetic rate (%)
Insulin	Hepatocytes	Control	100	9	100
		Dexamethasone[b]	517	9.4	474
	Adipocytes	Control	100	10.2	100
		Dexamethasone[b]	213	18.2	115
		Insulin	43	4.2	110
Progesterone	Breast cancer cells	Control	100	21	100
		Estradiol	1000	21	700[c]
		R5020[d]	15	6	<10
Glucocorticoid	Pituitary cells	Control	100	22.5	100
		Triamcinolone[b]	51	10.2	100
1,25-DHCC	Fibroblasts and intestinal cells	Control	100	4	100[c]
		1,25-DHCC	300	8	100[c]

[a] Data are from references 1-8.
[b] Synthetic glucocorticoid.
[c] mRNA synthesis.
[d] Synthetic progestin.
1,25-DHCC, 1,25-dihydroxycholecalciferol.

decreases the PR by both inhibiting synthesis and shortening the receptor half-life. Homologous regulation of the GR and vitamin D receptor (VDR) are examples of steroid receptor regulation exclusively through receptor stability. Triamcinolone, another synthetic glucocorticoid, halves GR number in a pituitary cell line by reducing the half-life, whereas 1,25-dihydroxycholecalciferol (1,25-DHCC) triples its own receptor in fibroblasts and intestinal cells by prolonging the half-life.

These examples were deliberately chosen because of their straightforward data: either only one mechanism is used, or, if both receptor synthesis and stability are affected, the two mechanisms reinforce one another. It is possible to have both mechanisms activated but in opposite directions: epidermal growth factor (EGF) increases its own receptor mRNA 5-fold, but it also increases EGF receptor (EGFR) degradation. The net effect is determined by the balance between these two processes. Initially, degradation predominates as the EGFR is down-regulated; then synthesis becomes preeminent as the receptors are replenished. The β-adrenergic receptors (βARs) show a different time course: β-agonists initially stimulate βAR levels as a result of increased transcription, but continuous receptor activation leads to downregulation by decreasing the βAR mRNA half-life from 12 to 5 hours.

Mechanisms for affecting gene expression and mRNA stability are covered in Chapters 13 and 15. Protein half-life is often regulated by proteolytic mechanisms; one such pathway is ubiquitination. Ubiquitin is a 76-amino acid peptide whose carboxy terminus is covalently coupled to the ε-amino group of lysine in a target protein destined for degradation. Essentially, ubiquitin is a molecular contamination notice that targets the protein for destruction. The process consists of three proteins: E1 activates ubiquitin and transfers it to E2; E2 then attaches it to the substrate with the help of E3, which is the substrate recognition unit. E3 is often referred to as the *E3 ligase*. Either one or multiple ubiquitin chains can be coupled to a protein. The latter, called *polyubiquitination*, can occur through either the K-48 or K-63 of ubiquitin and may require a fourth ubiquitination component known as *E4*. K-48 chains are required for targeting proteins to the proteasome, a tubular complex of proteases. K-63 chains often serve other functions; for example, this modification of the proliferating cell nuclear antigen (PCNA) is necessary for the error-free branch of RAD6-dependent DNA repair. Labeling with a single ubiquitin molecule, called *monoubiquitination*, directs the protein to the lysosome. Monoubiquitination can also have activities other than degradation: it can act as a chaperone for RNA proteins, recruit regulatory proteins, and alter chromatin structure (see Chapter 14).

There are three other proteins, Sumo, Nedd, and ISG15 (interferon [IFN]–stimulated gene 15), which are homologous to ubiquitin and which are coupled to proteins by a similar process, called *sumoylation, neddylation,* and *ISGylation,* respectively. Sumoylation can (1) compete with ubiquitination and inhibit degradation, (2) allosterically affect transcription or DNA repair factors, (3) prevent telomere lengthening, (4) and affect localization to, from, or within the nucleus (see Part 4). Major targets of neddylation are the Cullin proteins, which are involved with protein degradation by facilitating ubiquitination. The activity of steroid receptors (SRs) and several other transcription factors are inhibited by neddylated Cullin proteins through increased ubquitination and degradation. ISGylation is the most recently evolved modification; it is not found in nonvertebrates. It is also the least characterized; however, it does appear to facilitate IFN signaling.

Ubiquitination of receptor tyrosine kinases (RTKs), nuclear receptors (NRs), G-protein coupled receptors (GPCRs), the tumor growth factor β receptor (TGFβR), and the cytokine receptors has been detected and is usually ligand dependent. The mechanisms for each receptor group differ and may include recruitment, phosphorylation, and oligomerization. For example, RTK autophosphorylation attracts Cbl, an E3 ligase, whereas arrestin recruits an E3 ligase for the β_2AR. AR downregulation by insulin-like growth factor 1 (IGF-I) or interleukin 6 (IL-6) involves the stimulation of protein kinase B (PKB); the subsequent PKB phosphorylation of both the E3 ligase and AR then lead to ubiquitination. Retinoic acid receptor (RAR) ubiquitination also requires serine phosphorylation, as well as dimerization, but not transcriptional activity. Growth hormone receptor (GHR) is another receptor requiring dimerization for ubiquitination.

An alternate degradatory pathway involves ectoproteases; this process has been described for RTKs, GPCRs, and guanylate cyclases. These enzymes release membrane-bound ligands, as well as the extracellular domains of receptors; because tumor necrosis factor α (TNFα) was the first substrate identified, the protease was called *TNF alpha converting enzyme*, or TACE. TACE can be activated by protein kinase C (PKC), mitogen-activated protein kinase (MAPK), calmodulin-dependent protein kinase II (CaMKII), and nitric oxide (NO). NO nitrosylates an inhibitory domain within TACE, but the mechanisms of the other factors are unknown. Cytoplasmic proteases can also degrade receptors. Calcium activates calpain (see Chapter 10), which then cleaves the intracellular domain of the common subunits of the cytokine receptors (β_c, γ_c, and gp130).

Finally, some receptor ligands can exist in either a soluble or membrane-bound form. The membrane-bound stem cell growth factor (SCF) decreases SCF receptor (SCFR) downregulation. It is suspected that the receptor cannot be internalized without the ligand, and the ligand, in this case, is affixed onto another cell. A similar trick has been used by pharmacologists to prolong the half-life of several hormones: when coupled with polyethylene glycol, the hormones cannot be internalized with their receptors and degraded.

Activation

Receptors are normally activated by the binding of their ligands. However, as mentioned in Chapter 7, some receptors may be nonfunctional; these are the *cryptic receptors*. The recruitment of these receptors into the responsive pool is another mechanism whereby the number of *functional* receptors can be increased. This mechanism can be very rapid because protein synthesis is not required.

Another way of increasing the number of functioning receptors without ligand binding is by promiscuous dimerization leading to heterologous activation. RTKs are activated by autophosphorylation; however, some RTKs can phosphorylate and activate closely related, but unoccupied, RTKs. For example, the insulin receptor can phosphorylate and activate empty insulin-like growth factor-I receptors (IGF-IRs). Transphosphorylation between receptor families has also been documented; examples include EGFR-PDGFR (platelet-derived growth factor receptor), EGFR-IGF-IR, EGFR-gp130, Trk-Ret, and SCFR-EPOR (erythropoietin receptor).

Transphosphorylation and activation can also occur without direct oligomerization; generally, there is a kinase intermediate. For example, GPCRs coupled to G_q elevate calcium, which stimulates a calcium-activated soluble tyrosine kinase (STK) that phosphorylates and activates RTKs. Activation can be very selective: GH, a cytokine, activates Janus kinase 2 (Jak2), which phosphorylates the EGFR only at the growth factor receptor-bound 2 (Grb2) site. As a result, GH uses the EGFR to stimulate MAPK without triggering any of the other EGF activities. Furthermore, phosphorylation can be achieved indirectly through phosphatases: ultraviolet irradiation and oxidative stress can modify the cysteine in the active site of protein tyrosine phosphatases (PTPs). The ensuing inhibition of PTP leads to an elevated level of phosphotyrosine (pY) on RTKs. Receptors other than RTKs can also be affected: the PR can be phosphorylated and activated by dopamine stimulation.

One other factor can affect activation: priming. Integrin enhances RTK signaling by preclustering these receptors, thereby facilitating dimerization after ligand binding (see Chapter 12).

Compartmentalization

Finally, receptors may be physically present and functional, but not accessible to their ligands. Because the ligands for membrane receptors are hydrophilic, they cannot traverse the plasmalemma; therefore receptors located on internal membrane structures cannot bind their ligands. Translocation of the receptors from these structures to the plasma membrane would be another way of increasing the number of receptors. The IGF-IIR undergoes translocation to the plasmalemma in response to insulin, EGF, IGF-I, or IGF-II. This mechanism can increase the number of IGF-IIRs at the cell surface by 50% to 100% within 10 to 15 minutes. The *N*-methyl-D-aspartate (NMDA), nerve growth factor receptor (TrkB), thrombin, and prolactin receptors are also recruited to the plasmalemma.

There are several possible mechanisms for this phenomenon. First, tyrosine phosphorylation and phosphatidylinositol-3 kinase (PI3K) activation are required; the former can activate the latter, which is involved in the exocytotic process (see Chapter 10). Glycosylation is known to target receptors to the plasma membrane; some receptors are trapped intracellularly in a partially glycosylated state. For example, both PDGF and NO stimulate the completion of glycosylation in the insulin-like growth factor receptor (IGFR) and the prolactin receptor (PRLR), respectively, resulting in their migration to the cell surface. Conversely, the oncostatin M receptor (OSMR) boxes 1 and 2 block glycosylation; Jak binding masks this region and allows glycosylation and migration to the plasma membrane to occur. A third mechanism involves receptor binding to the cytoskeleton. The sodium (Na$^+$)-hydrogen exchanger regulatory factor (NHERF) is actually an adaptor between receptors and the cytoskeleton (see Chapter 12). Agonist-induced serine phosphorylation of the βAR prevents NHERF binding, and the βAR is destroyed in the lysosome. Dephosphorylation of βAR permits NHERF binding and attachment to the cytoskeleton. The βAR eludes the lysosome and is eventually recycled to the plasma membrane. Fourth, protein kinase A (PKA) is required for TrkB and EGFR recruitment, although it inhibits insulin-induced migration of the IGF-IIR. The target of this kinase is unknown. Finally, TGFβ recruits GDNF (glial-cell line-derived neurotropic factor) family receptor 1 (GFRα1) to the cell surface via a MAPK pathway; GFRα1 is the coreceptor of Ret (rearranged in transformation) and forms the receptor complex for GDNF. In summary, many recruitment mechanisms appear to involve the exocytotic machinery (tyrosine phosphorylation and the PI3K pathway), receptor processing (glycosylation), and interactions with the cytoskeleton, but other mechanisms are not as well understood.

If moving receptors to the cell surface increases the sensitivity of the cell to hormones, then moving the receptors back inside should render the cells less responsive. Indeed, this mechanism was thought to play an important role in the desensitization of cells to β-agonists: after binding β-agonists, the βAR internalizes in a process called sequestration, and this shift is associated with the loss of cell responsiveness. However, the association appears to be fortuitous: first, careful temporal studies have shown that desensitization precedes internalization. Second, mutant βARs incapable of internalization can still be desensitized. Therefore sequestration merely appears to be a means for dephosphorylating the βAR during recycling, rather than a mechanism for reducing the hormonal responsiveness of a cell. However, internalization may still be an important mechanism for regulating the sensitivity of the cell to other hormones; for example, insulin and IGF-I can induce GHR internalization by blocking recycling.

In respiratory epithelium there is an interesting variation on this type of regulation where the ligand is membrane-bound, and it and its receptor are segregated in different regions of the plasma membrane. Specifically, heregulin-α is restricted to the apical surface, whereas its receptor, the HER2-HER4 dimer, is found exclusively in the basolateral membrane. Under basal conditions, there is no ligand-receptor interaction, and the receptor remains inactive. However, any damage that disrupts epithelial integrity will allow the two components to come together and initiate pathways leading to mitosis and repair.

Finally, many hormones can also be compartmentalized by binding components of the extracellular matrix. The release of these hormones following injury can elicit inflammatory responses; in addition, they can establish gradients in morphogenesis.

Receptor Affinity and Specificity

The other major receptor characteristic that determines the responsiveness of a cell to hormones is affinity. Many of the factors that affect receptor affinity also affect its ligand specificity; therefore, both topics are covered in this section.

Covalent Modification

Phosphorylation is a major means of regulating the substrate affinity of enzymes, so it should not be surprising that it also affects receptor affinity. EGF- or PDGF-induced tyrosine phosphorylation of the EGFR reduces its affinity for EGF, whereas a similar modification of the gonadotropin-releasing hormone (GnRH) receptor by the EGFR enhances receptor affinity. Glycosylation is another post-translational modification that can occur on receptors, and it has been shown to affect the binding specificity of Notch1. Notch1 is a receptor for two morphogens, Delta and Jagged1; the receptor has several EGF motifs containing the sugar fucose. Fringe is an enzyme that attaches an *N*-acetylglucosamine to the fucoses; this modification occurs in the ligand-binding site and increases the affinity of Notch1 for Delta, while decreasing the signal for Jagged1.

Covalent modifications of ligands can also affect receptor-binding specificity. Osteopontin is a cytokine that is chemotactic and antiapoptotic for inflammatory cells; it also stimulates tissue remodeling during inflammation. It binds to several integrins and to CD44; the binding specificity is regulated by cleavage and phosphorylation. First, it binds some integrins via an arginine-glycine-aspartate (RGD) motif, but binds the integrin $\alpha_v\beta_3$ via a Gly-Arg-Gly-Asp-Ser (GRGDS) motif. The latter requires thrombin cleavage of osteopontin to expose the GRGDS sequence. Second, integrin binding requires osteopontin phosphorylation, but CD44 binding does not. Casein kinase 1 (CK1), CK2, and protein kinase G (PKG) have been implicated in this phosphorylation.

Hormone modification can also affect receptor kinetics. TNF is synthesized as a membrane-bound protein; cleavage yields a soluble hormone. The former slowly dissociates from TNF receptor 2 (TNFR2), whereas the latter dissociates very rapidly.

Allosterism

Another mechanism for altering receptor affinity is allosterism; this can occur either with small regulatory molecules or subunit oligomerization. An example of the former is the effect of adenosine 5'-triphophate (ATP) on the membrane guanylate cyclase; the association of atrial natriuretic peptide (ANP) with its receptor leads to the binding of ATP to the membrane proximal kinase-like domain. The ATP binding is necessary for cyclase activity, and it is also responsible for inducing a low-affinity state of the receptor.

Oligomerization can either be homologous or heterogenous. For example, EGFR dimerization represents homologous aggregation and results in a higher affinity for EGF as compared with the EGFR monomer. Homologous oligomerization may also decrease the receptor affinity in a process called *negative cooperativity*. It was originally thought that the insulin receptor exhibited this phenomenon because insulin-binding data generate a curvilinear Scatchard plot (see Chapter 5). Such a plot could represent multiple isoreceptors having different affinities or negative cooperativity; because only one insulin receptor gene has been found, negative cooperativity appeared to be the best interpretation. This conclusion was further supported by dissociation experiments: the dissociation of labeled insulin from its receptor was faster in the presence of excess unlabeled hormone than in its absence. The excess of unlabeled hormone kept the receptor occupancy at a high level, thereby maintaining a low affinity and promoting rapid dissociation. In the absence of this excess hormone, the receptor affinity would increase as the insulin dissociated and receptor occupancy fell; this higher affinity would retard further dissociation.

However, this concept has been challenged and is still controversial. For example, the dissociation experiment may be explained by the *mobile receptor* hypothesis. This idea has been developed mathematically, but only a conceptual presentation is given here. The hypothesis is that a hormone may bind either to its receptor alone or to the receptor-effector complex; furthermore, the affinity of the latter for the hormone is greater. In fact, this has been shown for several of GPCRs, which can be precoupled to G proteins. Hormone dissociation, therefore, would be slower from the receptor-effector complex than from the receptor alone. If an excess of unlabeled hormone were present, it would bind to the free receptors and because of its abundance would outcompete the labeled hormone-receptor complex for effector binding. Without an attached effector, the labeled hormone would now dissociate more rapidly. Again, this is exactly what is observed for the βAR, despite the fact that this receptor yields a linear Scatchard plot, indicating no allosterism. Alternatively, the curvilinear plot for insulin may be a result of hormone degradation, which gives rise to fragments having a spectrum of receptor affinities. This occurs in the liver; when the data are corrected for this degradation, the Scatchard plot becomes linear.

Oligomerization can also be heterologous. Most cytokine receptors consist of at least two nonidentical, although homologous, subunits (Chapter 7); either subunit can only bind its ligand with low to intermediate affinity, whereas both components bind hormone with high affinity. The IL-6 receptor represents an example of how such a system can be regulated. The IL-6 receptor consists of gp130 and a soluble α-subunit, sIL-6Rα. Endothelial cells have gp130 and constitutively produce their own IL-6; however, because they cannot make their own sIL-6Rα and the affinity of gp130 for IL-6 is so low, no signal is generated. Infiltrating macrophages and lymphocytes secrete sIL-6Rα to control locally the hormone sensitivity of the tissue. In the presence of sIL-6Rα, the receptor affinity increases and a signal can be transduced.

Receptor affinity may also be affected by interacting with a nonbinding subunit; as just noted, many GPCRs are precoupled to G proteins, and this association increases the affinity of the attached receptor for its ligand. Coreceptors, like proteoglycans, can also enhance ligand binding; for example, the requirement of fibroblast growth factor receptors (FGFRs) for heparin. The affinity of the neural nicotinic acetylcholine receptor (nAChR) is increased by calcium through the visinin-like protein 1 (VILIP-1), a myristoylated protein with three EF, calcium-binding domains. Activation of the nAChRs results in an influx of calcium, which binds VILIP-1; the ensuing change in conformation exposes the myristic acid, which targets VILIP-1 to the plasma membrane and nAChRs. VILIP-1 then allosterically increases the affinity of the nAChR for acetylcholine by threefold. This represents a rare positive feedback loop in biology.

Finally, different receptor subunits can associate and generate new binding specificities. The human EGF receptor homologue 1 (HER1) and HER2 can heterodimerize to produce a hybrid receptor with very high affinity toward EGF, but HER2-HER3 dimers prefer the neuregulins. Similarly, calcitonin receptor-like receptor (CRLR) binds calcitonin gene-related peptide (CGRP) when coupled to receptor activity modifying protein 1 (RAMP1) but binds adrenomedullin when coupled to RAMP2.

RNA Processing

Alternative splicing of the receptor mRNA is another way of affecting receptor affinity and specificity. Exon 11 of the insulin receptor codes for 12 amino acids in the extracellular domain near to the membrane. The intact receptor (IR-B) has half the insulin affinity of the one (IR-A) lacking these amino acids. This variation may have physiological significance because the long form is found in the liver, which is exposed to high concentrations of insulin secreted from the pancreas. The lower affinity may protect the liver from overstimulation. This splicing also affects ligand selectivity in heterodimers between the insulin receptor and IGF-IR. IR-A/IGF-IR hybrids will bind IGF-I, IGF-II, and insulin, and all of these ligands are active. However, IR-B/IGF-IR hybrids will bind IGF-I with high affinity, bind IGF-II with low affinity, and not bind insulin at all.

The FGFR2 can also undergo alternate splicing involving the carboxy-half of the third immunoglobulin loop: FGFR2 binds only FGF, whereas the variant binds both FGF and keratinocyte growth factor (KGF). Finally, some receptors are subject to RNA editing, where the sequence of the RNA transcript is changed after transcription. All editing of the $5-HT_{2C}$ (5-hydroxytryptomine 2C) decreases ligand affinity.

Endogenous Antagonists

Ligand Antagonists

With a fixed receptor number, affinity, and location, receptor function can be influenced by the presence of hormone antagonists. Antagonists are usually thought of as synthetic compounds originating from pharmacology laboratories, but many organisms produce their own hormone antagonists to further regulate receptor activity. Such antagonists can be generated in several ways. First, they may arise from a unique gene; for example, the IL-1 antagonist is homologous to IL-1α and IL-1β and binds to, but does not activate, the IL-1R. Second, antagonists can be generated by alternate splicing from the same gene as the agonist. For example, an alternately spliced hepatocyte growth factor (HGF) is truncated in the second kringle domain; sufficient structure is preserved to allow it to bind to the HGF receptor (HGFR), but it cannot activate the receptor. Angiopoietin is another truncated hormone that acts as an antagonist. Third, antagonists can be agonists that have undergone post-translational modifications. PKC phosphorylation of prourokinase converts it to an antagonist at the urokinase receptor, and mono(ADP-ribosylation) of α defensin-1 makes it an antagonist. Finally, subunit composition can affect hormone activity; for example, IL-12 composed of a 35- to 40-kDa dimer is an agonist, but the 40-kDa homodimer is an antagonist.

In addition to antagonists, some hormones have binding molecules that can inhibit transduction by sequestering them from their receptors. Noggin binds bone morphogenetic protein (BMP), and follistatin binds either activin or BMP.

Receptor Antagonists

Receptor antagonists are of two kinds: (1) soluble receptors and (2) membrane-bound receptors. Most receptor antagonists are modifications of normal receptors, although a few, like IL-1RII and osteoprotegrin, are dedicated antagonists transcribed from unique genes. The soluble receptors represent the extracellular domains of membrane-bound receptors; they are usually generated by alternate splicing, although some may be released by proteolysis. These proteins dampen hormone effects, presumably by sequestering hormones and preventing them from reaching their intact receptors. Note that not all soluble receptors are antagonists; the α-subunits of cytokine receptors frequently function solely to increase hormone binding affinity. As a result, only their extracellular domains are required, and soluble forms are still active.

Other receptor antagonists may be truncated beyond the membrane domain or lack critical autophosphorylation sites as a result of alternate splicing; such proteins are still membrane-bound and can dimerize but have no transducing activity. Because the biological activity of RTKs and cytokine receptors is usually dependent on the dimerization of functional proteins, a heterodimer containing an intact and a truncated receptor is inactive. Essentially, the truncated receptor binds an intact receptor and removes it from the pool of potentially active receptors. As with ligand antagonists, not all truncated or kinase-dead variants are antagonists: neither EphB6 nor ErbB3 have kinase activity, but they both provide phosphorylation sites for their active partners. Essentially, these subunits have become converted to scaffolds.

Other Receptor Inhibitors

There are also several inhibitors that are not homologous to either hormones or receptors. The most intensively studied group was first identified as inhibitors of cytokine signaling, which is reflected in the various names that have been given to them: cytokine inducible SH2-containing protein (CIS), Stat-induced Stat inhibitor (SSI), Janus binding protein (JAB), and suppressor of cytokine signaling (SOCS). All of these proteins exhibit some homology and have a central SH2, but they have different molecular mechanisms. Virtually all of them are pseudosubstrates; that is, they resemble Jak substrates but cannot be phosphorylated. In addition, by occupying the active site of Jak, they prevent any other substrate from being phosphorylated. Some of these inhibitors may also bind cytokine receptors and block Stat access to Jak. Others may target Jak to proteasomes; SOCS and CIS are E3 ligases. SOCS is not limited to the cytokine pathway; it can bind a variety of STKs and RTKs. In addition, because SOCS can be tyrosine phosphorylated, it can bind transducers possessing SH2 domains. These inhibitors are usually induced by cytokines in an apparent negative feedback loop.

Transduction Uncoupling

Receptors without any intrinsic activity must couple to a transducer; these include the GPCRs that activate G proteins and the cytokine receptors that activate STKs. The former have been more intensively studied and are the major subject matter for this section. The GPCR can be uncoupled by changes in either the receptor or the G protein; the former is discussed here, whereas the latter is covered in Chapter 9.

Receptor Phosphorylation

The major mechanism for rapid, reversible uncoupling is phosphorylation. Homologous uncoupling of the βAR can be mediated by two different kinases depending on the concentration of the agonist. In the presence of low agonist concentrations, G protein activation leads to the elevation of $3'5'$-cyclic AMP (cAMP) and the stimulation of PKA (Fig. 8-1, *A*). PKA will then phosphorylate the βAR in the carboxy-half of the third cytoplasmic loop and in the juxtamembrane portion of the carboxy terminus. Any hormone elevating cAMP can stimulate the phosphorylation of any other receptor similarly coupled to G_s, even if the hormone for the other receptor is not present; this is known as *class desensitization*. However, phosphorylation is more rapid when the βAR is occupied. The PKA effect is slow, requiring about 3.5 minutes, and does not lead to receptor internalization.

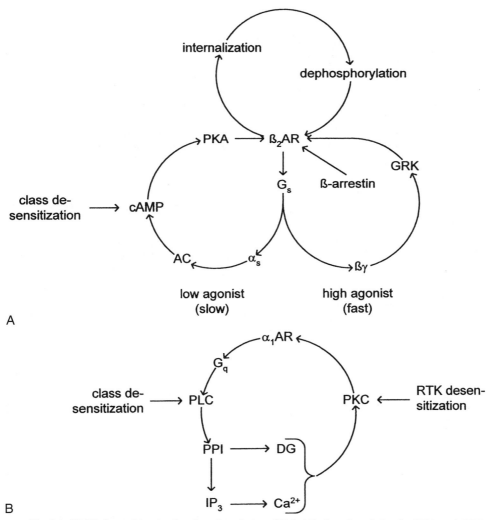

Fig. 8-1. GPCR desensitization by phosphorylation: (A) β_2AR phosphorylation by PKA and GRK and (B) α_1AR phosphorylation by PKC. *AC*, Adenylate cyclase; *DG*, diacylglycerol; *GPCR*, G protein-coupled receptor; *GRK*, GPCR kinase; *IP₃*, inositol 1,4,5-trisphosphate; *PLC*, phospholipase C; *PPI*, polyphosphoinositide.

High agonist concentrations activate a second kinase. Because the first identified substrate for this kinase was the βAR, it was originally called the *β-adrenergic receptor kinase* (βARK). However, it is now known that most GPCRs are substrates, so the kinase has been renamed the *GPCR kinase* (GRK). Actually, there are seven kinases in this family, but they exhibit overlapping substrate specificity. Rather, they are divided into three groups based on their location and regulation: GRK1 and GRK7, GRK2 and GRK3, and GRK4-6. GRK1 and GRK7 are retinal and function in light perception. The βγ subunits of G proteins activate GRK2 and GRK3 by direct binding and by recruiting them to the membrane where the GPCRs are located; this latter effect also requires polyphosphoinositides (PPI). GRK2 can be further stimulated by phosphorylation by PKA, PKC, or Src (an STK), and it can be inhibited by MAPK phosphorylation at the βγ–binding site. GRK4-6 also bind the plasmalemma through PPI, and they are stimulated by polyamines, which are basic second messengers (see Chapter 11). CaM binding inhibits both polyamine and membrane binding. GRK5 can be further inhibited by PKC phosphorylation.

GRK phosphorylates the S-T-rich region in either the carboxy terminus or the third cytoplasmic loop, depending on the GPCR. The phosphorylation per se does not uncouple the GPCR from G proteins; rather, the modified site binds another molecule, β-arrestin, which competes with G_s for GPCR binding. This phosphorylation occurs rapidly, requiring less than 15 seconds, and leads to receptor internalization. As previously noted, sequestration is not part of the desensitization process but is necessary for dephosphorylation and recycling. GRK2 and GRK3 can inhibit GPCRs by three other mechanisms. First, they bind and sequester βγ; second, they bind and inhibit PKCβ, a downstream effector of several GPCRs; and third, they stimulate the guanosine 5'-triphosphate (GTP) hydrolytic activity of G_q. Such proteins are called *GTPase activating proteins* (GAPs). Because GTP activates G proteins, its hydrolysis would terminate G protein stimulation (see Chapters 4 and 9). In addition, GRK can have effects unrelated to G proteins. For example, it can phosphorylate the PDGF receptor and both trigger its desensitization and enhance its ubiquitination. In this example, GRK is activated by the βγ subunits of G_i, which is stimulated by the PDGF receptor. GRK can also phosphorylate synuclein; because synuclein inhibits phospholipase C (PLC), its inhibition would activate PLC. Finally, GRK can phosphorylate ribosomal protein P2 and stimulate protein synthesis. P2 forms part of the elongation factor-binding site in the 60S ribosomal subunit.

As just mentioned, β-arrestin competes with G_s for binding to GPCRs. However, β-arrestin has other functions, too. First, it recruits cAMP phosphdiesterase; that is, in addition to inhibiting the generation of cAMP, it promotes the hydrolysis of existing cAMP. Second, it is an adaptor for clathrin and adaptor protein 2 (AP-2), which is required for GPCR internalization. Third, as an adaptor for the ATPase required for intracellular trafficking and for ArfGAP, it is involved with GPCR recycling. Arf is a small G protein required for internalization and dephosphorylation (Chapter 9). Finally, it serves as a scaffold for Arf, MAPK activation, and several STKs.

The duration of the association between β-arrestin and the GPCR determines the speed of resensitization. For example, β-arrestin dissociates from Class A receptors, like βAR, at the plasma membrane, and the GPCRs resensitize quickly. However, β-arrestin is internalized with Class B receptors, like the vasopressin receptor 2, and the GPCRs resensitize slowly.

Although the pathway just outlined is considered the classic one, there are several variations. For example, CK2, not GRK, phosphorylates the carboxy terminus of TRHR for the β-arrestin-mediated internalization via coated pits, and CK2 phosphorylation is a prerequisite for GRK phosphorylation of the leukotriene B_4 (LTB_4) receptor. β-Arrestin binding to the luteinizing hormone receptor (LHR) only requires that the GPCR be active, not phosphorylated, and the recycling of the parathormone (PTH) receptor requires neither dephosphorylation nor the dissociation of β-arrestin. Finally, some GPCRs, like those for angiotensin (AT) and secretin, are targets for GRK and β-arrestin but are internalized via endocytotic vesicles rather than coated pits.

The $\alpha_1 AR$ uses a different transduction system, which elevates calcium and phospholipid metabolites (see Chapter 10); one of its major outputs is PKC (Fig. 8-1, *B*). As one might expect, PKC can phosphorylate the $\alpha_1 AR$ and uncouple it from its transducer. As is true of the $\beta_2 AR$ system, this modification proceeds twice as rapidly in the presence of an agonist as in its absence. Heterologous regulation by bradykinin, which uses the same transducer, has the same effect; this is another example of class desensitization. However, bradykinin does not induce internalization because the $\alpha_1 AR$ is not occupied. The $\alpha_1 AR$ is internalized when it is stimulated by its own agonist, norepinephrine.

This negative feedback can be very specific. The metabotropic glutamate receptor (mGluR) can activate both G_q and G_s, which elevate calcium and cAMP, respectively. Calcium stimulates PKC, which phosphorylates mGluR and disrupts coupling to G_q but not to G_s; that is, G_q activation feeds back to inhibit only its own signaling pathway. A similar phenomenon is seen with the $\alpha_{2A} AR$, which stimulates both G_q and G_i. Again, PKC phosphorylation uncouples G_q but not G_i.

Desensitization can also cross different transduction pathways; for example, PKC increases fivefold the concentration of β-agonist necessary to activate adenylate cyclase. However, the phosphorylation state of the $\beta_2 AR$ is unchanged. In this case, PKC appears to act at the G protein or adenylate cyclase. Conversely, PKA phosphorylates a site in the third cytoplasmic loop of the muscarinic receptor type 1 ($M_1 R$), leading to its desensitization. RTKs can also affect GPCR-G protein coupling; insulin can induce the tyrosine phosphorylation of the $\beta_2 AR$ and reduce coupling efficiency by 50%. Because direct stimulation of the transducers was normal, it was assumed that the receptor modification was the cause for the uncoupling. Finally, NO can modify the cysteine in the membrane proximal region of the carboxy terminus and block palmitoylation, leading to decreased coupling between the βAR and G_s.

The uncoupling triggered by GRK is physical because β-arrestin competes with the G protein. However, the actual mechanism of uncoupling induced by other kinases is unknown except that it appears to be a functional, rather than a physical, uncoupling. For example, the desensitized muscarinic acetylcholine receptor is internalized with its G protein. Furthermore, the affinity of the uncoupled muscarinic receptor is still increased in the presence of its G protein, indicating that the two are still physically interacting.

Other Mechanisms

Another mechanism that can affect coupling to transducers is alternate splicing; this mechanism is slower and would have more long-term effects than phosphorylation. There are two dopamine 2 receptors; the D_{2A} has 29 more amino acids in the third intracellular loop than the D_{2B}. Both isoreceptors elevate calcium, but the D_{2B} inhibits cAMP levels twice as effectively as the D_{2A}. In other words, the additional amino acids appear to interfere with the coupling of the D_2 to G_i.

In a second example, alternate splicing of the PGE receptor, EP3, gives rise to a protein with four different carboxy termini, which influence G protein coupling. Isoform A activates G_i or G_o; isoforms B and C, G_s; and isoform D, G_i, G_s, and G_q. RNA editing is a different form of RNA processing that can affect coupling: the unedited 5-HT$_{2C}$ couples to both G_q and G_{13}, but the edited receptor only couples to G_q.

The cytokine receptors represent a different transduction system; receptor activation leads to its phosphorylation by Jak. The PRLR has three forms generated by progressively larger carboxy-terminal truncations arising from alternate RNA splicing. All three forms stimulate mitosis because the transducers for this effect bind to proximal phosphorylation sites. However, the shortest PRLR cannot induce milk proteins because this activity requires the distal phosphorylation sites.

Receptor-effector coupling can also affect and be affected by the lipid composition of the membrane. β-Adrenergic agonists increase the phosphatidylcholine content of membranes, thereby increasing their fluidity. This in turn is thought to facilitate the coupling of the receptor to G_s (see Chapter 10). This hypothesis is further supported by studies in which the fluidity of the membrane was experimentally altered; lipids that increase fluidity enhance β-agonist–induced adenylate cyclase activity, whereas lipids that decrease fluidity have the opposite effect. In addition, gangliosides, sphingolipids, and cholesterol have long been known to influence receptor function; for example, gangliosides can impair PDGFR dimerization, increase Trk autophosphorylation, and decrease EGF binding to its receptor. The discovery of detergent-insoluble, ganglioside-enriched complexes (DIGs) (see Chapter 7) led to the realization that these lipids were probably markers for rafts and has resulted in a reinterpretation of these studies. Lipid facilitation of transduction likely represented increased coupling efficiency in a membrane domain rich in signaling molecules. Inhibition may be based on the physical properties of DIGs; for example, their rigidity would impair lateral diffusion and oligomerization. The net effect on any given system would depend on what factors were particularly important for transduction in that pathway.

Finally, hormone modifications can affect coupling; apparently, different hormone isoforms induce different receptor conformations, leading to altered activity. For example, fully glycosylated follicle-stimulating hormone (FSH) activates G_s, but the incompletely glycosylated hormone activates G_i; core fucosylated thyroid-stimulating hormone (TSH) elevates both cAMP and inositol 1,4,5-trisphosphate (IP$_3$), but unfucosylated TSH only elevates cAMP. Hormone phosphorylation can also influence signaling; unphosphorylated PRL induces its receptor to couple to Jak2 and mitogenesis, but PKA-phosphorylated PRL induces coupling to a different STK that still activates Stat but stimulates differentiation instead of mitogenesis. Lastly, peptide cleavage can affect signaling; intact neurokinin A generates a rapid rise in calcium followed by an increase in cAMP, whereas the truncated hormone only elevates calcium.

Developmental Regulation

Many cellular processes are so complex that several hormones are involved, in which case the regulation of their receptors is frequently coordinated. The mammary gland is an excellent example. In the nonpregnant mouse, the mammary epithelium is poorly developed; although the ducts extend throughout the fat pad, they are widely spaced and there are no alveoli. During pregnancy, the epithelium rapidly proliferates under the influence of estrogens and progesterone; EGF may play an ancillary role. After parturition, growth ceases and lactation begins; in the mouse, milk production requires insulin, cortisol, and PRL. Growth and differentiation are generally mutually exclusive events; for example, progesterone and EGF stimulate epithelial proliferation but inhibit milk protein synthesis. Figure 8-2 schematically shows the tissue concentration of these receptors in the mammary gland during pregnancy and lactation; all of the data are from studies of mice except those for the PRLR, which are from rat studies. Data for other hormone receptors in the mammary gland are less complete but follow the same general pattern; for example, the GHR, IGF-IR, and FGFR are elevated during pregnancy and decrease with lactation. It is clear that the concentrations of receptors for the growth promoters are high during pregnancy but fall near parturition. The levels of receptors for the hormones that include differentiation, however, do not rise until late pregnancy and are maximal during lactation.

Many of the mechanisms discussed earlier in this chapter can also be used for the developmental regulation of receptors. First, because developmental changes take a relatively long time to occur, gene induction is often fast enough to produce the necessary modulations in receptor levels. Alternate splicing is another mechanism; although a truncated LHR mRNA is transcribed in the immature rat ovary, it cannot be translated. The intact mRNA is not synthesized and translated until 7 days postpartum. Finally, compartmentalization has also been documented in a developmental system. The EPOR is present in many myeloid precursors but these cells are unresponsive to EPO because the EPORs are intracellular. These receptors migrate to the plasma membrane in the precursors destined to enter the erythrocyte lineage.

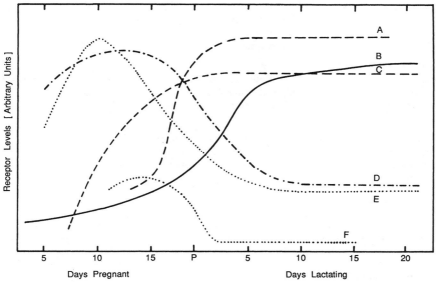

Fig. 8-2. Developmental regulation of the mammary gland receptors for cortisol (A), prolactin (PRL) (B), insulin (C), estrogen (D), epidermal growth factor (E), and progesterone (F). All data are from studies of mice except those for PRL receptors, which are from studies of rats.

Summary

The types of receptor regulation discussed in this chapter are summarized in Table 8-2. Actual receptor number can be changed by altering either synthesis or degradation, whereas the number of functional receptors can be changed by cross-activation or migration between cellular compartments. Receptor affinity and selectivity can be changed acutely by phosphorylation, glycosylation, or allosterism; a slower but longer effect can be achieved by changing the structure of the receptor, such as by alternate splicing or RNA editing. One could also view endogenous antagonists as a means of altering receptor affinity because their presence increases the concentration of hormone required for a given response. A ligand antagonist can either be a pure antagonist transcribed from a unique gene or an alternately spliced agonist that can still bind the receptor but cannot activate it. Receptor antagonists are either soluble receptors that sequester the hormone or kinase-defective molecules that bind with intact receptors to produce a nonfunctional dimer. Finally, the receptor may switch transducers or be uncoupled from all second messengers. These effects can be accomplished by

Table 8-2
General Mechanisms for Hormone Receptor Regulation

Mechanisms	Examples
Change Number By	
Synthesis	Insulin receptor induction by glucocorticoids in hepatocytes
Stability	Insulin receptor induction by glucocorticoids in adipocytes; membrane-bound SCF inhibition of SCFR downregulation
Compartmentalization	Shift of IGF-IIR to plasmalemma by insulin
Activation	IGF-IR phosphorylation by the insulin receptor; PDGFR and EGFR phosphorylation by calcium-activated STK
Change Affinity By	
Phosphorylation	EGFR autophosphorylation
Glycosylation	Fucosylation of Notch1
Allosterism	Guanylate cyclase by ATP; EGFR dimerization; sIL-6Rα secreted locally by infiltrating macrophages and lymphocytes; proteoglycan coreceptors
RNA processing	Insulin receptor $\pm$ exon 11
Antagonists	
Ligand	IL-1 antagonist; HGF alternate splicing; phosphorylated prourokinase
Receptor	Soluble TNFR; membrane-bound, truncated PDGFR
Other inhibitors	SOCS (suppressor of cytokine signaling)
Alternate Coupling to Transducer	
Receptor phosphorylation	β_2AR phosphorylation by PKA or GRK
Receptor S-nitrosylation	Blocks repalmitoylation in βAR and G_s coupling
Alternate splicing	Alternate carboxy termini of prostaglandin E_3 isoreceptor (EP3)
RNA editing	Unmodified 5-HT$_{2C}$ couples to G_q and G_{13} but the edited receptor only couples to G_q
Membrane properties	Lipid composition and the βAR
Hormone modification	Glycosylated FSH and TSH; phosphorylated PRL; truncated neurokinin A

ATP, Adenosine 5′triphosphate; *EGFR,* epidermal growth factor receptor; *FSH,* follicle-stimulating hormone; *GRK,* G protein-coupled receptor kinase; *HGF,* hepatocyte growth factor; 5-*HT$_2$,* 5-hydroxytryptamine receptor 2C; *IGF-IR,* insulin-like growth factor I receptor; *IGF-IIR,* insulin-like growth factor II receptor; *PDGFR,* platelet-derived growth factor receptor; *PKA,* protein kinase A; *SCF,* stem cell growth factor; *SCFR,* SCF receptor; *STK,* soluble tyrosine kinase; *TNFR,* tumor necrosis factor receptor; *TSH,* thyroid-stimulating hormone.

receptor phosphorylation, *S*-nitrosylation, or RNA processing; by hormone phosphorylation, glycosylation, or cleavage; or by altering the properties of the plasma membrane.

References

Receptor Number and Affinity

Haltiwanger, R.S., and Stanley, P. (2002) Modulation of receptor signaling by glycosylation: Fringe is an *O*-fucose-β1,3-*N*-acetylglucosaminyltransferase. *Biochim. Biophys. Acta* 1573, 328-335.

Hicke, L. (2001). Protein regulation by monoubiquitin. *Nature Rev. Mol. Cell Biol.* 2, 195-201.

Kim, K.I., Baek, S.H., and Chung, C.H. (2002). Versatile protein tag, SUMO: Its enzymology and biological function. *J. Cell. Physiol.* 191, 257-268.

Kornitzer, D., and Ciechanover, A. (1999). Modes of regulation of ubiquitin-mediated protein degradation. *J. Cell. Physiol.* 182, 1-11.

Peters, M., Müller, A.M., and Rose-John, S. (1998). Interleukin-6 and soluble interleukin-6 receptor: Direct stimulation of gp130 and hematopoiesis. *Blood* 92, 3495-3504.

Primakoff, P., and Myles, D.G. (2000). The ADAM gene family: Surface proteins with adhesion and protease activity. *Trends Genet.* 16, 83-87.

Endogenous Antagonists

Kile, B.T., Schulman, B.A., Alexander, W.S., Nicola, N.A., Martin, H.M.E., and Hilton, D.J. (2002). The SOCS box: A tale of destruction and degradation. *Trends Biochem. Sci.* 27, 235-241.

Kroiher, M., Miller, M.A., and Steele, R.E. (2001). Deceiving appearances: Signaling by "dead" and "fractured" receptor protein-tyrosine kinases. *Bioessays* 23, 69-76.

Naka, T., Fujimoto, M., and Kishimoto, T. (1999). Negative regulation of cytokine signaling: STAT-induced STAT inhibitor. *Trends Biochem. Sci.* 24, 394-398.

Jones, S.A., and Rose-John, S. (2002). The role of soluble receptors in cytokine biology: The agonistic properties of the sIL-6R/IL-6 complex. *Biochim. Biophys. Acta* 1592, 251-263.

Transduction Uncoupling

Bünemann, M., and Hosey, M.M. (1999). G-protein coupled receptor kinases as modulators of G-protein signalling. *J. Physiol. (Camb.)* 517, 5-23.

Burger, K., Gimpl, G., and Fahrenholz, F. (2000). Regulation of receptor function by cholesterol. *Cell. Mol. Life Sci.* 57, 1577-1592.

Ferguson, S.S.G. (2001). Evolving concepts in G protein-coupled receptor endocytosis: The role in receptor desensitization and signaling. *Pharmacol. Rev.* 53, 1-24.

Kohout, T.A., and Lefkowitz, R.J. (2003). Regulation of G protein-coupled receptor kinases and arrestins during receptor desensitization. *Mol. Pharmacol.* 63, 9-18.

Morris, A.J., and Malbon, C.C. (1999). Physiological regulation of G protein-linked signaling. *Physiol. Rev.* 79, 1373-1430.

Tsao, P., Cao, T., and von Zastrow, M. (2001). Role of endocytosis in mediating down-regulation of G-protein-coupled receptors. *Trends Pharmacol. Sci.* 22, 91-96.

Williams, S., Meij, J.T.A., and Panagia, V. (1995). Membrane phospholipids and adrenergic receptor function. *Mol. Cell. Biochem.* 149/150, 217-221.

Cited References

1. Salhanick, A.L., Krupp, M.N., and Amatruda, J.M. (1983). Dexamethasone stimulates insulin receptor synthesis in cultured rat hepatocytes. *J. Biol. Chem.* 258, 14130-14135.
2. Knutson, V.P., Ronnett, G.V., and Lane, M.D. (1982). Control of insulin receptor level in 3T3 cells: Effect of insulin-induced down-regulation and dexamethasone-induced up-regulation on rate of receptor inactivation. *Proc. Natl. Acad. Sci. U.S.A.* 79, 2822-2826.
3. Hyde, B.A., Blaustein, J.D., and Black, D.L. (1989). Differential regulation of progestin receptor immunoreactivity in the rabbit oviduct. *Endocrinology (Baltimore)* 125, 1479-1483.
4. Frick, G.P., Leonard, J.L., and Goodman, H.M. (1990). Effect of hypophysectomy on growth hormone receptor gene expression in rat tissues. *Endocrinology (Baltimore)* 126, 3076-3082.
5. Nardulli, A.M., Greene, G.L., O'Malley, B.W., and Katzenellenbogen, B.S. (1988). Regulation of progesterone receptor messenger ribonucleic acid and protein levels in MCF-7 cells by estradiol: Analysis of estrogen's effect on progesterone receptor synthesis and degradation. *Endocrinology (Baltimore)* 122, 935-944.
6. Nardulli, A.M., and Katzenellenbogen, B.S. (1988). Progesterone receptor regulation in T47D human breast cancer cells: Analysis by density labeling of progesterone receptor synthesis and degradation and their modulation by progesterone. *Endocrinology (Baltimore)* 122, 1532-1540.
7. McIntyre, W.R., and Samuels, H.H. (1983). Triamcinolone acetonide regulates glucocorticoid receptor levels by decreasing the half-life of the activated nuclear receptor form. *Program Abstr., 65th Annu. Meet. Endocr. Soc.*, p. 228.
8. Wiese, R.J., Uhland-Smith, A., Ross, T.K., Prahl, J.M., and DeLuca, H.F. (1992). Up-regulation of the vitamin D receptor in response to 1,25-dihydroxyvitamin D3 results from ligand-induced stabilization. *J. Biol. Chem.* 267, 20082-20086.

PART **3**

Transduction

CHAPTER **9**

G Proteins and
Cyclic Nucleotides

CHAPTER OUTLINE

Because peptide and certain other hormones cannot cross the plasma membrane, they must interact with their receptors on the cell surface. Therefore if the hormone is to have an effect on cellular processes, a signal must be generated from the other side of the membrane; this signal can then carry out the actions of the hormone. If the hormone itself is considered to be the primary, or first, messenger, then the signal becomes the second messenger.

Part 3 discusses all of the second messengers. This chapter begins with a brief discussion of the criteria for these mediators and then presents an overview of the guanosine 5′triphosphate (GTP)–binding proteins. Finally, the cyclic nucleotides are covered. Chapter 10 describes the role of phospholipids in signal transduction: phosphatidylinositol is involved with the elevation of intracellular calcium and the activation of several protein kinases, whereas phosphatidylcholine may affect membrane fluidity and generate the precursor for the eicosanoids. Arachidonic acid, sphingomyelin, and the lysophospholipids are also mentioned. Chapter 11 describes a number of mediators that are not as well defined as the cyclic nucleotides and calcium; they include the polyamines, oligosaccharides, cellular pH, magnesium, reactive oxygen intermediates, and the cytoskeleton. Chapter 12 is a general synthesis of the preceding three chapters and has a strong emphasis on protein phosphorylation. Several model systems are then presented to demonstrate how second messengers can directly affect cellular processes.

Criteria for Second Messengers

How does one identify a second messenger for a hormone; that is, what characteristics should a molecule have to be called a second messenger? Basically, there are five criteria that a compound must satisfy, and they are discussed with glycogen breakdown as an example; in the liver, glucagon stimulates glycogenolysis through 3′5′-cyclic AMP (cAMP).

1. The mediator, or its analog, must mimic the action of the hormone. Because cAMP does not readily cross the membrane, dibutyryl cAMP is frequently used on intact cells; once inside the cell, the butyryl side chains are removed. Unfortunately, most investigators do not realize that this hydrolysis results in a significant accumulation of butyric acid, which can affect cellular metabolism in its own right (see Chapter 14). As such, all experiments with dibutyryl cAMP should have butyric acid controls. Nonetheless, this analog does induce glycogenolysis in liver cells.

2. The hormone must induce elevated levels of the mediator. Glucagon does elevate cAMP levels in liver cells.

3. The hormone must appropriately affect the enzymes of synthesis and/or degradation of the mediator. cAMP is synthesized from adenosine 5′-triphosphate (ATP) by adenylate cyclase and is hydrolytically inactivated by a cyclic nucleotide phosphodiesterase (PDE) (Fig. 9-1). cAMP concentrations can be elevated by stimulating the adenylate cyclase or inhibiting the PDE. In the liver, glucagon does the former.

4. An appropriate temporal relationship must exist among the hormone, mediator, and hormonal effect. It would be difficult to argue that cAMP is the second messenger of glucagon in the liver if cAMP levels did not rise

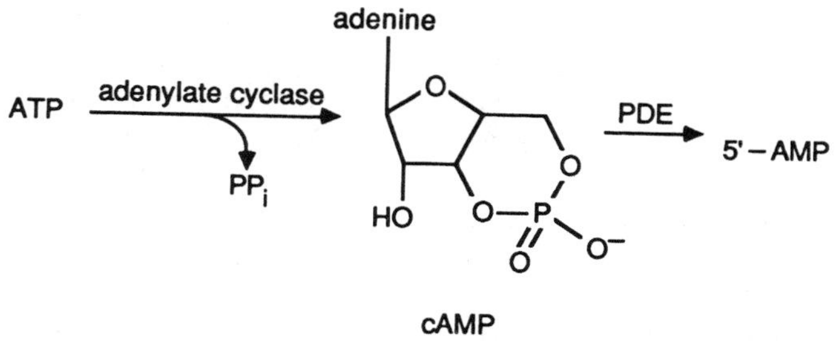

Fig. 9-1. The formation and degradation of cAMP. *PDE,* Cyclic nucleotide phosphodiesterase.

until after glycogenolysis had already started. In fact, the elevation of cAMP concentrations precedes glycogenolysis in this system.

5. Finally, if drugs are available to modulate the endogenous level of the mediator, they should also mimic or inhibit, as appropriate, the effects of the hormone. For example, cAMP levels can be raised by (1) stimulating the adenylate cyclase with either forskolin or aluminum fluoride or (2) inhibiting the PDE with methylxanthines. cAMP levels can be lowered by inhibiting the adenylate cyclase with 2′,3′-dideoxyadenosine or 9-(tetrahydro-2-furyl)adenine. Forskolin, aluminum fluoride, and the methylxanthines can stimulate glycogenolysis in liver cells.

Therefore one can reasonably conclude that cAMP is indeed the second messenger for glucagon in the liver. However, care must be taken not to over-interpret these data; for example, they do not argue that cAMP is the only mediator for glucagon in the liver. In fact, there is evidence that glucagon may activate another transduction system in this organ. The data also do not state that cAMP will be the mediator for glucagon in all tissues. As far as is known, this statement is true for glucagon; however, the effect of vasopressin is mediated by cAMP in the kidney but by calcium in the liver.

G Proteins

Components and Cycle

Several major transduction systems use GTP-binding proteins, or simply G proteins, to couple a hormone signal to an effector. Basically, a G protein is a molecular switch with a built-in automatic timer; when bound to guanosine diphosphate (GDP), the switch is in the off position, but when bound to GTP, the switch is on. Because the G protein itself is an inefficient GTPase, it will eventually hydrolyze the bound GTP and turn itself off. The G proteins are all homologous at the catalytic site, but they can vary considerably elsewhere. Indeed, the size variation alone is substantial: 20 to 100 kDa.

The GTPase activity of the small G proteins is so low that it requires an accessory factor, the GTPase activating protein (GAP). In addition to augmenting the intrinsic GTPase activity of small G proteins and turning them off, it has been proposed that GAP may serve as an effector. This possibility is discussed later. GAP can also affect the cellular distribution of G protein function: in several cell lines, Ras and Rap1 are activated throughout the cell, but the uneven distribution of GAP results in the restricted localization of Ras and Rap1 activities. Although GAPs show some overlap in G protein specificity, there appears to be a separate GAP that is primarily directed toward each small G protein. Large G proteins have a greater GTPase activity and originally were not thought to need GAPs. However, such proteins do exist for the G protein trimers; they are called *negative regulators of G protein signaling* (RGS).

External signals can be fed into the system through a component that triggers the exchange of GDP for GTP, thereby flipping the switch on. There have been several names given to this component; the most commonly used one is the GNP (GDP-GTP) exchange facilitator (GEF), although the term *GDP dissociation stimulator* (GDS) is still occasionally seen in the scientific literature. This text uses GEF.

The GEF is opposed by a GDP dissociation inhibitor (GDI) that not only inhibits GNP exchange but also represses the intrinsic and GAP-stimulated GTPase activity. The GDI appears to be preassociated with the small G proteins and probably keeps basal activity at a low level; the GDI also influences the localization of these molecules. Because the βγ-dimer performs similar functions on the large G proteins, some researchers have suggested that GDI is the small G protein equivalent of the βγ-dimer. The system is also dampened by a GTPase inhibiting protein (GIP). It binds the G protein in either the GDP- or GTP-bound form and inhibits GAP.

The general cycle is presented in Figure 9-2. In response to some stimulus, GEF facilitates the displacement of GDP by GTP and activates the G protein. The G protein by itself or with GAP, which subsequently binds to the G protein, triggers a series of biological effects. Eventually, the GTP is hydrolyzed, GAP dissociates, and the system is inactivated. Both GDI and GIP act to check the cycle. Table 9-1 adds flesh to this cycle by giving specific examples. The elongation factor Tu/elongation factor Ts (EF-T/u/EF-Ts) cycle in

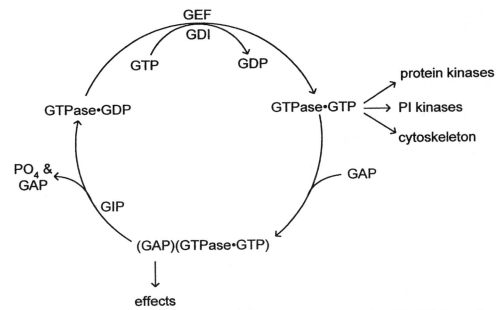

Fig. 9-2. The general cycle for G proteins. *GAP*, GTPase activating protein; *GDI*, GDP dissociation inhibitor; *GEF*, GNP exchange facilitator; *GIP*, GTPase inhibiting protein; *GTP*, guanosine 5′triphosphate; *PI*, phosphoinositide; *PO₄*, phosphate.

Table 9-1
Components of the GTPase Cycle for Several Systems

GTPase	GNP exchange facilitator (GEF)	GTPase activating protein	Function
EF-Tu	EF-Ts	Ribosome	Translation proofreading
Yeast Ras	Cdc25	IRA 1 and 2	Nutrient transduction
Mammalian Ras	mSOS	RasGAP	Proliferation
G protein trimer	Hormone receptor	RGS	Hormone signal transduction

RGS, Negative regulators of G protein signaling.

translation elongation is probably one of the most familiar examples to biology students.

Small G Proteins

The first G proteins discovered were involved with the signal transduction from G protein-coupled receptors (GPCRs); they are heterotrimers, whose GTPase subunit is about 40 kDa. These molecules are the ones most closely associated with the term *G protein,* although in this chapter they are referred to as *G protein trimers* to clearly distinguish them from the small G proteins. These latter proteins have molecular weights around 20 kDa and exist as either monomers or identical oligomers.

Lipid Modification of Proteins

A classification scheme for these small G proteins is presented in Table 9-2. Many of them are related to Ras both in amino acid sequence and in posttranslational modifications. Because some of these modifications are unusual, it is worthwhile to digress briefly and discuss them (Table 9-3, Fig. 9-3). In eukaryotes, there are four common ways in which lipids can be covalently attached to proteins; the simplest involves the attachment of a fatty acid. Myristic acid is coupled to the amino terminus through an amide linkage. This occurs during protein synthesis, and the modification is quite stable. Although many myristoylated proteins are membrane associated, others are cytoplasmic. This biphasic distribution is possible because myristic acid is too short to dictate membrane binding by itself. Rather, adjacent to the amino terminus is a cluster of basic amino acids that interact with the negatively charged membrane phospholipids. Phosphorylation in the basic domain will neutralize the charge and weaken membrane binding. Membrane binding of myristoylated proteins can be regulated in two other ways. First, in some peptides, like MARCKS (myristolated alanine-rich C kinase substrate), hydrophobic amino acids facilitate membrane binding; calmodulin (CaM) can bind and mask these amino acids. Second, in other proteins the myristic acid is tucked into a hydrophobic groove; ligand binding will cause the fatty acid tail to flip out and interact with the membrane. Examples of this latter mechanism include calcium binding to recoverin and GTP binding to Arf.

Palmitic acid is longer and is usually attached to cysteines by a thioester bond, although there are a few proteins known to be *N*-palmitoylated at the amino terminus. Palmitic acid is sufficiently hydrophobic, so it can bind to the membrane without the cooperation of any other factor. Furthermore, because it is a saturated fatty acid, it has a penchant to localize to DIGs (detergent-insoluble,

Table 9-2

Components of the GTPase Cycle for Several Systems

Group	Members	Function	Regulation (selected examples)[a]
Ras Superfamily **Ras family**	Ras	Mating and nutrient transduction (yeast); mitogenesis (mammals)	+GTPase: $p21^{ras}$ (NO) −GTPase: R-Ras (pY by Eph); K-Ras (Ca^{2+} − CaM) +GEF: RasGRF1/cdc25^{Mm} (Ca^{2+}-CaM, PKA, pY by Lck, calpain); RasGRP (Ca^{2+}-DG); CNrasGEF (cAMP-cGMP); mSos (Grb2 binding, PH-PIP$_n$); RasGDS1 (PKA) −GEF: mSos (MAPK); RasGRF1/cdc25^{Mm} (PKA) +GAP: Gap1^{IP4BP} (PH-IP$_4$) −GAP: RasGAP (PIP$_n$, p62^{dok} binding); p120-GAP (pY by Src)
	Ral	Endocytosis; vesicular trafficking; filapodia	+GTPase: RalA (CaM) +GEF: SHEP1 (SH2-pY); RalGDS (TC21, β-arrestin, PDK^b) −GEF: RalGDS (PKC)
	Rap	Cell adhesion	+GTPase: Rap1B (PKA) −GTPase: Rap1A (PKA) +GEF: Epac (cAMP); CalDAG-GEFI (Ca^{2+}-DG) −GEF: MR-GEF (Ras) +GAP: Rap1GAP (Gα$_z$ recruits) −GAP: Rap1GAP (Gα$_o$ sequesters)
	Rad, Gem, Kir, Rit, Rin	Cytoskeletal organization; secretion; ROCK inhibition	+GTPase: PKC and CaMKII by inhibiting CaM
	Rheb	Activate S6KII; inhibit Ras oncogenesis	−GTPase: Ca^{2+}-CaM
	κB-Ras	Bind PEST in IκB and decrease degradation	
	PIKE	Activate nuclear PI3K	+GEF: PLCγ (SH2-pY)
	AGS1	Activate G$_i$	
Rho family	Rho, Rac Cdc42, Rnd	Cytoskeletal organization; NADPH oxidase activation	+GTPase: RhoA (G$_h$ through transamidation, βγ via membrane recruitment, Etk through GDI dissociation^b); Cdc42Hs (PIP$_n$) −GTPase: RhoA (PKA); Rac1 (PKB)

Family	Members	Function	Regulation
			+GEF: Vav (SH2-pY, pY by Src); Cdc24p (PH-PIP$_n$); Dbl (α_{13}); Lbc (α_q); AKAP-Lbc (α_{12}); Tiam1 (CaMKII, Ras, CD44 binding); Pix (Nck, paxillin binding; Pak2); PDZ-RhoGEF (IGFI-R-IGFI and plexin B1-semaphorin 4D); ephexin (EphA-ephrin-A); P-Rex1 (PIP$_n$, $\beta\gamma$); SWAP-70 (PIP$_n$) +GAP: p190RhoGAP (pY by Src); ARAP (PH-PIP$_n$); Grit (recruitment by Trk); n-chimaerin (phosphatidylserine, PA) −GAP: BNIP-2 (pY by FGFR); n-chimaerin (lysophosphatidic acid, PI, AA) +GDI: RhoGDI (PKCα, fatty acids, phospholipids) −GDI: RhoGDI (ezrin-radixin-moesin binding)
Rab family	Rab, Ram, SEC4, YPT1	Vesicular trafficking; mitochondrial fission (Rab32)	−GTPase: Rab4 (MAPK); Rab3A (CaM) −GEF: GRAB (IP$_6$ kinase); Rab3-GEF (DD) +GAP: RN-tre (recruitment by EGFR-Eps8) ?GAP: AS160 (shift to cytosol by PKB) ±GDI: RabGDI (MAPK)^c
Ran family	Ran	Chromosomal condensation; nucleo-cytoplasmic transport; microtubule assembly (spindle)	+GEF: ARNO/cytohesin-1 (PH-PIP$_n$)
Arf family	ARF, CIN4, Sar	ER-Golgi-vesicle interactions; PLD; PI(4)P5 kinase	−GEF: ARNO/cytohesin-1 (PKA, PKC) +GAP: ArflGAP (PIP$_n$, DG); ARAP (PH-PIP$_n$)
G protein trimer	G$_s$, G$_i$, G$_q$, G$_{12/13}$	Hormone signal transduction	+GTPase: α_q (AGS1) −GTPase: α_{13} (PKA); α_i (PKC) +GEF: GPCR (ligand binding); mGluR5 (Homer) −GEF: GCR (GRK-arrestin binding) +GAP: RGS3 (PKG); RGS4 (PIP$_n$); RGS16 (pY by EGFR) −GAP: RGS2 (PKC); RGS7 (14-3-3 binding) +GDI: RGS14 (PKA)
Elongation factor	EF-Tu	Translation proofreading	
Signal recognition particle	SRP54	Transfer of proteins across the endoplasmic reticulum	

continued

continued

Group	Members	Function	Regulation (selected examples)[a]
Dynamin	Dynamin	Assembly of clathrin-coated vesicles; endocytotic neck constriction; vesicle movement	+GTPase: dynamin (PH-PIP$_n$) −GTPase: dynamin (PH-βγ; PKC)
ODN	Obg, DRG, NOG	Sensor for GTP/GDP ratios in stress and rRNA synthesis	

[a] The sign in front of the GTPase cycle component denotes stimulation (+) or inhibition (−) of that component. The regulation is given in parentheses. For example, p21^{ras} (a Ras GTPase) is stimulated by S-nitrosylation (NO); R-Ras (another Ras GTPase) is inhibited by tyrosine phosphorylation (pY) by the RTK, Eph; and K-Ras is inhibited by calcium-CaM binding.

[b] Allosteric activation.

[c] MAPK phosphorylation activates RabGDI; however, under some circumstances, this actually enhances Rab5 activity by speeding up recycling.

Table 9-3

Characteristics of the Covalent Lipid Modifications of Proteins

Characteristic	Modification			
	Myristoylation	Palmitoylation	Polyprenylation	Phosphatidylinositol glycan
Linkage	Amino-terminal amide	Thioester to cysteine[a]	Thioether to carboxy-terminal cysteine	Carboxy-terminal amide through phosphoethanolamine
Site of synthesis	Endoplasmic reticulum	Golgi complex and cytosol	Cytosol[b]	Endoplasmic reticulum[b]
Protein location	Plasmalemma and cytosol	Inner leaflet of plasma membrane (especially DIGs)	Plasmalemma and cytosol	Outer leaflet of plasma membrane
Turnover	Low	High	None except methylation	None except cleavage
Function	Amino-terminal blocker; reversible membrane association; protein association and folding; prequisite for palmitoylation	Membrane anchorage; receptor cycling; GRK activity	Membrane association; protein stability	Membrane anchorage (at cell apex and DIGs)

[a]Amino-terminal amide in a few proteins.

[b]Location of core modification. Additional modifications occur elsewhere.

DIGs, Detergent-insoluble, ganglioside-enriched complexes; GRK, G protein-coupled receptor kinase.

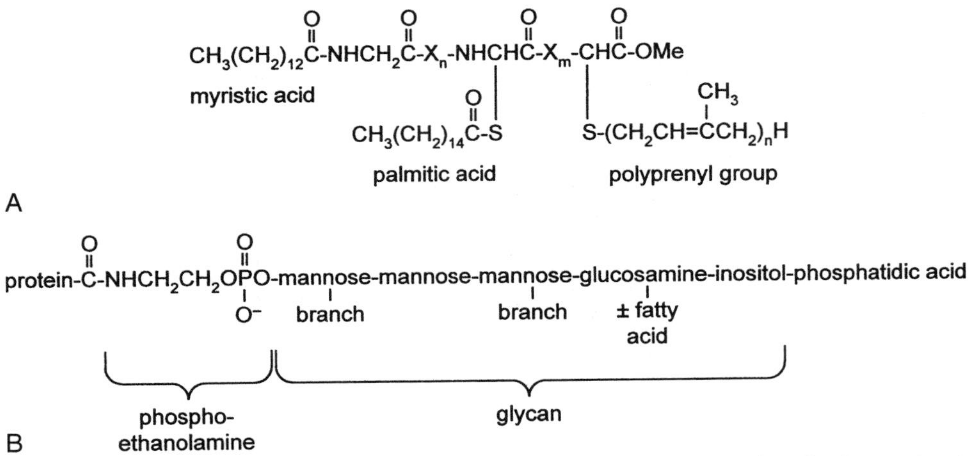

$$CH_3(CH_2)_{12}\overset{O}{\underset{\|}{C}}-NHCH_2\overset{O}{\underset{\|}{C}}-X_n-NHCHC-\overset{O}{\underset{\|}{C}}X_m-\overset{O}{\underset{\|}{C}}HC-OMe$$

myristic acid

$$CH_3(CH_2)_{14}\overset{O}{\underset{\|}{C}}-S$$

palmitic acid

$$S-(CH_2CH=\overset{CH_3}{\underset{}{C}}CH_2)_nH$$

polyprenyl group

A

$$protein-\overset{O}{\underset{\|}{C}}-NHCH_2CH_2O\overset{O}{\underset{\|}{P}}O-mannose-mannose-mannose-glucosamine-inositol-phosphatidic\ acid$$

O⁻   branch   branch   ± fatty
                                       acid

phospho-
ethanolamine

glycan

B

Fig. 9-3. Lipid modification of proteins. (A) a myristoylated, palmitoylated, and polyprenylated protein; (B) a protein attached to phosphatidylinositol (PI)-glycan.

ganglioside-enriched complexes). Unlike myristoylation, this modification has a high turnover, suggesting a regulatory role. Indeed, palmitic acid turnover is stimulated during GPCR desensitization (see Chapter 7), and this modification may determine the accessibility of adjacent serine-rich regions to kinases. Myristoylation of the amino terminus is usually a prerequisite for palmitoylation.

The other two lipid modifications are more complex. *Polyprenylation* refers to the attachment of a polyprenyl polymer, usually a farnesyl or a geranylgeranyl group. The former contains three isopentenyl subunits, whereas the latter has four. These polymers are generated by the polyprenyl or isoprenoid pathway, which also produces sterols; the latter is synthesized from a six-unit precursor that cyclizes to form the cyclopentanoperhydrophenanthrene nucleus. Proteins targeted for polyprenylation have the sequence C-A-A-X, where A is an aliphatic amino acid and X is any amino acid. If X is leucine, the cysteine receives a geranylgeranyl group; if it is serine, methionine, or asparagine, it will be farnesylated. Initially, the three carboxy-terminal amino acids are removed and the exposed cysteine is methylated. Then, the polyprenyl group is attached to the cysteine through a thioether linkage. In the small G proteins, there is often an adjacent cysteine that is palmitoylated; although the polyprenylation is not reversible, the methylation and associated palmitoylation do undergo turnover. For example, some small G proteins are maintained in the cytosol as an inactive complex with GDI; hormones that stimulate nucleotide exchange cause the complex to dissociate, and the newly exposed carboxy terminus of the G protein is carboxymethylated. Because the length of the polyprenyl chain is too short to induce membrane localization alone, the neutralization of the carboxy terminus by methylation provides the additional hydrophobicity to trigger the association of the G protein with the membrane, where many of its effectors reside.

The other modification is called *glypiation* and involves the coupling of phosphoinositide with the carboxy terminus of a protein via an oligosaccharide-phosphoethanolamine bridge (Fig. 9-3, *B*). First, the sugars and phosphoethanolamine in the main chain are added to phosphatidylinositol to form PI-glycan. All proteins that will be coupled with this anchor have a hydrophobic carboxy terminus preceded by a triplet of small amino acids. In the endoplasmic reticulum, this extension is cleaved and the preformed PI-glycan is attached via a pseudopeptidation reaction. In the Golgi complex, the oligosaccharide branches are added and the inositol may be acylated. The core sugar sequence shown in Fig. 9-3, *B*, is identical for all known PI-glycan anchors, but the sequence of the branches can vary considerably. This modification is used for membrane anchorage, and it is particularly prone to localize to DIGs. The modification can only be reversed by cleavage. This process is discussed further in Chapter 11.

Classification

Ras is a 21-kDa protein that is polyprenylated and palmitoylated near the carboxy terminus and myristoylated at the amino terminus. In yeast, Ras activates adenylate cyclase and mediates mating and nutrient transduction; in mammals, it is associated with mitogenesis (Fig. 9-4). First, Ras stimulates the MAPK (mitogen-activated protein kinase) cascade, which phosphorylates and activates several transcription factors essential for mitosis. Under basal conditions, Raf is phosphorylated on S-259 and S-621; the scaffold 14-3-3 exists as a dimer that can bind these two phosphorylated sites and keep Raf inactivated. Ras binds Raf, exposes S-259 to phosphatases, and localizes Raf to the plasma membrane. The dephosphorylation frees one arm of 14-3-3 to bind a yet unidentified activating kinase. Protein kinase C (PKC), p21-activated protein kinase (Pak), and soluble tyrosine kinases (STKs) have all been proposed to be the activating kinase. Indeed, even another Raf is a possibility because it is capable of autophosphorylation. Some of the confusion over identifying the responsible kinase may be a result of the need for multiple phosphorylations; for example, additional phosphorylations at both S-338 and Y-341 are required for Raf to bind MAPKK (mitogen-activated protein kinase kinase). This binding is facilitated by KSR, another scaffold, which binds Raf, 14-3-3, and MAPKK; the activated Raf then phosphorylates and activates MAPKK, which in turn phosphorylates and activates MAPK.

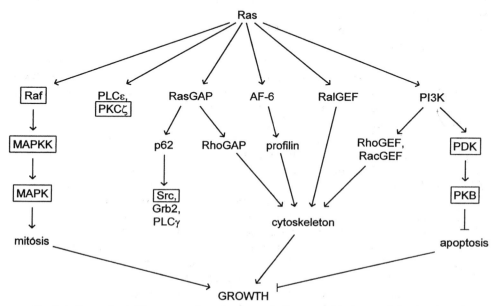

Fig. 9-4. Ras output. Kinases are boxed, arrows indicate stimulation, and flat heads indicate inhibition.

Growth also requires cytoskeletal remodeling. Ras directly binds RasGAP, AF-6, and RalGEF, which disrupt focal contacts and initiate migration. Finally, Ras directly binds and stimulates phosphatidylinositol-3 kinase (PI3K); the terminal kinase in this pathway, protein kinase B (PKB), is a major antiapoptotic factor (see Chapter 12). It is not surprising that constitutively active mutants of Ras are found in many cancers; it stimulates growth unresponsive to contact inhibition and the cells are resistant to apoptosis.

Normally, Ras is regulated at several sites (see Table 9-2). Ras itself may be the target of regulation: $p21^{ras}$ is stimulated by nitric oxide (NO) via S-nitrosylation of a cysteine, R-Ras is inhibited by tyrosine phosphorylation by a receptor tyrosine kinase (RTK), and K-Ras is inhibited by calcium-CaM. Alternately, a RasGEF may be under control; the major pathway for growth factors was outlined in Chapter 4 (Fig. 4-4). An activated RTK phosphorylates a docking protein that attracts the adaptor Grb2 via its SH2 domains. The SH3 domain of Grb2 then binds the polyproline helix in mSos, a RasGEF, and stimulates it. CNrasGEF has binding sites for cyclic nucleotides and can be stimulated by either cAMP or $3'5'$ cyclic GMP (cGMP); RasGRP is activated by calcium and DG; and RasGRF1/cdc25^{Mm} can be activated by calcium-CaM, tyrosine phosphorylation, or partial cleavage by calpain, a protease. Conversely, the latter can be inhibited by protein kinase A (PKA) phosphorylation, and mSos is inhibited by MAPK phosphorylation. Finally, Ras may be regulated through its GAP; Gap1^{IP4BP} is stimulated when its pleckstrin homology (PH) domain binds the phosphoinositol IP_4 (inositol 1,3,4,5-tetrakisphosphate); p120-GAP is inhibited by tyrosine phosphorylation; and RasGAP is inhibited when it binds polyphosphinositides (PPI).

An obvious question that arises is why does Ras have so many different GEFs and GAPs with so many different controls? First, the small G proteins are vital elements in cellular physiology and must be responsive to a large number of different inputs. Second, some components are tissue specific; for example, control by calcium-CaM is particularly common in nervous tissue because elevated calcium levels occur with electrical activity. Finally, different regulatory mechanisms have different effects with respect to time course and reversibility; for example, calpain cleavage of RasGRF1/cdc25^{Mm} is irreversible.

The other members of the Ras family have a 50% to 60% homology to Ras and are polyprenylated, except for the Rad and κB-Ras subfamilies. Details of the function and regulation of these small G proteins are given in Table 9-2; however, two features are worth mentioning here. First, this group illustrates some of the cross talk that occurs among the G proteins. TC21, a member of the Ras family, binds and activates a RalGEF, whereas Ras inhibits a RapGEF. Such interactions include the large G proteins; the α-subunit of a G protein trimer can bind and sequester Rap1GAP. Under basal conditions, β-arrestin, which is associated with G protein trimer signaling (see later) binds and sequesters RalGDS in the cytosol. After fMLP (formylmethionylleucylphenylalanine) stimulation, β-arrestin brings RalGDS to the plasma membrane and releases it. Second, some of these factors have CaM-binding sites, whose consensus sequence has a PKC phosphorylation site within it; phosphorylation at this site precludes CaM binding. This PKC-CaM antagonism is extremely common and is discussed further in Chapter 10. Here it can be seen in the Rad subfamily, which is inhibited by CaM; however, PKC phosphorylation stimulates this group by blocking CaM binding.

The Rho family is more distantly related to Ras and has a 30% homology to Ras. It is also polyprenylated and includes Rho, Rac, and Cdc42. This family is involved with cytoskeletal organization (Fig. 9-5); components of the cytoskeleton

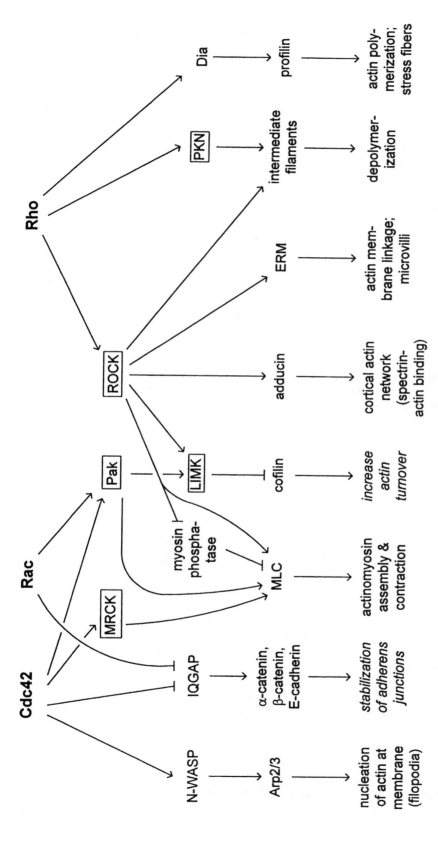

Fig. 9-5. Rho family output. Kinases are boxed, arrows indicate stimulation, flat heads indicate inhibition, and italics refer to net actions. For example, cofilin *stabilizes* actin, but because it is inhibited by LIMK, the net effect is an *increased* actin turnover. *ERM*, Ezrin-radixin-moesin; *IQGAP*, IQ (CaM-binding) and RasGAP domain containing protein; *LIMK*, LIM motif-containing protein kinase; *MLC*, myosin light chain; *MRCK*, myotonic dystrophy kinase related Cdc42-binding kinase; *Pak*, p21-activated protein kinase; *PKN*, novel PKC-like protein kinase; *ROCK*, Rho-associated coiled coil protein kinase; *N-WASP*, neural Wiskott-Aldrich syndrome protein.

are covered in Chapter 11. Rho activates Dia, which through profilin organizes actin and microtubules that are parallel to the cell's long axis and that end in focal adhesions. Rho also binds and stimulates the novel PKC-like protein kinase (PKN), which phosphorylates and dissembles intermediate filaments. Rho stimulation of the kinase ROCK (Rho-associated coiled coil protein kinase) facilitates formation of the submembrane actin network and actin membrane linkage. Rac activates Pak, which with ROCK can phosphorylate the myosin light chain and LIMK (LIM motif-containing protein kinase). The former is involved with actinomyosin assembly and contraction and the latter phosphorylates and inhibits cofilin. Cofilin stabilizes actin; its inhibition results in increased actin turnover. Cdc42 also activates Pak and a kinase that phosphorylates the myosin light chain. In addition, IQGAP (IQ and RasGAP domain-containing protein) binds β-catenin and E-cadherin, leading to the disruption of cell-cell adhesion. Cdc42 and Rac stabilize cell-cell adhesion by binding and inhibiting IQGAP. Finally, Cdc42 binds WASP (Wiskott-Aldrich syndrome protein), a scaffold, and exposes its carboxy terminus. This region of WASP binds Arp2/3, an actin polymerizer. In summary, the Rho family stimulates actin stress fibers, focal adhesion formation, and cytoskeletal turnover.

The Rho family regulation involves many of the second messengers just described. However, there is also a unique control mechanism: several RhoGEFs bind RTKs and are activated by ligand binding. Examples include PDZ-RhoGEF binding to insulin-like growth factor-I receptor (IGF-IR) and plexin B, and ephexin binding to EphA. In addition, this group has GDIs that can be regulated.

Like the Rho family, the Rab family has a 30% homology to Ras and is polyprenylated. However, its carboxy terminus may vary from the classic C-A-A-X motif, and this variation requires a unique set of enzymes for polyprenylation. Both the yeast (YPT1 and SEC4) and mammalian members (Rab and Ram) are involved with vescular trafficking between cell organelles and the plasmalemma. An important control seen in this group involves the partitioning of its components between the membrane and cytosol. For example, MAPK phosphorylation inhibits Rap4 by releasing it from membranes. Other unusual controls include the death domain in Rab3-GEF and the inositol hexakisphosphate kinase (IP$_6$K) inhibition of Rab3A. IP$_6$K competes with Rab3A for binding to its GEF, GRAB (guanosine nucleotide exchange factor for Rab3A).

The Ran family is unique in that it lacks any amino- or carboxy-terminal motifs associated with lipid modification. Ran prevents premature chromosomal condensation during mitosis.

The Arf family was initially discovered as accessory factors required for adenosine diphosphate (ADP)–ribosylation by various bacterial toxins; as such, they were called *ADP-ribosylation factors* (Arfs). They do not affect eukaryotic ADP-ribosylating enzymes and do not appear to have any GTPase activity without an ArfGAP. They are not polyprenylated but Arf is myristoylated, and they are intermediate in homology between Ras and the G protein trimers. They are major coat proteins of the Golgi complex stacks and the endoplasmic reticulum, where they play a role in budding and uncoating of transfer vesicles. The Arfs also activate several phosphoinositide kinases, and ArfGAP is required for GPCR internalization. The components of this group are primarily regulated by phosphoinositide binding.

Large G Proteins

Classically, the G protein trimers couple the GPCRs to their effectors; however, they can interact with many receptor groups (Chapter 7). The G protein is actually a heterotrimer with a subunit structure of $\alpha\beta\gamma$. The α-subunit is the GTPase and was originally thought to be primarily responsible for the biological activity of the G protein. As such, the classification of the G proteins parallels that of the α-subunits; that is, G proteins containing α_s are designated G_s and so on. There are four basic families. (1) α_s Stimulates adenylate cyclase, some ion channels, and several STKs. This group also includes α_{olf}, a G protein in the olfactory epithelium that is responsible for the sensory transduction of smell. (2) α_i Inhibits adenylate cyclase, stimulates Rap1-GTPase activating protein (RapGAP) and Src, and opens potassium channels. Other members of this family include α_z, α_o, α_t, and α_g. The latter two transmit sensory signals involved with light and taste perception, respectively; α_t is coupled to a cGMP-specific phosphodiesterase and α_g elevates cAMP in sweet receptors and calcium in bitter receptors. (3) α_q Activates phospholipase Cβ (PLCβ), the STK Bruton's tyrosine kinase (Btk), and some channels; α_{11}, α_{14}, α_{15}, and α_{16} are also in this group. (4) Finally, $\alpha_{12/13}$ is involved with the cytoskeleton, phosphorylation, and several miscellaneous functions. For example, $\alpha_{12/13}$ directly binds and activates radixin and RhoGEF, and it dissociates β-catenin from cadherin; these actions stabilize adherens junctions, induce actin attachment to membranes, and generate stress fibers. In addition, they bind and stimulate several STKs and at least one protein phosphatase (PP5). Finally, they can also stimulate the sodium-hydrogen exchanger 1 (NHE1), and α_{12} can activate phospholipase Cε (PLCε) and RasGAP. NHE1 regulates cell pH, which can function as a second messenger (see Chapter 11).

The protein α_s is 45-52 kDa; the larger variant is a result of alternate splicing of a single mRNA. The α-subunit is bilobed: one lobe resembles Ras, whereas the other lobe is unique. The GNP binds in the crevice between the two lobes. The carboxy terminus binds the hormone receptor and the effector, while the amino terminus binds the $\beta\gamma$-dimer. In α_i and α_o, the amino terminus is myristoylated; this modification enhances membrane binding, increases the affinity of the α-subunit for the $\beta\gamma$-dimer, and is required for coupling between the α_i and adenylate cyclase. Although it is not myristoylated, the α_s subunit is palmitoylated on the third residue; this fatty acid serves the same functions of membrane binding and coupling for α_s and α_q, as myristic acid does for α_i and α_o. Also, α_s binds aluminum fluoride (an exogenous activator) and is modified by cholera toxin. Cholera toxin removes nicotinamide from NAD^+ and attaches the rest of the molecule to an arginine (Fig. 9-6); this is called *mono(ADP-ribosyl)ation*. This amino acid is in the catalytic site, and its modification inactivates the GTPase. Because it is the hydrolysis of GTP that turns the system off, mono-(ADP-ribosyl)ation results in a persistently activated state. All of the members of

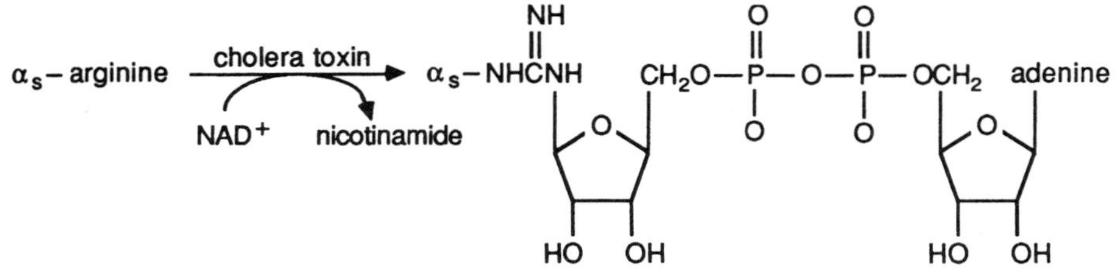

Fig. 9-6. Biochemical mechanism by which cholera toxin activates α_s.

the α_i family except α_z can undergo mono(ADP-ribosyl)ation by means of another bacterial toxin, pertussis toxin.

The β-subunit is a 35-kDa protein containing seven WD-40 repeats. These repeats are wedge shaped and form a torus known structurally as a *β-propeller*. The γ-subunit is 8.4 kDa and is polyprenylated; its amino terminus forms a coiled coil with the amino terminus of the β-subunit. The carboxy-terminal end of the third intracellular loop (IC3) in the GPCR binds the α-subunit at the carboxy terminus, which is in the larger lobe. IC3 also binds wedge 7 of the β-subunit, whose flat side lies against the smaller lobe of the α-subunit; that is, the GPCR is attached directly to the large lobe of α and indirectly through $\beta\gamma$ to the small lobe. Ligand binding to the GPCR results in rotation of the transmembrane helices, which in turn pry open the GNP site (similar to fingers prying open a clam) (Fig. 9-7). Because the concentration of GTP in the cell normally exceeds that of GDP, the former out-competes the latter, and the G protein will become activated.

Initially, it was thought that there was only a single $\beta\gamma$-dimer for all of the different α-subunits and that the dimer was not directly involved with mediating biological activity. However, both assumptions were proven incorrect. There are several isoforms for both of these subunits, and there appears to be some specificity in their association; for example, β_1 will bind γ_1 or γ_2, β_2 will only bind γ_2, and β_3 will bind γ_4. This coupling can also be receptor specific; the $\alpha o_1 \beta_3 \gamma_4$ mediates the inhibition of voltage-gated calcium channels by M4, whereas the $\alpha o_2 \beta_1 \gamma_3$ mediates the inhibition of these same channels by the somatotropin release-inhibiting factor (SRIF) receptor. In addition, certain $\beta\gamma$-isoforms can couple to specific effectors; for example, β_1 selectively couples to PLCβ.

The $\beta\gamma$-dimer serves several functions independent of the α-subunit. First, it facilitates membrane binding; for example, it helps to localize G protein-coupled receptor kinase (GRK), PLCβ, and STKs to the plasmalemma. Second, it is important in macromolecular assembly. It is obviously involved with the coupling of Gα and GPCR, but it also couples GRK to GPCRs and can act as a scaffold for Src and PLCγ. Furthermore, $\beta\gamma$ can facilitate microtubule assembly, activate a RasGEF, and open a potassium channel. The actions of the $\beta\gamma$-dimer are discussed in more detail later.

There is one final G protein that does not fit anywhere. G$_h$ is a multifunctional protein: it is a G protein, an NO reservoir, and a transglutaminase. There are

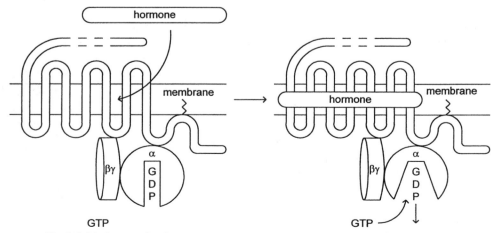

Fig. 9-7. Mechanism for G protein-coupled receptor (GPCR) activation of a G protein.

actually eight transglutaminases, but only the type 2 enzyme, also called G_h, is widely distributed and possesses GTPase activity. G_h is cysteine rich and can bind as many as eight NO molecules under basal conditions; this modification does not affect the enzymatic activity of G_h. Calcium elevation to about $1\,mM$ alters the conformation of G_h, leading to the release of NO. Calcium is also required for the transglutaminase activity. Transglutaminases catalyze the reversible formation of isopeptide bonds between glutamine and lysine, as well as the incorporation of amines into proteins. G_h can cross-link phospholipase A2 (PLA_2), a phospholipase used in signaling (see Chapters 4 and 10); Sp1, a transcription factor; and the hormone midkine. This modification increases the activity of all three proteins. It can also cross-link the latent transforming growth factor β (TGFβ)–binding protein and activate TGFβ. The transamidation function of G_h results in the incorporation of polyamines into eukaryotic initiation factor 5A (eIF-5A) and RhoA. The former changes the mRNA binding specificity of eIF-5A and the latter increases the ability of RhoA to bind and activate ROCK.

Nontransglutaminase activity is presumed to be a result of the G protein activity and includes the stimulation of PLCδ, a calcium-activated potassium channel, and protein synthesis; it also inhibits adenylate cyclase. The role of GTP is controversial; data from some studies suggest that GTP stimulates the function of G_h, whereas the results from other studies support an inhibitory role. This controversy may be the result of an indirect action of GTP. For example, G_h is reported to inhibit PLCδ and GTP causes the dissociation of G_h from this enzyme; that is, by *inhibiting* G_h, GTP *stimulates* PLCδ. A similar effect is seen with protein synthesis: G_h binds and sequesters the translation factor eIF-5A and GTP causes its dissociation.

Cyclic AMP Pathway

Components and Cycle

In the cAMP pathway, hormones stimulate or inhibit adenylate cyclase (AC) through G_s and G_i, respectively (the G proteins were previously discussed). In this section, the enzymes that produce and degrade cAMP are covered. At least nine ACs have been sequenced; they all have the same overall topology (Fig. 9-8, *A*). They can be divided into two highly homologous halves: each begins with six transmembrane helices, which are followed by the catalytic domain. Both catalytic sites are required for activity, although they do not have to be contiguous; that is, the molecule can be physically split in half and still be active as long as both halves are present. The cyclases fall into three groups: the first group, which includes AC I, AC III, and AC VIII, are CaM dependent (Fig. 9-9). Because AC I is also stimulated by α_s but inhibited by βγ, some researchers think that any G protein input cancels itself out and that calcium-CaM is the major activator of this group. AC III was originally identified in the olfactory epithelium, but it has subsequently been found in other tissues. AC II, AC IV, and AC VII form the second group, have shorter carboxy termini, and are primarily stimulated by α_s. This stimulation can be enhanced by βγ; AC II can also be activated by phosphorylation by PKC. Both of these groups can be inhibited by CaM-dependent protein kinase II (CaMKII) phosphorylation. AC V and AC VI, which comprise the third group, also have shorter carboxy termini but longer amino termini. They are primarily activated by α_s and inhibited by calcium. Furthermore, this group is

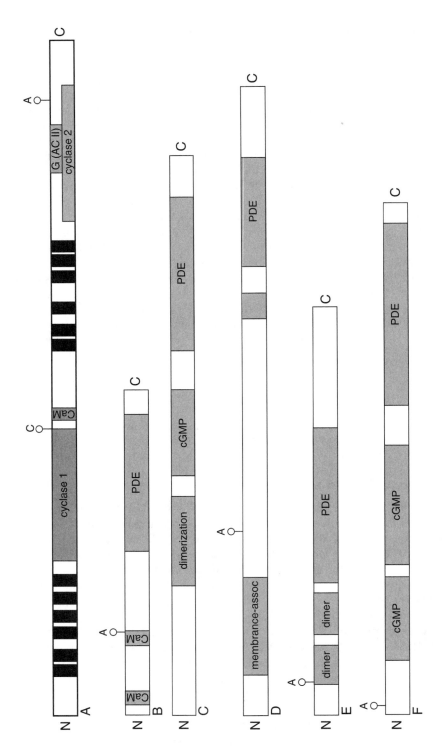

Fig. 9-8. Components of the cAMP cycle. (A) Adenylate cyclase (AC) and (B-F) various isozymes of the cyclic nucleotide phosphodiesterases. The calmodulin (CaM) domain would be found in the type I AC, while the G protein-binding domain would be found in the type II enzyme.

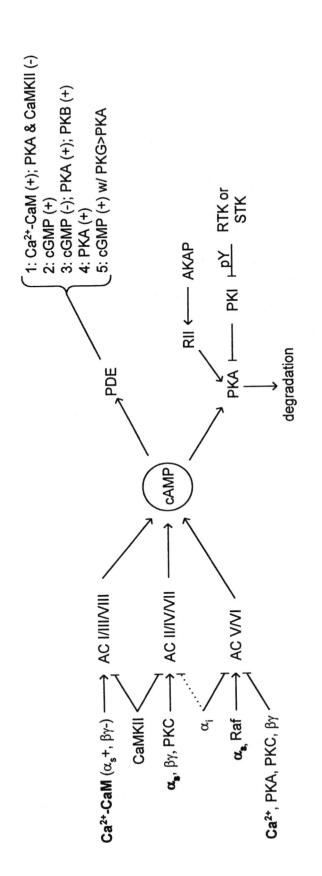

Fig. 9-9. cAMP metabolism. Adenylate cyclases (AC) are shown on the left; the factors in bold type are the major activators or inhibitors of the different cyclase families. The numbers in the upper right-hand corner refer to the phosphodiesterase (PDE) isozymes and their regulation (see also Table 9-4). Arrows indicate stimulation and flat heads indicate inhibition. *AKAP*, A kinase anchoring protein; *PKI*, protein kinase inhibitor; *RII*, type II regulatory subunit of PKA.

affected by phosphorylation; Raf stimulates and both PKA and PKC inhibit these cyclases. In addition, this last group is the most sensitive to the inhibitory effects of α_i, AC II from the second group is modestly suppressed by α_i, and AC I from the first group is barely affected.

The cyclic nucleotides are degraded by phosphodiesterases (PDEs). To date, there are 11 families, but only PDE1-6 are well characterized (Table 9-4). All PDEs have a homologous catalytic domain in the center or carboxy terminus of the molecule, and a regulatory site is usually located in the amino terminus (Fig. 9-8, B-F). PDE1 is stimulated by calcium-CaM but inhibited by phosphorylation by either PKA or the CaMKII. Because the CaM masks an autoinhibitory domain, proteolytic removal of this domain by calpain will also activate PDE1. PDE2 is allosterically activated by cGMP. This represents a rare direct action of cGMP; as noted later, most of the actions of cGMP are mediated by protein kinase G (PKG). PDE3 is stimulated by phosphorylation by either PKA or PKB. These 3 PDEs will use either cAMP or cGMP as a substrate. However, because the V_{max} of PDE3 for cGMP is so low, cGMP essentially becomes a competitive inhibitor of this family. Phosphodiesterase 4 (PDE4) degrades only cAMP; it is activated by PKA phosphorylation and by phosphatidic acid. MAPK phosphorylation inhibits the D3 isoform of PDE4. In addition, arrestin recruits PDE4 to the βAR to limit the generation of cAMP by this receptor. The cGMP-specific PDE uses only cGMP as a substrate and was originally thought to be restricted to the retina, where it is activated by transducin (G_t). However, it has recently been found in smooth muscle and platelets. This latter PDE is now known as PDE5, and the retinal form is called PDE6. PDE5 has an allosteric cGMP-binding site that acts as a costimulator with PKA or PKG phosphorylation.

Table 9-4
Cyclic Nucleotide Phosphodiesterases

Isoform[a]	Substrate	Regulation	Location
Calcium-CaM PDE (PDE1)	cGMP > cAMP	Ca^{2+}-CaM (+); calpain (+); PKA (−); CaMKII (−)	Nervous system
cGMP-stimulated PDE (PDE2)	cAMP > cGMP	cGMP (allosteric, +)	Adrenal cortex and brain
cGMP-inhibited PDE (PDE3)	cAMP > cGMP	PKA (+); PKB (+); cGMP (low V_{max} makes it a competitive inhibitor)	Smooth muscle, platelets, and heart
cAMP-specific PDE (PDE4)	cAMP	PKA (+); PDE4D3 by MAPK (−); PA (+); alternate splicing creates an SH3 binding site for STKs; recruited to βAR by arrestin	Ubiquitous
cGMP-specific PDE (PDE5)	cGMP	PKG > PKA (+); cGMP (allosteric costimulator)[b]	Smooth muscle and platelets
cGMP-specific PDE (PDE 6)	cGMP	α_t(+); cGMP (allosteric; increases GTPase activity of α_t)	Retina

[a] There are 11 families in all, but PDE7-11 are less well-known.

[b] PDE5 has a large cGMP-binding capacity; it may act to sequester cGMP or function as a cGMP reservoir.

$\beta_2 AR$, β_2 adrenergic receptor; *CaM*, calmodulin; *CaMKII*, CaM-dependent protein kinase II; *cAMP*, 3'5'-cyclic AMP; *cGMP*, 3'5'-cyclic GMP; *MAPK*, mitogen-activated protein kinase; *PA*, phosphatidic acid; *PDE*, phosphodiesterase; *PKA*, protein kinase A; *PKB*, protein kinase B; *PKG*, protein kinase G; *STK*, soluble tyrosine kinase.

Although PDEs have traditionally been considered a mechanism to terminate cAMP signaling, they can also act as effectors. In the juxtaglomerular cells of the kidney, NO stimulates renin secretion through PDE3: NO elevates cGMP, which then inhibits PDE3. As a result cAMP accumulates and triggers secretion; that is, PDE can actually be used to elevate cAMP in some systems.

These components interact in the following manner (Fig. 9-10). Hormones whose actions are mediated by elevated cAMP levels bind to their receptors. The hormone-receptor complex then binds to an inactive G_s; that is, a G_s whose α_s subunit is binding GDP instead of GTP. In some systems, the receptor may be precoupled to both G_s and AC; for example, 50% of the βARs are precoupled to G_s. This precoupling increases the affinity of the receptor for its ligand. As previously mentioned, when the hormone binds to its GPCR, the transmembrane helices rotate and pry open the GNP-binding crevice in α_s. Because of its higher cellular concentration, GTP will then displace GDP from α_s.

The next steps are controversial: according to conventional wisdom, α_s dissociates from the rest of the complex; however, a minority of researchers claim that the dissociation seen under experimental conditions is really an artifact. Regardless of the route, the α_s will now stimulate AC activity, but only as long as the GTP remains intact. Because α_s contains low GTPase activity, the GTP is slowly hydrolyzed to GDP, and α_s recombines with the β- and γ-subunits. Because of this automatic shut off system, cAMP generation is quickly terminated when hormones are removed. There are some analogs of GTP that have unusual bonds between the phosphates and that cannot be hydrolyzed by GTPases such as α_s. Such analogs permanently activate AC because the system cannot turn itself off. Cholera toxin has a similar effect by inactivating the GTPase (see preceding paragraphs).

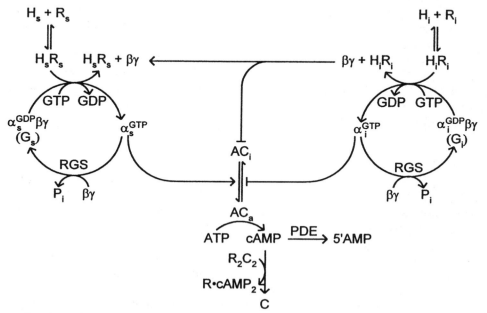

Fig. 9-10. The activation and inhibition of adenylate cyclase. AC_a and AC_i, Active and inactive adenylate cyclase, respectively; C, the catalytic subunit of the cAMP-dependent protein kinase; H_i inhibitory hormone (and its receptor, R_i); H_s, stimulatory hormone (and its receptor, R_s); R, the regulatory subunit of the cAMP-dependent protein kinase; RGS, negative regulators of G protein signaling. Other abbreviations are defined in the text and the legend to Fig. 9-1.

The α-subunit has a much higher intrinsic GTPase activity than the small G proteins do. For that reason, it was assumed that these proteins had no GAP or that a GAP-like domain had been incorporated into the larger α-protein. However, GAPs do exist for the α-subunit; they are called *negative regulators of G protein signaling* (RGS) (their control is discussed later). In addition, several effectors are known to accelerate GTP hydrolysis when they are bound to α. For example, PLCβ increases the GTPase activity of α_q, and a cGMP-specific phosphodiesterase will enhance GTP hydrolysis by α_t. Occasionally, this activity itself is regulated: the GAP activity in the amino terminus of Kir3, a G protein-regulated potassium channel, is only manifest after TrkB has phosphorylated Kir3. However, not all effectors have GAP activity; adenylate cyclase lacks such a function.

Hormones whose actions are mediated by lowering cAMP levels act in a similar manner, except that their hormone-receptor complexes bind G_i, which is then activated when GTP displaces the GDP on α_i. However, it is not clear how G_i inhibits AC; if the α_i system were analogous to the G_s system, α_i should directly inhibit AC, but the activity of α_i is very weak. The βγ-dimer is also an effective inhibitor, and this fact has led to the speculation that the released βγ may force α_s to reassociate into G_s, thereby terminating the AC stimulation. In other words, the activation of AC requires that G_s dissociate to liberate α_s:

$$\alpha_s\beta\gamma \rightleftharpoons \alpha_s + \beta\gamma$$

If the concentration of βγ can be increased from another source, such as G_i, then the laws of mass action dictate that the equilibrium would shift back to the left. However, there is one experiment in which data conflict with this hypothesis; the mono(ADP-ribosyl)ation of α_s by cholera toxin not only permanently activates α_s, but it also renders α_s unable to reassociate with βγ. Despite this, somatostatin-stimulated G_i is can still inhibit AC in cholera toxin–treated pituitary cells. This inhibition suggests that G_i does not act by promoting the reassociation of G_s in all systems. Another mechanism of inhibition involves the ability of βγ to inhibit directly two of the three cyclase families.

The α_i also has GTPase activity that is responsible for the termination of its action and that is augmented by separate RGS. This subunit is inhibited by pertussis toxin, which mono(ADP-ribosyl)ates α_i at a carboxy-terminal cysteine. In most tissues, G_i is present in a 4- to 10-fold excess over G_s. This excess is important in maintaining a low basal activity of adenylate cyclase.

Outputs

The best-known output of this system is the allosteric regulator, cAMP, which is generated by AC in response to α_s (Fig. 9-11). This modulator can activate cyclic nucleotide-gated (CNG) channels (see Chapter 7); several enzymes, including Ras-GEF and RapGEF; and cAMP-GEFII, which triggers exocytosis. It can also directly inhibit connexin, which is a component of the gap junctions between cells. However, the major target of this molecule is a cAMP-dependent protein kinase or PKA.

The PKA is a heterotetramer (R_2C_2), consisting of two regulatory subunits (R) that inhibit the two catalytic subunits (C). The amino terminus of R contains a dimerization domain, as well as a pseudosubstrate site, that occupies and inhibits the catalytic site of C (see Fig. 12-1, *A*). This region is then followed by two tandem, cAMP-binding sites. When cAMP binds the regulatory subunits, the tetramer dissociates and the liberated catalytic subunits become active:

$$R_2C_2 \text{ (inactive)} + 4cAMP \rightleftharpoons (R \cdot cAMP_2)_2 + 2C \text{ (active)}$$

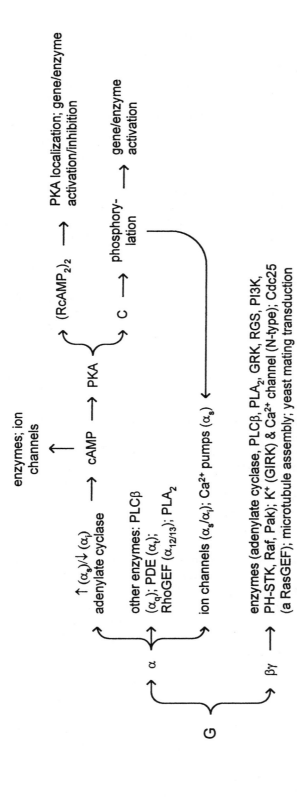

Fig. 9-11. Schematic representation of the various outputs of the cAMP pathway. *PLA₂*, phospholipase A₂; *PLCβ*, Phospholipase Cβ; *GRK*, GPCR kinase; *PH-STK*, STKs containing PH domains. Other abbreviations are defined in the text and in the legend to Fig. 9-10.

Activation also requires the autophosphorylation of C in the activation loop, turn motif, and the hydrophobic motif; these phosphate groups stabilize the kinase and properly orient the substrates. These effects lower the Michaelis-Menten rate constant (K_m) for ATP and the substrate. C is myristoylated; although this modification is unnecessary for either the localization or activation of PKA, it does provide greater thermal stability to the kinase. There are three isoforms of C; there are no known functional differences between Cα and Cβ, except for their tissue distribution. All three phosphorylate serines or threonines at the following consensus sequence: $(+)$-$(+)$-X-S/T, where $(+)$ is a basic amino acid, X is any amino acid, and S/T is either serine or threonine. PKA can phosphorylate an enormous range of substrates. However, Cγ differs slightly in substrate specificity and is less sensitive to a protein kinase inhibitor. The PKA phosphorylation of receptors has already been discussed in preceding chapters; other substrates include enzymes, the cytoskeleton, channels, and transcription factors.

There are two major regulatory subunits: RI and RII. RI has a high affinity for cAMP; this results in the rapid dissociation of the tetramer and a slow reassociation. It predominates in undifferentiated or embryonic tissues. RII has a lower affinity for cAMP; therefore dissociation is slow and reassociation is rapid. However, PKA II can autophosphorylate its RII subunit; this modification increases the affinity of RII for cAMP and decreases subunit reassociation. Conversely, cell cycle-dependent kinase 1 (Cdk1) can phosphorylate RII in the first cAMP-binding site; this modification decreases the affinity of RII for cAMP and favors the formation of the inactive tetramer. PKA II is more abundant than PKA I in differentiated tissues.

In addition to their structural differences, these two isoenzymes are regulated differently: in rat anterior pituitary cells, PKA I is activated more rapidly and with less growth hormone-releasing hormone (GHRH) than PKA II. In rat liver, a similar phenomenon is seen with glucagon. Data suggest that PKA I may be involved with acute responses, such as secretion or enzyme activation, whereas PKA II may trigger more prolonged effects, such as gene transcription. These types of regulations may be tissue specific; in normal osteoblasts, parathormone (PTH) activates PKA I and PKA II for the same length of time and with the same concentration dependency.

A final mechanism of regulation is localization. There are a large number of A kinase anchorage proteins (AKAPs) that tether PKA to specific sites within the cell (Table 9-5). This geographical distribution would restrict the kinds of substrates to which PKA could have access. AKAPs interact with PKA through R, and this association can be affected by phosphorylation. For example, Cdk2 phosphorylation of RII is required for binding to AKAP95 and AKAP450 but reduces the affinity of RII for microtubule associated protein (MAP2). Some AKAPs are also scaffolds; for example, AKAP250 appears to be involved with GPCR desensitization and internalization because it binds the β_2 adrenergic receptor (β_2AR), PKA, PKC, G protein-coupled receptor kinase 2 (GRK2), β-arrestin, clathrin, and actin. AKAP450, which anchors PKA to ryanodine receptors, also seems to be involved with signal termination because it binds PDE4D3, which hydrolyzes cAMP, and PP2A, which dephosphorylates PKA substrates.

Until recently, protein phosphorylation by the catalytic subunit was thought to be the only action exerted by PKA, but there is some evidence that the regulatory subunits may have functions independent of the catalytic subunit.

Table 9-5
Some A Kinase Anchorage Proteins and Their Anchorage Sites

Anchorage site	AKAP
Cytoskeleton	
Actin	Ezrin-AKAP78; AKAP200
Cortical actin	AKAP-KL
Microtubules	MAP2
Centrosome	AKAP350; pericentrin
Flagella	
Cytoskeleton	AKAP80
Fibrous sheath	AKAP82
Flagella (sperm)	AKAP110
Filopodia	Gravin-AKAP250
Membranes	
Sarcoplasmic reticulum	AKAP100 (to the ryanodine receptor)
Nuclear membrane	AKAP450
Plasma membrane	
Lateral membrane	AKAP18α
Apical membrane	AKAP18β
Postsynaptic densities	AKAP75/79 (to the GluR)
Calcium and sodium channels	AKAP15
NMDA	AKAP450; alternately spliced AKAP350
G protein	AKAP2
Organelles	
Golgi complex	AKAP85; AKAP350 (to the chloride channel)
Mitochondria	AKAP84; AKAP121/149[a]; Rab32
Peroxisomes	AKAP220
Acrosome	AKAP110
Nucleus	AKAP95

[a]AKAP121 is the murine homolog of the human AKAP149.

AKAP, A kinase anchorage protein; *GluR*, glutamate receptor; *MAP2*, microtubule-associated protein 2; *NMDA*, N-methyl-D-aspartate.

For example, RII has also been reported to inhibit the phosphoprotein phosphatase, an enzyme responsible for terminating the glycogenolysis induced by cAMP-stimulated phosphorylation (see Chapter 12). Consequently, RII would complement the kinase by preventing the dephosphorylation of its substrates. In addition, RI has been reported to bind and inhibit cytochrome c oxidase. More controversial is the possible role of RII in transcription. It is clear from several studies that phosphorylation by PKA is essential for cAMP-induced gene transcription (see Chapter 13), but whether RII contributes to this effect is unknown. RII can bind and inhibit the cAMP response element binding protein (CREB), which is also the target of PKA phosphorylation. However, the physiological relevance of RII binding is still unclear.

Until now, the section has been dominated by cAMP and PKA; this emphasis is only appropriate because PKA phosphorylation is an output of major importance. However, the α-subunit of G proteins can affect other activities (Fig. 9-11); for example, it can modulate the cGMP-specific PDE (see previous discussion) and the PLCβ and PLA$_2$ (see Chapter 10). The α-subunit can also directly affect many ion channels, including the calcium and potassium channels. Actually, these channels can often be regulated by the same pathway through different mechanisms; in endocrine and smooth muscle cells, the calcium channel is directly modulated by α, and in skeletal and cardiac muscle it is phosphorylated by PKA, which can also induce the gene for this channel.

Until recently, all of the output for the G proteins was thought to occur through the α-subunit, but there is now considerable evidence that the $\beta\gamma$-dimer can also have biological effects (Fig. 9-11). This dimer inhibits AC I and stimulates AC II and AC IV (see previous discussion) and activates RGS (see later), PI3K, PLCβ, and PLA$_2$ (Chapter 10). Activation of PLA$_2$ occurs in the retina. It can open potassium (GIRK) and calcium (Ca$_V$2.2) channels and activates several kinases, including GRK (see Chapter 7), Raf, Pak, and STKs containing PH domains. Although PH domains are more closely associated with PPI binding, a subset of these motifs binds $\beta\gamma$. Although $\beta\gamma$ stimulates Pak in yeasts, it inhibits it in mammals. Finally, the $\beta\gamma$-complex mediates both the pheromone induction of the mating response in yeast and the 1-methyladenine stimulation of oocyte meiosis in starfish. Some authorities have questioned the significance of these effects because of the high concentrations of $\beta\gamma$ required. In response, those favoring a physiological role for $\beta\gamma$ claim that this dimer can arise from many different G proteins and can, therefore, accumulate to high levels. In addition, it has been argued that the difference in effective concentrations between the α- and $\beta\gamma$-subunits is artifactual: many G protein effectors accelerate GTP hydrolysis and shorten the half-life of the active α-subunit. However, because α effects are usually measured in the presence of nonhydrolyzable analogs of GTP, the α activity is artificially prolonged, rendering the experimental system more sensitive to α. As such, the true effective concentrations of α and $\beta\gamma$ *in vivo* may not be that different.

Regulation

The regulation of the ACs and PDEs was discussed previously. In this section, control of the other components of this system is covered. G proteins can be regulated by (1) inhibitory proteins, (2) compartmentalization, (3) synthesis and degradation, and (4) covalent modification. Although α has a higher intrinsic GTPase activity than Ras alone, it is still less than that of Ras-GAP complex; to augment this activity, G protein trimers also have a GAP, which is called *RGS*. There are more than 30 different RGS that fall into six families. However, for simplicity, most RGS can be placed into three broad groups: those that only possess an RGS domain; those that contain both an RGS domain and one or more other motifs, such as a PDZ, DEP (disheveled/EGL-10 pleckstrin-related domain), Gγ-like (GGL), or phox homology (PX) domain; and those that are hybrid proteins. Examples of the latter include p115RhoGEF, GRK2, and PLCβ, which have RGS-like domains, but which also possess other activities.

RGS is the best-studied class of inhibitory proteins for G protein trimers. There are several ways that RGS can be regulated. First, RGS is inhibited by the palmitoylation of the G protein α subunit. Because the turnover of this fatty acid is accelerated after G protein activation, only activated G proteins would be susceptible to RGS inhibition. Second, RGS can be phosphorylated; PKG phosphorylation of RGS3 and RGS4 shifts them to the plasma membrane, where they have access to α, and tyrosine phosphorylation of RGS16 by epidermal growth factor receptor (EGFR) stimulates GAP activity. However, phosphorylated RGS7 is inhibited by its binding to 14-3-3, and tumor necrosis factor (TNF) activates RGS7 by stimulating its dephosphorylation. PKC phosphorylation of RGS2 and PKA phosphorylation of RGS10 are also inhibitory. RGS12 and RGS14 are unique among RGS in that they also have separate GDI activities; PKA phosphorylation near the GDI domain of RGS14 selectively enhances this GDI activity. Third, RGS can be regulated by allosterism; for example, RGS4 is inhibited by phosphatidic acid and PPI binding; the latter effect is reversed by calcium-CaM. Finally, RGS can be regulated by localization; RGS3 has a PDZ domain through which it binds EphB2. Ligand binding induces EphB2 clustering in DIGs, where RGS3 inhibits G proteins activated by chemotactic GPCRs.

The most recognizable functions of RGS are the maintenance of low basal G protein activity in the absence of stimulation and the termination of activity after stimulation. Although it has GAP activity, there are two other mechanisms it can use to terminate signaling. First, by binding G proteins, it can sequester them from effectors; for example, it can inhibit PLC by competing with the phospholipase for α_q. Second, RGS2 can directly inhibit AC III in olfactory epithelium. Paradoxically, RGS can also stimulate or modify G protein signaling. For example, by increasing the turnover of α, RGS allows for a more rapid response to the next signal. Those RGS with GGL domains can bind β and function as $\beta\gamma$-dimers; for example, the GGL domain can combine with β_5 to activate G protein-activated inwardly rectifying potassium (GIRK) channels. The GGL domain can mediate interactions with other proteins, as well; for example, it binds SCG10 in neurons. RGS can affect the spatial distribution of signaling: only those G proteins near receptors would become reactivated, whereas the activity of those G proteins further away would be suppressed. This mechanism of spatial focusing is particularly noticeable where receptors are highly concentrated, such as at synapses. Finally, one RGS family is also a RhoGEF; for example, the RGS domain in Lbc is used as an adaptor to bind α_q, which activates the GEF. Similarly, its splice variant, AKAP-Lbc, is activated by α_{12}. In addition to G protein-related functions, several other activities have been associated with RGS; some of those with multiple binding domains can act as scaffolds, and RGS12TS-S inhibits DNA synthesis and transcription by a repressor domain. Finally, RGD-PH1 and RGS19 are involved with intracellular trafficking.

While RGS targets the α-subunit, phosducin binds the $\beta\gamma$-dimer. In the retina and pineal gland, phosducin inhibits $\beta\gamma$ activity both by blocking effector binding and by interfering with $\beta\gamma$ recycling. Either PKA or CaMKII phosphorylation of phosducin can remove this inhibition. The phosducin-like protein is more widespread and serves a similar function. Its regulation by phosphorylation has not been determined.

Compartmentalization of G protein trimers is another regulatory mechanism. Early in the history of cyclic nucleotide research, the cell membrane was considered to be uniform with virtually no barriers to diffusion. G proteins were thought to form a common pool from which all receptors could compete. This view turns out to be overly simplistic; first, not all receptors have equal access to all G proteins. G_s activates both the plasmalemma calcium pump and an AC in liver; β-agonists can activate the G_s coupled to the cyclase but not that coupled to the pump. In addition, G proteins can be differentially localized to DIGs by palmitoylation or by caveolin binding; for example, caveolin binds α_q but not α_i.

Feedback regulation of G proteins often occurs by decreasing G protein number. For example, β-agonists, which activate G_s, decrease α_s 25% in lymphoma cells, while elevating α_i threefold; in other words, not only does the α_s pathway get dampened, but its antagonist, α_i, is also stimulated. The muscarinic acetylcholine receptor uses G_q in Chinese hamster ovarian (CHO) cells, and these agonists will downregulate G_q 40%. Heterologous regulation is also usually accomplished through altering G protein number. Many of the symptoms of hyperthyroidism are related to the overactivity of the βARs: rapid heart rate and elevated body temperature, for example. This relationship is partially a result of the induction of α_s by triiodothyronine (T_3); conversely, hypothyroidism increases α_i. Homologous regulation is often due to changes in G protein half-life, whereas heterologous regulation occurs through transcription.

Three types of covalent modifications have been described: (1) S-T phosphorylation, (2) tyrosine phosphorylation, and (3) mono(ADP-ribosyl)ation. The phosphorylation of α_i by either PKA or PKC impairs its ability to inhibit AC, and PKA phosphorylation of α_{13} inhibits its activation of Rho. On the other hand, PKC phosphorylation stimulates α_z by preventing its reassociation with βγ and its association with RGS; PKC also stimulates the activity of $\alpha_{15/16}$, a member of the α_q family. There are conflicting reports concerning the phosphorylation of α_s; this problem probably arises from the fact that there are two α_s isoforms. The longer form has an additional short sequence containing a PKC consensus site. RTKs and STKs can phosphorylate α_s, α_i, α_q, and α_o; the effects of this modification are also controversial. EGFR activates α_s in the heart but inhibits it in A431 cells. EGFR phosphorylates as many as two sites in α_s, and the differing effects may be due to the site of phosphorylation. Finally, other subunits can be phosphorylated: PKC phosphorylation of γ_{12} increases its affinity for α_o and AC but not for PLCβ.

As previously noted, several bacterial toxins can remove ADP-ribose from NAD^+ and attach it to G proteins. An endogenous enzyme with a similar catalytic activity and protein structure has been isolated. Because several of the bacterial toxins modify the G proteins of the cAMP system, it was suggested that the endogenous enzyme might function to modulate this system physiologically. This hypothesis was tested for human chorionic gonadotropin (hCG) in a Leydig tumor cell line, for luteinizing hormone (LH) in Leydig cells, and for thyroid-stimulating hormone (TSH) in a thyroid cell line; each stimulates AC in their respective target cells. However, in no system was the stimulation nicotinamide adenine dinucleotide (NAD)–dependent, and no mono(ADP-ribosyl)ation of G_s could be found. The conclusion was that none of these hormones acted by means of mono(ADP-ribosyl)ation.

In other systems, inhibitors of mono(ADP-ribosyl)ation blocked the effects of T_3 on oxidative phosphorylation and histamine on the production of eicosanoids. However, no data were given on the effects of T_3 or histamine on actual ADP-ribosylation levels or on enzymes involved in this modification. These crucial data were obtained in platelets, where prostacyclin treatment did result in the mono(ADP-ribosyl)ation of α_s, and this modification was associated with enhanced cAMP production. Furthermore, both prostaglandin E_2(PGE₂) and angiotensin II were shown to stimulate mono(ADP-ribosyl)ation activity in smooth muscle, and the selective inhibition of this modification blocked the effects of both hormones on DNA and RNA synthesis. Although the exact substrate of this enzyme was not identified, it appeared to be a component of the Rho pathway. Other substrates that have been identified for this enzyme include elements of the cytoskeleton, a calcium pump, and glutamate dehydrogenase; modifications of these proteins are usually inhibitory.

However, in most systems mono(ADP-ribosyl)ation does not appear to be the primary mechanism by which receptors activate G proteins; instead, it functions in secondary gain. This conclusion also comes from work in platelets, where thrombin mono(ADP-ribosyl)ates a G protein via PKC, an effector of thrombin that is downstream of the G protein. On the other hand, it has been shown that the βγ-dimer can be inhibited by mono(ADP-ribosyl)ation in a GPCR-dependent manner, and it was proposed that this modification was actually a mechanism for the termination of signaling.

Little work has been done on the regulation of mono(ADP-ribosyl)ation. Because the modification induced by thrombin requires PKC, one might speculate that the mono(ADP-ribosyl)transferase may be activated by PKC phosphorylation. In rods, NO can stimulate this transferase to modify α_s.

There are a few brief reports of other regulatory mechanisms of G proteins: α_s can be cleaved and activated by calpain; CaM can bind the amino terminus of β and block the reconstitution of the trimer, thereby prolonging activity; and hydrogen peroxide can covalently modify α_i and α_o, causing βγ to dissociate and activate PI3K, PKB, and MAPK. The modification is unknown but suspected to involve cysteine oxidation. The regulation of cAMP by PDEs was previously discussed; however, the role of PDE in compartmentalizing cAMP was not. In embryonic kidney cells, cAMP levels near the plasma membrane rise and fall rapidly because of a membrane-associated PDE. However, some cAMP escapes and diffuses to the cytosol where it can accumulate to produce a sustained elevation. The converse occurs in cardiac myocytes, where PDE keeps cAMP levels low outside the T-tubule area. Other factors affecting cyclic nucleotide levels are the pumps known as the multidrug resistance proteins (MRP): cAMP is pumped out of cell by MRP4, while cGMP is removed by MRP5.

PKA is primarily controlled by cAMP, the type of regulatory subunit, and anchoring proteins (see preceding text). However, there are two other regulatory mechanisms: (1) the protein kinase inhibitor (PKI) and (2) downregulation. PKI is a PKA pseudosubstrate whose levels can be regulated. For example, in the kidney, PTH, which acts through cAMP, elevates PKI 30% to 60%; this probably represents negative feedback. Conversely, both 1,25 dihydroxycholecalciferol (1,25-DHCC) and calcium lower PKI 90% to 95% in the kidney. In addition to the alteration of PKI levels, EGFR can phosphorylate PKI on tyrosine; this modification impairs PKI activity.

cAMP causes RI and RII to dissociate from C. Although this activates C, it also facilitates the downregulation of C because the free form is much more susceptible to degradation. Long-term exposure to cAMP can also elevate RI and RII; this drives the equilibrium back to the inactive tetramer.

Specificity

One final problem needs to be addressed: specificity. For simplicity, the presentation in this chapter implied that there was a unique and clear pathway from hormone receptor to G protein to effector. In truth, there is considerable overlap among these routes (Fig. 9-12). For example, many receptors can activate the same G protein; glucagon, vasopressin, and angiotensin all activate G_q in the liver. Conversely, one receptor can activate many G proteins; LH and PTH activate both G_s and G_q, M2 agonists activate G_q and G_i, and D2 is coupled to calcium channels by α_o and to potassium channels through α_{i3}. One G protein can induce many effectors; α_s can activate both AC and calcium channels. Finally, an effector can be stimulated by many G proteins; PLCβ can be activated by any member of the G_q family. An additional complication arises from the fact that some effectors, such as AC and PLCβ, can be affected by the βγ-dimer, as well as the α-subunit. How is signaling fidelity achieved?

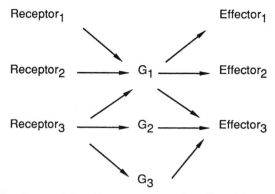

Fig. 9-12. Promiscuous interactions among receptors, G proteins, and effectors.

First, specificity can be receptor based; that is, it can be dependent on receptor concentration, isoform, phosphorylation, or subcellular localization. In the case where a receptor is coupled to multiple G proteins, there is often a preferred order related to the relative affinities among the various components. At low receptor concentrations, only the G protein having the highest affinity couples to the receptor. However, high receptor concentrations can compensate for low affinity. For example, at low concentrations, luteinizing hormone receptor (LHR), the vasopressin receptor 2 (V2), and βAR only couple to G_s, but at high concentrations they couple to both G_s and G_q. As discussed in Chapter 8, many receptors can undergo alternate splicing to generate multiple isoforms, each with unique coupling specificities. The reader is referred to the PGE receptor example therein. Receptor phosphorylation can also affect coupling. For example, $β_2AR$ is linked to both G_s and G_i, but PKA phosphorylation will impair signaling through G_s, while enhancing the coupling to G_i. The βγ from G_i will activate a STK-Ras-MAPK pathway, so that receptor phosphorylation results in a switch from PKA to MAPK. PKC phosphorylation of neurokinin A receptor induces a switch from G_q to G_s, and GRK phosphorylation of the platelet-activating factor (PAF) receptor shifts coupling from G_i and leukotriene C_4 (LTC_4) production to G_q and chemokine production. In some cases, phosphorylation simply expands the coupling rather than switching it: the prostacyclin receptor is coupled to G_s, but PKA phosphorylation allows coupling to G_i and G_q as well.

As previously noted, G proteins can be compartmentalized; as such, receptor localization can influence specificity. In renal cortical cells, parathyroid receptor (PTHR) in the basolateral membranes are coupled to G_s, but receptors in the apex are not. Localization in DIGs can also affect coupling choice.

Second, specificity can be hormone based; that is, it can be dependent on hormone concentration or structure. As with the receptor, the higher the hormone concentration, the more likely that coupling to low affinity G proteins will occur. For example, at the melanocortin receptor 3 (MCR3) the melanocyte-stimulating hormone (MSH) EC_{50} (effective concentration necessary to give 50% of the maximal effect) for G_q is $10^{-12}M$ and for G_s is $10^{-11}M$. Below $10^{-12}M$ MSH, MCR3 is coupled to G_q, but above $10^{-11}M$, it activates both G proteins. In another example, low PTH concentrations stimulate PKC, higher concentrations switch to PKA, and at still higher concentrations both kinases are activated.

In addition, different hormone structures can induce different receptor conformations; for example, fully glycosylated follicle-stimulating hormone (FSH) activates G_s, but incompletely glycosylated FSH activates G_i. Full-length neurokinin A produces a rapid elevation of calcium followed by a slow rise in cAMP; however, the truncated hormone only elicits the rapid elevation of calcium. Finally, wild-type LH is more potent at generating cAMP, whereas a variant LH with only one amino acid difference is more potent at stimulating PLC.

Third, specificity can be cell based; that is, it can be dependent on the cell cycle or cell type. Calcitonin (CT) stimulates the (Na^+, K^+)ATPase in G_2 via cAMP (G_s), but it inhibits this same enzyme in the S phase of mitosis via PKC (G_q). G_s is still present in S and can be activated by other agonists, but it is somehow uncoupled from the CT receptor. Similarly, vasopressin receptor 1a (V_{1a}) is coupled to G_i at interphase and to G_q during proliferation. Cell type is also important; in osteoblasts, PTH stimulates proliferation through G_q, but in osteoclasts, it stimulates bone resorption through G_s.

Fourth, specificity can be affected by accessory proteins, such as scaffolds. NHERF2 (sodium-hydrogen exchanger regulatory factor 2) has two PDZ domains: one binds the PTHR and the other binds PLCβ. When NHERF2 binds PTHR, it displaces G_s and favors coupling to G_i. The βγ from G_i then activates PLCβ. Essentially, NHERF2 switches PTHR from cAMP to PLCβ-calcium.

Finally, specificity can be affected by a variety of other factors, including combinatorial logic, temporal patterns, and feedback loops. Combinatorial logic refers to the fact that an effect is determined by the combination of pathways stimulated. For example, as noted previously, PTH can stimulate both calcium and cAMP; however, the simultaneous presence of insulin or low serum phosphate will selectively inhibit the calcium arm, whereas high serum phosphate will inhibit the cAMP branch. The duration of stimulation can also affect specificity; for example, the effects of PKA (G_s) are usually more prolonged than PKC (G_q). This appears to be a result of differences in GTPase kinetics, both intrinsic and RGS stimulated. In another example, endothelin uses G_{i3} for the initial, transient contraction of smooth muscle but G_{i1-2} for sustained contractions.

When a receptor is coupled to multiple G proteins, the products of one pathway can feedback and selectively uncouple that pathway. The metabotropic glutamate receptor (mGluR) can activate both G_q and G_s; PKC, a product of the G_q pathway, phosphorylates a threonine in the second intracellular loop (IC2). This phosphorylation disrupts coupling to the G_q but not G_s. In another example, α_{2A}AR is coupled to both G_q and G_i; again, PKC, a product of the former signaling pathway, phosphorylates a serine in the amino terminus of IC3, which selectively uncouples α_{2A}AR from G_q. Finally, β_2AR is coupled to both G_s and G_i. Receptor activation attracts β-arrestin, which then recruits PDE4; the resulting hydrolysis of cAMP shifts the output from G_s to G_i.

A final few comments are necessary about methodology and interpretation. Although multiple couplings appear to be real in many systems, they may be artifacts in others. These artifacts can arise from the techniques used to study signaling phenomena. First, a common method for studying transduction is to use a naive cell line and transfect it with genes coding for the signaling components of interest. It is often difficult to control accurately expression levels; indeed, it is frequently desirable to overexpress the signaling components so that they are easier to measure. However, overexpression can overcome low affinity (see discussion on receptor concentration); can overcome compartmentalization, for example, by saturating anchoring proteins; and does not mimic temporal organization. Second, nonprimary cell lines are not normal initially and can evolve during propagation. For example, PTH stimulation of AC in SaOS-2 cells increases with cell passage without a change in the amount or activity of either G_s or the AC; rather, there appears to be enhanced coupling between the PTH receptor and G_s. Third, *in vitro* experiments may omit *in vivo* modulators that add specificity, such as cofactors and scaffolds. Fourth, polymerase chain reaction, another common technique, can detect components not present at physiological levels. Finally, G protein-GPCR binding affinity does not always correlate with biological output. Therefore when the interaction among transduction elements is studied, it is important to mimic the physiological situation as closely as possible.

Cyclic GMP Pathway

The generation of cGMP has already been covered (see Chapter 7). Briefly, there are two synthetic pathways: one uses the membrane-bound guanylate cyclases and the other involves the soluble enzymes. The former cyclases are directly activated by ligand binding, whereas the latter are secondarily stimulated by hormones that elevate calcium, which with CaM triggers the synthesis of NO. It is NO that is the proximate activator of the soluble guanylate cyclases.

There are three major targets for cGMP: (1) two PDEs, (2) a protein kinase, and (3) ion channels. The effects of cGMP on several of the PDEs were previously discussed (see Table 9-4). Basically, the PDEs are one of the mechanisms through which cross talk between cGMP and cAMP takes place: cGMP stimulates PDE2, which would lower cAMP level, and it inhibits PDE3, which would have the opposite effect. There are two cGMP-dependent protein kinases, or protein kinase G (PKG): a ubiquitous, soluble form (PKG I) and a membrane-bound form (PKG II). They are very homologous to PKA; there is a dimerization domain in the amino terminus followed by two cGMP-binding sites and finally the catalytic site at the carboxy terminus (see Fig. 12-1). The consensus phosphorylation site for the two kinases is also the same: (+)-(+)-X-S/T. The major difference is that the protein is contiguous rather than being split into R and C subunits, as occurs with PKA. Another difference is that Src can phosphorylate a tyrosine in a cGMP binding site and activate PKG in a cGMP-independent manner. Palmitoylation of PKG II is responsible for its membrane association.

Alternate splicing of the amino terminus of PKG I gives rise to two forms: PKG Iα and PKG Iβ. PKG Iβ has a reduced affinity toward cGMP and may be activated by physiological concentrations of cAMP. Cross-stimulation can also occur after PKG autophosphorylation, which increases the affinity of PKG for cAMP. Conversely, high concentrations of cGMP can stimulate PKA. This is the second major mechanism for cross talk between the two pathways. Because both PKA and PKG have the identical phosphorylation consensus site and each can be stimulated by either cyclic nucleotide, attempts to elucidate a unique role for PKG have been difficult. It is most abundant in smooth muscle, where it induces relaxation by lowering intracellular calcium levels, and in certain parts of the central nervous system, where it is required for axon guidance and synapse formation.

Finally, cGMP can directly interact with several ion channels: the CNG channels. They are related to the voltage-gated ion channels (see Fig. 7-15, *A*); they are tetramers, whose subunits each have six transmembrane helices and a single transmembrane loop. There is a single cGMP-binding site in the carboxy terminus. The first CNG channel to be characterized was the cation channel of the photoreceptor; however, similar channels have been characterized from other tissues, although they differ in their ion specificity and cyclic nucleotide preference.

Cyclic CMP Pathway

Work on other cyclic nucleotides has been less fruitful. Cyclic CMP (cCMP) has been detected in tissues and is elevated in rapidly proliferating tissues; for example, levels in regenerating liver are 130 times those in normal liver. Furthermore, dibutyryl cCMP is mitogenic in both leukemic cells and in normal mammary epithelium. A cytidylate cyclase has also been identified and is similarly elevated in rapidly dividing cells. Finally, a very specific cCMP PDE has been partially purified; it does not hydrolyze cAMP or cGMP and is not inhibited by the methylxanthines. Its levels are low in rapidly proliferating tissues, and it can be inhibited by polyamines. Polyamines are associated with rapid growth (see Chapter 11); therefore, their inhibition of this enzyme would allow cCMP, a postulated mitogen, to accumulate and synergize with the polyamines and possibly other growth signals.

Transcellular Signaling

Second messenger signals can be propagated throughout a tissue in the absence of direct hormonal stimulation of each cell. With the use of radioautography to detect ^{125}I-labeled hCG binding and cytochemistry to detect activated PKA, receptor-kinase coupling could be studied in granulosa cells. Every cell showing hCG binding had activated PKA; every *isolated* cell without hCG binding did not have activated PKA. However, if a cell without hCG binding was in physical contact with one that did, then *both* cells had activated PKA. The hypothesis is that the cAMP generated in cells showing hCG binding is diffusing through gap junctions into the surrounding cells and activating their PKA.

This method of propagation also occurs for the other major transduction system, calcium. For example, in follicular cells, angiotensin II initiates calcium waves that are propagated to the adjacent oocyte. However, calcium has a very short range of diffusion, and it is not this cation that is passing through the gap junctions; instead, it is the second messenger IP_3 that diffuses through these pores and opens calcium reservoirs within each cell (see Chapter 10).

Gap junctions are formed when two hemichannels from adjacent cells come together. Each hemichannel is composed of six connexin proteins whose function is regulated by phosphorylation. In general, PKA phosphorylation and CaM binding increase permeability, whereas PKC, MAPK, and tyrosine phosphorylation decrease gating. Casein kinase 1 phosphorylation is required for gap junction assembly.

There are three other mechanisms for transcellular activation; the first involves the shedding of second messenger-laden packets. Activated platelets release microparticles rich in arachidonic acid; these particles are taken up by adjacent platelets, which use the arachidonic acid to synthesize thromboxane B (TXB) and prostaglandin I_2 (PGI$_2$). The second mechanism involves calcium receptors. One way to terminate a calcium signal is to pump the calcium outside of the cell (see Chapter 10). Adjacent cells possessing calcium receptors can detect this calcium efflux and also become activated. The third mechanism is called *transendocytosis*. Receptors are internalized into endosomes. They then form secondary vesicles within the endosome to create multivesicular bodies (see Fig. 7-17). However, rather than fuse with lysosomes and be destroyed, these exosomes are shed by exocytosis and internalized by adjacent cells where they can exert biological activity and/or be passed on to other cells. These mechanisms help to coordinate all of the cells of a particular tissue.

Summary

GTP-binding proteins, or G proteins, are molecular switches with built-in or closely associated automatic timers. They are activated when GTP replaces the GDP and are inactivated when the intrinsic GTPase activity hydrolyzes the GTP. Most of the small G proteins are involved with vesicular trafficking and cytoskeletal organization, but Ras has been strongly implicated in the mitogenic signal transduction of several growth factors. Hormones can control the small G proteins through second messengers acting directly on the G protein itself or on their GEFs, GAPs, or occasionally other components of their cycle.

The large G proteins are trimers. The α-subunit is the GTPase and designates which signal pathway will be affected; α_s and α_i alter AC activity and ion channels, α_q activates the phospholipid-calcium pathway, and $\alpha_{12/13}$ is involved with the cytoskeleton. The $\beta\gamma$-dimer has its own set of effectors, including enzymes, ion channels, and the cytoskeleton. PKA is a major target of cAMP and can have both acute effects by phosphorylating enzymes and ion channels and long-term effects by phosphorylating transcription factors and inducing genes. There is considerable overlap in the coupling of receptors, G proteins, and effectors; the exact biological activity produced is influenced by the receptor and agonist properties, cell type and proliferative state, scaffolds, duration of stimulation, presence of other signals, and feedback loops.

cGMP is produced by membrane-bound or soluble guanylate cyclases; the former are hormone receptors, whereas the latter are second messenger receptors for NO and carbon monoxide. cGMP can affect PDEs, ion channels, and PKG. cCMP is the newest and least studied member of the cyclic nucleotides. Although it is associated with proliferation, an unequivocal role in hormone action has not been demonstrated.

Finally, signals can diffuse through gap junctions and spread throughout a tissue without the hormone having to contact each and every cell. This phenomenon has been demonstrated for both cAMP and inositol 1,4,5-trisphosphate, a calcium-releasing second messenger.

References

G Protein Components and Cycle

Bernards, A. (2003). GAPs galore! A survey of putative Ras superfamily GTPase activating proteins in man and *Drosophila*. *Biochim. Biophys. Acta* 1603, 47-82.

Donovan, S., Shannon, K.M., and Bollag, G. (2002). GTPase activating proteins: Critical regulators of intracellular signaling. *Biochim. Biophys. Acta* 1602, 23-45.

Jackson, C.L., and Casanova, J.E. (2000). Turning on ARF: The Sec7 family of guanine-nucleotide-exchange factors. *Trends Cell Biol.* 10, 60-67.

Moon, S.Y., and Zheng, Y. (2003). Rho GTPase-activating proteins in cell regulation. *Trends Cell Biol.* 13, 13-22.

Peck, J., Douglas, G., Wu, C.H., and Burbelo, P.D. (2002) Human RhoGAP domain-containing proteins: Structure, function and evolutionary relationships. *FEBS Lett.* 528, 27-34.

Quilliam, L.A., Rebhun, J.F., and Castro, A.F. (2002) A growing family of guanine nucleotide exchange factors is responsible for activation of Ras-family GTPases. *Prog. Nucleic Acid Res. Mol. Biol.* 71, 391-444.

Schmidt, A., and Hall, A. (2002). Guanine nucleotide exchange factors for Rho GTPase: Turning on the switch. *Genes Dev.* 16, 1587-1609.

Spang, A. (2002). ARF1 regulatory factors and COPI vesicle formation. *Curr. Opin. Cell Biol.* 14, 423-427.

Sprang, S. (2001). GEFs: Master regulators of G-protein activation. *Trends Biochem. Sci.* 26, 266-267.

Small G Proteins

Bishop, A.L., and Hall, A. (2000). Rho GTPases and their effector proteins. *Biochem. J.* 348, 241-255.

Dasso, M. (2002). The Ran GTPase: Theme and variations. *Curr. Biol.* 12, R502-R508.

Kaibuchi, K., Kuroda, S., and Amano, M. (1999). Regulation of the cytoskeleton and cell adhesion by the Rho family GTPases in mammalian cells. *Annu. Rev. Biochem.* 68, 459-486.

Rebollo, A., and Martinez-A, C. (1999). Ras proteins: Recent advances and new functions. *Blood* 94, 2971-2980.

Ridley, A.J. (2001). Rho GTPases and cell migration. *J. Cell Sci.* 114, 2713-2722.

Rodman, J.S., and Wandinger-Ness, A. (2000). Rab GTPases coordinate endocytosis. *J. Cell Sci.* 113, 183-192.

Sever, S. (2002). Dynamin and endocytosis. *Curr. Opin. Cell Biol.* 14, 463-467.

Takai, Y., Sasaki, T., and Matozaki, T. (2001). Small GTP-binding proteins. *Physiol. Rev.* 81, 153-208.

Takuwa, N., and Takuwa, Y. (2001). Regulation of cell cycle molecules by the Ras effector system. *Mol. Cell. Endocrinol.* 177, 25-33.

Yamamoto, T., Taya, S., and Kaibuchi, K. (1999). Ras-induced transformation and signaling pathway. *J. Biochem. (Tokyo)* 126, 799-803.

Zwartkruis, F.J.T., and Bos, J.L. (1999). Ras and Rap1: Two highly related small GTPases with distinct function. *Exp. Cell Res.* 253, 157-165.

Large G Proteins

Fesus, L., and Piacentini, M. (2002) Transglutaminase 2: An enigmatic enzyme with diverse functions. *Trends Biochem. Sci.* 27, 534-538.

Lorand, L., and Graham, R.M. (2003). Transglutaminases: Crosslinking enzymes with pleiotropic functions. *Nature Rev. Mol. Cell Biol.* 4, 140-156.

Schwindinger, W.F., and Robishaw, J.D. (2001). Heterotrimeric G-protein βγ-dimers in growth and differentiation. *Oncogene* 20, 1653-1660.

Vanderbeld, B., and Kelly, G.M. (2000). New thoughts on the role of the βγ subunit in G protein signal transduction. *Biochem. Cell Biol.* 78, 537-550.

cAMP Components and Cycle

Francis, S.H., Turko, I.V., and Corbin, J.D. (2001). Cyclic nucleotide phosphodiesterases: Relating structure and function. *Prog. Nucleic Acid Res. Mol. Biol.* 65, 1-52.

Hanoune, J., and Defer, N. (2001). Regulation and role of adenylyl cyclase isoforms. *Annu. Rev. Pharmacol. Toxicol.* 41, 145-174.

Mehats, C., Andersen, C.B., Filopanti, M., Jin, S.L.C., and Conti, M. (2002). Cyclic nucleotide phosphodiesterases and their role in endocrine cell signaling. *Trends Endocrinol. Metab.* 13, 29-35.

cAMP Outputs

Antoni, F.A. (2000). Molecular diversity of cyclic AMP signaling. *Front. Neuroendocrinol.* 21, 103-132.

Diviani, D., and Scott, J.D. (2001). AKAP signaling complexes at the cytoskeleton. *J. Cell Sci.* 114, 1431-1437.

Feliciello, A., Gottesman, M.E., and Avvedimento, E.V. (2001). The biological functions of A-kinase anchor proteins. *J. Mol. Biol.* 308, 99-114.

Kaupp, U.B., and Seifert, R. (2002). Cyclic nucleotide-gated ion channels. *Physiol. Rev.* 82, 769-824.

Mark, M.D., and Herlitze, S. (2000). G-protein mediated gating of inward-rectifier K^+ channels. *Eur. J. Biochem.* 267, 5830-5836.

Michel, J.J.C., and Scott, J.D. (2002). AKAP mediated signal transduction. *Annu. Rev. Pharmacol. Toxicol.* 42, 235-257.

cAMP Regulation

Chen, C.A., and Manning, D.R. (2001). Regulation of G proteins by covalent modification. *Oncogene* 20, 1643-1652.

Hollinger, S., and Hepler, J.R. (2002). Cellular regulation of RGS proteins: Modulators and integrators of G protein signaling. *Pharmacol. Rev.* 54, 527-559.

Ross, E.M., and Wilkie, T.M. (2000). GTPase-activating proteins for heterodimeric G proteins: Regulators of G protein signaling (RGS) and RGS-like proteins. *Annu. Rev. Biochem.* 69, 795-827.

Sierra, D.A., Popov, S., and Wilkie, T.M. (2000). Regulators of G-protein signaling in receptor complexes. *Trends Cardiovasc. Med.* 10, 263-268.

Steinberg, S.F., and Brunton, L.L. (2001). Compartmentation of G protein-coupled signaling pathways in cardiac myocytes. *Annu. Rev. Pharmacol. Toxicol.* 41, 751-773.

G Protein Specificity

Albert, P.R., and Robillard, L. (2002). G protein specificity: Traffic direction required. *Cell. Signal.* 14, 407-418.

Dumont, J.E., Dremier, S., Pirson, I., and Maenhaut, C. (2002). Cross signaling, cell specificity, and physiology. *Am. J. Physiol.* 283, C2-C28.

Hur, E.M., and Kim, K.T. (2002). G protein-coupled receptor signalling and cross-talk: Achieving rapidity and specificity. *Cell. Signal.* 14, 397-405.

Tucek, S., Michal, P., and Vlachov, V. (2002). Modelling the consequences of receptor-G-protein promiscuity. *Trends Pharmacol. Sci.* 23, 171-176.

Cyclic GMP Pathway

Hofmann, F., Ammendola, A., and Schlossmann, J. (2000). Rising behind NO: cGMP-dependent protein kinases. *J. Cell Sci.* 113, 1671-1676.

Transcellular Signaling

Cruciani, V., and Mikalsen, S.O. (2002). Connexins, gap junctional intercellular communication, and kinases. *Biol. Cell* 94, 433-443.

Evans, W.H., and Martin, P.E.M. (2002). Gap junctions: Structure and function. *Mol. Membr. Biol.* 19, 121-136.

González-Gaitán, M. (2003). Signal dispersal and transduction through the endocytic pathway. *Nature Rev. Mol. Cell Biol.* 4, 213-224.

Jongsma, H.J., van Rijen, H.V.M., Kwak, B.R., and Chanson, M. (2000). Phosphorylation of connexins: Consequences for permeability, conductance, and kinetics of gap junction channels. *Curr. Top. Membr.* 49, 131-144.

Lampe, P.D., and Lau, A.F. (2000). Regulation of gap junctions by phosphorylation of connexins. *Arch. Biochem. Biophys.* 384, 205-215.

CHAPTER 10

Calcium, Calmodulin, and Phospholipids

The second major transduction system is the polyphosphoinositide (PPI) pathway. Very briefly, phosphoinositide (PI) is phosphorylated on positions 4 and 5 to become PPI (Fig. 10-1); this in turn is hydrolyzed by a hormone-dependent phospholipase C (PLC). Two second messengers are produced: a phosphorylated head group (inositol 1,4,5-trisphosphate or IP_3) and diacylglycerol (DG). Inositol 1,4,5-trisphosphate elevates cellular calcium levels, and DG stimulates a calcium-activated, phospholipid-dependent protein kinase (protein kinase C or PKC). In addition, phosphatidylinositol-3 kinase (PI3K) can phosphorylate the 3 position. This phospholipid cannot be hydrolyzed by any known PLC; rather, the intact molecule activates another protein kinase, PKB, as well as several GEFs (GNP exchange facilitators). Finally, phospholipase A_2 (PLA_2) can release the second fatty acid, which is predominantly arachidonic acid, a precursor to the eicosanoids (see Chapter 3). This overall scheme is depicted in Figs. 4-2 and 4-3.

Criteria for Second Messengers

Satisfying Sutherland's postulates for the PPI pathway is much more difficult than for the 3'5'-cyclic AMP (cAMP) pathway for several reasons: (1) there are many more mediators generated; (2) some of these mediators are produced at the end of a long chain of reactions; and (3) one of these mediators is an ion that is more difficult to manipulate than a small organic molecule. Nonetheless, with care and diligence it can be done for at least some of the second messengers in this pathway, such as calcium:

1. The mediator, or its analog, must mimic the action of the hormone. There are several ways that calcium can be introduced into the cell; the simplest uses the calcium ionophore, A23187 (Fig. 10-2). This compound is a hydrophobic calcium chelator; it binds these divalent cations and shuttles them across the plasmalemma from the medium into the cytosol. If one needs more temporal or spatial control, "caged" calcium can be used. This is another calcium chelator that is sensitive to ultraviolet light. The calcium remains tightly bound to the chelator until it is exposed to light, which isomerizes the molecule, thereby disorienting the chelating groups and releasing the calcium. When used with a laser beam, calcium can be elevated in very discrete locations within the cell. Finally, calcium can

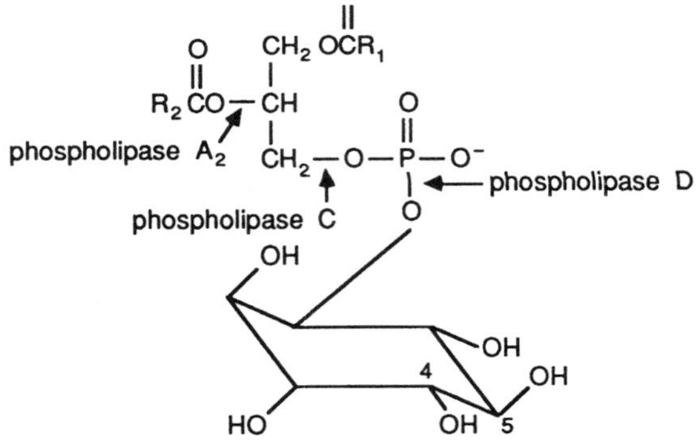

Fig. 10-1. Chemical structure of phosphoinositide showing the bonds hydrolyzed by the different phospholipases.

be released from internal stores by the natural mediator IP_3. Because of the size and charge of this compound, plasma membranes must be permeabilized with detergents to introduce it into cells. However, it also comes in a hydrophobic, caged form that readily passes through the plasmalemma and releases IP_3 after photolysis.

2. The hormone must induce elevated levels of the mediator. There are four methods for measuring calcium levels in response to hormones: (1) calcium-specific electrodes, (2) fluxes of calcium radioisotopes, (3) fluorescent calcium-sensitive proteins, and (4) light-emitting, calcium-sensitive proteins. Both electrodes and fluxes measure cellular calcium changes indirectly. As is discussed later, initial alterations in cellular calcium levels are usually due to an internal redistribution; exchanges with the extracellular environment, which are detected by the electrodes and fluxes, are secondary events. The other two methods directly determine intracellular calcium concentrations. There are several compounds, such as quin 2, that fluoresce when their acidic groups chelate calcium; the degree of fluorescence is proportional to the calcium concentration. These binding groups are esterified so that the compounds can traverse the plasma membrane; once inside, nonspecific esterases remove the blocking groups and the compounds become trapped (Fig. 10-2). The light-producing proteins, such as aequorin, act in a related manner. Aequorin, which is responsible for bioluminescence in certain jellyfish, is a 20-kDa protein containing an organic prosthetic group called *luciferin*. This complex emits light in response to elevated calcium levels; again, the intensity is proportional to the calcium concentration. The disadvantage of this method is that either the protein has to be microinjected or its gene has to be transfected into cells.

3. The hormone must appropriately affect the enzymes of synthesis and/or degradation of the mediator. Obviously, calcium is an element: it cannot be "synthesized" or "degraded." However, hormones do activate a PLC that releases IP_3; as noted in Chapter 7, the IP_3 receptor (IP_3R) is a calcium channel that releases internal stores of this ion. As such, PLC activity can be used as a surrogate for "calcium synthesis." In addition, calcium is removed by adenosine triphosphatase (ATPase) pumps, some of which are hormone regulated.

4. An appropriate temporal relationship must exist among the hormone, mediator, and hormonal effect. The normal course of events is to observe an increase in PLC activity, elevated IP_3, elevated calcium, and then activation of some downstream effector.

5. Finally, if drugs are available to modulate the endogenous level of the mediator pharmacologically, they should also mimic or inhibit, as appropriate, the effects of the hormone. Again, calcium is not synthesized or degraded; rather, it is shifted from one compartment to another through channels and pumps. Therefore these membrane proteins become the targets for pharmacological manipulations. As noted previously, calcium ionophores can be used to elevate intracellular calcium levels. On the other hand, calcium levels may be lowered by EGTA (ethylene glycol-bis[β-amino ethyl]-N, N, N', N-tetraacetic acid), a chelating agent relatively specific for calcium. Cellular depletion of calcium is often accelerated with a combination of both EGTA and an ionophore; because of the

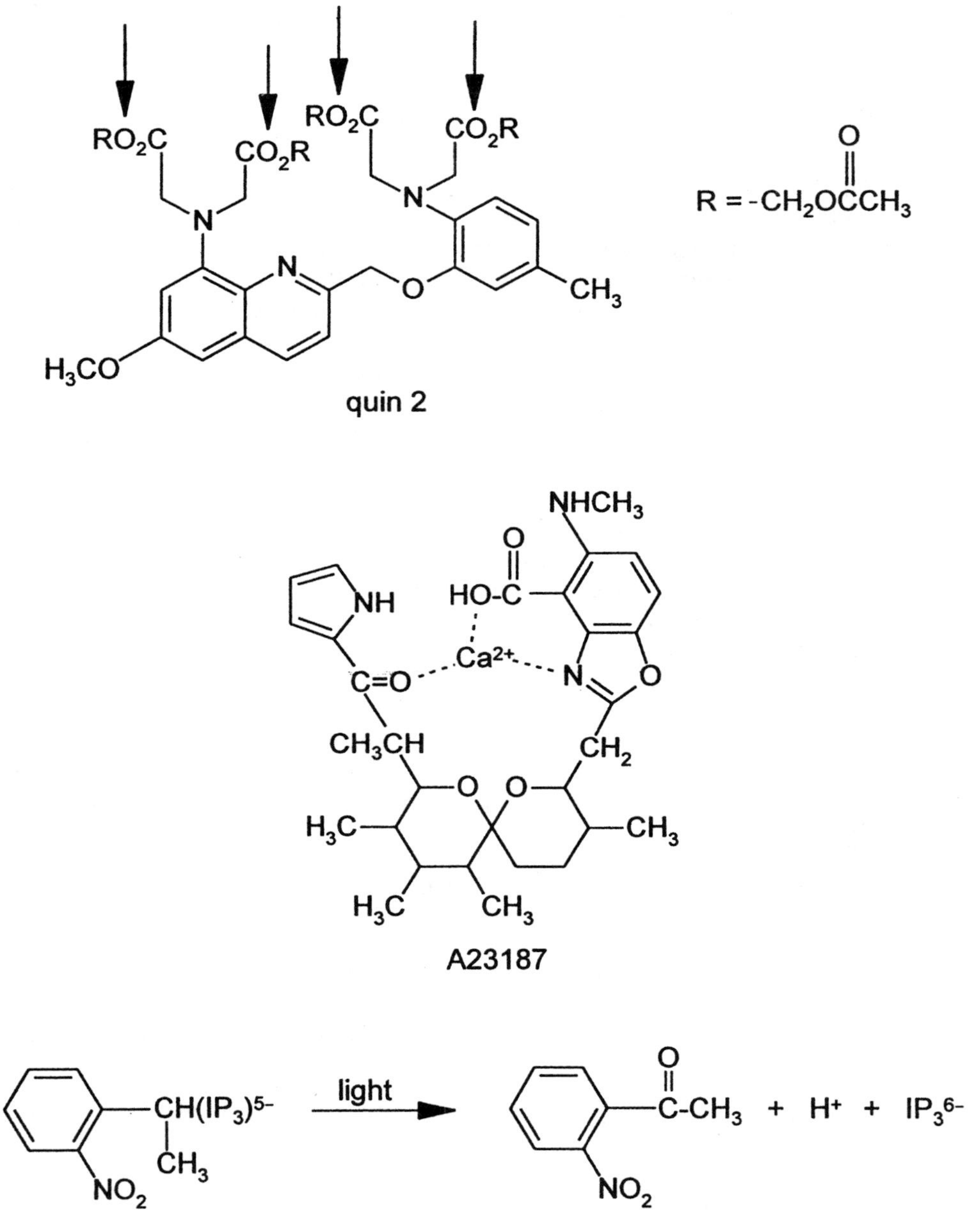

quin 2

$$R = -CH_2O\overset{\displaystyle O}{\overset{\displaystyle \|}{C}}CH_3$$

A23187

Fig. 10-2. Chemical structures for quin 2, a calcium-triggered fluorescent compound; A23187, a calcium ionophore shown with bound calcium; and caged IP₃, which releases IP₃ after photolysis. The arrows indicate the quin 2 calcium binding sites, which are exposed after the blocking groups have been removed.

EGTA in the medium, the calcium concentration is now reversed and the ionophore will carry calcium out of the cell. Caged chelators that are activated by light are also available for greater temporal and spatial control. Finally, one can deplete cells of calcium by blocking the calcium channel. There are four pharmacological classes of antagonists, each of which binds to one of several sites on the channel; the prototypical drugs are verapamil, nifedipine, diltiazem, and lidoflazine. The first drug appears to be the most popular for studies on the PPI pathway.

PLC can be inhibited by either neomycin or U73122. The mechanism of neomycin is actually indirect: it binds PPI and prevents PLC access to its substrate. Unfortunately, it also inhibits another phospholipase, PLA_2. U73122 is a steroid derivative with no reported extraneous effects; however, all drugs have side effects, and care should be exercised whenever any of them are used in experiments.

Two enzymes involved with recycling are also targets of drugs. Once the IP_3 is released, it is progressively dephosphorylated before being recycled; lithium salts block the final dephosphorylation, thereby inhibiting the pathway at the recycling step. Unfortunately, lithium also inhibits the glycogen synthase kinase and interferes with the binding of guanosine 5'-triphosphate (GTP) to the G proteins. Finally, DG kinase phosphorylates DG in preparation for its recycling. It can be inhibited by R59949, but this compound also inhibits PKC and PLA_2.

There are three other mediators that have received considerable attention: (1) calmodulin, a calcium-binding protein that can affect the activity of many enzymes, (2) PKC, and (3) PI3K, which generates the second messenger $PI(3)P_n$. Calmodulin (CaM) presents an additional problem to satisfying Sutherland's postulates: hormones do not acutely change total CaM levels. Rather, hormones elevate calcium that binds to and activates CaM. However, CaM can be covalently tagged by calcium-sensitive fluorescent probes that will emit light only when the calcium-binding sites are occupied and the CaM is active. CaM can be inhibited by compounds that bind to its central hydrophobic helix; these drugs include the phenothiazines and the naphthalene sulfonamides.

PKC can be specifically activated by TPA (12-*O*-tetradecanoylphorbol-13-acetate) and other related phorbol esters. However, there are at least seven PKC isoforms (see later), and their sensitivities to TPA differ. PKCε is the most sensitive, being stimulated by only 0.01 to 0.1 nM TPA; PKCα, PKCβ, and PKCγ are the least sensitive (10 to 1000 nM TPA); and PKCδ, PKCη, and PKCζ are intermediate in sensitivity (1 to 10 nM TPA). Because prolonged treatment with TPA causes PKC downregulation, phorbol esters can also be used as specific inhibitors; however, the inhibitory ability of TPA does not always correlate with the sensitivity of the PKC isozymes to TPA stimulation. In addition, the inhibitory effect occasionally shows tissue specificity. For example, in adipocytes PKCα, which is the most resistant to TPA, is downregulated the most, whereas PKCε, which is the most sensitive to TPA, is not repressed at all. In neuroblastoma cells, PKCα is downregulated, but in leukemia cells, it is upregulated after an 18-hour exposure to TPA. Finally, TPA can have effects on ion channels and hormone secretion that are not mediated by PKC and appear to be nonspecific actions. Other protein kinase inhibitors have been postulated to be relatively specific for PKC, but they are limited: rottlerin appears to be relatively specific for PLCδ and LY333531, for PLCβ. Alternatively, PKC mRNA can be neutralized with antisense RNA.

PI3K can be inhibited by either wortmannin or LY294002. However, wortmannin can also inhibit the p120-GAP that terminates Rab5 signaling; as a result, wortmannin stimulates Rab5 and endocytosis independent of its effect on PI3K. LY294002 has side effects, as well; it binds the estrogen receptor (ER) where it acts as a potent estradiol antagonist.

The Phosphoinositide Effect

Cellular Calcium

Perhaps the very first question that should be addressed is simply "Why calcium?" Initially, it seems an unusual choice for a second messenger; however, there are three facts that make calcium an excellent transducer. First, the cell uses phosphate for energy and regulation. Because of the insolubility of calcium phosphate, elevated cellular phosphate levels require low calcium concentrations. This creates a concentration gradient that can be exploited for rapid responses. Second, calcium is a metal ion with six to eight coordination sites that allow calcium to interact with proteins to modify their function. Finally, this binding buffers calcium and restricts its diffusion, which permits more precise signaling.

Calcium Channels

Although the free calcium concentration in extracellular fluid is about 1 mM, cytoplasmic levels are only 0.01 μM at rest and even during hormonal stimulation only reach 10 μM. As such, the simplest way to elevate cytosolic calcium levels would be to open plasma membrane calcium channels. There are two broad groups of such channels (Fig. 10-3): the receptor-operated calcium channel (ROCC) and the voltage-operated calcium channel (VOCC); both were briefly described in Chapter 7. The ROCCs are ligand-gated calcium channels; examples include the glutamate receptors (GluRs). There are six VOCC families, but only four are known to be hormonally regulated (Table 10-1).

The VOCC families were originally given letter designation based on their electrical properties, but they are currently classified by their α-subunit. For example, the skeletal Ca_V1 (originally L) channel consists of a single subunit of each of the following components: α_1, α_2-δ, β, and γ. The α_1 subunit conforms to the basic motif of the voltage-gated channels and can form a functional channel by itself (see Fig. 7-15, A). The α_2 and δ subunits are transcribed from the same gene, cleaved after translation, and remain attached by disulfide bonds. It increases current amplitude and confers valinomycin sensitivity on the channel. The β-subunit is a peripheral protein that accelerates the rates of both activation and inactivation. The γ subunit is unique to the skeletal muscle isoform and may be responsible for the high density of this channel in skeletal muscle; it may also determine inactivation properties.

The Ca_V1 channel requires a large depolarization (above −10 mV) for activation and produces a large, prolonged conductance. Because dihydropyridine (DHP) is a common pharmacological agonist for this channel, it is sometimes known as the *DHP receptor*. The Ca_V3 (previously T) channel requires very little depolarization (above −70 mV) for stimulation, but it only produces a tiny, transient conductance. The $Ca_V2.2$ (previously N) channels are so named because they are only found in neurons. They require intermediate degrees of depolarization (above −30 mV) and produce a moderate, transient conductance.

Although named for their regulation by membrane voltage, hormones can also control their activity to affect cytoplasmic calcium levels. The $Ca_V2.2$ channels are regulated by G proteins; the effect is determined by the β-subunit of the channel. VOCCs are also subject to phosphorylation: Ca_V1 channels are stimulated by protein kinase A (PKA), PKC, calmodulin-dependent protein kinase II (CaMKII), or tyrosine phosphorylation. Tyrosine phosphorylation inhibits the $Ca_V2.1$, $Ca_V2.2$, and Ca_V3 channels; Ca_V3 channels are actually stimulated by

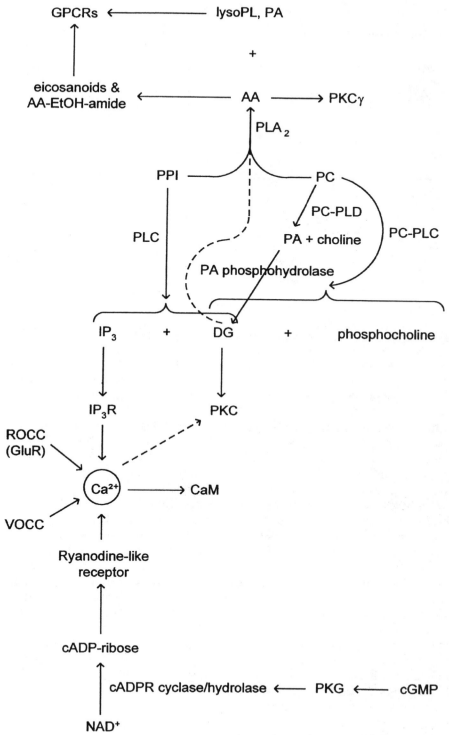

Fig. 10-3. Phospholipid signaling pathway. *AA,* Arachidonic acid; *AA-EtOH-amide,* arachidony-lethanolamide (anandamide); *cADPR,* cyclic ADP-ribose; *lysoPL,* lysophospholipid; *PA,* phosphatidic acid; *PC,* phosphatidylcholine; *PC-PLC,* PC-dependent PLC; *PC-PLD,* PC-dependent PLD; *ROCC,* receptor-operated calcium channel; *VOCC,* voltage-operated calcium channel. Other abbreviations are identified in the text.

Table 10-1
Voltage-Operated Calcium Channels

	Ca_v1 (L, DHP)	$Ca_v2.2$ (N)	$Ca_v2.1b$ (P)	$Ca_v2.1a$ (Q)	$Ca_v2.3$ (R)	Ca_v3 (T)
Activation voltage	High	High	High	High	High	Low
Inactivation						
Speed	Slow	Intermediate	Very slow	Intermediate	Fast	Fast
Ca^{+2}-dependence	Yes	Yes	No	Unknown	Unknown	No
Cloned α_1	C,D,F	B	A-b[a]	A-a[a]	E	G,H,I
Regulation						
G protein		G_s, G_o, G_i				$\beta\gamma(-)$
$PI(4,5)P_2$[b]		(+)	(+)	(+)		
Phosphorylation						
S-T[b]	PKA, PKC, CaMKII (+)	PKA[c]	PKA[c]	PKA[c]		
Tyrosine[b]	(+)	(−)	(−)	(−)		(−)
Calcium-CaM[b]	(−)	(−)	(−)	(−)	(−)	

[a] A-a and A-b are splice variants of α_{1A}.
[b] (+), factor stimulates channel opening; (−), factor inhibits channel opening.
[c] PKA phosphorylation overcomes $PI(4,5)P_2$ deficiencies.

phosphotyrosine (pY) dephosphorylation. In $Ca_v2.2$ channels, inhibition arises from the fact that pY recruits RGS12, which then inactivates the G protein. Calcium-CaM inhibits all VOCCs. $PI(4,5)P_2$ is required for $Ca_v2.1$ and $Ca_v2.2$ activity; hydrolysis of this phospholipid by PLC will inhibit these channels. However, PKA phosphorylation of the channels will eliminate the need for $PI(4,5)P_2$.

In addition to second messengers, some hormones can directly bind VOCCs; for example, anandamide, which is a ligand for the cannabinoid G protein-coupled receptor (GPCR), can also bind and inhibit the Ca_v3 channel. Finally, hormones can open VOCC by altering the membrane voltage; for example, adrenocorticotropic hormone (ACTH) and angiotensin II (AT) stimulate potassium efflux through PKA; the resulting depolarization activates Ca_v3 in adrenal cells. The resulting elevated calcium stimulates steroidogenesis.

Intracellular calcium can also be elevated by releasing this cation from internal sources; the major channels here are the IP_3, ryanodine, and nicotinic acid ADP (NAADP) receptors (see Chapter 7). There is still considerable controversy over exactly where these calcium stores are located. Researchers agree that one component is a special subcompartment of the smooth endoplasmic reticulum that is found near the plasma membrane. These vesicles have been reported to contain calcium-binding proteins that facilitate calcium concentration. IP_3R can also be found on nuclear membranes; however, the presence of IP_3Rs on plasmalemma or nuclear membranes is still unsettled. The former appears to occur in specialized cells; for example, plasma membrane IP_3Rs have been reported in cellular appendages that lack endoplasmic reticulum and whose geometry precludes rapid IP_3-calcium diffusion, such as the olfactory cilia and the outer segment of photoreceptors.

The IP_3R is a ligand-gated calcium channel, but IP_3 is not the only means of regulating this protein. The receptor has an allosteric adenosine 5'-triphosphate (ATP) site, whose occupancy increases IP_3-induced calcium fluxes by 50%. Alkanization can also enhance cation release by increasing the affinity of the receptor for IP_3. The role of calcium itself is complex: luminal calcium is required for channel opening but the effects of cytosolic calcium are concentration dependent and mediated by an accessory protein, calmedin. At low cytosolic calcium levels (0.1 to $0.3\,\mu M$), cation flux is augmented, but at 0.3 to $1\,\mu M$ calcium, ion release is inhibited. The effect of CaM is also biphasic; it inhibits calcium release at sub-maximal IP_3 but not at maximal levels. IP_3R phosphorylation provides another level of control; participating kinases include PKA, PKC, and CaMKII. The effects of PKA are dependent on the PKA concentration and the IP_3R isoform, but, in general, PKA facilitates calcium release by IP_3. PKC increases the rate of release and maximal release of calcium, and tyrosine phosphorylation augments the effect of IP_3. However, not all IP_3R isoforms have these phosphorylation sites. The immunophilin FKBP12 (FK506 binding protein 12) is another allosteric modulator that binds the IP_3R; it increases the open time. In addition to various ligands, many calcium channels can be selectively coupled to hormone receptors to enhance specificity. For example, in neurons, the actin cytoskeleton allows a more efficient coupling of the IP_3R with the bradykinin receptor than with the muscarinic receptor. Finally, the differential expression of the various isoforms of the IP_3R represents a final regulatory mechanism; there are three major isoforms of the receptor with additional forms generated by alternative splicing. These forms have different affinities for IP_3.

The ryanodine receptor is homologous to the IP_3R and is organized into the same square tetramer (see Chapter 7). It was originally thought to be restricted to muscle cells, where its body was imbedded in the sarcoplasmic reticulum and its large amino terminus formed the footlike structure that coupled it to the Ca_V1 channel in the T-tubule. However, a second isoform exists in cardiac muscle, and a third is widely distributed; although they often colocalize in tissues with IP_3Rs, the two regulate distinctly separate calcium pools. The ryanodine receptor is regulated in much the same manner as the IP_3R; it has CaM and ATP allosteric binding sites and both effects are biphasic. However, the calcium concentration curve is shifted to the right; activation occurs at 0.1 to $1\,\mu M$ (vs 0.1 to $0.3\,\mu M$ for IP_3), whereas inhibition requires 100 to $1000\,\mu M$. The ryanodine receptor can also be activated by PKA and CaMKII phosphorylation. Also, like the IP_3R, the ryanodine receptor binds FKBP12. Two functions have been suggested for this immunophilin: first, it may act as an adaptor for PP2B, which would dephos-phorylate and inactivate the receptor. Second, it appears to be required for coupling to Ca_V1 channels in the plasmalemma.

Finally, the ryanodine receptor can be activated by ligands, although there is still controversy over the physiological relevance of each. Obviously, it is acti-vated by ryanodine, but this is a pharmacological tool and not an endogenous ligand. Many authorities claim that the most important ligand is calcium; that is, rising calcium levels from open IP_3Rs trigger the opening of ryanodine receptors. Because of their larger pores and longer open time, ryanodine receptors rapidly empty the contents of their pools to produce an explosive spike. This phenom-enon has been called *calcium-induced calcium release* (CICR). Other ligands are simply thought to enhance this effect; for example, ryanodine increases the recep-tor sensitivity to calcium by a 1000-fold.

However, others suggest that cyclic ADP-ribose (cADPR) is the endogenous ligand. cADPR is synthesized by CD38 from NAD^+ (Fig. 10-4). CD38 actually acts as both a cyclase and a hydrolase; these two activities are differentially regulated; for example, zinc, nicotinamide adenine dinucleotide (NAD^+), or ATP can selectively inhibit the hydrolase activity. CD38 was first identified in

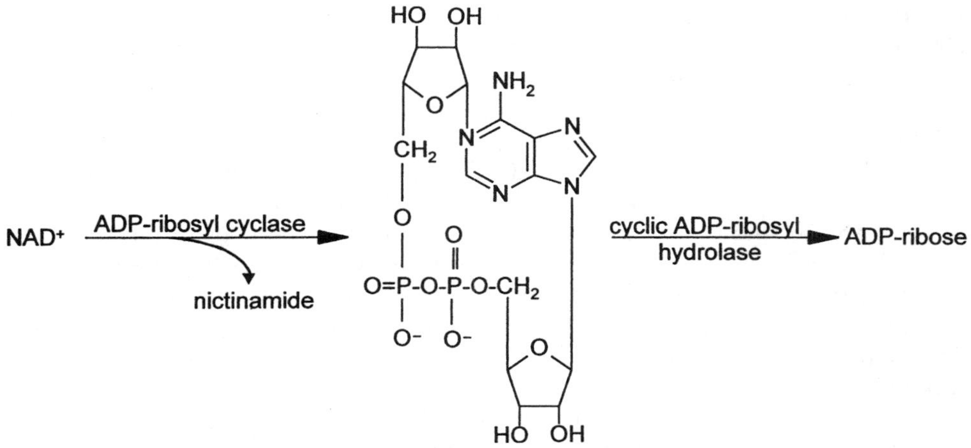

Fig. 10-4. Synthesis and degradation of cyclic ADP-ribose.

plasma membranes where its catalytic site faced the extracellular space. This presented several problems. First, NAD$^+$ would have to be exported, thereby depleting cellular NAD$^+$. Second, cADPR would then have to reenter the cell; however, CD38 does have a central channel through which cADPR could be imported. Alternatively, it has been suggested that the polarity of CD38 may be flexible. Finally, both the equilibrative and concentrative nucleoside transporters have been shown to import cADPR from outside the cell. However, these speculations may all be moot because CD38 has now been identified in both the endoplasmic reticulum and nuclear membranes; in addition, a yet uncharacterized soluble cyclase appears to exist. In the endoplasmic reticulum, CD38 is associated with the ryanodine receptor. These internal sites would relegate the plasma membrane CD38 to a paracrine function, which has been documented in some systems.

There is not much known about the regulation of cADPR cyclases. PKG stimulates the soluble cyclase; in addition, G proteins have been implicated in an unidentified cyclase. The cyclase can be activated by M1 and M3 and this effect is mimicked by cholera toxin; the cyclase is inhibited by M4 and this effect is blocked by pertussis toxin. Nitric oxide (NO) selectively inhibits CD38 cyclase activity without affecting hydrolase activity; this effect is cGMP (3'5'-cyclic GMP) independent and appears to be a result of S-nitrosylation. Finally, CD38 in lymphocytes is activated by antibodies that cross-link the cyclase, and cyclase oligomers are more active than monomers. Therefore there is the possibility that membrane-bound cyclases have endogenous ligands that activate the enzyme by oligomerization, like receptor tyrosine kinases (RTKs).

The question of relevance remains unresolved: is cADPR an important ligand in its own right, or does it simply modulate the effect of calcium? The answer may depend on the system; for example, it makes a significant contribution to the sperm-induced elevations in *Xenopus* oocytes, but plays no role in the fertilization of hamster oocytes.

There is yet a third second messenger that regulates calcium release from a third pool. NAADP can be synthesized by the same enzyme that generates cADPR, but it is unclear whether this is the physiological pathway. Low pH, cAMP, and high nicotinic acid favor the production of NAADP over cADPR. The former regulation is particularly interesting considering that the calcium pools controlled by NAADP reside in a lysosomal-like organelle, where the pH is acidic. Its NAADP receptor has not been purified but the receptor properties resemble that of Ca_V1 channels. Like the ryanodine receptor, it generates rapid, large calcium elevations, although they are not sustained. A comparison? of these three calcium release pathways is presented in Table 10-2.

Calcium Pools

There are three other calcium stores within the cell: the mitochondria, the nucleus, and the cytoplasmic surface of the plasma membrane. Calcium binds to the latter through the negatively charged head groups of the membrane phospholipids. Although the mitochondria can accumulate large amounts of calcium, it was thought that they did not release the cation in response to hormones or second messengers. Rather, the mitochondria were believed to act as sinks for pathologically elevated calcium, thereby preventing potential cell damage. However, it is now known that mitochondria play a more active role in calcium signaling. First, mitochondria can amplify the IP_3 signal. Calcium released by IP_3 enters the mitochondria because of the organelle's negative charge (Fig. 10-5). The charge compensation raises the pH, which triggers the permeability transition pore (PTP) and the Na^+-Ca^{2+} exchanger. Because the mitochondrial calcium release is synchronized with the cytosolic oscillations, it amplifies the calcium response. Calcium uptake by mitochondria has other effects, as well; calcium can stimulate ATP production for use by calcium pumps. There are other pathways that can evoke calcium release from mitochondria; for example, AMPA (α-amino-3-hydroxy-5-methyl-4-isoxazole propionic acid receptor) can affect the Na^+-Ca^{2+} exchanger through G_i.

A second function of mitochondria is to buffer high calcium at the mouth of IP_3R and other calcium channels to prevent negative feedback and promote sustained calcium levels. The calcium is later recycled to the adjacent endoplasmic reticulum. A third function is to restrict calcium signaling. In pancreatic acinar cells, mitochondria surround secretory vesicles and the nucleus. Rapid calcium uptake by mitochondria prevents the calcium from spreading outside of these areas, thereby increasing response specificity. Fourth, mitochondrial uptake of calcium helps to empty the endoplasmic reticulum, which in turn activates the uptake of external calcium through capacitative channels (see later). Finally, mitochondria play an important role in apoptosis. Ceramide, a lipid second

Table 10-2
Second Messenger-Gated Calcium Channels

	IP_3	cADPR	NAADP
Synthesis	PLCγ by pY; PLCβ by α_q and βγ; PLCε by Ras	PM cyclase by cGMP	Soluble cADPR cyclase (by cAMP?)
Calcium pool[a]	Cytosol and apex	Cortical and basal	Lysosome-like
Function	General calcium signaling (internal)	CICR, exocytosis, ion currents (internal and paracrine)	One-time events (for example, fertilization), secretion, supply calcium to other pools
Calcium profile	Puffs resulting in waves	Spikes	Brief, large elevation[b]
Calcium release	Slow	Rapid	Rapid
Receptor	Ryanodine family	Ryanodine family	Ca_V1-like

[a] Depends on cell type; for example, NAADP pools are apical in pancreatic acinar cells.
[b] Results in rapid, profound desensitization; may produce oscillations when interacting with the other two pathways.

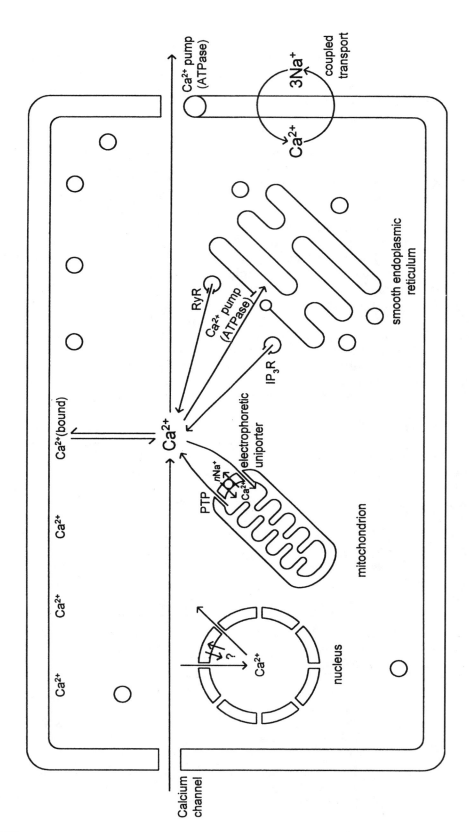

Fig. 10-5. Cellular distribution of calcium and its interchange. *PTP*, Permeability transition pore; *RyR*, ryanodine receptor.

messenger that induces apoptosis (see later), sensitizes the PTP to calcium so that the mitochondria generate slower, larger calcium waves. These waves coordinate mitochondria to release cytochrome c, which initiates apoptosis. NO can also release mitochondrial calcium through the PTP.

The last pool is in the nuclear membrane. There is no question that calcium exists in the nuclear membrane, that it can be released by IP_3, and that all of the components of the PPI pathway are present in the nucleus. However, a controversy does exist over whether nuclear calcium release is independent or dependent of cytosolic calcium. Under most conditions, nuclear calcium fluxes simply mirror those in the cytoplasm, but during the cell cycle, nuclear calcium fluctuates independently of cytosolic calcium. Apart from the question of autonomy, there are several interesting characteristics of calcium propagation through the nucleus. First, calcium waves move more quickly through the nucleus and can reach the other side of the cell faster than waves going around the nucleus. Second, the nuclear membrane possesses type II IP_3Rs that are more sensitive to IP_3 than those in the endoplasmic reticulum. As a result, nuclear calcium levels often rise more quickly and to a greater magnitude that cytosolic calcium levels do. Third, the nucleus has a slower relaxation, which allows calcium spikes to staircase. This phenomenon, in turn, permits the nucleus to integrate calcium signals. Finally, the nucleus appears to have a kinetic barrier that dampens high frequencies without affecting sustained elevations of calcium.

Phospholipase C

There are four major groups of phosphatidylinositol-specific PLCs: PLCβ, PLCγ, PLCδ, and PLCε. PLCα was an artifact, a degradation product of PLCδ. All PLCs share several domains (Fig. 10-6); the split catalytic domains, called X and Y before their function was known; one or more pleckstrin homology (PH) domains that bind phosphoinositide polyphosphate (PIP_n), Gβγ, or Rac; and EF and C2 (PKC conserved region 2) domains. EF and C2 domains have usually been associated with calcium binding, but their function in PLC has not been determined except for the PLCδ C2 domain, which binds G_h.

PLCβ activation is most closely associated with G proteins; α_q binds the carboxy terminus and neutralizes its autoinhibitory activity (Fig. 10-6, *A*). Alternatively, cleavage by the protease calpain can activate PLCβ by removing this site. The PH domain facilitates activity by membrane localization; this can be accomplished by either Rac or PI(3)P (polyphosphoinositide phosphorylated at the 3′ position) binding. The βγ dimer binds to the catalytic site; its activation of PLCβ requires high concentrations. However, because the βγ dimers can originate from any G protein, they can accumulate from several different pathways to reach the necessary levels. Other G proteins stimulating PLCβ include Rab3 and Cdc42. Finally, PKA and PKC phosphorylation inhibit activity, but mitogen-activated protein kinase (MAPK) phosphorylation stimulates the enzyme.

PLCγ activation is most closely associated with activation by RTKs. Its two SH2 (Src homology domain 2) domains bind pYs, which both allosterically activate the enzyme and localize it to the plasma membrane. The amino-terminal PH domain facilitates this shift to the membrane by binding $PI(3, 4, 5)P_3$. The split PH can bind either PIP_2 (phosphatidylinositol 4,5-bisphosphate) or the translation elongator factor EF-1α, which acts as a cofactor. It is interesting that EF-1α also stimulates PI4K (phosphatidylinositol-4 kinase), which synthesizes the substrate for PLCγ. The SH3 domain binds several proteins, including transient receptor potential channel (TrpC), a plasma membrane calcium channel (see

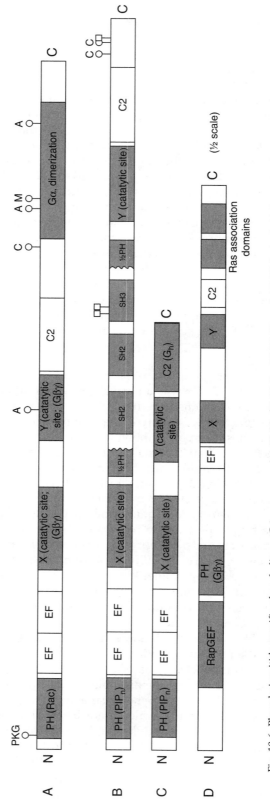

Fig. 10-6. Phosphoinositide-specific phospholipase C isoenzymes. (A) PLCβ, (B) PLCγ, (C) PLCδ, and (D) PLCε (at half the scale of A-C). Phosphorylation sites are represented by either circles (serine or threonine) or squares (tyrosine). Shaded regions represent domains of known function. *A*, PKA; *C*, PKC; *EF*, EF-handed calcium-binding domain; *M*, MAPK; *PH*, pleckstrin homology domain; *SH*, Src homology domain; *X and Y*, catalytic site.

later). Recently, it has been shown that PLCγ can stimulate calcium influx independent of its lipase activity, suggesting that it can allosterically activate calcium channels like TrpC. Tyrosine phosphorylation stimulates the enzyme, whereas PKC phosphorylation induces downregulation.

PLCδ can be regulated by many factors, although it is still not clear which ones are physiological. It is activated by $PI(4,5)P_2$ and calcium. The former effect is mediated by the PH domain, but the site of the latter effect has not been localized, although the EF motifs would be prime candidates. Other activators are spermine (see Chapter 11) and sphingosine. PLCδ is inhibited by two G proteins, Rho and G_h; this inhibition is relieved by RhoGAP and GTP, respectively. It is also inhibited by sphingomyelin.

PLCε is the newest and most unusual PLC. It is most similar in sequence to PLCβ, and it should be no surprise that it is stimulated by G proteins, specifically G_{12} and βγ. It also has a Ras association domain and can be activated by Ras acutely and by Rap2B in a more prolonged manner. Finally, it has a RapGEF domain whose regulation has not yet been described.

Phosphoinositide Cycle

In 1953 Hokin and Hokin reported that acetylcholine stimulated the turnover of PI in the pancreas; this phenomenon became known as the *PI effect*. In 1975, Mitchell published several other important observations:

1. Hormones that stimulate PI breakdown also stimulate calcium influx.

2. PI hydrolysis precedes calcium influx.

3. PI turnover does not require elevated calcium levels.

4. Calcium ionophores could mimic the overall effect of the hormones but do not stimulate PI breakdown.

It is now known that this latter finding is not true in all systems (see later); however, the data were sufficient for Mitchell to propose that these hormones acted by stimulating the breakdown of PI, which then evoked calcium fluxes. Finally, it has been shown that PPI is the actual substrate being hydrolyzed; increased PI turnover in most systems is a result of the recycling of the PPI breakdown products.

The hormone-receptor complexes are primarily coupled to PPI hydrolysis in two ways (see earlier text): G proteins and tyrosine phosphorylation. The RTKs and soluble tyrosine kinases (STKs) that are coupled to cytokine receptors to activate PLCγ. On the other hand, GPCRs use G proteins to activate PLCβ, PLCδ, and PLCε. These PLCs may also be secondarily stimulated by other second messengers; as noted previously, both PLCβ and PLCγ can be phosphorylated by second messenger–activated kinases, and PLCδ can be stimulated by polyamines.

Initially, the source of calcium for elevating cytosolic concentrations was believed to be internal because removal of this cation from the medium did not acutely block cytoplasmic fluctuations. However, repeated stimulations did lead to successively smaller fluctuations in the absence of external calcium, suggesting that external calcium is necessary to replenish internal stores. Cell fractionation studies show that hormone stimulation depleted calcium stores in the microsomes but not in the mitochondria or elsewhere. The major component of the microsomes was the smooth endoplasmic reticulum, which is known to sequester calcium in muscle cells. However, there is some controversy over whether the pool is a simple subcompartment of the smooth endoplasmic reticulum or whether it is a more highly specialized structure. Although these internal vesicles are the sole source of calcium in many systems, it is now known that other systems also take advantage of the high external calcium concentrations by activating calcium plasmalemma channels to augment the fluxes from internal stores. Calcium release from internal stores is mediated by the IP$_3$R, whereas fluxes across the plasma membrane use the VOCC. In addition, there is evidence that there are ligand-gated channels at the cell surface; some of these latter channels are activated by PPI metabolites and are involved with calcium replenishment (see later).

Calcium Profiles

Initial studies with quin 2 showed that agonists evoked a rise in cytoplasmic calcium levels that were smooth and static as long as the agonists were present. However, studies with aequorin in single cells revealed that calcium is released in pulses 7 seconds in duration and 20 to 100 seconds apart. The latency, frequency, and amplitude often depend on the agonist concentration; at high concentrations, the pulses fuse to form a bona fide plateau. The earlier studies with quin 2 consisted of monitoring entire cell populations, and these individual pulses were lost in the resulting statistical summation. These pulses are initiated from the subplasmalemma vesicles immediately underneath the stimulus and spread outward in waves. These waves are generated by IP$_3$; local calcium is released as the IP$_3$ diffuses throughout the cell. Calcium itself is sequestered or buffered too quickly to diffuse very far.

These pulses do not result from fluctuations in IP$_3$ generation because exogenously administered IP$_3$ evokes this same pattern, nor are the pulsations intrinsic to the IP$_3$R. Two models have been proposed to explain this pattern of calcium release. In the *Two Pool Model*, calcium is stored in an IP$_3$-sensitive pool and in an IP$_3$-insensitive, calcium-activated pool; the latter has a large pore allowing for the rapid and massive release of calcium. This phenomenon is known as *CICR* and may correspond to the ryanodine-sensitive pool. Calcium released by IP$_3$ triggers the explosive release of calcium from the CICR pool. The pools must be replenished before the next pulse can occur. In the *Calcium Agonist Model*, the pulses are explained in terms of the biphasic effects of calcium on the IP$_3$R. IP$_3$ releases calcium, which synergizes with IP$_3$ to accelerate the cation release. However, as calcium levels rise, the calcium eventually becomes inhibitory and further release is attenuated. The major criticism of the second model is that it is probably too slow to generate spikes, although it may be responsible for sinusoidal oscillations. In either model, calcium elevations can spread from the point of origin throughout the cell. The calcium itself has a very limited diffusion range; rather, it is the IP$_3$ that diffuses and releases calcium along the way.

Oscillations are slower than spikes, are more symmetrical in their fluctuations, and transmit information through amplitude rather than frequency (Table 10-3). In addition to the calcium feedback described in the calcium agonist model, they may be generated by two other mechanisms. In astrocytes, glutamate elevates calcium, which initially activates IP$_3$R and PLC. However, as calcium continues to rise, IP$_3$R is inhibited and PKC is stimulated. PKC then phosphorylates and inhibits both PLC and GluR; calcium levels fall and the cycle repeats

Table 10-3
Characteristics of Calcium Spikes and Oscillations

Property	Baseline spikes	Sinusoid oscillations
Frequency	<1/min	>1/min
Symmetry	Rapid rise with slower decay	Symmetrical
Baseline	Unchanged	Elevated
Duration	Sustained	Declines with time
Signaling	Increased frequency	Increased amplitude
Mechanism	Calcium positive-negative feedback vs CICR	PKC negative feedback on PLC or RGS on G protein
Function	Localized or waves	Waves distributed throughout cell

itself. Oscillations can also be produced by GPCRs, which use G proteins to activate PLCβ, and elevate IP_3 and calcium. Calcium-CaM then reverses the inhibitory effect of PIP_n on RGS4 (regulator of G-protein-signaling protein type 4); the GTP bound to the G proteins is hydrolyzed and PLCβ stimulation ends. Again, the cycle restarts.

In addition to these global fluctuations, changes in calcium concentrations can be much more restricted. For example, in exocrine cells, high affinity IP_3Rs exist on the luminal surface; IP_3, generated at the basal surface in response to secretagogues, diffuse to and specifically activate these high-affinity IP_3Rs. The resulting spikes are local and do not spread but are sufficient to induce secretion. Such an arrangement would not only be energetically efficient but would also be less likely to disrupt the rest of the cellular machinery. In fact, some investigators have suggested that this mechanism, whereby high affinity IP_3Rs are located only where they are needed, may represent the physiological situation and that global changes only occur in response to pharmacological stimulation.

As noted previously, many systems activate plasma membrane channels to allow extracellular calcium to supplement the internal release. This calcium influx is not pulsed, although it does increase the frequency and velocity of the IP_3 induced waves. Therefore extracellular calcium does not overwhelm the endogenously produced calcium waves that are so important for frequency signaling.

There are several advantages to coding information in patterns rather than simple amplitude. First, the system has greater fidelity, especially at low hormone concentrations. Brief pulses of calcium also prevent calcium overload and toxicity, conserve energy, and prevent desensitization. Finally, patterns can increase the repertoire of responses a single hormone can have. For example, during the follicular phase, gonadotropin-releasing hormone (GnRH) levels are low and induce calcium oscillations that stimulate luteinizing hormone (LH) transcription in pituitary cells but inhibit LH release. At ovulation, high GnRH levels produce calcium pulses or a plateau that inhibits LH transcription but releases the pool of LH that has accumulated during the follicular phase. In another example, large, transient spikes activate the transcription factors, NF-κB (nuclear factor that stimulates the κ chain in B cells) and AP-1 (activator protein 1), whereas a low, sustained calcium plateau stimulates the transcription factor NFAT (nuclear factor of activated T cells) (see Chapter 13). Finally, small or short-term calcium elevations cause PLA_2 to migrate to the Golgi complex, but higher or long-term elevations shift PLA_2 to the endoplasmic reticulum and perinuclear membranes in addition to the Golgi complex.

In addition to the profile and duration of calcium elevations, location is also important in determining the response. For example, calcium spikes restricted to the plasmalemma open large capacitance potassium channels, hyperpolarize the myocyte, and inhibit smooth muscle contraction. However, global calcium elevations promote actin-myosin interactions and contractions. The transcription factor CREB (cAMP response element binding protein) requires elevated nuclear calcium, but serum response factor (SRF) only requires elevated cytosolic calcium.

Termination

Signal termination involves two processes: the removal of calcium from the cytosol and the degradation of IP$_3$. The former is performed by two families of calcium pumps: the smooth endoplasmic reticulum calcium ATPases (SERCAs) and the plasma membrane calcium ATPases (PMCAs). The SERCA has an auto-inhibitory-dimerization site that can be masked by either CaM binding or dimerization; it can also be removed by calpain cleavage. Both CaM and calpain are activated by elevated calcium; their stimulatory effects would represent feedback regulation. SERCA is inhibited by two proteins: phospholamban and calnexin. PKA or CaMKII phosphorylation of the former causes it to dissociate from the pump. Phosphorylation of the latter is actually required for binding; elevated calcium activates a protein phosphatase, probably PP2B (see Chapter 12), which removes the phosphate and disrupts binding.

The PMCA can be substrate driven; calcium sharply stimulates the pump in the range 0.1 to 1 μM. CaM activates the pump by increasing its affinity for calcium 30-fold. In addition, because CaM dissociates slowly, PMCA reacts faster when it has been previously exposed to elevated calcium levels; that is, the pump has a memory. PMCA is also stimulated by PKA but inhibited by STK phosphorylation. Finally, long-term regulation is possible; glucocorticoids down-regulate PMCA1, which decreases calcium clearance and prolongs calcium effects.

Inositol phosphatases are classified according to the position they dephosphorylate (Fig. 10-7). Inositol polyphosphate 5-phosphatases are divided into four groups. Group I prefers soluble substrates, like IP$_3$, and is primarily responsible for terminating calcium signaling. This group includes the SH2-containing inositol phosphatase (SHIP); pY binding by the SH2 domain shifts SHIP to the plasma membrane but does not directly stimulate catalysis. Some members of this group can also be regulated by calcium. Calcium-activated PKC phosphorylates pleckstrin, which then stimulates the phosphatase. However, CaMKII phosphorylation of the phosphatase has the opposite effect and inhibits the enzyme. Group II

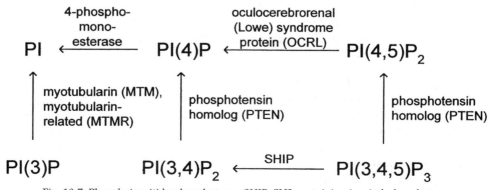

Fig. 10-7. Phosphoinositide phosphatases. *SHIP,* SH2-containing inositol phosphatase.

includes the oculocerebrorenal (Lowe) syndrome protein (OCRL). These phosphatases are isoprenylated, membrane-associated, and probably involved with vesicular trafficking. Group III requires a phosphate in the 3′ position and terminates $PI(3)P_n$ signaling. Group IV is poorly characterized.

The inositol polyphosphate 3-phosphatases include the phosphatase and tensin homologue (PTEN), myotubularin (MTM), and myotubularin-related protein (MTMR). PTEN prefers $PI(3, 4, 5)P_3$ and $PI(3, 4)P_2$ as substrates. PKB phosphorylation decreases PTEN degradation; this effect would represent negative feedback because it would remove the PPIs that activate PKB. On the other hand, phosphorylation by CK2 increases the stability of PTEN, while simultaneously reducing its activity. Dephosphorylation at the CK2 sites increases PTEN activity but renders it susceptible to proteolysis. Tyrosine phosphorylation by Src also inhibits PTEN activity. MTM and MTMR prefer $PI(3)P$ and $PI(3, 5)P_2$ as substrates; the product of the latter reaction, $PI(5)P$, allosterically activates these phosphatases in a positive feedback loop. Finally, inositol polyphosphate 1-phosphatase is best known as the target of lithium therapy in manic-depressive illness; lithium inhibits this phosphatase.

Little is known about the other second messengers. There is an NAADP 2′-phosphatase that is activated by calcium; this would represent feedback inhibition.

Replenishment

Maintenance of the system would require replenishment of the calcium stores and resynthesis of the PPI. Both SERCA and PMCA remove calcium from the cytosol. The former would return it to internal pools, but the latter would remove some calcium from the cell with each stimulation. This effect is easily seen when cells are repeatedly stimulated in the absence of extracellular calcium: the calcium spikes become progressively smaller and eventually stop. As such, internal stores must be refilled from the extracellular fluid. The channels through which this importation takes place are called *capacitative* or *store-operated channels* (SOC) because they open when they sense that internal stores are low. Recently, these channels have been identified as belonging to the TrpC family within the voltage-gated superfamily. This structural classification provides a clue as to the mechanism of sensing. In skeletal muscle, the action potential in the T-tubule is sensed by the Ca_V1 channel, which conveys this information to the ryanodine receptor in the lateral sacs of the sarcoplasmic reticulum. The information is transmitted through the long amino terminus of the ryanodine receptor; this region forms a footlike process connecting the T-tubule and lateral sacs. In a similar fashion, the long amino terminus of the IP_3R binds TrpC3. This physical association could provide an information route for TrpC3 to sense the calcium status of internal pools. It would also guarantee that calcium influx would occur very close to these internal calcium pools, so that replenishment could occur without significantly affecting cytoplasmic calcium concentrations. CaM can disrupt this association by displacing the amino terminus of IP_3R. Furthermore, TrpC3 can be stimulated by IP_3, G proteins, and PKC phosphorylation. Other members of this family can also be regulated by second messengers; TrpC4 is activated by G_q; TrpC3 and TrpC6, by DG, independent of PKC; TrpC6, by calcium-CaM; and TrpC7, by ADPR (a metabolite of cADPR).

The synthesis of PI can only occur in the endoplasmic reticulum and involves four processes: translocation of the phospholipids between the two membrane systems, the activation of DG, the recoupling of CDP-DG (cytidine

5'-diphosphate–diacylglycerol) and inositol, and the phosphorylation of the head group.

The translocation is accomplished by two proteins: the PI transfer protein (PITP) and the sterol carrier protein-2 (SCP-2). Because PI is limiting, long-term stimulation cannot be sustained without PITP and SCP-2. Little is known about the regulation of SCP-2, but PITP can be regulated by phosphorylation and localization. PLC activation hydrolyzes PPI to DG, which activates PKC. PKC, in turn, phosphorylates PITP, which induces its migration to the Golgi complex and changes its binding specificity from phosphatidylcholine (PC) to PI. Finally, PITP returns to the plasmalemma with PI. In addition to the transfer of PI between membranes, PITP is required for the epidermal growth factor (EGF) activation of PLCγ and PI4 kinase. Because these components all coprecipitate, they probably form a complex wherein PITP may present PPI to the kinase or PLC.

DG kinase (DGK) serves four functions: first, it terminates the DG signal. DG activates PKC, among other actions, and its phosphorylation lowers DG levels. Second, in the basal state, it maintains DG at very low levels. Indeed, DGKζ is precoupled to PKC to prevent spurious activation. Third, DGK converts DG to phosphatidic acid (PA), another second messenger with effects distinct from those of DG. Finally, DGK is involved with replenishing PI because PA will be conjugated with cytidine 5'triphosphate (CTP) to form CDP-DG. Phosphoinositide synthase is then responsible for joining the inositol and CDP-DG. Time course studies have shown that, early after PPI hydrolysis, 80% of PI is resynthesized from DG (salvage pathway); however, after several hours, PI is produced by the acylation of glycerol phosphate (*de novo* pathway). Interestingly, the PLC and DGK appear to interact in such a way as to channel the DG from the former to the latter to increase the efficiency of the replenishment of PI.

There are five groups of DGK; all enzymes have a catalytic site and C1 (PKC conserved region 1) domains. Type I is activated by calcium and tyrosine phosphorylation. DGKδ is a type II DGK; its carboxy-terminal SAM domain mediates oligomerization and association with the endoplasmic reticulum, whereas the amino-terminal PH domain binds PIP_n. PKC stimulates DGKδ phosphorylation, disaggregation, and translocation to the plasma membrane; it is not known if PKC directly phosphorylates DGKδ or initiates a kinase cascade that targets DGKδ. Type III is the simplest DGK because it possesses only the catalytic site and a C1 domain. DGKζ, a member of the type IV group, is associated with the carboxy terminus of the leptin receptor; leptin binding to its receptor is believed to activate and dissociate the DGKζ. Conversely, PKC phosphorylation of DGKζ inhibits its activity. Finally, RhoA directly binds and inhibits DGKθ, the only known type V DGK. DGKθ also has a central PH domain of unknown function.

The last step is the progressive phosphorylation of the inositol. It is this process that is being examined when the turnover of PI is being measured by the incorporation of radioactive phosphate into PI. However, because this occurs before hydrolysis, radioactive incorporation is really a reflection of recycling. In fact, increased incorporation could be seen in the complete absence of PPI hydrolysis, if the hormone simply stimulated a total increase in PI content through the *de novo* pathway.

A summary of the kinases and their activators is shown in Figure 10-8. PI3K is discussed later because its product has unique functions. Type II PI4 kinases bind

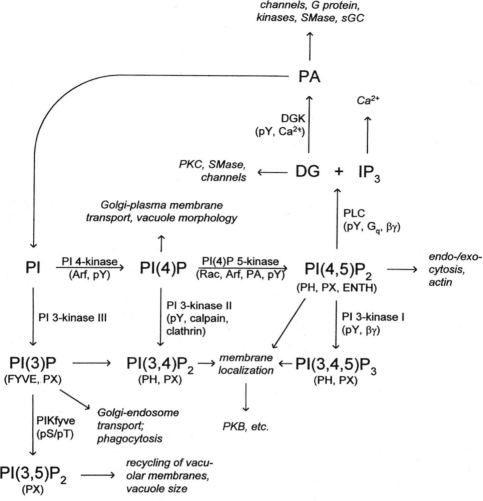

Fig. 10-8. Synthesis of second messengers from phosphoinositol. Lipids and lipid components are in large type, protein domains that bind each messenger are in parentheses, effects of each messenger are italicized, and regulation is printed along the arrows. *DGK,* DG kinase; *PA,* phosphatidic acid; *sGC,* soluble guanylate cyclase; *SMase,* sphingomyelinase.

the EGF receptor in the juxtamembrane region and are activated by tyrosine phosphorylation. Type III PI4 kinases are involved with Golgi-plasma membrane transport; as such, they are recruited to the Golgi complex and are stimulated by Arf. The PI(4)P5 kinase can also be activated by Arf, as well as Rac and PA, and it is inhibited by PKA phosphorylation. The γ-isoform of PI(4)P5 kinase occurs in nervous tissue; it binds talin, which localizes the kinase at focal adhesions. There the focal adhesion kinase (FAK) tyrosine phosphorylates and activates the PI(4)P5 kinase. The type III PI(3)P5 kinase, called *PIKfyve,* shifts from the cytosol to microsomes after being S-T phosphorylated in response to insulin stimulation. Amazingly, each PPI in this pathway has its own functions mediated by proteins with domains relatively specific for it: PI(4)P is involved with Golgi-plasma membrane transport; PI(3)P, Golgi-endosome transport; PI(3, 5)P_2, with recycling of vacuolar membranes; and PI(3, 4)P_2, PI(4, 5)P_2, and PI(3, 4, 5)P_3, with membrane localization and signaling.

Output

Calcium and Calmodulin

Calcium is a major output of this transduction system, and there are several enzymes whose activities are affected by concentrations of this cation; examples include the calcium activation of glyceraldehyde phosphate dehydrogenase, pyruvate dehydrogenase, and α-ketoglutarate dehydrogenase. More frequently, however, calcium acts in concert with calcium-binding peptides such as CaM.

CaM is a 148-amino acid peptide (16.7 kDa); it is heat stable; ubiquitous, both histologically and phylogenetically; and highly conserved evolutionarily. The peptide structurally resembles a dumbbell: it has two globular ends separated by a seven-turn helix (Fig. 10-9, *A*). Each globular end contains two calcium-binding sites, which have a helix-loop-helix configuration wrapped around the cation; the innermost site in each end uses the long, connecting helix as one of the sides. The two binding sites in the carboxy-terminal half have a slightly higher affinity than those in the amino-terminal half, but the difference is within an order of magnitude. Most of the dissociation constants reported for these sites range between 10^5 and 10^6 M. Each globular end also has a hydrophobic groove.

Calcium binding to the globular domains exposes the hydrophobic groove, which binds either end of the target α-helix. The intervening α-helix in CaM is quite flexible and allows the globular ends to properly position themselves (Fig. 10-9, *B*). The central helix in CaM also provides additional interactions with the target helix; the amino-terminal half of the CaM helix is hydrophobic, whereas the carboxy-terminal half is acidic. The target sites in many CaM-binding proteins are complementary to this long helix; that is, they are half hydrophobic and half basic. Although this CaM-protein interaction is considered to be proto-typical, other interactions have also been found. In addition, not all CaM activity requires calcium binding; in the absence of calcium, the amino-terminal groove is completely closed, but the carboxy-terminal one is partially open, allowing it to bind some targets.

CaM usually affects its targets in one of two ways. First, it can bind and block autoinhibitory domains in substrates, such as the CaMK (see later). Second, it can span molecules and induce dimerization. A head-to-head CaM dimer can encircle the α-helical coiled coil domain of several basic helix-loop-helix transcription factors (see Chapter 13) and inhibit DNA binding. A similar effect is seen with the petunia glutamate decarboxylase and with the small potassium conductance (SK) channel, which is a calcium-activated channel.

The major mechanism by which hormones control CaM activity is by altering the cytoplasmic calcium concentration. CaM can also be covalently modified; for example, CK2 can phosphorylate CaM and inhibit its activity. The effects of tyrosine phosphorylation by RTKs and STKs are both substrate and site specific. There are two potential phosphorylation sites: Y-99 and Y-138. In general, phosphorylation of both tyrosines has no effect or is inhibitory, whereas phosphorylation of Y-99 alone is stimulatory. Modification of the amino-terminal site enhances the activation of phosphodiesterase 1 (PDE1), calcium ATPase, CaMKII, nitric oxide synthase (NOS), and sodium-hydrogen exchanger 1 (NHE1), but activation of PP2B is unaffected. In addition, CaM binding is regulated by phosphorylation of its targets; the consensus CaM-binding domain has a PKC phosphorylation site within it. Phosphorylation of this site blocks CaM binding. Finally, one of the lysines in CaM is constitutively methylated; this does not affect biological activity, but it does prevent the ubiquitination and rapid degradation of CaM.

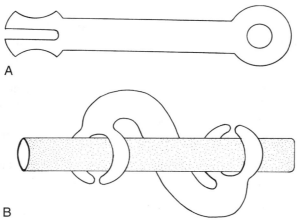

Fig. 10-9. Schematic representation of the calmodulin molecule and its interactions with substrates. (A) Each globular domain has two calcium-binding sites and a groove. The two globular domains are turned 90° with respect to each other and are separated by a seven-turn α-helix. (B) α-Helical substrates bind to CaM in the grooves of the globular domains.

Sequestration is another mechanism by which hormones can regulate CaM. The proteins, neurogranin and neuromodulin, act as reservoirs for CaM; they have multiple binding sites for CaM. Elevated calcium levels activate PKC (see later), which phosphorylates these sites and releases CaM. The third intracellular loop (IC3) of opiate receptors also binds CaM that is released by agonist binding.

Long-term regulation involves the transcription and translation of CaM and can be influenced by hormones. CaM levels decline in the fat pads of diabetic rats and in the myometrium of oophorectomized rabbits; levels are restored by insulin and estrogen replacement therapy, respectively. Furthermore, the CaM content of rat mammary glands rises twofold to threefold during pregnancy. Finally, CaM can be regulated by cellular localization; mammals have three CaM alleles that code for five CaM mRNAs, although the CaM translated from all of them are identical in sequence. However, the various mRNA isoforms are targeted to different sites within the cell where they are translated; for example, in neurons, CaM1 and CaM2 are localized to dendrites, and CaM3 migrates to the cell body.

There are many enzymes regulated by CaM, but only some that are more important or that are better understood are discussed. Some enzymes, such as the cyclic nucleotide phosphodiesterase 1 and some adenylate cyclases, have already been covered (see Chapter 9), whereas others like the myosin light chain kinase are discussed later (see Chapter 12). In these examples CaM binds to the enzyme as a transient, extrinsic subunit. However, in phosphorylase kinase CaM is actually a permanent part of the enzyme complex; this enzyme is critical in glycogenolysis, where it phosphorylates and activates the glycogen phosphorylase (see Chapter 12). The phosphorylase kinase has the following quaternary structure: $(\alpha\beta\gamma\delta)_4$. The γ subunit contains the catalytic site, and the δ subunit is CaM. The α- and β-subunits are additional regulatory components; their phosphorylation increases the affinity of the enzyme for calcium, thereby making activation easier.

Because phosphorylation is such a common mechanism to control cellular processes, it is not surprising that CaM can activate more general protein kinases as well. The CaM-dependent protein kinase, type II (CaMKII), is very large (600 to 650 kDa) and composed of 12 subunits arranged in a 2-layered hexamer. The kinase contains several homologous subunits; γ and δ are ubiquitous, whereas α

and β are restricted to the nervous system. All of the subunits are catalytic, bind CaM, and can be phosphorylated. Both homooligomers and heterooligomers exist, although heterooligomers having α- and β-subunits in a 3:1 ratio predominant. The amino-terminal kinase and CaM-binding domains are located above and below the hexameric layers; the hexameric core consists of the aggregated carboxy termini. The regulatory site is a pseudosubstrate for both the ATP-binding site and the catalytic site (see Fig. 12-1, *D*); CaM activates the kinase by binding to and masking the carboxy-terminal region of the regulatory domain.

Autophosphorylation in the amino terminus of the regulatory domain then activates the kinase. A second round of autophosphorylation within the CaM domain has two effects: first, it increases the affinity of the regulatory domain for CaM 1000-fold, and second, it can preserve 40% of the kinase activity even in the absence of CaM. The higher CaM affinity maintains CaM binding to the kinase long after calcium levels have begun to fall. This phenomenon enables the CaMKII to act as a calcium sensor; if the calcium pulses are occurring fast enough, the CaMKII will remain fully active despite the fluctuating ion levels.

The existence of CaMKII implies the existence of CaMKI; indeed, there are four CaM-dependent protein kinases. Originally, only CaMKII was thought to have a broad enough specificity to have widespread effects. However, CaMKI and CaMKIV are now known to have important general functions. On the other hand, CaMKIII is highly specific for the eukaryotic elongation factor (eEF)–2 translation factor; indeed, it is often referred to as the *eEF-2 kinase* and is discussed in Chapter 15. CaMKI and CaMKIV differ from CaMKII in several ways (Table 10-4); they are both monomers, they require a CaM-dependent kinase kinase (CaMKK) for the initial phosphorylation, and neither becomes independent of calcium-CaM. There are two CaMKKs: α is nuclear and β is cytosolic. The CaMKs and CaMKKs are homologous and are activated in a similar fashion; calcium-CaM binds CaMKK and induces autophosphorylation and activation. CaMKKα and CaMKKβ then phosphorylate CaMKIV and CaMKI, respectively. CaMKIV can autophosphorylate itself in the CaM domain to render itself partially independent of calcium-CaM. There is some cross talk between the calcium-CaM and cAMP pathways at this site in that PKA can phosphorylate and inhibit CaMKKα.

Another CaM-regulated enzyme, calcineurin or protein phosphatase 2B (PP2B), reverses phosphorylation (see Chapter 12); it is a widespread, multifunctional phosphatase. This enzyme is composed of two subunits, α and β. The α-subunit is 61 kDa and binds both zinc and ferric ions; it also contains the catalytic and CaM-binding sites. Interestingly, the β-subunit (19 kDa) is another member of the CaM family and has four calcium binding sites. Therefore this phosphatase is controlled by two calcium-binding peptides, CaM and the β-subunit.

Protein Kinase C

The other major second messenger produced by the PPI pathway is DG, which activates another protein kinase, PKC. This kinase is divided into four conserved (C) regions and the intervening variable (V) regions (see Fig. 12-1, *C*). C1 contains the TPA(12-*O*-tetradecanoylphorbol-13-acetate)-DG–binding site. It is subdivided into two cysteine-rich sheets, which each form a half-site that binds TPA with low affinity; together they form a single, high-affinity binding site. In PKCε, C1 also binds actin. C2 contains four zinc ions, probably in a binucleate cluster, and binds calcium at the edge of an eight-stranded β-sandwich. C2 also binds to receptors for activated C kinase (RACKs), which are responsible for linking PKC to other molecules, such as membrane receptors, PI3K, PTP, and

Table 10-4
Properties of General Purpose Calmodulin-Dependent Kinase Isoforms

Property	CaMKI	CaMKII	CaMIV
Tissue Distribution	Ubiquitous	Ubiquitous	Limited (primarily nervous system)
Subcellular Distribution	Cytosol	Mostly cytosol	Nuclear and cytosol
Subunit	Monomer	Dodecamer	Monomer
Activation			
Calcium-CaM	Yes	Yes	Yes
First phosphorylation	CaMKK	Autophosphorylation	CaMKK
Second phosphorylation	No	Autophosphorylation	Autophosphorylation
Calcium-CaM independence after autophosphorylation	No	Yes (80%)	Yes (20%)

Src. Finally, the C2 in PKCδ binds actin, and the C2 in PKCμ contains the plasma membrane binding site. C3 and C4 are the ATP-binding and catalytic sites, respectively. V1 is at the extreme amino terminus and contains a pseudosubstrate site that inhibits kinase activity; it is thought that the binding of calcium and phospholipids alters the conformation of PKC in such a way that this region is displaced from the catalytic site. In PKCζ, this region also binds tubulin. There are several autophosphorylation sites; although the unmodified PKC can bind substrate and cofactors, it is not active until it has been autophosphorylated in C4. The atypical PKCs also contain a C4 tyrosine phosphorylation site that is required for nuclear localization.

There are three major PKC families (Table 10-5). Group A is also called the *classical* or *conventional PKCs* (cPKCs) and includes the PKC isoforms α, β, and γ. They possess the C2 domain and are calcium dependent, and they have good activity toward the classic kinase substrates, such as histones, myelin basic protein, and protamine. Group B, also known as the *new* or *novel PKCs* (nPKCs), includes PKC isoenzymes δ, ε, η, and θ. They do not have a C2 domain and are calcium independent, although they still require DG and phosphatidylserine. They do not phosphorylate the classic kinase substrates well. Group C, or the atypical PKCs (aPKCs), includes PKCζ, PKCι/λ, and possibly PKCμ; they lack both the C2 and half of the C1 domains. They are calcium independent and DG independent but still require phosphatidylserine. PKCμ, also called *protein kinase D1* (PKD1), is unusual in that it possesses a hydrophobic amino terminus, and its sequence is very divergent. Indeed, some authorities question its inclusion in the aPKC; rather, it has been suggested that PKCμ, PKCυ (also called PKD3), and PKN constitute a fourth PKC group.

Many of these isoforms also differ in which signaling pathways they affect. This appears to be a result not only of substrate specificity but also of cellular localization and tissue distribution. Many isoforms are cytoplasmic; upon activation, they may be translocated to the nucleus (PKCα and PKCζ); to the Golgi complex (PKCη); to the cytoskeleton (PKCβ, PKCε, and PKCζ); or to regions within the plasmalemma, such as focal adhesions (PKCα), ruffles (PKCβ), or cell junctions (PKCδ). This redistribution is facilitated by various anchoring proteins: RACK1 binds the C2 and V5 of PKCβII and localizes it to the perinuclear region; RACK2 binds the V1 of PKCε and localizes it to the Golgi complex; PICK1 localizes PKCα to the nucleus; and F-actin localizes PKCβII and PKCε to the cytoskeleton. Tissue distribution also differs among the isoforms; PKCα, PKCδ, PKCε, and PKCζ are ubiquitous, whereas other isozymes have a

Table 10-5
Protein Kinase C Isozymes

	cPKC	nPKC	aPKC
Members	α, β, γ	δ, ε, η, θ	ζ, μ (PKD), λ/ι
Regulation			
Ca^{2+}-dependent	+	0 (no C2)	0 (no C2)
DG-dependent	+	+	0 (missing half of C1)
PS-dependent	+	+	+
Other[a]	All: 14-3-3	ε: DHCC; PIP$_2$; 14-3-3	All: PI(3)P$_n$; pY required for nuclear translocation
	α: 20-HETE	δ: pY (R/STK) prerequisite for activity	ζ: Ceramide; Ras
	α, γ: DHCC	η: Phosphatidic acid	λ/ι: LIP[b]
	γ: Arachidonic acid		μ: DG; βγ; PKC
Location[c]	α: Cytosol to plasma membrane (focal adhesion) and nucleus	δ: Nucleus to plasma membrane (cell junctions)	ζ: Cytosol to nucleus, plasma membrane, and cytoskeleton
	β: Cytosol to plasma membrane (ruffles) and cytoskeleton	ε: Nucleus (pore complex) to plasma membrane and cytoskeleton	μ: Golgi complex
	γ: Cytosol to Golgi complex and plasma membrane	η: Cytosol to Golgi complex, plasma membrane, and nuclear pores	

[a] Stimulation or requirement for activation.
[b] Lambda interacting protein.
[c] "To" refers to the redistribution that takes place after PKC activation.

more restricted tissue distribution. For example, PKCβ is abundant in the brain and spleen, PKCη is mostly located in skin and lungs, and PKCθ is found predominantly in skeletal muscle and hematopoietic stem cells. Finally, the different isoforms vary in their length of activation; PKCα is transiently stimulated, whereas PKCβ retains its activity longer.

There are several mechanisms to regulate PKC. Calcium is a major activator for the Group A PKCs, and phospholipids affect all PKCs. The most important phospholipids are DG and phosphatidylserine; the latter appears to coordinate calcium binding with the C2 region. However, other lipids can also affect PKC activity (Table 10-5); PIP$_n$ stimulates aPKCs and PKCε; DHCC stimulates PKCα, PKCγ, and PKCε; lipoxygenase metabolites stimulate PKCα; arachidonate, PKCγ; phosphatidic acid, PKCη; and ceramide, PKCζ, although other sphingolipids are generally inhibitory. PIP$_n$ acts as a membrane anchor; it is not known if the other lipids are acting as DG or phosphatidylserine substitutes or if they have separate effects.

PKCs are subject to phosphorylation; the tyrosine phosphorylation of PKCδ has been well studied, but similar phosphorylation of PKCα, PKCβ, PKCγ, PKCε, and the aPKCs has been described. Tyrosine phosphorylation does not affect activity per se but renders PKCδ independent of calcium and lipids. Tyrosine phosphorylation of the aPKCs is required for nuclear localization. Phosphorylation of PKCμ by PKCη or PKCε can also induce its nuclear translocation.

PKC can be activated by cleavage in V3 by the proteases calpain and caspase 3. Removal of the pseudosubstrate site in the amino terminus renders PKC permanently active, even in the absence of calcium and phospholipids. In addition to being constitutively active, the clipped PKC, also called PKM, would no longer be bound to the membrane; the resolubilized PKC could have a greater access to its substrates or even be able to phosphorylate a different spectrum of substrates. Cleavage by caspase 3 during apoptosis appears to be physiological, but there is conflicting evidence as to whether calpain cleavage actually occurs in intact cells. Interestingly, PKMζ, which is generated from PKC? mRNA and not proteolysis, is responsible for long-term potentiation in the hippocampus; that is; its constitutive activity is the basis for memory in this part of the brain.

Finally, PKC activity can be influenced by other proteins. The effect of the scaffold, 14-3-3, is confusing. Some authorities report that it inhibits PKC by acting as a pseudosubstrate; others claim it activates PKC by acting as an adaptor to membrane phospholipids. In all likelihood, 14-3-3 maintains PKC in an inactive but primed state for stimulation; this mechanism would be similar to its role in the Ras activation of MAPK (see Chapter 9). PKCμ is unique in having a PH domain; βγ activates PKCμ by binding to this domain. PKCλ/ι can be stimulated by the Rho family member Cdc42 and by a lambda interacting protein (LIP).

Calpain

Calpain is an acronym for *cal*cium-activated, *papain*-like protease. There are two isozymes that have homologous catalytic subunits and identical regulatory subunits; both subunits bind calcium. The regulatory subunit also has a phospho-lipid-binding site that is responsible for membrane association. Calpain is usually not a degradatory protease; rather it has very discrete cleavage sites. In particular, it prefers to cleave at interfaces between hydrophobic and hydrophilic domains; such sites exist in many CaM-binding proteins (see earlier mention) and in PEST (proline, glutamic acid, serine and threonine) regions.

Proteins with CaM-binding sites often become CaM-independent after cleavage; they include phosphorylase kinase, calcium ATPase, PDE1, calcineurin, and a CaM-activated Ras. Cleavage activates other proteins by removing autoinhibitory domains; they include PKC, PLCβ, and PI3K. Finally, calpain can influence the cellular localization of transducers; usually, cleavage removes the anchoring domain and solubilizes the protein. For example, PTP dissociates from the plasma membrane and may be stimulated or inhibited depending on the isoform; Fak, an STK, dissociates from the cytoskeleton and is inhibited; and AMPA, a GluR, dissociates from the synapse.

Calpain is regulated by several mechanisms: calcium, phospholipids, calpastatin, and phosphorylation. Calcium activates both isozymes, but the concentration requirement differs: calpain I only needs micromolar amounts of calcium, whereas calpain II requires millimolar concentrations. However, phospholipids can reduce this requirement. Calpain can be inhibited by calpastatin, a protein composed of four internal repeats. PKA phosphorylation of calpastatin results in its aggregation and sequestration near the nucleus. On the other hand, calcium elevation activates PP2B, which dephosphorylates, solubilizes, and activates calpastatin. Calpastatin is also subject to proteolysis by calpain and caspase; this cleavage releases the A and C repeats, which are actually calpain activators, rather than inhibitors. Finally, both calpain and its substrates can be phosphorylated. PKA phosphorylation of calpain stabilizes its inactive conformation, whereas tyrosine phosphorylation near the cleavage site of its substrates inhibits proteolysis.

Other Outputs

There are several other potential outputs. First, the PPI, DG, or both may be subject to further hydrolysis to release arachidonic acid, which either can be active by itself or serve as a precursor to the eicosanoids. These compounds are covered in more detail later in this chapter. PIP_n is recognized by several proteins modules, such as the PH, phox homology (PX), and other domains (see Table 7-2); many of the proteins containing these modules control either vesicular trafficking (see Fig. 10-8) or cytoskeletal organization (see Chapter 12). In addition to opening the IP_3R calcium channel, IP_3 stimulates a calcium ATPase and the PP1, while inhibiting protein tyrosine phosphatase IA, aldolase, and the (Na^+, Ca^{2+})exchanger. IP_3 can be dephosphorylated to IP_2, which stimulates phosphofructokinase, DNA polymerase, and a calcium ATPase.

IP_3 can be further phosphorylated to produce additional transducers. IP_4 prolongs the activity of IP_3 by inhibiting the 5-phosphatase that converts IP_3 to IP_2 and by inhibiting the calcium ATPase that clears calcium from the cytosol. At low concentrations, it can enhance IP_3 by activating calcium channels, but at high concentrations, it is an IP_3R antagonist, which sharpens calcium oscillations. It can also stimulate AMP deaminase, PP1, and a RasGAP, but it inhibits chloride channels. Finally, IP_4 and IP_5 inhibit the chromatin-remodeling factor ISWI (imitation of switch); indeed, Ipk2, a yeast kinase that converts IP_3 to IP_4 and IP_5, is a component of a transcription complex.

Conversely, IP_6 stimulates another chromatin-remodeling factor, SWI/SNF (switch and sucrose nonfermentors genes). IP_6 also blocks the recruitment of AP-2 and clathrin-dependent endocytosis, but it promotes dynamin I-mediated endocytosis. Furthermore, it stimulates an unidentified kinase that phosphorylates pacsin-syndapin I and increases its binding to dynamin I. It binds to the Ku subunit of the DNA-dependent protein kinase (DNA-PK) and activates the end-joining activity of this enzyme in the repair of nicked DNA.

The reader should be warned that there are three serious methodological problems with studying IP_3 metabolites; as such, investigations on these molecules should include important controls. First, the highly negative charge of IP_3 metabolites can produce nonspecific electrostatic attraction. This possibility should be evaluated by testing unnatural isomers of the metabolites and by using physiological concentrations of calcium, which will partially neutralize the charge. Second, the reported effects may be due to the sequestration of calcium by these metabolites. One can rule out this possibility by showing that the effect is not mimicked by calcium chelators. Finally, preparations may be contaminated with other PIs.

Although DG is best known as an activator of PKC, it too can have other functions; for example, it can activate the acid sphingomyelinase, NADPH (reduced nicotinamide adenine dinucleotide phosphate) oxidase, a RasGEF and RapGEF, and TrpC3 and TrpC6, and it inhibits the cGMP-gated channel in the retina. Indeed, it could potentially regulate any protein possessing a C1 domain, which is the DG-binding site in PKC but which also exists in many other proteins.

Regulation

Interrelationships of the Branches

The calcium and PKC branches of the PPI pathway can interact in four different ways: (1) synergistically, (2) antagonistically, (3) independently, (4) and temporally separately. In most systems, the two branches reinforce each other.

For example, in platelets thrombin-induced protein phosphorylation is only partially mimicked by PKC activators or calcium ionophores alone, but it is fully mimicked by a combination of the two. Examples of antagonism are rarer; in neuronal cells, calcium elicits hyperpolarization by increasing potassium currents, whereas PKC activators depolarize the cells by decreasing potassium currents. In lymphocytes, the two branches act independently; TPA stimulates mitosis without any apparent changes in intracellular calcium concentrations. Finally, the two branches can interact temporally; IP_3 and calcium trigger the acute responses, whereas PKC and its phosphorylated substrates have more prolonged effects. Glucose-provoked insulin secretion from the pancreas is biphasic; the initial peak is mimicked by A23187, whereas the later, sustained release is mimicked by TPA. A similar phenomenon occurs with aldosterone production in adrenal glomerulosa cells.

Such an array of interactions may at first seem contradictory: how can two branches antagonize, or be independent of, one another, when they are both part of the same pathway? Recent studies suggest that these two branches may, in fact, be independently regulated. For example, hormones can stimulate a phosphatidyl-choline-specific phospholipase C, which hydrolyzes phosphatidylcholine (PC) into DG and phosphocholine (see later). The phosphocholine does not affect intracellular calcium levels, but the DG can still activate those PKC isozymes that are calcium independent. Conversely, calcium levels can be regulated independently of PKC activation; for example, calcium can be recruited from IP_3-independent pools or from the external medium through plasma membrane channels. In these cases, no DG is produced, and PKC is not activated. Differential tissue distribution results in another form of independent control. In the brain, IP_3 receptors and PKC are not always localized in the same regions. Although both are present in the molecular layer of the cerebellum, hippocampus, corpus striatum, and cerebral cortex, IP_3 receptors are immeasurable in the external plexiform layer of the olfactory bulb and in the substantia gelatinosa, even though PKC can still be detected in concentrations comparable to that in other brain regions. Therefore there are mechanisms for regulating these outputs separately.

Homologous Regulation

Homologous regulation consists primarily of negative feedback by one of the two major products of this pathway: calcium and PKC (Fig. 10-10). Negative feedback mediated by PKC takes three forms: (1) inhibition of phospholipase C, (2) removal of active products of the pathway, and (3) decreasing the hormonal sensitivity of the cell to subsequent stimulation. First, PKC phosphorylates both PLCβ and PLCγ and impairs their activation. Next, PKC removes IP_3 by phosphorylating and activating the IP_3 5'-phosphomonoesterase, which then dephosphorylates IP_3. PKC also phosphorylates DG kinase, which converts DG to phosphatidic acid. Although this modification does not directly affect enzyme activity, it does stabilize the kinase. Calcium is removed by stimulation of the calcium transport ATPases, which pump the cation out of the cell; the sarcoplasmic calcium ATPase is phosphorylated by PKC, whereas the (Ca^{2+}, H^+)ATPase is activated by calcium and CaM. This effect is complemented by the PKC phosphorylation of IP_3R, which reduces additional calcium release. Finally, PKC can reduce its own levels (downregulate itself); however, these reports must be cautiously interpreted. Whenever TPA is used to activate PKC, downregulation of several PKC isozymes is inevitable and rapid; this appears to be a result of enzyme degradation. However, when PKC is naturally stimulated, its

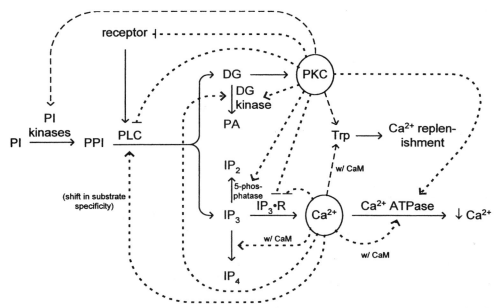

Fig. 10-10. Homologous regulation of the polyphosphoinositide pathway. The major feedback control (dotted lines) is affected by the principal outputs, PKC and calcium (circled); arrows indicate stimulation, and flat heads indicate inhibition. Solid lines represent the primary signaling pathway, and the replenishment steps are represented by dashed lines. *PA*, Phosphatidic acid.

activity persists. As a result, PKC downregulation, as induced by TPA, may be either an artifact or an exaggerated response due to either the great potency of TPA or its long half-life. An exaggerated response would not occur normally because DG is metabolized within minutes of its formation; as such, the natural control of PKC may be through levels of its DG activator rather than downregulation. A similar situation occurs with progesterone receptors; progesterone agonists downregulate the receptor, whereas the natural steroid does not. The reason is that progesterone has such a short-half life that it does not occupy the receptor long enough to downregulate it; the agonists all have much longer half-lives. Another factor influencing effects of TPA is the level of hsp 70. This chaperone binds PKC and is known to favor PKC recycling rather than downregulation. The final mechanism by which PKC exerts negative feedback is by phosphorylation of the hormone receptors. For example, this modification of the EGFR lowers its affinity for EGF and reduces the responsiveness of the cell toward further stimulation by EGF.

Calcium is the other major product, and it also contributes to this negative feedback. First of all, elevated calcium levels in combination with calcimedin interfere with the binding of IP_3 to its receptor, thereby decreasing further calcium release. In combination with CaM, it activates the calcium pump. It also facilitates the clearance of other second messengers by stimulating DG kinase and the IP_3 3-phosphokinase. Another way that calcium is involved with negative feedback is through altering the substrate specificity of the phospholipase C; at low calcium concentrations, the enzyme prefers the fully phosphorylated PPI and IP_3 is liberated. However, at higher concentrations, the phospholipase switches to phosphatidylinositol 4,5-bisphosphate and IP_2 is freed. Because IP_2 does not bind to the IP_3 receptor, the internal calcium release ceases; however, DG is still generated and can continue to activate PKC.

In addition to their involvement with negative feedback, PKC and calcium prepare the cell for the next round of stimulation. First, both PKC and calcium-CaM activate TrpC channels, which are responsible for refilling internal calcium pools (see previous text). Other metabolites, like IP_4, may also stimulate these channels. Second, both stimulate DG kinase; although this terminates DG signaling, it is also the first step in recycling DG back to PI. Finally, PKC activates the kinases that convert PI to PPI.

Heterologous Regulation

The relationship between the PPI and the cAMP pathways can either be synergistic or antagonistic (Fig. 10-11); the former occurs during glycogenolysis in the liver (see Chapter 12), whereas the latter occurs in platelets. The PPI pathway can antagonize the cAMP system at several levels; for example, PKC can phosphorylate the β-adrenergic receptor (βAR) and human chorionic gonadotropin (hCG) receptors (see Chapter 8). Unlike the phosphorylation of the EGFR, the phosphorylation of these receptors does not change their affinities but rather uncouples them from the adenylate cyclase. Another possible mechanism is suggested by the fact that both systems use G proteins. Activation of PLCβ by G_q liberates βγ subunits, which could then tie up the α_s from the cAMP pathway. This mechanism would be analogous to the one postulated for G_i (see Chapter 9). Finally, the calcium generated by the PPI pathway can bind to CaM to stimulate PDE1 that hydrolyzes cAMP, thereby destroying its biological activity. Conversely, cAMP can inhibit both the PI kinases and the phospholipase C; the former does not require phosphorylation but appears to be a direct effect of cAMP. In addition, PKA phosphorylates and activates the PMCA, which removes calcium from the cytoplasm.

In other systems, the two pathways complement one another. Indeed, glucagon and ACTH each activate both pathways. Synergism can occur in several ways. First, PKC can enhance cAMP production and reduce its elimination. For example, PKC can phosphorylate and inhibit R_i, G_i, and PDE, while stimulating several adenylate cyclase isozymes (see Chapter 9). Second, many of the eicosanoids produced by the PPI pathway have receptors that stimulate adenylate cyclase; this results in signal amplification. Conversely, PKA can phosphorylate IP_3R and increase its affinity for IP_3. The effect of calcium-CaM on adenylate

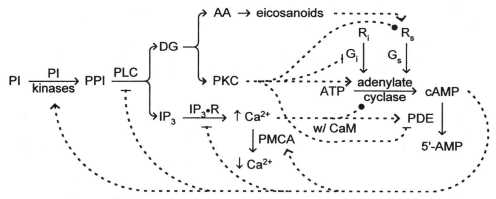

Fig. 10-11. Heterologous regulation of the polyphosphoinositide pathway. The primary signaling pathway is indicated by solid lines, and regulatory controls are represented by dotted lines; arrows indicate stimulation, flat heads indicate inhibition, and circled heads may be either depending on the system. *AA,* Arachidonic acid; *PMCA,* plasma membrane calcium ATPase.

cyclase activity is biphasic. The catalytic subunit of the cyclase binds CaM, and both basal and hormone-stimulated activity requires calcium concentrations of 0.01 to 0.1 μM; higher levels (1 μM or greater) inhibit the enzyme. Therefore although basal levels of calcium are required for cyclase activity, the elevated concentrations observed during activation of the PPI pathway would both inhibit the cyclase and stimulate the PDE, resulting in lower cAMP levels.

Other Phospholipids

Other Phospholipases

Phosphatidylcholine-Specific Phospholipase C

Phosphatidylcholine Hydrolysis. As work on the PPI pathway became more sophisticated and quantitative, certain discrepancies began to arise. In particular, the data showed that there was far more DG produced than PIP$_2$ hydrolyzed; for example, in hepatocytes, vasopressin hydrolyzed 9 ng of PIP$_2$ per milligram of tissue but generated 400 ng of DG per milligram. Furthermore, the discrepancies were greatest at later time points. With phospholipids tagged with different fatty acids, it was determined that PIP$_2$ supplied the DG acutely, but that PC was the source of DG after the first few minutes of stimulation. There are two ways that PC can release DG: (1) a PC-specific PLC (PC-PLC) can hydrolyze PC in a manner analogous to the PPI pathway, or (2) PC could be acted on by a PC-specific phospholipase D (PC-PLD). The latter reaction would actually produce phosphatidic acid (PA) and choline (see Fig. 10-3); the former would have to be degraded further by a PA phosphohydrolase to produce DG. Both pathways exist; the contribution by each will vary depending on the specific system.

Phosphocholine, the other product of the first hydrolytic pathway, has not been well studied. However, there are a few reports that suggest that it can transduce signals. For example, phosphocholine triggers mitosis, activates the transcription factor NF-κB, and synergizes with sphingosine-1-phosphate in the stimulation of Raf.

PC-PLC is stimulated by MAPK phosphorylation and inhibited by G$_o$. The effects of this enzyme can also be potentiated by factors that interfere with PC recycling. CTP:phosphocholine cytidylyltransferase is involved in the reutilization of DG and can be regulated by hormone mediators. For example, cholecystokinin (CCK), α-adrenergic agonists, and bombesin inhibit this enzyme in the pancreas by elevating calcium and activating CaM. When cytidylyltransferase activity is suppressed, DG recycling is delayed and DG accumulates.

Phosphatidylcholine Synthesis. In addition to being an alternate source for DG, the accumulation of PC can alter membrane properties in a way that enhances cAMP production. PC can be synthesized from phosphatidylserine through a decarboxylation to phosphatidylethanolamine followed by successive methylations (Fig. 10-12). Although two different phospholipid methyltransferases were originally postulated, recent data suggest that there may be only one. Because phosphatidylserine is predominantly located on the cytoplasmic face of the plasmalemma and PC is usually on the extracellular side, this conversion to PC is accompanied by a transverse migration across the membrane.

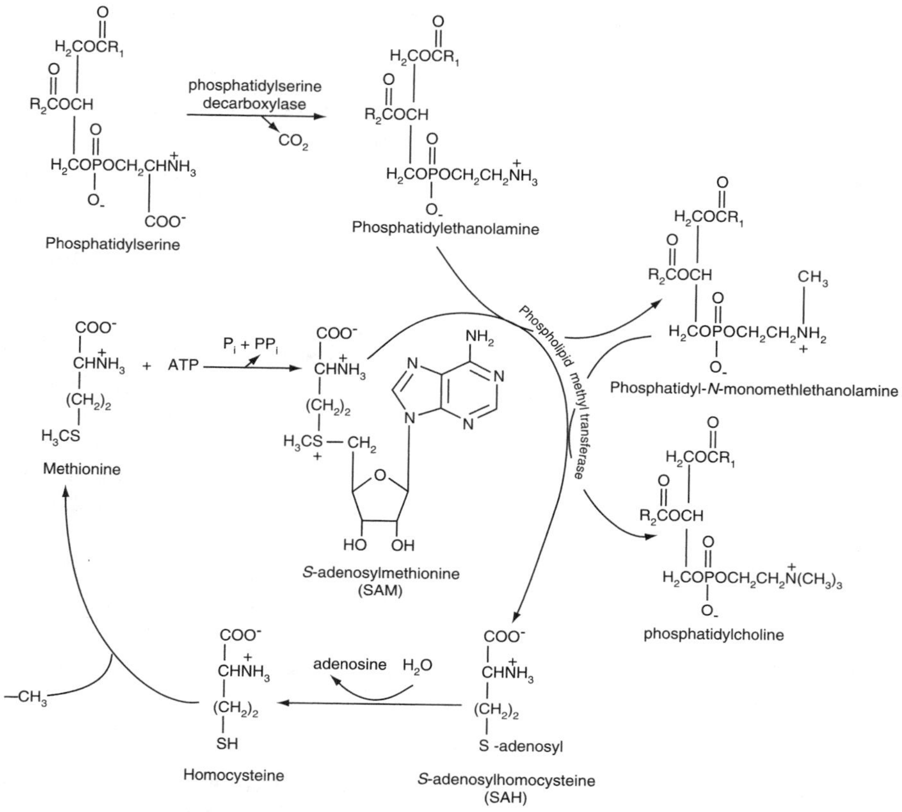

Fig. 10-12. Pathway for the synthesis of phosphatidylcholine.

The source of the methyl groups is *S*-adenosylmethionine, which is converted to *S*-adenosylhomocysteine (SAH). The methyltransferase is subject to product inhibition by SAH, and several drugs have been developed to take advantage of this fact. For example, 3-deazaadenosine and its structural variants either are or can be metabolized to SAH analogs and are potent inhibitors of the methyltransferase.

Phospholipid methylation has been closely associated with cAMP production in several systems. In fibroblasts, bradykinin stimulates phospholipid methylation before cAMP content rises; in most systems, an elevation in cAMP concentrations is the most common response to PC synthesis. However, in *Xenopus* oocytes, progesterone stimulation of phospholipid methylation is associated with a decline in cAMP levels. In both systems, the methylation peaks at 15 seconds, although the changes in cAMP content require 2 to 5 minutes, suggesting a cause-and-effect relationship. This is supported by the use of methyltransferase inhibitors, which also inhibit the changes in cAMP concentrations. Finally, cholera toxin and fluoride can stimulate the adenylate cyclase through G_s without affecting the PC levels; this finding, along with the time courses, would eliminate the possibility that the changes in the PC metabolism are a secondary event.

How might phospholipid methylation influence cAMP production? The first argument is that it increases membrane fluidity, thereby facilitating the coupling of receptor, G proteins, and adenylate cyclase. The ability of isoproterenol, a β-adrenergic agonist, to stimulate adenylate cyclase in turkey erythrocytes is influenced by membrane fluidity; loading the membranes with cholesterol decreases membrane fluidity and dampens isoproterenol-induced cyclase activity. Conversely, loading the membranes with vaccenic acid increases fluidity and enhances cyclase activity. Likewise, increasing the PC content of these membranes increases their fluidity and coupling efficiency.

Second, phospholipid methylation may act through calcium. Calcium influxes are stimulated after phospholipid methylation but before changes in cAMP content are observed; these fluxes usually occur in 0.5 to 2 minutes depending on the system. Furthermore, methylation inhibitors also inhibit these fluxes, suggesting that they were evoked by the methylation. Because bradykinin has been shown to stimulate the PPI pathway, it is possible that the changes in membrane fluidity induced by changes in PC metabolism could also have facilitated PPI hydrolysis, which then led to the calcium fluxes. Regardless of the exact mechanism, the resulting calcium fluctuations would alter adenylate cyclase activity (see later).

Third, the increase in PC may stimulate a PC-specific phospholipase A_2(PC-PLA$_2$), which would release arachidonic acid for eicosanoid synthesis (see later); many receptors for the eicosanoids are coupled to adenylate cyclase. Methyltransferase inhibitors block the release of arachidonic acid and cAMP elevation; mepacrine (also called quinacrine) is an inhibitor of PLA$_2$ and has the same effect. Alternatively, the active agent may not be arachidonic acid or its metabolites but the other hydrolytic product, lysophosphatidylcholine. Lysophosphatidylcholine, like the eicosanoids, is a parahormone that binds a GPCR. In addition, lysophospholipids, which lack a fatty acid in the second position, are strong detergents and, in sufficiently large concentrations, can lyse cells (see later). Indeed, the active ingredient in several snake toxins is a PLA$_2$, and the toxicity of the venom can be directly attributed to this lytic effect. In smaller amounts, lysophosphatidylcholine might act as a membrane fusogen and aid in secretion; phospholipid methylation has been implicated in a number of secretory systems. Lysophospholipids have also been implicated in the regulation of the (Na$^+$, H$^+$)antiport system that influences cellular pH.

This scheme is not without its critics; there are, in fact, three basic problems with this hypothesis. First, in many systems the elevation in cAMP precedes phospholipid methylation, and the methylation can be induced by cAMP. These systems include glucagon in the liver, ACTH in adipocytes, and hCG in Leydig cells. This is in contrast to the findings in turkey erythrocytes and *Xenopus* oocytes noted earlier. Second, the various inhibitors used all have side effects; the methyltransferase inhibitors can also inhibit other types of methylation reactions, and mepacrine can bind to chromatin and inhibits both oxidative phosphorylation and the mitochondrial ATPase. Furthermore, the effects of these drugs can be inconsistent; some methyltransferase inhibitors will block the actions of certain hormones in a particular system, whereas other inhibitors cannot, even though both effectively suppress methyltransferase activity.

Third, methyltransferase activity does not necessarily correspond to PC content. Phosphatidylcholine can be synthesized by two separate pathways: the methylation pathway (Fig. 10-12) and the salvage pathway. The latter pathway activates phosphocholine with CTP to form CDP-choline, which is then coupled to DG to form PC. In the liver, the two pathways are reciprocally controlled so as to maintain a constant PC content. For example, glucagon, β-agonists, and vasopressin stimulate the methyltransferase but inhibit the salvage pathway; 3-deazaadenosine inhibits the methyltransferase but stimulates the salvage pathway. Similarly, exogenous choline activates the salvage pathway and suppresses methylation; choline deficiency has the opposite effect. If this type of control operated in all tissues, PC content would remain constant regardless of the methyltransferase activity, and membrane fluidity would not be altered. However, PC could still be a significant source of arachidonic acid, DG, and lysophosphatidylcholine, even if total PC content and membrane fluidity does not fluctuate.

Phospholipase D

As noted previously, there is an alternate way to generate DG from PC: hydrolysis by PLD to produce choline and PA. PA is then dephosphorylated by PA phosphohydrolase. There are two PLDs, and both are involved with vesicular trafficking: PLD1 translocates granules to cell periphery, whereas PLD2 regulates calcium-dependent fusion of granules with the plasma membrane.

PLD is highly regulated (Fig. 10-13); *in vitro* it can be stimulated by G proteins (Rho, Arf, and Ral), PI(3)P$_n$, and several kinases (MAPK, PKC, and PKN). Although MAPK phosphorylates PLD, the activation by PKC and PKN does not require this modification. PKC is thought to act like a PLD anchoring protein. PKC can also phosphorylate PLD, but the effects are conflicting. GRK (G protein-coupled receptor kinase) stimulates PLD indirectly; it phosphorylates and neutralizes synuclein, a PLD inhibitor. However, when PLD activation is studied *in vivo*, there are several discrepancies with the regulation of the enzyme *in vitro*. Less work has been done on the PA phosphohydrolase. It does bind the EGFR, where it is sequestered; it dissociates after EGF binding.

The PLD-PA phosphohydrolase pathway is not simply a backup for the PC-PLC pathway; there are distinct differences in the two. First, the effects of PLD are

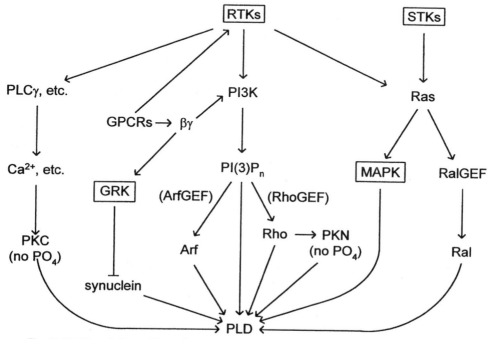

Fig. 10-13. Phospholipase D regulation. Kinases are boxed, arrows indicate stimulation, and flat heads indicate inhibition.

longer lasting. This is due to the preference of PLD for PC with saturated and monounsaturated fatty acids; the resulting DG is cleared more slowly that DG containing polyunsaturated fatty acids. Second, PLD is predominantly nuclear in some cells and may be primarily responsible for the production of nuclear PA and DG. Finally, there is a shift in output; although both produce DG, the species produced by PLD is not as potent in stimulating PKC as the polyunsaturated DG produced by PC-PLC. In addition, there is a new second messenger: PA. PA has many actions including the recruitment of proteins for receptor endocytosis, the activation of kinases (PKCη, PKCζ, TOR [target of rapamycin], and Raf), the stimulation of PTP1C and Ras, and the inhibition of PP1. PA can also synergize with Arf to activate PI(4)P5 kinase. Unfortunately, many of these actions have only been demonstrated *in vitro*, and so their physiological relevance is unknown.

Phospholipase A$_2$

PLA$_2$ hydrolyzes the fatty acid in the second position of phosphoglycerides (see Fig. 10-1). There are 11 PLA$_2$ groups, but only those associated with trans-duction are discussed here. The cytosolic PLA$_2$ (cPLA$_2$) belongs to Group IV and is calcium dependent. Calcium or ceramide binding is used for plasma membrane localization, and vimentin binding is used for perinuclear localization; the latter is induced by CaMKII phosphorylation of the enzyme. cPLA$_2$ is activated by PIP$_n$ through a PH domain, by α$_i$ and G$_h$, and by phosphorylation by PKC, CaMKII, MAPK, STKs, and RTKs. It is inhibited by PKA or PKG phosphorylation. Finally, cPLA$_2$ is inhibited by lipocortin. The EGFR and hepatocyte growth factor receptor (HGFR) can phosphorylate lipocortin, resulting in the release of this inhibitor and the stimulation of cPLA$_2$.

The calcium-independent PLA$_2$ (iPLA$_2$) belongs to Group VI. It is inhibited by CaM and stimulated when calcium levels fall and CaM dissociates. It is also stimulated by caspase cleavage during apoptosis. It functions in phospholipid remodeling. In addition, it mediates the effects of catecholamines and vasopressin in smooth muscle; however, the exact mechanism by which these hormones stimulate iPLA$_2$ is unknown.

The secreted PLA$_2$s (sPLA$_2$) belong to Groups I, II, and V. Because they are located extracellularly, one might not expect them to play a role in signaling. Indeed, Group I PLA$_2$s function exclusively in phospholipid digestion. However, Group II PLA$_2$s can bind heparin sulfate, be internalized, and migrate to the nucleus where they generate arachidonic acid (AA). This distribution has import-ance for the ultimate fate of the AA. In the nucleus, sPLA$_2$ associates with the 5-lipoxygenase and feeds AA into the leukotriene pathway. On the other hand, cPLA$_2$ is cytosolic and feeds AA into the prostaglandin pathway. Group V PLA$_2$s use a different mechanism; they remain extracellular to generate lysophosphati-dylcholine and AA, which then bind to GPCRs, elevate intracellular calcium, and activate cPLA$_2$.

In addition to acting as a precursor for the eicosanoids (see Chapter 3), this acid can be converted to arachidonylethanolamide, a ligand for the cannabinoid receptor. Finally, AA can exert several activities by itself; it activates PKCγ, Ras, the soluble guanylate cyclase, and the neutral sphingomyelinase. It also directly stimulates potassium channels and potentiates the glutamate (NMDA [*N*-methyl-D-aspartate) channels. On the other hand, it inhibits CaMKII, the myosin light chain phosphatase, and estradiol binding to its receptor. The imme-diate source of the AA is phosphatidylethanolamine (PE), which is located in the inner leaflet of the plasmalemma and has a high content of this fatty acid.

However, PC indirectly makes a major contribution by transferring its AA to the lysophosphatidylethanolamine to regenerate PE. PI is also rich in AA and may be responsible for as much as one-third of the liberated AA in some systems. AA can be released from either intact PI by PLA_2 or from DG by DG lipase after PLC hydrolysis; the latter pathway provides 12% to 15% of the AA released in platelets.

The other product of PLA_2 is a lysophospholipid. In addition to their possible roles as mediators of endocytosis or exocytosis (see previous), lysophospholipids can be released as parahormones that act through either GPCRs or the peroxisome proliferator activated receptor γ (PPARγ). These lipids may also serve as precursors to the platelet-activating factor (PAF); an alkylether PC can be converted to PAF by PLA_2 and lyso-PAF:acetyl CoA acetyltransferase. The latter enzyme can be activated by PKA, PKC, or CaMKII, depending on the system. PAF is degraded by PAF-acetylhydrolase, and this enzyme can be regulated by steroids; serum levels of the hydrolase are induced by progesterone and glucocorticoids and repressed by estradiol. The elevation occurs through gene induction because protein synthesis is required. The PAF would be secreted and act in an autocrine-paracrine fashion (see Chapter 3).

One last mediator in this family is lysophosphatidic acid. This is a lysophospholipid whose head group has been removed and is generated by the sequential action of PLD and either PLA_1 or PLA_2. The PLD-PLA sequence is possible because PLA does not require a head group for its function. However, PLD does not act on lysophospholipids; therefore the reverse sequence, PLA-PLD, requires a special PLD, called *autotaxin*. Like lysophospholipids, lysophosphatidic acid is a parahormone that acts through GPCRs. It is primarily involved with chemotaxis and related phenomena.

Phosphatidylinositol-3 Kinase

As noted at the beginning of this chapter, the central phospholipid in the PPI pathway is phosphatidylinositol 1,4,5-trisphosphate (PIP_2), which is hydrolyzed to yield IP_3 and DG. However, other isomers also exist; for example, PI-3′ kinase (PI3K) generates a phosphatidylinositol phosphorylated on the 3′ position. The PI3K is activated by several mitogens, and the resulting phospholipid has been closely associated with mitogenesis. There are, in fact, three classes of PI3K. All of the classes are homologous to each other and to a group of protein kinases that include the DNA-PK and TOR. Indeed, these protein kinases are sometimes referred to as *Class IV PI3Ks*, although they do not possess any lipid kinase activity. However, some PI3Ks have been shown to have protein kinase activity under special circumstances (see later), but the number of known substrates is very small. Class I is divided into IA (PI3Kα, PI3Kβ, and PI3Kδ) and IB (PI3Kγ). In the Class IA PI3Ks, the catalytic subunit, also known as p110, has a Ras binding domain at the amino terminus followed by a C2 domain, a HEAT (huntingtin-EF3-A subunit of PP2A-TOR1) domain, and finally the catalytic site. The HEAT domain, also called the *helical domain*, acts as an internal scaffold. The Class IB PI3K has an additional carboxy-terminal C2 and PX domain, which results in this enzyme being constitutively membrane bound.

The Class IA PI3Ks also have a regulatory subunit; the best-known one is called p85 and consists of two SH2, one SH3, and a BH (breakpoint cluster homology) domain. p85 is an inhibitory subunit; in addition, it is an adaptor that not only links p110 to specific activators but can alter the activity of the catalytic subunit. For example, p85 binding to pY through its SH2 domains will relieve the inhibition, but the resulting effect will depend on the RTK or docking protein to which it is bound; when bound to the insulin receptor, PI3K acts like a protein kinase that phosphorylates and activates PDE3B, but when bound to insulin receptor substrate (IRS), it is a lipid kinase. Other factors acting through p85 include CaM, which binds the SH2 domain and stimulates p110, and Ruk, a proline-rich scaffold that binds the SH3 domain and inhibits p110. Tyrosine phosphorylation of p85 results in intramolecular binding to the SH2 and also relieves the inhibition. Finally, PKA phosphorylation stabilizes the PI3K-Ras complex.

The p110 subunit can also be directly regulated. As noted previously, there is a Ras-binding domain in the amino terminus, and Ras activates PI3K. Rab and PIKE (PI3K enhancer) are other small G proteins affecting PI3K; for example, Rab5 recruits both the Class I and Class III PI3Ks. PIKE is a nuclear G protein that is stimulated by PIP_n through its PH domain; it then activates nuclear PI3K. Tyrosine phosphorylation of p110 is inhibitory. The $\beta\gamma$ subunits of G proteins activate PI3Kγ both directly and indirectly; they recruit PI3Kγ to the plasma membrane through its regulatory subunit, p101, and they directly stimulate both the lipid and protein kinase activities.

p85 has several functions independent of p110. In addition to regulating p110 activity, p85 acts as an adaptor for a variety of transducers; it mediates the binding of IRS and $p62^{Dok}$, both docking proteins, to the insulin receptor; Gab1, another docking protein, to RTKs; and TOR to S6KI, so that the former kinase can phosphorylate and activate the latter kinase. p85 also binds Rac and mediates some of its effects, such as the formation of lamellipodia and the activation of NFAT, a transcription factor. P55γ is another regulatory subunit for p110; it binds and sequesters the antioncogene Rb (see Chapter 16). The activation of PI3K induces the dissociation of Rb.

In general, the Class I catalytic subunits are involved with signaling and with the initial phases of phagocytosis; these effects are a result of the synthesis of $PI(3)P_n$. First, this phospholipid activates PKB by means of a three-step process. Initially, PKB is recruited to the plasma membrane by the interaction between $PI(3)P_n$ and the PH domain of PKB. Then PKB is phosphorylated on S-473; the identity of this kinase is still very controversial. Finally, a PIP_n-dependent kinase 1 (PDK1), which is homologous to PKB and also has a PH domain, is recruited to the plasmalemma, is activated, and phosphorylates PKB at S-308. PKB is a critical kinase in insulin transduction and apoptosis inhibition (see Chapter 12). $PI(3)P_n$ can also activate several PKC isoforms (Table 10-5).

Second, $PI(3)P_n$ acts as a membrane anchor for proteins containing PH, PX, or FYVE (Fab1p-YOTP-Vac1p-EEA1) domains; these proteins include RhoGEFs, the Tec family of STKs, and proteins involved with the cytoskeleton and endosomal trafficking. Finally, $PI(3)P_n$ binds and activates dynamin, which is responsible for neck stricture of coated pits in the internalization of receptors.

In addition to $PI(3)P_n$, Class I PI3K can phosphorylate a limited number of proteins. Its activation of PDE3B was previously noted. Autophosphorylation of PI3Kβ, but not PI3Kγ, leads to inhibition, whereas IRS phosphorylation inhibits subsequent tyrosine phosphorylation. Both of these effects probably represent negative feedback. Finally, PI3K can phosphorylate the inhibitory translation factor, 4E-BP1; although the effect was not reported, modification of this factor usually neutralizes its inhibition of protein synthesis.

Class II PI3Ks include C2β (PI3KIIβ) and Cpk/p170. The SH3 domain of Grb2 binds the amino-terminal polyproline region in these PI3Ks and mediates their recruitment to RTKs, where they are activated by clathrin and participate in endocytosis. C2β is also stimulated by calpain cleavage. Class III PI3Ks include the VPS34-like proteins and have a structure similar to Class IB except that there is no Ras-binding domain. These PI3Ks are involved with membrane trafficking and phagolysosome formation.

Sphingolipids

Sphingolipids are membrane lipids with a structure similar to phospholipids: a backbone containing two fatty acids and a head group. However, the backbone is serine instead of glycerol. In sphingosine, the carboxylic acid of serine is attached to palmitic acid in a carbon-carbon bond, rather than the ester bond seen in phospholipids. When the second fatty acid is coupled through an amide bond, the molecule becomes ceramide. The head group defines three different classes of sphingolipids: the cerebrosides contain a single sugar; gangliosides, complex sugars; and sphingomyelin, phosphocholine. Sphingomyelin breakdown occurs during stress. If the damage is repairable, a stress response is activated; however, if the damage is irreparable, apoptosis is triggered to prevent cancerous mutations. These opposite reactions are executed by different branches of the pathway: sphingosine-1-phosphate stimulates the stress response, whereas ceramide initiates apoptosis (Fig. 10-14).

This pathway is initiated by the activation of sphingomyelinase (SMase). There are two enzymes that are hormonally regulated: an acid SMase (A-SMase) and a neutral SMase (N-SMase). The former is located in lysosomes, which immediately present a problem: How do hormones on the cell surface stimulate an enzyme in a cell organelle? The first solution is to generate a soluble second messenger that can migrate to the lysosome. For example, several hormones activate PC-PLC to produce DG, which then stimulates A-SMase. A second solution is to internalize the hormone-receptor complex; endocytosis of the IL-18R is required for it to activate A-SMase. Conversely, the enzyme could be brought to the surface; the CD40 ligand induces A-SMase containing vesicles to migrate to the plasma membrane with which they fuse. The N-SMase has a death domain through which it can couple to receptors through adaptors, like Fan. It can also be activated by arachidonic acid released by PLA_2. In addition to differences in their control, these enzymes appear to function at different time; in many cells, A-SMase is responsible for the acute, transient rise in ceramide; and the N-SMase is responsible for the sustained elevation of ceramide. However, in colon cells, the reverse is true.

The direct regulation of these enzymes is not the only way to control this pathway; substrate availability is important. Sphingomyelin is usually located on the outer leaflet of the plasmalemma, although a small intracellular pool does exist. During apoptosis and calcium elevations, membrane lipids can be flipped between leaflets by scramblase. Some sphingolipids, like sphingosine-1-phosphate, can also be transported by an ABC transporter. Yet another way to regulate this pathway is at the level of product metabolism. SMase hydrolyzes sphingomyelin into ceramide and phosphocholine. The former is the active product and can be cleared either by sphingomyelin synthase, which simply reverses the SMase reaction, or by glucosylceramide synthase, which feeds ceramide into the cerebroside-ganglioside pathway. Tumor necrosis factor (TNF)

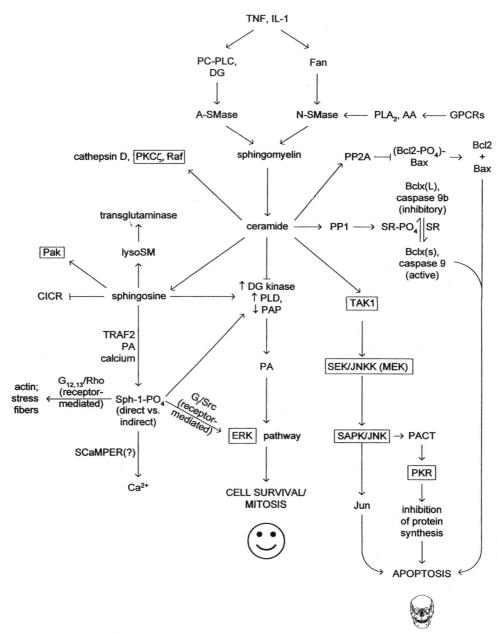

Fig. 10-14. Sphingomyelin signaling pathway. Kinases are boxed, arrows indicate stimulation, and flat heads indicate inhibition. *AA,* Arachidonic acid; *A-SMase,* acid sphingomyelinase; *Bcl,* B cell leukemia-lymphoma (an apoptotic factor family); *CICR,* calcium-induced calcium release channel; *ERK,* extracellular-signal regulated kinase; *Fan,* factor-associated with N-SMase (an adaptor); *lysoSM,* sphingosylphosphocholine; *N-SMase,* neutral sphingomyelinase; *PA,* phosphatidic acid; *PACT,* PKR activating protein; *PAP,* PA phosphohydrolase; *PC-PLC,* PC-specific PLC; *PKR,* dsRNA-activated protein kinase; *SAPK/JNK,* stress-activated protein kinase/Jun amino-terminal kinase; *SCaMPER,* sphingolipid calcium release-mediating protein of the endoplasmic reticulum; *SEK/JNKK (MEK),* SAPK-ERK kinase/Jun amino-terminal kinase kinase (a MAPK-ERK-activating kinase); *SR,* a splicing factor; *TAK,* TGFβ-activated kinase.

stimulates caspases, which are apoptotic proteases and which inactivate these enzymes, allowing ceramide to accumulate.

Ceramide induces apoptosis via several routes. One involves protein phosphatases and the Bcl family. The latter include both proapoptotic and antiapoptotic members; in particular, phosphorylated Bcl2 binds and inhibits the apoptotic actions of Bax. Ceramide activates PP2A, which dephosphorylates Bcl2 and releases Bax (Fig. 10-14). Another family member, Bclx, and a caspase have multiple isoforms; when phosphorylated, the SR (serine [S]-arginine [R]-rich) splicing factors generate the apoptotic inhibitory forms, Bclx(L) and caspase 9b. However, ceramide also activates PP1, which dephosphorylates SR. Unmodified SR generates the active apoptotic forms, Bclx(s) and caspase 9. Ceramide can also stimulate several kinases. The stress-activated protein kinase (SAPK), also called the *Jun amino-terminal kinase* (JNK), phosphorylates the transcription factor, Jun, which can induce apoptotic factors. In addition, it phosphorylates PACT, an activator of protein kinase R (PKR). PKR is also known as the *dsRNA-activated protein kinase* because it is stimulated during viral infections; its phosphorylation of translation factors turns off protein synthesis to abort viral reproduction. The same effect occurs in apoptosis: in this case, PACT activates PKR to inhibit translation. Finally, the physiochemical properties of ceramide may contribute to its activity. For example, its small head group, hydrophobicity, and tendency to aggregate facilitate the fusion of vesicles and the coalescence of DIGs (detergent-insoluble, ganglioside-enriched complexes). The latter could promote transduction by assembling the various signaling components at one site.

Ceramidase converts ceramide to sphingosine. There are three enzymes: the acid ceramidase is a lysosomal enzyme and does not participate in signal transduction. The neutral ceramidase is mitochondrial, and the alkaline enzyme is in the endoplasmic reticulum; both are stimulated by some cytokines. Sphingosine can be converted to sphingosylphosphocholine, also called *lysosphingomyelin,* which complements the actions of ceramide. First, it directly stimulates a transglutaminase by lowering the calcium requirement; this enzyme cross-links proteins during apoptosis. In addition, sphingosylphosphocholine is a parahormone that is secreted from cells and binds GPCRs to activate Jun.

Alternatively, sphingosine can be phosphorylated by sphingosine-1-phosphate kinase. This kinase can be regulated in several ways; first, it can bind and be activated by TRAF2, an adaptor involved with the TNF signaling pathway. PA released by PLD stimulates the kinase, and calcium-CaM facilitates kinase activity by shifting the kinase to the plasma membrane. Finally, because sphingosine is primarily extracellular, some authorities suggest that substrate availability is a mechanism for controlling this enzyme. For example, PKC induces the phosphorylation of the kinase, its translocation to the plasma membrane, and even its secretion. The latter would give sphingosine-1-phosphate kinase access to its substrate. Although PKC is suspected to phosphorylate this lipid kinase directly, an indirect effect through a kinase cascade cannot be ruled out because direct modification by PKC has been difficult to demonstrate *in vitro*. Sphingosine-1-phosphate is degraded through sphingosine-1-phosphate phosphohydrolase. Interestingly, this phosphohydrolase does not just terminate sphingosine-1-phosphate signaling; it also elevates ceramide levels by recycling the sphingosine.

Sphingosine-1-phosphate opposes ceramide by promoting cell survival and proliferation; this is accomplished, in part, through the ERK (extracellular-signal regulated kinase) pathway. ERK is stimulated in two ways. First, sphingosine-1-phosphate is secreted and activates a GPCR that stimulates ERK through G_i and Src. Second, it elevates PA levels; PA can activate ERK through Raf. Both sphingosine-1-phosphate and sphingosine stimulate PLD, which generates PA from phospholipids; stimulate DGK, which generates PA from DG; and inhibit PA phosphohydrolase, which breaks down PA into DG. This is a site for cross talk with ceramide, which lowers PA levels by inhibiting PLD and DGK and by stimulating PA phosphohydrolase. Sphingosine-1-phosphate also inhibits the A-Smase through its activation of PKB, an antiapoptotic kinase; this effect is independent of the G protein or ERK pathways. In addition, it induces stress fiber formation via its receptor. Finally, sphingosine-1-phosphate elevates calcium levels, although the mechanism for the calcium release is not known. Originally, it was thought to activate a calcium channel, SCaMPER, but recent data suggest that SCaMPER may not be a channel, although it may still be a regulatory subunit for such a channel.

Lipoxygenase Pathway

The role of several lipoxygenase products in hormone action has been investigated. In many systems, lipoxygenase inhibitors specifically block the action of the hormone, and this inhibition can be overcome by a particular product of this pathway. Finally, the hormone can elevate the levels of these molecules. These systems include ACTH and EGF in steroidogenesis; dopamine and GnRH in SRIF (somatotropin release-inhibiting factor) and LH release, respectively; and angiotensin in aldosterone synthesis and the inhibition of renin release (Table 10-6). The first step in leukotriene synthesis is 5-lipoxygenase, and it is primarily regulated by substrate availability. For example, calcium induces its localization to membranes, where the AA is generated. The enzyme also has a SH3 domain, which binds Grb2, α-actinin, and actin, and which contributes to the membrane localization. Finally, it is stimulated by FLAP, a protein that facilitates the transfer of arachidonic acid to the enzyme. In addition to substrate availability, the activity of 5-lipoxygenase can be stimulated by phosphorylation by members of the MAPK family; ERK directly phosphorylates the enzyme, whereas p38 acts through MAPK activated protein kinase 2 (MAPKAP2).

Leukotrienes are parahormones that use GPCRs. However, there is evidence that at least a few actions of these lipids may be direct, such as the inhibition of CaMKII and the regulation of potassium channels.

Table 10-6
Role of the Lipoxygenase Pathway in Hormone Action

Hormone	Activity	Presumed mediator
Angiotensin	Inhibit renin release; steroidogenesis	12-Hydroxy-6,8,11,14-eicosa-tetraenoic acid (12-HETE)
Dopamine	SRIF release	8,9-Epoxyeicosatrienoic acid
GnRH	LH release	Leukotriene C_4
ACTH	Steroidogenesis	5-Hydroperoxy-6,8,11,14-eico-satetraenoic acid (5-HPETE)
EGF	Steroidogenesis	12-Hydroxy-6,8,11,14-eicosa-tetraenoic acid (12-HETE)

Summary

Hormones can use several phospholipid metabolites as second messengers (see Figs. 10-3 and 10-14). Hormones can activate PI-specific PLCβ through G_q, PLCγ through tyrosine phosphorylation, PLCε through several G proteins (G_{12}, βγ, Ras, and Rap2B), and PLCδ through PI(4,5)P_2, calcium, and the inhibition of Rho and G_h. The head group, IP_3, triggers the release of calcium from internal stores; the calcium is released in pulses whose frequency contains information about the intensity of the signal. Hormones can also elevate calcium through ROCC and mitochondria. However, mitochondria function primarily in signal amplification and in restricting calcium elevation to specified regions of the cell. Although calcium can modulate some enzymes, this cation usually acts in combination with CaM. Calcium-CaM binds many target proteins, but the greatest effects are seen with a multifunctional protein kinase, CaMKII, and the protein phosphatase, calcineurin.

DG is an allosteric activator of another kinase, PKC; this is actually a large family of kinases, some of which also require calcium and phospholipids. The hydrolysis of PC generates DG for PKC stimulation without elevating calcium because the phosphocholine head group is inactive. This pathway appears to be more important at later periods after hormone stimulation. PC hydrolysis can be accomplished either by a PC-PLC or by a combination of PC-PLD and PA phosphohydrolase. PC-PLC can be activated by MAPK phosphorylation, Ras, and G proteins, and PLD is stimulated by G proteins, PI(3)P_n, and several kinases.

Another phospholipase, PLA_2, produces AA, which can directly affect protein kinases and ion channels or be converted to various eicosanoids. PLA_2 releases AA primarily from PE and PC and can be hormonally regulated through G proteins, calcium, and phosphorylation. A small amount of AA may come from DG through DG lipase. PLA_2 also generates a lysophospholipid as the other product; this metabolite may affect endocytosis and exocytosis, mitogenesis, cAMP synthesis, and calcium release. Finally, sphingomyelinase can initiate the breakdown of sphingomyelin into ceramide and sphingosine-1-phosphate. The former initiates apoptosis, whereas the latter promotes cell survival and proliferation.

Homologous regulation of the PPI pathway revolves around feedback inhibition by the two major products, PKC and calcium. They stimulate the removal of their activators and suppress further PPI hydrolysis. However, they also prime the system for the next round of stimulation by replenishing calcium and PPI stores. Heterologous regulation by cAMP can be either synergistic or antagonistic, depending on the system being studied.

References

Methodology

Brose, N., and Rosenmund, C. (2002). Move over protein kinase C, you've got company: Alternative cellular effectors of diacylglycerol and phorbol esters. *J. Cell Sci.* 115, 4399-4411.

Kazanietz, M.G. (2002). Novel "nonkinase" phorbol ester receptors: The C1 domain connection. *Mol. Pharmacol.* 61, 759-767.

Takahashi, A., Camacho, P., Lechleiter, J.D., and Herman, B. (1999). Measurement of intracellular calcium. *Physiol. Rev.* 79, 1089-1125.

Calcium Channels

Carafoli, E., Santella, L., Branca, D., and Brini, M. (2001). Generation, control, and processing of cellular calcium signals. *Crit. Rev. Biochem. Mol. Biol.* 36, 107-260.

Catterall, W.A. (2000). Structure and regulation of voltage-gated Ca^{2+} channels. *Annu. Rev. Cell Dev. Biol.* 16, 521-555.

Chini, E.N., and de Toledo, F.G.S. (2002). Nicotinic acid adenine dinucleotide phosphate: A new intracellular second messenger? *Am. J. Physiol.* 282, C1191-C1198.

da Silva, C.P., and Guse, A.H. (2000). Intracellular Ca^{2+} release mechanisms: Multiple pathways having multiple functions within the same cell type? *Biochim. Biophys. Acta* 1498, 122-133.

Fill, M., and Copello, J.A. (2002). Ryanodine receptor calcium release channels. *Physiol. Rev.* 82, 893-922.

Higashida, H., Hashii, M., Yokoyama, S., Hoshi, N., Chen, X.L., Egorova, A., Noda, M., and Zhang, J.S. (2001). Cyclic ADP-ribose as a second messenger revisited from a new aspect of signal transduction from receptors to ADP-ribosyl cyclase. *Pharmacol. Ther.* 90, 283-296.

Keef, K.D., Hume, J.R., and Zhong, J. (2001). Regulation of cardiac and smooth muscle Ca^{2+} channels ($Ca_V1.2a, b$) by protein kinases. *Am. J. Physiol.* 281, C1743-C1756.

Kiselyov, K., Shin, D.M., and Muallem, S. (2003). Signalling specificity in GPCR-dependent Ca^{2+} signalling. *Cell. Signal.* 15, 243-253.

Lee, H.C. (2001). Physiological functions of cyclic ADP-ribose and NAADP as calcium messengers. *Annu. Rev. Pharmacol. Toxicol.* 41, 317-345.

Meissner, G. (2002). Regulation of mammalian ryanodine receptors. *Front. Biosci.* 7, d2072-d2080.

Patel, S., Churchill, G.C., and Galione, A. (2001). Coordination of Ca^{2+} signalling by NAADP. *Trends Biochem. Sci.* 26, 482-489.

Venkatachalam, K., van Rossum, D.B., Patterson, R.L., Ma, H.T., and Gill, D.L. (2002) The cellular and molecular basis of store-operated calcium entry. *Nature Cell Biol.* 4, E263-E272.

Calcium Pools

Bootman, M.D., Thomas, D., Tovey, S.C., Berridge, M.J., and Lipp, P. (2000). Nuclear calcium signalling. *Cell. Mol. Life Sci.* 57, 371-378.

Cocco, L., Martelli, A.M., Gilmour, R.S., Rhee, S.G., and Manzoli, F.A. (2001). Nuclear phospholipase C and signaling. *Biochim. Biophys. Acta* 1530, 1-14.

Martelli, A.M., Tabellini, G., Borgatti, P., Bortul, R., Capitani, S., and Neri, L.M. (2003). Nuclear lipids: New functions for old molecules? *J. Cell. Biochem.* 88, 455-461.

Rutter, G.A., and Rizzuto, R. (2000). Regulation of mitochondrial metabolism by ER Ca^{2+} release: An intimate connection. *Trends Biochem. Sci.* 25, 215-221.

Phospholipase C

Carpenter, G., and Ji, Q. (1999). Phospholipase C-γ as a signal-transducing element. *Exp. Cell Res.* 253, 15-24.

Fukami, K. (2002). Structure, regulation, and function of phospholipase C isozymes. *J. Biochem. (Tokyo)* 131, 293-299.

Rebecchi, M.J., and Pentyala, S.N. (2000). Structure, function, and control of phosphoinositide-specific phospholipase C. *Physiol. Rev.* 80, 1291-1335.

Rhee, S.G. (2001). Regulation of phosphoinositide-specific phospholipase C. *Annu. Rev. Biochem.* 70, 281-312.

Williams, R.L. (1999). Mammalian phosphoinositide-specific phospholipase C. *Biochim. Biophys. Acta* 1441, 255-267.

Calcium Profiles

Bootman, M.D., Lipp, P., and Berridge, M.J. (2001). The organisation and functions of local Ca^{2+} signals. *J. Cell Sci.* 114, 2213-2222.

Dupont, G., Swillens, S., Clair, Tordjmann, T., and Combettes, L. (2000). Hierarchical organization of calcium signals in hepatocytes: From experiments to models. *Biochim. Biophys. Acta* 1498, 134-152.

Termination

Kanoh, H., Yamada, K., and Sakane, F. (2002). Diacylglycerol kinases: Emerging downstream regulators in cell signaling systems. *J. Biochem. (Tokyo)* 131, 629-633.

Laporte, J., Blondeau, F., Buj-Bello, A., and Mandel, J.L. (2001). The myotubularin family: From genetic disease to phosphoinositide metabolism. *Trends Genet.* 17, 221-228.

Leslie, N.R., and Downes, C.P. (2002). PTEN: The down side of PI 3-kinase signalling. *Cell. Signal.* 14, 285-295.

Rohrschneider, L.R., Fuller, J.F., Wolf, I., Liu, Y., and Lucas, D.M. (2000). Structure, function, and biology of SHIP proteins. *Genes Dev.* 14, 505-520.

Waite, K.A., and Eng, C. (2002). Protean PTEN: Form and function. *Am. J. Hum. Genet.* 70, 829-844.

Wishart, M.J., Taylor, G.S., Slama, J.T., and Dixon, J.E. (2001). PTEN and myotubularin phosphoinositide phosphatases: Bringing bioinformatics to the lab bench. *Curr. Opin. Cell Biol.* 13, 172-181.

Replenishment

Clapham, D.E., Runnels, L.W., and Strübing, C. (2001). The TRP ion channel family. *Nature Rev. Neurosci.* 2, 387-396.

De Matteis, M.A., Godi, A., and Corda, D. (2002). Phosphoinositides and the Golgi complex. *Curr. Opin. Cell Biol.* 14, 434-447.

Hardie, R.C. (2003). Regulation of TRP channels via lipid second messengers. *Annu. Rev. Physiol.* 65, 735-759.

Harteneck, C., Plant, T.D., and Schultz, G. (2000). From worm to man: Three subfamilies of TRP channels. *Trends Neurosci.* 23, 159-166.

Kanaho, Y., and Suzuki, T. (2002). Phosphoinositide kinases as enzymes that produce versatile signaling lipids, phosphoinositides. *J. Biochem. (Tokyo)* 131, 503-509.

Putney, J.W. (1999). TRP, inositol 1,4,5-trisphosphate receptors, and capacitative calcium entry. *Proc. Natl. Acad. Sci. U.S.A.* 96, 14669-14671.

Topham, M.K., and Prescott, S.M. (2002). Diacylglycerol kinases: Regulation and signaling roles. *Thromb. Haemost.* 88, 912-918.

Vennekens, R., Voets, T., Bindels, R.J.M., Droogmans, G., and Nilius, B. (2002). Current understanding of mammalian TRP homologues. *Cell Calcium* 31, 253-264.

Zitt, C., Halaszovich, C.R., and Lückhoff, A. (2002). The TRP family of cation channels: Probing and advancing the concepts on receptor-activated calcium entry. *Prog. Neurobiol.* 66, 243-264.

Calcium and Calmodulin

Bayer, K.U., and Schulman, H. (2001). Regulation of signal transduction by protein targeting: The case for CaMKII. *Biochem. Biophys. Res. Commun.* 289, 917-923.

Benaim, G., and Villalobo, A. (2002) Phosphorylation of calmodulin: Functional implications. *Eur. J. Biochem.* 269, 3619-3631.

Chin, D., and Means, A.R. (2000). Calmodulin: A prototypical calcium sensor. *Trends Cell Biol.* 10, 321-328.

Fujisawa, H. (2001). Regulation of the activities of multifunctional Ca^{2+}/calmodulin-dependent protein kinases. *J. Biochem. (Tokyo)* 129, 193-199.

Hoeflich, K.P., and Ikura, M. (2002). Calmodulin in action: Diversity in target recognition and activation mechanisms. *Cell (Cambridge, Mass.)* 108, 739-742.

Hook, S.S., and Means, A.R. (2001). Ca^{2+}/CaM-dependent kinases: From activation to function. *Annu. Rev. Pharmacol. Toxicol.* 41, 471-505.

Hudmon, A., and Schulman, H. (2002). Structure-function of the multifunctional Ca^{2+}/calmodulin-dependent protein kinase II. *Biochem. J.* 364, 593-611.

Rusnak, F., and Mertz, P. (2000). Calcineurin: Form and function. *Physiol. Rev.* 80, 1483-1521.

Soderling, T.R., Chang, B., and Brickey, D. (2001). Cellular signaling through multifunctional Ca^{2+}/calmodulin-dependent protein kinase II. *J. Biol. Chem.* 276, 3719-3722.

Toutenhoofd, S.L., and Strehler, E.E. (2000). The calmodulin multigene family as a unique case of genetic redundancy: Multiple levels of regulation to provide spatial and temporal control of calmodulin pools? *Cell Calcium* 28, 83-96.

Protein Kinase C

McCahill, A., Warwicker, J., Bolger, G.B., Houslay, M.D., and Yarwood, S.J. (2002). The RACK1 scaffold protein: A dynamic cog in cell response mechanisms. *Mol. Pharmacol.* 62, 1261-1273.

Ron, D., and Kazanietz, M.G. (1999). New insights into the regulation of protein kinase C and novel phorbol ester receptors. *FASEB J.* 13, 1658-1676.

Van Lint, J., Rykx, A., Maeda, Y., Vantus, T., Sturany, S., Malhotra, V., Vandenheede, J.R., and Seufferlein, T. (2002). Protein kinase D: An intracellular traffic regulator on the move. *Trends Cell Biol.* 12, 193-200.

Calpain

Sato, K., and Kawashima, S. (2001). Calpain function in the modulation of signal transduction molecules. *Biol. Chem.* 382, 743-751.

Other Phosphoinositide Outputs

Cockcroft, S., and De Matteis, M.A. (2001). Inositol lipids as spatial regulators of membrane traffic. *J. Membr. Biol.* 180, 187-194.

Payrastre, B., Missy, K., Giuriato, S., Bodin, S., Plantavid, M., and Gratacap, M.P. (2001). Phosphoinositides: Key players in cell signalling, in time and space. *Cell. Signal.* 13, 377-387.

Shears, S.B. (2001). Assessing the omnipotence of inositol hexakisphosphate. *Cell. Signal.* 13, 151-158.

Toker, A. (2002). Phosphoinositides and signal transduction. *Cell. Mol. Life Sci.* 59, 761-779.

Phospholipase D

Banno, Y. (2002). Regulation and possible role of mammalian phospholipase D in cellular functions. *J. Biochem. (Tokyo)* 131, 301-306.

Exton, J.H. (2002) Regulation of phospholipase D. *FEBS Lett.* 531, 58-61.

Liscovitch, M., Czarny, M., Fiucci, G., and Tang, X. (2000). Phospholipase D: Molecular and cell biology of a novel gene family. *Biochem. J.* 345, 401-415.

Phospholipase A_2

Capper, E.A., and Marshall, L.A. (2001). Mammalian phospholipases A_2: Mediators of inflammation, proliferation and apoptosis. *Prog. Lipid Res.* 40, 167-197.

Hernández, M., Nieto, J.L., and Crespo, M.S. (2000). Cytosolic phospholipase A_2 and the distinct transcriptional programs of astrocytoma cells. *Trends Neurosci.* 23, 259-264.

Murakami, M., and Kudo, I. (2002). Phospholipase A_2. *J. Biochem. (Tokyo)* 131, 285-292.

Six, D.A., and Dennis, E.A. (2000). The expanding superfamily of phospholipase A_2 enzymes: classification and characterization. *Biochim. Biophys. Acta* 1488, 1-19.

Phosphatidylinositol-3 Kinase

Anderson, K.E., and Jackson, S.P. (2003). Class I phosphoinositide 3-kinases. *Int. J. Biochem. Cell Biol.* 35, 1028-1033.

Backer, J.M. (2000). Phosphoinositide 3-kinases and the regulation of vesicular trafficking. *Mol. Cell Biol. Res. Commun.* 3, 193-204.

Cantrell, D.A. (2001). Phosphoinositide 3-kinase signalling pathways. *J. Cell Sci.* 114, 1439-1445.

Corvera, S. (2001). Phosphatidylinositol 3-kinase and the control of endosome dynamics: New players defined by structural motifs. *Traffic* 2, 859-866.

Djordjevic, S., and Driscoll, P.C. (2002). Structural insight into substrate specificity and regulatory mechanisms of phosphoinositide 3-kinases. *Trends Biochem. Sci.* 27, 426-431.

Sphingolipids

Pettus, B.J., Chalfant, C.E., and Hannun, Y.A. (2002). Ceramide in apoptosis: An overview and current perspectives. *Biochim. Biophys. Acta* 1585, 114-125.

Huwiler, A., Kolter, T., Pfeilschifter, J., and Sandhoff, K. (2000). Physiology and pathophysiology of sphingolipid metabolism and signaling. *Biochim. Biophys. Acta* 1485, 63-99.

Merrill, A.H. (2002). De novo sphingolipid biosynthesis: A necessary, but dangerous, pathway. *J. Biol. Chem.* 277, 25843-25846.

Ohanian, J., and Ohanian, V. (2001). Sphingolipids in mammalian cell signalling. *Cell. Mol. Life Sci.* 58, 2053-2068.

Spiegel, S., and Milstien, S. (2002). Sphingosine 1-phosphate, a key cell signaling molecule. *J. Biol. Chem.* 277, 25851-25854.

van Blitterswijk, W.J., van der Luit, A.H., Veldman, R.J., Verheij,M., and Borst, J. (2003). Ceramide: Second messenger or modulator of membrane structure and dynamics? *Biochem. J.* 369, 199-211.

Young, K.W., and Nahorski, S.R. (2001). Intracellular sphingosine 1-phosphate production: A novel pathway for Ca^{2+} release. *Semin. Cell Dev. Biol.* 12, 19-25.

Lipoxygenase Pathway

Rådmark, O. (2002). Arachidonate 5-lipoxygenase. *Prostaglandins Lipid Med.* 68-69, 211-234.

Soberman, R.J., and Christmas, P. (2003). The organization and consequences of eicosanoid signaling. *J. Clin. Invest.* 111, 1107-1113.

CHAPTER *11*

Miscellaneous Second Messengers

CHAPTER OUTLINE

Polyamines

An inevitable concomitance of growth and differentiation in almost any system is the increase in polyamines. These small molecules are straight-chain organic compounds that have two or more amino groups. They are positively charged and are required for DNA synthesis, transcription, and translation. Therefore it is not surprising that they have been implicated in hormone action.

Polyamine Synthesis and Degradation

In eukaryotes, polyamine synthesis begins with ornithine, a product from the urea cycle (Fig. 11-1). Its decarboxylation by ornithine decarboxylase (ODC) is the committed step and leads to the first polyamine, putrescine. Spermidine and spermine are then synthesized by the sequential addition of aminopropyl groups to each end of the putrescine. The donor is *S*-adenosylmethionine (SAM); the attachment of the adenosine to the sulfur renders the thioether bonds labile. Normally, it is the methyl group that is donated in biosynthetic pathways; however, in polyamine synthesis, SAM is decarboxylated and the other side chain, the aminopropyl group, is transferred to putrescine to form spermidine. A second transfer to the other end of spermidine yields spermine. The regulation of ODC occurs primarily through altering enzyme levels. Putrescine, the initial product of this pathway, feeds back to inhibit the translation of ODC mRNA; decreases the half-life of the enzyme; and induces an enzyme inhibitor, the ODC-antizyme (Fig. 11-2, *A*). The enzyme can be stimulated by these same mechanisms; for example, protein kinase C (PKC) increases ODC transcription, whereas prolactin (PRL) decreases the antizyme.

Polyamine synthesis is also regulated by substrate availability; lysophosphatidylcholine increases polyamine levels by stimulating arginine transport. In addition, nitric oxide synthase (NOS) competes with ODC for limited arginine levels (see Chapter 7); nitric oxide (NO) can also *S*-nitrosylate and inhibit ODC. Finally, polyamines can be imported from the extracellular fluid. In astrocytes polyamine transport is increased by tyrosine kinases and decreased by PKC and the ODC antizyme; the targets of the kinases are unknown.

There are several useful inhibitors for the synthetic pathway. ODC can be inhibited by the substrate analogs, HAVA (α-hydrazino-δ-amino-valeric acid) and DFMO (α-difluoromethylornithine). The former is a competitive inhibitor, but the latter is a suicidal inhibitor, which irreversibly reacts with the enzyme. *S*-adenosylmethionine decarboxylase can be inhibited by MGBG [methylglyoxal bis(guanylhydrazone)], a polyamine analog, which binds to an allosteric site on the enzyme. SAM decarboxylase has an absolute requirement for putrescine; this ensures that SAM will not be decarboxylated unless there is putrescine available to accept the aminopropyl group. A second allosteric site binds spermine and inhibits the enzyme; this represents simple negative feedback (Fig. 11-1). MGBG is either a putrescine antagonist or a spermine agonist; in either case, enzyme activity is suppressed.

Polyamines are inactivated by acetylation. This modification has three major effects: first, it blocks the cationic site, which is critical to the biological activity of polyamines. Second, it accelerates their excretion because the polyamine uptake system in the kidneys does not recognize the acetylated form. Finally, polyamines are substrates for polyamine oxidase, whose product can either be recycled or further degraded by diamine oxidase. Polyamines

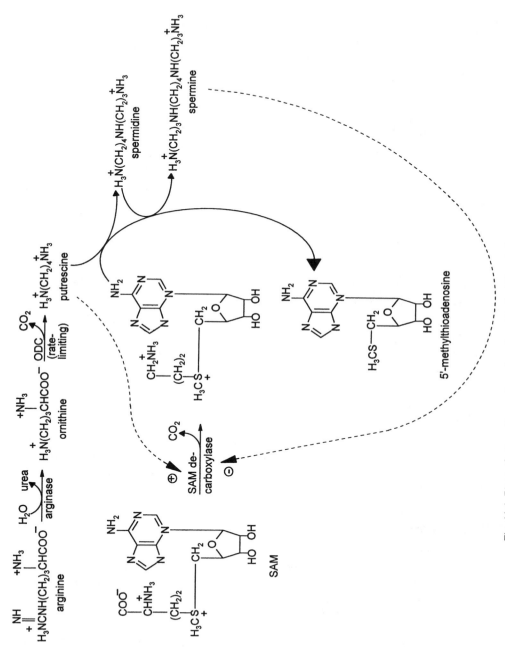

Fig. 11-1. Biosynthetic pathway for polyamines. Dashed lines indicate regulatory pathways.

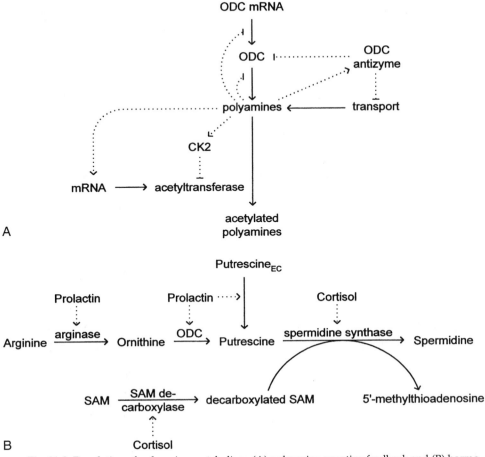

Fig. 11-2. Regulation of polyamine metabolism. (A) polyamine negative feedback and (B) hormonal regulation in the mammary gland.

induce the N^1-acetyltransferase by increasing the half-life of its mRNA; polyamines can also bind the N^1-acetyltransferase and block its ubiquitination and degradation. On the other hand, casein kinase 2 (CK2) phosphorylates the enzyme and triggers its proteasomal degradation.

Model Systems

Mammary Gland

Mammary gland differentiation is one system in which polyamines may mediate hormone actions (Fig. 11-2, *B*). Insulin, cortisol, and PRL are all required for mammary gland differentiation *in vitro*. Cortisol and PRL also elevate polyamine levels and stimulate the enzymes in the biosynthetic pathway: PRL stimulates arginase and ODC, whereas cortisol stimulates SAM decarboxylase and spermidine synthase. Furthermore, PRL can also elevate intracellular polyamine levels by increasing their transport. Because of their charge, polyamines do not readily cross membranes but require a transport system. The V_{max} for this system is stimulated 2.5-fold by PRL; the Michaelis-Menten constant (K_m) does not change. In addition, the stimulation during the first 12 hours does not depend on transcription or translation, suggesting that this effect is an early event in PRL action. This transport could be very important in altering intracellular polyamine concentrations because polyamine levels in the blood increase 3-fold during pregnancy when the mammary epithelium is undergoing rapid proliferation.

Figure 11-3 shows the time course of DNA synthesis, differentiation (that is, casein and α-lactalbumin accumulation), spermidine, and the enzyme activities for polyamine synthesis; this time course study was performed in mouse mammary gland explants cultured with insulin, cortisol, and PRL. Clearly, increases in ODC activity and spermidine levels precede differentiation. Furthermore, although the activities of SAM decarboxylase, spermidine synthase, and the second ODC peak require transcription and translation, the first ODC peak does not, again suggesting that this initial stimulation is a primary event.

Additional support for the role of polyamines in hormone-induced mammary differentiation comes from inhibitor studies. MGBG inhibits DNA synthesis, elevations in spermidine levels, and the production of milk proteins. This is not a nonspecific inhibition because it can be reversed by exogenous spermidine. However, the acid test for any postulated second messenger is whether the putative mediator can replace the hormone; in the mouse mammary gland, spermidine can replace cortisol, but not PRL, and it cannot replace any of the hormones in rat or rabbit mammary glands. In summary, polyamines appear to be involved in the hormonal induction of milk proteins, but they cannot be the sole mediators of hormone action in this system.

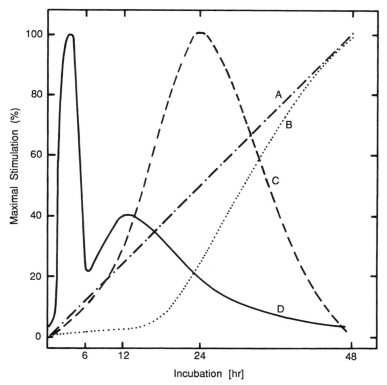

Fig. 11-3. Time course of differentiation and the activation of polyamine synthesis in mouse mammary gland explants cultured with insulin, cortisol, and PRL. Spermidine levels and the enzymatic activities of arginase, SAM decarboxylase, and spermidine synthase all rise linearly (A). The other lines represent casein and α-lactalbumin accumulation (B), DNA synthesis (C), and ornithine decarboxylase activity (D). Hormones were present throughout the culture period.

Kidney

Another system in which polyamines may mediate hormone effects is the kidney. In brief, administration of testosterone *in vivo* stimulates ODC activity within 30 seconds, the levels of all polyamines within 2 minutes, and both endocytosis and the uptake of amino acid and glucose analogs within 5 minutes. All of these effects are blocked by DFMO and reversed by exogenous putrescine. Finally, putrescine alone can mimic all of the actions of testosterone. Similar results are obtained with β-agonists. Unfortunately, both the supraphysiological dose of testosterone and the use of intact animals complicate the interpretation of these data; furthermore, at least one group has not been able to duplicate the result with testosterone. However, if the original results are eventually validated and shown to be physiologically relevant, it would characterize a hormone system in which polyamines have satisfied all of the criteria for a second messenger.

Output

Casein Kinase 2

Given that polyamines are at least involved in hormone action, how do they affect cellular functions? As already mentioned, they are required for DNA synthesis, transcription, and translation, but their mechanism of action in these processes is unknown. In part, this requirement may be due to their ability to stabilize nucleic acid structure and neutralize the negative charges; they accomplish this function by lying in the minor groove of the helix and coordinating the phosphates with their amino groups. This also facilitates Z-DNA formation (see Chapter 14) and increases the helical twist. Because polyamines have a preference for binding the TATA box, these structural alterations may keep the promoter open for transcription factors.

Other major effects are mediated by a polyamine-dependent protein kinase, CK2. This is a ubiquitous protein kinase that can use either adenosine 5′-triphosphate (ATP) or guanosine 5′-triphosphate (GTP), although it has higher affinity for the former. It is a heterotetramer and has two 38-kDa α- and two 27-kDa β-subunits. The α-subunit possesses the catalytic site, but the β-subunit is still required for full enzymatic activity. The α-subunit also has the binding site for heparin, which inactivates the kinase.

The β-subunit is the regulatory subunit. It is responsible for tetramer formation through its dimerization motif (see Fig. 12-1), and it determines substrate specificity. The β-subunit also contains the binding site for polyamines. CK2 is activated by spermidine in a two-step process: initially, the positively charged polyamines bind to the negatively charged heparin or other proteoglycan inhibitor and cause it to dissociate from the kinase; polyamines then directly bind to the β-subunit. Spermidine increases the V_{max} of the kinase fivefold and shifts CK2 into the nucleus, but it does not affect the K_m. Some authorities have questioned the physiological role of polyamines in the regulation of CK2; however, inhibitors of polyamine synthesis inhibit this kinase, and CK2 activity is correlated with endogenous polyamine levels. For example, ODC transfection increases polyamines and CK2 activity. Other basic molecules, like histones, can also activate CK2 under certain circumstances. The β-subunit may also serve as a general regulatory subunit for other kinases; for example, its binding activates A-Raf but inhibits Mos. The β-subunit can be modified by several kinases; however, the phosphorylation status of CK2 does not greatly affect kinase activity because CK2 has a relatively high basal activity. CK2 can be autophosphorylated at the extreme amino terminus, which is highly conserved. Autophosphorylation has no effect on tetramer formation or response to polyamines, but it does increase the half-life of the kinase and decreases its activity toward some substrates. PKC can phosphorylate CK2, but this only stimulates the kinase 50%. Cdk2, a proline-directed kinase associated with the cell cycle (see Chapter 12), phosphorylates CK2 at the carboxy terminus near the site of tetramer formation, but the effects are controversial. Some researchers claim that this modification results in as much as a fivefold stimulation in kinase activity, whereas others report that deleting this site has no effect on activity or tetramer formation.

There are other mechanisms for CK2 regulation. As noted previously, polyamines shift CK2 into the nucleus, where CK2 is brought closer to many of its substrates. This translocation may be an important regulatory mechanism in those systems where total CK2 activity does not appear to change in response to hormones. Localization may also be controlled by polyphosphoinositide (PPI); CKIP-1 (CK2 interacting protein) is a CK2 anchor that possesses a pleckstrin homology (PH) domain. As such, its membrane anchorage could be influenced by PPI synthesis or hydrolysis. Another way CK2 can localize to the plasma membrane is through receptor binding: CK2 binds the intracellular domains of CD5, an antigen receptor, and CD163, a scavenger receptor. Subsequent aggregation stimulates the kinase. The fibroblast growth factor receptor (FGFR) can also bind CK2, but the effect of this association is unknown. Because the FGFR is a receptor tyrosine kinase (RTK), it may act by tyrosine phosphorylation and/or aggregation. For example, the soluble tyrosine kinase (STK) Abl can phosphorylate and inhibit CK2.

Finally, CK2 is allosterically regulated by several factors. It is activated by sphingosine, and modestly inhibited by calcium. Mitogen-activated protein kinase (MAPK) allosterically stimulates CK2 autophosphorylation, whereas the binding of the βγ subunit of the translation elongation factor, EF-2, shifts the substrate specificity of CK2 toward translation factors. Lastly, the β-subunit of CK2 has a cyclin-like sequence. Cyclins are regulatory subunits of cell cycle dependent kinases and can be inhibited by the binding of p21$^{WAF1/CIP1}$. This protein also binds and inhibits CK2.

Several substrates for this kinase have been identified, but the effects of phosphorylation are not always clear because CK2 is involved in a number of "silent phosphorylations." Silent phosphorylations are those that have no direct effect on a substrate but that are required for subsequent phosphorylations by other kinases. For example, CK2 modifies glycogen synthase; although this has no immediate effect on synthase activity, the phosphorylation facilitates the subsequent modification of this synthase by glycogen synthase kinase 3. Phosphorylation by this latter kinase does inhibit the synthase. Conversely, the modification of acetyl coenzyme A carboxylase by CK2 facilitates the dephosphorylation of this enzyme at the protein kinase A (PKA) site.

CK2 is a major nuclear kinase, and many of its substrates are involved with nucleic acids. CK2 phosphorylation increases the activities of DNA ligase, topoisomerase II, and several transcription factors (see Chapter 13). The translation initiation factor, eIF-2, is also phosphorylated. This factor is responsible for bringing the first amino acid to the ribosome. Phosphorylation results in a slight increase in affinity for Met-tRNA$_f^{Met}$, but the significance of this is unknown. Finally, CK2 can phosphorylate the 20S protease complex; IRS-1, a major mediator of insulin action; and other kinases like Cdk2. The effects of these latter modifications are unknown.

Other Outputs

Polyamines also affect enzymes other than kinases. Polyamines can stimulate DNA and RNA polymerases, DNA gyrases, DNA methylases, and topoisomerases I and II. However, in many of these enzymes, divalent cations can substitute for the polyamines, suggesting that the effects of polyamines in these systems are nonspecific and due simply to their charges.

Another way that polyamines can affect enzyme activity is by binding to substrates. For example, the ability of polyamines to stimulate phosphoinositide (PI) kinases has been attributed to its binding to negative phospholipids and making them better substrates. In a similar manner, polyamines inhibit phospholipase C (PLC), phospholipase A_2, (PLA$_2$), and diacylglycerol (DG) kinase; in this case, polyamine binding protects the substrates.

Polyamines can also bind enzyme modulators. The central helix in calmodulin (CaM) is partially acidic and can bind polyamines; this binding blocks the activation of calcineurin and phosphodiesterase 1 (PDE1). Finally, the polyamine pathway can affect enzymes by competing for limiting precursors. For example, in plants both the polyamine and the ethylene pathways use SAM. Because this metabolite is limiting, the levels of polyamines and ethylene vary in a reciprocal manner. Similarly, reciprocity exists between polyamines and NO.

In addition to enzymes, polyamines can affect channels and hormone receptors. They frequently inhibit cation channels by blocking them. Conversely, they enhance the binding of growth hormone (GH), follicle-stimulating hormone (FSH), and insulin to their membrane receptors; promote the binding of glycine to the NMDA (*N*-methyl-D-aspartate) glutamate receptor; increase the stability of the estrogen receptor; and promote the activation of the progesterone receptor and facilitate its binding to DNA.

Finally, polyamines themselves may be used to modify proteins. A translation initiation factor, eIF-5A, contains an unusual amino acid, *hypusine*:

It is formed when the butylamine group from spermidine is transferred to lysine by a transglutaminase; the resulting deoxyhypusine is then hydroxylated. This modification is not required for general translation. However, it allows eIF-5A to recognize a specific mRNA sequence that is associated with mRNA for cell cycle proteins, so that it is required for cell division. Transglutaminase can also transamidate RhoA and increase its binding to, and activation of, ROCK (Rho-associated coiled coil protein kinase). Another type of protein modification catalyzed by transglutaminases is the cross-linkage of proteins through isopeptide bonds. These bonds occur between the ε-amino group of lysine and the γ-carbonyl of glutamine. Polyamines increase the enzyme activity 30-fold. Transglutaminases can cross-link PLA$_2$, the latent transforming growth factor β (TGFβ) binding protein, the hormones midkine and interleukin 2 (IL-2), and the transcription factor Sp1; this modification increases the activity of all 5 proteins. This enzyme also plays a critical role in cross-linking proteins during apoptosis.

Oligosaccharides

The fact that the insulin receptor is a tyrosine kinase does not exclude its use of other signaling pathways. In an attempt to identify another mediator, researchers incubated insulin with liver plasma membranes, which contain insulin receptors, and the supernatent was examined for insulin-like activity. Indeed, the supernatant stimulated mitochondrial pyruvate dehydrogenase, PDE3, acetyl coenzyme A carboxylase, steroidogenesis, and the phosphatases in glycogen metabolism; it also inhibited adenylate cyclase, PKA, and pyruvate kinase. The supernatent from untreated membranes was inactive. The activity has not been definitively identified, but preliminary characterization suggests that it is a 1- to 2-kDa oligosaccharide.

The source of this carbohydrate appears to be a PI-glycan; that is, a phosphoinositide in which additional sugars are attached to the inositol. There are several lines of evidence supporting this hypothesis. First of all, such a PI-glycan has been purified from membranes, and its polar head group can be removed by a PLC specific for this glycolipid. This head group has all of the activity of the insulin mediator. Second, its sugar composition is similar to that for the natural messenger; this composition includes a non-*N*-acetylated glucosamine, which is unusual in eukaryotic systems. Finally, this PI-glycan can be labeled in intact cells and, following insulin stimulation, this label appears in the putative mediator.

Synthesis

This PI-glycan is strikingly similar to one that anchors proteins in membranes. There are many ways in which fatty acids can be attached to proteins (see Chapter 9). The PI-glycan anchor involves the coupling of phosphoinositide to the carboxy terminus of a protein via an oligosaccharide-phosphoethanolamine bridge (see Fig. 9-3, *B*). The PI-glycan is synthesized first; the sugars are added sequentially followed by the phosphoethanolamine. The glucosamine is actually added as *N*-acetylglucosamine, which is then deacetylated. The inositol is initially acylated, but this fatty acid may be removed or reattached as the structure undergoes remodeling. This modification is important because the presence of a fatty acid on inositol inhibits the activity of PLCs. The fatty acids attached to the glycerol backbone of the PI can also be exchanged during this period.

All proteins that will be coupled to this anchor have a hydrophobic carboxy terminus preceded by a triplet of small amino acids. In the endoplasmic reticulum, this extension is cleaved and the preformed PI-glycan is attached by means of a pseudopeptidation reaction. In the Golgi complex, the oligosaccharide branches are added. The core sugar sequence shown in Fig. 9-3, *B*, is identical for all known PI-glycan anchors, but the sequence of the branches can vary considerably. In addition, there are other documented variations; for example, there can be additional phosphoethanolamines or phosphoceramide can replace the phosphatidic acid.

Function

PI-glycan modification can have several functions. First, it targets certain proteins to the apical surface. Second, it can increase the lateral mobility of membrane proteins. Most integral membrane proteins have diffusion rates slower than lipids, but proteins with PI-glycan anchors are only held in the plasmalemma by a phospholipid; therefore their lateral diffusion rate approaches that of lipids. Third, it allows for the quick release of the intact protein from the cell surface through a phospholipase. Several hormones can trigger this release; for example, insulin stimulates the release of alkaline phosphatase, heparin, and lipoprotein lipase. However, it is not clear if this liberation is part of the biological action of insulin or simply represents protein turnover. Fourth, the PI-glycan may allow the protein to couple with other transducers; several antigenic markers in immune cells are anchored by PI-glycan and are coupled to the soluble tyrosine kinase Lck. This is a specific interaction because replacement of the PI-glycan with a transmembrane helix uncouples these markers from Lck.

However, the most controversial aspect of this modification is its proposed role in signal transduction. There are two potential mediators generated by the breakdown of the PI-glycan. The first is DG, which can activate PKC. However, it should be noted that the fatty acids in the PI-glycan are usually saturated and therefore differ from those in PPI. Because the fatty acid composition of DG has been shown to influence its ability to activate the different isozymes of PKC, the effects of the PI-glycan–generated DG may differ from its PPI-generated counterpart.

The other product is the oligosaccharide, or glycan. As noted previously, insulin stimulates the appearance of an oligosaccharide that is capable of acutely affecting the activities of several enzymes. In addition, the glycan is release by, and can mimic some of the effects of, epidermal growth factor (EGF), nerve growth factor (NGF), brain-derived neurotrophic factor (BDNF), neurotropin 3 (NT-3), TGFβ, thyroid-stimulating hormone (TSH), IL-2, PRL, and erythropoietin (EPO). These effects appear to be mediated by PI3K (phosphatidylinositol-3 kinase). It has been hypothesized that PI-glycans disrupt DIGs (detergent-insoluble, ganglioside-enriched complex) and release activated STKs, which then phosphorylate IRS (insulin receptor substrate). Finally, this docking protein binds and stimulates PI3K.

However, there are problems with this scheme, not the least of which is the fact that PI-glycan anchors are located in the outer leaflet of plasma membranes. Because extracellular oligosaccharides and PI-glycan–specific PLC mimic some hormone activity, the glycan either must be taken up into the cells or bind to an external receptor. In addition, if the oligosaccharide originates from a membrane-anchored protein, two bonds must be hydrolyzed: the one between inositol and glycerol and the one between the glycan and the protein. The first bond can be hydrolyzed by a PI-glycan PLC or phospholipase D (PLD), whereas the latter would require a protease. Alternatively, the substrate may be the PI-glycan precursor in the endoplasmic reticulum. In this case, only one bond needs to be broken; however, the insulin receptor in the plasma membrane must somehow be coupled to the PI-glycan in the endoplasmic reticulum through some other mediator. Because the release of the glycan requires G_q, PLCβ may provide this link.

Cellular pH

All mitogens increase the cytoplasmic pH by 0.1 to 0.3 units by activating a (Na^+, H^+) antiporter. Antiporter mutants or inhibitors will block growth, and this effect can be reversed, if cellular alkanization is achieved by adding bicarbonate to the medium. These experiments suggest that the elevation of cellular pH is obligatory for mitogenesis; however, alkanization by itself will not induce growth. It is interesting to compare H^+ to another ionic second messenger, Ca^{2+}. First the change in H^+ concentrations is much less than that for Ca^{2+}; the former also has greater mobility, whereas the latter is quickly buffered. As a result, H^+ does not have responses that are localized. Finally, because H^+ transporters are slow, H^+-regulated processes are activated more slowly.

There are several forms of the antiporter; NHE1 is ubiquitous and is the one most closely associated with growth factors. It is a typical transporter with 10 transmembrane helices; it also has a long carboxy terminus containing several potential phosphorylating sites that are important for growth factor regulation. For example, PKC, ROCK, MAPKs, and the Nck interacting kinase (NIK) all phosphorylate and stimulate NHE1 (Fig. 11-4).

Calcium activates NHE1 by several mechanisms. First, NHE1 has an auto-inhibitory domain in the carboxy terminus; calcium-CaM binds and blocks this domain. Second, calcium activates NOS to produce NO (see Chapter 7), which can *S*-nitrosylate and stimulate NHE1. Third, calcium can activate PKC. Finally, phosphorylated protein phosphatase 2B (PP2B) binds and inhibits NHE1; calcium activates this phosphatase, which autodephosphorylates and dissociates from the exchanger. Cortisol, aldosterone, and triiodothyronine (T_3) elevate NHE1 activity by inducing the gene for this antiporter, whereas parathormone (PTH) stimulates its degradation. Finally, hyperpolarization can activate NHE1.

NHE3 is another isoform that is regulated by signaling pathways. It has two PPI-binding sites in the carboxy terminus, and PI3K activation causes NHE3 to shift from intracellular sites to the plasma membrane. This redistribution is facilitated by calcium, which induces the binding of NHE3 to the cytoskeleton and inhibits its internalization. Cortisol can induce the serum and glucocorticoid-regulated kinase (SGK), which phosphorylates and activates NHE3. Finally, NHE3 can be inhibited by an NHE regulatory factor (NHERF). NHERF is a multi-functional protein: (1) it acts as a scaffold, (2) it is involved in the oligomerization of some receptors, and (3) it participates in the membrane targeting, trafficking,

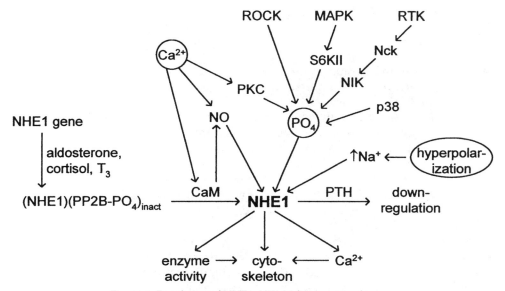

Fig. 11-4. Regulation of NHE1. *NIK*, Nck interacting kinase.

and sorting of some membrane proteins. PKA phosphorylation of NHERF enhances its inhibition of NHE3, whereas GRK (G protein-coupled receptor kinase) phosphorylation increases its oligomerization function. β-Adrenergic receptor (βAR) activation shifts NHERF from NHE3 to the carboxy terminus of this G protein-coupled receptor (GPCR), thereby relieving the inhibition of NHE3.

The NHE is opposed by the (Cl^-, HCO_3^-) antiporter, which lowers the pH. This transporter is localized at focal adhesions and associates with Fak, an STK. The subsequent tyrosine phosphorylation and activation of this transporter lowers the pH and counters the action of mitogens. This may be the basis of contact inhibition, where the establishment of cell-cell contacts inhibits cell proliferation.

There are three major targets of an elevated pH: calcium levels and profile, enzymes, and the cytoskeleton. The extrusion of H^+ is accompanied by the import of Na^+; this, in turn, is rectified by a (Na^+, Ca^{2+}) exchanger, which elevates intracellular calcium concentrations with its consequences (see Chapter 10). Alkalinization also converts calcium spikes to waves.

All enzymes have pH optima. Elevated pH inhibits the acid endonuclease and apoptosis, which are antigrowth. It also inhibits protein phosphatases, thereby prolonging the effect of protein kinases. Alkalinization alone can translocate $PKC\alpha, PKC\delta$, and $PKC\epsilon$ to the plasma membrane. Finally, an elevated pH stimulates oxidative phosphorylation in muscles.

pH has multiple effects on the cytoskeleton. First, it increases actin turnover to supply monomers for elongation during motility; this is accomplished by scinderin and cofilin (see later). Scinderin is inhibited by polyphosphoinositide (PIP_n); elevated pH shifts scinderin from PIP_n to actin, which it severs. This process is facilitated by calcium, which favors cytoskeletal breakdown. Alkalinization also activates cofilin, which uses the actin monomers to reform actin during motility. An elevated pH is required for Rho-mediated stress fiber formation and shifts EF-1α from actin to the ribosome, where it increases protein synthesis.

Magnesium

At first glance, magnesium appears to be an unlikely candidate for a second messenger. First, magnesium gradients across the plasma membrane and organelles are small, so that transport is required for a change in cell content to occur. Second, hormones can only alter total cell magnesium 10% to 15%. Normal intracellular concentrations are 0.5 to 1 mM with 90% to 95% of magnesium bound. Nonetheless, hormones can affect magnesium levels. In the heart, βAR agonists increase, and insulin decreases, magnesium concentrations; the mitochondria appears to be the source. Arachidonic acid stimulates magnesium release from mitochondria through a (Mg^{2+}, H^+) exchanger. Finally, calcium and magnesium appear to be inversely regulated.

Most important, there are clear biological effects associated with magnesium. Many enzymes require magnesium, and magnesium can alter the function of ion channels. However, the most convincing effects are on transcription. The 3'5'-cyclic AMP (cAMP) response element binding protein (CREB) binds to a DNA sequence known as the *cAMP response element* (CRE). In reality, the CRE is not an inviolate sequence but possesses some variability. Magnesium can alter the affinity of CREB for some of these sequences by 1000-fold.

Reactive Oxygen Intermediates

Higher organisms are often invaded by pathogens. One of the mechanisms for fighting infections is for macrophages to generate reactive oxygen intermediates (ROIs), also known as *reactive oxygen species* (ROS); in large concentrations, these are highly toxic molecules that can be used to kill pathogens. Interestingly, the NADPH (reduced nicotinamide-adenine dinucleotide phosphate) oxidase used to produce ROIs in defense cells has homologues in many tissues. These enzymes produce lower levels of ROIs that are used as signaling molecules for stress. In some cases, ROI is converted to hydrogen peroxide, which can mediate some of the oxidative effects of ROI. The reaction can occur spontaneously or be catalyzed by enzymes like the superoxide dismutase:

$$2 \cdot O^{2-} + 2\,H^+ \rightleftharpoons H_2O_2 + O_2$$

The NADPH oxidase is activated by Rac (see Chapter 13), and the resulting ROIs can affect many processes. First, they can facilitate tyrosine kinase stimulation: the oxidation of cysteine in the activation loop of RTKs increases the accessibility of the loop to autophosphorylation. The stimulation of STK occurs either through Ras activation or through the inactivation of calcium pumps. The latter would elevate calcium levels and stimulate calcium-activated STKs, like Src and Pyk. Furthermore, ROI can inactivate protein tyrosine phosphatase (PTP) by oxidizing the cysteine in the active site; this would prolong the effects of tyrosine phosphorylation. ROIs can also stimulate MAPK and NF-κB (nuclear factor that stimulates κ chain in B cells); the latter is a transcription factor that induces stress-related genes and is activated by ROI-stimulated STKs. Finally, ROI oxidizes two cysteines in both G_i and G_o; the oxidation of the former induces subunit dissociation and oxidation of the latter increases GTP binding.

Cytoskeleton and Substratum

Cell shape often reflects the metabolic and developmental activities of the cell; for example, the change from squamous to columnar epithelium when secretory processes are stimulated. Cell shape, in turn, is determined by elements of the cytoskeleton: microtubules, microfilaments, and intermediate filaments. It is also determined by contacts with other cells and with the substratum. First, these components are discussed individually; then their roles in hormone action are examined.

Components

Microtubules are hollow tubes, 20 to 30 nm in diameter and 4.0 to 7.0 nm thick. They are composed of two nearly identical globular proteins: α- and β-tubulin. These subunits are 50 to 60 kDa and alternate as they spiral along the length of the tube like a collapsed spring. Each subunit has a GTP binding site, and this nucleotide is hydrolyzed during the noncovalent polymerization process:

$$n(\text{tubulin-GTP}) \rightarrow (\text{tubulin-GDP})_n + nP_i$$

Colchicine, vinblastine, and vincristine disrupt microtubules by binding to the GTP site and preventing assembly. As a consequence, these compounds are very useful in studying the role of microtubules in various cellular processes; unfortunately, they also have several side effects. For example, colchicine can (1) bind to cell membranes and inhibit fluid and nucleoside transport, (2) suppress protein synthesis and secretion, (3) alter intermediate filament organization, and (4) affect cell shape at concentrations that do not disrupt microtubules.

In contrast to these drugs, paclitaxel (Taxol) promotes microtubule formation. In particular, it (1) stabilizes microtubules against the disruptive actions of low temperatures and calcium, (2) lowers the critical tubulin concentration required for polymerization, and (3) decreases the lag time for assembly. Unfortunately, the microtubules formed under the influence of paclitaxel (Taxol) are randomly located and disorganized.

In addition to α- and β-tubulin, there are several other proteins involved with the formation of and structure of microtubules. The microtubule-associated proteins 1 and 2 (MAP 1 and 2) are 270- to 300-kDa proteins that form knobs along the length of the microtubules. Their axial periodicity of 32 nm corresponds to about 1 MAP molecule per 9 tubulin dimers. MAP 1 and 2 also (1) cross-link microtubules with themselves and other cytoskeletal elements; (2) stiffen and stabilize microtubules, leading to longer microtubules; (3) act as anchorage sites for PKA and CaM; and (4) can generate force. The latter effect is analogous to the action of dynein in cilia and flagella. The tau proteins are another group of cytoskeletal proteins that range in size from 58 to 65 kDa. They are homologous to MAP in the carboxy terminus and serve a similar function: they cross-link and stiffen microtubules. On the other hand, stathmin destabilizes microtubules by sequestering $\alpha\beta$-dimers. Phosphorylation by PKA, Pak (p21-activated protein kinase), or Cdk interferes with the action of stathmin and stabilizes microtubules.

Microfilaments are structurally identical to the thin filaments of muscle; indeed, their subunits, both called *actin*, are homologous to each other. Actin is a 42-kDa, globular protein, which binds ATP; the ATP is hydrolyzed when the subunits noncovalently polymerize:

$$n(\text{actin-ATP}) \rightarrow (\text{actin-ADP})_n + nP_i$$

The actin forms two rows that twist such that a complete turn occurs every 13–14 monomers; the resulting filament is 6 to 7 nm in diameter. As is the case with microtubules, microfilament assembly can be affected by drugs. Cytochalasin B disrupts microfilaments by capping the barbed ends and blocking elongation. Like all drugs, it has side affects; these include (1) binding to cell membranes and inhibiting both hexose transport and, in plant cells, photopolarization; (2) suppressing protein synthesis and secretion; (3) inducing nuclear extrusion; (4) inhibiting cell movement and phagocytosis; and (5) blocking cytokinesis in concentrations that do not disrupt microfilaments. Cytochalasin D does not inhibit hexose transport but does have many of the other undesirable effects. Therefore all experiments with any of these compounds should include a search for these effects and, if appropriate, controls should be established for them. Other drugs include latrunculin A, which binds actin monomers; misakinolide A, which binds actin dimers and caps the barbed ends; and swinholide, which binds actin dimers. All three of these drugs block the nucleation and elongation of microfilaments.

Like microtubules, microfilaments have many accessory proteins. These are primarily concerned with organizing the actin filaments, controlling their polymerization, or facilitating their interaction with membrane proteins. Some of the more common ones are listed in Table 11-1.

Table 11-1
Actin Accessory Proteins

Protein	Function	Regulation
Bundling and Cross-Linking Proteins		
Actin-binding protein	Organizes actin into orthogonal nets	
α-Actinin	Organizes actin into loose, parallel bundles	(+) by PIP_n binding (−) by PIP_n hydrolysis (PLC)
Annexin	Actin bundling near plasma membrane	
Fimbrin	Organizes actin into tight, parallel bundles	
Filamin	Cross-links actin; attaches receptors to actin; scaffold for signaling molecules	
Synapsin I	Actin bundling	(−) MAP kinase phosphorylation
Actin Disruption Proteins		
Cap100	Blocks barbed end but does not sever actin	(−) by PIP_n binding (+) by PIP_n hydrolysis (PLC)
Cofilin	Increases actin growth during motility by increasing GTP·G actin concentration by increasing actin turnover	(−) LIM kinase phosphorylation
Depactin	Nibbles at actin	
Destrin	Severs actin and binds monomer	
Fragmin	Blocks barbed end	
Gelsolin/villin	Severs actin; caps barbed end; but fragments may act as polymer nucleus	(−) by PIP_n binding (+) by PIP_n hydrolysis (PLC) and calcium
Profilin	Binds and sequesters monomer (lower eukaryotes); catalyzes ADP-ATP exchange to promote turnover	With WASP, polymerization With VASP, depolymerization
Scinderin	Severs actin	(+) by calcium and elevated pH
Thymosinβ4	Inhibits ADP-ATP exchange	
Interfacing Proteins		
Adducin	Binds actin to spectrin	(+) PKC phosphorylation (−) CaM
BPAG1n	Binds actin to neurofilaments	
ERM family (ezrin-radixin-moesin)	Binds actin to plasma membrane via CD44	(+) ROCK phosphorylation; PIP_n
MARCKS	Binds actin to membrane	(−) PKC and ROCK phosphorylation (+) PIP_n and CaM block PKC
Paxillin	Binds vinculin and FAK; adaptor for FAK	(+) pY and Pak phosphorylation
Talin	Binds vinculin to membrane (for example, through fibronectin receptor)	(+) pY
Vinculin	Binds actin to talin	(+) by PIP_n binding (−) by PIP_n hydrolysis (PLC)
Actin Filament Nucleation and Stabilization		
Arp2/3	Promotes actin nucleation and branching	
Caldesmon	Prevents actin reorganization	pY, disassemble; CK2 phosphorylation blocks myosin binding
Nebulin	Promotes actin nucleation and stabilizes the filament in skeletal muscle	
Tropomyosin	Stabilizes actin	

Both microtubules and microfilaments are involved with cell motility and structure. The microtubules form the mitotic spindle, cilia, and flagella, whereas microfilaments are essential for pinocytosis and phagocytosis. Both have structural functions as well: microtubules form the marginal band in erythrocytes, whereas microfilaments form the terminal web beneath the plasma membrane and within the microvilli.

After the discovery of microtubules and microfilaments, another group of filaments was found. Because their sizes (8 to 10 nm in diameter) were between those of the other two, these filaments were called *intermediate filaments*. The molecules have a central α-helix; the helices from two molecules align themselves in a parallel fashion and twist around each other in a coiled coil motif. Two such dimers then associate in a staggered, antiparallel arrangement to form a tetramer. This aggregation continues until the intermediate filament is complete. A detailed discussion of these filaments is beyond the scope of this chapter; however, the basic properties of some of the major groups are summarized in Table 11-2. Intermediate filaments can be disrupted by acrylamide.

Thus far, this text has only considered the signals generated by hormone-receptor interactions. However, similar signals can be produced by receptors that recognize the substratum or other cells. A summary of these adhesion molecules is presented in Table 11-3. Integrins can bind either the substratum (e.g., collagen, laminin, and fibronectin) or other cells (e.g., cadherin). They are αβ-heterodimers: the β-subunit is the transducer and activates the focal adhesion kinase (FAK), and the α-subunit inhibits the β-subunit prior to ligand binding. Both the immunoglobulin (Ig)-CAM family and the selectins are involved with cell-cell interactions and couple to STKs; the former functions in morphogenesis, and the latter, in inflammatory responses. The collagen receptor is an RTK and was discussed in Chapter 7.

Cadherins mediate cell-cell adhesion in tissue organization and embryogenesis. In the classical pathway, cadherins on separate cells exhibit calcium-dependent homotypic binding at the adherens junctions. The intracellular domains connect with actin by means of a β-catenin (or plakoglobin)—α-catenin—α-actinin bridge. In the nonclassical pathway, the cadherins are located at desmosomes and bind intermediate filaments.

Model Systems

To begin with, the cytoskeleton is not involved in the actions of all hormones. For example, in rat adipocytes, colchicine, vinblastine, or vincristine does not have any effect on insulin-induced glucose oxidation. Cytochalasin B has no effect on insulin-induced proteins or on insulin inhibition of lipolysis, but at high concentrations, it does suppress glucose oxidation. This suppression is not observed with cytochalasin D, indicating that it is a result of the inhibition of glucose transport. Therefore neither microtubules nor microfilaments play any role in the action of insulin in adipocytes. Similar results have been reported for MSH action in melanoma cells.

Steroidogenesis, however, does involve the cytoskeleton. In rat luteal cells, cytochalasin B inhibits the progesterone synthesis induced by either human chorionic gonadotropin (hCG) or cAMP. The drug does not suppress protein synthesis, lower hCG receptor number or affinity, or affect cAMP production. Furthermore, the inhibition of steroid synthesis could not be reversed by adding

Table 11-2

Basic Properties for Some of the More Common Intermediate Filaments

Property	Keratin filaments (tonofilaments)	Desmin filaments	Vimentin filaments	Neurofilaments	Lamins	Nestin
Source	Epithelium	Skeletal, cardiac, and smooth muscle	Mesenchyme and derivatives	Neurons (central and peripheral nervous systems)	Nucleated cells	Neuroepithelial stem cells
Subunits and organization	Acidic and basic keratins (40–65 kDa); dimers containing one of each	Desmin (50 kDa)	Vimentin (52 kDa)	Core (68 kDa) and two peripheral proteins (150 and 200 kDa)	Lamins (60–70 kDa)	Nestin
Location	Junctional complexes (desmosomes)	Z and M lines; intercalating discs	Perinuclear	Cellular appendages	Between heterochromatin and inner nuclear membrane	Like vimentin but does not extend as far into cytoplasm
Possible functions	Cell adhesion	Framework for myofibrils; align Z lines; biogenesis of T-SR system	Framework for membrane; role in nuclear transport	Tensile strength for axon	Forms nuclear lamina	Early neural differentiation

T-SR, T-tubule-sarcoplasmic reticulum.

Table 11-3
Adhesion Molecules

Property	Integrins	Ig-CAM Family	Cadherins	Selectins	Collagen receptors
Structure	αβ-heterodimers	Immunoglobulin domain with or without FNIII; some have enzymatic activity (e.g., tyrosine kinase in Eph; phosphotyrosine phosphatase in RPTP)	Dimer	Lectin-, EGF- and complement-like domains	RTK with lectin-like extra-cellular domain
Ligands	Collagen, laminin, vitronectin, fibronectin, Ig-CAM, cadherin, thrombospondin, Von Willebrand factor	Homotypic: NCAM Heterotypic: DCC; Eph	Calcium-dependent homotypic cell-cell at adherens junctions and desmosomes	Heterotypic cell-cell through Ca^{2+}-dependent recognition of sialyated glycans; FGFR for E-selectin	Collagen
Function	β localizes to focal adhesions and transduces α inhibits β without ligand	Morphogenesis	Classical: Connected to actin through β-catenin (or plakoglobin) at adherens junctions Nonclassical: Connected to intermediate filaments at desmosomes Tissue organogenesis embryogenesis	Inflammatory cell adhesion	Morphogenesis
Transduction	FAK, Src, and Shc	Fyn, Src, dephosphorylation (RPTP)	Wnt, tyrosine phosphorylation, Rac, Rho, PI3K	STK, MAP kinase	RTK

DCC, Deleted in colon cancer (a chemoattractant for commissural axons in the spinal cord); *FAK*, focal adhesion kinase; *NCAM*, neural cell adhesion molecule; *RPTP*, receptor protein tyrosine phosphatase.

glucose to the medium or by using cytochalasin D, indicating that glucose transport is not a problem. Finally, colchicine has no effect at all.

Similar effects are observed in adrenal tumor cells. Again, cytochalasin B suppresses the stimulation of side-chain cleavage by adrenocorticotropic hormone (ACTH) and cAMP. However, it does not inhibit protein synthesis, cholesterol transport into the cell, or ATP levels. Furthermore, it has no effect on cleavage either in isolated mitochondria or by the purified enzyme. Subsequent studies have shown that the microfilaments are required to transport the cholesterol from the plasma membrane to the mitochondria, where cleavage takes place.

There also appears to be a role for intermediate filaments in steroidogenesis: some cholesterol in the adrenal gland is stored as droplets attached to vimentin. Acrylamide stimulates steroid synthesis by disrupting the intermediate filaments and liberating the cholesterol for transport to the mitochondria. Therefore both microfilaments and intermediate filaments have been implicated in steroidogenesis; the latter is involved with cholesterol storage and the former transports this sterol to the mitochondria.

The mammary gland represents another example of a system in which the cytoskeleton and substratum influence hormone action. In the rabbit, PRL converts the squamous-cuboidal epithelium to a cuboidal-columnar epithelium; it also increases the number and length of microvilli and develops the Golgi complex and rough endoplasmic reticulum. Finally, PRL induces milk protein synthesis. Are the changes in the cytoarchitecture and gene induction linked, or are they independent events? This question can be answered by isolating rabbit mammary epithelial cells and culturing them in Petri dishes coated with a collagen gel. The cells will form a flat monolayer; because the substratum is fixed and the cells form junctional complexes, the cells are forced to maintain a squamous configuration. In the presence of PRL, the cellular shape does not change, and there is no stimulation of milk proteins or their mRNAs. However, if the collagen gel is loosened from the Petri dish, it will float and shrink, allowing the cells to change shape. Now PRL induces both the previously described morphological changes and milk protein synthesis. Similar results have been reported for the mouse mammary epithelium, although the effect is less complete. On attached cells, some casein mRNA does accumulate in the presence of hormones, but it is only 40% of that in floating gels. Furthermore, casein phosphorylation and secretion are impaired on attached gels. Other data suggest that epithelial cell–substratum interactions may be more important than cell shape; if basement membrane components are provided, the cell does not have to be in a secretory conformation. However, if the cell must make its own basement membrane, cell shape plays a greater role. These effects appear to be mediated by tyrosine phosphorylation. First, the basement membrane inhibits PTP; this allows PRL-induced phosphorylation of Jak and Stat to reach levels capable of triggering milk protein expression. Second, substratum binding to integrin activates C/EBP (CCAAT/enhancer-binding protein), another transcription factor required for casein gene induction.

Regulation

There are several ways that hormones can affect the cytoskeleton (Fig. 11-5). Hormones can alter the cytoskeleton by direct interactions, by covalent modification, or through second messengers. The intracellular domain of many hormone receptors binds the cytoskeleton, which it can then modify. For example, EGF

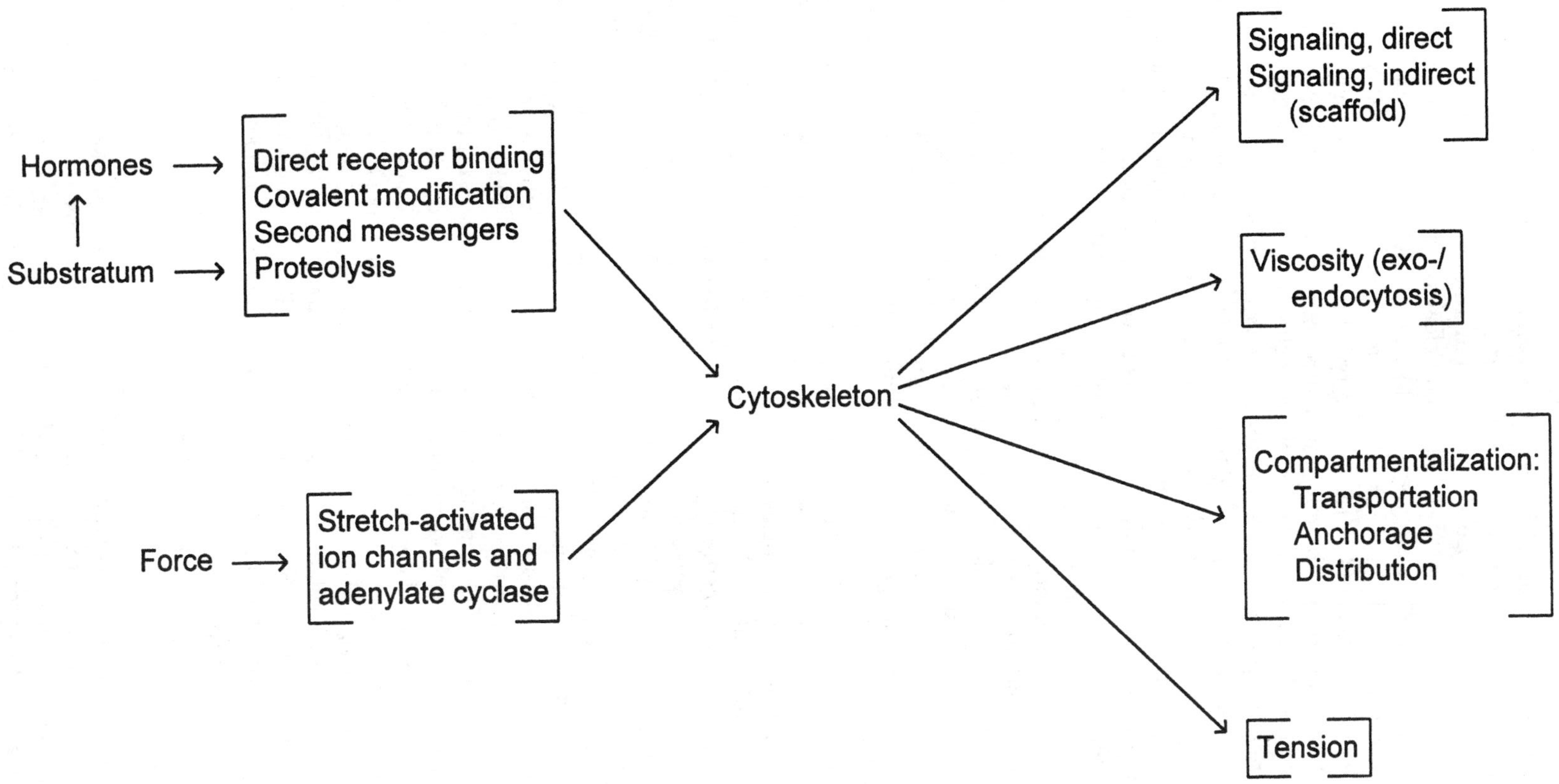

Fig. 11-5. Regulation of cellular activity through the cytoskeleton.

binding to its receptor induces an EGFR-WASP (epidermal growth factor receptor–Wiskott-Aldrich syndrome protein) complex that depolymerizes actin.

Phosphorylation is the major form of covalent modification. In general, S-T phosphorylation disrupts the cytoskeleton; it decreases polymerization and impairs the binding of the cytoskeleton to other components and to the plasmalemma. This result is independent of the type of kinase. Tyrosine phosphorylation has a similar effect: PDGFR (platelet-derived growth factor receptor) phosphorylation of talin causes it to dissociate from vinculin, thereby severing the connection between microfilaments and the plasma membrane. Trk (nerve growth factor receptor) and EGFR phosphorylation of β-catenin causes it to dissociate from cadherin, which also severs its connection to actin. Finally, Src phosphorylation of cortactin and tensin inhibit actin cross-linking.

Tubulin undergoes several unusual modifications. Tyrosine can be reversibly attached to the carboxy terminus. PKC phosphorylates and inhibits the tubulin tyrosine ligase; the detyrosinated tubulin is more stable and can recruit intermediate filaments through kinesin. Acetylation also stabilizes microtubules; the hormonal regulation of this modification is described in Chapter 14.

Deamination is another type of covalent modification; it refers to the conversion of arginine within a protein to citrulline. This modification leads to the disassembly of vimentin, desmin, and glial fibrillary acidic protein (GFAP) in the presence of micromolar calcium. Mono(ADP-ribosyl)ation induces the depolymerization of desmin and actin. Finally, the cytoskeleton can be degraded by calpain; talin, filamin, and fodrin are particularly susceptible to calpain, a calcium-regulated protease. Fodrin is the nonerythrocyte isoform of spectrin, a long rodlike protein that forms a cytoskeletal network with actin just underneath the plasma membrane. Many of these effects are interrelated; for example, the dephosphorylation of neurofilaments renders them more susceptible to proteolysis by calpain, whereas the polyamination of tau by a transglutaminase makes them less susceptible.

Several second messengers can bind components of the cytoskeleton. Calcium can act through PKC, CaMKII (calmodulin-dependent protein kinase II), calpain (see preceding paragraph), or CaM (see later). However, it can also act alone. In the presence of calcium, gelsolin fragments actin and caps the barbed ends, but as calcium falls, gelsolin loses the ability to sever actin and the fragments can then act as nucleation sites. In combination with CaM, it also regulates caldesmon. Caldesmon suppresses the actin-activated adenosine triphosphatase (ATPase) of smooth muscle myosin; this prevents muscle relaxation and maintains muscle tension. Calcium-CaM binds at the carboxy terminus of caldesmon and removes this inhibition. Phosphorylation by PKC at the carboxy terminus or by CaMKII or CK2 at the amino terminus also eliminates the inhibition. MARCKS is another target of CaM; this acronym stands for *m*yristoylated, *a*lanine-*r*ich *C* *k*inase *s*ubstrate. MARCKS cross-links actin and secures it to the plasma membrane. CaM binding to MARCKS impairs actin cross-linking but not membrane binding: as such, CaM increases the plasticity of the network without disrupting its membrane attachment. On the other hand, PKC phosphorylation of MARCKS impairs both actin cross-linking and membrane binding, resulting in the solubilization of the actin. Calcium can also affect many kinases and calpain (see previous discussion).

Phospholipids are other mediators that affect the cytoskeleton. PIP_n binds to gelsolin, profilin, scinderin, and Cap100 and prevents these elements from disrupting actin. When hormones activate PLC, the PIP_n is hydrolyzed and gelsolin is free to sever actin and cap the barbed ends, while profilin sequesters the monomers; as a result, actin depolymerizes. When the signal is terminated, PIP_n is regenerated and binds gelsolin and profilin; actin rapidly repolymerizes from primed fragments. On the other hand, PIP_n can be masked by GAP43, MARCKS, or CAP23; these inhibitors will dissociate from PIP_n after PKC phosphorylation or calcium-CaM binding to these proteins. Finally, PIP_n can bind vinculin to expose the talin and actin binding sites, which will lead to the formation of focal adhesions.

G proteins and pH can also directly affect the cytoskeleton; α_s, α_i, and α_o bind tubulin and activate GTPase, which inhibits microtubular assembly. Scinderin is the best-known target of pH. Scinderin is an actin-severing protein that is inhibited by binding to PIP_n. Calcium and alkanization shift scinderin from PIP_n to actin near the plasma membrane; the PIP_n and actin compete for the same site on scinderin. Scinderin then severs actin in the vicinity of the plasmalemma. This local disruption is thought to play a role in exocytosis.

The substratum acts in a manner similar to hormones; various components in the extracellular matrix bind to plasma membrane receptors to generate second messengers that can then affect the cytoskeleton. The mechanism by which mechanical force alters the cytoskeleton is less well characterized; however, stretch-activated ion channels and adenylate cyclase have been reported.

Output

How are these changes in the cytoskeleton converted into biological activity? First, many elements of the cytoskeleton can augment or initiate classical signaling pathways. For example, integrin can precluster many RTKs and prime them for hormonal stimulation. In addition, many tyrosine phosphorylated cytoskeletal components can act as adaptors, dockers, and scaffolds. For example, the pY on villin binds and activates PLCγ, the pY on ezrin binds and activates PI3K and PKB, and phosphorylated paxillin is an intermediate in the integrin-Ras pathway. Because many cytoskeletal molecules bind PIP_n, some of them can regulate the availability of this phospholipid; for example, MARCKS binds and sequesters PIP_n to inhibit PLC, and MARCKS releases PIP_n after PKC phosphorylation. Finally, some parts of the cytoskeleton can directly affect transducers: gelsolin stimulates PLD, whereas fodrin inhibits it; profilin stimulates PI3K; and ezrin-radixin-moesin (ERM) inactivates RhoGDI, thereby stimulating Rho.

Second, the depolymerization or proteolysis of components of the cytoskeleton can decrease cytosol viscosity and facilitate exocytosis and endocytosis; for example, actin disassembly is required for PRL secretion. Third, the cytoskeleton is involved with compartmentalization; for example, it is involved with transport by acting as a guidewire for signaling molecules greater than 500 Da. Microtubules are used for long distances and use kinesin and dynein motors; actin is used for short distances and uses myosin-based motors. The role of the cytoskeleton in the transport of cholesterol from its stores into the mitochondria has already been mentioned. The movement of glucocorticoid receptors (GRs) to the nucleus and glucose transporters to the plasma membrane represent other examples. The cytoskeleton also plays a role in the partitioning of mRNA between ribosomes and ribonucleoprotein (RNP) particles. RNPs sequester certain mRNAs and render them translationally inactive. The degree of translation is determined by a balance between the mRNA in the two pools. For example, dynein binds the 3'-untranslated end of the PTH mRNA and moves it along microtubules.

Paxillin binds PABP1 (poly(A) binding protein)-mRNA complex and recruits it to actin; it is displaced by eIF-4G at the ribosome. This redistribution may also affect the half-life of mRNA. For example, laminin, another component of the extracellular matrix with a cellular receptor, increases the casein mRNA half-life in mammary cells from 10 to 41 hours, whereas the half-life of total mRNA is only slightly altered. This stabilization is blocked by either cytochalasin D or colchicine. The cytoskeleton can affect metabolic enzymes as well; glycolytic enzymes are more efficient when they are bound to the cytoskeleton. Glucagon elevates calcium, which dissociates the enzymes and inhibits glycolysis. Finally, the cytoskeleton can act as anchorage sites; indeed, some of its components double as AKAPs (A kinase anchoring protein), and their binding to PKA affects substrate availability.

Fourth, the cytoskeleton can affect transcription; the best-studied example is β-catenin, which is sequestered in cytosol by cadherin. Its tyrosine phosphorylation by RTKs results in its release and migration to the nucleus, where it acts as a cofactor for other transcription factors. Hic-5 is another cofactor distributed between focal adhesions and the nucleus. Signals from cell attachment sites release Hic-5, which translocates to the nucleus and enhances steroid receptor transcription.

Finally, it has been proposed that the tension generated by changes in cell shape can be translated into cellular signals. For example, there are tension-gated ion channels that act in mechanotransduction. In addition, tension may be transmitted along the components of the cytoskeleton all the way to the nuclear matrix, where these forces can affect gene induction. This transduction from physical to chemical information is postulated to occur through changes in thermodynamic variables. For example, tension could alter free energy to promote the unraveling of chromatin; this change in chromatin structure might then facilitate transcription (see Chapter 14).

Summary

In addition to the cyclic nucleotides and phospholipids, there are several other systems that can mediate or affect signal transduction across the plasmalemma. These include polyamines, oligosaccharides, pH, magnesium, ROIs, and the cytoskeleton. Polyamines are essential for DNA replication, transcription, and translation; they also affect the activity of many enzymes, including a multifunctional protein kinase, CK2. Both the synthesis of polyamines and the activity of this kinase are under hormonal control. The oligosaccharides are released from a PI-glycan by a specific PLC, which liberates the polysaccharide moiety; the sugar residues, in turn, affect enzyme activity and gene expression, whereas the DG may activate certain isozymes of PKC.

Alkanization is an inevitable consequence of mitogenesis, but it is not sufficient by itself to induce cell division. Mitogens raise the pH by activating the (Na^+, H^+) antiporter through calcium and phosphorylation. The resulting alkanization further elevates calcium and affects enzyme activity and the cytoskeleton. Magnesium concentration is changed by hormone-activated transporters and affects enzymes and transcription factors. ROIs are generated by the Rac-NADPH oxidase pathway; they can elevate tyrosine phosphorylation and activate MAPK and some G proteins.

The cytoskeleton is also involved in hormone action; for example, microfilaments are required to transport cholesterol into the mitochondria for the hCG- or ACTH-induced side-chain cleavage. In addition to cytoplasmic transport, the cytoskeleton is important in mediating hormone-induced changes in cell shape required for specific functions. Hormones can affect the cytoskeleton by covalent modifications, such as phosphorylation or proteolysis, or by second messengers. The altered cytoskeleton, in turn, can affect biological activity by stimulating signaling pathways, by decreasing the cytoplasmic viscosity, by transporting or distributing materials within the cell, or by affecting transcription through the release of coactivators from the plasma membrane.

References

Polyamines

Faust, M., and Montenarh, M. (2000). Subcellular localization of protein kinase CK2: A key to its function? *Cell Tissue Res.* 301, 329-340.

Igarashi, K., and Kashiwagi, K. (2000). Polyamines: Mysterious modulators of cellular functions. *Biochem. Biophys. Res. Commun.* 271, 559-564.

Litchfield, D.W. (2003). Protein kinase CK2: Structure, regulation and role in cellular decisions of life and death. *Biochem. J.* 369, 1-15.

Medina, M.Á., Urdiales, J.L., Rodríguez-Caso, C., Ramírez, F.J., and Sánchez-Jiménez, F. (2003). Biogenic amines and polyamines: Similar biochemistry for different physiological missions and biomedical applications. *Crit. Rev. Biochem. Mol. Biol.* 38, 23-59.

Thomas, T., and Thomas, T.J. (2001). Polyamines in cell growth and cell death: Molecular mechanisms and therapeutic applications. *Cell. Mol. Life Sci.* 58, 244-258.

Oligosaccharides

Jones, D.R., and Varela-Nieto, I. (1998). The role of glycosyl-phosphatidylinositol in signal transduction. *Int. J. Biochem. Cell Biol.* 30, 313-326.

Müller, G., and Frick, W. (1999). Signalling via caveolin: Involvement in the cross-talk between phosphoinositolglycans and insulin. *Cell. Mol. Life Sci.* 56, 945-970.

Cellular pH

Counillon, L., and Pouysségur, J. (2000). The expanding family of eucaryotic Na^+/H^+ exchangers. *J. Biol. Chem.* 275, 1-4.

Putney, L.K., Denker, S.P., and Barber, D.L. (2002). The changing face of the Na^+/H^+ exchanger, NHE1: Structure, regulation, and cellular actions. *Annu. Rev. Pharmacol. Toxicol.* 42, 527-552.

Voltz, J.W., Weinman, E.J., and Shenolikar, S. (2001). Expanding the role of NHERF, a PDZ-domain containing protein adapter, to growth regulation. *Oncogene* 20, 6309-6314.

Magnesium

Murphy, E. (2000). Mysteries of magnesium homeostasis. *Circ. Res.* 86, 245-248.

Reactive Oxygen Intermediates

Cooper, C.E., Patel, R.P., Brookes, P.S., and Darley-Usmar, V.M. (2002) Nanotransducers in cellular redox signaling: Modification of thiols by reactive oxygen and nitrogen species. *Trends Biochem. Sci.* 27, 489-492.

Dröge, W. (2002). Free radicals in the physiological control of cell function. *Physiol. Rev.* 82, 47-95.

Finkel, T. (2003). Oxidant signals and oxidative stress. *Curr. Opin. Cell Biol.* 15, 247-254.

Haddad, J.J. (2002). Antioxidant and prooxidant mechanisms in the regulation of redox(y)-sensitive transcription factors. *Cell. Signal.* 14, 879-897.

Reth, M. (2002). Hydrogen peroxide as second messenger in lymphocyte activation. *Nature Immunol.* 3, 1129-1134.

Sauer, H., Wartenberg, M., and Hescheler, J. (2001). Reactive oxygen species as intracellular messengers during cell growth and differentiation. *Cell. Physiol. Biochem.* 11, 173-186.

Thannickal, V.J., and Fanburg, B.L. (2000). Reactive oxygen species in cell signaling. *Am. J. Physiol.* 279, L1005-L1028.

Cytoskeleton and Substratum

Braga, V.M.M. (2002). Cell-cell adhesion and signalling. *Curr. Opin. Cell Biol.* 14, 546-556.

Calderwood, D.A., Shattil, S.J., and Ginsburg, M.H. (2000). Integrins and actin filaments: Reciprocal regulation of cell adhesion and signaling. *J. Biol. Chem.* 275, 22607-22610.

Coulombe, P.A., Ma, L., Yamada, S., and Wawersik, M. (2001). Intermediate filaments at a glance. *J. Cell Sci.* 114, 4345-4347.

Fuchs, E., and Karakesisoglou, I. (2001). Bridging cytoskeletal interactions. *Genes Dev.* 15, 1-14.

Hynes, R.O. (2002) Integrins: Bidirectional, allosteric signaling machines. *Cell* (Cambridge, Mass.) 110, 673-687.

Juliano, R.L. (2002). Signal transduction by cell adhesion receptors and the cytoskeleton: Functions of integrins, cadherins, selectins, and immunoglobulin-superfamily members. *Annu. Rev. Pharmacol. Toxicol.* 42, 283-323.

Schwartz, M.A., and Shattil, S.J. (2000). Signaling networks linking integrins and Rho family GTPases. *Trends Biochem. Sci.* 25, 388-391.

Yin, H.L., and Janmey, P.A. (2003). Phosphoinositide regulation of the actin cytoskeleton. *Annu. Rev. Physiol.* 65, 761-789.

CHAPTER **12**

Phosphorylation and Other Nontranscriptional Effects of Hormones

CHAPTER OUTLINE

The rise of molecular biology has resulted in an undue emphasis on gene regulation. Certainly transcriptional control is a major mechanism by which hormones affect cellular processes, and it is discussed in Part 4. However, there are other mechanisms that do not involve gene regulation but are no less important. The major portion of this chapter discusses the regulation of phosphorylation by protein kinases and phosphatases and their effects on certain model systems.

General Properties

Criteria and Characteristics

The major nontranscriptional mechanism for cellular regulation is phosphorylation. Why is phosphorylation the premier mechanism for regulating eukaryotic proteins? First, its strong charge favors the formation of ionic bonds, which can readily alter protein conformation. Second, kinase recognition sites are fairly simple and can be easily inserted into proteins. Indeed, multiple phosphorylation sites can be inserted into proteins to provide for signal integration. Finally, phosphorylation is easily reversible by a simple hydrolytic reaction.

A number of serine-threonine protein kinases have already been discussed in previous chapters, and their structure and control are summarized in Table 12-1. Although many of these protein kinases have been extensively studied, the roles of other protein kinases are less well characterized because many of their major substrates are still unidentified. Such identification is complicated by the question of physiological relevance; almost any protein of sufficient size will have consensus sequences for many kinases and can, in fact, be phosphorylated under the appropriate *in vitro* conditions. Therefore how does one determine if a particular modification is a genuine regulatory mechanism? Four criteria have been proposed as requirements for a physiologically relevant phosphorylation.

1. The substrate should be phosphorylated *in vitro* at a reasonable rate and in a stoichiometric manner.

2. This phosphorylation should appropriately alter the function of the substrate. For example, if the substrate is an enzyme, its activity should be modulated. Determining whether this criterion has been satisfied can be complicated by two phenomena. First, phosphorylation may have unexpected effects; for example, rather that affecting enzyme activity, it may alter its cellular localization. Second, in hierarchical phosphorylation, the first phosphorylation is frequently silent (see the following).

3. The phosphorylation and dephosphorylation should occur in intact cells or tissues and should be correlated with the functional changes in the substrate. Furthermore, substrates mutated at these sites should not be phosphorylated or have their activity changed. Finally, specific kinase inhibitors, if available, should lead to dephosphorylation of the substrate and reversal of the activity.

4. Lastly, the mediators controlling the responsible kinases and/or phosphatases should fluctuate in a manner correlated with the degree of phosphorylation. The levels of the kinases rarely change; therefore the correlation must be made between the kinase effector and the modification. For

Table 12-1
Summary of Multipurpose Serine-Threonine Protein Kinases Involved with Hormone Action

Kinase	Subunits	Regulation	Substrate specificity
AGC family			
PKA	R_2C_2	cAMP	$(+)-(+)-X-S/T^a$
PKB	Dimer	PI(3)P_n and PDK1 phosphorylation; arachidonic acid-ROI	R-X-R-Sm-Sm-S/T-HΦ^b
PKC	Monomer	Ca^{2+} and/or phospholipid	$(+)_{1-3}X_{1-2}(S/T)X_{1-2}(+)_{1-3}{}^c$
PKG	Homodimer	cGMP	$(+)_{2-3}-X-S/T^c$
PKN	Monomer	Rho; arachidonic acid	
S6KI	Monomer	S-T phosphorylation by MAPK and PDK1	R-(R)-R-X-X-S
S6KII	Monomer	Phosphorylation by MAPK	R-(R)-R-X-X-S
CaM kinase family			
CaMKII	Hexamer bilayer	Ca^{2+} and CaM; autophosphorylation	R-X-X-S/T^d
STE family			
Pak	Monomer	Rac, Cdc42, PKB, Etk, pY	(K/R)-R-(−)-S
IKK	$(IKK\alpha)_4(IKK\beta)_4$ $(IKK\gamma)_4(hsp90)_2$ $(Cdc37)_{2-3}$	Phosphorylation by NIK	D-S-HΦ-X-S
CMGC family			
CK2	$\alpha_2\beta_2$	Polyamines; pS/pT	S/T-X/ (−)-X/(−)-(−)
GSK3	Monomer	Inhibited by PKC, PKB, or S6KII phosphorylation	S/T-P-X$_2$-(−)^e
MAP kinase	Monomer$_{inact}$/ dimer$_{act}$	pT and pY by MAPKK	P-X$_{1-2}$-S/T-P
Cycle-dependent kinases	Cdk2 + cyclin A, for example	pT and pY on cyclin	S/T-P or P-L-S/T-P
H1 kinase	Cdk2 + cyclin B + Suc1	pT and pY on cyclin	S/T-P-X-(+)
Tyrosine kinase-like family			
Raf kinase	Monomer	pS and pY(?);Ras	HΦ-HΦ-X-S/T-HΦ-HΦ^f
Mixed Lineage kinase family			
ROCK	Monomer	Rho	$(+)-X-(+/H\Phi)-S/T$
TOR family			
DNA-PK	Trimer^g	Requires DNA binding	(−)-S/T-Q or Q-S/T-Q

^{a} Arginine is the preferred basic amino acid.
^{b} Sm, small, nonglycine amino acid; HΦ, especially bulky, hydrophobic amino acid.
^{c} Either arginine or lysine may be the basic amino acid.
^{d} Valine often follows S-T.
^{e} Prefers a proline-rich environment but positions are not absolute; (−) represents a phosphoserine or phosphothreonine.
^{f} Especially aliphatic amino acids.
^{g} Catalytic subunit + Ku heterodimer (targeting subunit, helicase and stimulates DNA ligase).

example, if a particular phosphorylation is catalyzed by protein kinase A (PKA), then 3′5′-cyclic AMP (cAMP) levels should rise as the degree of phosphorylation increases.

Both glycogenolysis and smooth muscle contraction, which are described later, meet these criteria.

Kinases frequently operate in cascades. As just noted, kinase levels are usually constant and activation rarely increases enzyme activity more than 20-fold.

Therefore for an increase in the gain of this system, several kinases are arranged in a cascade that can greatly amplify the initial response. Cascades can also create a steeper response curve, increase the range of substrates phosphorylated, and introduce regulatory sites for better integration of the response.

Another common characteristic of protein phosphorylation is its hierarchal nature. *Hierarchy* refers to the phenomenon whereby phosphorylation at one site influences the phosphorylation at subsequent sites. Usually, it is the first modification that is regulated by second messengers, although this phosphorylation by itself is without effect. Its only purpose is to facilitate or inhibit the subsequent modifications. Such an arrangement permits graded or threshold effects and allows for regulatory integration.

There are several examples of hierarchal phosphorylation; in glycogen synthase, there are four pairs of sites where the modification of the first is required for phosphorylation of the second. At the amino terminus, PKA phosphorylation must precede that of casein kinase 1 (CK1). At the carboxy terminus, there are two additional sites where CK1 is the secondary kinase; in one either PKA or protein kinase C (PKC) must act first, whereas the other site requires initial phosphorylation by PKA or CaMKII (calmodulin–dependent protein kinase II). Finally, in the middle of glycogen synthase, there is a site that must be modified by CK2 before GSK3 (glycogen synthase kinase 3) can phosphorylate it.

There are also examples where the first phosphorylation inhibits the second. The hormone-sensitive lipase can be phosphorylated at either S-563 or S-565, but not both. Normally, PKA would phosphorylate the former and activate the lipase, but prior modification at the latter site will block PKA action.

Another characteristic of kinases is their substrate specificity. Obviously, the catalytic site exhibits specificity for the sequence around the phosphorylation site (Table 12-1); however, many kinases also recognize other sites on the substrate. Indeed, these kinases are often preassociated with their substrates at these docking sites. This arrangement can have several consequences. First, it allows for a faster response because the kinase does not have to find its substrate. Second, preassociation creates a high local substrate concentration, which can compensate for a phosphorylation site in a sequence that lacks perfect identity to the consensus site. Third, it can introduce an additional level of regulation; PKC phosphorylation of the NMDA (*N*-methyl-D-aspartate) receptor at the docking site for CaMKII prevents the binding of CaMKII and its modification of the receptor. Finally, the docking site can orient the kinase so that it will only phosphorylate selected sites. For example, the transcription factor Elk has two docking sites for MAPK (mitogen-activated protein kinase); MAPK bound to the FQFP site specifically phosphorylates S-383, whereas MAPK bound to the D-domain specifically modifies other sites.

Specificity is also affected by subcellular localization, which determines the access a kinase has to any given substrate. Scaffolds and anchoring proteins were discussed in Chapter 7. However, it should be noted that some scaffolds can sequester and inhibit kinases, rather than facilitate kinase cascades. For example, the Raf kinase inhibitory protein (RKIP) can bind either Raf or MEK (MAPK/ERK-activating kinase) in a mutually exclusive manner, thereby keeping the two kinases separated and interrupting the MAPK cascade.

Finally, specificity can be determined by cell-type or duration of stimulation. Not all cells express all kinases; as noted in Chapter 10, many of the PKC isozymes have cell-specific distributions. The duration of stimulation can also affect the substrates modified and their ultimate effects. For example, the transient activation of ERK (extracellular-signal regulated kinase) stimulates the proliferation of PC12 cells, while sustained activation stimulates differentiation.

Nonkinase Activity

Protein kinases are extremely versatile molecules; although named for their protein kinase activity, some of them exhibit other enzymatic activity. Some members of the TOR (target of rapamycin) family, like PI3K (phosphatidylinositol-3 kinase), have both protein and lipid kinase activities, and the PHD of MEKK1 (MAPK/ERK kinase kinase 1) is an E3 ligase for both itself and ERK2. In addition, kinases can act as anchoring proteins, adaptors, and scaffolds. For example, PKCμ binds and localizes PI4K (phosphatidylinositol-4 kinase), and PI4,5K (phosphatidylinositol-4,5 kinase) to the plasma membrane; Tra1, another member of the TOR family, acts as an adaptor between acidic TADs (transcriptional activation domains) and histone acetyltransferases; and the integrin-linked kinase (ILK) recruits vinculin, actin, myosin, integrin, and perlecan to integrin foci. Indeed, some kinases have lost their enzymatic activity and now function solely as scaffolds: for example, KSR (kinase suppressor of Raf) is homologous to Raf but lacks any kinase activity; rather, it is a scaffold in the Ras-MAPK pathway. The kinase homology domain (KHD) of membrane-bound guanylate cyclases (GCs) represents another nonfunctional kinase that now serves a different function: it inhibits the GC in the absence of adenosine 5'-triphosphate (ATP) (see Chapter 7).

Finally, some kinases can allosterically activate or inhibit other molecules independent of direct phosphorylation. The activation of phospholipase D (PLD) by PKC (see Chapter 10) and the stimulation of PLCγ and PI3K by binding to pYs (phosphotyrosines) on RTKs (receptor tyrosine kinases) and STKs (soluble tyrosine kinases) (see Chapter 7) have already been mentioned. S6KII (S6 kinase II) binding to the CREB (cAMP response element binding protein) binding protein (CBP) inhibits it by blocking its access to a helicase. The phosphorylation of S6KII induces its dissociation from CBP, thereby activating CBP. Conversely, S6KII allosterically activates the ER TAD2.

Frequently, kinase binding activates a protein by blocking or displacing an autoinhibitory domain; for example, the TOR activation of S6KI and the PDK (PIP$_2$-dependent kinase) activation of RalGDS occur through this mechanism. Alternately, the kinase may displace an inhibitor; Etk, an STK in the Btk family, binds the PH domain of RhoA and dislodges RhoGDI. In many cases, these additional activities augment the primary effects of the kinases; Raf1 and MEK1 are both involved in the MAPK mitogenic cascade. Raf1 also binds and inhibits the apoptotic signal-regulated kinase 1 (ASK1), and MEK1 binds and inhibits MyoD, a muscle differentiation transcription factor. Both apoptosis and differentiation oppose mitosis. In other cases, the kinase is activating a negative feedback loop; ERK binds and stimulates the MAPK phosphatase (MKP), which will dephosphorylate and inactivate ERK.

Regulation

Protein kinases can be regulated in several ways. First, all kinases must be phosphorylated in their activation loops, and many have other regulatory phosphorylation sites. Second, kinases can be allosterically activated by second messengers: for example, cAMP and PKA, 3'5'-cyclic GMP (cGMP) and protein kinase G (PKG), and calmodulin (CaM) and CaMKII. In some cases, the kinase can faithfully reflect the pattern of second messenger release; CaMKII is initially activated by calcium-CaM, but subsequent autophosphorylation renders the kinase partially active in the absence of calcium-CaM. This modification allows CaMKII to remain partially active for several seconds after calcium levels decline. Depending on the pulse frequency, the CaMKII may still be active when the next calcium elevation occurs, allowing the signals to be summed: the greater the frequency, the greater the summation. This phenomenon converts the signal frequency into a graded response by the kinase.

Third, kinases can be regulated by compartmentalization; the role of anchoring proteins for PKA and PKC was discussed in Chapter 9. Fourth, because kinases have autoinhibitory sites to maintain a low basal activity, they can be activated by proteases that remove these domains. This appears to occur physiologically during apoptosis, where caspases cleave and stimulate ROCK (Rho-associated coiled coil protein kinase) and PKR (dsRNA-activated protein kinase). Fifth, many proteins have specific inhibitors: PKI for PKA and the Raf kinase inhibitory protein (RKIP) for several kinases, including Raf, NIK (Nck interacting kinase), the transforming growth factor β (TGFβ) activated kinase 1 (TAK1), and the IκB (inhibitor of NF-κB) kinase α and β (IKKα and IKKβ). Sixth, in the long term, the kinases can be regulated by induction and turnover. Finally, phosphatases can control both the kinases themselves and their substrates because both are phosphorylated.

Protein Kinases

General

There are many S-T kinase families, but eight are of special interest for their roles in hormone signal transduction (Table 12-1). The AGC family derives its name from the first three members: PKA, PKG, and PKC. It also includes PKB, PKN, and PDK, which phosphorylate many AGC kinases in their activation loop; GRK (G protein-coupled receptor kinase); ribosomal S6 kinase I (S6KI or p70^{s6k}); ribosomal S6 kinase II (S6KII; also called RSK or p90^{s6k}); and the serum and glucocorticoid-regulated kinase (SGK), also called the *cytokine-independent survival kinase* (CISK). The CaM kinase family includes the CaM-dependent kinases, the myosin 'light chain kinase (MLCK), and the death-associated protein kinase (DAPK). This family is closely related to the AGC family, and both have a preference for sites near basic amino acids.

The Ste20 family is also known as the stress-related kinases and includes Pak (p21-activated protein kinase), NIK, IKK, the hematopoietic progenitor kinase (HPK), and several of the kinases upstream of MAPK (MAPK kinase and MAPK kinase kinase). The CMGC family is named for its best-known members: Cdk (cell-cycle dependent kinase), MAPK, GSK, and the Cdk-like kinases (Clk). It also includes CK2 and ASK (apoptosis stimulating kinase). The family is divided into two groups: CK2 prefers phosphorylation sites adjacent to acidic groups, whereas the other group prefers sites near prolines.

Tyrosine and S-T kinases are highly homologous in their catalytic domains; in fact, the change of only a few amino acids can change the specificity between tyrosine and S-T. Therefore it should not be too surprising to find several S-T kinases in the tyrosine kinase family. The dsRNA-activated kinase (PKR) is clearly within this family; Raf and Mos are probably members, too. The MLK (mixed lineage kinase) family also has some structural similarities to tyrosine kinases. Known members have a coiled coil motif known as a *leucine zipper* (see Chapter 13) and include ROCK and some kinases upstream of JNK (Jun ami-no[N]-terminal kinase) and p38.

The TOR family derived its name from the fact that its premier member is the *target of rapamycin* inhibition. Some members of this family, like PI3K, have both lipid and protein kinase activity. The DNA-PK is another member of this family. The TGFβR family includes the only known group of receptor S-T kinases, as well as the LIM motif-containing protein kinase (LIMK). Finally, although the tyrosine and S-T kinases are closely related, it should be noted for completeness that the histidine kinases, which play major roles in signal transduction in plants, fungi, and bacteria, are totally unrelated to other protein kinases.

Many of these kinases have already been covered in previous chapters: GRK (Chapter 8); PKA, PKG, Pak, ROCK, and LIMK (Chapter 9); PKC and CaMKII (Chapter 10); and CK2 (Chapter 11). Other kinases, like IKK, are so closely associated with transcription that they are discussed with these factors (see Chapter 13). Many of the others are described in this chapter. All of these kinases have similar structures: a homologous kinase domain, a regulatory domain, and a (pseudo)autophosphorylation site (Fig. 12-1). The latter two elements tonically inhibit the kinase in the absence of a positive modulator. The proline-directed kinases and several other miscellaneous kinases are discussed next.

Proline-Directed Protein Kinases

One of the first kinases in this class was initially discovered as a kinase that phosphorylated the microtubule associated protein 2 (MAP2); it became known as the *MAP2 kinase* or, more simply, MAPK. Later, it was discovered to have more widespread effects and be stimulated by growth factors. Not wanting to change an entrenched acronym, the scientific community merely redefined MAPK to mean mitogen-activated protein kinase. The MAPKs exist in several isoforms and must be phosphorylated on both threonine and tyrosine for activation. Both modifications are performed by a single group of homologous kinases, the MAPK kinases (MAPKK). The MAPKKs are also activated by phosphorylation (Fig. 12-2); these kinases form a heterogeneous, nonhomologous group known as the *MAPKK kinases* (MAPKKK).

Several MAPK isoforms are shown in Table 12-2. The first three are the best-known MAPKs. Members of the first group are closely associated with mitogen-esis and are called the *extracellular-signal regulated kinases* (ERKs); their MAPKKs and MAPKKKs are known as the MEK and *MEK kinases* (MEKK), respectively. Classically, this pathway is activated by the Ras-Raf pathway as discussed in Chapter 9 and shown in Fig. 12-2. However, there are other factors that can influence this basic mechanism. For example, Pak phosphorylates MEK1 and primes it for activation by Raf, and PKC can phosphorylate a Raf kinase inhibitory protein (RKIP) and induce its dissociation from Raf. cAMP can stimulate or inhibit the pathway, depending on the cell type. Inhibition can occur via two routes: first, PKA can phosphorylate and inhibit c-Raf. Second, cAMP can bind and stimulate Epac, a RapGEF; Rap1 then blocks the binding of Ras to c-Raf. This inhibition can be converted to stimulation in those cells with B-Raf, which is activated by Rap1. Stimulation can also occur when PKA phosphorylates protein

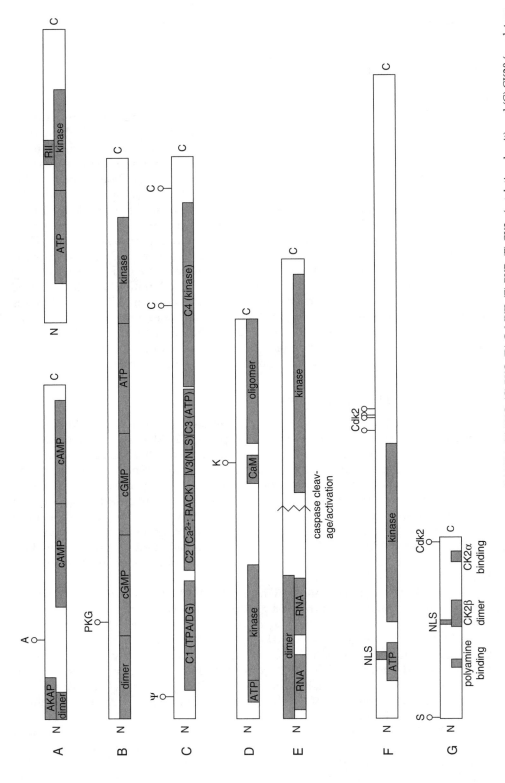

Fig. 12-1. The major kinases regulated by second messengers: (A) PKA, (B) PKG, (C) PKC, (D) CaMKII, (E) PKR, (F) CK2α (catalytic subunit), and (G) CK2β (regulatory subunit). Phosphorylation sites are designated by circles and the responsible kinase is identified by the following letters: *A*, PKA; *C*, PKC; *Cdk2*, cell cycle-dependent kinase 2; *K*, CaMKII; *ψ*, pseudosubstrate site; *S*, CK2. Other abbreviations are as in the text.

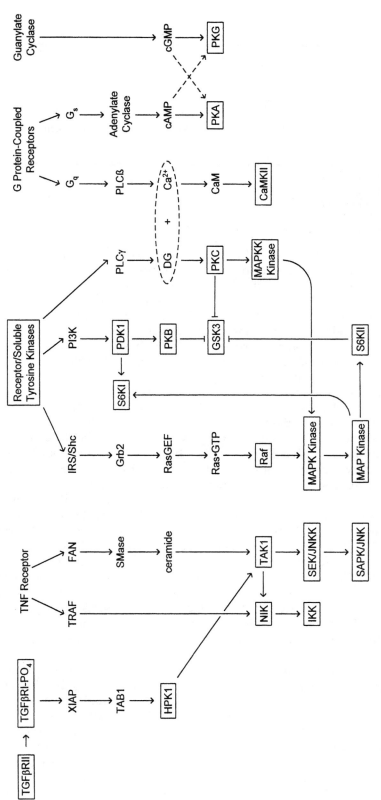

Fig. 12-2. Flowchart of some hormonally regulated protein kinases. Kinases are boxed, arrows indicate stimulation, flat heads indicate inhibition, and abbreviations are defined in the text.

Table 12-2
The Mitogen-Activated Protein Kinase Family

MAPKKK[a]	MAPKK[b]	MAPK[b]	Function
Raf, Mos, Tpl2	MEK1,2	ERK1,2	Mitogenesis
MLK3, MEKK1-4	MEK4,7	JNK1-3	Stress, AP-1[c]
ASK, TAK, MEKK1-4, MLTKα	MEK3,6	p38[d], ERK6	Stress, immunity, differentiation
MEKK3	MEK5	ERK5	MEF2[c]
TAK	?	NLK	Negative regulation of Wnt
?[e]	ERK3 kinase	ERK3	
?	?	ERK7,8	Growth inhibition

[a] Nonhomologous.
[b] Homologous.
[c] Transcription factor.
[d] Can also be directly activated by TAB1 binding.
[e] Question marks signify that the kinases are still unidentified.

tyrosine phosphatase SL (PTP-SL). PTP-SL dephosphorylates ERK and sequesters it in the cytoplasm; PKA phosphorylation causes PTP-SL to release ERK. Finally, there is an interesting relationship between Tpl2, an MAPKKK for the ERK pathway, and p105, a precusor to the p50 subunit of the NF-κB transcription factor. p105 binds and inhibits Tpl2; upon activation, and presumably cleavage of p105, Tpl2 is released and initiates the ERK cascade.

The next two isoforms are associated with stress responses; indeed, one of them is known as the *stress-activated protein kinase* (SAPK). Because one of its substrates is Jun, a transcription factor that regulates several stress-related genes, it is also known as the *Jun amino-terminal kinase* (JNK). The upstream kinases are known as the *stress-ERK kinase* (SEK) or *JNK kinase* (JNKK), although many authorities simply use the MEK designation for all MAPKKs. MEKK4 has a long amino terminus with an autoinhibitory domain; GADD45 is a protein that is rapidly induced after DNA damage, binds next to this domain, and blocks its inhibitory activity. MLK3, another MAPKKK for JNK, is activated by Rac and Cdc42 in a similar manner; the small G proteins bind near the auto-inhibitory SH3 (Src homology domain 3) of MLK3. Phosphorylation of MAPKKKs is another regulatory mechanism; PKN phosphorylation of MLTK stimulates this p38 MAPKKK. On the other hand, Pak phosphorylation of MEKK1 blocks the binding and subsequent activation of JNK. In addition, some MAPKs can be directly phosphorylated by kinases that are not part of the MAPKKK-MAPKK cascade; for example, the DNA-PK phosphorylation and inhibition of JNK. This effect may in part be responsible for the suspension of the cell cycle during DNA damage. The other stress MAPK, p38, is also stimulated by a kinase cascade. In addition, it can be allosterically activated: p38 is stimulated by TAB1, a scaffold protein in the signaling pathway of TGFβ and IL-1.

Less is known about the other MAPK isoforms. ERK5 activates the myocyte enhancer factor 2 (MEF2), a transcription factor that induces the differentiation of muscles and neurons. The Nemo-like kinase (NLK) inhibits Wnt, a morphogen that binds a G protein-coupled receptor (GPCR) (see Chapter 7). Finally, ERK7 constitutively inhibits growth; presumably it is controlled by inhibition rather than stimulation.

The MAPK cascade presented in Fig. 12-2 is simplified; in fact, there are many routes by which MAPKs can be activated. First, there is some cross talk between the different MAPK pathways listed in Table 12-2. Second, other kinases can activate the cascade. As previously noted the MAPKKKs are a very heterogeneous group and only the major ones are listed in Table 12-2.

GSK3 was first discovered as a component of glycogen metabolism, from which it derives its name. It is unusual in several respects: first, it has a high constitutive activity and is regulated by inhibition. Second, it usually inhibits its substrates, and third, substrates must be primed by a previous phosphorylation near the GSK3 phosphorylation site. It is a major kinase in the insulin pathway (see later) where it inhibits several critical metabolic enzymes. Insulin activates these enzymes by stimulating PKB to phosphorylate GSK3. This modification creates a pseudosubstrate site within the amino terminus; it is called a *pseudosubstrate* because there is no adjacent phosphorylatable residue. As a result, the GSK3 remains locked in a nonfunctional enzyme–pseudosubstrate complex. Primed substrates are not required in the other major pathway that involves GSK3. In the Wnt pathway, GSK3 and its substrates are held together on scaffolds; the high local concentration compensates for the normally low activity GSK3 shows toward unprimed substrates.

The constitutive activity of GSK3 is further stimulated by tyrosine phosphorylation by Pyk2 (proline-rich tyrosine kinase 2). However, it is inhibited by many S-T kinases, including PKA, PKB, PKC, SGK, S6KII, and ILK. S6KII is one of the few substrates activated by GSK3 phosphorylation. Its ability to phosphorylate and inhibit GSK3 probably represents negative feedback.

The cell cycle kinases have had several acronyms: MPF for maturation promoting factor or M phase promoting factor, Cdc for cell-division-cycle kinase, and Cdk for cell cycle-dependent kinase. The latter term is the one currently favored and the one that is used in this text. These kinases are inactive alone; they require a regulatory subunit, called *cyclin*. As the name implies, these kinases are important in cell division. However, the Cdk2-cyclin A complex is a multipurpose protein kinase that is active on many substrates during interphase. It can be stimulated by epidermal growth factor (EGF)–and nerve growth factor (NGF)–induced threonine phosphorylation on the cyclin subunit. Complexed with cyclin B and suc1, Cdk2 is active only during mitosis. The presence of suc1 shifts the substrate specificity of the complex to proteins involved with cell division; for example, Cdk2 will phosphorylate (1) histone H1 that is involved with chromosomal condensation, (2) lamins that result in nuclear envelope breakdown, (3) single-stranded DNA-binding proteins that are required for the initiation and elongation in DNA synthesis, and (4) the retinoblastoma gene product, which inhibits DNA synthesis in its underphosphorylated form (see Chapter 16). Because of the foremost effect, this complex is also called the *H1 kinase*.

The cell cycle begins when the cyclin B is induced. Initially, it loosely associates with Cdk2, but the binding becomes tighter after T-161 of Cdk2 is phosphorylated. This modification is very quickly followed by phosphorylation at T-14 and Y-15; these amino acids are in the ATP-binding site and inhibit the kinase. When DNA synthesis is complete, the latter two sites are dephosphorylated by Cdc25, a threonine-tyrosine phosphatase, and the now activated Cdk2 initiates mitosis. At the end of mitosis, cyclin B is degraded, and Cdk2 once again becomes inactive.

Other Protein Kinases

Raf is the first kinase in the Ras-MAPK cascade, and its regulation by Ras was described in Chapter 9. Raf can also be controlled by phosphorylation. Many cytokines enhance Raf activity via tyrosine phosphorylation by STKs. Initially this modification was controversial because it was not seen in all systems. Then it was realized that some Raf isoforms have an aspartic or glutamic acid at this site; the negative charge mimics phosphorylation, thereby rendering tyrosine phosphorylation unnecessary in systems where these isoforms exist. In addition, Raf can be phosphorylated by S-T kinases; PKC and Pak stimulate Raf activity, whereas PKA inhibits it.

The S6Ks were first identified as kinases that phosphorylated ribosomal protein S6 in response to mitogens. Two major groups of S6Ks were then isolated and characterized. S6KI was originally purified from rat liver, has a single kinase domain, has a molecular weight of 70 kDa, and acts more slowly than S6KII; its activity is maximal 5 to 20 minutes after stimulation, whereas S6KII is fully activated in only 2 to 5 minutes. Nonetheless, it is probably the true S6K; phosphorylation of ribosomal protein S6 results in a modest increase in translation. S6KIα is activated by PKB and TOR, but S6KIβ is less dependent on phosphorylation for activity. Rac and Cdc42 bind and translocate S6KI to the ribosome but do not activate it.

S6KII was originally purified from frog oocytes, contains two kinase domains, and has a larger molecular weight of 90 kDa. The amino-terminal kinase sequence is most similar to the PKA family and to S6KI; for this reason, the S6K activity is thought to reside in this domain. The carboxy-terminal kinase sequence most closely resembles phosphorylase kinase, although phosphorylase *b* is not a substrate for S6KII. In fact, the only known substrate for the carboxy terminus is the hinge region between the two kinase domains. Briefly, S6KII is regulated as follows: first, MAPK phosphorylates and activates the carboxy-terminal kinase. Second, the latter phosphorylates the hinge to create a docking site for PDK1. Finally, PDK1 phosphorylates and activates the amino-terminal kinase. PKC can further phosphorylate the β isozyme; this modification does not affect kinase activity, but it does mask the nuclear location signal (NLS) and results in the kinase being retained in the cytoplasm. S6KII has broad substrate specificity, including GSK3 and several transcription factors; however, the ribosomal protein S6 does not appear to be a physiological substrate.

PKB has an autoinhibitory PH domain. The generation of $PI(3)P_n$ (polyphosphoinositide phosphorylated at the 3′ position) by PI3K recruits PKB to the plasma membrane through the PH domain. Final activation is achieved by two phosphorylations. The identity of the first kinase, which modifies S-473, is controversial; some authorities have proposed that PKB actually autophosphorylates itself, whereas others suggest the process is performed by a kinase downstream of p38. In contrast to this dispute, PDK1 is clearly responsible for the second phosphorylation, which occurs on T-308. PDK1 is related to PKB, has a PH domain, and is also activated by $PI(3)P_n$ (see later). In addition to the PI3K pathway, PKB can be phosphorylated and stimulated by PKC, ILK, and CaMKK; the latter occurs in neurons. EGF activation of PKB occurs by means of tyrosine phosphorylation of Y-474 by Src.

PKB has three major activities (Fig. 12-3): antiapoptotic, mitogenic, and insulinomimetic. Apoptosis is a noninflammatory form of induced cell death that occurs during development and defense. Bcl2 is an antiapoptotic factor that is neutralized by binding to Bad, a proapoptotic member of the same family. PKB-phosphorylated Bad is bound and sequestered by 14-3-3, and Bcl2 is released to bind and neutralize other proapoptotic family members. FoxO3a (formerly called FKHRL1) and Nur77 are transcription factors that induce several apoptotic

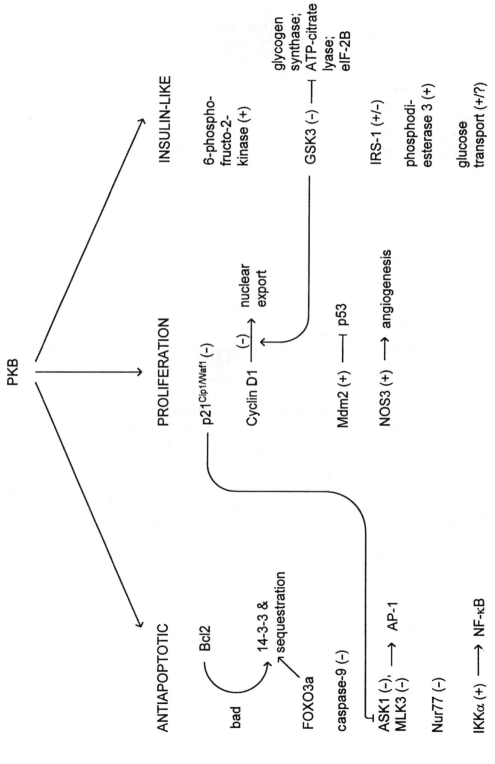

Fig. 12-3. The biological activities of PKB. Plus and minus signs indicate stimulation and inhibition, respectively, after PKB phosphorylation; arrows indicate stimulation; flat heads indicate inhibition; and abbreviations are defined in the text.

factors. In addition, FoxO3a can bind and inhibit HNF-4, a transcription factor that mediates some of the actions of insulin. Both FoxO3a and Nur77 are inhibited by PKB phosphorylation; in particular, phosphorylation of FoxO3a targets it for either sequestration by 14-3-3 or for ubiquitination and destruction. Caspase-9 is a protease stimulated during apoptosis to degrade cellular proteins. PKB phosphorylation inhibits autocleavage and activation. ASK, MLK3, and the IL-1 receptor associated kinase (IRAK) are kinases that mediate apoptosis and all are inhibited by PKB phosphorylation. Finally, IκB inhibits NFκB, a transcription factor associated with the stress response that opposes apoptosis. IKKα activates NF-κB by targeting IκB for degradation; and PKB phosphorylation stimulates IKKα.

PKB is a major mediator of insulin activity. Because this aspect of PKB is fully discussed in the following section about insulin, only a brief description is provided here. Insulin stimulates glucose transport; glycolysis; and glycogen, lipid, and protein synthesis. PKB phosphorylates and stimulates 6-phospho-fructo-2-kinase, the rate-limiting step in glycolysis. GSK3 phosphorylates and inhibits glycogen synthase; ATP-citrate lyase, which shifts acetate away from the TCA cycle and into fatty acid synthesis; and eIF-2B, a translation factor. PKB inhibits GSK3, thereby indirectly activating these three pathways. Glucose transport is a complex process requiring several steps. PI3K is clearly necessary and its effect appears to be mediated by PKB, but its exact role is still unknown. IRS-1 (insulin receptor substrate 1) is a primary docking protein for insulin. PKB phosphorylation of IRS-1 retards its tyrosine dephosphorylation, thereby prolonging its activity. However, it also inhibits the recruitment of PI3K, so that not all of its activities are enhanced. PKB can also inhibit PTP-1B (protein tyrosine phosphatase 1B) to retard the dephosphorylation of the insulin receptor and prolong its activity. Finally, insulin is opposed by cAMP, and the PKB phosphorylation and activation of PDE3 serve to reduce cAMP levels.

The last major role of PKB is in promoting proliferation. One Cdk and cell cycle inhibitor is p21$^{Cip1/Waf1}$. PKB phosphorylation shifts it from the nucleus into the cytosol, where it can no longer inhibit Cdk and where it binds and inhibits ASK1. Cyclin D1 is a required regulatory subunit of Cdk; GSK3 phosphorylation of cyclin D1 promotes its nuclear export and degradation. As previously noted, PKB inhibits GSK3; this effect stabilizes cyclin D1 and keeps it in the nucleus where is stimulates mitosis. p53 is an antioncogene that inhibits growth (see Chapter 16); Mdm2 is an E3 ligase that ubiquinates p53. PKB phosphorylation of Mdm2 shifts it into the nucleus where it can block the antigrowth effect of p53. Finally, PKB phosphorylates and stimulates nitric oxide synthase 3 (NOS3), which promotes angiogenesis to support growth. Other activities of PKB include the stimulation of TOR and Pak and the inhibition of SEK1 and Raf.

As already noted, PDK is related to PKB, and it has a PH domain that can bind PI(3)P$_n$. It is a general activator for the AGC kinase family. Specifically, it phosphorylates the activation loop in these kinases. With one exception, this does not require PI(3)P$_n$ and therefore appears to be a constitutive activity. The exception is PKB, whose modification does require PI(3)P$_n$; as such, PDK phosphorylation of PKB serves a regulatory function. In addition to the AGC family, PKD can phosphorylate and stimulate Pak.

PDK is activated by PI(3)P$_n$, but tyrosine phosphorylation by Trk (tropomyosin-related kinase) bypasses this requirement. On the other hand, tyrosine phosphorylation by the insulin receptor does not alter its activity, but the modification does shift the kinase to the plasma membrane. Sphingosine has also been reported to stimulate PDK.

SGK is another kinase related to PKB. Although its name is derived from the ability of glucocorticoids to induce it, SGK can also be regulated acutely. For example, insulin can stimulate SGK through PI3K and PDK1 in much the same manner as it stimulates PKB; however, PI(3)P$_n$ acts through a PX (phox homology), and not a PH, domain. In addition, it can be activated by CaMKK phosphorylation. The functions of SGK include the phosphorylation and inhibition of GSK3 and FoxO3a; the activation of NHE3 (Na$^+$-hydrogen exchanger 3), probably by direct phosphorylation; and the stimulation of ENaC (epithelial Na$^+$ channel) by phosphorylating and inhibiting the NEDD (neural precursor cell-expressed developmentally downregulated) ligase, thereby decreasing the degradation of ENaC.

TOR is homologous to PI3K, although it lacks any lipid kinase activity. It is also called *FRAP* (FK506 binding protein-rapamycin associated protein) or *RAFT1* (rapamycin and FK506 target 1). These names arose from the way this kinase was discovered: rapamycin and FK506 are immunosuppressants, whose mechanism of action, in part, depends on its ability to bind and inhibit this kinase. In lower organisms, TOR is a nutrient sensor that requires the presence of amino acids for activity. Because its activity is required for protein synthesis (see the section on Model System), an amino acid deficiency would suppress translation. Because it has a high K$_m$ for ATP, it is an ATP sensor as well; that is, when ATP levels are high, TOR is activated and energy is shifted into stored forms like protein. In higher-order organisms, TOR appears to have acquired an input from anabolic hormones, but the nature of this control is still highly controversial (see later for further discussion). Both the PI3K pathway and the PLD-phosphatidic acid pathway have been proposed as regulators of TOR. In addition to the translation machinery, TOR can also phosphorylate and activate PKCδ.

DNA-PK is another member of the TOR-PI3K family. It consists of a large catalytic subunit and a small heterodimeric regulatory subunit, called *Ku*. Ku targets DNA-PK to free DNA ends; it also has helicase activity and stimulates DNA ligase. DNA-PK is required in recombination and may inhibit DNA replication until DNA repair has been completed. However, it probably has additional activities because it modifies many nuclear proteins, including nuclear receptors, transcription factors, DNA topoisomerases, and RNA polymerases. Its regulation is poorly understood; it does require DNA binding for activity. However, binding to at least one coactivator (TR [thyroid receptor] binding protein or TRBP) stimulates DNA-PK in the absence of DNA; TRBP also changes the substrate specificity of DNA-PK. In addition, the kinase has been reported to bind the EGFR; EGF binding releases DNA-PK, which then translocates to the nucleus. Finally, DNA binding and kinase activity is inhibited by autophosphorylation.

PKR is another nucleic acid-binding protein kinase. It is a dimer containing two RNA binding sites in the amino terminus (see Fig. 12-1, *E*). These sites are autoinhibitory domains that are neutralized by RNA binding or autophosphorylation. A PKR activating protein (PACT), a phosphorylated regulatory protein, can stimulate PKR allosterically, and during apoptosis, caspases can activate PKR by removing the RNA binding domains. A 58-kDa inhibitor of protein kinase (p58^{IPK}), another regulatory protein, acts by blocking PKR dimerization. Classically, PKR is induced by interferon and modifies a component of the translation machinery; translation is inhibited and viral replication is blocked (see Chapter 15). However, it also has other functions. For example, it can phosphorylate and activate IKK, and it can serve as a scaffold for the PDGFR (platelet-derived growth factor receptor); in the latter case, it recruits STK and MAPK to phosphorylate Stat.

Protein Phosphatases

If phosphorylation is important, so must dephosphorylation. Indeed, phosphatases are 100 to 1000 times more active than kinases, and it is unlikely that the stimulation of any kinase would result in the net increase in phosphorylation of a substrate without concomitant inhibition of phosphatase activity. However, phosphatases do not simply terminate the signaling; rather, they also play a specific role in shaping the signal. Although kinases determine the signal amplitude, phosphatases determine the speed and duration of signaling. Like kinases, many phosphatases are prebound to substrates and to various scaffolds; in fact, some phosphatases can act as scaffolds themselves. There are four phosphatase families: the PPP, PPM, PTP, and FCP families. The PPP family members are non-metal-dependent S-T phosphatases and comprise PP1; PP2A, including the closely related PP4 and PP6; PP2B; PP5-RdgC, including the closely related PPEF and PP7; and PPKL, including the closely related prokaryotic forms. PPM family members are metal-dependent S-T phosphatases and include PP2C. The PTP family contains the protein tyrosine and dual specificity phosphatases. Finally, the FCP family dephosphorylates the carboxy-terminal domain (CTD) of RNA polymerase II and includes FCP1.

Serine-Threonine Phosphatases

The major S-T phosphatases are listed in Table 12-3. Many of them show the same variety of regulation by second messengers that the protein kinases exhibit. Protein phosphatase 1 (PP1) is a dimer consisting of a catalytic and a localization subunit. Regulation occurs primarily by localization through the targeting subunits, by inhibitor proteins, and by phosphorylation of both of these regulatory proteins. There are more than 50 targeting subunits; for example, the G subunit attaches PP1 to glycogen particles; M, to myosin; NI, to nuclear spliceosomes; PNUTS (protein phosphatase 1 nuclear targeting subunit), to nuclear nucleic acid complexes; NF-L, to neurofilaments; and neurabin I, to actin. Inhibitory proteins include Inhibitor 1 and Inhibitor 2 (glycogen), which are thought to act like pseudosubstrates; CPI17 (myosin); and DARRP-32 (neurons). Phosphorylation of the G subunit by PKA or PKG causes PP1 to dissociate from glycogen; the free form is much more likely to bind and be inhibited by Inhibitor 1. This binding also requires the phosphorylation of Inhibitor 1 by PKA. PP1 is stimulated by the insulin-induced S6KII phosphorylation of the G subunit and by the phosphorylation of Inhibitor 2 by MAPK, Cdk2, or GSK3. These kinases modify the same site and lead to the dissociation of Inhibitor 2 from PP1. On the other hand, tyrosine phosphorylation by the insulin receptor activates Inhibitor 2. Phosphorylation can also affect other PP1 inhibitors; PKA, CK2, and tyrosine phosphorylation inhibits NI, resulting in the stimulation of PP1; PKA phosphorylation of PNUTS releases and activates PP1; PKA activates DARPP-32, which inhibits PP1; and PKC stimulates PI17, which inhibits PP1.

Protein phosphatase 2A (PP2A) is a trimer. C is the catalytic subunit; B, the regulatory subunit that inhibits C; and A, a scaffold for B and C. Like PP1, PP2A is

Table 12-3
Summary of Several Widespread, Multifunctional Phosphatases

Phosphatase	Location	Function	Structure	Regulation[a]
PP1	Glycogen particle; microsomes; ribosomes; myofibrils	Glycogen metabolism; cholesterol and protein synthesis; muscle relaxation	Dimer: catalytic (37 kDa) and organelle-binding subunits (103 kDa); different binding subunits for each organelle	Localization via targeting subunits (>50): for example, G (glycogen), M (myosin), NI (nucleus), NF-L (neurofilaments) Inhibitors: Inh1 & Inh2 (glycogen), CPI17 (myosin), DARPP-32 (neurons) Both targeting subunits and inhibitors subject to phosphorylation
PP2A	Cytosol	Glycolysis and gluconeogenesis; fatty acid synthesis and amino acid catabolism; protein kinases	Trimer: catalytic (C, 36 kDa), regulatory (B, 65 kDa) and scaffold (A, highly variable)	Localization via adaptors: SG2NA (nucleus), cyclin G (Mdm2) Allosterism: polyamines, ATP, ceramide Phosphorylation Carboxymethylation
PP2B (calcineurin)	Cytosol	Dephosphorylation of RII (PKA), Inh1, CaM-dependent PDE, and microtubules; immune response	Dimer: catalytic (61 kDa) and regulatory subunit (19 kDa); latter is a member of CaM family and binds calcium	Allosterism: calcium, CaM, phospholipids Phosphorylation Inhibitors: immunophilins, Cain, SOCS3
PP2C	Cytosol	Cholesterol synthesis	Monomer of 45 kDa	Calcium binding site in plant enzyme
PP5	Nuclear	RNA biosynthesis; cell growth	Monomer of 58 kDa	Cleavage of TPRs Allosterism: arachidonic acid, G_{12}, G_{13}

[a] See text for details.

regulated by its recruitment to specific areas or substrates; for example, striatin localizes PP2A to postsynaptic densities; SG2NA, to the nucleus; PR130, to ryanodine receptors; and cyclin G, to Mdm2, which is an E3 ligase for the antioncogene, p53. PP2A can also be allosterically regulated; ATP and a protein modulator (protein tyrosine phosphatase activator or PTPA) convert PP2A to a tyrosine phosphatase. In addition, polycations, such as spermine, stimulate PP2A *in vitro*, but the physiological role of polyamines in the regulation of this phosphatase remains to be determined. Finally, PP2A can be stimulated by ceramide.

In addition to allosteric regulation, PP2A can be phosphorylated. PP2A is stimulated by PKG and PKR, and it is inhibited by TOR and tyrosine phosphorylation by EGFR or Jak. The effect of CK2 phosphorylation depends on the substrate; it stimulates PP2A activity toward MAPKK but decreases its activity toward the transcription factor, Sp1. Finally, PP2A is the target of an unusual modification: carboxymethylation. This modification can be stimulated by either cAMP or calcium, depending on the cell type; it is not known if cAMP acts through PKA phosphorylation. Carboxymethylation alters the substrate specificity of PP2A and is required for the association of the subunits.

Calcineurin (PP2B) is a dimer consisting of a catalytic and a regulatory subunit. The latter is a member of the CaM family and binds calcium; this binding activates PP2B activity. PP2B can be further stimulated by the binding of exogenous calcium-CaM or phospholipids. It can be inhibited by CaMKII and PKC phosphorylation of the CaM-binding domain. However, the most intriguing regulation is by the immunosuppressants, cyclosporin A and FK506. These drugs are used clinically to suppress tissue rejection. The mechanism of action of these drugs involves the inhibition of PP2B after complexing with endogenous proteins; the cyclosporin A binds cyclophilin, FK506, and the FK506 binding protein (FKBP). Apparently, activation of the T lymphocytes requires the dephosphorylation of a transcription factor by PP2B; only the dephoshorylated form can enter the nucleus and bind DNA. This dephosphorylation by PP2B is blocked by the immunosuppressants. It is suspected that cyclophilin and FKBP may have endogenous ligands that are involved with the physiological regulation of PP2B. There are also other natural inhibitors, including the suppressor of cytokine signaling 3 (SOCS3), calcineurin inhibitor (Cain), and the modulatory PP2B interacting protein 1 (MCIP1). Cain is a noncompetitive inhibitor that blocks the binding of CaM to PP2B, and MCIP1 is a pseudosubstrate that is activated by either MAPK or GSK3. Finally, like PP1 and PP2A, PP2B binds to scaffolds and anchoring proteins, which localize the phosphatase to specific substrates or subcellular sites.

Less is known about the other S-T phosphatases. Protein phosphatase 2C (PP2C) is a monomer that is either distinct from or only distantly related to PP1, PP2A, and PP2B. In animals, there is no identifiable regulatory domain, but plant PP2C possesses a calcium binding motif. Protein phosphatase 5 (PP5) is a monomeric nuclear phosphatase that has four amino-terminal tetratricopeptide repeats (TPRs), which inhibit catalytic activity. PP5 is stimulated when this inhibitory domain is removed by proteolysis, although it is not known if this mechanism is physiological. Unsaturated fatty acids, such as arachidonic acid, can bind the TPR and suppress its inhibition. G_{12} and G_{13} can also bind and activate PP5. Because the effect of these G proteins is additive with arachidonic acid, they presumably bind at another site on the phosphatase. Finally, there are several nonspecific protein phosphatases, like the acid and alkaline phosphatases; these enzymes are active in catabolic pathways, where they recover phosphate for recycling.

Phosphotyrosine Phosphatases

The other major group of protein phosphatases remove phosphate from phosphotyrosine: the phosphotyrosine phosphatases (PTPs). They are regulated by one or more of six basic mechanisms. First, all PTPs have a cysteine in their catalytic sites; oxidation or *S*-nitrosylation will inactivate the enzyme. Second, many soluble PTPs have localization domains; for example, PTP1B has a hydrophobic carboxy terminus that associates with the endoplasmic reticulum; PTPS has a carboxy terminus that binds DNA; PTPε has an NLS in its amino terminus; and the amino terminus of PTP1B is homologous to protein 4.1, which is known to interact with the cytoskeleton (Fig. 12-4, *A*). SH2, SH3, and PDZ (PSD-95/DLG/ZO-1) domains also function in localization (Fig. 12-4, *B*). Third, other PTPs have PEST (proline, glutamic acid, serine, threonine-rich) domains that regulate enzyme turnover (Fig. 12-4, *D*). Fourth, PTPs can be phosphorylated; S-T phosphorylation is usually inhibitory, but the effect of tyrosine phosphorylation depends on the site. In addition to direct phosphorylation, pY can allosterically activate those PTPs with SH2 domains. Fifth, membrane-bound PTPs are inhibited by dimerization. The mechanism for this effect is unknown but may be a result of transdephosphorylation in those PTPs activated by tyrosine phosphorylation. Alternately, the amino-terminal end of the intracellular domain of one PTP has been proposed to block the catalytic site of its partner. Finally, many PTPs bind ligands, which can affect enzymatic activity; this phenomenon is particularly common for the extracellular domain of membrane-bound PTPs (Table 12-4), where natural ligands usually inhibit activity by inducing oligomerization. Soluble PTPs may bind second messengers; for example, MEG2 has an amino terminus that is homologous to retinaldehyde-binding protein and PI transfer protein (Fig. 12-4, *C*). Both of these latter proteins bind small molecules that are active in signal transduction, and the homology with this PTP suggests that the PTP may also bind and be regulated by retinoids or IP$_3$ (inositol 1,4,5-trisphosphate).

Originally, the PTPs were divided into three classes: soluble (class I) and membrane bound (classes II and III). The soluble PTPs are structurally simple: they all have a homologous catalytic domain of about 250 amino acids and either an amino-terminal or carboxy-terminal extension possessing some regulatory domain. The membrane-bound PTPs have either one (class II) or two cytoplasmic catalytic domains (class III). It is universally agreed that the membrane proximal domain is active, but the status of the distal domain is uncertain. Because it also has some, albeit low, phosphatase activity, some authorities claim it either has a narrow substrate specificity or functions only under special conditions. Alternately, it may serve as a pY binding domain, like SH2. Although it has also been suggested that the distal domain may facilitate substrate binding for the proximal site, this possibly seems unlikely for two reasons. First, the three-dimensional structure of the class III phosphatases shows the binding site of the two domains facing opposite directions. Second, the proximal site is active when cloned alone, suggesting that it needs no help from the distal domain. Another possibility is that the distal site serves as an autoinhibitory domain, similar to the KHD in the membrane-bound GCs. Finally, the distal site may act as a dimerization domain.

A few specific PTPs are worth discussing in detail. Because of its two SH2 domains, SHP (SH2-containing protein tyrosine phosphatase) has been extensively studied for its role in the signaling of tyrosine kinases. The amino terminus is an autoinhibitory domain; pY binding relieves this inhibition and activates the

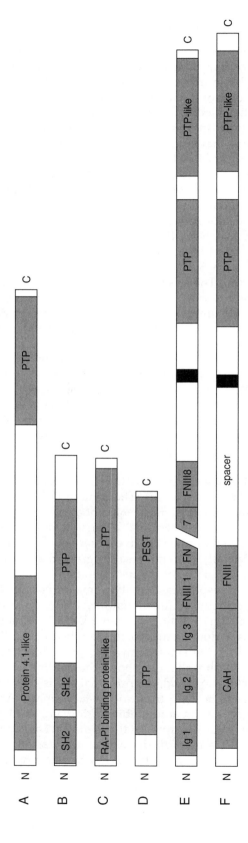

Fig. 12-4. Several phosphotyrosine phosphatases: (A) PTP1B, (B), SHP, (C) MEG2, (D) PTP-PEST, (E) PTPδ, and (F) PTPζ. A-D are soluble PTPs (class I) each having a different type of regulatory site. E and F are membrane-bound (class III). The black area represents the transmembrane domain; the middle six FNIII domains in PTPδ are ellipsed. *CAH*, Carbonic anhydrase-like; *FN III*, fibronectin type III domain; *Ig*, immunoglobulin-like; *RA-PI*, retinaldehyde-phosphoinositide. Other abbreviations are defined in the text.

Table 12-4
Summary of Known Ligands for Receptor Protein Tyrosine Phosphatases

RPTP	Ligand	Response
CD45	Galectin-1	Inhibits
DEP-1	Matrigel	Stimulates
LAR	Laminin, nidogen	
PTPα	Contactin	
PTPσ	Heparin (agrin and collagen XVIII)	
PTPζ	Pleiotrophin, midkine, contactin, tenascin	Inhibits
VE-PTP	VE-cadherin	Inhibits

phosphatase. In addition to pY, the SH2 domains can bind βγ, which recruit SHP to the plasma membrane. SHP can also be phosphorylated on tyrosines, but the effects vary depending on the site. SHP is a substrate for S-T kinases; it is inhibited by MAPK or PKCδ phosphorylation, but it is stimulated by PKA phosphorylation. Finally, SHP binds to the angiotensin receptor by a mechanism that requires α_s; ligand binding results in the dissociation and activation of SHP.

As one might expect, SHP desensitizes some RTKs by dephosphorylating them. Surprisingly, however, it can also mediate the effects of some RTKs. Most of these actions are related to the fact that SHP can act as an adaptor; for example, it serves as an adaptor between the PDGFR and Grb2 (growth factor receptor-bound protein 2), between the insulin receptor and IRS, and between IRS and Jak.

As already noted, PTP1B has a hydrophobic carboxy terminus that associates with the endoplasmic reticulum; cleavage of this domain stimulates the phosphatase. PTP1B can also be activated by tyrosine phosphorylation by either the insulin receptor or EGFR, but it is inhibited by PKA, PKB, or CK2 phosphorylation. Calcium causes it to dissociate from at least one of its substrates, band 3 of the red blood cell. As with SHP, it can act as an adaptor for Grb2.

PTP-PEST is inhibited by PKA or PKC phosphorylation. In addition, it is sensitive to the phosphorylation status of some of its substrates; for example, PKC phosphorylation of ShcA (SH2-containing protein A) is required for PTP-PEST binding and tyrosine dephosphorylation. Finally, PTP-PEST can also act as an adaptor; the NPLH sequence binds the PTB (phosphotyrosine binding) domain of Shc, and the polyproline region binds the SH3 domain of p130^{cas}, a docking protein. In addition to affecting turnover, the PEST domain can be used to precouple PTP-PEST to some of its substrates; for example, this region can bind paxillin.

The membrane-bound PTPs are best known by their large, often complex extracellular domain; PTPζ will serve as an example. The amino terminus of the extracellular domain is homologous to carbonic anhydrase. Although it is not enzymatically active, it does bind contactin. The following FNIII (fibronectin type III) domain binds tenascin. The spacer between the FNIII and the transmembrane domains binds CAM, pleiotrophin (PTN), and midkine (MK), all of which dimerize and inhibit the phosphatase. PTN and MK actually share two receptor subunits: PTPζ and the RTK, ALK (anaplastic lymphoma kinase). The third receptor subunit differs between the two hormones and provides the signaling specificity; MK binds LRP (LDL receptor-related protein), whereas PTN binds syndecan. This represents the only known example of hormone binding to, and regulation of, PTP; rather, most membrane-bound PTPs appear to bind elements in the substratum or are involved with cell adhesion.

In addition to ligand binding, membrane-bound PTPs can be regulated by second messengers and phosphorylation. CaM binds and inhibits the distal phosphatase domain in PTPα. HePTP (hematopoietic protein tyrosine phosphatase) dephosphorylates MAPKs; PKA phosphorylation of HePTP interferes with MAPK binding and releases the active kinase.

Two other PTP groups are also briefly mentioned. First, the cysteine phosphatases, also called the *low molecular weight* PTPs (LMW-PTPs), are 18-kDa proteins with no homology to the preceding PTPs. One of their distinguishing characteristics is their activation by purines, especially cGMP. They are also stimulated by Y-131 phosphorylation by several STKs. As with the other PTPs, nitric oxide (NO) *S*-nitrosylates the active site cysteine and inhibits the LMW-PTP. These phosphatases are known to dephosphorylate RTKs and p190RhoGAP, which results in the inhibition of growth and the stimulation of adhesion. In addition, phosphorylated Y-131 binds Grb2 and inhibits the Grb2-MAPK pathway, either by sequestering Grb2 or by dephosphorylating MAPK. The second group of phosphates consists of the mixed function phosphatases that are capable of removing phosphates from serines, threonines, and tyrosines. PP2A is one such enzyme; its phosphotyrosine phosphatase activity is differentially enhanced by ATP and by the small t and middle T antigens from various viruses (see Chapter 16). Cdc25, which dephosphorylates and activates Cdk2, is another dual function phosphatase. Finally, there are several dedicated MAPK phosphatases (MKPs); MKP-1 is specific for ERK, and MKP-3 is specific for JNK.

Several functions have been attributed to PTPs. The first and most obvious function is their regulatory role in tyrosine phosphorylation. They can oppose the actions of tyrosine kinases by dephosphorylating their substrates. In addition, they can have more direct effects on RTKs; these kinases are activated by autophosphorylation and can be inactivated by dephosphorylation. Second, dephosphorylation can modify the output of RTKs; for example, DEP-1 removes some, but not all, of the phosphates in PDGFR, thereby altering the nature of the response. In another example, MAPK induces MKP, which switches the MAPK response from a sustained stimulation to a proportional response and eventually to an acute stimulation. Third, PTPs can stimulate tyrosine kinases; many soluble tyrosine kinases are inhibited by autophosphorylation sites and can be stimulated by their dephosphorylation. For example, the PTP, CD45, can dephosphorylate and activate Lck, and Cdc25 can do the same with Cdk2. Finally, PTPs can act as adaptors and docking proteins. Although this often results in a positive response, it can also suppress a pathway by sequestering mediators; for example, the sequestration of Grb2 by PTPα or PTPγ.

Model Systems

At this point, all of the major signaling pathways have been presented. This chapter concludes with a general synthesis of these pathways. First, three hormones are analyzed to show how these transducers mediate their actions: TGFβ, a growth-inhibiting hormone that uses an S-T kinase as a receptor; insulin, an anabolic hormone that uses an RTK; and TNF (tumor necrosis factor), a stress hormone that uses a class 2, group 1 cytokine receptor. Then, glycogen metabolism and smooth muscle contraction-relaxation are discussed to show how these pathways can synergize and antagonize each other.

Individual Hormone Action

Transforming Growth Factor β

TGFβ is an important hormone in embryogenesis, where it often limits the growth of tissues and organs (see Chapter 3). It binds a receptor S-T kinase, TGFβRII, which then phosphorylates and activates TGFβRI; the latter appears to be the major signal transducer (Fig. 12-5). First, TGFβ binding to TGFβRII stimulates the association of cyclin B to TGFβRII. Normally, cyclin B serves as a regulatory subunit for Cdk2, a kinase critical in mitosis. However, in this case, cyclin B actually functions as an adaptor to bring Cdk2 to TGFβRII so that it can be phosphorylated and inhibited. This is one mechanism by which TGFβ inhibits growth. Another mechanism involves the suppression of protein synthe-

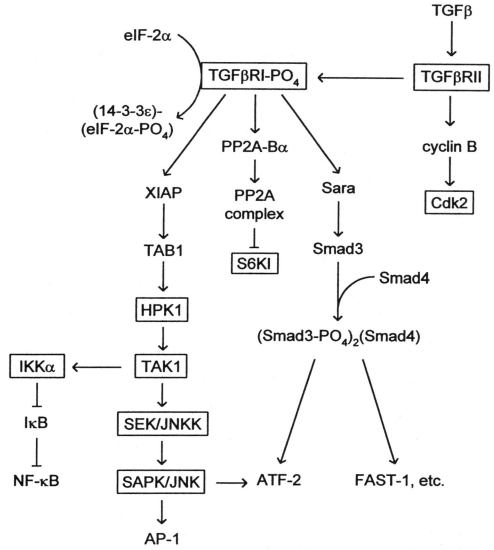

Fig. 12-5. The signaling pathways for TGFβ. Kinases are boxed, arrows indicate stimulation, and flat heads indicate inhibition. *HPK1*, Hematopoietic progenitor kinase 1; *TAB1*, TAK binding protein 1; *TAK1*, TGFβ activated kinase 1; *XIAP*, *Xenopus* inhibitor of apoptosis protein.

sis; TGFβ also induces PP2A-Bα binding to TGFβRI. PP2A-Bα is a regulatory subunit for PP2A-C, which it recruits to TGFβRI along with a second regulatory subunit, PP2A-Aβ. The intact, active PP2A trimer then dissociates, dephosphorylates, and inactivates S6KI. S6KI mediates the stimulation of translation by mitogens by phosphorylating ribosomal protein S6. Finally, the translation factor eIF-2α is phosphorylated by either TGFβRI or RII; the modified factor then binds 14-3-3ε. Although eIF-2α activity was not measured, phosphorylation of this factor is usually inhibitory (see Chapter 15), and 14-3-3 normally sequesters the phosphoproteins that bind it. This would be expected to further suppress protein synthesis.

Some of the effects of TGFβ are mediated by transcription. The activation of Smad was discussed in Chapter 7. Briefly, the adaptor Sara recruits Smad3 to TGFβRI. Upon activation, TGFβRI phosphorylates Smad3, which then trimerizes with another modified Smad3 and the Co-Smad, Smad4. Because Sara only binds the monomer, the trimer dissociates and migrates to the nucleus to initiate transcription. Although it has its own DNA binding domain and TAD, it most commonly acts as a coactivator or corepressor with other transcription factors (see Chapter 13). TGFβ can also activate transcription factors associated with the stress response, such as AP-1 and NF-κB. These factors are stimulated by means of a protein kinase cascade that begins with the adaptor XIAP (X chromosome-linked inhibitor of apoptosis protein) and scaffold TAB1 (TAK binding protein 1). TAB1 assembles HPK1 and TAK1 (TGFβ–activated kinase 1) so that the former can phosphorylate and activate the latter. TAK1 is an MAPKKK for the SAPK-JNK pathway that leads to AP-1. The other transcription factor, NF-κB, is tonically inhibited by IκB (inhibitor of NF-κB). TAK1 phosphorylates and stimulates IKKα (IκB kinase α), whose phosphorylation of IκB triggers its degradation (see Chapter 13). NF-κB then migrates to the nucleus and initiates transcription at stress-related genes.

Insulin

Insulin is an anabolic hormone; that is, it stores energy (see Chapter 2). As such, it promotes the uptake of glucose and amino acids into cells where they are stored as protein, lipid, and glycogen. In adipocytes, it also stimulates glycolysis to generate the acetates for fatty acid synthesis. Finally, many catabolic hormones use cAMP as a second messenger; insulin opposes the actions of these hormones by suppressing the cAMP pathway.

Metabolism. Central to many of these actions is the PI3K pathway and the antagonism between PKB and GSK3 (Fig. 12-6). GSK3 phosphorylates and inhibits glycogen synthase; ATP-citrate lyase, which shifts acetate from the tricarboxylic acid cycle to lipogenesis; and eIF-2B. The insulin receptor is an RTK that phosphorylates the docking protein IRS, which in turn binds and activates PI3K. The product of this lipid kinase, $PI(3)P_n$, resembles glucose-6-phosphate; as such, it is a competitive inhibitor of glucose-6-phosphatase. After uptake, glucose is trapped within the cell by phosphorylation; dephosphorylation is required before glucose can be exported. Through inhibition of glucose-6-phosphatase, $PI(3)P_n$ ensures that glucose will not leave the cell. $PI(3)P_n$ also stimulates PKB, which phosphorylates and inhibits GSK3; this results in the reactivation of glycogen synthase, ATP-citrate lyase, and eIF-2B with the concomitant stimulation of glycogen, lipid, and protein synthesis, respectively. PKB also phosphorylates and stimulates phosphofructokinase 2, a critical regulatory step in glycolysis

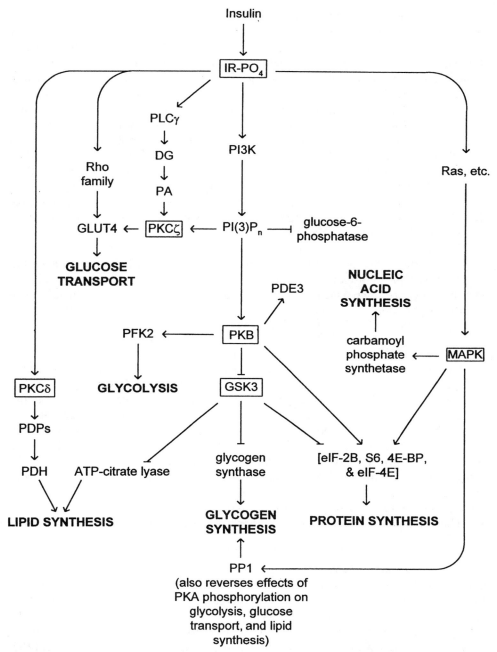

Fig. 12-6. The signaling pathways for insulin; note that not all pathways are present within a single tissue. Kinases are boxed, arrows indicate stimulation, and flat heads indicate inhibition. *IR,* Insulin receptor; *PA,* phosphatidic acid; *PDH,* pyruvate dehydrogenase; *PDPs,* PDH phosphatases 1 and 2; *PFK2,* phosphofructokinase 2.

that supplies acetate for lipogenesis. Insulin also stimulates lipogenesis through another route: the insulin receptor phosphorylates PKCδ which then phosphorylates and activates pyruvate dehydrogenase phosphatases (PDPs). The latter dephosphorylate and stimulate pyruvate dehydrogenase (PDH), which converts pyruvate to acetyl-CoA.

Insulin inhibits the cAMP pathway in several ways. First, PKB can phosphorylate and activate PDE3B, which hydrolyzes cAMP. PDE3B can also be phosphorylated by PI3K, which functions as a protein kinase when it is directly bound to the insulin receptor. The β-adrenergic receptor (βAR) is a major source of cAMP generation; the insulin receptor phosphorylates the βAR and blocks G_s binding. Third, the insulin receptor can directly bind and activate G_i, which inhibits the adenylate cyclase. The fourth mechanism involves the MAPK pathway. Insulin stimulation of MAPK involves the insulin receptor phosphorylation of another docking protein, Shc. Phosphorylated Shc then recruits Grb2, which initiates a cascade that includes mSOS, Ras, Raf, MAPKK, and MAPK (see Fig. 12-2). S6KII, a kinase downstream of MAPK, phosphorylates the G subunit of PP1, which then becomes activated. Finally, PP1 reverses the PKA-induced phosphorylation. In summary, insulin inhibits βAR and adenylate cyclase from stimulating the production of cAMP, hydrolyzes the cAMP that has already been synthesized, and reverses PKA phosphorylation.

The MAPK pathway is also important in supporting growth. Its effects on protein synthesis are described later. In addition, MAPK phosphorylates carbamoyl phosphate synthetase, the rate-limiting enzyme in pyrimidine synthesis. This modification increases the sensitivity of the enzyme toward 5-phosphoribosyl-α-pyrophosphate, an enzyme activator, while decreasing its sensitivity toward UTP, an inhibitor; that is, it stimulates the enzyme and nucleic acid synthesis.

Glucose Transport.   The regulation of glucose transport is more complex. In most cells, glucose transport occurs by facilitated diffusion; that is, it is transported by a carrier down its concentration gradient and no energy input is required. The transporter is symmetrical in that glucose can travel in either direction with equal ease. However, glucose transport does require that the hexose be properly oriented; the reducing end always points toward the cytoplasm. The carrier is 55 kDa, contains 15% carbohydrate, and binds cytochalasin B stoichiometrically. It crosses the plasma membrane 12 times, although only the last 6 transmembrane helices are believed to be directly involved with glucose transport. There are actually a group of homologous glucose carriers; the one most closely associated with insulin regulation is GLUT4.

The transport of glucose by its carrier is analogous to an enzyme reaction:

$$\text{glucose}_{out} + \text{carrier} \; \rightleftharpoons \; \text{glucose-carrier} \; \rightarrow \; \text{glucose}_{in} + \text{carrier}$$

Such an analysis yields values for K_m and V_{max}; although they are not exactly equivalent to the enzymatic parameters, they do convey similar information. Insulin stimulates glucose transport by increasing both the V_{max} (transport capacity) and the K_m (transporter affinity).

In initial studies cytochalasin B was used to assay glucose carriers in the plasma membrane and in the low-density microsomal fraction. It was found that insulin increases the content of transporters in the plasma membrane, although it decreases the content in microsomes; this correlates with the observed stimulation of glucose transport. One can also assay for the carrier by measuring transport activity in the different cellular fractions, and the same results are obtained. Therefore insulin increases V_{max} by inducing the migration of glucose carriers from the microsomes to the plasma membrane. In fact, there appears to be a small amount of basal cycling; and insulin both increases GLUT4 translocation to the plasmalemma, while reducing its internalization. In addition, there appear to be multiple pools that are regulated separately; insulin recruits transporters from

GLUT4 storage vesicles via microtubules, whereas exercise and G protein trimers recruit them from endosomes via Arf6 and actin.

However, other data have not supported this migration hypothesis as the sole mechanism for insulin-induced glucose uptake. First of all, cycloheximide, an inhibitor of protein synthesis, also blocks the translocation of the glucose transporter from the microsomes to the plasma membrane but does not inhibit insulin-stimulated glucose uptake. Second, there is a quantitative discrepancy between transporter migration and glucose transport: the former only increases 2.6-fold, whereas the latter is stimulated 11-fold. Finally, there is a temporal discrepancy: the carrier reaches the cell surface several minutes before glucose uptake is increased, suggesting the existence of an activation step.

The molecular mechanism for the translocation appears to involve PI3K, PKCζ, and microtubules. PI3K is required but not sufficient; inhibition of PI3K blocks glucose uptake, but other hormones, like PDGF, can also stimulate PI3K without affecting glucose transport. The magnitude and location of the PI3K response appear to be important. For example, PDGF only stimulates PI3K in the plasma membrane and the effect is small, wherease insulin stimulates PI3K in both the plasmalemma and the microsomes, where the transporters are located and the effect is greater. The pattern of $PI(3)P_n$ production is another important parameter: small, continuous levels or large, brief pulses are ineffective in inducing GLUT4 translocation. Insulin also stimulates PKCζ, which is likewise required for glucose transport but not sufficient. Insulin appears to stimulate PKCζ via two pathways: $PI(3)P_n$ and phosphatidic acid. The former is a product of PI3K, and the latter is derived from PLCγ-generated DG, which is then phosphorylated by DGK.

The docking and fusion of transporter-laden vesicles to the plasma membrane are regulated through a member of the Rho family. A complex of APS, CAP, and Cbl bind the phosphorylated insulin receptor. After the receptor phosphorylates Cbl, the complex dissociates and binds the adaptor CRKII. CRKII is analogous to Grb2 in the Ras pathway; it binds C3G, a RhoGEF, which activates TC10, a member of the Rho family.

The current hypothesis has PKB and PKCζ releasing GLUT4 vesicles from intracellularly tethered sites. For example, PKCζ is known to phosphorylate several proteins associated with the GLUT4 vesicles. TC10 then regulates the docking and fusion of these vesicles with the plasma membrane. Finally, insulin inhibits Rab5 and dynein via the PI3K pathway; the former is accomplished by PKB phosphorylation and activation of RabGAP. Dynein are molecular motors attached to microtubules and transport material along the microtubules like freight trains along a railroad track. The inhibition of Rab5 is believed to prevent the internalization of GLUT4.

Protein Synthesis.   Protein synthesis is a complex process which is affected by hormones at several sites (Fig. 12-7). This process is more fully covered in Chapter 15; however, the four sites influenced by insulin are discussed here: eIF-2B, eIF-4E, 4E-BP, and S6. eIF-2 binds the initiator methionine-tRNA, and eIF-2B is the GEF (GNP exchange facilitator) for eIF-2. The regulation of eIF-2B is simple and has been described. Briefly, it is phosphorylated and inhibited by GSK3; this prevents eIF-2 from being recycled and suppresses translation. Insulin activates eIF-2B by stimulating PKB to phosphorylate and inactivate GSK3 (see Fig. 12-6). In addition, the insulin receptor phosphorylates the adaptor, Nck, which induces its translocation to the ribosomal compartment where it binds eIF-2β and stimu-

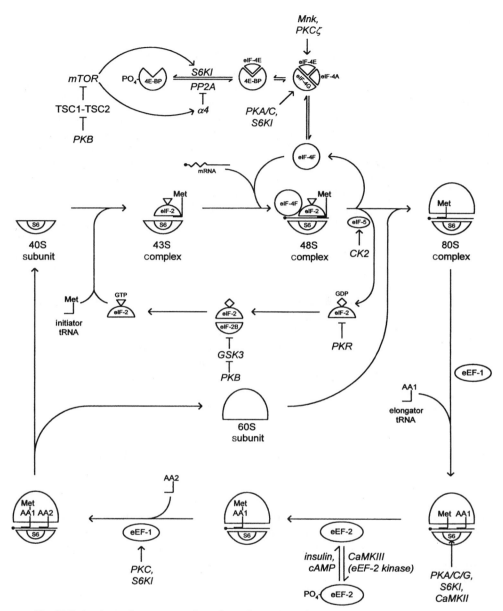

Fig. 12-7. A schematic representation of translation. Regulatory factors are italicized. *AA,* Amino acid; *Mnk,* MAPK interacting kinase; *TSC1,* tuberous sclerosis 1 (also called *hamartin*); *TSC2,* tuberous sclerosis 2 (also called *tuberin*).

lates translation. Although the exact mechanism is unknown, Nck may form a complex with Sam68, an RNA-binding protein that is also phosphorylated by the insulin receptor. This complex could target specific mRNA to the ribosome.

eIF-4F is a complex that is involved with binding mRNA, removing secondary structure, locating the initiation start site, and bringing the mRNA to the small ribosomal subunit. Among other components, it includes eIF-4E, which binds the mRNA cap; eIF-4A, a helicase; and eIF-4G, a scaffold that holds the complex together. eIF-4E can be directly phosphorylated and stimulated by Mnk, a kinase downstream of MAPK; there is some evidence that PKCζ can also target this translation factor. 4E-BP is an inhibitor that binds and sequesters eIF-4E. Phosphorylation of 4E-BP results in the release of eIF-4E and the stimulation of protein synthesis. This modification can be reversed by PP2A. However, mTOR phosphorylation of α4, a PP2A subunit, inhibits the protein phosphatase and prolongs the phosphorylated state of 4E-BP.

How insulin stimulates the phosphorylation of 4E-BP is still unresolved; the major candidates are mTOR and S6KI. In part, this confusion is the result of multiple levels of phosphorylation and discrepancies between *in vivo* and *in vitro* experiments; for example, rapamycin inhibits 4E-BP phosphorylation *in vivo* but not *in vitro*. This would suggest that S6KI directly modifies 4E-BP and that mTOR is upstream. However, several protein kinases can affect S6KI because it possesses multiple phosphorylation sites. MAPK phosphorylation primes S6KI for other kinases by exposing the carboxy terminus; PDK modifies the activation loop, and either PDK or mTOR can phosphorylate the carboxy terminus. There are additional complications because there are several isoforms of S6KI; S6KIβ is less dependent on PKB and mTOR than is S6KIα. Additional confusion arises when rapamycin is used to probe mTOR function; it appears that this compound only inhibits the phosphorylation of some substrates. Therefore phosphorylation of a protein in the presence of rapamycin does not necessarily eliminate mTOR as the responsible kinase. In any event, it is not clear how mTOR is activated by PDK-PKB. The most recent data suggest that a complex of the antioncogenes TSC1 and TSC2 inhibits mTOR and that PKB phosphorylation of TSC2 inhibits the TSC1-TSC2 complex, thereby activating mTOR. Therefore the most likely pathway would be PDK → PKB → TSC1-TSC2 → mTOR → S6KI → 4E-BP with MAPK and PDK playing supporting roles in S6KI activation (Fig. 12-7). On the other hand, some authorities have suggested that these are parallel pathways; nutrient availability regulates translation through the mTOR pathway, whereas growth factors use PDK.

Insulin has two other effects on protein synthesis. First, S6KI phosphorylates ribosomal S6 protein; this modification increases AUG binding and translation fourfold. Second, insulin inhibits eEF-2 kinase. Because eEF-2 kinase inactivates eEF-2, its inhibition by insulin would reactivate eEF-2 and stimulate translation. The eEF-2 kinase can be phosphorylated and inhibited by either S6KI (via mTOR) or S6KII (via MAPK).

Feedback Loops.   Within the preceding scheme there are several feedback loops. Amplification of the insulin signal can be achieved in many ways. First, PKB phosphorylation of IRS protects it from the actions of PTPs. PKB also phosphorylates and inhibits PTP-1B, which reduces the dephosphorylation of the insulin receptor. Finally, Ras synergizes with IRS to stimulate PI3K.

On the other hand, there are also negative feedback loops. First, mTOR, PKCζ, MAPK, and GSK3 phosphorylation of IRS inhibits its tyrosine phosphorylation. PKC phosphorylation of the insulin receptor triggers its desensitization. and MAPK phosphorylation of mSOS causes it to dissociate from Grb2.

Tumor Necrosis Factor

TNF is a stress hormone. As with most such hormones, it has the seemingly paradoxical effects of facilitating either apoptosis or survival. In fact, both responses are interrelated; when faced with an insult, the cell must make a decision. If the damage is great enough to induce malignant transformation, apoptosis is triggered; it is better for the cell to die than for the whole organism to die. On the other hand, if the damage is minor, stress and defense responses are induced to facilitate cell survival. This section concentrates on four specific responses: caspase activation, sphingolipid hydrolysis, PKB stimulation, and transcription of stress-related genes (Fig. 12-8).

The activation of caspases is relatively simple. First, caspases are proteases involved with apoptosis; they exist as zymogens with low basal activity. They are associated with the tumor necrosis factor receptor (TNFR) through adaptors. When TNF induces TNFR trimerization, the caspases are brought into proximity. Cross-cleavage and activation ensue; there is some evidence that aggregation can also allosterically activate the caspases. The major function of these initial caspases is to cleave other caspases in a protease cascade that quickly amplifies the signal.

TNF can stimulate both sphingomyelinases by recruiting them to the plasma membrane through adaptors. In addition, one of these adaptors, TRAF (TNFR-associated factor) binds and activates sphingosine kinase so that the ceramide produced is converted to sphingosine-1-phosphate; the former favors apoptosis, whereas the latter favors survival. TRAF actually plays a central role in the survival pathway of TNF. First, it binds and stimulates Src, an STK. The generation of pY activates the PI3K pathway and ultimately PKB. PKB is a major antiapoptotic signal (see Fig. 12-3). Second, TRAF binds and activates inhibitor of apoptosis protein (IAP), which ubiquinates the caspases and targets them for degradation.

TNFR initiates several protein kinase cascades, which lead to transcription factors (see Chapter 13 for a more thorough discussion of these transcription factors). These pathways are often critical in determining whether apoptosis or stress responses will be induced. For example, AP-1 (also called *Jun-Fos*) and NF-κB favor inflammation. However, the subunit composition is important; c-Jun is antiapoptotic, whereas JunB has the opposite effect. The duration of stimulation is also critical; delayed, persistent activation of AP-1 favors apoptosis, but rapid, transient stimulation enhances survival. The intensity of stimulation is another factor; strong ERK activation favors survival, whereas weak activity leads to apoptosis. Finally, the inhibition of RNA or protein synthesis favors apoptosis.

There are three kinase cascades leading to three transcription factors. First, TRAF directly binds and activates the apoptosis signal-regulating kinase (ASK), which initiates the JNK cascade. JNK not only phosphorylates and activates the Jun subunit of AP-1, but it also phosphorylates Bcl2, which binds and sequesters proapoptotic Bcl2 family members. In association with TANK (TRAF-associated NF-κB activator), TRAF binds and stimulates the NF-κB inducing kinase (NIK). This action initiates another kinase cascade leading to NF-κB (see Chapter 13 for details). Finally, TNF activates the ERK cascade. It begins with the binding of Grb2 to a proline-rich domain in the TNFR; binding is also facilitated by Fan (factor associated with neutral sphingomyelinase). The cascade then proceeds as has been previously described except that the kinase RIP (receptor interacting protein) replaces MEK. In truth, many different routes to ERK have been proposed; the one just given is presented because of its simplicity and clarity. The target of ERK is Ets, an accessory transcription factor that facilitates many other factors, like NF-κB.

Hormone Synergism and Antagonism

Glycogen Metabolism

Glycogen metabolism is an excellent example of how hormone transducers can synergize and antagonize with each other in the regulation of a metabolic

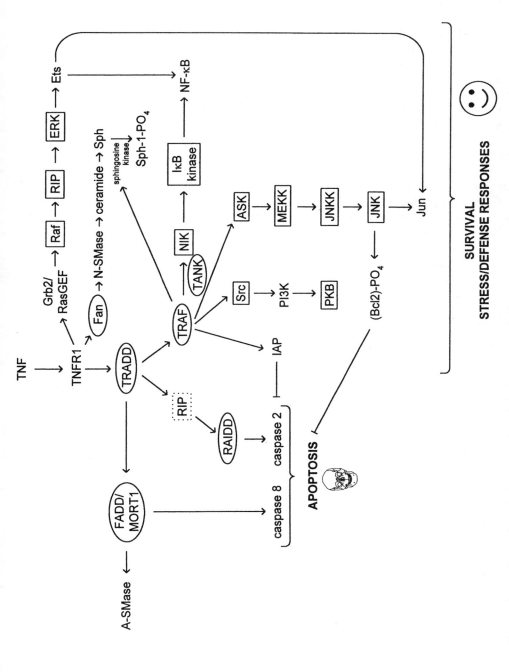

Fig. 12-8. The signaling pathways for TNF. Kinases are boxed; the boxes have solid lines if kinase activity is required for the indicated action, but the boxes have dotted lines if the kinase activity is not required. In the latter case, the kinase is probably acting like a scaffold. Arrows indicate stimulation and flat heads indicate inhibition. *ASK*, Apoptosis signal-regulating kinase; *A-SMase*, acid sphingomyelinase; *FADD*, Fas-associated protein with a death domain; *Fan*, factor associated with neutral sphingomyelinase; *IAP*, inhibitor of apoptosis protein; *N-SMase*, neutral sphingomyelinase; *NIK*, NF-κB inducing kinase; *RAIDD*, RIP-associated ICE-like death domain protein; *RIP*, receptor interacting protein; *TANK*, TRAF-associated NF-κB activator; *TRADD*, TNFR-associated death domain protein; *TRAF*, TNFR-associated factor.

377

pathway (Fig. 12-9). Three separate hormone groups are considered: cAMP-dependent hormones (glucagon in the liver and epinephrine in muscle); calcium-dependent hormones (vasopressin in the liver, which acts through the PPI pathway, and acetylcholine in the muscle, which releases calcium via action potentials); and insulin, which uses a variety of second messengers. For accuracy, it should be noted that in the liver glucagon has been shown to activate both adenylate cyclase and the PPI pathway; this is not surprising because the two mediators reinforce one another in this system. However, for simplicity the pathways will be considered separately.

Glucagon (liver) and epinephrine (muscle) inhibit glycogen synthesis and stimulate glycogenolysis; all of these effects are mediated by phosphorylation by PKA. Glycogen synthase in muscle has a substrate-binding site for UDP-glucose and allosteric sites for ADP and glucose-6-phosphate. The former inhibits the enzyme (if ADP is high, then energy levels are low and the cell should not be storing glucose but burning it), and the latter stimulates it (if glucose-6-phosphate levels are high, it should be stored). The enzyme has seven phosphorylation sites. Phosphorylation at five of these sites results in an increased K_m for UDP-glucose and K_a for glucose-6-phosphate and a decreased K_i for ADP; that is, it binds to its inhibitor more efficiently but binds to its substrate and activator less efficiently. Essentially, the enzyme is inhibited. These phosphorylation sites are modified by

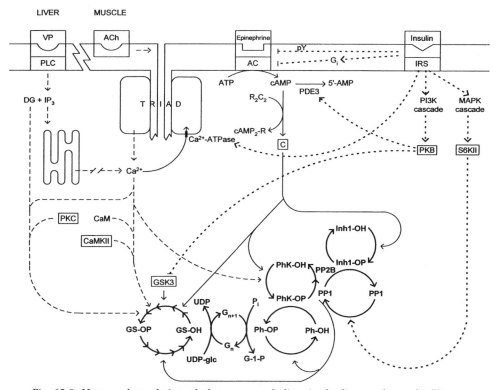

Fig. 12-9. Hormonal regulation of glycogen metabolism in the liver and muscle. Kinases are boxed, arrows indicate stimulation, and flat heads indicate inhibition. Solid lines depict the cAMP pathway; dashed lines, the calcium pathway; and dotted lines, the insulin pathways. *AC,* Adenylate cyclase; *GS,* glycogen synthase; G_n, glycogen; *G-1-P,* glucose-1-phosphate; *Ph,* glycogen phosphorylase; *PhK,* glycogen phosphorylase kinase; R_2C_2, the PKA tetramer; *UDP-glc,* UDP-glucose. Other abbreviations are as in the text.

different protein kinases, and the phosphorylations are additive. As a result, glycogen synthase is not regulated in a strictly on-off fashion but is controlled more like a rheostat. This enzyme is inhibited, in part, by PKA-mediated phosphorylation.

Glycogen breakdown is accomplished by glycogen phosphorylase, which exists in two forms: phosphorylase *b* is a dimer, which is subject to allosteric control, but phosphorylase *a* is a tetramer, which is fully active and insensitive to allosteric modulators. The conversion from phosphorylase *b* to phosphorylase *a* is a result of phosphorylation by phosphorylase kinase. This kinase has a subunit composition of $(\alpha\beta\gamma\delta)_4$; γ contains the catalytic site, δ is CaM, and α and β are regulatory subunits that are both phosphorylated. This phosphorylation, which is also performed by PKA, leads to a 15- to 20-fold increase in activity; it also decreases the K_a of the δ subunit for calcium, another activator.

Finally, the these phosphorylations can be reversed by PP1. Therefore for glycogenolysis to be maintained, this enzyme must be inactivated. Once again, this is done by PKA, which phosphorylates the G subunit of PP1; this modification causes PP1 to dissociate from the glycogen particle and bind Inhibitor 1. This binding also requires that Inhibitor 1 be phosphorylated by PKA.

Vasopressin can also stimulate glycogenolysis in the liver, and this effect is mediated by the PPI pathway. The free calcium binds to the δ subunit (CaM) of phosphorylase kinase, resulting in further activation. The phosphorylation of the glycogen phosphorylase will then stimulate the enzyme to break down glycogen. Calcium can also activate CaMKII and PKC; both of these kinases can phosphorylate and further inactivate glycogen synthase. The elevation of calcium by the PPI pathway can be augmented by cAMP, which activates CNG calcium channels.

In the muscle, calcium elevations are a result of action potentials induced by acetylcholine. The calcium released from the sarcoplasmic reticulum not only stimulates muscle contraction but also glycogen breakdown; the latter supplies the energy for the former.

Insulin inhibits glycogenolysis through several pathways. First of all, phosphorylation by PKA is stopped by (1) tyrosine phosphorylation of the βAR and uncoupling it from G_s, (2) stimulating G_i to suppress adenylate cyclase, and (3) activating PDE3 through the PI3K-PKB pathway to hydrolyze cAMP. PKB also phosphorylates and inhibits GSK3. Insulin terminates calcium signaling by phosphorylating IRS, which binds and stimulates the sarcoplasmic calcium pump. This pump will then lower cytoplasmic calcium levels. Finally, insulin reverses the phosphorylation of glycogen synthase, glycogen phosphorylase, and phosphorylase kinase; that is, the synthase is reactivated, and the phosphorylase returns to tight allosteric control. Insulin accomplishes this dephosphorylation by inducing S6KII to phosphorylate and activate PP1.

Smooth Muscle Contraction

In glycogen metabolism, cAMP and calcium synergized with each other and were antagonized by the PI3K-PKB and MAPK pathways. In smooth muscle contraction, cyclic nucleotides and calcium antagonize each other. Contraction occurs when the thick filaments (primarily myosin) and thin filaments (actin) in muscle slide past one another. This requires the stimulation of an actin-activated Mg^{2+}-ATPase that resides in the head region of the heavy chain of myosin. Also associated with the head are two light chains. Contraction is initiated when intracellular calcium levels rise (Fig. 12-10); because the

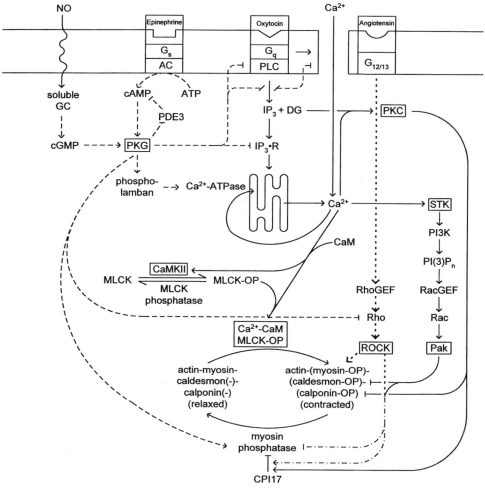

Fig. 12-10. Hormonal regulation of smooth muscle contraction. Kinases are boxed, arrows indicate stimulation, and flat heads indicate inhibition. Solid lines depict the calcium pathway; dashed lines, the PKG pathway; dotted lines, the Rho pathway; and alternating dots and dashes, the overlap between the Rho and Rac pathways. *GC*, Guanylate cyclase; *MLCK*, myosin light chain kinase. Other abbreviations are as in the text.

sarcoplasmic reticulum in smooth muscle is not as well developed as it is in skeletal muscle, the calcium is thought to originate from both intracellular and extracellular sources. The calcium and CaM activate CaMKII, which in turn phosphorylates and activates a kinase that modifies one of the myosin light chains. This myosin light chain kinase (MLCK) phosphorylates the P-light chain, which then stimulates the Mg^{2+}-ATPase and triggers contraction. In addition, there are two contraction inhibitors that must be neutralized: calponin and caldesmon. The elevated calcium activates PKC, which phosphorylates and inhibits calponin. Furthermore, calcium stimulates a calcium-dependent STK that initiates the PI3K pathway. A $PI(3)P_n$-activated RacGEF leads to the stimulation of Pak, which phosphorylates and inhibits caldesmon. These phosphorylations can be reversed by myosin light chain phosphatase, which is really PP1 with the M localizing subunit. Both PKC and Pak phosphorylate CPI17 and enhance its inhibition of PP1.

Oxytocin is a major hormone inducing smooth muscle contraction. Its receptor is coupled to G_q, which activates the PPI pathway to release calcium from internal stores; G_q can also directly open plasma membrane calcium channels. On the other hand, angiotensin is coupled to $G_{12/13}$, which triggers the Rho pathway that activates ROCK. Like CaMKII, ROCK can phosphorylate and stimulate myosin. In addition, ROCK inhibits myosin phosphatase in two ways; first, like PKC and Pak, it phosphorylates CPI17. Second, both ROCK and Pak phosphorylate M; this modification causes PP1 to dissociate from the myofilaments. In summary, the PPI, Rho, and Rac pathways stimulate smooth muscle contraction by activating actin-myosin interactions, blocking the effects of contraction inhibitors, and inhibiting the myosin phosphatase.

PKG opposes these pathways and induces smooth muscle relaxation. PKG can be activated in two ways. First, NO can diffuse through the plasma membrane, bind soluble GC, and generate cGMP. Second, epinephrine is coupled to G_s, which elevates cAMP to levels that result in the cross-activation of PKG. PKG facilitates cAMP accumulation by allosterically inhibiting PDE3. PKG has three effects: it inhibits the PPI and Rho pathways, and it stimulates myosin phosphatase. Suppression of the PPI pathway involves the phosphorylation and inhibition of PLCβ, and the inhibition of the IP_3 receptor via phosphorylation of an associated regulatory protein. PKG can also phosphorylate and activate several potassium channels; this hyperpolarizes the muscle cell and inhibits the voltage-gated calcium channels. PKG lowers existing calcium levels by phosphorylating and stimulating phospholamban, an activator of the calcium pump in the sarcoplasmic reticulum. Finally, PKG phosphorylates and inhibits RhoA. Although PKG also stimulates the plasma membrane calcium pump and the myosin phosphatase, the molecular mechanisms for these effects are unknown.

Summary

Phosphorylation is a major mechanism by which hormones directly affect cellular metabolism, the cytoskeleton, and membrane transport. The activities of many enzymes are influenced by phosphorylation, and the responsible kinases are frequently under the control of hormone mediators: PKA by cAMP, PKG by cGMP, PKC by calcium and phospholipids, CaM-dependent protein kinase by CaM, PKB by $PI(3)P_n$, and CK2 by polyamines. In addition, there are many other kinases that may not be directly regulated by second messengers but that form part of protein kinase cascades initiated by hormones. These cascades amplify the signal and introduce additional sites for regulation.

The S-T phosphatases, which regulate the reverse process, are subject to similar controls. PP1 is primarily regulated by targeting subunits and inhibitors, both of which are affected by phosphorylation. PP2A is allosterically regulated by polyamines and ceramide; PP2B, by calcium and CaM; and PP5, by G proteins and arachidonic acid. The soluble phosphotyrosine phosphatases can be regulated by phosphorylation or allosterism; an example of the latter is the activation of SHP by pY. The membrane-bound PTPs may mediate hormone or cell-matrix transduction.

Hormones use many of these pathways to mediate their biological activities. Insulin primarily uses the PI3K-PKB pathway for its metabolic effects, whereas TNF uses the sphingomyelinase and SAPK pathways for stress responses. Finally, many of these transducers can interact with one another; for example, cyclic nucleotides and calcium synergize with each other in glycogenolysis but antagonize each other in smooth muscle contraction.

References

Kinase Regulation

Bauman, A.L., and Scott, J.D. (2002). Kinase- and phosphatase-anchoring proteins: Harnessing the dynamic duo. *Nature Cell Biol.* 4, E203-E206.

Kolch, W. (2000). Meaningful relationships: The regulation of the Ras/Raf/MEK/ERK pathway by protein interactions. *Biochem. J.* 351, 289-305.

Sharrocks, A.D., Yang, S.H., and Galanis, A. (2000). Docking domains and substrate-specificity determination for MAP kinases. *Trends Biochem. Sci.* 25, 448-453.

Tanoue, T., and Nishida, E. (2003). Molecular recognitions in the MAP kinase cascades. *Cell. Signal.* 15, 455-462.

Proline-Directed Protein Kinases

Barr, R.K., and Bogoyevitch, M.A. (2001). The c-Jun N-terminal protein kinase family of mitogen-activated protein kinases (JNK MAPKs). *Int. J. Biochem. Cell Biol.* 33, 1047-1063.

Chang, L., and Karin, M. (2001). Mammalian MAP kinase signalling cascades. *Nature (London)* 410, 37-40.

Chong, H., and Vikis, H.G., and Guan, K.L. (2003). Mechanisms of regulating the Raf kinase family. *Cell. Signal.* 15, 463-469.

Doble, B.W., and Woodgett, J.R. (2003). GSK-3: tricks of the trade for a multi-tasking kinase. *J. Cell Sci.* 116, 1175-1186.

Frame, S., and Cohen, P. (2001). GSK3 takes centre stage more than 20 years after its discovery. *Biochem. J.* 359, 1-16.

Harwood, A.J. (2001). Regulation of GSK-3: A cellular multiprocessor. *Cell (Cambridge, Mass.)* 105, 821-824.

Other Protein Kinases

Brazil, D.P., and Hemmings, B.A. (2001). Ten years of protein kinase B signalling: A hard Akt to follow. *Trends Biochem. Sci.* 26, 657-664.

Datta, S.R., Brunet, A., and Greenberg, M.E. (1999). Cellular survival: A play in three Akts. *Genes Dev.* 13, 2905-2927.

Kandel, E.S., and Hay, N. (1999). The regulation and activities of the multifunctional serine/threonine kinase Akt/PKB. *Exp. Cell Res.* 253, 210-229.

Lawlor, M.A., and Alessi, D.R. (2001). PKB/Akt: a key mediator of cell proliferation, survival and insulin responses? *J. Cell Sci.* 114, 2903-2910.

Rohde, J., Heitman, J., and Cardenas, M.E. (2001). The TOR kinases link nutrient sensing to cell growth. *J. Biol. Chem.* 276, 9583-9586.

Serine-Threonine Phosphatases

Cohen, P.T.W. (2002). Protein phosphatase 1—targeted in many directions. *J. Cell Sci.* 115, 241-256.

Janssens, V., and Goris, J. (2001). Protein phosphatase 2A: A highly regulated family of serine/threonine phosphatases implicated in cell growth and signalling. *Biochem. J.* 353, 417-439.

Klee, C.B., Ren, H., and Wang, X. (1998). Regulation of the calmodulin-stimulated protein phosphatase, calcineurin. *J. Biol. Chem.* 273, 13367-13370.

Zolnierowicz, S., and Bollen, M. (2000). Protein phosphorylation and protein phosphatases. *EMBO J.* 19, 483-488.

Phosphotyrosine Phosphatases

Beltran, P.J., and Bixby, J.L. (2003). Receptor protein tyrosine phosphatases as mediators of cellular adhesion. *Front. Biosci.* 8, D87-D99.

Camps, M., Nichols, A., and Arkinstall, S. (2000). Dual specificity phosphatases: A gene family for control of MAP kinase function. *FASEB J.* 14, 6-16.

Petrone, A., and Sap, J. (2000). Emerging issues in receptor protein tyrosine phosphatase function: Lifting fog or simply shifting? *J. Cell Sci.* 113, 2345-2354.

Raugei, G., Ramponi, G., and Chiarugi, P. (2002). Low molecular weight protein tyrosine phosphatases: Small, but smart. *Cell. Mol. Life Sci.* 59, 941-949.

Transforming Growth Factor β

Miyazono, K. (2000). Positive and negative regulation of TGF-β signaling. *J. Cell Sci.* 113, 1101-1109.

Piek, E., Heldin, C.H., and ten Dijke, P. (1999). Specificity, diversity, and regulation in TGF-β superfamily signaling. *FASEB J.* 13, 2105-2124.

Insulin

Baumann, C.A., and Saltiel, A.R. (2001). Spatial compartmentalization of signal transduction in insulin action. *Bioessays* 23, 215-222.

Bickel, P.E. (2002). Lipid rafts and insulin signaling. *Am. J. Physiol.* 282, E1-E10.

Elmendorf, J.S. (2002). Signals that regulate GLUT4 translocation. *J. Membr. Biol.* 190, 167-174.

Farese, R.V. (2002). Function and dysfunction of aPKC isoforms for glucose transport in insulin-sensitive and insulin-resistant states. *Am. J. Physiol.* 283, E1-E11.

Hajduch, E., Litherland, G.J., and Hundal, H.S. (2001). Protein kinase B (PKB/Akt)—a key regulator of glucose transport? *FEBS Lett.* 492, 199-203.

Khan, A.H., and Pessin, J.E. (2002). Insulin regulation of glucose uptake: A complex interplay of intracellular signalling pathways. *Diabetologia* 45:1475-1483.

Kido, Y., Nakae, J., and Accili, D. (2001). The insulin receptor and its cellular targets. *J. Clin. Endocrinol. Metab.* 86, 972-979.

Simpson, F., Whitehead, J.P., and James, D.E. (2001). GLUT4—At the cross roads between membrane trafficking and signal transduction. *Traffic* 2, 2-11.

Whitehead, J.P., Clark, S.F., Ursø, B., and James, D.E. (2000). Signalling through the insulin receptor. *Curr. Opin. Cell Biol.* 12, 222-228.

Whiteman, E.L., Cho, H., and Birnbaum, M.J. (2002). Role of Akt/protein kinase B in metabolism. *Trends Endocrinol. Metab.* 13, 444-451.

Tumor Necrosis Factor

Baker, S.J., and Reddy, E.P. (1998). Modulation of life and death by the TNF receptor superfamily. *Oncogene* 17, 3261-3270.

Baud, V., and Karin, M. (2001). Signal transduction by tumor necrosis factor and its relatives. *Trends Cell Biol.* 11, 372-377.

Chung, J.Y., Park, Y.C., Ye, H., and Wu, H. (2002). All TRAFs are not created equal: common and distinct molecular mechanisms of TRAF-mediated signal transduction. *J. Cell Sci.* 115, 679-688.

Denecker, G., Vercammen, D., Declercq, W., and Vandenabeele, P. (2001). Apoptotic and necrotic cell death induced by death domain receptors. *Cell. Mol. Life Sci.* 58, 356-370.

Kumar, S., and Colussi, P.A. (1999). Prodomains—adaptors—oligomerization: The pursuit of caspase activation in apoptosis. *Trends Biochem. Sci.* 24, 1-4.

Matsuzawa, A., and Ichijo, H. (2001). Molecular mechanisms of the decision between life and death: Regulation of apoptosis by apoptosis signal-regulating kinase 1. *J. Biochem. (Tokyo)* 130, 1-8.

Wajant, H., and Scheurich, P. (2001). Tumor necrosis factor receptor-associated factor (TRAF) 2 and its role in TNF signaling. *Int. J. Biochem. Cell Biol.* 33, 19-32.

Smooth Muscle Contraction

Abdel-Latif, A.A. (2001). Cross talk between cyclic nucleotides and polyphosphoinositide hydrolysis, protein kinases, and contraction in smooth muscle. *Exp. Biol. Med.* 226, 153-163.

Fukata, Y., Amano, M., and Kaibuchi, K. (2001). Rho-Rho-kinase pathway in smooth muscle contraction and cytoskeletal reorganization of non-muscle cells. *Trends Pharmacol. Sci.* 22, 32-39.

Pfitzer, G. (2001). Regulation of myosin phosphorylation in smooth muscle. *J. Appl. Physiol.* 91, 497-503.

PART 4

Gene Regulation by Hormones

Hormonally Regulated Transcription Factors

CHAPTER OUTLINE

In the preceding unit, mechanisms by which hormones and their second messengers directly affect cellular processes were discussed. This unit describes mechanisms by which these molecules affect gene expression. Chapter 13 discusses the hormonal control of specific transcription factors and their mechanisms of action. Chapter 14 then examines the role of DNA organization and conformation on transcription; in particular, nucleosomes and their covalent modification, DNA methylation, and DNA conformation are discussed. Finally, the hormonal control of various posttranscriptional events is summarized in Chapter 15; such events include RNA stability, processing, transport, and translation, as well as the posttranslational modification of proteins.

Specific Transcription Factors: Structure

For completeness, this section will briefly cover each major structural class of specific transcription factors. This discussion is then followed by a more detailed examination of those groups that are hormonally regulated. Transcription factors can be divided into two major groups: specific and general. General transcription factors are those proteins responsible for the actual transcription of genes: for example, RNA polymerases, the TATA box binding protein (TBP), and related factors. They are continuously active on housekeeping genes, but induced genes are transcribed only under certain circumstances. The general transcriptional machinery has no intrinsic mechanism to tell when these latter genes need to be expressed. It is the function of the specific transcription factors to locate sets of genes that are only expressed intermittently. First, the endocrine system senses the condition and triggers the secretion of hormones that activate specific transcription factors; second, these factors then search for specific DNA sequences associated with the various gene sets and recruit the general transcription machinery to them. As such, specific transcription factors form a critical link between hormones and gene expression.

In a review of the known factors, it is important to keep in mind several basic principles. First, the α-helix is the most common secondary structure used to bind DNA; presumably this frequency is a result of the dimensions of the α-helix, which fits well in the major groove of DNA. Most hormone response element (HRE) recognition helices in eukaryotic transcription factors have the same basic structure: an amino-terminal half that binds the DNA bases and a carboxy-terminal half that interacts with the deoxyribose-phosphate backbone. Second, the factors primarily bind in the major groove because the nucleotides are more accessible, especially to an α-helix. Minor groove contacts are usually made by an isolated peptide strand, which augments the major contacts in the wider groove. Third, the DNA backbone constitutes about half of all the contacts. Although these contacts are to the deoxyribose-phosphate backbone, they are not necessarily nonspecific because the base sequence can produce local alterations in the DNA structure sufficient to add to the overall specificity of binding. Finally, there is no unique correspondence between amino acids and the bases they contact; that is, there is no simple code for determining to what nucleotide sequence a given protein sequence will bind.

Families

Multiple α-Helical Family

This rather large and diverse family includes transcription factors having one α-helix in the major groove and one or more α-helices overlying the first helix. One of the simplest and evolutionarily earliest members of this family is the helix-turn-helix (HTH) motif. The bacterial prototype consists of two α-helices separated by a β-turn (Fig. 13-1, *A*). The second, or carboxy-terminal, helix is the recognition helix and lies in the major groove to make specific contacts with the bases. The first helix lies on top of the second at a right angle and makes nonspecific contacts with the backbone. The amino-terminal tail also binds in an adjacent major groove. Transcription factors with HTH motifs bind as dimers to DNA sequences with dyad symmetry.

Higher-order organisms have a variation on this structure called the *home-odomain* (Fig. 13-1, *B* and *C*). This family has three helices; as with the HTH proteins, the carboxy-terminal helix lies in the major groove to make specific contacts. The first two helices lie on top, nearly perpendicular to the third. Because of the additional helix, the amino-terminal tail of the homeodomain is displaced and lies in the 5′ minor groove. Although the binding of this tail is weak ($K_d \sim 10^{-7}\,M$), it adds to the overall DNA affinity of this family. Alone, the third helix has a K_d of $\sim 1\text{-}2 \times 10^{-9}\,M$. These transcription factors are often involved with developmental programs. Examples of this family include the homeotic proteins of *Drosophila*, the MAT factors in yeast, and TTF1 (formerly called thyroid transcription factor 1 [TTF1]) in mammals.

The POU family represents a variation of the homeodomain. The name is an acronym derived from the first three groups of transcription factors in which the motif was recognized: Pit-1, the Oct subfamily, and Unc-86. This motif contains a standard homeodomain, called *POU$_H$*, preceded by a conserved 75 to 82 amino acid sequence called *POU$_S$*. This latter region is composed of four α-helices; the second and third form a classic HTH structure with the third helix lying in the major groove (Fig. 13-1, *D*). The first and fourth helices are antiparallel to each other and lie on top of the HTH structure and perpendicular to the second helix. There is a hydrophobic core between these helices. The POU$_S$ domain is homologous to several bacterial repressors, such as the λ and Cro repressors. These inhibitors have an absolute requirement for dimerization for DNA binding; this is accomplished by a fifth helix that acts as a dimerization interface. This helix does not exist in POU$_S$; rather, it is replaced by a variable linker (14-26 amino acids) that couples POU$_S$ to POU$_H$. This bridge lies in the minor groove and determines the spacing between POU$_S$ and POU$_H$. POU$_H$ binds the 3′ adenine-thymine (AT)–rich half of the HRE, whereas POU$_S$ binds the 5′ ATGC sequence. Interestingly, the spacer is long enough to allow POU$_S$ to bind its half of the HRE in either orientation.

Dimerization depends on the factor and the HRE. Members of the Oct subfamily dimerize weakly in solution, although their binding does exhibit cooperativity when binding to duplicated sequences. Pit-1 forms a much stronger dimer (Fig. 13-1, *D*); this phenomenon is probably related to the fact that its DNA binding site is also usually dimeric. The dimerization is thought to be mediated by the first two helices of POU$_H$.

The winged HTH (wHTH) has three α-helices; the second and third form a classic HTH motif, whereas a β-sheet (the wing) lies overhead (Fig. 13-1, *E*). This

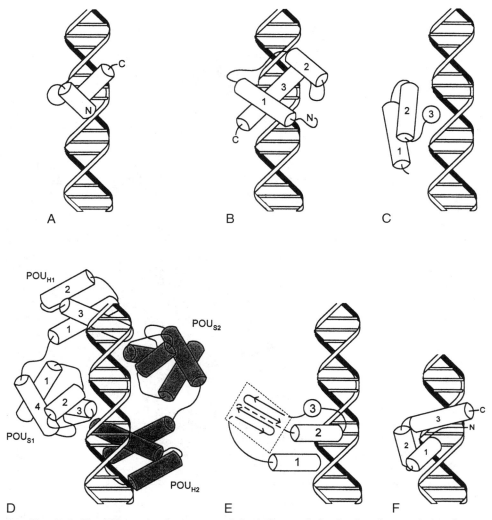

Fig. 13-1. Three-dimensional structure of the helix-turn-helix family of transcription factors. (A) Helix-turn-helix (HTH) domain; (B) homeodomain (viewed from above); (C) homeodomain (viewed from the side); (D) POU domain; (E) winged HTH; and (F) the high mobility group (HMG) domain.

family binds DNA as monomers. Each helix has a conserved tryptophan; the tryptophans in the second and third helices contact the DNA. However, the tryptophan in the first helix actually intercalates between the sixth and seventh base pair (bp) in the 3′ HRE and induces a 120° bend in the DNA. This family includes the Ets (E26 transformation specific), IRF (interferon regulatory factor), and Fox subfamilies.

The high mobility group (HMG) domain was first discovered in nuclear proteins that migrated rapidly in polyacrylamide gels, and this characteristic gave rise to their name. The HMG domain also contains three α-helices; the second and third helices are perpendicular and form a twisted "L" (Fig. 13-1, *F*). The first helix along with the amino terminus fit into the minor groove; this binding widens the minor groove. The third helix makes contact with the deoxyribose-phosphate backbone. Aromatic amino acids from all three helices form a hydrophobic core that stabilizes the structure. Members of this family include the T cell factor/lymphoid enhancer factor (TCF/LEF) and Sox.

Zinc-Containing Family

There are several transcription factor families that bind zinc, and none have any structural similarity to the others. The tetrahedral zinc is simply used to stabilize the three-dimensional structure. The first group was discovered in transcription factor IIIA (TFIIIA) and is referred to as the true zinc fingers, or simply zinc fingers. This motif consists of 12 to 14 amino acids bounded by a cysteine pair at the amino terminus and a histidine pair at the carboxy terminus; this arrangement is often abbreviated as C_2H_2 (Fig. 13-2, *A*). These amino acids force the intervening sequence into a loop; the ascending limb is part of a β-sheet, whereas the descending limb forms an α-helix. The helix sits in the major groove at about a 45° angle with its amino-terminal end contacting the DNA. Because only the tip of the finger touches the DNA, each finger can only contact 3 to 4 bases and many fingers are required to recognize a specific sequence. These fingers are very periodic; each unit rotates 32° along the DNA axis. The ascending limb of the β-structure makes contact with the deoxyribose-phosphate backbone. The Sp1 factor is an example of this family.

The nuclear receptors are a second group containing zinc. As discussed in Chapter 6, all four coordinating amino acids are cysteines (C_2C_2) and the overall conformation is that of a loop-helix-extended region (Fig. 13-2, *B*). There are two "fingers" and the α-helix begins in the carboxy-terminal cysteine pair of each and

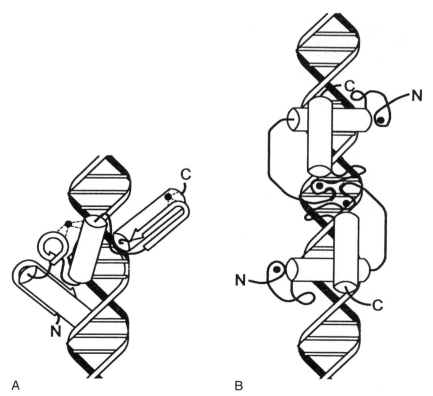

A   B

Fig. 13-2. Three-dimensional structure of several zinc-containing transcription factors. (A) Zinc finger; (B) zinc twist (nuclear receptors).

extends into the following sequence. The amino-terminal helix is the recognition helix and lies in the major groove, whereas the preceding loop binds the deoxyribose-phosphate backbone; the second helix lies on top. Both helices are amphipathic, and their hydrophobic sides form the core of their association. For this reason, this structure is more accurately referred to as a zinc twist; others have called it a *double loop zinc helix*. Essentially, it is a fancy HTH stabilized by zinc coordination. The nuclear receptors bind DNA as dimers and this association is mediated, in part, by the amino-terminal knuckle of the second zinc twist.

Two other members of this group should be mentioned. The GATA family resembles the NRs by having a C_2C_2 arrangement and includes the GATA and white collar-1 factors. The last group is the zinc cluster. Because it has a coiled coil dimerization motif, it is discussed below.

Coiled Coil Family

A coiled coil is an oligomerization motif first identified in intermediate filaments. Basically, two α-helices line up in parallel and twist into a left-handed supercoil. This interaction is stabilized by hydrophobic binding; indeed, a coiled coil motif can be recognized from its primary sequence, which shows a heptad repeat. Every fourth and seventh amino acid is hydrophobic. In Figure 13-3, *A*, these positions appear to line up opposite each other; in fact, the amino acids are slightly offset. For example, the leucine at position d is bounded both in front and behind by the a' amino acids from consecutive repeats on the opposite strand. This leucine is also bounded below by the leucine at d' and above by the amino acid at e'. This four-point enclosure can be constructed around every hydrophobic amino acid in the a/a' and d/d' positions; this arrangement is often referred to as a "knobs-in-holes" structure. In dimers, the amino acids at e and g form ionic bonds that determine the specificity of dimerization. In tetramers, these residues are hydrophobic and form lipophilic troughs for two additional helices (Fig. 13-3, *B*).

The coiled coils of transcription factors are unique in that position d is almost exclusively leucine. This fact led to speculation that the dimer was formed when

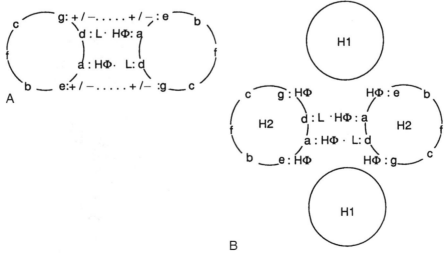

Fig. 13-3. A cross-sectional representation of a coiled coil dimer (A) and tetramer (B). *a-g*, The amino acid positions in a heptad repeat; *HΦ*, any hydrophobic amino acid; *L*, leucine.

the forked side chains of these amino acids interlocked, like a zipper, and the motif became known as a leucine zipper. In truth, leucine is the preferred amino acid because its dimensions produce the best and closest packing. Then why do intermediate filaments not have a strict requirement for leucine at d? The reason lies in the pitch, which is very shallow; it takes 16 repeats for 1 full turn of the supercoil. Transcription factors usually have only 4 to 6 repeats; this represents between only a quarter and a half a turn. This very limited entwining facilitates interchange between subunits, but it also reduces the dimer stability; the use of leucine maximizes the dimer interaction over this short distance. The extraordinary length of intermediate filaments results in several supercoil turns; the additional contacts compensate for the use of slightly less optimal hydrophobic amino acids.

The simplest transcription factors in this group have a basic, amino-terminal extension of the α-helix (bZIP). The carboxy terminus dimerizes, whereas the amino terminus bows out to straddle the DNA in the major grooves; the intervening hinge determines the spacing of the $\frac{1}{2}$ HRE (Fig. 13-4, *A*). This figure depicts the factor GCN4 (general control nonderepressible 4); the spacing of HRE for Jun-Fos is slightly different in that the contacts between the basic end of the helix, and the DNA cannot be made without bending the DNA. Jun bends the DNA toward the dimer interface, whereas Fos bends it away. The CCAAT/enhancer-binding protein (C/EBP) is another member of the bZIP family.

The basic helix-loop-helix (bHLH) factors have a second coiled coil carboxy-terminal to the first; the two are separated by an Ω loop. The second pair of helices fit into the hydrophobic troughs formed by dimerization of the first pair (Fig. 13-4, *B*). Essentially, bHLH is a bZIP where the carboxy terminus forms a tetramer instead of a dimer. The loop is required to flip the second pair so that they are parallel to the first set; a parallel coiled coil is more stable. In addition, the loop contacts the deoxyribose-phosphate backbone to further increase stability. The bHLH family includes MyoD, Ah, and HIF-1α (hypoxia-inducible factor 1α). The helix-span-helix factors are identical to the bHLH except that the loop is longer. AP-2 is an example of such a factor. Finally, the second pair of α-helices can have a carboxy-terminal extension beyond the tetramer (Fig. 13-4, *C*). These factors have two long helices; the amino-terminal pair forms a classic bZIP. The carboxy-terminal pair also forms a coiled coil at their carboxy-terminal end; however, the amino terminus flares out to fit in the grooves of the first zipper to form a tetramer. Essentially, these factors are bHLH proteins followed by a zipper; as such, they are often abbreviated bHLH-ZIP and include Myc, Max, Mad, and SREBP-2.

The last group possesses characteristics from several classes. The zinc cluster has two zinc atoms in a binuclear cluster that stabilizes two short α-helices at right angles (Fig. 13-4, *D*). The end of the amino-terminal helix is basic and points into the major groove. An extended linker originating from the carboxy-terminal helix traces the DNA backbone for three-fourths of a turn, where the peptide forms a coiled coil with its dimerization partner. There is evidence for another dimerization region carboxy-terminal to the first; this second oligomerization domain may fold back to form a HLH motif. Basically, these factors are bZIP or bHLH proteins without the DNA-binding, amino-terminal extension of the first pair of helices. Instead, the DNA-binding region is a separate module attached to the coiled coil by a tether. This structure allows the components of the HRE to be widely separated; for example, GAL4 recognizes the CCG sequence at the ends of a 17-bp HRE. This distance represents one and a half turns of the DNA helix.

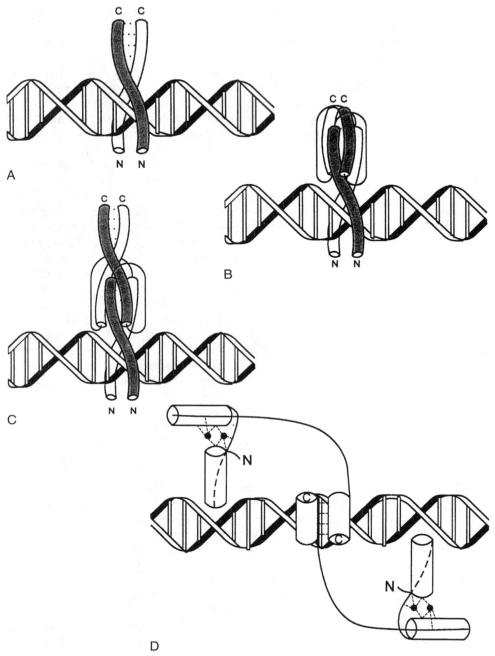

Fig. 13-4. Three-dimensional structure of the coiled coil family of transcription factors. (A) Basic-zipper (bZIP); (B) basic helix-loop-helix (bHLH); (C) bZIP-HLH; (D) zinc cluster.

Interestingly, the intervening sequence is also recognized by another transcription factor that presumably forms a supercomplex with GAL4.

Immunoglobulin-like Family

The immunoglobulin (Ig)–like family is characterized by a DBD that has an Ig-like β-barrel structure. The best-characterized example in this group is the Rel subfamily, which possesses two β-barrels connected by a hinge similar to the cytokine receptor motif (Fig. 13-5, *A*). The AB loop in the amino-terminal barrel extends into the major groove in order to recognize specific base sequences. The carboxy-terminal barrel consists of eight strands: three from each monomer form the dimerization interface, whereas the remaining five bind DNA. The carboxy-terminal end of this barrel flares apart to accommodate IκB, an inhibitor of one member of this family (NF-κB). The gap between the amino-terminal barrels is occupied by HMGA, a nucleosomal protein (see Chapter 14) that facilitates binding of the transcription factor. Members of this subfamily include NF-κB and NFAT (nuclear factor of activated T cells). Stat has a similar DBD but lacks the second β-barrel; it dimerizes by means of an alternate mechanism.

The T-box transcription factors represent another member of this family. In these transcription factors, the Ig fold is concave and arches over the DNA (Fig. 13-5, *B*). The apex forms the dimerization domain, while the lower level forms a scaffold for the DNA-binding helices. In this case, it is an α-helix, not the Ig loop that binds the DNA. The fourth helix binds in the minor groove and causes it to widen; the third helix lies on top to secure the former helix in place.

MADS-Box Family

Like POU, MADS is an acronym derived from the first letter of the initial four transcription factors identified with this motif: MCM1, AG, DEFA, and SRF (serum response factor). This group is also known as the MEF2-box family after another member: the myocyte enhancer factor 2 (MEF2). The MADS-box is a three-tiered structure (Fig. 13-6, *A*). An α-helix from each monomer lies in the major groove in an antiparallel orientation. The amino termini extend into the minor grooves, causing the DNA to bend 72°. The second tier is an overlying β-sheet formed from two strands from each monomer. In addition to forming the dimerization domain, this layer interfaces with Ets, an accessory transcription factor, via the loop between the sheet and the third tier. The third layer consists of two antiparallel α-helices, which forms the TAD (transcriptional activation domain).

β-Family

The last family is grouped more for convenience than structural homology; in fact, its only similarity with the other families is that all bind DNA through β-strands. The transforming growth factor β (TGFβ) receptor phosphorylates and activates the Smad family of transcription factors. The MH1 ([S]mad homology domain 2) domain is the DBD (DNA-binding domain), which consists of four α-helices and three pairs of β-strands (Fig. 13-6, *B*). The hydrophobic fourth helix occupies a central position; the other three helices are located on one side and the β-strands on the other side. The β-loop between strands three and four binds flat in the major groove. Although MH1 can bind as a dimer to palindromic HREs, there is no synergism between the subunits.

The GCC-box binding domain is found in the ethylene-responsive element binding proteins and consists of a three-stranded antiparallel β-sheet with an overlying α-helix (Fig. 13-6, *C*). The β-sheet lies in the major groove. The Y-box is another β-structure and can bind either RNA or DNA. It is found in the frog Y-box protein 2 (FRGY2, also called *messenger ribonucleoprotein 4* [mRNP4]), the DNA binding protein B (dpbB), and the ZO-1 associated nucleic acid binding protein (ZONAB). The nucleotide-binding domain is a five-stranded β-barrel; however, the structural basis for nucleotide binding is unknown. The

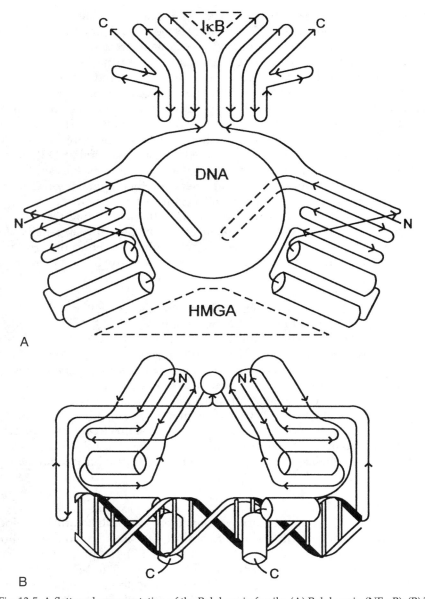

Fig. 13-5. A flattened representation of the Rel domain family. (A) Rel domain (NF-κB); (B) T-box. Cylinders represent α-helices, arrows are β strands, and the central circle in (A) is the DNA in cross-section. Accessory proteins in (A) are depicted with dashed lines.

DBD is followed by an arginine-proline-rich tail, which is responsible for oligomerization. At low concentrations, FRGY2 destabilizes RNA in preparation for translation, but at high concentrations, it can sequester RNA in ribonucleoprotein particles. ZONAB binds an inverted repeat of CCAAT to promote the formation of single-stranded DNA; it can either stimulate or repress transcription.

The Tubby family possesses a 12-stranded β-barrel with a long central α-helix; there are four smaller helices located in the loops. There is a basic groove that runs halfway around the barrel; this furrow is presumed to be the DBD.

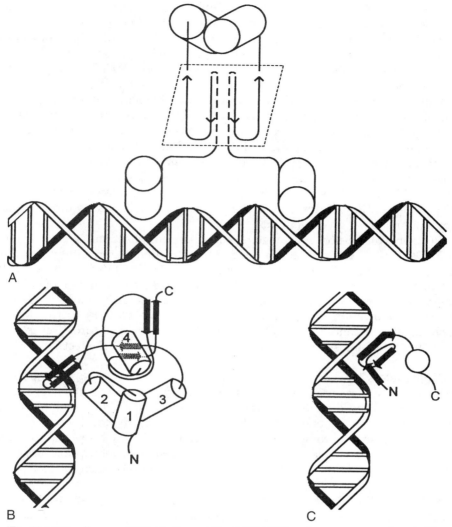

Fig. 13-6. Miscellaneous DNA-binding motifs. (A) MADS-box (serum response factor); (B) MH1 (Smad); (C) Y-box (ethylene-responsive element binding protein). Cylinders represent α-helices and arrows are β strands.

Transcription Activation Domains

Motifs and Molecular Mechanisms

Transcription activation domains have been mapped in several factors; they are often characterized by an abundance of a particular amino acid rather than a specific sequence. The most studied motif is one enriched in acidic amino acids, known as an *acid blob, negative noodle,* or *acid noodle;* the names refer to both the charge and the original belief that these sequences formed random structures. Later, they were postulated to form amphipathetic α-helices that interacted with other transcription factors having basic helices. Most recently, they have been predicted to form a hydrophobic β-sheet, and the hydrophobic amino acids have been shown to be critical for transcription. The negative charges may only be necessary to make the lipophilic sheet more accessible. Whatever the secondary structure, acidic domains are known to stimulate RNA polymerase II. They also bind to the TATA-box factors TFIIB and TFIID; to SWI/SNF (switch and sucrose nonfermentors gene products), a chromatin-remodeling complex; and to the replication protein A. The latter protein is involved with DNA unwinding. In addition to direct contacts with the transcriptional machinery, acidic domains can bind coactivators, which form bridges between the TAD and general transcription factors.

Other domains rich in glutamine, proline, isoleucine, or serine-threonine have also been documented. Like the acidic motif, the activity of the glutamine TADs is dependent on the hydrophobic amino acids and not the glutamine. The glutamine domain can bind the TBP. The proline-rich regions probably form polyproline helices (see Chapter 7); they can bind TFIIB and several transcriptional coactivators. With the exception of the S-T TAD, these TADs are all hydrophobic; this property makes them "sticky" and allows them to recruit other proteins. The S-T domains probably act as regulatory sites through phosphorylation and/or glycosylation.

From this discussion, several general mechanisms of action can be envisioned for TADs. First, TADs can recruit general transcription factors either directly or through coactivators; for example, binding to TBP, TFIIB, TFIID, TFIIF, and TFIIH have all been documented. Second, TADs may alter the conformation of either the general transcription factor and/or the DNA. Finally, TADs may stimulate the covalent modification of proteins; they have been shown to recruit kinases, acetyltransferases, and methyltransferases. These mechanisms are discussed further in the following section on mechanisms for transcription effects.

Coactivators

Coactivators are bridges that link the specific transcription factors to the general ones. They directly bind the TAD of the former; this binding is often contingent on a hormone signal. For example, coactivator association may depend on ligand binding in nuclear receptors (NRs) or on phosphorylation within the TAD. Coactivators also interact with general transcription factors. As bridges, they usually do not enhance basal transcription by themselves.

CREB-Binding Protein. One of the most common and best studied coactivator is CREB-binding protein (CBP) and its close homologue, p300. The name for this coactivator originated from the fact that it was first found associated with the kinase interacting domain (KID) of the cAMP (3'5'-cyclic AMP)–binding protein (CREB), but it is now known to bind to virtually every class of specific transcription factor, including members from the NRs, bZIP, bHLH, POU, wHTH, Rel, MADS, and Smad families (Fig. 13-7). It also binds several elements of the general transcription machinery, including the TBP, TFIIB, TFIID, and the RNA polymerase II. Other proteins it binds include p53, an antioncogene; PCAF (p300/CBP associated factor), a protein acetyltransferase; SWI/SNF, a chromatin remodeling complex; RNA helicase A (RHA); and NAP-1, which acts as a histone H2A-H2B shuttle in chromatin assembly. Indeed, its numerous simultaneous interactions facilitate synergy among transcription factors. Finally, in addition to being a general bridge and organizer, it possesses histone acetyltransferase activity (HAT).

Both CBP and p300 have very similar structures, function, and regulation, but there are some differences. First, CBP is recruited to TATA start sites, whereas p300 is recruited far from the start site. Second, CBP is required for hematopoietic stem cell replication, whereas p300 is required for stem cell differentiation. Finally, CBP mediates the proteolysis of some transcription factors, like p53,

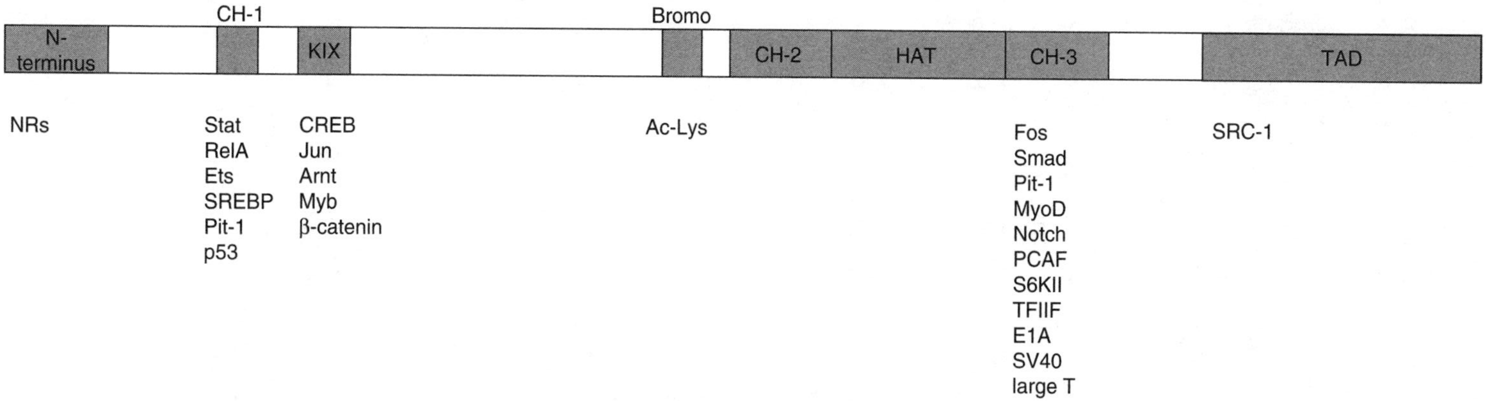

Fig. 13-7. Schematic structure of CBP. The transcription factors that bind the individual domains are listed below those domains. *CH*, Cysteine-histidine-rich domain; *HAT*, histone acetyltransferase; *KIX*, KID interacting domain. Other abbreviations are as in the text and List of Abbreviations at the end of the book.

through its E4 polyubiquitinating activity, whereas p300 facilitates this modification by coupling the transcription factor and its E3 ligase through hHR23A, a DNA repair protein. Both processes lead to degradation in the proteasome.

CBP is regulated in several ways. First, the HAT activity can be stimulated by either mitogen-activated protein kinase (MAPK) or cyclin E-Cdk2 (cell-cycle dependent kinase 2) phosphorylation in the carboxyl terminus. Phosphorylation can also affect recruitment. As discussed in Chapter 6, the TAD2 of NRs binds the sequence LXXLL; protein kinase C (PKC) phosphorylation near this sequence in p300 prevents its recruitment to NRs. Protein kinase B (PKB) phosphorylation in the carboxy terminus inhibits binding to C/EBPβ. On the other hand, PKC phosphorylation in the amino terminus of CBP is required for recruitment to AP-1 and Pit-1. Phosphorylation of the transcription factor also affects CBP interactions; for example, the KID domain of CREB must be phosphorylated to attract CBP. In addition to phosphorylation, CBP can be methylated on an arginine in the KIX domain by protein arginine methyltransferase 4 (PRMT4). This modification prevents its binding to the KID domain of CREB but not to the TAD2 of NRs. On the other hand, methylation at an arginine carboxy-terminal to KIX is required for NR transcription. Finally, CBP can be allosterically regulated; S6KII binding to CBP blocks its association with RHA, and the kinase activity of S6KII is not required for this effect. Allosteric regulation of transcription factors can also affect CBP binding; ligand binding to NRs creates the TAD2 that binds the LXXLL motif in CBP.

The p160 Family.   Members of the p160 family were first discovered as coactivators of NRs; as such they were named *steroid receptor coactivator* (SRC) or *NR coactivator* (NCoA). However, they have been shown to interact with many classes of other transcription factors, including members of the bZIP, bHLH, Rel, and MADS families. They have a bHLH-PAS domain at the amino terminus. PAS is a domain that mediates the interaction between two proteins or between proteins and small signaling molecules and is named after the first three proteins in which it was first found: Per, Arnt, and Sim. They also possess three nuclear receptor interaction domains (NIDs) or LXXLL motifs used to bind NRs. The NIDs are not equivalent; the carboxy-terminal NID binds TAD1, whereas the NID in the linker region binds TAD2. Finally, the carboxy terminus contains a glutamine-rich region overlapping a HAT domain. The HAT activity is considerably less than that associated with CBP and some authorities question whether it is physiologically relevant. The glutamine-rich domain is an alternate binding site for NRs; for example, it binds TAD1b of the androgen receptor.

There are three members of this family: SRC-1 (NCoA-1); SRC-2 (NCoA-2), which is also called GRIP1; and SRC-3 (NCoA-3), which is also called p/CIP, TIF2, RAC3, ACTR, or TRAM-1. All three are phosphoproteins; SRC-1 and SRC-2 are stimulated by MAPK phosphorylation, and IKK (IκB kinase) phosphorylation of SRC-3 shifts it into the nucleus. All three have consensus sites for sumoylation, although only SRC-1 and SRC-2 have been shown to be modified. Sumoylation occurs in the NID and interferes with NR binding. Finally, all three can form higher-ordered structures; for example, they can also bind CBP, PCAF, PRMT4, and SWI/SNF.

Vitamin D Receptor Interacting Protein.   Vitamin D receptor interacting protein (DRIP) is a coactivator initially associated with the thyroid family; indeed, DRIP is also known as the *TR associated protein* (TRAP). DRIP can bind members of the zinc finger, bHLH-ZIP, and Rel family, too. In the case of NRs, binding is ligand dependent. DRIP is usually part of a preformed complex of 14-16 proteins. Such complexes are called *enhanceosomes*, which are generally defined as stable multimeric protein complexes that recruit coactivators, general transcription factors, and RNA polymerase II to the promoter. The DRIP complex does not contain SRC-1, CBP, p300, or any HAT activity. In addition, this complex only stimulates transcription on chromatin, not naked DNA.

The transformation-transcription domain associated protein (TRRAP) also forms an enhanceosome. The TRRAP complex contains GCN5, a HAT; TRRAP, which is an adaptor for NRs; and the transcription factor $TAF_{II}30$. It forms at about the same time as DRIP and after CBP is recruited (see section on Integration later for a discussion of the timing of enhanceosome formation).

Corepressors

Corepressors directly bind specific transcription factors and inhibit transcription. They usually bind these factors when the latter are in the unstimulated state: for example, unliganded NRs. There are several mechanisms for this inhibition. First, corepressors can recruit deacetylases (see Chapter 14). Second, they can bind and mask TADs in specific transcription factors. Finally, they can bind and directly inhibit general transcription factors.

Regulation.   There are several general mechanisms that regulate corepressors. First, ligands can affect corepressor binding; for example, the empty TR binds the NR corepressor (NCoR) but triiodothyronine (T_3) displaces it. Also, either calmodulin (CaM) or 14-3-3 can bind histone deacetylase (HDAC) and prevent its binding to MEF2. Second, the HRE can affect corepressors; for example, RAR (retinoic acid receptor) is always associated with corepressors on DR1 whether retinoic acid is bound, but it is only associated with corepressors on DR5 when it is unoccupied. Third, phosphorylation of either corepressors or transcription factors can affect their interactions. MAPKKK phosphorylation of the silencing mediator of RAR and TR (SMRT) decreases its binding to specific transcription factors and shifts SMRT to the cytosol. On the other hand, casein kinase 2 (CK2) phosphorylation of SMRT stabilizes its binding to NRs. The transcription factor can also be modified; MAPK phosphorylation of TR induces the dissociation of SMRT and the cyclin A-Cdk2 phosphorylation of B-Myb releases NCoR. Finally, CaMKIV (CaM-dependent protein kinase IV) phosphorylation of p65, a subunit of NF-κB, triggers the exchange of a corepressor (SMRT) for a coactivator (CBP).

Fourth, corepressors can be regulated by subcellular localization. Phosphorylation of either HDAC or the 140-kDa receptor interacting protein (RIP140) results in their binding to 14-3-3 and their sequestration in the cytoplasm. Fifth, competition for cellular components can affect corepressor recruitment; for example, Sin3 is an adaptor that recruits corepressors. c-Ski (Sloan Kettering virus isolate) recruits Sin3 to Mad, a bHLH-ZIP transcription factor, but v-Ski does not. Both isoforms compete for binding to Mad, and the proportion of each bound to Mad will determine the degree of corepressor recruitment. Finally, corepressors can be targeted for proteosomal degradation.

Families.   SMRT is an alternately spliced SRC-1. Although named for its effect on the TR family, it also represses transcription by other families, including the bZIP, POU, Rel, and MADS families. It is negatively and positively affected by

MAPKKK and CK2 phosphorylation, respectively, as described earlier. It has two major mechanisms of action: first, it recruits class I HDACs via Sin3 and directly binds class II HDACs. Second, it binds and sequesters components of the general transcriptional machinery, such as TFIIB.

NCoR directly binds the D and E domains of NRs and represses their transcriptional activity. There is some specificity between NCoR and SMRT; the former prefers TR, whereas the latter prefers RAR. Like SMRT, NCoR recruits class I HDACs via Sin3 and directly binds class II HDACs.

RIP140 represses NRs and the aromatic hydrocarbon receptor (Ah); this effect is enhanced by ligand binding. Repression is the result of RIP140 competing with SRC-1 for binding to transcription factors. It is regulated by phosphorylation and acetylation. The former results in RIP140 being sequestered in the cytosol by 14-3-3. RIP140 is acetylated by CBP at a site used to bind another corepressor, the carboxy-terminal binding protein (CtBP); acetylation disrupts this interaction.

PIAS was first described as an inhibitor of Stat signaling, but it has both positive and negative effects on a variety of transcription factors. PIAS is also a Sumo E3 ligase (see Chapter 8), and many of its effects are due to this modification. For example, PIAS-induced sumoylation of TCF/LEF or Arnt sequesters them in promyelocytic leukemia (PML) nuclear bodies. Similar modifications of AR, AP-2, Sp3, IRF-1, and SRC-2 are also inhibitory. On the other hand, sumoylation of CREB stabilizes this factor and promotes its nuclear localization; Sumo conjugation of C/EBPε suppresses an autoinhibitory domain, and modification of the glucocorticoid receptor (GR) facilitates its synergy. The effect on the progesterone receptor (PR) is more complex; although PIAS inhibits transcription activation by PR, it is required for PR-induced repression. A recent hypothesis proposes that sumoylation localizes transcription factors and coactivators to PML (promyelocytic leukemia oncogene) bodies for assembly into transcription complexes; removal of Sumo would then activate the complexes and allow them to leave. Little is known about the hormonal regulation of this modification, but sumoylation of both androgen receptor (AR) and GR is enhanced by ligand binding. In the heat shock transcription factor 1, MAPK phosphorylation of a serine near the conjugation site enhances sumoylation.

The cyclins have long been known as regulators of Cdks; however, they can also mediate cross talk between cell cycling and transcription by acting as either coactivators or corepressors. Cyclin 1 binds to and enhances the activation of estrogen receptor (ER) by recruiting SRC-1 and PCAF, and cyclin 3 facilitates complex formation between CRABPII (cellular retinoic acid binding protein II) and RAR. CRABPII promotes the transfer of retinoic acid to RAR. On the other hand, cyclin 1 inhibits AR by competing with PCAF and inhibits TR by recruiting HDAC.

Chromatin Remodeling

As is discussed in Chapter 14, DNA is packaged into nucleosomes, which must be remodeled to make the resident genes accessible to transcription. One of the functions of specific transcription factors is to recruit the proteins responsible for this remodeling.

SWI/SNF Superfamily

There are three families in this group. The SWI/SNF family gets its name from the fact that it is required for the expression of mating type *switch* and sucrose *non*fermentors genes. It is effective in inducing both nucleosome sliding and displacement. The ISWI family (*im*itiation of *sw*itch) only triggers nucleosomal sliding. In addition, its function requires the presence of histone tails that protrude beyond the edge of the nucleosome; the SWI/SNF family does not require the presence of these tails. The chromo-helicase DNA binding (CHD) family, also called the *Mi-2 family*, combines chromatin remodeling with HDACs and is covered in more detail later.

Their mechanism, which requires ATP (adenosine 5'-triphosphate), involves peeling back the ends of the DNA from the nucleosome and enabling the nucleosome to slide or be displaced. This is particularly important in the promoter region, which generally must be clear for the transcription initiation complex to assemble. However, these family members can catalyze nucleosomal sliding in either direction. As such, they can be associated with corepressors, as well as coactivators. They are regulated by allosterism, phosphorylation, and recruitment. For example, SWI/SNF is stimulated by IP_4 and IP_5 but inhibited by MAPK phosphorylation; ISWI is inhibited by IP_6. In addition, many specific transcription factors can bind to these chromatin remodelers either directly or through coactivators. Once they have been recruited to the promoter, they can be retained by binding to acetylated histones via their bromodomains.

Other Complexes

BRG/Brm is the *Drosophila* homologue of SWI/SNF but includes actin and actin-like components. Nuclear PIP_n (phosphoinositide polyphosphate) localizes the complex to the chromatin and nuclear matrix and stimulates the ATPase (adenosine triphosphatase) activity. SUG1, also called the SR binding factor, is a $3'-5'$ DNA helicase that facilitates DNA unwinding. Finally, heterochromatin protein 1 (HP1) binds to heterochromatin, which is the condensed, transcriptionally inactive form of chromatin, and induces its assembly and silencing. CK2 phosphorylation of HP1 is required for these functions.

Integration

These factors do not act in a haphazard fashion; rather, their activity is highly orchestrated. A typical sequence involves three phases (Fig. 13-8). First, a specific transcription factor recruits a SWI/SNF family member to displace nucleosomes in the regulatory region. For example, both TAD1 and TAD2 in NRs bind SWI/SNF. The p160 family arrives next and they recruit CBP. Histone acetylation by CBP will lock the nucleosomes in the open conformation. CBP also acetylates SRC-3 and induces its dissociation in preparation for the third phase. Acetylation may also release polyamines from chromatin; these liberated polyamines will decrease p160 binding to NRs, while increasing DRIP binding. The recruitment of DRIP represents the third stage; this complex will act as a bridge to the general transcription factors. In addition, DRIP is important in reinitiation for subsequent rounds of transcription.

In fact, there are many factors that can affect this sequence. First, the initial state of the chromatin is important; inert, condensed chromatin will require more remodeling than rRNA genes, which are never inactive and bind RNA polymerase I even during mitosis. Second, the complexity of gene organization plays a role; genes with complex, extended regulatory region will probably require more remodeling than those with simple promoters. Finally, when transfected cells are

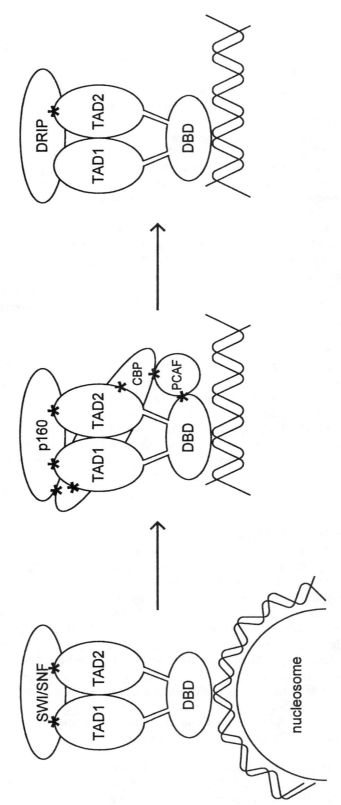

Fig. 13-8. A typical temporal sequence for the action of coactivators. Asterisks indicate interaction sites.

being used, the level of transfected transcription factors affects cooperativity: high levels can eliminate the need for coactivator synergism.

Specific Transcription Factors: Regulation

Leucine Zipper

Activating Transcription Factor Family

The best-known member of the activating transcription factor (ATF) family is the cAMP response element binding protein (CREB). CREB (Fig. 13-9, *A*) is regulated by protein kinase A (PKA) phosphorylation of S-133 in the KID domain. This modification triggers the subsequent phosphorylation of several adjacent sites by CK2 and one nearby site by GSK3 (glycogen synthase kinase 3). This cluster of negative charges recruits CBP, but it is not required for DNA binding. Kinases other than PKA can also phosphorylate S-133; for example, in response to elevated calcium levels in neurons, CaMKIV phosphorylates the same serine as PKA and activates transcription by CREB (Fig. 13-10, *A*). However, PKA still plays a role; it phosphorylates and activates the PP1 inhibitor, DARPP-32. DARPP-32 prolongs the effect of CREB by delaying dephosphorylation. Epidermal growth factor (EGF) stimulates CREB through S6KII; the components of this pathway are more stable than those of PKA, resulting in a longer effect. Finally, stress and inflammatory cytokines activate CREB via the p38/MAPKAP-2 pathway.

There are other ways that CREB is regulated. CREB is sumoylated at two sites near the nuclear location signal (NLS) during hypoxia; this stabilizes the transcription factor and promotes its nuclear localization. CREB2 (also called ATF4) is a homologous protein that lacks both a TAD and a PKA site. It binds CRE but cannot activate transcription; as such, it acts as a repressor. Indeed, there are several truncated forms of CREB, called inducible cAMP early repressors (ICERs) that suppress CREB activity by means of the same mechanism. DREAM (downstream regulator element antagonist modulator) binds KID and blocks CBP recruitment; it also binds a DREAM response element (DRE), which is a negative HRE. DREAM is a calcium-binding protein related to recoverin; calcium binding decreases the affinity of DREAM for the DRE. Several members of the CREM family (see later) neutralize DREAM by displacing it from the DRE and competing with CREB for binding to this repressor.

There are five members of the CREM subfamily; all of these forms are generated by alternate splicing. The gene itself is fascinating in that several exons are duplicated; therefore alternating splicing involves not only omitted exons but also choices among two DNA-binding exons and two coiled coil exons. The short forms are all inhibitors. CREMα and CREMβ differ in the DNA-binding exon used, and CREMγ lacks a small exon coding for 12 amino acids. All three are missing the pair of glutamine-rich TADs that bracket the KID; therefore they cannot directly activate transcription. Inhibition of intact CREB and CREM requires both dimerization and CRE binding. Interestingly, the short forms still possess the PKA site; this is important because these forms can also facilitate CREB activation by binding and sequestering DREAM through their KIDs. Even in the absence of DREAM, the short forms can attract coactivators that reduce their repression. CREMγ is the intact form. It is partially active even without PKA-induced phosphorylation; basal activity is about three times that

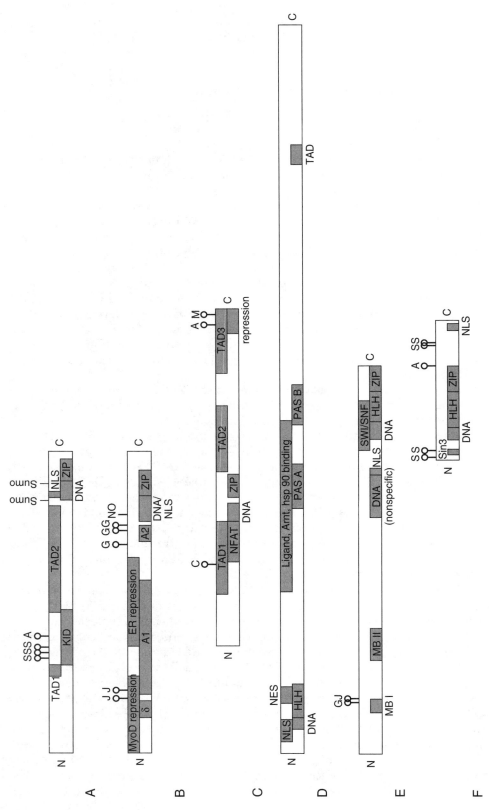

Fig. 13-9. Schematic structure of several transcription factors. (A) cAMP response element binding protein (CREB); (B) Jun and (C) Fos, which dimerize to form the anterior pituitary factor 1 (AP-1); (D) the aromatic hydrocarbon (Ah) receptor; and (E) Myc and (F) Max, which dimerize. Circles represent pS or pT. A, PKA; C, PKC; G, GSK3; J, JNK; M, MAPK; NO, S-nitrosylated cysteine; S, CK2; Sumo, sumoylation site. Other abbreviations are as in the text and List of Abbreviations at the end of the book.

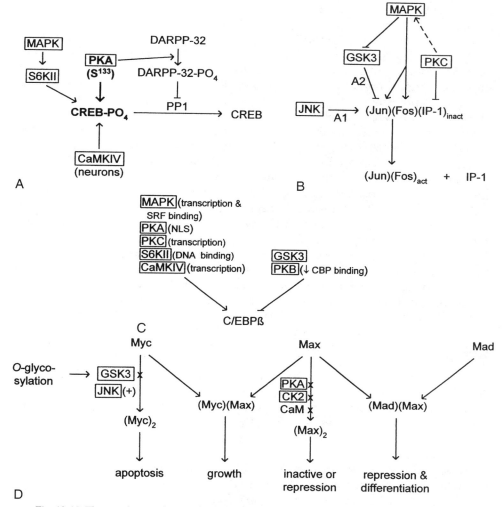

Fig. 13-10. The regulation of several transcription factors by phosphorylation. (A) CREB; (B) AP-1; (C) C/EBPβ; and (D) Myc-Max. Kinases are boxed, arrows indicate stimulation, and flat heads indicate inhibition; bold lines depict central pathways. *DARPP*, Dopamine and cAMP regulated phosphoprotein; *IP-1*, inhibitor of AP-1.

of CREB. However, transcription is further increased by either PKA or Cdk2 modification of KID. S-CREM possesses an alternate start site such that the first TAD and the P-box are omitted. This omission results in S-CREM being a repressor, although it is weaker than the short forms that lack both TADs.

There are several other members of the ATF family. ATF2 is activated by JNK (Jun amino [N]-terminal kinase) phosphorylation. ATF6 has a transmembrane domain that is embedded in the endoplasmic reticulum. Stress results in the cleavage of this domain and the migration of ATF6 to the nucleus. As such, it resembles SREBP-2; indeed, it uses the same protease (see later).

Jun-Fos (AP-1) Family

AP-1 was originally identified as an anterior pituitary transcription factor that appeared to be primarily regulated by PKC. This assumption arose from the finding that TPA, a phorbol ester that activates PKC, also stimulates AP-1.

Subsequent studies have shown that the regulation is much more complex. Although the most active form of this factor is a heterodimer between Jun and Fos, each protein is first described separately. Jun has a classic bZIP motif in the carboxy terminus and two TADs, A1 and A2, in the amino terminus (see Fig. 13-9, *B*). A1 is acidic and A2 is glutamine rich and proline rich; both have important phosphorylation sites. Transcriptional activity is associated with an increase in phosphorylation in A1 but a decrease in A2. These changes are induced by mitogens, Ras, and activators of PKC; however, PKC does not directly modify Jun. These factors are believed to act at A1 through JNK (Fig. 13-10, *B*), which directly phosphorylates two critical residues in this region. In contrast to A1, A2 is an inhibitory domain when phosphorylated. GSK3 is the major kinase responsible for constitutively modifying the A2 sites. MAPK can phosphorylate and inhibit GSK3; because phosphate groups are constantly turning over, the suppression of GSK3 activity would eventually lead to a reduction in the level of A2 phosphorylation. It has also been proposed that MAPK directly phosphorylates Jun and increases its half-life.

Jun activates transcription by several mechanisms. A1 primarily affects transcription by recruiting CBP in much the same way as KID in CREB. On the other hand, the negatively charged A2 adjacent to the DBD inhibits DNA binding; its dephosphorylation increases the binding of Jun to DNA. Jun also recruits TFIID and TAF1. The amino terminus of TAF1 binds and inhibits TBP; its sequestration by Jun would activate TBP. Finally, AP-1 forms a complex with Ets. Ets is a general accessory transcription factor that facilitates the activity of many specific transcription factors.

There are several other regulatory sites on Jun. The δ domain is amino-terminal to A1 and binds Ras, which facilitates phosphorylation. Several domains are involved with transcription repression; an amino-terminal domain that overlaps δ suppresses MyoD activity, whereas a carboxy-terminal domain that overlaps the zipper region suppresses ER-induced transcription. Interactions between the bZIP family and the nuclear receptor family are discussed in greater detail later. In addition to phosphorylation, AP-1 can be regulated by ROI (reactive oxygen intermediates); their target is a conserved cysteine in the DBD of Jun. Either *S*-nitrosylation or glutathionylation of this cysteine inhibits AP-1.

There are several isoforms of Jun. JunB and JunD bind the same HRE as Jun and can dimerize with Fos, but JunB requires multiple copies of this HRE for activity. At singlets, JunB acts as a repressor. These isoforms also target different genes to give rise to different activities; for example, c-Jun favors cell proliferation and antiapoptosis, whereas JunB has the opposite effect.

Fos is related to Jun, but it has its bZIP motif more centrally located in the protein. It has two TADs, which bracket the bZIP structure, and a third TAD in the carboxy terminus (see Fig. 13-9, *C*). Like Jun, Fos is a substrate for many protein kinases; a PKA site in the extreme carboxy terminus is required for the transcriptional repression mediated by this domain. MAPK phosphorylation near the same region increases Fos half-life and transcriptional activity. Also like Jun, Fos can interact with other transcription factors; for example, it synergizes with NFAT, which binds the carboxy-terminal region of TAD1. Fos has several isoforms: FosB, Fra-1, and Fra-2. ΔFosB is a splice variant of FosB that lacks the carboxy-terminal 101 amino acids. Because ΔFosB does not have the carboxy-terminal repression domain, it cannot inhibit transcription; however, its transcriptional activity is also reduced.

Jun can homodimerize, bind its HRE, and activate transcription. The Fos zipper sequence is incompatible with homodimerization and this deficiency prevents DNA binding. If Fos is given a Jun zipper, it can homodimerize and bind DNA. Fos also has a shorter half-life than Jun; this difference in stability means that Jun-Jun homodimers are predominant in the unstimulated state. After Fos induction, Jun-Fos appears. This change in dimer composition is reflected by a switch in HRE preference and gene induction; Jun-Jun prefers the HRE, TGACATCA, whereas Jun-Fos prefers TGACTCA. It also results in a greater stimulation of transcription because Jun-Fos is a more potent transcription factor than Jun-Jun.

Finally, AP-1 can be regulated by an inhibitor, IP-1. This protein loosely binds to AP-1 through the zipper region and blocks DNA binding. It can be phosphorylated and inactivated by either PKA or PKC.

C/EBP Family

Members of the C/EBP family all act at the DNA sequence, CCAAT, from which their name is derived: CCAAT/*enhancer-b*inding *p*rotein. The one most closely associated with hormonal regulation is called NF-IL6 in human beings and C/EBPβ in mice. This transcription factor has an unusual variation of the bZIP motif; there is a cysteine that immediately follows the zipper and is involved with disulfide bond formation. Therefore, the coiled coil dimer is reinforced by a covalent link. The amino terminus recruits SWI/SNF. It is regulated by several kinases (Fig. 13-10, C); MAPK, PKC, and CaMKIV enhance transcription. The former also facilitates synergism with SRF. PKA and S6KII increase DNA binding; PKA also induces nuclear localization. GSK3 and PKB inhibit C/EBPβ; the latter does so by inhibiting CBP binding.

There are several isoforms and all can heterodimerize with each other; the heterodimers have greater transcriptional activity than homodimers. As with virtually all other specific transcription factors, truncated forms can act as inhibitors.

Helix-Loop-Helix and Variations

Basic Helix-Loop-Helix

The aromatic hydrocarbon (Ah) receptor binds several pollutants, the most famous of which is dioxan. The response element for the Ah receptor is found in genes coding for enzymes that metabolize such compounds. In addition to detoxification, the Ah receptor inhibits adipose differentiation and is required for hepatocyte synthesis of albumin. Because it is a ligand-regulated transcription factor and is retained in the cytoplasm by heat shock protein (hsp) 90 and immunophilins until activation, it was originally thought to be a member of the NR family, perhaps a variation of the PPAR (peroxisome proliferator activated receptor) that bound aromatic instead of aliphatic hydrocarbons. However, cloning of the receptor components revealed them to be bHLH transcription factors. It may still be related to the endocrine system because the receptor can also be affected by several endogenous compounds; at physiological concentrations, a condensation product of tryptophan and cysteine activates the Ah receptor, whereas 7-ketocholesterol inhibits it. Ligand binding occurs in the PAS domains. Furthermore, tyrosine phosphorylation is required for DNA binding.

The Ah receptor is a dimer. Arnt (Ah receptor nuclear translocator) is responsible for transporting the LBS (ligand-binding subunit) into the nucleus;

both are bHLH proteins, although they share only two small regions of homology (see Fig. 13-9, *D*). Arnt appears to be a common subunit for several bHLH-PAS transcription factors. Hsp 90 binds the ligand binding domain (LBD) and DBD of the LBS; it stabilizes the conformation of the LBS and inhibits dimerization, DNA binding, and nuclear translocation. After ligand binding, hsp 90 dissociates; and the Ah receptor migrates to the nucleus and binds DNA. The LBS binds the 5′ end of the HRE, whereas Arnt binds the 3′ end. Neither subunit alone can bind DNA.

The hypoxia-inducible factor 1 (HIF-1) is also a bHLH-PAS transcription factor; the HIF-1α subunit senses oxygen, and the HIF-1β is Arnt. In normal oxygen tension, HIF-1α is hydroxylated at two sites: hydoxyasparagine in the TAD blocks p300 binding and hydroxyproline triggers ubiquitination and destruction of HIF-1α. In low oxygen tension, proline is not hydroxylated and there is no ubiquination. HIF-1α then accumulates in the nucleus, dimerizes with Arnt, and induces genes related to hypoxia: for example, glycolytic enzymes, vascular endothelial growth factor (VEGF), and erythropoietin (EPO). The absence of hydroxylated asparagine allows the p300 coactivator to bind the TAD. In the absence of hypoxia, MAPK phosphorylates the inhibitory domain and derepresses HIF-1, and nitric oxide (NO) can stabilize and activate HIF-1α by *S*-nitrosylation of the factor. Ras, soluble tyrosine kinase (STK), and PKB can also stabilize HIF-1α; the first two inhibit proline hydroxylation, whereas PKB stimulates HIF-1α synthesis.

The PAS domain is extremely versatile; during evolution it has been adapted to bind many different ligands, some of which are used to sense light or gases. In several bacterial proteins, the PAS domain binds heme, which acts as an oxygen sensor. In *Neurospora*, the PAS domain in a GATA transcription factor binds FAD, which is activated by light. NPAS2 is a bHLH-PAS transcription factor involved with the circadian rhythm; both of its PAS domains bind heme. In this case, DNA binding is inhibited by the binding of carbon monoxide to the hemes.

Another bHLH transcription factor is MyoD, which is involved with muscle differentiation. MyoD does not have a PAS domain. Growth factors, such as fibroblast growth factor (FGF), inhibit differentiation so that proliferation can continue; these mitogens inhibit MyoD through two PKC-mediated pathways. First of all, PKC directly phosphorylates MyoD in the DBD and abolishes DNA binding. Second, PKC indirectly activa' Jun, whose zipper can heterodimerize with the HLH region of MyoD; this heterodimer is inactive.

Helix-Span-Helix

Only one member of the helix-span-helix family is known to be hormonally regulated. Like AP-1, AP-2 is found in the anterior pituitary, where it stimulates the transcription of several pituitary hormones. There is nothing unusual about its structure. It is stimulated by either PKA or PKC; both sites are located in the DBD, but PKA phosphorylation does not affect DNA binding. Sumoylation decreases transcriptional activity.

Basic Helix-Loop-Helix-ZIP

Myc-Max. The Myc-Max heterodimer mediates the effects of several mitogens. Myc is an unusual bHLH-ZIP factor in that it has a second DBD amino-terminal to the one associated with the HLH (see Fig. 13-9, *E*). This second domain binds DNA nonspecifically and is thought to enable Myc to slide along the DNA to locate its HRE. Transcriptional activity resides in two Myc boxes: MB

I mediates transcription, whereas MB II induces apoptosis, blocks differentiation, and represses transcription. Myc homodimerizes poorly and its DNA affinity is low. JNK phosphorylation on S-62 and S-71 stimulates the activity of Myc dimers, which induce apoptosis (Fig. 13-10, *D*). Conversely, GSK3 phosphorylation on T-58 inhibits transcriptional activity. The attachment of *N*-acetylglucosamine (NAG) to T-58 blocks its phosphorylation. Modification by a single NAG is being reported on an increasing number of transcription factors. As such, it is worth briefly digressing to discuss this type of glycosylation.

The attachment of NAG has many functions. First, in Myc and in the carboxy-terminal domain (CTD) of RNA polymerase II, glycosylation inhibits phosphorylation. This is a result of the two modifications occurring on the same amino acids; as such, phosphorylation and glycosylation are mutually exclusive events. Second, it may increase stability. This effect may be related to the first function, since some of these sites are located in PEST (proline, glutamic acid, serine, and threonine-rich) domains, which trigger degradation if they are phosphorylated. By inhibiting phosphorylation in these regions, glycosylation also inhibits degradation. Third, they can either increase or decrease transcription. An example of the former is the antioncogene p53; glycosylation masks the inhibitory carboxy terminus and stimulates transcription. Inhibition is achieved by affecting macromolecular assembly; it blocks ER binding to the ERE, and it prevents Sp1 oligomerization and binding to general transcription factors. In addition, glycosylation can facilitate the formation of repressor complexes. Finally, glycosylation is involved with organelle targeting; for example, NAG is believed to be a nuclear localization signal.

The enzyme responsible for the attachment of NAG, *O*-NAG transferase (OGT), has been cloned. There are several potential sites for posttranslational modifications, but their physiological relevance is unknown. However, one major regulatory mechanism appears to be recruitment: GRIF-1 (GABA$_A$ receptor interacting factor 1) recruits OGT to the GABA$_A$ receptor; OIP106 (OGT interacting protein 106), to the RNA polymerase II; and Sin3, to transcription complexes.

Max is a dimerization partner of Myc. Max, called Myn in mice, is also a bHLH-ZIP factor, but it is much smaller than Myc because it lacks any known TAD (see Fig. 13-9, *F*). Although both the HLH and the zipper regions are required for dimerization, they exhibit different specificities: the HLH is a major determinant in Max homodimer formation, whereas the zipper is primarily involved with heterodimerization. Because Max does not have a TAD, homodimers do not activate transcription. In fact, they act as repressors, because they can still bind DNA and block its access to other factors. There are several isoforms of Max. Max(s) lacks the first 9 amino acids, which are located next to the DBD; this isoform tolerates greater variability at the 5′ end of the HRE. ΔMax is a product of alternate splicing that results in a frameshift at amino acid 98 with termination at amino acid 103. Because the zipper is disrupted, dimerization with Myc is less stable and dimerization with Max or ΔMax does not occur at all.

Max is constitutively synthesized, whereas Myc is induced. Because Myc homodimerizes poorly and Max lacks a TAD, the Myc-Max heterodimers are the dominant transcriptionally active form when both subunits are present; they mediate growth. On the basis of gel migration studies, Myc-Max bends DNA in a manner similar to Jun-Fos; Max bends DNA away from the HLH-ZIP domain, whereas Myc bends DNA toward this region. Surprisingly, this bending is not seen in the three-dimensional structure of Max. It is possible that the process of crystalization straightens the DNA to force all of the molecules into a more regular structure. The x-ray structure also shows that the loop between the α-helices contacts the deoxyribose-phosphate backbone of the minor groove. These added contacts may be necessary to compensate for the reduced flexibility of the bHLH-ZIP; the tetramer is a more rigid structure than the simple dimer of the bZIP factors.

Myc-Max can be regulated either by phosphorylation or by inhibitors. PKA impairs Max homodimer formation, whereas CK2 decreases the DNA-binding affinity of Max homodimers, but not Myc-Max dimers. Essentially, Max is activated by modifications that favor heterodimer formation. Both JNK and GSK3 phosphorylate the TAD of Myc; the former is required for Myc-mediated apoptosis, whereas the latter inhibits it. In addition to phosphorylation, several homologous inhibitor proteins have been characterized; examples would include Mad (Max dimerization) and Mxi1 (Max interactor 1). They heterodimerize with Max, bind DNA, and repress transcription. It is believed that this alternate heterodimerization acts like a switch; in the uninduced state, Max, which is constitutively synthesized, exists predominantly as homodimers. During growth, Myc is induced, and Myc-Max forms and stimulates the expression of growth-related genes. During differentiation, Mad is induced, and Mad-Max forms and represses the transcription of these same genes (Fig. 13-10, *D*).

Sterol Response Element Binding Protein 2.   Sterol response element binding protein 2 (SREBP-2) is a bHLH-ZIP transcription factor with a membrane anchor in its amino terminus. The SREBP-2 cleavage activating protein (SCAP) is a transport protein that acts as a sterol sensor; high sterol levels decrease the half-life of SCAP and no cleavage occurs. Low sterol levels stabilize SCAP, which then transports SREBP-2 from the endoplasmic reticulum to the Golgi complex, where a protease cleaves off the amino terminus. SREBP-2 is released to migrate to the nucleus to induce the genes for sterol synthesizing enzymes. As previously noted, this same regulatory mechanism is used by ATF6, whose membrane anchor is cleaved in response to stress. MAPK phosphorylation of SREBP-2 enhances its transcriptional activity.

Winged Helix-Turn-Helix

The Fox subfamily belongs to the wHTH family. FoxH1 (formerly called *FAST-1*) is a major target of the TGFβ-Smad pathway (see Chapter 7). Briefly, the activin receptor phosphorylates Smad2 on a serine in the carboxy terminus and induces trimer formation with Smad4. The trimer then migrates to the nucleus and binds FoxH1. The regulation of this complex is considered further under the Smad transcription factors that follow.

FoxO3a (formerly called *FKHRL1*) is a transcription factor that mediates apoptosis. Insulin is frequently added to cell cultures to increase viability; one of the mechanisms for this effect is the inhibition of FoxO3a. Insulin activates PKB (see Chapter 12) which phosphorylates FoxO3a. This modification reduces DNA binding and releases FoxO3a from the DNA; FoxO3a is then sequestered in the cytoplasm by binding to 14-3-3.

There are also two major groups of accessory transcription factors in the wHTH family. An accessory transcription factor is a specific transcription factor that possesses a DBD and TAD and that can be regulated by external stimuli, but it rarely acts on its own. Rather it forms complexes with other specific transcription factors and recognizes nucleotides in the 5' end of the HRE. For example, the

Ets group facilitates the activity of transcription factors belonging to the NR, bZIP, bHLH-ZIP, POU, Rel, MADS, and zinc finger families. The IRF group is most closely associated with the Stat transcription factors and will be covered with them. In some cases, it has been shown that the accessory factor is dispensable if the primary transcription factor is overexpressed.

Homeodomains and Variations

Homeodomains. Homeodomain transcription factors are usually involved in developmental programs; as such, few of them are acutely hormonally regulated. However, a few do persist after development to regulate organ function. For example, the thyroid transcription factor, TTF1, is required for the development of the brain, lungs, and thyroid gland. After organogenesis, TTF1 induces thyroglobulin and thyroid peroxidase in response to thyroid-stimulating hormone (TSH). TSH stimulates FoxE1 through PKA phosphorylation and disulfide bond oxidation. On the other hand, PKC phosphorylation inhibits TTF1 by trapping it in the cytoplasm.

POU Factors. Pit-1 is a pituitary transcription factor that belongs to the POU family. POU_S is responsible for high-affinity DNA binding, whereas POU_H only binds DNA with low affinity (Fig. 13-11, *A*). It is unusual in that it forms dimers more readily than other members of this family and this tendency corresponds to its HRE, which consists of imperfect repeats. The transcription complex also contains Ets as an accessory factor. PKA phosphorylation of POU_H inhibits Ets binding. The effects of PKC phosphorylation are more complex because PKC appears to act both directly and indirectly. Direct PKC phosphorylation increases the DNA affinity at low Pit-1 concentrations. PKC phosphorylation also alters the HRE preference of Pit-1; the unmodified factor will bind the TATTCAT core with or without the AT-rich sequence at the 5' end, but phosphorylated Pit-1 shows a requirement for this 5' sequence. In addition, PKC can stimulate Pit-1 function by activating the MAPK pathway, which targets Ets.

Pit-1 can also be regulated by alternate dimerization. At high Pit-1 concentrations, Pit-1 homodimerizes; but at low concentrations, Pit-1 will heterodimerize with Oct-1, another member of the POU family. Dimerization occurs through the POU_H of Pit-1; Oct-1 replaces POU_H by binding the AT-rich half of the HRE. Certain HREs have a dimer preference, which allows the gene to respond to the concentration of activated Pit-1. For example, the growth hormone (GH) gene prefers the Pit-1 homodimer; this preference means that GH transcription requires high levels of Pit-1.

Finally, Pit-1 function can be regulated by alternate splicing. Pit-1 induces both GH and prolactin genes. Pit-1T, which has an additional 14 amino acids in the TAD, transcribes TSHβ, and Pit-1β, which possesses 26 extra amino acids in the TAD, only induces GH.

High Mobility Group

TCF/LEF is an HMG transcription factor, which is the target of β-catenin. Cadherins mediate cell-cell adhesion and are connected to the cytoskeleton by β-catenin. However, free β-catenin can also act as an accessory transcription factor for TCF/LEF. In the unstimulated state, GSK3 phosphorylates APC and axin, which are scaffolds (Fig. 13-12). These modifications increase their binding to free β-catenin, which is then sequentially phosphorylated by casein kinase 1 (CK1) and GSK3. The phosphorylated β-catenin is now targeted for destruction.

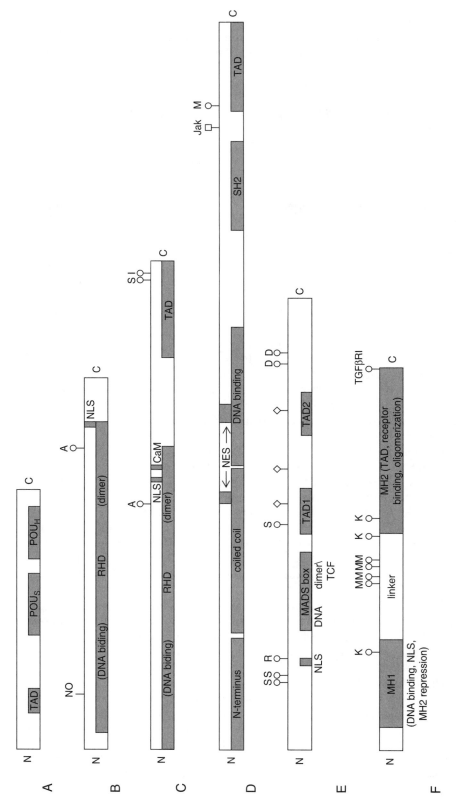

Fig. 13-11. Schematic structure of the other miscellaneous transcription factors. (A) Pit-1; (B) p50 and (C) p65, which dimerize to form the NF-κB factor; (D) Stat; (E) serum response factor (SRF); and (F) Smad. Circles represent pS or pT; squares, pY; and diamonds, glycosylation sites. *D*, DNA-PK; *I*, IKKβ; *K*, CaMKII. Other abbreviations are as in the text and Fig. 13-9.

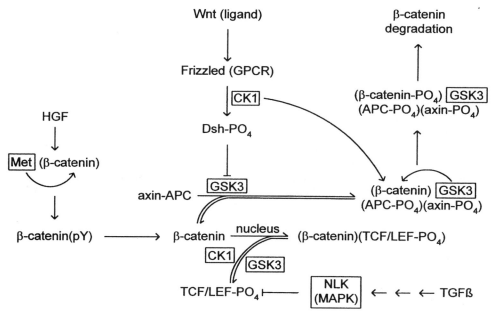

Fig. 13-12. The regulation of β-catenin. Kinases are boxed, arrows indicate stimulation, and flat heads indicate inhibition. *APC*, Adenomatous polyposis of the colon; *HGF*, hepatocyte growth factor; *Met*, HGF receptor; *NLK*, Nemo-like kinase (a MAPK).

There are several ways that hormones can affect this pathway. In the classic pathway, the morphogen Wnt binds its GPCR (G protein-coupled receptor) Frizzled. Frizzled then stimulates CK1ε by a still unknown mechanism. CK1δ and CK1ε have long carboxy termini whose autophosphorylation inhibits the kinase. The mGluR (metabotropic glutamate receptor) has recently been shown to activate CK1ε by dephosphorylating the carboxy terminus through PP2B with the following pathway: PLCβ → IP_3 → IP_3R →↑ calcium → PP2B. Both Frizzled and mGluR are GPCRs, and they may also share the same mechanism for stimulating CK1.

Once activated, CK1 phosphorylates Disheveled (Dsh). Dsh then inhibits GSK, but it does not do so directly. Rather, it is thought that Dsh may change the conformation of axin in a way that blocks the access of GSK to its substrates. Because the APC-axin complex includes PP2A, the complex will be dephosphorylated and release β-catenin. Finally, β-catenin migrates to the nucleus where it activates TCF/LEF by displacing HDAC1. TCF/LEF can also be regulated by phosphorylation; CK1 and GSK3 phosphorylation can increase or decrease, respectively, β-catenin binding to TCF/LEF. TGFβ inhibits TCF/LEF via the MAPK pathway, which inhibits its nuclear transport, and PIAS-induced sumoylation of TCF/LEF sequesters it in PML nuclear bodies. β-Catenin can undergo other modifications; acetylation of β-catenin is required for the repression of the Myc gene. In addition to TCF/LEF, β-catenin is a coactivator for AR, which also acts as a shuttle for the nuclear translocation of β-catenin. No other NR is known to transport β-catenin.

There are alternate activation pathways for β-catenin. For example, RTKs like Met and EGFR bind β-catenin. Indeed, in the rat liver, as much as 30% to 40% of all β-catenin is bound to Met. Upon ligand binding, β-catenin is tyrosine phosphorylated, released, and migrates to the nucleus.

Immunoglobulin-like Family

NF-κB

NF-κB is a nuclear factor that binds to a response element in the κ chain gene of the B lymphocyte (see Chapter 3). However, this response element has since been shown to be more widespread; it is closely associated with genes involved with defense mechanisms. NF-κB is a dimer of identical or partially homologous subunits; the most studied form has subunits of 50 and 65 kDa. All NF-κB subunits are homologous in the DNA-binding and dimerization domains, which usually form the amino terminus and is called the Rel domain (see Fig. 13-11, *B* and *C*). Indeed, this group of transcription factors is called the Rel family. p50 has no recognizable TAD, and homodimers were thought to be transcriptionally inactive; however, it does activate a variant HRE (Table 13-1). In a closely related subunit, p49 (also called p50B), TADs are located in the amino-terminal 100 and carboxy-terminal 150 amino acids. The p50 precursor has about a 550 amino acid, carboxy-terminal extension that acts as an autoinhibitory domain. This extension has seven ankyrin-like repeats, and the inhibitory site has been mapped to the last two repeats and the intervening acidic region, which appears to bind the basic nuclear localization signal (NLS) and mask it. This carboxy terminus can be generated as a separate molecule by alternate splicing and can still act as an inhibitor; in this form it is referred to as IκBγ. Ankyrin domains are also known to bind the cytoskeleton, and it has been speculated that this binding may be a second mechanism by which the carboxy terminus of the precursor prevents nuclear migration.

Unlike p50, p65 has a clearly defined TAD located in the carboxy terminus (see Fig. 13-11, *C*). However, it binds DNA poorly as a homodimer; this appears to be due to steric hindrance by the carboxy terminus because DNA binding is markedly increased after carboxy-terminal truncation. Binding can also be increased by the presence of multiple HREs or by another subunit, like p50, that appears to displace the carboxy terminus. p50 and p65 are truly a matched set; p50 binds DNA well but is a weak transcription activator, whereas p65 binds DNA poorly but has a strong TAD. p65 binds the 3', less conserved end of the HRE, and p50 binds the 5', more conserved half.

There are several other members of this family. p49 has already been mentioned; it is 60% identical to p50. Rel is more closely related to p65 and

Table 13-1
DNA Sequence Specificity for Subunits of NF-κB

	NF-κB response elements[a]			
NF-κB composition	GGGGACTTTCC	GGGGAATCTCC	GGGGAATCCC	GGGAAAGTAC
Homodimers				
p49	+	0	0	
p50	B	B	+	
Rel	+	B/0	+	
p65	Weak	0	0	
Heterodimers				
p49-p65	++	B/0	0	
p49-Rel	++	0	0	
p65-Rel	0		B	B

[a]+, Stimulates transcription; *0*, does not stimulate transcription; *B*, binds but transcriptional activation has not been demonstrated; *B/0*, binds but does not activate transcription.

appears to bind to a wide range of HRE elements. RelB was originally called I-Rel because no transcriptional activity could be detected and the dimer was thought to be inhibitory. However, others have demonstrated activity; the difference may be a result of promoter usage or cell specificity. Herein probably lies the explanation for so many subunit variants: each possible combination has a different HRE preference and may also exhibit differences in promoter or cell specificity (Table 13-1). Alternate splicing generates even more possibilities; Δp65 lacks a segment in the DBD. Although it can still homodimerize with p65, it cannot bind DNA; therefore it acts as an inhibitor.

Altering subunit composition is only one way that NF-κB can be regulated; the most common form of regulation is through an inhibitory subunit, IκB and its variants. All of these inhibitors have 3-7 ankyrin repeats in the middle of the peptide and function in the same manner as IκBγ; they mask the NLS and trap the NF-κB in the cytoplasm until activation. They also have a carboxy-terminal acidic and PEST region. NF-κB activation occurs when IκB is phosphorylated on S-32 and S-36, resulting in its ubiquitination and proteolysis with the subsequent release of NF-κB. One of the most difficult problems in deciphering this pathway was determining the responsible kinase, especially since so many kinases appeared capable of activating the system. In truth, there was only a single kinase complex that directly phosphorylated IκB; the confusion arose from the fact that this kinase complex also acted as an integration site for many other kinases (Fig. 13-13, A). The IκB kinase (IKK) is an extremely large complex. First, it consists of two heterodimers of the homologous kinases, known as IKKα and IKKβ. In addition, there are 4 IKKγ subunits; these are regulatory subunits and adaptors that interface the kinase complex with IκB. Finally, it includes an hsp 90 dimer with 2-3 of its targeting subunits, Cdc37. Hsp 90 is a molecular chaperone that is usually associated with intrinsically unstable proteins, like the SRs. IKKα represents the major input site for other kinases, although at least one kinase, PKC, has been reported to activate the IKK complex through the phosphorylation of IKKβ. Although IκB is the main target of IKK, there are at least five other known substrates: p100, p105, p65, SRC-3, and IRS-1. p100 and p105 are the precursors to p52 and p50, respectively; IKK phosphorylation triggers their ubiquitination and cleavage. It is one of the few examples where the proteasome makes a single cleavage rather than totally destroying a protein. IKKβ phosphorylates p65 in the TAD and enhances its activity, whereas IKK phosphorylation of SRC-3 shifts it to the nucleus where it acts as a coactivator for NF-κB. Finally, IKK phosphorylation of IRS-1 is responsible, in part, for the insulin resistance induced by many inflammatory signals.

IκB can also be inhibited by ROIs. In response to stress, RacGDI dissociates from Rac, which then shifts to the plasma membrane and is activated by RacGEF (Fig. 13-13, A). Rac • GTP acts as an adaptor that recruits p67*phox* and cytochrome b_{558} to the plasmalemma. In the meantime, PKC phosphorylates p47*phox* and p22*phox*; this exposes the PX domain on both proteins, resulting in their migration to the plasma membrane. In addition, the phosphorylation of p47*phox* exposes a SH3 domain that binds p22*phox*. At this point the assembly of the NADPH (reduced nicotinamide adenine dinucleotide phosphate) oxidase is complete and it begins to generate ROIs, which then stimulate a STK (see Chapter 11). There are two targets for the STK. First, direct phosphorylation of IκB on Y-42 will induce its dissociation from NF-κB without proteolysis; second, the STK can also phosphorylate and activate IKKα.

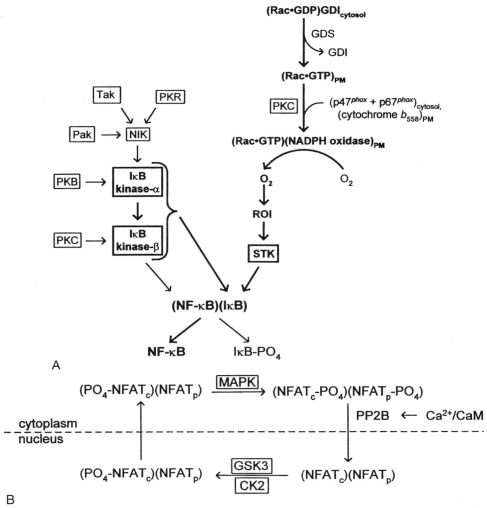

Fig. 13-13. The regulation of NF-κB (A) and NFAT (B). Kinases are boxed; bold lines indicate core pathways. *PM*, Plasma membrane.

In addition to IKK and ROIs, IκB is subject to regulation by several other factors. First, CK2 phosphorylation in the carboxy-terminal PEST domain promotes calpain cleavage; however, this represents basal turnover and not induced destruction. Finally, κB-Ras is a Ras superfamily member that binds the PEST domain and decreases IKK-induced degradation.

The reason for several IκBs lies in their specificity for the different NF-κB subunits and in their regulation (Table 13-2). For example, IκB is specific for p65; IκBα, for p50; IκBγ, for both, and the effects of IκBβ are dose dependent. The structural determinants for subunit specificity appear to reside within the first ankyrin repeat. However, this scenario is not as simple as it first appears. As already noted, p50 homodimers are not very active transcriptionally, although they can still bind DNA well; as a result, they can act like inhibitors. If IκBα inhibits an inhibitor, there will be stimulation. In fact, TADs have been located at both ends the IκBα, and some authorities claim that the IκBα-p50 complex directly stimulates transcription.

Table 13-2
Specificity and Regulation of NF-κB Inhibitors

Inhibitor	Alternate name	NF-κB subunit inhibited[a]		Duration
		p50	p65	
IκB		0	+	
IκBα	Bcl3 variant	+	+/0	Transient
IκBβ	p40, MAD-3	+[b]	+	Persistent
IκBγ	Carboxy terminus of p50	+	+	

[a] +, Inhibits subunit; 0, does not inhibit subunit.
[b] Requires high concentrations.

Direct phosphorylation of the NF-κB subunits also occurs. Phosphorylation of p65 by IKKβ has already been mentioned. PKA phosphorylation induces the dissociation of the amino and carboxy termini to expose the CBP binding site. CaMKIV phosphorylation then triggers the release of the repressor SMRT and recruits the coactivator CBP. PKC and CK2 phosphorylation of the TAD stimulates activity. Finally, p65 can be acetylated; modification of K-221 blocks IκB binding, while K-310 acetylation is required for maximal transcriptional activity. p50 can also be modified. PKC phosphorylation increases DNA binding. In addition, p50 is sensitive to the cell redox; C-62 can be *S*-nitrosylated or undergo *S*-glutathionylation, both of which inhibit activity.

Finally, several NF-kB subunits can be regulated by CaM and by the poly-(ADP-ribose) polymerase (PARP), an enzyme involved with histone modification (see Chapter 14). Although CaM binding has no known effect on p65, it does inhibit the nuclear localization of Rel. As such, CaM binding can determine the composition of the active nuclear transcription factor. PARP binds, stabilizes, and inhibits both p50 and p65. Upon activation by DNA strand breaks, PARP automodifies itself and dissociates, thereby activating NF-κB. Essentially, this is another way that stress can stimulate this pathway.

Nuclear Factor of Activated T Cells

NFAT is a tetramer of $NFAT_c$, $NFAT_p$, and AP-1. AP-1 occupies a position similar to HMG for NF-κB (see Fig. 13-5, *A*). AP-1 recognizes the 3' half of the HRE and may also serve as an adaptor to JNK. Basically, NFAT is inhibited by phosphorylation and activated by dephosphorylation. The negatively charged pS in NFAT binds the basic NLS and blocks nuclear translocation. Hormones that elevate calcium activate PP2B to dephosphorylate NFAT, which then migrates to the nucleus and initiates transcription (Fig. 13-13, *B*). This signal is terminated by PKA and GSK3 phosphorylation of $NFAT_c$. PKA phosphorylation primes $NFAT_c$ for the GSK3 phosphorylation, whereas the latter modification facilitates the nuclear export of $NFAT_c$. CK2 phosphorylation of NFAT4, another isoform, has the same effect. Phosphorylation by members of the MAPK family blocks nuclear reentry by masking the NLS.

Stat

Like all members of the Ig-like family, the Stat transcription factors use an Ig-like domain for DNA binding. However, there is no second Ig-like domain; rather, dimerization is mediated by an SH2-pY interaction in the carboxy terminus (see Fig. 13-11, *D*). The close position of these two structures prevents intramolecular binding. Amino-terminal to the DBD is a coiled coil that interacts with the IRF family, accessory factors that occupy the gap between the dimers as

HMG does for NF-κB and AP-1 does for NFAT (see Fig. 13-5, *A*). The extreme amino terminus mediates receptor binding and tetramer formation; the latter is responsible for synergism between dimers bound to multiple HREs. Finally, the extreme carboxy terminus contains a TAD.

Stats are prebound to membrane receptors; the sequence surrounding the tyrosine selects for specific Stat isoforms (Fig. 13-14, *A*). Upon hormone binding, the RTK or STK becomes activated and phosphorylates Stat. The SH2 domain has a higher affinity for the pY on another Stat than it does for the receptor; therefore after phosphorylation the Stats dissociate from the receptor, dimerize, enter the nucleus, and bind DNA. Although it is not universally required, members of the IRF family, like IRF9 (also called ISGF3γ or p48), often facilitate Stat activity by binding to the 5′ end of the HRE and by recruiting coactivators. The activities of these Stat dimers were discovered before their structures were determined; as a result, the older scientific literature contains unique names for these complexes. For example, the Stat1 homodimer was called GAF (IFN-γ activated factor) and its HRE, GAS (IFN-γ activated site); the Stat1-Stat3 heterodimer was called SIF (*sis*-inducible factor) and its HRE, SIE (*sis*-inducible element); and the Stat1-Stat2-IRF9

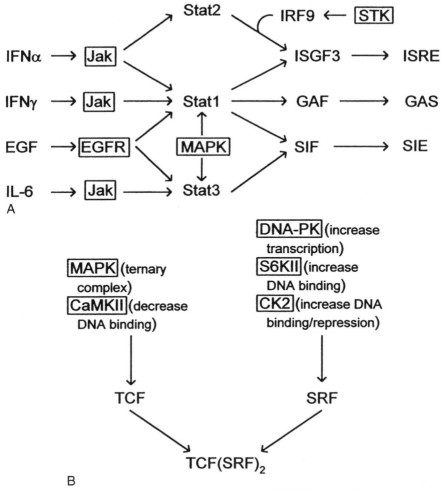

Fig. 13-14. The regulation of Stat (A) and SRF (B). Kinases are boxed.

complex was called ISGF3 (IFN-stimulated gene factor 3) and its HRE, ISRE (IFN-stimulated response element).

In addition to tyrosine phosphorylation, the carboxy-terminal TAD of Stat1, Stat3, Stat4, and Stat5 is serine phosphorylated. Several potential kinases have been identified, including PKCδ, CaMKII, MAPK, and kinases downstream of MAPK or PI3K. The effects of pS on DNA binding are controversial. It is not required for nuclear localization, although it does inhibit tyrosine phosphorylation and therefore may be involved with downregulation. It may also fine-tune transcriptional activity or partner choice. For example, pS is required for Stat1 to bind to BRCA1 (antioncogene deleted in breast cancer 1), a coactivator, and it inhibits Stat5-induced casein gene expression in the absence, but not presence, of cortisol. In the latter case, pS may inhibit Stat5 until all the coactivators are present. Finally, the effect of pS is promoter dependent; for example, although pS inhibits Stat5-induced casein expression, it stimulates luciferase induction, and serine phosphorylation of Stat4 is required for T cell differentiation and IL-12 secretion but not for T cell proliferation.

Stats are also subject to inhibition by alternately spliced Stats and by PIAS1. Methylation at R-31 blocks PIAS1 binding and activates Stat in the absence of tyrosine phosphorylation. Both p21$^{Cip1/Waf1}$, a cell cycle inhibitor, and cyclin D, a regulatory subunit for Cdk4 and Cdk6, inhibit Stat3. Cyclin D can also bind and inhibit several other transcription factors, including the AR. Because it binds PCAF and p300, cyclin D may interfere with coactivator recruitment. PKR allosterically inhibits Stat1, but it dissociates after Stat1 activation; this effect is not kinase dependent. Finally, tyrosine phosphorylation of Stats can be blocked by nitration by NO metabolites. Because these two modifications occur on the same tyrosine, they are mutually exclusive.

Regulation can also be exerted through IRFs. For example, tyrosine phosphorylation of IRF blocks DNA binding of IRF unless it is associated with Stat; acetylation of IRF also inhibits DNA binding.

Runt

The best-known member of this group is CSL, which is the target of Notch signaling. CSL (CBF1/Su(H)/LAG-1) has several components; CBFA (core binding factor A) has an Ig-like DBD that is inhibited by the carboxy terminus. CBFB activates CBFA by displacing the carboxy terminus and stabilizing the exposed DNA binding loops. Notch is a membrane receptor for the DSL family of ligands (delta/serrate/LAG-2). Stimulation of this pathway delays differentiation until the cell fate has been determined. Ligand binding triggers the cleavage of the extracellular domain by ADAM, a metalloprotease. This allows γ-secretase, another protease, access to the transmembrane region to release the intracellular domain, which primarily consists of ankyrin repeats. This Notch fragment migrates to the nucleus, binds CSL, displaces SMRT, and recruits HAT. CSL can also be regulated by phosphorylation: G-CSF stimulates differentiation by inhibiting Notch through MAPK phosphorylation. On the other hand, GSK3 phosphorylation increases the half-life of Notch.

MADS-box Family

Serum Response Factor

The SRF is a dimer with a MADS domain and two TADs. Both TADs contain multiple sites that may be reciprocally phosphorylated or *O*-glycosylated (see

Fig. 13-11, *E*). It is associated with an accessory transcription factor, TCF (ternary complex factor). TCF is also called Elk1 and is a member of the Ets family. TCF stabilizes the SRF-SRE complex; the half-life of this complex increases from 1.6 to 2.5 minutes to 20 to 30 minutes. It also binds to the CAGGA sequence just 5′ to the SRE.

Both SRF and TCF are regulated by phosphorylation (Fig. 13-14, *B*). MAPK modifies TCF in the carboxy terminus; this modification enhances ternary complex formation between TCF and SCF dimers and increases transcriptional activity. CaMKII phosphorylation of TCF stabilizes the autoinhibitor domain and decreases DNA binding. DNA-PK phosphorylation of SRF increases transcription, while S6KII phosphorylation increases DNA binding. CK2 phosphorylation in the amino terminus of SRF also increases DNA binding, whereas modification of the carboxy terminus enhances repression by SRF.

Myocyte Enhancer Factor 2

In the inactive state, MEF2 is complexed with Cain (Fig. 13-15, *A*), which inhibits MEF2 in multiple ways. First, it recruits HDAC through the adaptor Sin3; second, it competes with the coactivator p300; and third, it blocks the phosphorylation of the carboxy-terminal TAD by MAPK. Finally, Cain is a PP2B inhibitor; this blocks the stimulation of NFAT, which synergizes with MEF2 in myocyte differentiation. MEF2 is activated by CaM. First, CaM displaces Cain from the carboxy terminus; second, CaM stimulates CaMKIV, which phosphorylates HDAC, resulting in its sequestration in the cytoplasm by 14-3-3. The dissociation of HDAC allows MAPK to phosphorylate the TAD and activate transcription. Finally, CaM activates NFAT via PP2B.

β-Family

The transcription factors under this heading are structurally heterogeneous; their only common connection is that they bind DNA through β-structures other than Ig-like domains.

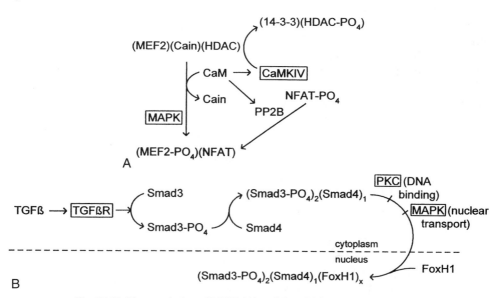

Fig. 13-15. The regulation of MEF2 (A) and Smad (B). Kinases are boxed.

Smad Family

Smads are a major output for TGFβ. The structure is dominated by two regions: the MH1 and MH2 domains. MH1 contains the NLS and the DBD that recognizes CAGA or CGGCnCGC sequences. It also represses the TAD in MH2. MH2 binds the TGFβ receptor, is involved with trimer formation, and has a TAD that recruits coactivators and general transcription factors. Activation is primarily through phosphorylation by the TGFβ receptor; this modification induces the trimerization between two R-Smads and a Co-Smad. The trimer migrates to the nucleus and acts as an accessory factor for many transcription factors. Although most commonly envisioned as a partner for FoxH1, a wHTH factor, it can also facilitate transcription by factors belonging to the NR, zinc finger, bZIP, bHLH, homeodomain, and Ig-like families. There are additional mechanisms of regulation; MAPK phosphorylation stabilizes Smad but blocks nuclear translocation (Fig. 13-15, B). CaM binding to the amino terminus blocks MAPK phosphorylation. CaMKII phosphorylation also prevents migration to the nucleus and PKC phosphorylation prevents DNA binding.

Y-Box Family

Members of this family can bind RNA or DNA and can function either transcriptionally or posttranscriptionally, depending on the particular factor. For example, FRGY2 sequesters mRNA and inhibits translation. CK2 phosphorylation promotes the assembly of this oligomer, whereas dephosphorylation triggers the release of the mRNA and its translation. Both dbpB and ZONAB are transcription factors that are sequestered outside the nucleus. Full-length dbpB binds mRNA in the cytosol; thrombin stimulates the cleavage of dbpB and its migration to the nucleus. ZONAB binds to ZO-1 at tight junctions where it is released when changes occur in cell-cell connections.

Tubby Family

Tubby is a β-barrel with a basic groove along one side. One end of this groove also binds $PI(4,5)P_2$, which holds it at the plasma membrane. Activation of α_q and PLCβ cleaves $PI(4,5)P_2$ and releases Tubby to migrate to the nucleus and activate transcription.

Summary of Regulation

As the various transcription factors were just described, several mechanisms of regulation were encountered (Table 13-3). It may be useful to summarize them now. First of all, transcription factors can be induced; this mechanism is particularly important for short-lived factors like Fos. For example, basal AP-1 activity is due to Jun homodimers, because Fos has a much shorter half-life than Jun. With induction, Fos increases, forms heterodimers with Jun, increases its activity, and slightly shifts the HRE preference. Although important, this mechanism requires RNA and protein synthesis and, therefore, is not a primary response.

Another site of regulation occurs at the level of posttranscriptional modification. Many examples of alternate splicing were given above. The most common form is the omission of an exon; this deletion usually cripples the factor, resulting in the creation of an inhibitor. Alternatively, the effects may be more selective; the insertion of an extra 26 amino acids in Pit-1 eliminates its ability to induce transcription of the PRL gene, although its stimulation of GH expression is

Table 13-3
General Mechanisms for the Regulation of Transcription Factors

Mechanisms	Examples	Comments
Transcription	Fos; Myc	
Posttranscriptional Processing		
Deleted exons	ΔFosB; Δp65; ΔMax	Unusally inhibitory
Alternate exons	CREM; Pit-1	
Posttranslational Modifications		
Phosphorylation	See Figs. 13-9 to 13-15	
O-Glycosylation	Myc; SRF	Often reciprocal with phosphorylation
S-Glutathionylation	NF-κB (p50); Jun	Inhibitory
S-Nitrosylation	NF-κB (p50); Jun	Inhibitory
S-PGJ$_2$	NF-κB (p50)	Inhibitory
Acetylation	Nuclear receptors; NF-κB (p65)	Blocks IκB binding
Methylation	Stat1	Blocks PIAS1 binding
Hydroxylation	HIF-1α	Inhibitory
Ubiquitination	IκB; β-catenin	Proteolysis
Sumoylation	IκB; AR; GR; p53	Inhibits AR; stimulates GR; inhibits IκB ubiquitination
Oxidation of disulfide bonds	TTF1	Stimulatory
Cleavage	p50 precursor	Removes inhibitory carboxy terminus
Protein-Protein Interactions		
Inhibitors	IP-1; IκB; Cain	
Dimers	Jun-Jun vs Jun-Fos	
Nonhomologous partners	TCF for SRF; IRFs for Stats	
Cofactors	CBP-CREB	Interaction requires phosphorylation
Compartmentalization		
Transcription factor-inhibitor phosphorylation status	Rel; C/EBPβ; NFAT; Smad; Stat; FoxE1 (TTF1); β-catenin; FoxO3a	
Coactivator	Hic-5; NRs	
NLS-masking subunit	IκB; hsp 90	
Cleavage	SREBP-2; ATF6; Notch	
Anchor	Tubby sequestered at plasma membrane via PI(4,5)P$_2$	Released by hydrolysis
Allosterism		
Hormones	Nuclear receptors; NPAS2 (CO)	
Second messengers	Smad, MEF2, and Rel (CaM); DREAM (Ca^{2+}); κB-Ras and Jun (G protein)	
Other ligands	PPAR; Ah receptor; NF-κB (PARP)	NF-κB stimulated by PARP release
Other		
Light	White collar-1	Stimulated

unaltered. In more elaborate examples, there may be multiple copies of one or more exons, from which a single set is chosen; CREM represents such a paradigm.

However, the regulation most closely linked with the endocrine system is phosphorylation. Figures 13-9 to 13-15 depict phosphorylation sites and regulatory functions for several of the systems previously discussed; although many kinases may act on any given transcription factor, for clarity only the most prominent kinases are shown for each factor. Glycosylation is another posttranslational modification: it may (1) be reciprocal with phosphorylation (Myc); (2) block hydrophobic interactions (Sp1 with TF110); (3) mask inhibitory domains (carboxy terminus of p53); and (4) protect proteins from the proteasome (inhibit phosphorylation of PEST domains). The modification of cysteines by glutathione and NO signal the redox potential of the cell and are often inhibitory. A cysteine in p50 can also be conjugated with prostaglandin J_2; the physiological relevance of this curious modification is unknown.

Acetylation and methylation are usually associated with histones, but increasing numbers of transcription factors have also been discovered to be modified. Acetylation and methylation are covered in more detail in Chapter 14. Ubiquitination, sumoylation, and neddylation are all related (see Chapter 8); the former usually involves protein destruction by the proteasome, whereas sumoylation affects localization within the nucleus. Finally, p50 can be activated by the proteolytic removal of its inhibitory carboxy terminus; this modification pathway is obviously irreversible.

Protein-protein interactions represent a fourth mechanism. There are three types of interactions: (1) inhibitors, (2) oligomerization partners, and (3) cofactors. Inhibitors would include IP-1 and IκB, which bind, and block the DNA binding of, AP-1 and NF-κB, respectively; inhibition is relieved by phosphorylation of the former and destruction of the latter. As previously noted, most transcription factors are dimers whose partners can come from any member within a particular family. The transcriptional activity, HRE preference, and regulation are often influenced by the kinds of partners present in the dimer; for example, Myc-Max stimulates transcription, whereas Mad-Max inhibits it, and Jun-Jun prefers to bind the sequence TGACATCA, whereas Jun-Fos binds TGACTCA. In addition to binding family members, transcription factors may bind accessory factors that increase DNA binding specificity and transcriptional activity; examples include the IRF, Ets, and Smad families. Finally, transcription factors recruit coactivators. These proteins can influence gene expression because different promoters require different coactivators. In addition, many of these coactivators can be regulated themselves.

Fifth, transcription factors can be regulated by compartmentalization. Subcellular localization can be controlled by several mechanisms. First of all, nuclear migration can be induced by the phosphorylation of either the DNA-binding component or some other regulatory subunit; Stat and Smad are examples of the former, and IRF9 is an example of the latter. Nuclear localization can also be stimulated by dephosphorylation: for example, NFAT. Second, coactivators can be compartmentalized; Hic-5 has two TADs and is distributed between focal adhesions and the nucleus. Signals from the cell attachment site release Hic-5, which goes to the nucleus and enhances SR transcription. Third, nuclear localization can be achieved by unmasking the NLS; in NF-κB and GR, a masking subunit is removed by phosphorylation or ligand binding, respectively. Fourth, transcription factors may be tethered to extranuclear membranes and released by cleavage of the transmembrane domain; SREBP-2 and ATF6 are integral membrane proteins in the endoplasmic reticulum and Notch is located in the plasma membrane. Finally, Tubby is anchored in the plasmalemma by PIP_2 (phosphatidylinositol 4, 5-bisphosphate) and liberated by phospholipid hydrolysis.

Sixth, transcription factors can be regulated allosterically. The regulation of the nuclear receptors for steroids, retinoids, and thyroid hormones has already been extensively discussed (see Chapter 6). PPAR is activated by fatty acids, especially arachidonic acid, which is a second messenger in the PPI pathway, and NPAS2 is inhibited by carbon monoxide (CO) binding to the hemes in its PAS domains. The Ah receptor has no known endogenous ligand, but it is regulated by aromatic hydrocarbons in the environment. In addition, second messengers can bind and affect transcription factors or their regulators: CaM stimulates Smad and MEF2; calcium inhibits DREAM, which is still stimulatory because DREAM is a repressor; Ras binds Jun to facilitate phosphorylation and κB-Ras binds IκB to decrease degradation; and G_z and G_i2 bind Eya2, a coactivator for the Six family of homeodomain transcription factors, and sequester Eya2 in the cytosol. Finally, transcription factors can even respond directly to environmental stimuli. White collar-1 is a *Neurospora* transcription factor belonging to the GATA family. The GATA family has zinc fingers similar to the NRs. It also has three PAS domains, the first of which is modified to bind FAD and detect blue light.

Specificity

Problems

It should be apparent from the preceding discussion that there is extensive overlap in the regulation of these proteins. This overlap manifests itself in four ways: (1) many second messengers act through the same transcription factor; (2) many different transcription factors can heterodimerize; (3) many different factors can bind the same HRE; and (4) one factor may bind to many different HREs.

CREB is an example from the first category; PKA is considered to be the major activator of CREB, but S6KII and CaMKIV can also phosphorylate CREB. C/EBP (see Fig. 13-10, *C*) and NF-κB (see Fig. 13-13, *A*) also have a large number of inputs from different kinase pathways. This phenomenon can be further complicated by the promiscuous dimerization of transcription factors that are regulated differently, thereby leading to even greater convergence. For example, there is considerable cross-dimerization within the thyroid hormone receptor family, and most of the factors can bind to at least one common HRE. Although a single ligand can usually activate the receptor dimer, ligands for both components act synergistically, thereby merging the actions of two different hormones. Such a phenomenon is also seen in transcription factors stimulated by different kinases; a dimer between Jun, activated by PKC, and CREB, activated by PKA, can bind the CRE.

Additional loss of specificity occurs when multiple transcription factors bind the same HRE. The classic example is the glucocorticoid receptor family; the glucocorticoid, progesterone, androgen, and mineralocorticoid receptors all bind the identical HRE despite having very distinct biological activities. In addition, several transcription factors can bind multiple HREs; for example, Jun-ATF3 binds both the cAMP response element (CRE) and the AP-1 HRE; and the retinoic acid X receptor (RXR)–PPAR heterodimer can bind and activate both the PPAR response element and the ERE.

Possible Solutions

Jigsaw Puzzle Hypothesis

Several possible solutions to this specificity problem have been proposed; none are mutually exclusive. The first is the Jigsaw Puzzle Hypothesis; simply stated, this hypothesis claims that the action of a given hormone is determined by the entire constellation of transcription factors that it activates, as well as the degree, timing, and duration of stimulation. For example, transient activation of MAPK stimulates Fos, but sustained activation of this kinase stimulates Jun and Fra. In the liver, GH activates Stat1, Stat2, and Stat5 but generates a unique profile; Stat1 stimulation requires higher GH concentrations than the other two, and Stat1 and Stat3 desensitize more quickly than Stat5 after the first GH pulse.

Receptor Hypothesis

The Receptor Hypothesis states that hormone specificity lies in the tissue distribution of the receptor. For example, although GR and PR bind the same HRE, progesterone does not induce the glucocorticoid-stimulated genes in liver. This lack of a response is due to the absence of PRs in this tissue; progesterone will induce these genes, if the liver cells are transfected with PRs. This hypothesis can be expanded to include the distribution of steroid metabolizing enzymes, as well as receptors; the specificity of aldosterone has been shown to be due to the tissue distribution of enzymes that degrade cortisol (see Chapter 5). Finally, CaMKIV is restricted to the nervous tissue; therefore, its activation of CREB only occurs in neurons. Unfortunately, this hypothesis does not hold for all systems; in the liver T_3 induces malic enzyme and the enzymes for fatty acid metabolism. In fibroblasts transfected with TR, T_3 only induces the malic enzyme, and in transfected QT6 cells T_3 has no effect.

DNA Conformation Hypothesis

The DNA Conformation Hypothesis states hat two transcription factors may bind the same HRE but induce different conformation resulting in different effects. For example, Jun and Myc bend DNA toward the dimer, whereas Fos and Max bend DNA away from the dimer (Fig. 13-16). As such, Jun-Jun and Jun-Fos may bind the same HRE but the resulting DNA conformation is very different. A similar situation occurs with Max homodimers and Myc-Max dimers. This effect resides in the charge at the amino terminus of the bZIP; the more positive the charge, the more positive the DNA bend.

Ancillary Factor Hypothesis

The Ancillary Factor Hypothesis states that the transcription factor in question is only one of many proteins involved in the expression of a gene and that these other proteins can impart additional specificity. Such ancillary factors can include (1) subunit partners, (2) accessory transcription factors, (3) coactivators, (4) repressors, and (5) matrix localizing factors. An example of the first item is members of the TR family; vitamin D receptor (VDR) homodimers can bind perfect HREs, whereas RXR-VDR prefers imperfect ones. Second, accessory transcription factors usually bind in the 5' flanking region, which increases the

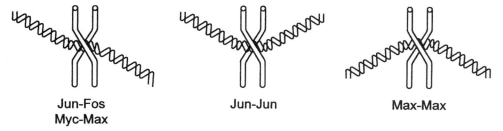

Jun-Fos　　　　　**Jun-Jun**　　　　　**Max-Max**
Myc-Max

Fig. 13-16. The direction of DNA bending in response to different bZIP dimer compositions.

effective length and specificity of the HRE. Third, all the members of the GR family bind the same HRE, but their TADs are different, leading to the recruitment of different coactivators; GR requires NFI and p160, but PR does not; AR requires the retinoblastoma protein (Rb), but GR does not; and the requirement for HMGB1 is greater for GR than PR, whereas neither AR nor MR need it at all. Fourth, members of the GR family also respond differently to repressors; NCoR and SMRT increase PR sensitivity but have little or no effect on GR; AR is less susceptible to repressors than GR on certain genes, and BAG-1M is a chaperone that inhibits GR DNA binding and transcriptional activity but does not affect MR. Finally, there are nuclear proteins that bind both the nuclear matrix and SRs. Such proteins may facilitate the localization of certain SRs near specific HREs.

Position and Sequence Effects

The Position and Sequence Effects hypothesis states that the binding to an HRE may be influenced by the surrounding sequence, HRE sequence variation, and/or spacing with other HREs. First, it is now known that the "consensus" HREs are actually only the core recognition unit; flanking sequences are also important and impart additional specificity. For example, while the P box of NRs bind the core HRE, the second zinc finger and hinge region of GR and AR bind to flanking sequences that can distinguish between a GRE (glucocorticoid response element) and an ARE. Such interactions with adjacent sequences have also been demonstrated for ER, TR, Myc-Max, and RXR-PPAR. Second, DNA can have different topologies and the conformation of a HRE can influence what factor recognizes it. For example, PR requires the HRE be on negatively supercoiled DNA, but GR does not. On the other hand, GR is more active on intact chromatin, whereas PR is more active on naked DNA.

Finally, the sequence, integrity, polarity, and spacing of the HREs can impart additional specificity on a system. As discussed later, natural HREs are rarely perfect; they usually exhibit slight variations. AR is more tolerant to imperfections in the 5′ end of the HRE than GR is. Furthermore, these variations can impose different conformations on the factors that bind them. ER, RXR-VDR, and NF-κB have been shown to have different conformations on different HREs. In the former case, this difference translates into a different spectrum of coactivator recruitment; ER binds SRC-1 and SRC-2 equally well on perfect and imperfect EREs, but it binds much less SRC-3 on the imperfect ERE. The occurrence of full versus half HREs is also important; the HRE for NRs usually occurs as direct repeats or palindromes. In either case, this arrangement engages NR dimers, whereas half sites bind monomers. NCoR is a corepressor that has two NIDs; it is only active when both NIDs have bound a NR, which would only occur on a full HRE. In fact, many coactivators and corepressors have multiple binding sites for transcription factors and can distinguish not only the number, but also the polarity, of HREs. NCoR can distinguish between direct repeats (DRs) and palindromes. In addition, heterodimeric transcription factors will assume different orientations on HREs with different polarity; this, in turn, can affect which other transcription factors are recruited and what the final effect on transcription will be. For example, Jun-Fos polarity determines whether it will interact with NFAT. In other cases, polarity can affect binding specificity; AR binds more strongly to single and tandem DRs than GR, which prefers palindromes.

HRE spacing represents a final mechanism to add specificity. All members of the TR family bind the same half HRE: AGGTCA (Table 13-4). However, the

Table 13-4
The Effect of Hormone Response Element Spacing on Nuclear Receptor Binding Specificity

Number of nucleotides between successive AGGTCA	Nuclear receptor binding specificity
1	RXR; COUP; PPAR; RXR-PPAR
2	RXR; TR-RXR
3	VDR-RXR
4	TR; TR-RXR
5	RAR; VDR; RXR-RAR; RXR-TR
6	VDR

spacing between these DRs determines which NR will bind. Like polarity and sequence variations, spacing can also affect cofactor recruitment. In the Pit-1 HRE of the GH gene there are two extra bp versus the prolactin gene. This arrangement puts POU$_S$ and POU$_H$ on the same side, allowing NCoR to bind and repress transcription. On the prolactin gene, these two domains are rotationally displaced, NCoR cannot bind, and the gene is transcribed.

Mechanisms for Transcription Effects

Positive Effects

DNA Allostery

Once the transcription factor has bound DNA, what happens next? Three major mechanisms for how these factors work have been proposed: (1) they alter DNA structure; and/or (2) they interact with other transcription factors, particularly those directly involved with transcription initiation; and/or (3) they block the effect of corepressors. Hormonally regulated transcription factors can change chromatin structure in four ways: (1) DNA unwinding or relaxation, (2) DNA bending, (3) DNA looping, or (4) nucleosomal sliding or displacement.

DNA Relaxation. GR induces both relaxation in negatively supercoiled DNA and induces positive supercoils in relaxed DNA; ER can also produce positive supercoiling, as well as stabilize single-stranded DNA, although there is no evidence that it actually induces melting. The effects of both GR and ER are specific because they require the presence of the appropriate HRE. There are four mechanisms by which transcription factors can affect DNA topology. First, it is suspected that the changes in supercoiling are a direct effect of the binding of the transcription factor to its HRE. However, there are at least three other mechanisms by which transcription factors can affect DNA topology. For example, intercalation would be a second mechanism; the conserved tryptophan in Ets is inserted between DNA bases, and the first helix in HMG lies in the minor groove. Both of these interactions distort DNA structure. Third, NRs can recruit helicases; DEAD box proteins are helicases that bind the ER TAD1, especially when it has been phosphorylated by MAPK. SUG1 and RNA helicase A are other NR cofactors; the latter binds through CBP.

Finally, some of these effects may be mediated through topoisomerases; topoisomerases catalyze changes in the topology of DNA. Topoisomerase I nicks one strand of the DNA and allows it to relax before ligating the free ends. Topoisomerase II nicks both strands, can introduce negative supercoils, and can knot and unknot DNA (catenate and decatenate, respectively). The latter enzyme is stimulated 2- to 3-fold by either CK2 or PKC phosphorylation, although the former kinase appears to be the more physiological one. Topoisomerase I is also activated by both kinases. Finally, the decatenation activity of topoisomerase II has been reported to be stimulated by a direct interaction with several hormonally regulated transcription factors, including CREB, ATF2, and Jun.

DNA Bending.   Many of these factors can also bend DNA; these proteins include TR, ER, RXR, CREB, Jun-Fos, Myc-Max, NF-κB, Stat, SRF, and POU factors. Jun-Fos and Myc-Max bend DNA in a subunit specific manner. Jun and Myc bend DNA toward the dimer interface; Fos and Max, away from the interface. As such, Jun homodimers produce a simple bend with the factor bisecting the acute angle, whereas Jun-Fos generates a DNA that roughly maintains a straight structure because the two bends nearly cancel each other out. There is a slight kink at the point that AP-1 binds (Fig. 13-16).

In the case of CREB, phosphorylation of CREB increases DNA bending. There is a correlation between DNA bending and transcriptional activity for some of these factors; for example, variant TR HREs that are not bent by TR do not support transcription. In addition, for PR isoforms and ER mutants, DNA bending correlates with the ability of the SR to bind coactivators and stimulate transcription. However, bending is not essential because not all transcription factors induce it. For example, if the function of DNA bending were to facilitate nucleosomal sliding to expose the promoter (see later), it would not be necessary on promoters that were already accessible.

DNA bending is determined by two major methods: x-ray crystallography and circular permutation. There are two problems with the former: first, bent DNA may be forced to straighten so as to facilitate packing in crystals. Second, because smaller molecules are easier to crystallize than large ones, investigators often use only the DBD and DNA. As previously noted, regions outside of the DBD may contribute to DNA binding: for example, the second zinc finger and hinge domain in SRs, and the TAD in Jun-Fos. These additional interactions may contribute to DNA bending but are not present in crystals containing only the DBD. For instance, the PR DBD bends DNA 25°, but the full length PR bends DNA 72° to 77°. Circular permutation involves circularizing a piece of DNA containing the HRE of interest, cutting the circle at different sites within the circle, and then running the DNA on gels. DNA that has the same length but is bent at different sites along this length will migrate differently on gels. These migration patterns can be used to measure approximately the DNA bending. However, this technique tends to overestimate the bending.

DNA Looping.   As discussed later, HREs often exist as multiple copies at the 5' end of genes; receptor-receptor interactions, either direct or through intermediates, can cause the intervening DNA to loop out. Such loops have been demonstrated for both PR bound to the uteroglobin gene and ER bound to the prolactin gene. Looping and bending serve several functions: (1) they increase contact between the receptor or transcription factor and DNA; (2) they bring separate transcription factors together; (3) bending may provide a recognition motif for transcription factors; and (4) they may induce changes in the nucleosomes.

Nucleosomal Sliding and Displacement.   Finally, the binding of transcription factors can result in the sliding or displacement of nucleosomes.

Nucleosomes are complexes of histones around which DNA is wrapped (see Chapter 14); they act as a physical barrier to transcription initiation. On the tyrosine aminotransferase gene, GR displaces two nucleosomes, and on the mouse mammary tumor virus (MMTV) genome, GR displaces a single nucleosome and allows NFI and TFIID to bind. A causal relationship is supported by the fact that hyperacetylation of the histones on the MMTV-integrated genes prevents nucleosomal displacement and blocks transcription. Furthermore, the MMTV genome can be derepressed by titrating out the nucleosome using competing DNA. However, although nucleosomal displacement appears necessary for transcription, it is not sufficient; MMTV on transient DNA does not have nucleosomes but still requires glucocorticoids for induction. Nor is this mechanism universal; ER binds to its ERE and induces changes in the DNA conformation without displacing any nucleosomes. Nucleosomal sliding and displacement is produced by SWI/SNF, which was discussed in the beginning of this chapter.

Interaction with Transcription Factors

The second mechanism by which transcription factors can stimulate gene expression is by interacting with components of the transcription initiation machinery. Some NRs can directly bind to general transcription factors and stabilize the preinitiation complex; this interaction is mediated by the TAD in the LBD. In addition, these factors can be recruited by coactivators, such as CBP, p160, and DRIP complexes. In fact, some NRs can act as bridges themselves; for example, estradiol stimulates the cyclin D1 gene through ATF2-Jun binding on a CRE. However, the DBD of ER is not required for this effect; rather, ER is acting as a bridge between ATF2-Jun and coactivators. Nonetheless, because of methodological problems with transfection, care must be taken when evaluating a claim that synergism is independent of DNA binding. Specifically, high copy numbers can lead to artifacts; for example, at high concentrations only the amino terminus of GR is required to synergize with Stat5 to induce casein gene expression. However, at lower physiological concentration, GR binding to its GRE is required.

Derepression

Finally, specific transcription factors can stimulate transcription by removing repressors. For example, Myc binds and inhibits YY1, a general repressor.

Negative Effects

Steric Occlusion

Transcription factors can also inhibit gene expression. The simplest mechanism is steric occlusion, where an HRE is located near or may even overlap sites for other, positively acting factors and sterically interferes with either the binding or activity of these latter factors. For example, the expression of the osteocalcin gene is stimulated by 1,25-DHCC (1,25-dihydroxycholecalciferol) and inhibited by both glucocorticoids and PKC. The GRE overlaps the TATA box, and the AP-1 site overlaps the VDRE; therefore the binding of GR would block the binding of preinitiation factors and that of AP-1 would interfere with the binding of VDR (see the osteocalcin VDRE in Table 13-5). In another example, both Sp1 and TR compete for the same site on the EGFR gene; because Sp1 is required for transcription, occupancy of this site by TR inhibits gene expression.

Table 13-5
Hormone Response Elements for Selected Transcription Factors

Hormone	DNA Sequence
Glucocorticoids, progesterone, androgens, and aldosterone (GRE)	GGTACAnnnTGTTCT CCATGTnnnACAAGA
Estradiol (ERE)	AGGTCAnnnTGACCT TCCAGTnnnACTGGA
Ecdysone (EcRE)	GGTCAnTGACC TCCAGTnACTGG
Triiodothyronine	AGGTCAnnnnAGGTCA TCCAGTnnnnTCCAGT
1,25-DHCC (Osteocalcin)[a]	TTGGTGACTCACCGGGTGAACGGGGGCAT AACCACTGAGTGGCCCACTTGCCCCCGTA
cAMP (CRE)	TGACGTCA ACTGCAGT
Jun-Fos (AP-1)	TGA$^{C}_{G}$TCAG ACT$^{G}_{C}$AGTC
C/EBP	RTTGCGYAAY[b] YAACGCRTTR
Ah Receptor (XRE or DRE)	T$^{A}_{T}$GCGTG A$^{T}_{A}$CGCAC
Myc/Max	CACGTG GTGCAC
SREBP-2	YCAYnYCAY[b] RGTRnRGTR
FoxO3a	TTGT TTAC AACAAATG
Pit-1	AnTnCATnn$^{G}_{T}$TATnCAT TnAnGTAnn$^{C}_{A}$ATAnGTA
NF-κB	GGGACT T TCC CCC TGAAAGG
NFAT	GGAAAA CCT TTT
Stat	TTCCX$_n$GGAA AAGGX$_n$CCTT
SRF (SRE)	AGATG$^{C}_{G}$CCATA TTTGG$^{G}_{C}$CATCT TCTAC$^{C}_{C}$GGTATAAACC$^{G}_{G}$GTAGA
MEF2	CTA($^{A}_{T}$)$_4$TAG GAT($^{T}_{A}$)$_4$ATC
Smad	CAGACAGT CAGACAGT GTCTGTCAGTCTGTCA GTCTAGAC CAGATCTG

[a] Dashed arrow represents a RARE; dotted arrow, a Jun-Fos HRE.
[b] R, Any purine; Y, any pyrimidine.

There are two variations on this theme. In the first, inhibition is ligand dependent; in the absence of ligand, the binding of the nuclear receptor results in inhibition, but in the presence of ligand, transcription is stimulated. This phenomenon is common in certain members of the TR family, including TR, RXR, and ecdysone. These receptors normally bind tightly to their HREs even without ligand; however, hormone binding is required to stimulate transcription. Originally, this situation was considered to be a form of autosteric occlusion because the receptor occupies its HRE without doing anything. However, it has recently been shown that the unoccupied TR is not just a passive obstruction but that it can actively inhibit formation of the preinitiation complex by recruiting corepressors that bind to the incomplete groove that forms TAD2. Upon ligand binding, helix 11/12 closes and completes the formation of TAD2. The additional determinants favor the binding of coactivators over corepressors.

A second variation is the negative HRE (nHRE). This DNA sequence is usually a variant HRE that mediates inhibition. It is thought that if the nuclear receptor does not bind correctly to the DNA, it will not assume a transcriptionally active comformation. For example, many nHRE are half-sites that only allow the monomer to bind, although the dimer is required for activity (see Chapter 6). Spacing is also important; the TRE and ERE only differ in the spacing between the half-sites. Each can bind the HRE of the other but cannot activate gene expression from it. Again, the factor occupies an HRE but is inactive. Finally, the variant sequence of the nHRE can influence partners. For example, the nHRE in the TSHβ gene only binds thyroid receptor (TR) homodimers, but not RXR-TR dimers. T_3 bound to TR homodimers will recruit HDAC and repress transcription.

Competition

Competition, or squelching, is a second inhibitory mechanism. In this form of repression, one specific transcription factor competes for scarce coactivators, dimerization partners, other synergistic transcription factors, or general transcription factors. It can be mimicked by TADs alone and does not require an HRE. Competition for the coactivator CBP is particularly common because it is so universally required and scarce. For example, Stat inhibits AP-1 and NF-κB by outcompeting them for CBP; antagonism between AR and AP-1 and between NR and NF-κB is also a result of competition for limiting amounts of CBP. Like CBP, RXR is present in the nucleus in limiting quantities, and because it is the common subunit for so many NRs, it is in heavy demand. For example, PPAR and VDR inhibit TR by binding and sequestering RXR. In insects, ultraspiracle (USP) serves the same function as RXR. JH binding to USP induces the formation of an USP homodimer; this decreases the amount of EcR-USP and inhibits ecdysone action. Third, other synergistic or accessory transcription factors can be the object of the competition; PR inhibits GR by competing for Stat5, and AR inhibits TCF/LEF by competing for β-catenin. Finally, TR can bind and sequester components of the general transcriptional machinery, such as TBP, TFIIA, and TFIIB. However, the physiological relevance of competition has recently been called into question; the amounts of receptor required to sequester enough factors to inhibit transcription far exceed levels normally found in tissues. For example, squelching by ER requires concentrations of ER 10 times the physiological levels.

Protein-Protein Antagonism

There are three types of protein-protein interactions that can result in inhibition: homologous and heterologous associations and specific inhibitors.

Homologous interactions refer to the binding of transcription factors to defective members from the same family; these inactive subunits include ΔCREB, ΔFosB, and CREMβ. They may fail to bind DNA or activate transcription, depending on the segment that is missing.

Heterologous inhibition refers to repression produced when one transcription factor binds another from a second family. The most intensively studied example is the interaction between members of the AP-1 and NR families. AP-1 can antagonize the action of GR, PR, AR, ER, RAR, and TR, but not RXR or MR. Although individual studies may yield slightly different results, the following basic picture emerges. First, a physical interaction between Jun and the nuclear receptor can be detected, usually by coimmunoprecipitation. This binding requires the coiled coil domain of Jun and possibly part of the amino terminus; both the DBD and the LBD of the nuclear receptor are also required. However, the presence of DNA is not necessary and the DBDs from several nuclear receptors are freely interchangeable; as such, the DBD is not being used to bind DNA. The relationship between Jun and TR is particularly interesting; without T_3, this interaction enhances transcription by AP-1, but in the presence of T_3 it becomes inhibitory. A related phenomenon occurs between AP-1 and GR, except that the composition of AP-1 determines the effect; Jun-Fos heterodimers inhibit GR, but Jun homodimers or Jun-Fra1 heterodimers are stimulatory. This and other data implicate Fos as the major inhibitory component of AP-1.

Interactions between other protein families also exist. Jun can bind and inhibit MyoD; the binding occurs between the zipper of Jun and the HLH domain of MyoD. In addition, Oct-1 can be suppressed by GR; this interaction requires the second helix in the POU_H domain. There are probably several mechanisms of this inhibition. First, the two transcription factors may be competing for the same coactivators; this is thought to underlie the antagonism between AP-1 and the NRs. Alternately, binding may block a site for a synergist, coactivator, or activating kinase.

The last type of protein-protein suppression of transcription is the result of specific inhibitors: for example, IP-1 and AP-1 or PIAS1 and Stat. As with heterologous inhibition, these inhibitors may obstruct critical binding sites. However, PIAS1 is also a Sumo E3 ligase, and some of its effects are probably due to sumoylation.

Silencing and Corepressors

Silencing is a more active type of inhibition. The interactions just discussed merely block the activation by other factors; however, corepressors actively suppress transcription, for example, by reversing stimulatory modifications, altering chromatin conformation, and displacing coactivators. For example, the RNA polymerase II requires phosphorylation on S-2 in the CTD for activity; GR inhibits NF-κB by recruiting an S-2 phosphatase to deactivate the RNA polymerase. However, most silencing is accomplished by several repressor complexes: Sin3, Mi-2/NuRD (nuclear remodeling and histone deacetylation), SMRT/TRAC, and NCoR.

Sin3 and Mi-2/NuRD bridge specific transcription factors and HDAC; HDAC reverses acetylation, which is associated with transcription (see Chapter 14). In general, Sin3 represses specific genes, whereas the inhibitory effect of NuRD extends over many genes. In addition, NuRD includes DNA remodeling subunits, which allows it to target less accessible genes. The Sin3 complex includes Sin3, an adaptor that binds the specific transcription factors; HDAC1 and HDAC2; RbAp46 and RbAp48, which are core histone binding subunits; an O-NAG transferase; and SAP18 and SAP30, whose functions are unknown. The NuRD complex also contains HDAC1, HDAC2, RbAp46, and RbAp48. In addition, it includes Mi-2, a DNA helicase that also recruits the KRAB repressor; MBD2, an mCpG binding protein that would target NuRD to methylated DNA; MBD3, which is probably a scaffold; and p66, MTA1, and MTA2, whose functions are not known.

SMRT/TRAC is an alternately spliced form of SRC-1 and forms a complex that is a hybrid between Sin3 and NuRD. It recruits several HDACs, the Sin3 adaptor, and the RbAp proteins, like the Sin3 complex. It also contains MBD3 and MTA proteins, such as occurs in NuRD. It does not include Mi-2. Although NCoR acts as a bridge between NRs and the Sin3 complex, it is not a static adaptor because it is sensitive to the state of the NR. For example, because it has two NIDs that must be occupied for activity, it represses NR dimers but not monomers. In addition, NCoR binding to TR homodimers is ligand dependent, but binding to RXR-TR dimers is not. In fact, T_3 will induce the dissociation of NCoR from the latter. Both of these phenomena are related to the HRE because the occurrence of full or half HREs determines dimer or monomer binding and DR spacing determines NR partners. This fact allows NRs to be transcription activators or repressors, depending on the HREs in the promoter region.

Hormone Response Elements

Previous sections of this chapter have referred to specific DNA sequences recognized by the individual transcription factors. In this section, the methods for identifying these sequences and some of their characteristics are covered.

Methodology

Several techniques have been used to identify HREs. One of the first methods to determine a consensus sequence involved the assembly of known sequences for all the genes induced by a particular hormone; one then searched for short sequences that are common to all the 5' regions. The logic is simple: if a certain hormone activates a half dozen or so genes, then its receptor must recognize some common structure in all of them. Unfortunately, this technique is the least intellectually gratifying because the results are not coupled to any binding or functional data. Another problem is that there is no single perfect HRE; rather, many HREs can have several positions that are variable, making their identification even more difficult. For example, following examination of the gene sequences for β casein, MMTV, human GH, and proopiomelanocortin, a 24-nucleotide consensus sequence for the glucocorticoid receptor-binding site was postulated. However, the sequence was located far upstream (−500 to −300 bp), and the sequence similarity among these sites was only 58% to 83%. Better methodologies (see later) have shown that the actual binding site consists of six nucleotides (TGTYCT, where Y is any pyrimidine) and is located closer than −200 bp; furthermore, the homology between sequences is 82% to 100% for any given nucleotide. However, this technique can be useful, if it is combined with another method that restricts the potential sequences to which the receptor binds.

Histology can be used to localize steroid receptors by immunofluoresence. The first use of this technique was on the insect hormone ecdysone because it has an α-β unsaturated carbonyl within the steroid nucleus. This can be activated by light having a wavelength of 320 nm, resulting in a covalent bond being formed between the hormone and the receptor; antibodies to ecdysone were then used to localize its receptor to the chromosomal puffs. With the purification of steroid receptors, monoclonal antibodies to the receptor itself are now available for these types of studies. With these antibodies, the estradiol receptor has been found on euchromatin. In another example, the unoccupied progesterone receptor is located in the condensed chromatin that becomes euchromatin after occupancy by the steroid. Although this technique is the most visually satisfying one, its major drawback is its lack of resolution.

Kinetic analysis measures the association or dissociation constants for the binding of hormone receptors to DNA fragments; binding to DNA reconstituted with histones and to intact chromatin has also been studied. The fragments with the highest affinity presumably contain the binding sequence. This technique is good for comparing the relative binding affinities among DNA preparations, but the absolute values are still too low to account for the specificity observed *in vivo*.

Functional analyses involve altering the 5' region of a gene, reintroducing the gene into cells, and examining the effects of this modification on the hormonal control of transcription. For example, one can make successively larger depletions in the 5' region until hormone inducibility is lost. Unfortunately, a simple deletion is actually two modifications in one; not only are certain sequences missing, but the spacing between the gene and other sequences upstream is altered. Such spacing may be critical for the proper functioning of the regulatory sequences. This problem is overcome with linker scanning mutants, which contain unrelated sequences in the place of those deleted so that the spacing is preserved. Another modification is insertion mutation, in which unrelated sequences are randomly inserted into the 5' region. The loss of hormone inducibility in a particular mutant would suggest that the receptor-binding site in that mutant has been interrupted. Finally, once the putative binding site has been restricted to a sufficiently small region of DNA, site-directed mutants can be generated. This technique involves the introduction of single base pair mutations in selected positions and results in the least perturbation of the surrounding DNA.

Footprint analysis involves binding the receptor to DNA and then enzymatically digesting the nucleic acid. The receptor-binding site should be spared because the bound protein interferes with the enzymatic digestion. Alternatively, selected guanine residues are methylated and receptor binding is measured; if a particular base interacts with the receptor, its methylation will interfere with binding. This latter technique is known as *methylation interference footprinting*.

When a putative HRE has been identified, it can be cocrystallized with the DBD of the transcription factor. However, crystallization may force unnatural contacts to improve packing. In addition, amino acids outside of the DBD may contribute to DNA binding or to the DBD conformation. Conversely, flanking nucleotides may be important in secondary contacts with the transcription factor. Finally, critical nucleotides may not be readily apparent in x-ray crystallography, if they are important in DNA conformation rather than DBD binding.

The putative HRE can also be cut out, spliced onto a reporter gene, placed into cells, and gene regulation examined. This is the functional acid test for validating an HRE; if a DNA sequence is the HRE for a particular hormone, then it should impart hormonal regulation to any gene to which it is attached.

Sequences

Table 13-5 lists the HREs for several transcription factors regulated by the endocrine system. Some general characteristics are apparent; first, most HREs occur as repeats, either direct or inverted. This corresponds to the dimeric nature of many transcriptional factors. Second, many HREs have similar or identical sequences; specificity is determined, in part, by spacing and orientation. The importance of spacing for direct repeats was previously discussed (see Table 13-4). However, it is also important for inverted repeats, also called *palindromes;* for example, the Stat family binds to the palindrome $TTCCX_n\ GGAA$. When n is zero, the HRE binds Stat3; when n is two, it binds Stat 6; and when n is one, it becomes a general binding site for the Stat family. It has been hypothesized that the spacing for the TR family is determined by the overhang in the A-T region of the NR; for example, the overhang for PPAR is 1 bp; RXR, 1 to 2 bp; TR, 4 bp; RAR, 5 bp; and VDR, 5 to 6 bp. On a palindrome, the NRs are head-to-head and there is no overhang: (A/T-*DBD*)(*DBD*-A/T). Therefore these NRs would bind palindromes without any spacing. On DRs, the overhang of the 3′ partner determines the spacing because its A-T region would occupy the gap between the two subunits: 5′(A/T-*DBD*)(A/T-*DBD*)3′. For example, the spacing for DRs would be 5 for RXR-RAR dimers but 1 for RAR-RXR dimers. For inverted palindromes, the spacing would be the sum of the overhangs because the A-T region from both subunits would be interposed: (*DBD*-A/T)(A/T-*DBD*). Because DNA is a spiral, HRE spacing has topological consequences. Pit-1 has a flexible linker between POU_S and POU_H, so that it can bind HREs with variable spacing. The addition of 2 extra bp to the consensus sequences puts POU_S and POU_H on the same side of the DNA; this arrangement allows NCoR to bind and repress transcription.

Many of the HREs for NRs in Table 13-5 are the consensus sequences that exhibit perfect symmetry. In nature, such sequences rarely exist; rather, the 5′ end is often imperfect. This asymmetry imparts polarity to the HRE; and polarity, in turn, can affect specificity, activity, and hormone sensitivity. First, imperfect HREs are possible because NRs bind DNA as dimers. Initially, one member binds the 3′, more perfect half, because it has a higher affinity than the imperfect half. Because the second member is tethered to the first, its local concentration is very high, and its binding to the 5′ half is favored, even though this half of the sequence is more variable. Essentially, the high concentration of the second partner compensates for the imperfect sequence of the HRE. The use of RXR as the second subunit also improves binding to imperfect HREs. RXR appears to be a general filler; it binds to the 5′, imperfect half, and its presence in a dimer allows its partner to bind to half sites with a greater variability in spacing (see Table 13-4). This is possible because RXR makes fewer contacts with the HRE; however, as a result it is also less active. As such, an imperfect-perfect sequence is more active than the reverse, and ligand binding to the 3′ subunit is more effective in stimulating transcription than binding to the 5′ subunit. For example, RXR-RAR binds to DR5 and activates transcription, but because of the overhang rule, DR1 selects for RAR-RXR dimers and transcription is inhibited. Likewise, RXR-TR is active on DR5 but inhibitory on DR1. Polarity can also affect cofactor binding; Jun-Fos polarity determines whether NFAT can be recruited. As such, polarity can increase the response repertoire of HREs.

In addition to imparting polarity, there are several other advantages to having imperfect HREs. First, they may alter the responsiveness of genes to transcription factors. For example, the CREB binds imperfect CREs with a lower affinity than perfect sequences; however, they give a greater induction because of their lower baseline. Second, imperfect HREs may also prevent strong inhibition by those nuclear receptors that repress transcription in the absence of their ligand. The perfect HRE may simply have too high an affinity to release the NR. Third, imperfect HREs can increase specificity. For example, the GR, AR, and PR all bind with maximal affinity to the same perfect HRE, but HREs with slight variations in the sequence, spacing, and/or flanking sequences show preferential binding. Fourth, imperfect HREs may represent composite HREs that may broaden specificity, or fine-tune the magnitude of the response. For example, the GRE and ERE are highly specific for GR and ER, respectively, but a hybrid ERE-GRE in the uteroglobin gene can bind and be activated by ER, GR, or PR. In addition, some composites are not as active and can lead to more moderate responses. Fifth, imperfect HREs can encourage synergism; a half GRE is insufficient to active transcription by itself and requires Stat. Finally, imperfect HREs may differentially affect the conformation of a transcription factor. RXR-VDR, NF-κB, and ER have all been shown to have different conformations on different HREs. In the case of NF-κB, different amino acid contacts with the HREs are responsible for these various structures. The different ER conformations lead to altered coactivator recruitment.

Table 13-5 presents a single HRE, but many genes have multiple copies; these copies frequently synergize with each other. For example, the DNA affinity for both GR and PR increases by as much as 100-fold when a second HRE is introduced. This synergism is more marked for imperfect HREs; the affinity of ER for an imperfect ERE increases 4- to 8-fold by the presence of another ERE, but there is no effect if both EREs are perfect. This synergism extends to transcription, as well; multiple HREs enhance the transcriptional effects of the factors that bind them. This transcriptional synergism is often greater in magnitude than the increase observed in DNA binding. It is believed that a single NR or transcription factor is not as effective as multiple copies in recruiting coactivators.

Summary

The structures of many transcription factors are known and they can be classified according to the structural motif they use to bind DNA. Many of them use an α-helix to make specific contacts with bases in the major groove. The HTH, wHTH, homeodomains, POU, HMG, and zinc twist factors use a short recognition helix overlaid with from one to three additional helices to secure the first in place. The MADS factors also use an α-helix to bind DNA, but in this group the helix is covered by a β-sheet. The coiled coil factors use a long α-helix, whose carboxy terminus forms a coiled coil and whose amino terminus straddles the major groove. The bZIP, bHLH, helix-span-helix, and bHLH-ZIP factors are all variations on this motif. β-Structures can also be used to bind DNA; the largest group uses an Ig-like β-barrel and includes NF-κB, NFAT, Stat, and Runt. Most factors form dimers and this is reflected in the DNA sequences that they recognize; most HREs are either direct or inverted repeats.

There is considerable overlap in the activities of these factors. Enhanced specificity is achieved by the selective distribution of hormone receptors, metabolizing enzymes, or second messengers; by the presence of ancillary factors; or by the position, polarity, sequence variation, and spacing of the HREs. Despite all of these mechanisms, some overlap still exists, and the final activity of any given hormone may simply reside in the spectrum of activities that it evokes.

Hormonal regulation of these factors involves (1) induction of the factors; (2) alternate splicing; (3) posttranslational modifications, including phosphorylation, glycosylation, acetylation, methylation, ubiquination, and cleavage; (4) protein-protein interactions, including variable dimerization partners, inhibitors, cofactors, and heat shock proteins; (5) compartmentalization; and (6) allosterism by ligands and second messengers. Once activated, these factors stimulate transcription by facilitating the assembly of the preinitiation complex in the promoter region. This may occur either directly by protein-protein binding or through changes in the DNA conformation, including (1) DNA unwinding or relaxation, (2) DNA bending, (3) DNA looping, or (4) nucleosomal sliding or displacement. Inhibition of transcription can be accomplished by (1) steric occlusion of stimulatory HREs, TATA boxes, or other critical sites; (2) competition for limiting coactivators, dimerization partners, other synergistic transcription factors, or general transcription factors; (3) protein-protein antagonism, where transcription factors bind and inhibit one another; and (4) silencing, where repressors directly reverse stimulatory modifications, alter chromatin conformation, or displace coactivators.

The DNA sequences recognized by these factors usually occur as direct or inverted repeats, reflecting the dimeric nature of the factors themselves. However, these HREs are rarely perfect; these variations may affect HRE selectivity, the responsiveness of the gene, reversibility, synergism, and the conformation of the factors. HREs exist in multiple copies, which allows the bound transcription factors to synergize with one another with respect to both DNA binding and transcription.

References

Zinc-Containing Family

Bouwman, P., and Philipsen, S. (2002). Regulation of the activity of Sp1-related transcription factor. *Mol. Cell. Endocrinol.* 195, 27-38.
Jenkins, B.D., Pullen, C.B., and Darimont, B.D. (2001). Novel glucocorticoid receptor coactivator effector mechanisms. *Trends Endocrinol. Metab.* 12, 122-126.
Rachez, C., and Freedman, L.P. (2000). Mechanisms of gene regulation by vitamin D_3 receptor: A network of coactivator interactions. *Gene* 246, 9-21.
Zhang, J., and Lazar, M.A. (2000). The mechanism of action of thyroid hormones. *Annu. Rev. Physiol.* 62, 439-466.

Leucine Zipper

De Cesare, D., and Sassone-Corsi, P. (2000). Transcriptional regulation by cyclic AMP-responsive factors. *Prog. Nucleic Acid Res. Mol. Biol.* 64, 343-369.
Lonze, B.E., and Ginty, D.D. (2002). Function and regulation of CREB family transcription factors in the nervous system. *Neuron* 35, 605-623.
Mayr, B., and Montminy, M. (2001). Transcriptional regulation by the phosphorylation-dependent factor CREB. *Nature Rev. Mol. Cell Biol.* 2, 599-609.
Ramji, D.P., and Foka, P. (2002). CCAAT/enhancer-binding proteins: Structure, function and regulation. *Biochem. J.* 365, 561-575.
Servillo, G., Fazia, M.A.D., and Sassone-Corsi, P. (2002). Coupling cAMP signaling to transcription in the liver: Pivotal role of CREB and CREM. *Exp. Cell Res.* 275, 143-154.

Shaywitz, A.J., and Greenberg, M.E. (1999). CREB: A stimulus-induced transcription factor activated by a diverse array of extracellular signals. *Annu. Rev. Biochem.* 68, 821-861.

Vo, N., and Goodman, R.H. (2001). CREB-binding protein and p300 in transcriptional regulation. *J. Biol. Chem.* 276, 13505-13508.

Helix-Loop-Helix and Variations

Andersen, B., and Rosenfeld, M.G. (2001). POU domain factors in the neuroendocrine system: Lessons from developmental biology provide insights into human disease. *Endocr. Rev.* 22, 2-35.

Baudino, T.A., and Cleveland, J.L. (2001). The Max network gone Mad. *Mol. Cell. Biol.* 21, 691-702.

Bilton, R.L., and Booker, G.W. (2003). The subtle side to hypoxia inducible factor (HIFα) regulation. *Eur. J. Biochem.* 270, 791-798.

Burgering, B.M.T., and Kops, G.J.P.L. (2002). Cell cycle and death control: Long live Forkheads. *Trends Biochem. Sci.* 27, 352-360.

Carlsson, P., and Mahlapuu, M. (2002). Forkhead transcription factors: Key players in development and metabolism. *Dev. Biol.* 250, 1-23.

Dang, C.V., Resar, L.M.S., Emison, E., Kim, S., Li, Q., Prescott, J.E., Wonsey, D., and Zeller, K. (1999). Function of the c-Myc oncogenic transcription factor. *Exp. Cell Res.* 253, 63-77.

Edwards, P.A., Tabor, D., Kast, H.R., and Venkateswaran, A. (2000). Regulation of gene expression by SREBP and SCAP. *Biochim. Biophys. Acta* 1529, 103-113.

Kikuchi, A. (2000). Regulation of β-catenin signaling in the Wnt pathway. *Biochem. Biophys. Res. Commun.* 268, 243-248.

Lando, D., Gorman, J.J., Whitelaw, M.L., and Peet, D.J. (2003). Oxygen-dependent regulation of hypoxia-inducible factors by prolyl and asparaginyl hydroxylation. *Eur. J. Biochem.* 270, 781-790.

Li, R., Pei, H., and Watson, D.K. (2000). Regulation of Ets function by protein-protein interactions. *Oncogene* 19, 6514-6523.

Lüscher, B. (2001). Function and regulation of the transcription factors of the Myc/Max/Mad network. *Gene* 277, 1-14.

Massari, M.E., and Murre, C. (2000). Helix-loop-helix proteins: Regulators of transcription in eucaryotic organisms. *Mol. Cell. Biol.* 20, 429-440.

Minet, E., Michel, G., Mottet, D., Raes, M., and Michiels, C. (2001). Transduction pathways involved in hypoxia-inducible factor-1 phosphorylation and activation. *Free Radical Biol. Med.* 31, 847-855.

Novak, A., and Dedhar, S. (1999). Signaling through β-catenin and Lef/Tcf. *Cell. Mol. Life Sci.* 56, 523-537.

Safran, M., and Kaelin, W.G. (2003). HIF hydroxylation and the mammalian oxygen-sensing pathway. *J. Clin. Invest.* 111, 779-783.

Seidensticker, M.J., and Behrens, J. (2000). Biochemical interactions in the wnt pathway. *Biochim. Biophys. Acta* 1495, 168-182.

Semenza, G.L. (2001). HIF-1, O_2, and the 3 PHDs: How animal cells signal hypoxia to the nucleus. *Cell (Cambridge, Mass.)* 107, 1-3.

Sharrocks, A.D. (2001). The ETS-domain transcription factor family. *Nature Rev. Mol. Cell Biol.* 2, 827-837.

Shimano, H. (2001). Sterol regulatory element-binding proteins (SREBPs): Transcriptional regulators of lipid synthetic genes. *Prog. Lipid Res.* 40, 439-452.

Yordy, J.S., and Muise-Helmericks, R.C. (2000). Signal transduction and the Ets family of transcription factors. *Oncogene* 19, 6503-6513.

Zhou, Z.Q., and Hurlin, P.J. (2001). The interplay between Mad and Myc in proliferation and differentiation. *Trends Cell Biol.* 11, S10-S14.

Immunoglobulin-like Family

Baron, M. (2003). An overview of the Notch signalling pathway. *Semin. Cell Dev. Biol.* 14, 113-119.

Chatterjee-Kishore, M., van den Akker, F., and Stark, G.R. (2000). Association of STATs with relatives and friends. *Trends Cell Biol.* 10, 106-111.

Decker, T., and Kovarik, P. (2000). Serine phosphorylation of STATs. *Oncogene* 19, 2628-2637.

Delfino, F., and Walker, W.H. (1999). Hormonal regulation of the NF-κB signaling pathway. *Mol. Cell. Endocrinol.* 157, 1-9.

Horvath, C.M. (2000). STAT proteins and transcriptional responses to extracellular signals. *Trends Biochem. Sci.* 25, 496-502.

Imada, K., and Leonard, W.J. (2000). The Jak-STAT pathway. *Mol. Immunol.* 37, 1-11.

Kadesch, T. (2000). Notch signaling: A dance of proteins changing partners. *Exp. Cell Res.* 260, 1-8.

Karin, M. (1999). How NF-κB is activated: The role of the IκB kinase (IKK) complex. *Oncogene* 18, 6867-6874.

Levy, D.E., and Darnell, J.E. (2002). STATs: Transcription control and biological impact. *Nature Rev. Mol. Cell Biol.* 3, 651-662.

Mumm, J.S., and Kopan, R. (2000). Notch signaling: From outside in. *Dev. Biol.* 228, 151-165.

Peters, R.T., and Maniatis, T. (2001). A new family of IKK-related kinases may function as IκB kinase kinases. *Biochim. Biophys. Acta* 1471, M57-M62.

Schmitz, M.L., Bacher, S., and Kracht, M. (2001). IκB-independent control of NF-κB activity by modulatory phosphorylations. *Trends Biochem. Sci.* 26, 186-190.

Zhu, J., and McKeon, F. (2000). Nucleocytoplasmic shuttling and the control of NF-AT signaling. *Cell. Mol. Life Sci.* 57, 411-420.

MADS-Box Family

McKinsey, T.A., Zhang, C.L., and Olson, E.N. (2002). MEF2: A calcium-dependent regulator of cell division, differentiation and death. *Trends Biochem. Sci.* 27, 40-47.

β-Family

Lutz, M., and Knaus, P. (2002). Integration of the TGF-β pathway into the cellular signalling network. *Cell. Signal.* 14, 977-988.

Massagué, J., and Wotton, D. (2000). Transcriptional control by the TGF-β/Smad signaling. *EMBO J.* 19, 1745-1754.

Moustakas, A., Souchelnytskyi, S., and Heldin, C.H. (2001). Smad regulation in TGF-β signal transduction. *J. Cell Sci.* 114, 4359-4369.

Zimmerman, C.M., and Padgett, R.W. (2000). Transforming growth factor β signaling mediators and modulators. *Gene* 249, 17-30.

General Regulation

Barolo, S., and Posakony, J.W. (2002). Three habits of highly effective signaling pathways: Principles of transcriptional control by developmental cell signaling. *Genes Dev.* 16, 1167-1181.

Bogdan, C. (2001). Nitric oxide and the regulation of gene expression. *Trends Cell Biol.* 11, 66-75.

Brivanlou, A.H., and Darnell, J.E. (2002). Signal transduction and the control of gene expression. *Science* 295, 813-818.

Cartwright, P., and Helin, K. (2000). Nucleocytoplasmic shuttling of transcription factors. *Cell. Mol. Life Sci.* 57, 1193-1206.

Conaway, R.C., Brower, C.S., and Conaway, J.W. (2002). Emerging roles of ubiquitin in transcription regulation. *Science* 296, 1254-1258.

Crabtree, G.R. (2001). Calcium, calcineurin, and the control of transcription. *J. Biol. Chem.* 276, 2313-2316.

Cyert, M.S. (2001). Regulation of nuclear localization during signaling. *J. Biol. Chem.* 276, 20805-20808.

Hanover, J.A. (2001). Glycan-dependent signaling: O-linked N-acetylglucosamine. *FASEB J.* 15, 1865-1876.

Holmberg, C.I., Tran, S.E.F., Eriksson, J.E., and Sistonen, L. (2002). Multisite phosphorylation provides sophisticated regulation of transcription factors. *Trends Biochem. Sci.* 27, 619-627.

Verger, A., Perdomo, J., and Crossley, M. (2003). Modification with SUMO: A role in transcriptional regulation. *EMBO Rep.* 4, 137-142.

Wells, L., Whalen, S.A., and Hart, G.W. (2003). O-GlcNAc: A regulatory post-translational modification. *Biochem. Biophys. Res. Commun.* 302, 435-441.

Whitmarsh, A.J., and Davis, R.J. (2000). Regulation of transcription factor function by phosphorylation. *Cell. Mol. Life Sci.* 57, 1172-1183.

Coactivators and Corepressors

Ahringer, J. (2000). NuRD and SIN3: histone deacetylase complexes in development. *Trends Genet.* 16, 351-356.

Bevan, C., and Parker, M. (1999). The role of coactivators in steroid hormone action. *Exp. Cell Res.* 253, 349-356.

Burke, L.J., and Baniahmad, A. (2000). Co-repressors 2000. *FASEB J.* 14, 1876-1888.

Hermanson, O., Glass, C.K., and Rosenfeld, M.G. (2002). Nuclear receptor coregulators: Multiple modes of modification. *Trends Endocrinol. Metab.* 13, 55-60.

Hu, X., and Lazar, M.A. (2000). Transcriptional repression by nuclear hormone receptors. *Trends Endocrinol. Metab.* 11, 6-10.

Ito, M., and Roeder, R.G. (2001). The TRAP/SMCC/Mediator complex and thyroid hormone receptor function. *Trends Endocrinol. Metab.* 12, 127-134.

Jepsen, K., and Rosenfeld, M.G. (2002). Biological roles and mechanistic actions of co-repressor complexes. *J. Cell Sci.* 115, 689-698.

Knoepfler, P.S., and Eisenman, R.N. (1999). Sin meets NuRD and other tails of repression. *Cell (Cambridge, Mass.)* 99, 447-450.

Latchman, D.S. (2001). Transcription factors: Bound to activate or repress. *Trends Biochem. Sci.* 26, 211-213.

Lee, K.C., and Kraus, W.L. (2001). Nuclear receptors, coactivators and chromatin: New approaches, new insights. *Trends Endocrinol. Metab.* 12, 191-197.

Leo, C., and Chen, J.D. (2000). The SRC family of nuclear receptor coactivators. *Gene* 245, 1-11.

McManus, K.J., and Hendzel, M.J. (2001). CBP, a transcriptional coactivator and acetyltransferase. *Biochem. Cell Biol.* 79, 253-266.

Näär, A.M., Lemon, B.D., and Tjian, R. (2001). Transcriptional coactivator complexes. *Annu. Rev. Biochem.* 70, 475-501.

Robyr, D., Wolffe, A.P., and Wahli, W. (2000). Nuclear hormone receptor coregulators in action: Diversity for shared tasks. *Mol. Endocrinol.* 14, 329-347.

Chromatin Remodeling

Becker, P.B., and Hörz, W. (2002). ATP-dependent nucleosome remodeling. *Annu. Rev. Biochem.* 71, 247-273.

Dilworth, F.J., and Chambon, P. (2001). Nuclear receptors coordinate the activities of chromatin remodeling complexes and coactivators to facilitate initiation of transcription. *Oncogene* 20, 3047-3054.

Fry, C.J., and Peterson, C.L. (2001). Chromatin remodeling enzymes: Who's on first? *Curr. Biol.* 11, R185-R197.

Havas, K., Whitehouse, I., and Owen-Hughes, T. (2001). ATP-dependent chromatin remodeling activities. *Cell. Mol. Life Sci.* 58, 673-682.

Muchardt, C., and Yaniv, M. (1999). ATP-dependent chromatin remodeling: SWI/SNF and Co. are on the job. *J. Mol. Biol.* 293, 187-198.

Sudarsanam, P., and Winston, F. (2000). The Swi/Snf family: Nucleosome-remodeling complexes and transcriptional control. *Trends Genet.* 16, 345-351.

Urnov, F.D., and Wolffe, A.P. (2001). Chromatin remodeling and transcriptional activation: The cast (in order of appearance). *Oncogene* 20, 2991-3006.

Vignali, M., Hassan, A.H., Neely, K.E., and Workman, J.L. (2000). ATP-dependent chromatin-remodeling complexes. *Mol. Cell. Biol.* 20, 1899-1910.

Zlatanova, J., Caiafa, P., and van Holde, K. (2000). Linker histone binding and displacement: Versatile mechanism for transcriptional regulation. *FASEB J.* 14, 1697-1704.

Integration

Cosma, M.P. (2002). Ordered recruitment: Gene-specific mechanism of transcription activation. *Mol. Cell* 10, 227-236.

McKenna, N.J., and O'Malley, B.W. (2002). Combinatorial control of gene expression by nuclear receptors and coregulators. *Cell (Cambridge, Mass.)* 108, 465-474.

Hormone Response Elements

Garvie, C.W., and Wolberger, C. (2001). Recognition of specific DNA sequences. *Mol. Cell* 8, 937-946.

Verrijdt, G., Haelens, A., and Claeseens, F. (2003). Selective DNA recognition by the androgen receptor as a mechanism for hormone-specific regulation of gene expression. *Mol. Genet. Metab.* 78, 175-185.

CHAPTER **14**

Modifications and Conformations of DNA and Nuclear Proteins

CHAPTER OUTLINE

Transcription involves more than simply activating a transcription factor (see Chapter 13); the DNA that it recognizes may not always be readily accessible. Because of its size and charge, DNA is extensively packaged by chromatin proteins and this complex can impede transcription. Furthermore, both these proteins and the DNA itself can be covalently modified; these alterations can change chromatin structure and affect transcription. The nature of the packaging and these modifications are now examined.

Nucleosomes

Components

DNA presents the cell with two packaging problems: (1) a very negatively charged molecule and (2) a very long polymer. The histones are designed to solve these problems; they are small, very basic proteins that are involved with DNA packaging. The histones are classified according to the abundance of their basic amino acids; histones H3 and H4 are the arginine-rich histones because they have slightly more arginine than lysine (Table 14-1). They are also among the most highly conserved proteins in nature; histone H4 sequences from cows and peas reveal only two amino acid differences, and both of these substitutions are conservative. A conservative substitution occurs when one amino acid is replaced by a structurally similar one so that the overall function of the protein remains unperturbed. Histones H2A and H2B are the moderately lysine-rich histones and are moderately conserved in evolution. All four of these proteins have the same charge distribution: a strongly basic amino terminus comprising one-third to one-half of the molecule, a basic carboxy terminus, and a neutral central region.

These four proteins are called the *core histones* because they form the nucleus around which the DNA winds. This nucleus, or nucleosome, has three parts: an $(H3)_2(H4)_2$ tetramer and two $(H2A)(H2B)$ dimers. The tetramer forms a shallow spiraling circle that resembles a locking washer; the DNA will run along the back of this tetramer, entering and exiting by the H3 amino-terminal tails. Its direct interaction with DNA explains why H3 and H4 are the most conserved of the histones. Each $(H2A)(H2B)$ dimer forms a wedge above and below the spiral to give the entire nucleosome a flatter, disklike shape. The disk is 55 Å by 110 Å and is encircled by 140 bp of DNA or about $1\frac{2}{3}$ turns (Fig. 14-1, *A*). Although they may interfere with transcription initiation, nucleosomes do not impede elongation. The $(H2A)(H2B)$ dimers may temporarily leave to open up the nucleosome and reduce any potential physical hindrance to the transcriptional machinery. The basic amino terminus of H4 binds the acidic $(H2A)(H2B)$ dimers to the $(H3)_2(H4)_2$ tetramer. This interaction is weak, and reduction of the basic charge by acetylation of the H4 tails further weakens the association and favors dissociation. The resulting hypothetical structure consisting of only the $(H3)_2(H4)_2$ tetramer is called a *lexosome*. The existence of the lexosome is supported by histone exchange rates within the nucleosome; 80% of H3 and H4 appear to be permanent residents of the nucleosome, whereas 40% of H2B has an exchange half-time of about 130 minutes. A smaller fraction of H2B has an even faster exchange rate.

Successive nucleosomes are connected by 20 to 100 base pairs (bp) of DNA, called *linker DNA*. The last histone, histone H1, is a lysine-rich histone that binds to the linker DNA and is essential to the further condensation of chromatin. H1 is known to inhibit the initiation, but not elongation, of

Table 14-1

Characteristics and Modifications of Histones and High-Mobility Group Proteins

	Molecular weight (kDa)	Lys/Arg	Nuclear protein modification[a]					
			Acetylation[b]	Phosphorylation	Poly-(ADP-ribosyl)ation	Methylation	Monoubiquitination	Glycosylation
Substrates								
Histone H1	21	63/3	0	++	++	0	+	0
Histone H2A	14	14/12	+	+	+	+	++	0
Histone H2B	13.8	20/8	+	0	0	+	++	0
Histone H3	15.3	13/18	++	+	0	++	++	0
Histone H4	11.3	11/14	++	+	0	++	+	0
HMGB (HMG 1-2)	31.7	50/10	+	0	++	+		+[c]
HMGN (HMG 14-17)	11.2	21/5	+	++/+[d]	+			+
HMGA (HMGI/Y/C)	12	16/10	+	+	+	+		
Control								
Turnover			Rapid	Rapid	Rapid	Slow (Lys); faster (Arg)	Unknown	
RNA/protein synthesis required			No	No	No	No	No	
Hormonally regulated			++	++	++	++	+	

[a] For intact normal cells or tissues. ++, Modification is common; +, modification occurs; 0, modification has not been detected.
[b] N^ε-Lysine only.
[c] Disputed for HMG 1.
[d] ++ for HMG 14 and + for HMG 17.

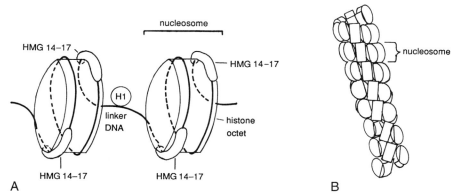

Fig. 14-1. Nucleosomes. (A) The histone octet with the intervening linker DNA and histone H1. (B) Aggregation of nucleosomes into a solenoid structure.

transcription; and this repression is due, in part, to the chromatin condensation induced by H1. In addition, H1 may displace some transcription factors from DNA. Originally, it was thought that H1 was itself displaced from linker DNA during transcription; it is now believed that it merely loosens its grip. H1 contains a central globular domain of four α helices in a barrel. In inactive chromatin, H1 is closely associated with DNA, but in transcriptionally active regions, only the amino- and carboxy-terminal tails touch the DNA. This reduced interaction apparently favors chromatin decondensation and does not interfere with transcription factors.

Another way to remove the inhibitory effects of H1 is to replace it with an H1 variant. Histone variants, called *isohistones*, have been identified for all histones except histone H4, the most evolutionarily conserved histone. Histone H1° binds to the linker DNA more tightly than histone H1 and, in fact, displaces it in terminally differentiated, but not in actively dividing, tissue. Histone H1° is also less effective in chromatin condensation than histone H1. Finally, histone H1° is hormonally regulated; this variant is lost in target organs following hormone deprivation and returns with hormone replacement. This effect is specific because histone H1 does not change in these same tissues and histone H1° in nontarget tissues is unaffected by the hormone status of the animal. Therefore some hormones may act, in part, by replacing histone H1 with histone H1° so that the chromatin will relax and the DNA can be transcribed.

Other histones also have variants; one is a protein conjugate between histone H2A and HMG 20. Because HMG 20 is identical with ubiquitin, the complex is presently called *uH2A*. As much as 10% of histone H2A and 1.5% of histone H2B are conjugated to ubiquitin; amazingly, the addition of this 76-amino acid peptide does not appear to alter the nucleosomal structure, although it may impede refolding once the nucleosome is opened.

Other isohistones include H2A.Z, H2AX, macroH2A (mH2A), and spH2B. H2A.Z inhibits chromatin condensation, reducing the need for SWI/SNF (switch and sucrose nonfermentors genes products) in transcription activation. It is associated with active promoters, recruits RNA polymerase II and TBP (TATA box binding protein), and destabilizes the nucleosomal core to allow the RNA polymerase II to pass through. H2AX also decondenses chromatin, but it is associated with DNA repair, not transcription. This more open chromatin structure allows greater access by DNA repair enzymes, which H2AX recruits. However, during meiosis H2AX is actually required for the formation of the sex body, which arises from the condensation of the X and Y chromosomes. mH2A is also associated with the condensed, inactive X chromosome, although it is not solely responsible for this process. Finally, spH2B forms part of the telomere-binding complex. The telomere is the structure at the ends of chromosomes.

Other nuclear proteins are also important in chromatin structure. One such class was initially extracted into 0.35 M sodium chloride (NaCl) and remained soluble in 2% trichloroacetic acid. The supernatent contained proteins with a high mobility in polyacrylamide gels; therefore they were named the high-mobility group (HMG) proteins. There are three classes: HMGN, HMGB, and HMGA. HMGN includes HMG 14 and HMG 17. HMGN proteins have an uneven charge distribution; the amino terminus is very basic, whereas the carboxy terminus is very negative. They bind as a homodimer to the nucleosome where they secure the DNA where it enters and exits the nucleosome; the basic amino terminus binds the DNA and the negative carboxy-terminus binds the histones. As a result, they stabilize the nucleosome, whether in the condensed or unraveled state. Indeed, in the presence of HMGN the amount of DNA wrapped around the nucleosome increases from $1\frac{2}{3}$ to $1\frac{3}{4}$ turns. They are also partially responsible for inducing regular spacing in nucleosomes, and they are thought to interact with the nuclear matrix through their sugar residues; unlike histones, HMG proteins are glycosylated (Table 14-1). Finally, they enhance transcription in several ways; first, they facilitate the DNA binding of some transcription factors. Second, they block the inhibitory effect of H1 on transcription and chromatin condensation by competing with H1 binding.

HMGB includes HMG 1 and HMG 2; they are larger than HMGN proteins and have a very different structure. HMGB proteins have a central hydrophobic domain containing duplicated DNA-binding motifs known as *HMG boxes* (see Chapter 13). The carboxy terminus is extremely acidic; in fact, the last 30 residues are either glutamic or aspartic acid. The physiological function for this carboxy terminus is unknown, but *in vitro* it is capable of unwinding the DNA helix. It is also very efficient in bending DNA. There are several facts implicating these proteins in gene activation:

1. They partially unwind the DNA helix; they can also bend it.

2. They enhance the ability of ISWI (imitation of switch) to slide nucleosomes during remodeling.

3. They bind specific DNA sequences, especially at cruciform structures, where they remove the transcriptional block that these structures create.

4. They form an integral part of some transcription complexes, such as NF-κB (nuclear factor that stimulates the κ chain of B cells) and ATF2 (activating transcription factor 2), and they increase the DNA affinity of SRs (serine-arginine-rich protein; a splicing factor).

5. They stabilize the initiation complex.

HMGA includes HMGI, HMGY, and HMGC. They bind the minor groove of adenine-thymine (AT)–rich DNA through an AT hook: (P)-R-G-R-P flanked by basic residues. The proline creates the bend or hook, the arginines bind thymine, and the flanking amino acids bind the deoxyribose-phosphate backbone. Like HMGB, these proteins can stimulate transcription.

1. They can connect distant AT-rich regions to create loops that facilitate transcription factor interactions.

2. They displace H1.

3. They directly bind several transcription factors and increase their DNA affinity.

Organization

One of the more controversial topics involving nucleosomes is whether they occupy specific, fixed positions along the DNA. There are certain mechanisms that would allow this positioning to occur. First of all, certain DNA sequences favor bending in one plane and binding to nucleosomes; for example, AT-rich sequences are preferred where the minor groove is compressed. Second, proteins that bind to specific DNA sequences can form boundaries, after which nucleosomes would line up in phase. In general, the shorter the linker DNA, the more persistent the phasing. If these bound proteins occurred periodically, they could shepherd the nucleosomes over a considerable distance. Z-DNA can also act as a boundary because nucleosomes cannot bind Z-DNA. Third, chromatin folding can influence nucleosomal spacing. Regardless of the mechanism, phasing is never rigid; rather nucleosomal positions are the result of statistical probabilities.

The existence of nucleosomal positioning has important regulatory implications. For example, in several genes they lie over the promoter and must be removed before transcription can begin (see Chapter 13). Nucleosomes do not impede elongation, but they do repress initiation. Nucleosomes may also favor the binding of certain proteins; for example, the progesterone receptor (PR) prefers to bind at the edge of nucleosomes.

The nucleosomes themselves can condense into a higher ordered structure, a solenoid (Fig. 14-1, *B*). This superstructure is 200 to 300 Å in diameter and has 6 to 8 nucleosomes per turn. About 50 kilobases (kb) of the solenoid, or about 35 turns, twist into a loop; these loops can then form an even larger solenoid, called a *coil*. Six loops form a single turn of the coil; each turn is called a *rosette* and 30 rosettes make up the entire coil. A chromatid contains about 10 coils. Originally, the solenoid was thought to represent the condensed, inactive form, whereas the active structure was completely unraveled, producing a beads-on-a-string appearance. However, some authorities suggest that the solenoid never totally unwinds but that it only needs to loosen slightly to support transcription.

Histone and High Motility Group Protein Modifications

Histones are susceptible to innumerable modifications, only five of which are considered in this chapter: acetylation, phosphorylation, poly(ADP-ribosyl)ation, methylation, and monoubiquitination. Which characteristics might indicate that a particular modification is likely to be involved in gene activation? First of all, because gene induction is usually rapid, the modification should have a fast turnover so as to allow for the rapid modulation of the modification. Second, if the modification is to be a primary event in hormone action, it should be independent of RNA and protein synthesis. Finally, there should be evidence for hormonal regulation. All of these modifications satisfy most, if not all, of these requirements (Table 14-1).

When examining the various posttranslational modifications that histones can undergo, it is important to identify the substrate positively; this may not be easy because many modifications alter either the size or the charge of the substrate. In addition, proteins other than histones may be modified. Second, it is important to identify the particular site on the substrate because different sites may be functionally distinct; this is especially true for acetylation, phosphorylation, and methylation. Third, it is important to determine the degree of modification; a single lysine may be monoacetylated, diacetylated, or triacetylated. The same phenomenon occurs with methylation. Fourth, no single modification appears to act on its own; rather, any given modification only has meaning in the context of which other modifications may be present. This is known as the *histone code*. For example, H3 phosphorylated on S-10 is associated with condensed, inactive chromatin called *heterochromatin*, but when associated with the acetylation of K-9 and K-14, phosphorylated S-10 is found in open, active chromatin called *euchromatin*. Fifth, pharmacological agents are frequently used to perturb these modifications, but all drugs have undesirable side effects that are not always controlled for in these investigations.

The final problem relates to the physiological interpretation of the data. First, an increase in the incorporation of a labeled precursor, such as radioactive phosphate or acetate, may only represent an increased turnover of this modification. In other words, a faster turnover would allow a greater percentage of the modified units to become labeled, even though the total level of the modification remains unchanged. Second, hormone-altered modifications may, in fact, only represent hormone-altered label uptake, equilibrium, or substrate turnover. Third, there are always problems with artifacts, especially during processing; homogenization may reveal previously unexposed sites that would normally never be modified *in vivo*. Finally, in studies with intact animals the question of direct versus indirect action always arises. Insulin has been reported to induce histone H1 phosphorylation *in vivo*, but in intact animals, insulin induces hypoglycemia, which triggers the release of glucagon. Glucagon also stimulates histone H1 phosphorylation. Is insulin acting directly on histone H1 or indirectly through elevated glucagon levels? These caveats are discussed in greater detail later, as appropriate.

Acetylation

Acetylation refers to the covalent addition of an acetate to a free amino group: CH_3CONHR. The free amino group may be either at the amino terminus (N^α) or on the side chain of lysine (N^ε). Acetylation is performed by several acetyltransferases, which exhibit the following histone substrate specificity: $H3 \approx H4 \gg H2A \approx H2B \gg H1$. The N^α-acetyltransferases belong to one of three major families and attach acetyl groups to proteins during synthesis (e.g., histone H1); such groups are very stable, do not appear to be associated with gene activation, and are not considered further. N^ε-acetylation, however, has a rapid turnover and can be performed by two histone acetyltransferases (HATs), A and B. Like the N^α-acetyltransferases, N^ε-acetyltransferase B occurs in the cytoplasm and acetylates nascent proteins; it prefers lysines adjacent to neutral amino acids, such as those in histone H4. In contrast, acetyltransferase A is tightly bound to chromatin, prefers lysines adjacent to other basic amino acids, and will modify either free or nucleosomal histones. Both enzymes require adenosine 5'-triphosphate (ATP), magnesium, and acetyl coenzyme A, which is the source of the acetate.

The class A HATs are very diverse and include coactivators and specific and general transcription factors (Table 14-2). Indeed, the catalytic mechanisms for the

Table 14-2
A Summary of Major Histone Acetyltransferases

Family	Example	Substrate	Function
CBP/p300	CBP	Core	Global transcription
GNAT	PCAF	H3, H4	Transcription
	ELP3	Core[a]	Transcription elongation
SRC	SRC-1	H3, H4	NR-dependent transcription
Transcription factors[b]	TAF1	H3, H4	Part of TFIID
	TFIIIC	H2A, H3, H4	Essential for RNA polymerase II transcription
MYST	ESA1	H2A, H4	Part of NuA4; binds acidic domains through Tra1p subunit; creates acetylation gradient near telomeres
ATF2	ATF2	H2B, H4	Transcription
CDY	CDY	H4	Histone displacement by protamines during spermatogenesis

[a] Modifies only H3 and H4 when part of Elongator complex.
[b] No homology; grouped for convenience.

GNAT (GCN5-related *N*-acetyltransferase) and MYST (MOZ/Ybf-Sas3/Sas2/Tip60) families are totally different, suggesting that some of these families had distinct evolutionary origins. The major regulatory mechanisms for HATs include phosphorylation, allosterism, and localization. MAPK (mitogen-activated protein kinase) phosphorylation of CBP (CREB binding protein), JNK (Jun amino(*N*)-terminal kinase) phosphorylation of ATF2, and Cdk2 phosphorylation of Tip60 (MYST family) increase HAT activity, whereas PKCδ phosphorylation of p300 and DNA-PK phosphorylation of GCN5 (GNAT family) inhibit HAT activity. Some transcription factors, like C/EBP, bind and allosterically activate CBP, whereas others, like HOX, inhibit CBP. Transcription factors can also recruit HATs; this localization usually involves phosphorylation of the transcription factor. For example, the PKA phosphorylation of CREB (cAMP response element binding protein) and the PKA or CaMKIV (CaM-dependent protein kinase IV) phosphorylation of the p65 subunit of NF-κB attracts CBP. This recruitment appears to be a prerequisite for p300, and possibly other HAT, phosphorylation. HATs can also be inhibited by having their target sites masked; for example, histone H1 can sterically hinder HATs. Finally, acetylation is only one of several modifications that are intertwined in a complex regulatory web known as the *histone code* (see later). For example, several lysines can either be acetylated or methylated; such lysines must be demethylated before acetylation can occur. In addition, prior phosphorylation of nearby amino acids can increase or decrease acetylation at certain sites.

These enzymes are complemented by histone deacetylases (HDACs), which remove acetyl groups. There are three HDAC groups. Class I includes HDAC1-3, –8, and –11, is almost exclusively located in the nucleus, and is the group most frequently involved with transcription. It can deacetylate all sites except K-16 on H4. They are part of the Sin3 and NuRD (nuclear remodeling and histone deacetylation) complexes and bind SMRT (silencing mediator of RAR and TR) and NCoR (nuclear receptor corepressor) through the Sin3 adaptor (see Chapter 13). In addition to deacetylation, HDAC1 can inhibit transcription by recruiting PP1 to dephosphorylate some transcription factors. Class II is subdivided into IIa (HDAC4, -5, -7, and –9) and IIb (HDAC6 and –10). They have a high molecular weight of 600 to 2000 kDa and can bind SMRT and NCoR directly. Some Class II

HATs have an amino-terminal repressor domain that functions independently of the deacetylase activity. The Class IIa HDACs shuttle between the cytosol and nucleus, and enzymatic activity appears to require interaction with SMRT, NCoR, and/or other proteins. Class IIb HDACs have duplicated catalytic domains and are predominantly located in the cytoplasm where they have several functions; for example, HDAC6 acts as a microtubule deacetylase. Class III includes SIR2 and has a unique catalytic mechanism that requires NAD^+:

$$\text{acetylated histone} + NAD^+ \rightarrow \text{histone} + \text{nicotinamide} + 1\text{-}O\text{-acetyl-ADP-ribose}$$

It can deacetylate p53 and K-16 of H4.

Different HDACs are targeted to different regions of the chromatin to serve different functions. For example, Hos, a Class II HDAC, acts on histones at ribosomal protein genes, and SIR2 is located at telomeres; whereas Rpd3, a Class I HDAC, is never found at the ends of chromosomes.

Like HATs, HDACs can be regulated by kinases, recruitment, and allosterism. CaMKII phosphorylation of Class II HDACs results in their sequestration in the cytoplasm by 14-3-3. On the other hand, MAPK phosphorylation of this group shifts them into the nucleus. CK2 phosphorylation of Class I HDACs increases their activity and their ability to form repressor complexes; for example, CK2 phosphorylation of HDAC2 enhances its binding to the repressor Sp3.

HDACs are recruited to many NRs through cyclin D, which acts as an adaptor in this situation. Heterochromatin protein 1 (HP1) recognizes methylated lysines and recruits HDAC5 to these sites. Calcineurin inhibitor (Cain) recruits class II HDACs to MEF2 (myocyte enhancer factor 2) to inhibit its activity; CaM activates MEF2, in part, by displacing Cain and HDAC. After endothelin binding, its receptor is internalized and migrates to the perinuclear region where it binds HDAC7, causing it to leave the nucleus. In addition, there are several other mechanisms for regulating HDAC activity. As noted previously, SMRT and NCoR recruit class I HDACs; however, they also allosterically stimulate these HDACs. Sumoylation of HDAC1 and HDAC4 by RanBP2 results in their localization to the nucleus; CaMKII phosphorylation of HDAC4 prevents sumoylation. Finally, SIRT3, a class III HDAC located in the mitochondria, is activated by cleavage. Because the class III HDACs are dependent on NAD^+, they are also coupled to metabolism. For example, starvation elevates NAD^+ and activates SIR2.

The study of acetylation is facilitated by the availability of inhibitors. First, there are bisubstrate analogs for p300 and PCAF (p300/CBP associated factor); for example, p300 is inhibited by lysine conjugated to coenzyme A at the ε-amino group. PCAF is inhibited by a peptide consisting of the amino-terminal 20 amino acids of H3; the peptide is conjugated to coenzyme A at K-14. Second, HDACs are noncompetitively inhibited by butyrate, resulting in the hyperacetylation of all histones except histone H1. Butyrate is inexpensive and easily traverses the cell membrane; unfortunately, it also has several undesirable effects. For example, butyrate can inhibit DNA methylases; furthermore, it can induce differentiation apart from its effects on acetylation. Newer HDAC inhibitors, like trapoxin and trichostatin A, are more specific. However, the former does not inhibit HDAC6, and some cell lines can rapidly metabolize the latter, rendering trichostatin A ineffective. In addition, all HDAC inhibitors are beset with at least two other problems: first, they usually increase acetylation levels above those observed under physiological conditions. Second, they induce a global acetylation, whereas natural acetylation is more selective. This difference is important because modifications at specific positions are associated with distinct functions.

There are several facts that argue for the importance of acetylation in gene expression. First, several coactivators have HAT activity; they include SRC-1 (steroid receptor coactivator 1), CBP, and CAF. Second, some general transcription factors, like TFIID, contain HATs. Conversely, repressors, such as Sin3 and YY1 (Yin Yang 1), bind HDACs. Finally, if SRC-1 is artificially coupled to a HDAC, PR activity is inhibited. If acetylation is important in transcription, what role does it play? First, acetylation in excess of 10 groups per nucleosome results in a more relaxed chromatin structure as evidenced by increased DNase I susceptibility and by data from gel electrophoresis, electric dichroism, and sedimentation studies. It also favors the dissociation of (H2A)(H2B) dimers to form lexosomes, and it inhibits the activity of SIR (silencing information regulator), which binds to H4 and facilitates heterochromatin formation. However, many authorities do not believe that acetylation by itself is capable of decondensing DNA; rather, it appears more likely that acetylation destabilizes chromatin for other relaxation factors and/or fixes the open state after SWI/SNF acts. Acetylation can also affect protein binding to chromatin. First, it can loosen histone tails to allow transcription factors, SWI/SNF, and/or kinases to bind. Second, it can create docking sites for proteins with bromodomains, which bind acetylated lysines. Third, it can disrupt repressor binding. Finally, acetylation of proteins can alter their DNA binding; for example, HMGB1 acetylated near the DBD (DNA-binding domain) exhibits an increased affinity for damaged DNA and four-way junctions.

To this point, the discussion has been restricted to the acetylation of histones and HMGs. However, many nuclear proteins can be acetylated; this function has been referred to as *factor acetyltransferase* (FAT) *activity*. This designation is somewhat misleading because there are no unique enzymes that use these substrates; rather, this reaction is accomplished by the same HATs that modify histones. Table 14-3 contains selected examples of nonhistone acetylations and the variety of functions that this modification can serve. For example, it can stimulate transcription by increasing nuclear factor binding, nuclear retention, transcriptional activity, half-life, and coactivator recruitment. It can also influence gene selection, induce corepressor dissociation, or block autoinhibitory domains. More difficult to understand are examples where acetylation inhibits transcription. Several explanations have been advanced to rationalize this effect. First, it may simply be a means to terminate transcription after a period of stimulation for the system to remain responsive. In the case of NRs, the dissociation of CBP is thought to facilitate its replacement by the DRIP (vitamin D receptor interacting protein) complex (see Chapter 13). In addition to the effects listed in Table 14-3, acetylation increases the histone exchange rate, is required for nucleosome repositioning, and is involved with DNA recombination, double strand break repair, and apoptosis.

Whenever studies on acetylation are evaluated, several questions need to be addressed. First, what is the substrate? As just noted, transcription factors, coactivators, and corepressors can be modified in addition to histones and HMG proteins. Second, what acetylated pool is being measured? There are two pools, and only the one that turns over rapidly is associated with transcription. Third, how extensively is the substrate modified? Under basal conditions, histones are either nonacetylated or monoacetylated; after stimulation, they are monoacetylated and diacetylated, whereas inhibitors frequently generate triacetylated forms. Fourth, what is the site modifying? Different sites serve different, even opposite, functions; HMGA can be acetylated at two sites. CBP acetylates HMGA at K-65; this decreases DNA binding and destabilizes enhanceosome

Table 14-3

Functions of Selected Transcription Factor Acetylation

Effect	Substrate	HAT	Comment
Activates Transcription			
Increases nuclear binding	MyoD	PCAF, p300	
	p50 (NF-κB)	CBP, p300	
Nuclear retention	HNF-4	CBP	
	p65 (NF-κB)	CBP, p300	Decreases IκB binding and nuclear export (K-221)
	E2A	CBP, p300, PCAF	Also increases transcriptional activity
Increases transcriptional activity	p53	CBP, p300, PCAF	K-310 (CBP) but modification of K-122 and K-123 (p300, PCAF) decreases DNA binding
	p65 (NF-κB)	CBP	
Gene selection	β-Catenin	CBP	Only inhibits Myc gene transcription
Increases coactivator binding and/ or recruitment	EKLF	CBP, p300	Recruits SWI/SNF
	HNF-4	CBP	Increases CBP binding
	p53	CBP	Increases TRRAP and CBP binding
Increases half-life	Smad7	p300	Acetylation and ubiquitination compete for the same lysine
	p53	CBP	Acetylation and ubiquitination compete for the same lysine
	SREBP	p300	Acetylation and ubiquitination compete for the same lysine
	E2F1	PCAF	
Derepression	AR	p300	Induces NCoR dissociation
	CDP/cut	CBP, p300, PCAF	A repressor whose acetylation induces its dissociation
	RIP140	CBP, p300	A repressor whose acetylation induces its dissociation
	BCL6	p300	A repressor inactivated by acetylation
	Myb	CBP	Neutralizes autoinhibitory domain
	TAL1/SCL	p300, PCAF	Induces Sin3 dissociation
Inhibits Transcription			
Inhibits coactivator binding	TCF	CBP	Decreases β-catenin binding
	HMGA	CBP	Disrupts binding to NF-κB
	NRs	p300	Induces CBP dissociation
Inhibits DNA binding	Fos (DBD)	CBP	Blocks binding to CRE but not AP-1 HRE
	IRF-7 (DBD)	PCAF	
Recruits repressors	YY1	p300, PCAF	Recruits HDACs and induces Rb dissociation
Blocks sumoylation	Sp3		Sumoylation is required for Sp3 activity

CDP/cut, CCAAT displacement protein/cut homologue; DBD, DNA-binding domain; EKLF, Erythroid Krüppel-like factor; TAL1/SCL, T cell acute leukemia/stem cell leukemia.

formation. Conversely, PCAF acetylates HMGA at K-71; this blocks acetylation at K-65 and promotes enhanceosome formation. Fifth, what other modifications are present? As already noted, a single modification at a single site rarely determines activity; rather, it is the constellation of modifications, called the histone code, that signals a particular function. Finally, is the involvement of HATs or HDACs actually due to their activity on acetylation? HDACs can inhibit transcription by mechanisms other than acetylation; for example, HDAC1 can recruit PP1 to dephosphorylate and inactivate transcription factors, and the amino termini of class II HDACs can bind the MADS (MCM1-AG-DEFA-SRF) box of MEF2 and repress transcription independent of their deacetylase activity. On the other hand, CBP and p160 can stimulate transcription apart from their HAT activity; for example, they can recruit chromatin remodelers and general transcription factors.

Phosphorylation

Phosphorylation is another modification that has a rapid turnover and is independent of protein and RNA synthesis; furthermore, it is hormonally regulated. Like acetylation, the phosphorylation of specific sites has different functions. For example, phosphorylation can be associated with both DNA condensation and relaxation, depending on the substrate and the site. Chromatin condensation during mitosis is a result of H1 phosphorylation by H1 kinase, which is composed of Cdk2, cyclin B, and Suc1. PKC phosphorylation of H3 facilitates condensation in apoptosis. On the other hand, PKA phosphorylation of H1 and H3 leads to relaxation. H3 can be phosphorylated at S-10 and S-28 by either the MAPK-S6KII pathway or by the Aurora kinase, which is a member of the PKA family and which can be activated *in vitro* by PKA. This modification recruits condensin, a complex that directs the supercoiling of DNA. However, when the S-10 phosphorylation is associated with acetylation at K-9 and K-14, DNA relaxation and transcription are induced. Finally, Mec1, a member of the DNA-PK family, phosphorylates H2A after DNA damage. This modification results in decondensation to facilitate DNA repair through nonhomologous end joining.

Phosphorylation can also affect localization and DNA binding. For example, PKC phosphorylation of HMGB1, HMGN14, and HMGN17 leads to their nuclear exclusion. CK2 phosphorylation of HMGN14 increases its DNA affinity, locking the nucleosome in either the open or closed state. On the other hand, CK2 phosphorylation of HMGB and PKC phosphorylation of HMGA decrease DNA binding. The former can also decrease the affinity of HMGB to some transcription factors.

Poly(ADP-ribosyl)ation

Poly(ADP-ribosyl)ation involves the attachment of a branched polymer of ADP-ribose units, which are derived from NAD (nicotinamide adenine dinucleotide). The nicotinamide is removed, and the now available 1' carbon of the ribose is coupled to a negative residue, such as the side chain of glutamic acid or the carboxy terminus (Fig. 14-2). The 1' carbon of the next subunit is then attached to the first by means of the 2' carbon of the other ribose.

There are four groups of poly(ADP-ribose) polymerases (PARPs). The type I PARPs include PARP-1 and PARP-2 and are single-chain, globular proteins containing just over a 1000 amino acids. They catalyze chain initiation, elongation, and branching. The enzymes are divided into three approximately equal sections: amino-terminal DNA-binding domain, a central automodification domain, and a carboxy-terminal NAD-binding site. The DBD contains two zinc fingers, which bind DNA strand breaks, and two HTH motifs, which bind intact DNA. The amino-terminal end of the automodification domain contains a leucine zipper, which mediates homodimerization. Type I PARPs have a basic isoelectric point and are located on the linker DNA. Type II includes PARP-3, which is much smaller and lacks a DBD. Type III contains tankyrase 1 and 2 (also called *PARP-4a* and *PARP-4b*, respectively); these enzymes are 150 times less active than PARP-1 and appear to synthesize only linear chains. They are negative regulators of the telomerase, which extends the telomeres at the ends of chromosomes. Tankyrases are also associated with the Golgi complex and the endocytic machinery; they may be involved with receptor and transporter sorting. They can be activated by the insulin receptor through MAPK. Finally, the vault PARP (also called *PARP-5*) is part of the ribonucleoprotein (RNP) vault complex and may be involved with transport between the cytoplasm and nucleus.

Type I PARPs are primarily involved with the modification of nuclear proteins. The major substrates for these enzymes vary from system to system. Histone H1 is generally preferred over histones H2A and H2B, although in wheat embryos, this is reversed. Histones H3 and H4 are never modified. The HMGB proteins are usually preferred over HMGN. However, the best substrate is the PARP itself. The enzyme has 15 attachment sites, each of which may have side chains over 80 units long; this would add a minimum of 650 kDa to a protein that was originally only 110 kDa. The RNA polymerase is another potential substrate. Both the synthetase and the RNA polymerase are inhibited by this modification.

The regulation of type I PARPs is still poorly understood; for example, the data on the roles of the polyamines are confusing. Spermidine and spermine can fulfill the requirement for histones and divalent cations, but in fibroblasts inhibitors of polyamine synthesis stimulate the enzyme 2- to 3-fold. Phorbol esters can stimulate the enzyme 10-fold in these same cells, but *in vitro* PKC stochiometrically phosphorylates and inhibits the enzyme. PARP is stimulated by NO, but this effect is probably a result of the generation of NO-induced DNA damage, which is a potent PARP activator. Indeed, PARP participates in DNA repair (see later) and can trigger apoptosis if the DNA damage is severe. Bcl2 inhibits apoptosis and can inhibit PARP; Bcl2 binds PARP through ribosomal protein S3a, which acts as an adaptor. Finally, the enzyme can be inhibited by analogs of NAD: for example, 3-aminobenzamide and 3-methoxybenzamide. These compounds are very useful for studying the role of poly(ADP-ribosyl)ation in tissues, but they also have potential toxicity, including the inhibition of thymidine incorporation, lactate dehydrogenase, and glucose oxidase. Fortunately, such untoward effects require inhibitor concentrations of 1 mM or more, and frequently the activity of poly(ADP-ribosyl)synthetase is adequately suppressed at lower concentrations.

The degradation of this polymer is accomplished by three enzymes (Fig. 14-2). The poly(ADP-ribose)glycohydrolase cleaves ribose-ribose bonds, including those at the branching points; as a consequence, it can remove all but the last subunit. This enzyme has two isoforms: type I is nuclear, more active, and prefers histone H1 and the RNA polymerase, and type II is cytoplasmic and inhibited by cAMP. The second enzyme, ADP-ribosyl histone hydrolase, cleaves the ribose-protein bond, thereby freeing the last subunit. The resulting ADP-ribose units are then hydrolyzed to AMP by a yet unidentified enzyme, which

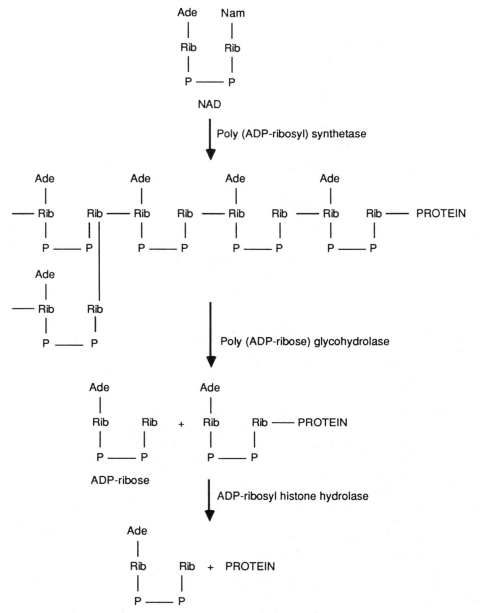

Fig. 14-2. Synthetic and degradatory pathway of poly(ADP-ribosyl)ated proteins. *Ade*, Adenine; *Nam*, nicotinamide; *P*, phosphate; *Rib*, ribose.

appears to be either a phosphodiesterase or an ADP-ribose pyrophosphatase. The first two degradatory enzymes vary reciprocally with the polymerizing activity.

Assays for PARP are fraught with difficulties. In general, cells or nuclei are exposed to a labeled precursor, such as NAD or adenosine, and radioactive incorporation into protein or histones is determined. The first problem involves the inability of NAD to enter intact cells. One can either permeabilize cells with a hypotonic solution or assay PARP in nuclei. In both cases, one must be very careful not to introduce nicks in the DNA during homogenization because DNA breaks will activate PARP. Labeled adenosine will enter intact cells, but it is a less specific precursor. This means that only a small fraction of the radioactive nucleoside becomes attached to protein; therefore more label must be used and the substrates must be carefully purified. Furthermore, it requires at least 18 hours for the adenosine to equilibrate with the intracellular NAD pool. The assay system chosen is of more than academic interest, labeling patterns differ between intact and broken cell preparations and even the concentrations of NAD used can affect chain lengths.

There are other problems that must also be solved: labile bonds, product identification, and competing reactions. Some of the polymer bonds are labile in alkali, so tissue processing should be performed rapidly at 4 °C in slightly acidic buffers. The charge and molecular weight of the product can be substantially altered; however, short incubation times will prevent the polymer from becoming so long that the product is no longer recognizable when compared with unmodified standards. Finally, there are other nonhistone-non-HMG substrates that can become heavily labeled: for example, the PARP itself. Adequate purification and identification of the reaction products will allow extraneous reactions to be segregated and analyzed separately. Finally, the function of PARP can be studied in knockout mice, but the existence of multiple forms with overlapping activity complicates this approach.

As previously noted, this enzyme is involved with DNA repair. It binds nicked DNA and protects the DNA from recombination or spurious transcription. It is also thought that the histones may be shuttled to the automodified PARP to expose the DNA to the repairing enzymes. The structure of poly(ADP-ribose) is helical, like other nucleic acids, and is capable of binding histones. After the damage is repaired, the polymers are digested and the histones return to the DNA. PARP also inhibits DNA synthesis and transcription until the repair has occurred. For example, the automodification domain binds the repressor YY1; after localizing to a DNA nick, PARP undergoes automodification and releases YY1 to inhibit transcription. In a severely damaged cell, PARP will be so active that it can deplete the cell of NAD and ATP, leading to apoptosis and elimination of a dangerously mutated cell. Finally, the PARP tails also facilitate nucleosome assembly during DNA replication; as in the case with DNA damage, nucleosomes are temporarily transferred to these tails during replication.

However, the role of PARP in gene expression is not as clear. PARP is found in RNP particles and modification of RNP releases mRNA for translation; it is also found in *Drosophila* chromosomal puffs and is required for their formation. Furthermore, in the absence of H1, PARP induces DNA relaxation, which is conducive to transcription. However, PARP can also cross-link H1 to stabilize chromatin condensation. Transglutaminase uses a similar mechanism: it cross-links H2A and H2B in the condensed chromatin of sperm and red blood cells. Finally, several genes cannot be transcribed in PARP knockout mice. As such, there appears to be a variable association between PARP activity and transcription.

There are several mechanisms by which PARP could facilitate transcription. First, as already noted, it can relax DNA in the absence of H1. Second, it inhibits DNA methylation, which is usually found in inactive chromatin (see later). It can modify and inhibit repressors, like YY1 and TEF1A (transcription enhancer factor 1A). It can also directly bind and activate several transcription factors. For example, the automodification domain of PARP binds Oct-1 and increases its DNA binding; furthermore, PARP binds p53 and has the same effect, while it suppresses the autoinhibitory domain of AP-2. Finally, it enhances the transcriptional activity of NF-κB. Many of these latter effects appear to be allosteric because they do not require enzymatic activity; alternatively, PARP may act as a scaffold and organizing factor for the preinitiation complex. PARP can inhibit transcription as well; in what may represent negative feedback, PARP binds the hairpin DNA in the promoter of its own gene and suppresses expression. This inhibition likewise does not require catalytic activity.

Methylation

Methylation refers to the attachment of from one to three methyl groups to the nitrogen of either lysine (N^ε) or arginine (N^η). The guanidinyl group in arginine has two nitrogens that can be methylated (Fig. 14-3); in symmetrical dimethylation, each nitrogen receives a methyl group, whereas in asymmetrical dimethylation both methyl groups are attached to only one nitrogen. This distinction divides protein arginine methyltransferases (PRMTs) into two groups: type I

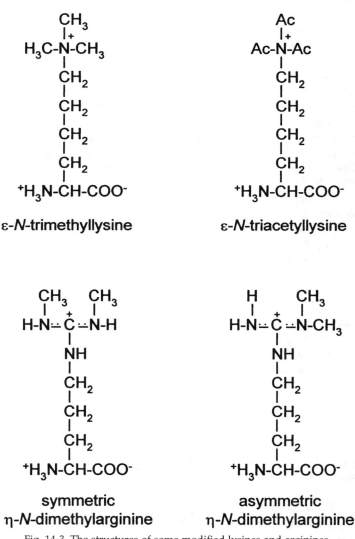

Fig. 14-3. The structures of some modified lysines and arginines.

Table 14-4
A Summary of Some Arginine Methyltransferases

PRMT	Class	Substrate	Comments
PRMT1	Type I	H4 (R-3), Stat1, fibrillarin, nucleolin, nRNPs, Sam68	Binds IFN-RI, SRC-1, and YY1
PRMT2	Type II	Ribosomal protein L12	Cofactor for NRs; SH3 in amino terminus
PRMT3	Type I		Zinc finger in amino terminus
PRMT4 (CARM1)	Type I	H3 (R-17 and R-26), CBP, PABP1, HuR	Cofactor for NRs; binds SRC-1; stabilizes labile mRNA
PRMT5 (JBP1)	Type II	Sm proteins, coilin	Assembles spliceosomes and RNPs; binds Jak
PRMT6	Type I	PRMT6	

PRMTs monomethylate and asymmetrically dimethylate arginines, whereas type II PRMTs monomethylate and symmetrically dimethylate these residues. Thus far, six PRMTs have been identified and their characteristics are given in Table 14-4. All lysine methyltransferases have a SET domain, whose name is an acronym for the first three proteins in which it was found: Su(var), E(z), and trithorax. Several lysine methyltransferases and their substrates are listed in Table 14-5. There are no known demethylases. However, this modification can be reversed by either histone replacement or by cleavage of the amino-terminal tail of H3.

Historically, histone methylation has been associated with chromatin condensation and mitosis. This association, along with a slow turnover and little or no evidence for direct hormonal regulation, suggested that methylation does not play a significant role in gene expression. However, a dramatic reevaluation of methylation has occurred recently. Results of earlier studies had come to erroneous conclusions as a result of several complications involved with studying this modification. First, it is important to distinguish between lysine and arginine methylation; the former is more stable and associated with chromatin condensation, whereas the latter is more labile and associated with transcription. Second, identification of the particular site modified is critical; histone H3 methylated on K-9 is localized to heterochromatin, whereas H3 methylated on K-4 is localized to euchromatin. In another example, methylation within the KIX (KID interacting domain in CBP) domain of CBP interferes with CREB binding but has no effect on NR activity; however, methylation carboxy-terminal to this domain is actually required for NR transcriptional activity. The latter example demonstrates another confounding factor; proteins other than histones can be methylated. Finally, the degree of modification influences its effects; H3 trimethylated on K-4 is only

Table 14-5
A Summary of Some Lysine Methyltransferases

Lysine methyltransferase	Substrate	
	Histone	Site(s)
SUV39H1	H3	K-9
G9a	H3	K-9 and K-27
Set2	H3	K-36
Set1 and Set9	H3	K-4
Dot1	H3	K-79
Set8 (PR-Set7)	H4	K-20

associated with active genes, but dimethylated K-4 may be associated with either active or inactive genes.

More recent studies have shown that methylation has a wide range of functions. First, methylation of K-9 in H3 is involved with silencing. Indeed, targeted methylation of this residue is sufficient by itself to repress transcription. This modification recruits HP1 and HPC2, which are adaptors for factors responsible for silencing and condensing chromatin. For example, HP1 can recruit HDAC5 and a DNA methyltransferase. Methylated K-9 also attracts Swi6 for chromatin remodeling and excludes the RNA polymerase II. Finally, methylation in the KIX domain of CBP prevents its interaction with CREB.

On the other hand, methylation can also facilitate transcription; this effect is usually the result of recruiting coactivators or excluding corepressors. For example, methylation at K-4 or K-79 in H3 displaces HDAC, and PRMT1 methylation on R-31 of Stat1 blocks the binding of PIAS1. In addition, methylated R-17 in H3 recruits SRC-2, CBP, and p300.

Methylation is involved with macromolecular assembly. The interaction between transcription factors and their coactivators and corepressors has already been mentioned. The Sm protein and coilin are part of the spliceosome, and their methylation is required for binding to survival motor neuron (SMN), another component of this complex. Finally, Sam68 is a polyproline domain-containing adaptor in the signal particle; methylation in the polyproline region prevents the binding of Sam68 to SH3, but not WW, domains. The last function of methylation is in cellular localization; methylation of hnRNP A1 and A2 facilitate their nuclear export.

Methyltransferases can be regulated in several ways. First, many NRs recruit these enzymes; for example, many PRMTs bind the TAD1, DBD, or ligand-binding domain (LBD) of NRs. SUV39H1 is recruited to the cyclin E promoter by Rb. In this case, recruitment is further regulated by phosphorylation; phosphorylation of Rb by cyclin D-Cdk4 or cyclin E-Cdk2 inhibits this association. PRMT5 binds Jak; indeed, PRMT5 was initially called *JAB1* (Jak binding protein 1) before it was known to be a methyltransferase. However, it is not know if Jak phosphorylates PRMT5. Finally, PRMT1 is associated with the intracellular domain of IFN-RI (interferon type 1 receptor); the significance of this interaction is unknown, but it may sequester PRMT1 in the cytoplasm until ligand stimulation.

Monoubiquitination

The basics of ubiquitination were covered in Chapter 8. The modification seen in histones is monoubiquitination; the major ubiquitin-conjugating enzyme is Rad6, and the major substrates are H2A, H2B, and H3. Several functions have been associated with this modification. First, it may facilitate the replacement of resident histones with other isohistones; although monoubiquitination does not target proteins for degradation in the proteasome, it can target them to lysosomes. Histones normally have very long half-lives: 9 to 15 days for histones H3 and H4, 4 to 6 days for histones H2A and H2B, and around 2 days for histone H1 in terminally differentiated kidney cells. Therefore what this modification may do is to increase the turnover of ordinary histones to facilitate the insertion of isohistones more conducive to transcription. Indeed, in *Drosophila* uH2A levels are highest in actively transcribed regions. Increased histone turnover may also be important for changing methylation patterns because no histone demethylases have yet been identified; for example, ubiquitination has been associated with decreased methylation at K-9 of H3. Second, it may act as a protein chaperone to maintain a slightly unfolded state of a single protein or protein complex. For example, ubiquitination at K-123 of H2B is required for the methylation of K-4 in H3; the former increases the accessibility of the latter to HATs. Ubiquitination also keeps the carboxy terminus of H2B exposed; this region is required for transcription. Third, this modification may recruit regulatory proteins; for example, the ubiquitination of XIAP-TRAP6 is necessary for attracting TAB1-TAK1 in the TGFβ signaling pathway. In addition, some HDACs have a ubiquitin-binding motif and may be recruited by ubiquitin to deacetylate the adjacent chromatin. Finally, this modification can affect chromatin structure; for example, deubiquitination of H2A is associated with chromatin condensation. Ubiquitination may favor an open conformation by disrupting histone-histone interactions; for example, the H2B attachment site for ubiquitin is near a tyrosine that binds the amino terminus of H2A.

The regulation of ubiquitination occurs at several levels. First, many proteins are targeted for ubiquitination by phosphorylation at key sites: for example, IKK phosphorylation of IκB (inhibitor of NF-κB) (see Chapter 13). Second, acetylation and ubiquitination often compete for the same sites (see Table 14-3); as such, acetylation can stabilize transcription factors by inhibiting their ubiquitination. Finally, the ubiquitination complex can itself be phosphorylated; Nedd (neural precursor cell-expressed developmentally downregulated) is involved in facilitating the ubiquitination of some substrates. The SGK (serum and glucocorticoid-regulated kinase) phosphorylation of Nedd causes it to dissociate from at least one target protein, the epithelial sodium channel, whose stability is then enhanced.

Interactions

Many of these modifications interact with each other in one of four ways: stimulatory, inhibitory, reciprocal, or sequential. There are many examples of a modification at one site promoting a second modification at another site; phosphorylation of S-10 in H3 increases the acetylation of K-14 10-fold (Fig. 14-4). Methylation at R-3 in H4 stimulates acetylation of K-8 and K-12; and the methylation of K-4 in H3 increases acetylation on both H3 and H4. The latter effect is a result of methylated K-3 displacing HDAC.

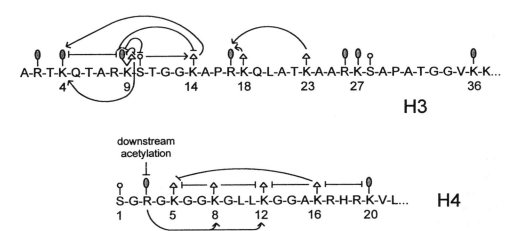

Fig. 14-4. Sites of H3 and H4 modification and their interrelationships. Arrows indicate stimulation; flat heads, inhibition; circles, phosphorylation sites; triangles, acetylation sites; and shaded ovals, methylation sites.

Several modifications inhibit each other; methylation of K-9 in H3 is inhibited by phosphorylation of S-10, acetylation of K-14, or methylation of K-4 (Fig. 14-4). This effect is mutual; that is, methylation of K-9 will also inhibit the other modifications. Acetylation of K-8 or K-16 in H4 inhibits the acetylation of K-5 and K-12, and methylation of K-20 and acetylation of K-16 in H4 are mutually inhibitory. Finally, the acetylation of H4 inhibits the methylation of R-3, and the methylation of K-9 in H3 inhibits the acetylation of both H3 and H4 by p300.

At least one amino acid, K-9 in H3, can be either methylated or acetylated; such modifications are obviously reciprocal. Finally, modifications can also be sequential; K-9 and K-14 in H3 must be deacetylated prior to methylation of K-9. Conversely, K-18 and K-23 in H3 must be acetylated before methylation at R-17, and ubiquitination at K-123 in H2B is required for the methylation of K-4 in H3.

These interrelationships ensure that certain modifications occur together; and it is these combinations, rather than any individual modification, that determine the effect of the modifications. This is known as the *histone code*. For example, euchromatin and transcription are associated with acetylation of K-9 and K-14 in H3, phosphorylation of S-10, and, in some systems, the methylation of R-17. The association of methylated K-4 and acetylated K-14 in H3 or methylated R-3 and acetylated K-5 in H4 serves the same function. On the other hand, heterochromatin and silencing are associated with any of the following individual modifications: phosphorylation of S-10 in H3, methylation of K-9 in H3, acetylation of K-12 in H4, or acetylation of K-16 in H4.

The requirement for a specific constellation of modifications to stimulate or repress transcription lies in the binding specificity of critical proteins. For example, the single bromo domain of BRG1, which is the ATPase subunit in SWI/SNF, will only bind acetylated K-8 of H4, whereas the double bromo domains of TAFII250, which is a subunit of TFIID, require the concomitant acetylation of both K-9 and K-14 of H3. The histone code can also stipulate which corepressor will be recruited; methylation of only K-9 in H3 will bind HP1, but methylation of both K-9 and K-27 will specifically bind the Polycomb group proteins.

Template Activity

DNase Activity

As noted previously, the modification of histones and HMG proteins can affect chromatin structure; these different conformations can influence how efficiently the resident genes are transcribed. This variable ability to be transcribed is referred to as *template activity*. For example, the tightly coiled solenoid is often referred to as *condensed DNA* and forms the transcriptionally inactive heterochromatin, whereas the unraveled structure is known as *relaxed DNA* and forms the transcriptionally active euchromatin. These two types of DNA conformations can be detected by limited digestion with various DNases; the more open the DNA structure, the more susceptible it is to digestion. Therefore DNase sensitivity can be used as a crude indicator of DNA structure. The following examples demonstrate the association between hormone-induced gene expression and DNA conformation.

Micrococcal nuclease is a DNase that specifically cleaves the linker DNA. In the oviduct of immature chicks, no ovalbumin mRNA is detectable, and the gene is insensitive to the nuclease. After primary stimulation with diethylstilbestrol, a synthetic estrogen, ovalbumin mRNA accumulates and the DNA is digested into single or small clusters of nucleosomes. Four or five days after hormone withdrawal, the gene once again becomes resistant to micrococcal nuclease, and no transcription takes place. Secondary induction restores both transcription and DNA susceptibility to digestion.

The vitellogenin genes in *Xenopus* liver show a similar, although not identical, effect. These genes are divided into two groups, known as the A genes and B genes. The B genes are always sensitive to digestion with DNase I, a nonspecific nuclease, and are always readily inducible. The A genes are resistant to DNase I in the untreated male frog, and primary induction with estradiol is slow. One month after primary stimulation and hormone withdrawal, the A genes exhibit DNase I sensitivity and are rapidly inducible following secondary stimulation. However, if 8 months are allowed to elapse between primary and secondary induction, the A genes return to a nuclease-resistant state, and the induction of transcription is slow.

These experiments suggest that transcription is associated with relaxed DNA and that if the DNA is already relaxed before induction, induction occurs more quickly. This latter effect may explain why secondary induction in many systems is faster and greater in magnitude than primary induction; that is, this memory effect may reside in the DNA structure. The second example also illustrates the point that, although a change in chromatin structure may facilitate transcription, it is not sufficient by itself.

DNA Methylation

Methodology

Another postulated mechanism for hormonally altering template activity is the methylation of cytidines at position 5. This methylation usually occurs in C_pG sequences, has been associated with gene repression, and can be inherited. The latter characteristic is due to DNA methyltransferase 1 (DNMT1), whose activity is linked to DNA synthesis and is relatively specific for hemimethylated DNA. After DNA replication, the parental strand is still methylated, but the daughter strand is not; this hemimethylated site is then fully methylated by DNMT1. If a site was not originally methylated in the parental strand, both strands will be unmethylated; such a site is a poor substrate for DNMT1. Therefore methylated sites remain methylated and unmethylated sites remain unmethylated; for this reason DNMT1 is often referred to as the *maintenance methyltransferase*. DNMT3 is responsible for *de novo* methylation and is important in the establishment of embryonic methylation patterns.

One can locate these methylation sites by restriction mapping. Bacteria are susceptible to infection by certain viruses called *bacteriophages*; resistant bacterial strains develop enzymes, called *restriction endonucleases*, which cleave viral DNA at highly specific sequences. Similar sequences in the host are protected by methylation; that is, methylation at these sites renders them resistant to endonuclease cleavage (Table 14-6). With these enzymes, cleavage sites can be mapped within any given piece of DNA. If this is done before and after hormone treatment and if additional sites appear after the treatment, then it is concluded that the hormone must have stimulated demethylation at these new sites, thereby

Table 14-6
Sequence Specificity for Selected Restriction
Endonucleases

Methylation effect	Restriction endonuclease[a]	
	HpaII	*HhaI*
Endonuclease-sensitive	$^{(m)}C^{\downarrow}CGG$	$GCG^{\downarrow}C$
Endonulcease-resistant	$^{(m)}C^{m}CGG$	$G^{m}CGC$

[a] Arrows indicate cleavage site.

rendering them susceptible to restriction endonucleases. The major caveat with this technique is that only a few sites are being sampled; there could be numerous sites at which methylation status is changing but which are not recognized by any of the enzymes that one may be using. For example, *HpaII* produces more fragments in the α-globin gene in erythrocytes than it does in brain or sperm, suggesting that the gene is hypomethylated in erythrocytes. This observation is consistent with the fact that the α-globin gene is active only in erythrocytes. However, no difference in cleavage patterns is noted when *HhaI* is used; if one had only tested this latter restriction endonuclease, one might have concluded that methylation patterns were identical in all tissues. The more enzymes one uses, the more complete picture one sees. Methylation sites can also be determined by direct sequencing; however, this technique is more laborious than the use of restriction endonucleases.

In addition to observing the natural changes in methylation patterns, these patterns can be experimentally manipulated. The simplest method is to use 5-azacytidine, a suicide inhibitor of the DNMT. To be effective, it must be incorporated into the DNA where it causes hypomethylation. For example, the metallothionein I gene is normally inducible by glucocorticoids or cadmium in several lines but not in mouse thymoma cells; however, it does become inducible following treatment with 5-azacytidine. These are fairly typical results; the drug, either by itself or in combination with other inducers, stimulates gene transcription in cells that normally do not express that gene. Such data have led to the hypothesis that methylation represses gene expression and demethylation is involved with their induction. Unfortunately, the drug is nonspecific and hypomethylates extensive regions of the genome; furthermore, it may also inhibit other methylation reactions, such as phospholipid methylation.

A more selective way of altering methylation patterns is to use cloned genes. For example, various portions of the γ-globin gene were methylated, and the gene was then transfected into cells. Methylation of the structural gene or the vector did not interfere with expression, but methylation of the 5′ flanking sequences totally blocked transcription. Other experiments have confirmed that repression is most effective when methylation occurs in the preinitiation domain of a gene.

Examples
Unfortunately, the association of demethylation with transcription is far from perfect; for example, the vitellogenin gene is still expressed in *Xenopus* liver, even though it remains fully methylated, and 5-azacytidine specifically inhibits milk protein synthesis in the mouse mammary gland. However, it is the chick vitellogenin gene that has been among the most extensively studied

methylation systems and that can provide some of the most useful insights into this process.

Estradiol treatment of immature chicks results in both the transcription of the vitellogenin gene and the appearance of DNase hypersensitive sites in the 5′ region; a hypersensitive site is a site of DNA cleavage induced by a brief incubation with low concentrations of DNase I. Another site in the 5′ region is demethylated, but this occurs after transcription begins. The coding strand is the first to be demethylated; the complementary, nonsense strand follows about 24 hours later. The demethylation does not require DNA synthesis, so it is not a result of repeated DNA replication combined with a failure to remethylate. This passive mechanism has been postulated by others, who suggest that transcription factors bound to active genes prevent methylation so that the original methylation is eventually diluted out. Instead, this system demonstrates that an active mechanism is required; one hypothesis is that the 5-methylcytidine is demethylated directly. Another is that gene induction is accompanied by DNA nicks, which stimulate repair-type synthesis and the removal of the methylated nucleotides.

After hormone withdrawal, the demethylated site is not remethylated, although one hypersensitive site is lost and transcription ceases. This might indicate that, in this system, some of the hypersensitive sites are related to transcription, whereas demethylation is involved in the memory effect. However, this latter hypothesis was not supported by experiments in chick embryos. When chick embryos are given a single dose of estradiol, the vitellogenin genes respond as just described. Twenty-five weeks later, the hypersensitive site is lost and the demethylated site has undergone variable remethylation. Despite this, the memory effect is still present in many of the birds and is not correlated with the degree of methylation. Furthermore, the other hypothesis linking the hypersensitive sites to transcription is not supported by other experiments in the immature chick; both the hypersensitive sites and demethylation appear in the vitellogenin gene in all estrogen-responsive tissues, such as the oviduct, even though the gene is only expressed in the liver. The gene is not altered in tissues unresponsive to estradiol. This is not an uncommon finding; the gene for SVS IV, an androgen-dependent protein, is hypomethylated in the seminal vesicles and the prostrate gland. Both tissues are androgen responsive. However, the SVS IV gene is only expressed in the seminal vesicles. Therefore both hypersensitive sites and demethylation appear to be generalized responses by tissues sensitive to a particular hormone. They apparently occur in all genes inducible by that hormone and in all the tissues responsive to the hormone, despite the fact that only a particular subset of these genes will actually be activated in any given tissue.

Vitellogenin, like all of the other proteins discussed, is a tissue-specific protein, which is made only when induced. For such genes, methylation and demethylation are discrete events at restricted locations in the 5′ region. This pattern does not hold for housekeeping genes; these are the genes that are constitutively transcribed in virtually all cells. These latter genes are associated with large islands (500 to 2000 bp), which are very rich in the sequence C_pG. These islands are located in the 5′ region and are inevitably nonmethylated, except for those genes on the inactive X chromosome. Further support for the association of island methylation and gene inactivation comes from transfection experiments with an artificially methylated gene for amidophosphoribosyl transferase; only methylation of the 5′ region inhibited gene expression. It has been postulated that, when these genes are active, the islands are protected from methylation by certain proteins, perhaps related to transcription. Furthermore, because this class of genes is almost always active, methylation is a rare event.

Function

What then is the role of DNA methylation? In all systems examined thus far, methylation in the 5' region inhibits gene expression, but demethylation occurs in all tissues responsive to a particular hormone, whether those genes are transcribed. Most authorities agree that changes in methylation patterns are not primary events in gene expression or repression. Rather, it has been suggested that methylation merely serves to imprint inactivity on a gene that was initially repressed by some other mechanism. Furthermore, because of its persistence and its relatively low level in somatic cells, methylation may be more important in developmental processes than in acute gene regulation.

How might methylation maintain gene repression? There are two major mechanisms. The first and simplest is that methylation of a hormone response element (HRE) blocks the binding of some transcription factors. For example, the HREs for CREB, Myc-Max, NF-κB, and Ets contain C_pG; if this sequence is methylated, these transcription factors will not bind. However, it is also possible that methylation may block the binding of a repressor and actually stimulate gene transcription. This possibility has been advanced to explain the association between the induction of the human prolactin (PRL) gene and its methylation. Not all transcription factors are affected by this modification; the methylation of the binding site for Sp1 does not block Sp1 binding.

A second mechanism involves the specific binding of methylated DNA by proteins that repress transcription. These proteins can be hormonally regulated. For example, in the immature hen, the ovalbumin gene is repressed by the binding of these proteins to methylated sites. Estrogen stimulation lowers the levels of these proteins and exposes the sites to demethylation; these changes permit the expression of ovalbumin. During molting or the egg-laying pause, the estrogen levels fall and the methylated DNA-binding proteins increase to repress ovalbumin once again. Although the sites in the ovalbumin gene are still unmethylated, the higher repressor concentration compensates for its lower affinity for the unmethylated sequences.

Several methylated DNA binding proteins have been identified. Phosphorylated MDBP-2-H1, a carboxy-terminally truncated H1 isohistone, binds methylated DNA and promotes chromatin condensation. Estradiol stimulates the dephosphorylation of MDBP-2-H1 and terminates its repression. Methylcytosine binding protein 2 (MeCP2) forms a complex with HDAC through Sin3, NCoR, and Ski; however, MeCP2 can also repress transcription independent of HDAC. This latter effect may arise from the ability of MeCP2 to recruit an H3 K-9 methyltransferase. MBD2, another methylcytosine binding protein, forms a complex with the Mi-2/NuRD deacetylase complex. DNMT3a itself can bind both HDAC1 and a H3 K-9 methyltransferase.

These processes are thought to occur in an ordered sequence. First, histones are deacetylated and the chromatin is remodeled to allow the SET (su[var]/E[z]/trithorax) methyltransferases access to histone H3. H3 is then methylated on K-9, which recruits HP1. HP1, in turn, attracts DNMT and SWI/SNF to methylate the DNA and further remodel the chromatin. Finally, the methylcytosine recruits Mi2/NuRD and probably other factors to assist in the chromatin remodeling and maintain deacetylation.

Demethylation can be active or passive. The latter occurs when DNA synthesis takes place without remethylation. The former is accomplished by demethylases. Normally, histone tails prevent the demethylases from gaining access to the DNA; however, histone acetylation exposes the DNA so that demethylation can occur.

Z-DNA

In the previous discussion, the relationship between template activity and both DNA packaging and methylation were explored. One other determinant of template activity has also been postulated: DNA helical structure. DNA can assume several conformations. B-DNA is the classic Watson-Crick model, whereas A-DNA resembles the structure of RNA and is found in DNA-RNA hybrids. Z-DNA is a very elongated, left-handed helix that can be found in DNA regions rich in guanine and cytosine. Such regions frequently occur in the 5′ regions of genes, particularly at transcription initiation sites. In addition, Z-DNA may occur in certain HREs; for example, the estrogen hormone response element (ERE) appears to assume a non-B conformation based on chemical reactivity and S1 nuclease digestion. This structure is intrinsic to the sequence because it does not require chromatin proteins and is not observed in related HREs. The formation of Z-DNA is also favored by polyamines, which are known to be elevated by many hormones (see Chapter 11). Finally, there is one example that suggests a relationship between Z-DNA and transcription; cAMP increases both Z-DNA content and transcription of the corticotropin-releasing hormone (CRH) gene, whereas dexamethasone decreases both.

How might Z-DNA affect gene expression? One answer, for which there is some supporting evidence, is that Z-DNA is recognized by specific transcription factors. For example, dsRNA adenosine deaminase type I, an RNA editing enzyme, recognizes Z-DNA through a wHTH (winged helix-turn-helix) domain. This illustration suggests that certain transcription factors could also recognize Z-DNA.

RNA Polymerases

RNA polymerases are the enzymes that transcribe the template. There are three such enzymes: RNA polymerase I synthesizes rRNA except for the 5S rRNA; RNA polymerase II synthesizes mRNA; and RNA polymerase III synthesizes tRNA, 5S rRNA, and other small RNAs. RNA polymerase II has a carboxy-terminal domain (CTD) consisting of repeats, where every second and fifth residue is serine. The unphosphorylated polymerase is recruited to the promoter by TBP for assembly into a preinitiation complex. Phosphorylation of S-5 in the CTD promotes transcription by causing RNA polymerase II to dissociate from the initiation complex and by recruiting the capping enzyme. The switch to phosphorylation on S-2 promotes elongation and processing by recruiting factors for splicing, poly(A) tail synthesis, and so on.

TFIIH is probably the major S-5 kinase and is recruited by NRs. In response to mitogens and stress, ERK can also phosphorylate S-5. Finally, S-5 can be O-glycosylated by NAG (N-acetylglucosamine) (see Chapter 13); this modification is reciprocal with phosphorylation and is involved with RNA polymerase II recycling. P-TEFb is probably the major S-2 kinase and is recruited by NF-κB and several NRs. However, glucocorticoids, which are anti-inflammatory, antagonize NF-κB, which triggers defense responses. One site of this antagonism is the S-2 of the CTD; GR inhibits S-2 phosphorylation either by recruiting an S-2 phosphatase or by inhibiting the S-2 kinase.

Only limited information about the hormonal regulation of the other two polymerases is known. For example, MAPK phosphorylation of the RNA polymerase I-specific initiation factor, TIF-IA, is essential for pre-rRNA synthesis. Myc, a hormonally regulated transcription factor (see Chapter 13), directly binds TFIIIB in the RNA polymerase III complex and stimulates transcription by the polymerase. This would suggest that hormones can stimulate RNA polymerase III the same way they stimulate RNA polymerase II; they activate specific transcription factors to recruit, directly or through coactivators, elements of the transcription complex. Whether the phosphorylation of RNA polymerase III by hormonally regulated kinases also plays a role is unknown.

Summary

The major effects of histone and HMG protein modifications appear to be on chromatin structure; relaxed chromatin is associated with active transcription, whereas condensed chromatin is not. However, single modifications rarely dictate which structure will be favored; rather, certain modifications tend to occur together, and it is this combination that determines chromatin structure. Such a profile is known as the histone code. In general, acetylation is more conducive to the open state, whereas lysine methylation favors condensation and gene repression. Phosphorylation can induce either conformation depending on the other modifications present. Many nonhistone proteins can also be modified. The acetylation of transcription factors can facilitate gene expression by increasing their nuclear binding, nuclear retention, half-life, transcriptional activity, coactivator binding and/or recruitment, or by inducing corepressor dissociation (see Table 14-3). Nonhistone protein methylation usually affects recruitment of coactivators, corepressors, or other components of macromolecular complexes. HATs, HDACs, and PRMTs can be regulated by phosphorylation, recruitment, and allosterism. Many kinases are controlled by second messengers.

Template activity can also be altered by DNA methylation, which is usually associated with gene inactivation. However, hypomethylation is not closely related to transcription; instead, it appears to be a general marker for potential hormone sensitivity. As such, DNA methylation appears to fix gene inactivity brought about by some other factor. Methylated sequences can repress transcription either directly by blocking the binding of certain transcription factors or indirectly by binding repressor proteins. Hormones can regulate DNMTs indirectly by regulating histone modifications; for example, H3 methylation recruits HP1-DNMT.

Finally, hormones can affect the enzymes that transcribe the template. RNA polymerase II is phosphorylated by kinases whose recruitment is regulated by NRs, and S-5 can be directly phosphorylated by ERK, a signal-regulated kinase. A MAPK can also modify and stimulate RNA polymerase I, whereas Myc can recruit RNA polymerase III.

References

Nucleosome Components

Ausió, J., Abbott, W., Wang, X., and Moore, S.C. (2001). Histone variants and histone modifications: A structural perspective. *Biochem. Cell Biol.* 79, 693-708.

Thomas, J.O., and Travers, A.A. (2001). HMG 1 and 2, and related 'architectural' DNA-binding proteins. *Trends Biochem. Sci.* 26, 167-174.

Acetylation

Bannister, A.J., and Miska, E.A. (2000). Regulation of gene expression by transcription factor acetylation. *Cell Mol. Life Sci.* 57, 1184-1192.

Bertos, N.R., Wang, A.H., and Yang, X.J. (2001). Class II histone deacetylases: Structure, function, and regulation. *Biochem. Cell Biol.* 79, 243-252.

Brown, C.E., Lechner, T., Howe, L., and Workman, J.L. (2000). The many HATs of transcription coactivators. *Trends Biochem. Sci.* 25, 15-19.

Chan, H.M., and La Thangue, N.B. (2001). p300/CBP proteins: HATs for transcriptional bridges and scaffolds. *J. Cell Sci.* 114, 2363-2373.

Chen, H., Tini, M., and Evans, R.M. (2001). HATs on and beyond chromatin. *Curr. Opin. Cell Biol.* 13, 218-224.

Cress, W.D., and Seto, E. (2000). Histone deacetylases, transcriptional control, and cancer. *J. Cell Physiol.* 184, 1-16.

Gray, S.G., and Ekström, T.J. (2001). The human histone deacetylase family. *Exp. Cell Res.* 262, 75-83.

Kouzarides, T. (2000). Acetylation: A regulatory modification to rival phosphorylation? *EMBO J.* 19, 1176-1179.

Marmorstein, R. (2001). Structure of histone acetyltransferases. *J. Mol. Biol.* 311, 433-444.

Ng, H.H., and Bird, A. (2000). Histone deacetylases: Silencers for hire. *Trends Biochem. Sci.* 25, 121-126.

Ogryzko, V.V. (2001). Mammalian histone acetyltransferases and their complexes. *Cell. Mol. Life Sci.* 58, 683-692.

Roth, S.Y., Denu, J.M., and Allis, C.D. (2001). Histone acetyltransferases. *Annu. Rev. Biochem.* 70, 81-120.

Verdin, E., Dequiedt, F., and Kasler, H.G. (2003). Class II histone deacetylases: Versatile regulators. *Trends Genet.* 19, 286-293.

Phosphorylation

Giet, R., and Prigent, C. (1999). Aurora/Ipl1p-related kinases, a new oncogenic family of mitotic serine-threonine kinases. *J. Cell Sci.* 112, 3591-3601.

Poly(ADP-ribosyl)ation

Chiarugi, A. (2002). Poly(ADP-ribose) polymerase: Killer or conspirator? The 'suicide hypothesis' revisited. *Trends Pharmacol. Sci.* 23, 122-129.

Smith, S. (2001). The world according to PARP. *Trends Biochem. Sci.* 26, 174-179.

Methylation

Kouzarides, T. (2002). Histone methylation in transcriptional control. *Curr. Opin. Genet. Dev.* 12, 198-209.

Lachner, M., and Jenuwein, T. (2002). The many faces of histone lysine methylation. *Curr. Opin. Cell Biol.* 14, 286-298.

McBride, A.E., and Silver, P.A. (2001). State of the Arg: Protein methylation at arginine comes of age. *Cell (Cambridge, Mass.)* 106, 5-8.

Stallcup, M.R. (2001). Role of protein methylation in chromatin remodeling and transcriptional regulation. *Oncogene* 20, 3014-3020.

Zhang, Y., and Reinberg, D. (2001). Transcription regulation by histone methylation: Interplay between different covalent modifications of the core histone tails. *Genes Dev.* 15, 2343-2360.

Monoubiquitination

Moore, S.C., Jason, L., and Ausió, J. (2002). The elusive structural role of ubiquitinated histones. *Biochem. Cell Biol.* 80, 311-319.

Ulrich, H.D. (2002). Degradation or maintenance: Actions of the ubiquitin system on eukaryotic chromatin. *Eukaryotic Cell* 1, 1-10.

Interactions

Berger, S.L. (2002). Histone modifications in transcriptional regulation. *Curr. Opin. Genet. Dev.* 12, 142-148.

Fischle, W., Wang, Y., and Allis, C.D. (2003). Histone and chromatin cross-talk. *Curr. Opin. Cell Biol.* 15, 172-183.

Jenuwein, T., and Allis, C.D. (2001). Translating the histone code. *Science* 293, 1074-1080.

Richards, E.J., and Elgin, S.C.R. (2002). Epigenetic codes for heterochromatin formation and silencing: Rounding up the usual suspects. *Cell (Cambridge, Mass.)* 108, 489-500.

Turner, B.M. (2002). Cellular memory and the histone code. *Cell (Cambridge, Mass.)* 111, 285-291.

DNA Methylation

Attwood, J.T., Yung, R.L., and Richardson, B.C. (2002). DNA methylation and the regulation of gene transcription. *Cell Mol. Life Sci.* 59, 241-257.

Dobosy, J.R., and Selker, E.U. (2001). Emerging connections between DNA methylation and histone acetylation. *Cell Mol. Life Sci.* 58, 721-727.

Geiman, T.M., and Robertson, K.D. (2002). Chromatin remodeling, histone modifications, and DNA methylation—How does it all fit together? *J. Cell Biochem.* 87, 117-125.

Robertson, K.D. (2002). DNA methylation and chromatin—unraveling the tangled web. *Oncogene* 21, 5361-5379.

RNA Polymerases

Kobor, M.S., and Greenblatt, J. (2002). Regulation of transcription elongation by phosphorylation. *Biochim. Biophys. Acta* 1577, 261-275.

Prelich, G. (2002). RNA polymerase II carboxy-terminal domain kinases: Emerging clues to their function. *Eukaryotic Cell* 1, 153-162.

CHAPTER **15**

Posttranscriptional Control

CHAPTER OUTLINE

The previous chapters in this part have primarily analyzed gene control from a transcriptional perspective; in this chapter other, posttranscriptional, control points are covered. The three major focal points are RNA processing, translational control, and posttranslational modifications.

RNA Processing

Stability

In the rat mammary gland, prolactin stimulates the accumulation of casein mRNA by 34-fold, but the absolute rate of transcription increases only 2-fold. The difference is made up by the change in half-life, which increases from 5.4 hours in cultures without prolactin to 96 hours in cultures with this hormone. In this system, gene expression, as evidenced by mRNA accumulation, is primarily evoked by altering mRNA half-life. This mechanism operates in many systems; in chick oviduct, estrogens increase the mRNA half-life of ovalbumin from around 4 to 5 hours to 24 hours and that of conalbumin from 3 to 8 hours; in *Xenopus* liver, these same steroids increased the half-life of vitellogenin mRNA from 16 hours to 3 weeks. Decreasing the half-life of mRNAs is also possible. In cultures of chick embryo hepatocytes, triiodothyronine (T_3) induces the malic enzyme, whereas glucagon inhibits it. T_3 elevated the malic enzyme mRNA 11- to 14-fold, although the transcriptional rate only doubled, suggesting an increased half-life. However, glucagon definitely affects the half-life by reducing it from 8 to 11 hours to only 1.5 hours; glucagon has no effect on the transcription of this enzyme. Finally, a single hormone can affect the half-lives of two different mRNAs in opposite ways: in a pituitary cell line, thyrotropin-releasing hormone (TRH) stimulates prolactin mRNA accumulation but inhibits growth hormone (GH) mRNA accumulation. TRH increases the half-life of the prolactin mRNA from 17 to 27 hours and reduces that of the GH mRNA from 24 to 15 hours. TRH does not affect mRNA processing in this system.

There are several factors that can affect mRNA stability; the first is the length of the poly(A) tail. There exists a group of poly(A) binding proteins that combine with this tail to form nucleosome-like structures; these particles are relatively resistant to nuclease attack. Protein kinase C (PKC) lengthens the poly(A) track on corticotropin-releasing hormone (CRH) mRNA by 100 nucleotides, and 3'5'-cyclic AMP (cAMP) does the same for vasopressin mRNA; on the other hand, T_3 decreases the length of the poly(A) tail on thyroid-stimulating hormone β (TSHβ) mRNA. The most obvious target for these hormones and second messengers is the poly(A) polymerase, and there are numerous reports of hormones affecting the activity of this enzyme. Unfortunately, most studies are done under nonsaturating conditions. To understand this problem, assume that under basal conditions only 25% of the enzyme is being used; after hormone exposure, RNA synthesis triples. Now 75% of the enzyme is being used, and its total endogenous activity will increase even though enzyme number and intrinsic activity are unchanged. In other words, the hormone is not directly stimulating enzyme activity but only increasing substrate levels. For one to determine if a hormone is actually altering enzyme number or specific activity, the enzyme should be assayed under conditions in which it is maximally used; that is, substrate should be saturating.

As alluded to previously, there are two mechanisms for increasing the activity of poly(A) polymerase. Steroids increase enzyme number as evidenced by the

Table 15-1
Summary of Hormonally Regulated RNA Processing

Site	Mechanism	Example
mRNA Stability		
Poly(A) polymerase	Induction	Glucocorticoid on GH mRNA
	Phosphorylation	PKC (indirect) in T cells
Poly(A) binding protein (p38)	Phosphorylation	CK2 increases RNA binding; Tyr kinases decrease it
A/U-binding proteins	Induction	β-Agonist for βAR mRNA
	Repression	Insulin for βAR mRNA
	Phosphorylation	PKA induced dissociation from PEPCK mRNA; p38/MAPKAP-2 sequesters tristetraprolin
	Methylation	Required for HuR stabilization of A/U mRNAs
	A/U splicing	Deletion in CREMγ mRNA by FSH
p190RhoGDS	Bind destabilizing element	Stabilizes light neurofilament subunit mRNA
IRF (3′)	Activation	NO and PKC (nuclease resistance)
Endoribonuclease	Induction and pY	E_2 in liver on albumin mRNA
Processing		
Capping, splicing and poly(A) synthesis	Recruitment	MAPK phosphorylation of CTD of Pol II recruits enzymes
Methylation: PIMT	Recruitment	To PPARγ and RXRα via adaptor
Splicing		
U5, PGC-1, PSF	Recruitment	Nuclear receptors
SR	Dephosphorylation	Ceramide via PP1 (switch from Bclx[L] and caspase 9b to Bclx[s] and caspase 9)
SRp40	Phosphorylation	Insulin via PI3K (PKB?) stimulates alternate splicing of PKCβII
CBP80	Phosphorylation	S6KI stimulates cap binding activity and cap-dependent splicing
Sam68	Phosphorylation	MAPK stimulates splicing
SF1	Phosphorylation	PKG blocks spliceosome assembly
hnRNPA1	Phosphorylation	PKCζ or PKA (inhibits hnRNPA1 to stimulate splicing); p38 shifts hnRNPA1 to cytosol
Interchromatin granule	PIP_2	Required for mRNA splicing clusters
Editing		
APOBEC-1	Phosphorylation	PKC stimulates
TRNA transglycosylase	Phosphorylation	PKC required
Nuclear Egression and Mitochondrial Import		
p110	Phosphorylation	PKC increases mRNA transport in liver
nucleolin	Phosphorylation	CK2 increases pre-rRNA transport and helicase activity
ATPase	Allosterism (?)	Stimulated by steroids and cAMP in isolated nuclei
Topoisomerase II	Phosphorylation	CK2, PKCα, -β or -γ stimulate activity; inhibited by PKCζ
AKAP121	Phosphorylation	PKA stimulates mRNA binding and mitochondrial uptake

requirement for protein synthesis; the effect of glucocorticoids on the poly(A) tail of GH mRNA would be an example (Table 15-1). On the other hand, PKC increases the specific activity of the polymerase in T lymphocytes. This action certainly suggests phosphorylation as a mechanism for regulating enzyme activity; however, *in vitro* the enzyme is a very poor substrate for PKC. Only an NI-like kinase has been shown to phosphorylate quantitatively and to activate significantly poly(A) polymerase. It is possible that PKC triggers *in vivo* a kinase cascade that ends with the NI-like kinase. Finally, poly(A) polymerase can be regulated by localization. As noted in Chapter 14, mitogen-activated protein kinase (MAPK) phosphorylation of the CTD (carboxy-terminal domain) of RNA polymerase II recruits enzymes involved with poly(A) synthesis, as well as capping and splicing. Poly(A) binding proteins can also be regulated by phosphorylation. Casein kinase 2 (CK2) phosphorylation of p38 increases its RNA binding, whereas tyrosine phosphorylation decreases it.

Conversely, mRNA can be destabilized by the presence of A/U-rich repeats in the 3′ untranslated region. Once again, these sequences are recognized by specific RNA-binding proteins; the difference is that many of these A/U-binding proteins promote the degradation of the mRNA. An exception is HuR, whose binding stabilizes these labile mRNAs. Hormones can affect this system in four ways; first of all, they can alter the level of the proteins that promote mRNA degradation. β-Agonists downregulate the β-adrenergic receptor (βAR), in part, by inducing these proteins, whereas insulin and dexamethasone elevate βAR mRNA by repressing the A/U-binding proteins. A second mechanism involves the phosphorylation of these proteins; protein kinase A (PKA) phosphorylation causes these proteins to dissociate from the mRNAs for phosphoenolpyruvate carboxykinase and insulin-like growth factor binding protein 3 (IGFBP-3) and extends their half-lives. In another example, tristetraprolin binds to the A/U-rich repeats in the tumor necrosis factor α (TNFα) mRNA and promotes its degradation. Phosphorylation of tristetraprolin by the p38-MAPKAP-2 (MAPK-activated protein kinase 2) pathway results in its cytoplasmic sequestration by 14-3-3 with an increase in the TNFα mRNA half-life. Third, PRMT4 can be recruited by NRs to methylate HuR; this modification is required for HuR to stabilize mRNAs with A/U-rich repeats. Finally, splicing can eliminate some of these sequences; in the testis, follicle-stimulating hormone (FSH) induces CREMγ (cAMP response element modulator) with an alternate poly(A) site, which eliminates 9 of the 10 A/U-rich regions in the 3′ tail.

The iron response factor (IRF) is another mRNA binding protein. IRF is actually homologous to aconitase, an enzyme in the tricarboxylic acid cycle. However, nitric oxide (NO) will displace the iron and convert the enzyme into a protein that binds a specific sequence, called the *iron response element* (IRE), that forms a hairpin loop in the 3′ end of mRNAs for proteins induced during iron deficiency; for example, transferrin, which is involved with the cellular uptake of iron. IRF stabilizes this loop and inhibits nucleases. IRF can be further activated by PKC phosphorylation, which increases mRNA binding. Essentially, IRF is an iron sensor; when iron is abundant, it recombines with IRF to reform the aconitase and these mRNAs are degraded.

As previously described, stability means protection from nucleases; therefore these enzymes represent a potential control point. Poly(A) binding proteins recruit αCP, an endoribonuclease inhibitor, to the poly(A) tail. G3BP, a ribonuclease, is activated by phosphorylation. However, binding to the SH3 (Src homology 3) domain of RasGAP leads to its dephosphorylation and inactivation. Hormones can also stimulate these enzymes; for example, estradiol preferentially lowers albumin mRNA levels in *Xenopus* liver by activating an endonuclease that favors the albumin mRNA as substrate; this stimulation occurs through tyrosine phosphorylation. In addition, a poly(A)-specific 3′ exonuclease can be activated by spermidine.

Finally, hormones may affect mRNA stability indirectly. Through stimulation of translation initiation (see later), the ribosome density increases and protects the mRNA. In addition, there may a shift in poly(A) binding proteins. As previously noted, estradiol increases vitellogenin mRNA and decreases albumin mRNA; the massive elevation of vitellogenin mRNA will outcompete the albumin mRNA for the limiting amount of poly(A) binding protein, leaving the latter exposed to ribonucleases (RNases).

Modifications

The addition of a poly(A) track is not the only modification that RNA undergoes after transcription; many RNA species are also subjected to cleavage, splicing, and base modifications. In liver nucleoli from Leghorn roosters, the conversion of rRNA from a 32S precursor to the 28S mature form can be followed in pulse-chase experiments. In the absence of estrogen, this processing requires 20 minutes, but following estrogen treatment the conversion only requires a few minutes. Another type of processing requires nucleotide modifications, such as methylation. Both estradiol and prolactin stimulate tRNA methylase 2- to 3-fold in immature rat uteri and mouse mammary explants, respectively. Unfortunately, much of these data were obtained under nonsaturating conditions; therefore it is not clear if the processing machinery was really being changed or just more effectively used.

There are several mechanisms by which hormones can affect splicing; for example, many nuclear receptors (NRs) recruit splicing factors directly or through coactivators. Transcriptional activation domain 1 (TAD1) of androgen receptor (AR) binds ANT-1, a splicing factor that binds the U5 component of snRNP (small nuclear ribonucleoprotein). AR also binds the polypyrimidine tract binding protein-associated splicing factor (PSF) through the adaptor, FHL-2 (four and a half LIM-only protein 2), and PPARγ (peroxisome proliferator activated receptor γ) binds the PPARγ coactivator 1 (PGC-1), another splicing factor.

Alternately, splicing factors can be phosphorylated. Insulin stimulates the alternate splicing of PKCβII through a PI3K (phosphatidylinositol-3 kinase)–dependent phosphorylation of SRp40, a factor required for splice selection. S6KI phosphorylation of the CBP80 subunit of the RNA cap-binding complex (CBC) stimulates cap-binding activity and cap-dependent RNA splicing. Cdk2 (cell cycle-dependent kinase) phosphorylation of splicing factor 2 (SF2/ASF) increases hnRNP (heterologous nuclear RNP) U binding and spliceosome assembly. Finally, MAPK phosphorylation of Sam68 stimulates the splicing of CD44 mRNA. On the other hand, heterologous RNP A1 (hnRNPA1) binds splice sequences and inhibits splicing. PKCζ and PKA phosphorylate and inhibit hnRNPA1, thereby stimulating splicing, and p38 phosphorylation shifts hnRNPA1 out of the nucleus and into the cytosol. Finally, protein kinase G (PKG) phosphorylation of SF1 blocks its binding to other splicing factors and prevents prespliceosome assembly.

Dephosphorylation is another mechanism for regulation. Phosphorylated serine-arginine-rich protein (SR) generates mRNA for Bclx(L) and caspase 9b, which inhibit apoptosis. Ceramide activates PP1 (protein phosphatase), which dephosphorylates SR; unmodified SR generates alternately spliced mRNAs that synthesize Bclx(s) and caspase 9, which stimulate apoptosis. Finally, it has been shown that another second messenger, PIP$_2$ (phosphatidylinositol 4,5-bisphosphate), is required for mRNA splicing, although its mechanism is not known. PIP$_2$ is present in interchromatin granule clusters and may be involved either in the transport of mRNA or as a substrate for nuclear PLC to elevate calcium levels.

Enzymes involved with base modification can also be regulated by localization and phosphorylation. PIMT (PRIP interacting protein with methyltransferase domain), an RNA methyltransferase, is recruited to PPARγ and RXRα through the adaptor, PRIP (PPAR interacting protein). APOBEC-1 (apolipoprotein B editing catalytic subunit 1) converts C to U, which changes the codon for glutamine to one for stop in the mRNA for apolipoprotein B; this alteration prematurely terminates translation. PKC phosphorylation of APOBEC-1 stimulates its activity eightfold. PKC is also required for the activity of tRNA:guanine transglycosylase, which exchanges G for queuine; however, direct phosphorylation of this enzyme has not yet been demonstrated.

Nuclear Egression

Eukaryotes have their genetic material enclosed within a nuclear membrane, which acts as a barrier between transcription and translation. Its effectiveness is attested to by the fact that most mRNAs never leave the nucleus and are eventually degraded. Therefore nuclear egression of mRNA is certainly a potential control point. Dexamethasone induces α_{2u}-globulin in rat liver; after adrenalectomy, the mRNA for this protein accumulates in the nucleus, whereas the mRNA content in polysomes falls. Within 2 hours of dexamethasone administration, nuclear mRNA levels decline, whereas those in the polysomes rise; it was assumed that this transfer was too fast to be mediated by transcription, although no transcription inhibitors were tested. Furthermore, the size of this mRNA is unchanged, suggesting that processing is not involved. The conclusion is that glucocorticoids could directly stimulate the nucleocytoplasmic transport of mRNA.

How might this nuclear egression be regulated? Research has implicated a nuclear nucleotide triphosphatase in the transfer of mRNA from the nucleus to the cytoplasm. Human chorionic gonadotropin stimulated this enzyme 2- to 3-fold in the isolated nuclear membranes of luteal cells. This stimulation does not occur in the nuclear membranes of nontarget organs or in the nonnuclear membranes of luteal cells. Similar results have been reported in other systems; estradiol (E_2) stimulates the adenosine triphosphatase (ATPase) almost 40-fold in isolated uterine nuclei, whereas insulin and cAMP have more modest effects in liver nuclei. Where they are examined, these hormones affect only the V_{max} of the enzyme; the K_m is not significantly altered. p110 is a nuclear envelope protein associated with mRNA transport, and its phosphorylation by PKC is correlated with mRNA egression. Finally, topoisomerase II facilitates the exit of mRNA; as noted in Chapter 13, this enzyme is stimulated by CK2, PKCα, PKCβ, and PKCγ, and it is inhibited by PKCζ.

In contrast to mRNA, the egression of pre-rRNA involves the protein nucleolin, which also possesses RNA helicase activity. Insulin, in picomolar concentrations, rapidly stimulates the phosphorylation of nucleolin through CK2 to stimulate both pre-rRNA efflux and helicase activity. Cdk2 phosphorylation also stimulates helicase activity.

The nucleus is not the only membrane barrier that mRNAs may encounter; for example, several mRNAs must enter the mitochondria to be translated. AKAP121 (A kinase anchoring protein 121) is a PKA anchoring protein located on the outer mitochondrial membrane. PKA phosphorylation of AKAP121 stimulates its ability to bind to these mRNAs and enhances their translocation into mitochondria.

Translation

The transport of mRNA into the cytoplasm does not automatically result in its translation. Prolactin injection into pseudopregnant rabbits induces both casein mRNAs and casein proteins, but in virgin rabbits, almost no casein is synthesized, although its mRNA is induced 58-fold and is found attached to ribosomes. In another example, the C57BL mouse strain is infected with the mouse mammary tumor virus but only exhibits a low mammary tumor incidence. During lactation, the viral RNA is produced in abundance and is transported to the cytoplasm, where it cosediments with polysomes. However, no viral peptides are produced. The RNA is normal because it can be purified from these glands and be translated in the reticulocyte lysate system.

Regulation of General Translation

Many hormones are known to facilitate protein synthesis by inducing the cellular machinery for translation. This is particularly notable in systems in which hormones stimulate large amounts of protein secretion, as in milk production. In the mammary gland, both cortisol and prolactin have been shown to induce tRNAs, 28S rRNA, eEF-2 (eukaryotic elongation factor 2), and the formation of rough endoplasmic reticulum.

In other systems this effect can be semiselective. For example, if the protein to be made has a biased amino acid composition, the hormone may stimulate only the tRNAs for those amino acids, thereby favoring the synthesis of that protein. Alanine, serine, and glycine represent 80% of all the amino acids in silk fibroin. In the silk glands of the fifth instar silkworm, the tRNAs for these three amino acids are induced along with their respective acyl tRNA synthetases. This alteration in tRNA abundance is required for the efficient translation of this protein (Table 15-2). The antifreeze protein in winter flounder represents a hormonally regulated system; 60% of the amino acids in the antifreeze protein are alanines. The onset of winter causes GH to decline; this fall leads to an increase in both the mRNA for the antifreeze protein and in a single alanine isoacceptor. This isoreceptor is presumed to be the one that recognizes the codon GCC, which codes for 70% to 75% of all of the alanines in the mRNA for the antifreeze protein.

Regulation of Selective Translation

In addition to tRNA abundance, there are other examples of even more selective control. The 1,25-DHCC (1,25-dihydroxycholecalciferol) doubles the rate of chain elongation of chromogranin A in the parathyroid gland without affecting the translation of parathormone. In another example, polyamines inhibit the enzymes responsible for their synthesis by decreasing enzyme number. However, because the mRNAs for these enzymes do not change, the effect must be posttranscriptional. Indeed, in a reticulocyte lysate system, polyamines selectively inhibit the translation of the mRNAs for ODC (ornithine decarboxylase) and SAM (*S*-adenosylmethionine) decarboxylase; the translation of total protein or serum albumin was unaffected. In other words, in some systems it is possible for a hormone to direct the protein synthetic machinery to translate one or a small group of mRNAs while ignoring others. A similar phenomenon occurs in plants, where jasmonic acid induces defense genes, while it inhibits the translation of preexisting mRNAs. These mRNAs appear normal, but chain initiation is reduced.

Table 15-2
Summary of Hormonally Regulated Translation

Site	Mechanism	Example
General Translation		
Selective tRNA induction	Induction	Silk and antifreeze proteins
Translation factors	Phosphorylation	See Table 15-3
	Hypusine	eIF-5A in mitogenesis
mRNA-Binding Components		
3′ UTR masking element binding protein	Phosphorylation	MAPK-like phosphorylation leads to dissociation
Polyamines	Bind to GC-rich sequences	Inhibition of SAM decarboxylase
	Frameshift	Not read STOP codon in ODC antizyme
IRF (5′)	Activation	NO or PKC (activate IRF to inhibit translation)
Calreticulin	Phosphorylation	Inhibits translation
Miscellaneous		
eEF1A2	Binding by M4	Increase GNP exchange
Mdm2	Binding to amino terminus	Stimulate p53 translation
Compartmentalization		
mRNP3 and mRNP4 masking proteins	Phosphorylation	CK2 (required for inhibition; recruits other proteins)
FRGY2	Phosphorylation	CK2 (inhibits translation by promoting mRNA binding)
EIF4E	Unknown	Integrin shifts eIF4E from membrane to cytoskeleton
Protein Folding	Phosphorylation	CK2 and MAPK recruit chaperone to endoplasmic reticulum-bound ribosomes
	Calcium	Effect of plant hormones on α-amylase folding

Where is this control exerted? The induction of heat shock proteins provides a clue. When *Drosophila* cells are incubated at 36° C, special heat shock genes are transcribed, and their mRNAs are translated; no other mRNA is translated. This latter mRNA is not degraded and is normally translated when the temperature returns to 25° C. This system can be dissected further by incubating cells at either 25° C or 36° C and separating the lysates, which contain the protein synthetic machinery, from the mRNAs; the fractions can then be recombined in different ways. The high-temperature mRNA contains both normal and heat shock mRNAs but, when it is recombined with the high-temperature lysate, only the heat shock mRNAs are translated. This result is expected based on the experiments in whole cells. The low-temperature mRNA contains only normal mRNA and, when recombined with the high-temperature lysate, it is still not translated. This eliminates the possibility that the normal mRNAs were reversibly inactivated at the elevated temperature. When high-temperature mRNA is mixed with a low-temperature lysate, both normal and heat shock proteins are synthesized. This result suggests that the selectivity resides in the translational machinery.

Molecular Mechanisms

What are the molecular bases for affecting the activity of the translational machinery? Covalent modifications of the tRNA synthetases, initiation factors, elongation factors, or the ribosome may explain this phenomenon (see Fig. 12-7 and Table 15-3). For example, PKC phosphorylation of the valyl tRNA synthetase increases its activity. Of the initiation factors, eIF-2 (eukaryotic initiation factor 2) has been one of the most extensively studied; this factor is responsible for putting the first tRNA in place. It is a heterotrimer; the α subunit binds guanosine 5′triphosphate (GTP), the β subunit is involved with the GTP exchange, and the γ subunit binds Met-tRNA$_f^{Met}$. The phosphorylation of eIF-2α constitutes one aspect of the physiological control of hemoglobin. Hemoglobin consists of an iron-containing porphyrin ring, heme; and a protein, globin. In iron-deficient anemia, there is little iron and heme; in the absence of iron and heme, it is senseless to synthesize the globin. Therefore low heme concentrations activate the heme-regulated inhibitor kinase (HRI) that phosphorylates the α subunit and inhibits all translation; in reticulocytes, globin is virtually the only protein being made. Three other stress-related kinases act in a similar manner: PKR (dsRNA-activated protein kinase) is stimulated by interferon to shut down all translation and abort a viral infection. GCN2 (general control nonderepressible 2) is activated by amino deficiency, and PERK (PKR-like endoplasmic reticulum kinase) is stimulated by unfolded proteins. Although these kinases usually inhibit translation, they preferentially affect mRNA that binds weakly. Other mRNA, such as that for ATF4, can actually be stimulated. Several other hormonally regulated kinases can also modify eIF-2α with similar, but less dramatic, effects (Table 15-3). However, because most of these latter kinases are activated by growth factors whose actions would need enhanced translation, their effects are paradoxical and their physiological role, uncertain. Rather, most hormones act on eIF-2 through its GEF (GNP exchange facilitator), eIF-2B. GSK3 (glycogen synthase kinase 3) phosphorylation of eIF-2B is inhibitory; insulin reactivates eIF-2B by stimulating protein kinase B (PKB) to phosphorylate and inhibit GSK3 (see Chapter 12).

Another intensively studied initiation factor is eIF-4E, which binds the mRNA cap structure. Its regulation was discussed in Chapter 12. Briefly, 4E-BP (eIF-4E binding protein) is an inhibitor that binds and sequesters eIF-4E. Phosphorylation of 4E-BP results in the release of eIF-4E and the stimulation of protein synthesis. However, the kinase responsible for this phosphorylation is hotly debated; S6KI and mTOR are the major candidates. One of the reasons for the interest in eIF-4Eα is its potential for selectivity; its function is to melt secondary structures in mRNA to facilitate translation. This effect would be greatest in those mRNAs with such structures. For example, the mRNA for ornithine decarboxylase, the rate-limiting enzyme in polyamine synthesis, has such secondary structures and is preferentially translated following insulin stimulation. If the 5′ end of this mRNA is spliced onto other mRNAs, they too will become insulin responsive.

Elongation factors and ribosomal proteins can also be phosphorylated. eEF-2 kinase (formerly called CaMKIII) modifies eEF-2 and blocks translation; however, PKA phosphorylation of eEF-2 renders it active independent of eEF-2 kinase phosphorylation. In addition, eEF-2 kinase itself can be phosphorylated and inhibited by either S6KI (through mTOR) or S6KII (through MAPK) depending on the hormone: for example, IGF-I inactivates eEF-2 kinase via the former pathway. Ribosomal protein S6 has been the focus of many studies both because it is the mRNA binding site on the ribosome and because its phosphorylation is stimulated by so many growth factors. However, such modifications enhance general translation only modestly; rather, the phosphorylation of ribosomal protein S6 may increase the translation initiation of specific classes of mRNA. For example, it has been suggested that TOP mRNAs may be the target of

Table 15-3

Occurrence and Effects of Phosphorylation on the Translation Machinery

Translation factor	Function	Phosphorylation[a]					
		PKC	S6KI	CK2	PKA	Other	
Initiation Factors							
eIF-2α	Complex with GTP and initiator Met-tRNA and binds to 40S preinitiation complex	+	+	+		HRI, PKR, GCN2, PERK (all inhibit); dephosphorylated by insulin	Generally, kinases decrease translation by decreasing GTP exchange and recycling (forms nonfunctional complex with eIF-2β); but some translation (e.g., ATF4) is stimulated and others begin at internal ribosomal sites; CK2 required for activity
eIF-2Bε	GTP exchange			+		GSK3 (inhibitory)	Increases GTP exchange fivefold (partially overcomes phosphorylation of eIF-2α)
eIF-4B	eIF-4 recycling and mRNA binding to 40S preinitiation complex	+	+	+	+	Insulin; EGF	Selective translation
eIF-4E	Binds m7G cap of mRNA	+	+		+	Insulin, EGF, PDGF and TNFα (through Mnk)	PKA and PKC increase binding to capped mRNA; Mnk decreases binding to allow recycling
4E-BP	Binds and inhibits eIF-4E binding to eIF-4γ		+			PKB (via S6KI?); mTOR (?); MAPK	Decreases affinity for (and inhibition of) eIF-4E
eIF-4G	Required for translation of capped mRNA (scaffolding protein)	+	+		+	S6KII	Stimulates binding of capped mRNA to preinitiation complex
eIF-5	Facilitates GTP hydrolysis on eIF-2α			+			Increases 80S assembly but only when eIF-5 is limiting

Elongation Factors	Function	[a]	Protein kinase	Effect
eEF-1Bα and -β[b]	Amino acyl tRNA binding	+	Insulin-activated multipotential S6K[c], Cdk2	S6K increases translation 3-fold; PKC increases GTP exchange
eEF-1Bγ[b]	Amino acyl tRNA binding		Cdk2	Inhibits GTP exchange
eEF-2	Translocation	+	eEF-2 kinase	eEF-2 kinase blocks translocation; S6K and p38 inhibit eEF-2 kinase; PKA renders eEF-2 independent of eEF-2 kinase
Ribosomal Proteins				
S6	mRNA binding site on ribosome	+	PKG and CaMKII	Increases AUG binding and translation modestly
S18	Initiation and elongation	+	CaMKII	Stimulates protein synthesis

[a] +, Indicates physiological phosphorylation detected.
[b] eEF-1Bα, -β, and -γ were formerly called eEF-1δ, -β, and -γ, respectively.
[c] A new S6K distinct from S6KI and S6KII; it also phosphorylates eEF-1A (formerly called eEF-1α).

phosphorylated S6 protein. TOP mRNAs have a 5′ terminal oligopyrimidine (TOP) tract, which is characteristic of proteins involved with translation. Although the translation of TOP mRNAs requires PKB, the role of S6KI is still controversial.

Not all modifications are phosphorylations; eIF-5A is modified by butylamine, which is derived from spermidine. The butylamine is attached to a lysine and then hydroxylated to form hypusine (see Chapter 11). Although this modification is not required for general translation, it is required for cell division. Therefore regulation of the synthesis of hypusine may control the translation of mRNAs for genes involved with mitosis. Because this reaction is dependent on the polyamine concentration, those hormones that elevate polyamine levels could trigger this modification and facilitate the synthesis of proteins needed for cell division.

Besides covalent modifications, translation can be affected by mRNA binding factors. As noted previously, polyamines suppress the translation of the mRNA for the polyamine synthetic enzymes; in the SAM decarboxylase gene, this effect is a result of polyamine binding to the guanine-cytosine (GC)–rich region in the 5′ end. The ODC antizyme represents a different type of control. Its mRNA has a stop signal at codon 36; polyamines cause a frameshift that results in the production of the full-length protein. This represents negative feedback because the ODC antizyme would halt any more polyamine synthesis. In zygotes, mRNA is masked by 3′ binding proteins; phosphorylation by proline-directed kinases, like Cdk2 and MAPK, induces dissociation and translation of the mRNA. Another regulatory mechanism used by hormones is allosterism; eEF1A2 binds the third intracellular loop of the M4 receptor, and M4 stimulates GNP exchange and speeds the recycling of this factor.

As already noted, genes involved with iron metabolism have a specific sequence, the IRE, that is bound by the IRF. When the IRE is at the 3′ end, IRF binding stabilizes the mRNA; these mRNAs code for proteins induced during iron deficiency, like transferrin. However, when the IRE is at the 5′ end of the mRNA, IRF binding blocks translation; 5′ IREs occur in mRNAs for protein repressed during iron deficiency, like ferritin. IRF is created when NO causes the iron to dissociate from an aconitase-like protein. Like IRF, calreticulin serves multiple functions within the cell: calreticulin normally acts as a chaperone in the endoplasmic reticulum (see later), but when phosphorylated by PKC, it binds a stem-loop structure in the 5′ region of the C/EBPβ mRNA and inhibits translation. In addition to binding mRNA, regulators can affect translation by binding nascent proteins. Mdm2 normally functions as an E3 ligase for the antioncogene p53 (see Chapter 16), but it can actually stimulate p53 synthesis when bound to the nascent antioncogene.

Finally, translation can be regulated by compartmentalization of the mRNA into ribonucleoprotein (RNP) particles. For example, mRNAs for ribosomal proteins are distributed between polysomes and RNP particles; the latter are translationally inactive. The degree of translation is, therefore, determined by a balance between the two pools. In lymphosarcoma cells dexamethasone inhibits the translation of these mRNAs by shifting them into RNP particles. Translation factors can also be compartmentalized; integrins shift eIF-4E from the plasma membrane to the cytoskeleton, where the ribosomes are located. CK2 plays a pivotal role in regulating these proteins. CK2 phosphorylation of mRNP3 and mRNP4 is required for them to mask mRNA. This modification does not affect their affinity for mRNA; rather, it probably recruits other accessory proteins. CK2 also phosphorylates the frog Y-box protein 2 (FRGY-2); phosphorylation promotes oligomerization, which is required for binding mRNA.

Posttranslational Regulation

Protein Folding

Proteins destined for secretion or insertion into the plasma membrane are synthesized on ribosomes bound to the endoplasmic reticulum (ER). These proteins are *N*-glycosylated with a core sequence of oligosaccharides that end in glucose. Calreticulin and calnexin are ER chaperone proteins that recognize proteins with these terminal glucoses and retain them in the ER. A glucosidase removes these glucoses to release the proteins and allow them to fold. If they do not fold properly within a reasonable time, another enzyme reattaches the glucoses, and the proteins rebind calreticulin and calnexin to keep them from being transferred to the Golgi complex prematurely. The cycle then repeats itself until folding is complete, at which point the glucoses are not reattached and the proteins move to the Golgi complex.

As described in Chapter 10, the ER is an important reservoir for calcium. The calcium concentration ranges from 240 to 400 μM when fully loaded to 18 to 40 μM after hormone-induced depletion. The higher calcium level is required for calreticulin to bind glycoproteins. Protein disulfide isomerase is another chaperone, whose function is to catalyze the formation of disulfide bonds. In low calcium concentrations, the isomerase binds calreticulin and is inhibited. Therefore elevated calcium levels induce the dissociation and activation of the isomerase and enable the calreticulin to bind glycoproteins. Although these calcium effects are real and occur during physiological fluxes, it is not known if this actually serves as a regulatory mechanism.

However, calcium does function in the hormonal control of protein folding of at least one plant protein: α-amylase. Seeds contain a starchy center surrounded by a cellular layer, the aleurone layer. During germination, the cells of the aleurone synthesize α-amylase, which breaks the starch down so that it can provide the energy for early growth. These events are hormonally regulated: abscisic acid maintains dormancy, while gibberellic acid triggers germination. Gibberellic acid stimulates the influx of calcium, which is required for the proper folding and stabilization of α-amylase. Abscisic acid blocks this elevation in calcium; α-amylase is misfolded and rapidly degraded.

In contrast to calreticulin, calnexin is regulated by phosphorylation. Both CK2 and MAPK phosphorylate the cytosolic tail of calnexin; these phosphorylations act synergistically to localize calnexin to ER-bound ribosomes.

Posttranslational Modifications

Once a protein is synthesized, there are many modifications that it may undergo. These alterations are performed in a very orderly sequence, as the protein traverses several cellular organelles on its way to being secreted (Table 15-4). *N*-linked glycosylation involves the attachment of sugar residues onto an asparagine, as opposed to *O*-linked glycosylation, in which monosaccharides are coupled to a serine or a threonine. In the former, an entire oligosaccharide side chain is synthesized on a carrier lipid, which then transfers this core structure

Table 15-4
Location of Some Post-translational Modifications

Location	Modification
Rough endoplasmic reticulum	Cleavage of the signal sequence, hydroxylation, some end-group blockers, cross-linking by disulfide bonds, core glycosylation, addition of core PI-glycan, myristoylation
Golgi apparatus	Terminal glycosylation, *O*-linked glycosylation, phosphorylation, palmitoylation, sulfation
Secretory vesicles	Cleavage (as of the C peptide in insulin)
Extracellular	Cleavage (as in zymogens), cross-linking (as in collagen and fibrin)

en bloc to the nascent polypeptide. This is called *core glycosylation*. Although the protein is still in the rough ER, some of the sugars at the ends are removed; this continues in the Golgi apparatus. Finally, new sugars are added back to the ends; each type of protein gets a unique sequence. This is called *terminal glycosylation*. As discussed later, core glycosylation appears to be automatic, but terminal glycosylation is frequently under hormonal regulation and is associated with secretion.

The hormonal control of posttranslational modification has been investigated in a number of systems, only two of which are discussed: the mouse mammary tumor virus and TSH. The virus is a standard retrovirus having six genes:

1. The *gag* gene codes for two structural polyproteins.

2. The *pol* gene codes for the reverse transcriptase.

3. The *env* gene codes for a single glycosylated polyprotein, which will be split into two envelope proteins.

4. The *pro* gene overlaps the *gag* and *pol* genes and codes for a protease used in processing some of the other gene products.

5. The *du* gene overlaps the *pro* gene but requires a frameshift to be read; it codes for a dUTPase (deoxyuridine 5'-triphosphate diphosphohydrolase).

6. The *sag* gene in the 3' end of the genome codes for a superantigen that interacts with the immune system of the mouse.

The processing of the polyproteins is somewhat variable, but the following schemata appear to represent the predominant pathways. One of the *gag* products, Pr74^{gag}, first must be phosphorylated to Pr76^{gag} before cleavage can take place; the numbers refer to the protein molecular masses in kilodaltons. This phosphorylation is stimulated by dexamethasone; indirectly, so is the subsequent processing by cleavage because the precursor, Pr74^{gag}, accumulates in the absence of dexamethasone. In a similar manner, the *env* product, Pr60^{env}, must be glycosylated before further processing can occur. The core glycosylation to Pr74^{env} is automatic, but the terminal glycosylation to gp78 requires dexamethasone.

In another example, TRH stimulates the secretion of TSH from rat pituitary cell cultures. TSH is, however, inhibited by T_3 or SRIF (somatotropin release-inhibiting factor). The former represents specific negative feedback inhibition, whereas the latter is a general inhibitor of hormone release. None of these factors affects TSH synthesis or core glycosylation; however, TRH stimulates, and T_3 and SRIF inhibit, terminal glycosylation. These effects are selective because terminal glycosylation of other proteins is not affected. Finally, if terminal glycosylation is pharmacologically inhibited by monensin, TRH-induced TSH release is also suppressed; inhibition of core glycosylation by tunicamycin has no effect on TRH-stimulated TSH secretion. The conclusions are that

1. TSH is constitutively synthesized and core glycosylated.
2. Factors affecting secretion act at the terminal glycosylation step.
3. Terminal glycosylation immediately leads to secretion.

The coupling between terminal glycosylation and secretion appears to be a general phenomenon; for example, in mouse mammary explants, T_3 stimulates the terminal glycosylation and release of α-lactalbumin.

There are three molecular mechanisms underlying these effects (Table 15-5). First of all, hormones can induce the enzymes involved in these modifications. For example, glucocorticoids induce the expression of mannosidase II, a trimming enzyme, and estradiol can induce several of these enzymes, including

Table 15-5
Summary of Hormonally Regulated Protein Modification and Stability

Site	Mechanism	Example
Posttranslational Modification		
Glycosylation	Induction	Glucocorticoids and E_2
	Phosphorylation	PKA stimulation of mannosylphoshoryldolichol synthase; PKG stimulation of NAG transferase I
	Calcium	α1,2-mannosidase
Phosphorylation	Per kinases	See Table 12-1
Nitrosylation	Ca^{2+}/CaM activation of NOS	Stimulates Ras (Cys); inhibits prostacyclin synthase (Tyr)
Farnesylation/ geranylgeranylation	MAP kinase phosphorylation of farnesyl-/ geranyltransferase I	Insulin induces farnesyltransferase 2X and modifies Ras
Cleavage	Phosphorylation	p105 by IKK triggers cleavage
	Ca^{2+}-calpain	α subunit of L channel (short form 4X more active)
Protein Stability		
Protease activation	Ca^{2+}-calpain	Degradation of iNOS, tau, and spectrin
	Oligomerization	Caspases
	Phosphorylation	PKC inhibits calpastatins
Protease inhibition	Phosphorylation	PKB inhibits caspase 9
	S-Nitrosylation	NO inhibits caspase 8
	Allosterism	Glucocorticoid blocks insulin binding and degradation to its protease
	Competition vs. allosterism	AT_4 inhibits insulin-regulated aminopeptidase
Increased substrate susceptibility	Phosphorylation	S6KII stimulates IκB proteolysis
	Allosterism	ODC antizyme binds ODC and exposes PEST region
Decreased substrate susceptibility	Phosphorylation	PKC inhibits cleavage of MARCKS; JNK inhibits Jun ubiquitination
	Polyamines	Bind spermidine-spermine N^1-acetyltransferase and block ubiquitination
	Sulfotransferases	E_2 induction (cap and protect like sialic acid)

mannosylphosphoryldolichol synthase, which charges the lipid carrier with mannose, and oligosaccharyltransferase, which transfers the oligosaccharide core from the lipid carrier to the nascent protein. Estradiol can also induce enzymes involved with O-glycosylation, such as N-acetylgalactosyltransferase, which is the first enzyme in the synthesis of this type of oligosaccharide.

Second, these enzymes could be more acutely regulated by phosphorylation. For example, the mannosylphosphodolichol synthase is stimulated by phosphorylation with PKA, and NAG (N-acetylglucosamine) transferase I is stimulated by PKG phosphorylation. Finally, these enzymes may be affected by allosteric modulators; for example, calcium is a second messenger and can activate $\alpha1,2$-mannosidase, another trimming enzyme.

Other posttranslational modifications can also be hormonally regulated. Chapter 12 was entirely dedicated to phosphorylation. Nitrosylation of cysteines and nitration of tyrosines were covered in Chapter 7. For example, S-nitrosylation stimulates Ras but inhibits NF-κB, whereas tyrosine nitration of prostacyclin synthase is inhibitory. NO is produced by nitric oxide synthase (NOS), which is stimulated by calcium and calmodulin (CaM).

Insulin increases the activities of farnesyltransferase and geranylgeranyltransferase I through MAPK phosphorylation. This stimulation results in a two-fold increase in Ras modification. Finally, cleavage can also be hormonally regulated. p50 is a component of NF-κB and is generated from p105, whose cleavage is triggered by IKK (IκB kinase) phosphorylation. In a second example, the α-subunit of the Ca$_V$1 channel can exist in either a short or a long form; the former is four times more active than the latter. Glutamate increases the amount of the short form by elevating calcium and activating calpain, which cleaves the α-subunit.

Protein Stability

Hormones can also affect protein stability. For example, in mouse mammary explants, cortisol will progressively stimulate casein accumulation over a concentration range of 10 ng/ml to 1 μg/ml; however, the actual rate of synthesis is no higher at 1 μg/ml than at 10 ng/ml. On the other hand, the half-life of the casein is markedly longer at the higher steroid concentration. At the lower hormone concentration, about half of all milk proteins are degraded in 40 hours; however, in the presence of 1 μg of cortisol per milliliter, only 10% of the casein is degraded, although the destruction of the other proteins remains at around 50%.

There are two major targets for controlling protein stability: the protease and the substrate. The role of calcium in activating the protease calpain was discussed in Chapter 10. In addition to calcium, PKC can activate calpains, albeit indirectly; PKC phosphorylation of calpastatins increases the calcium required for calpastatins to inhibit calpain. This elevated threshold decreases the effectiveness of calpastatins. Although calpain is usually associated with discrete cleavage and activation of proteins, it can also degrade proteins, such as iNOS (inducible NOS), tau, and spectrin. Caspases are proteases involved with apoptosis; they are activated by oligomerization through TNF receptors and related adaptors.

Hormones can also inhibit proteases. PKB phosphorylation of caspase 9 and S-nitrosylation of caspase 8 inhibit their activities. In some cases, inhibition can be selective; PKA phosphorylation of metalloendopeptidase decreases its activity toward gonadotropin-releasing hormone (GnRH) but not toward neurotensin. Allosterism is another mechanism to control proteases. The insulin-regulated aminopeptidase (IRAP) prefers peptides with an amino-terminal cysteine; these substrates include many small peptide hormones. AT_4 binds and inhibits IRAP, thereby prolonging the half-lives of the other peptide hormones.

Hormones can increase or decrease substrate susceptibility to proteases. Phosphorylation of IκB targets this factor for ubiquitination and degradation, whereas the phosphorylation of Jun by JNK (Jun amino-terminal kinase) inhibits its ubiquitination. Phosphorylation of substrates near cleavage sites can also prevent cleavage by altering the recognition site for the protease; PKC phosphorylation of MARCKS (myristolated alanine-rich C kinase substrate) and Src phosphorylation of NMDA (N-methyl-D-aspartate) receptor are examples of this mechanism. Substrates can be masked by sulfation; sulfates cap and protect oligosaccharides in glycoproteins in a manner similar to sialic acid. Estradiol can induce these sulfotransferases. Finally, polyamines can bind spermidine-spermine N^1-acetyltransferase and block ubiquitination. This represents a form of negative feedback because the polyamines are prolonging the half-life of an enzyme that inactivates them.

Summary

Hormones can have an impact on protein expression and function at multiple levels. Chapters 11 and 12 showed how transducers influenced enzyme activity and the cytoskeleton, and Chapters 13 and 14 demonstrated how second messengers affected transcription. This chapter completed the continuum by examining the involvement of hormones with posttranscriptional events. Specifically, hormones may positively and negatively affect RNA processing, nuclear egression, translation, posttranslational modifications, and protein stability. Phosphorylation of enzymes or translation factors is a major mechanism of regulation, but calcium and compartmentalization also contribute to the control of these processes.

For further illustration of these points, the effects of four hormones on prolactin induction of milk proteins are reviewed; most of the data have been presented separately in previous chapters (Table 15-6). At the receptor level, insulin and cortisol are required to maintain normal levels of the prolactin receptor; T_3 further elevates these levels and progesterone lowers them. Receptor

Table 15-6
The Interactions of Insulin, Cortisol, Progesterone, and Triiodothyronine on Prolactin Action in the Murine Mammary Gland

Parameter	Insulin[a]	Cortisol[a]	Progesterone[a]	T_3[a]
Prolactin Receptor	+	+	−	+
Second Messengers				
Polyamines		+		
Prostaglandins			−	
mRNA Accumulation				
Casein	+	+	−	
α-Lactalbumin	+	+	−	+
Translation	+	+	−	+[b]
Posttranslational Modification				+
Casein Half-Life		+		

[a] +, Stimulation; −, inhibition.
[b] Data from rabbits only.

concentrations are important because they will determine the sensitivity of the cell to any particular hormone. Progesterone also lowers the insulin receptor number and is capable of binding, but not activating, the cortisol receptor; essentially, progesterone at the elevated levels observed during pregnancy acts as a competitive antagonist. Additional sites of progesterone interference include (1) blocking prolactin stimulation of prostaglandin synthesis, (2) suppressing prolactin induction of casein gene transcription, and (3) inhibiting the translation of milk proteins. Both insulin and cortisol stimulate casein and α-lactalbumin gene transcription, and the latter hormone also stabilizes the resulting mRNAs. Insulin, cortisol, and T_3 also induce components of the translational machinery. Cortisol has the additional effects of increasing the half-life of casein and of augmenting the prolactin stimulation of polyamine synthesis by increasing both SAM decarboxylase and spermidine synthetase activities. Finally, T_3 selectively stimulates α-lactalbumin terminal glycosylation.

The rationale behind such a complex regulation is not known. In part, it may be related to the necessity of coordinating lactation with parturition, suckling, and maternal metabolism. In any event, one cannot entirely understand or appreciate the actions of a hormone by only examining one of its effects on gene expression in the absence of all other synergistic and antagonistic factors.

References

RNA Processing

Akker, S.A., Smith, P.J., and Chew, S.L. (2001). Nuclear post-transcriptional control of gene expression. *J. Mol. Endocrinol.* 27, 123-131.

Bevilacqua, A., Ceriani, M.C., Capaccioli, S., and Nicolin, A. (2003). Post-transcriptional regulation of gene expression by degradation of messenger RNAs. *J. Cell. Physiol.* 195, 356-372.

Bouton, C. (1999). Nitrosative and oxidative modulation of iron regulatory proteins. *Cell. Mol. Life Sci.* 55, 1043-1053.

Chen, C.Y.A., and Shyu, A.B. (1995). AU-rich elements: Characterization and importance in mRNA degradation. *Trends Biochem. Sci.* 20, 465-470.

Csermely, P., Schnaider, T., and Szántó, I. (1995). Signalling and transport through the nuclear membrane. *Biochim. Biophys. Acta* 1241, 425-452.

Eisenstein, R.S., and Blemings, K.P. (1998). Iron regulatory proteins, iron responsive elements and iron homeostasis. *J. Nutr.* 128, 2295-2298.

Ginisty, H., Sicard, H., Roger, B., and Bouvet, P. (1999). Structure and functions of nucleolin. *J. Cell Sci.* 112, 761-772.

Ross, J. (1996). Control of messenger RNA stability in higher eukaryotes. *Trends Genet.* 12, 171-175.

Shim, J., and Karin, M. (2002). The control of mRNA stability in response to extracellular stimuli. *Mol. Cells* 14, 323-331.

Translation

Dever, T.E. (2002). Gene-specific regulation by general translation factors. *Cell (Cambridge, Mass.)* 108, 545-556.

Dufner, A., and Thomas, G. (1999). Ribosomal S6 kinase signaling and the control of translation. *Exp. Cell Res.* 253, 100-109.

Gingras, A.C., Raught, B., and Sonenberg, N. (2001). Regulation of translation initiation by FRAP/mTOR. *Genes Dev.* 15, 807-826.

Kleijn, M., Scheper, G.C., Voorma, H.O., and Thomas, A.A.M. (1998). Regulation of translation initiation factors by signal transduction. *Eur. J. Biochem.* 253, 531-544.

Kobor, M.S., and Greenblatt, J. (2002). Regulation of transcription elongation by phosphorylation. *Biochim. Biophys. Acta* 1577, 261-275.

Martin, K.A., and Blenis, J. (2002) Coordinate regulation of translation by the PI 3-kinase and mTOR pathways. *Adv. Cancer Res.* 86, 1-39.

Meyuhas, O. (2000). Synthesis of the translational apparatus is regulated at the translational level. *Eur. J. Biochem.* 267, 6321-6330.

Rhoads, R.E. (1999). Signal transduction pathways that regulate eukaryotic protein synthesis. *J. Biol. Chem.* 274, 30337-30340.

Scheper, G.C., and Proud, C.G. (2002) Does phosphorylation of the cap-binding protein eIF4E play a role in translation initiation? *Eur. J. Biochem.* 269, 5350-5359.

Shah, O.J., Anthony, J.C., Kimball, S.R., and Jefferson, L.S. (2000). 4E-BP1 and S6K1: Translational integration sites for nutritional and hormonal information in muscle. *Am. J. Physiol.* 279, E715-E729.

Shantz, L.M., and Pegg, A.E. (1999). Translational regulation of ornithine decarboxylase and other enzymes of the polyamine pathway. *Int. J. Biochem. Cell Biol.* 31, 107-122.

Wilson, K.F., and Cerione, R.A. (2000). Signal transduction and post-transcriptional gene expression. *Biol. Chem.* 381, 357-365.

PART **5**

Special Topics

CHAPTER 16

Pathogen-Endocrine System Interactions

Hormones have long been associated with certain cancers; in human beings, estrogens have been linked to breast and uterine tumors, whereas prolactin is associated with mammary tumors in rodents. Furthermore, certain endocrine deficiencies are known to have protective effects; for example, untreated eunuchs do not have prostate cancer, and patients with untreated Turner's syndrome, which includes nonfunctional ovaries, do not have breast cancer. Still, there has always been the question of whether these effects are direct or indirect. Are these hormones actually carcinogenic, or do they merely induce a proliferative state, which renders the tissue more susceptible to other agents?

Although the role of hormones in tumor induction is not clear, they are involved in the growth of certain tumors, most notably estrogens in breast and uterine cancers and androgens in prostate cancer. Furthermore, the appropriate endocrinectomy leads to remission. Unfortunately, even this association is variable; not all tumors are responsive, and even those that initially respond frequently relapse. These discrepancies might be explained by proposing that hormones could have induced hormone-dependent tumors, which then developed the ability to synthesize their own hormones; at this point, the tumors became independent of exogenous hormones. Certainly, many tumors do produce peptide hormones not made in the original tissue, and although no nonadrenal, nongonadal tumor can make steroids *de novo*, some mammary tumors do possess steroid-metabolizing enzymes, which allow them to synthesize active steroids from circulating precursors. However, the presence of these ectopic hormones or enzymes does not always correlate with the hormone-independent state. Because of counterregulatory mechanisms in animals, the simple ectopic production of hormones may induce hyperplasia but does not appear to be sufficient for malignant transformation.

Rather, it appears necessary to modify these hormones, or some other component of their transduction pathway, such that they remain active even in the face of counterregulatory measures. The genes coding for the normal proteins are called *proto-oncogenes*, whereas those generating constitutively active signaling molecules are called *oncogenes*. However, one oncogene is still not enough; usually there must be at least two. The first oncogene is usually plasma membrane-associated and induces *transformation*; transformation is the process by which cells become independent of environmental checks on cell growth. Such cells, for example, will grow in soft agar or will ignore cell-cell contacts and form multilayered cultures. The second oncogene is often nuclear-associated and induces *immortalization*; this is the process by which cells overcome senescence and is correlated with premalignancy. Most malignancies are associated with such pairs of oncogenes.

Certain viruses can also cause tumors, and they do so by means of the same mechanisms. Viruses are thought to be derived from pieces of the host's genome that have acquired an independent existence. If these pieces contain a proto-oncogene, the potential exists for its mutation into an oncogene that can induce cancers. Indeed, all of the known viral oncogenes (v-*onc*) have cellular counterparts (c-*onc*). This is an excellent strategy; viruses only have limited space in their genome, but cell growth is a complex phenomenon requiring the coordination of many cellular processes. Rather than trying to control all of these processes individually, the virus merely plugs into a preexisting regulatory system: the mitogenic pathway.

This chapter considers these various topics. First, it explores how pathogens circumvent the immune system with particular reference to the endocrine control of immunity. Then it discusses the various ways that oncoviruses can hijack the mitogenic pathway and how viruses can coordinate their life cycle with the host's physiology. Finally, it discusses defense mechanisms that have been developed against pathogens, specifically, antioncogenes.

Viral and Bacterial Genes

Infection

Intracellular pathogens must successfully accomplish three tasks in order to survive and reproduce. First, they must gain entry to the cell; this process is called *infection*. Second, they must evade the host's immune system; this activity is called *virulence*. Finally, they must create an environment conducive to their own reproduction. Often, this involves activating the host's transcriptional and translational machinery: a process that frequently leads to *mitogenesis* and *oncogenesis*.

Infection occurs when a pathogen binds to the surface of the cell and either fuses with the plasma membrane or is internalized and fuses with the endosomal membrane. This is not a nonspecific interaction; most pathogens have surface structures that have a high affinity to some integral membrane protein of the host. Although pathogens can use any such protein, hormone receptors are particularly well suited for this function (Table 16-1). First, hormone specificity is determined by the distribution of its receptor; as such, the use of a hormone receptor allows the pathogen to target certain cell types for infection. Second, hormone receptors are highly regulated; an important part of this control involves internalization. This process allows viruses that require an acidic environment for fusion to gain access to the endosomal compartment. Finally, hormone receptors regulate cell function and metabolism. If the pathogen does not just bind the receptor but also activates or inhibits it, then it could initiate the processes of virulence and/or mitogenesis.

Virulence Genes and the Immune System

Many bacteria and viruses possess virulence genes whose products facilitate the infection process, usually by interfering with the immune system (Table 16-2). Several viral proteins are either homologues or analogs of hormones; such proteins are called *virokines*. Those that are inactive are antagonists that block immune function; for example, MC53L and MC148R are interleukin 18 (IL-18) and chemokine antagonists, respectively (Table 16-2). On the other hand, agonists may serve other functions. For example, the immune system is potentially very damaging and is physiologically regulated by both activators and suppressors. IL-10 is a suppressor that inhibits cytokine synthesis, and BCRF1 (Epsetin-Barr virus *Bam*HI C fragment, rightward reading *frame 1*) is a viral homologue that mimics this function. Other viral agonists may use chemotaxis to recruit new targets for infection and/or stimulate growth (see the following section on the mitotic pathway).

In addition, some pathogens affect cytokine and steroid production. For example, IL-1β is synthesized as a precursor that must be cleaved to its active form, like proinsulin and angiotensinogen (see Chapter 2). Cytokine response modifier protein A (crmA) is a serine protease inhibitor (serpin) that inhibits the

Table 16-1
Some Hormone Receptors Bound by Pathogen Proteins for Activation or Cell Entry

Pathogen	Hormone Receptor
Virus	
Adeno-associated virus 5	Platelet-derived growth factor
Avian leukosis-sarcoma virus	Tumor necrosis factor–like
Epstein-Barr Virus	Interferonα-C3d
Feline and murine leukemia virus	Tumor growth factor β
Fibropapillomavirus	Platelet-derived growth factor
Friend spleen focus-forming virus	Erythropoietin
Herpes simplex, type 1	Fibroblast growth factor[a], insulin growth factor II, nerve growth factor[a]
Human cytomegalovirus	Epidermal growth factor
Human immunodeficiency virus	CD4, nicotinic acetylcholine, glutamate[b], chemokines[c]
Poxviruses	Epidermal growth factor
Rabies	Nicotine acetylcholine, nerve growth factor[a]
Reovirus	β-Adrenergic[d]
Bacteria	
Neisseria gonorrhoeae	Luteinzing hormone
Salmonella	Epidermal growth factor
Streptococcus pneumoniae	Platelet-activating factor
Actinobacillus actinomycetemocomitans	Platelet-activating factor
Listeria monocytogenes	Hepatocyte growth factor
Protozoa	
Plasmodium vivax (malaria)	Chemokines

[a] Low affinity receptor only.
[b] Binds NMDA at the glycine site.
[c] Some *Herpesvirus* and *Capripoxvirus* also bind chemokine receptors.
[d] Also increases epidermal growth factor (EGF) binding and epidermal growth factor receptor (EGFR) autophosphorylation, although it does not compete with EGF; EGFR facilitates but is not required for infection.

IL-1β converting enzyme. In a second example, the vaccinia virus produces a homologue of the 3β-hydroxysteroid dehydrogenase; although its exact function is unknown, it would be reasonable to assume that it elevates glucocorticoids, which are known to be immunosuppressive. Finally, cytokines can also be neutralized by sequestration. Many viruses code for cytokine receptors, called *viroceptors*. The soluble forms are believed to act as sinks for the cytokines. These receptors include those for interleukins, tumor necrosis factor (TNF), and interferon γ (IFNγ). Bacteria can use a similar strategy; the *Yersinia* genome contains a gene that codes for a homologue of the platelet membrane protein platelet glycoprotein Ibα (GPIbα). This protein is a general receptor for several components of the blood clotting system, including thrombin and von Willebrand factor, and it mediates platelet activation. It is thought that the GPIbα homologue sequesters thrombin and blocks its role in inflammatory reactions.

However, like virokines, some viroceptors are active and appear to function by recruiting new targets and disseminating the infection. In other cases, these active receptors induce the heterologous desensitization of other immune receptors that use the same transducers.

Other pathogen products interfere with the signaling pathways; some bind C4b or C3b to block the complement pathway, whereas others prevent the plasma membrane translocation of β$_2$-microglobulin or the major histocompatibility complex class I. *Helicobacter pylori* produces an arginase that depletes arginine

Table 16-2
Some Pathogen-Immune Interactions

Protein	Pathogen	Action
Cytokine Synthesis		
crmA	Cowpox[a]	Serpin that inhibits IL-1β converting enzyme and apoptotic proteases
BCRF1	Epstein-Barr[b]	IL-10 homologue; inhibits cytokine synthesis
MC53L	Molluscum contagiosum	IL-18 antagonist
MC148R	Molluscum contagiosum	Chemokine antagonist
p15E	Leukemia[c]	TGFβ agonist
SalF7L	Vaccinia[a]	3β-Hydroxysteroid dehydrogenase
Gingipains	*Porphyromonas gingivalis*	TNF proteases
Cytokine Sequestration		
B15R	Vaccinia[a]	IL-1β-like receptors (soluble)
B18R	Vaccinia[a]	IFN-αβ-like receptors (soluble)
ECRF3	Herpes	α Chemokine receptor
T2	Myxoma[a]	TNF receptor, type I (soluble)
T7	Myxoma[a]	IFNγ receptor (soluble)
US28	Herpes	β Chemokine receptor
YOPM	*Yersinia*	GPIbα homologue; binds thrombin
Complement		
VCP	Vaccinia[a]	Binds C4b; inhibits classical pathway
CCPH	Herpes saimiri	Binds C4b; inhibits classical pathway
gC-1	Herpes simplex	Binds C3b; inhibits both pathways
gE-gI	Herpes simplex	Binds Fc of IgE
Antigen Recognition		
E3-gp19K	Adenovirus	Binds MHC-I and prevents translocation to surface
UL18	Cytomegalovirus[b]	MHC-I homologue; binds $β_2$-microglobulin and prevents translocation to surface
Transduction		
v-Akt	AKT8[c]	= PKB (antiapoptotic)
E4-ORF1	Adenovirus	Stimulates PI3K
rocF	*Helicobacter pyloris*	Arginase depletes arginine
VA[d]	Adenovirus	Inhibits phosphorylation of eIF-2
EBER	Epstein-Barr[b]	Inhibits phosphorylation of eIF-2
σ3[d]	Reovirus	Inhibits phosphorylation of eIF-2
TAR[d]	HIV[c]	Inhibits phosphorylation of eIF-2
YOPJ	*Yersinia*	Binds and inhibits MAPKK
LMP2A	Epstein-Barr[b]	Binds and inhibits STKs; blocks calcium mobilization
YOP2b, YOP51	*Yersinia*	Protein tyrosine phosphatases
E8	Herpes	Death domain blocks association of adaptors with TNFR and apoptosis
MC159, MC160	Molluscum contagiosum	Death domain blocks association of adaptors with TNFR and apoptosis
Fumonisins	*Fusarium monilforme*	Inhibits ceramide synthase (substrate analog)
Transcription		
Ad12-13S-E1A	Adenovirus	Blocks processing of p105 to p50
v-Rel	Reticuloendotheliosis virus[c]	NF-κB-like transcription factor (dominant inhibitor)
A238L	African swine fever virus[e]	IκB; also binds and inhibits PP2B, thereby inhibiting NFAT
V protein	SV 5[f]	Targets Stat1 to proteasome
vIRF	KSHV[b]	Binds and inhibits p300

[a] DNA poxviruses.
[b] DNA herpes viruses.
[c] Retroviruses.
[d] VA, and possibly TAR, RNA binds to and inactivates PKR. σ3 is a capsid protein that binds dsRNA.
[e] DNA iridovirus.
[f] DNA papovavirus.

in macrophages; this decreases nitric oxide (NO) production, which normally activates the immune response. Several viral products inhibit the IFN-induced protein kinase, which normally limits viral infection by phosphorylating eukaryotic initiation factor 2 (eIF-2) and freezing translation. Others bind soluble tyrosine kinases (STKs), like Lyn and Fyn, and prevent their activation of the polyphosphoinositide (PPI) pathway in B lymphocytes. STKs are critical components of cytokine receptor transduction (see Chapter 7), and they can be neutralized by protein tyrosine phosphatases, which several pathogenic bacteria produce. Finally, several pathogen products thwart apoptosis; apoptosis is an attempt by the immune system to control an infection by killing the infected cells. Several pathogen products have death domains that block the association of adaptors with the tumor necrosis factor receptor (TNFR). Fumonisin, a fungal toxin, inhibits ceramide synthase by acting as a substrate analog; E4-ORF1 from type 9 adenoviruses binds and activates phosphatidylinositol-3 kinase (PI3K), whose product stimulates protein kinase B (PKB). Other viruses make PKB homologues. PKB is a potent antiapoptotic kinase (see Chapter 12).

The major transduction pathways for immune cells often terminate with the transcription factor NF-κB (nuclear factor that stimulates κ chain gene in B cells). The hormone response element (HRE) for this factor appears in the 5′ end of many defense genes. However, the effects of pathogen proteins on NF-κB are variable. NF-κB induces a stress response that could be detrimental to viruses, although apoptosis could release and spread the virus. On the other hand, NF-κB is also antiapoptotic and mitogenic for certain cell types, which could be beneficial to viruses. The choice may depend on which other pathways are being activated by the virus, and/or the stage of infection, and/or the cell type. For example, tuberculous bacillus inhibits apoptosis in the host cell but induces it in macrophages, and *Chlamydia* inhibits host apoptosis early in the infection but induces it later after replication has occurred.

Oncogenes and the Mitotic Pathway

Viral and bacterial products that actually induce tumor formation are involved with the mitogenic pathway (Table 16-3). First of all, these oncogene products may be either mitogens, like fibroblast growth factor (FGF) or platelet-derived growth factor (PDGF), or enzymes that synthesize nonpeptide mitogens. The latter situation occurs in plants infected with *Agrobacterium tumefaciens*, which contains the T_i plasmid. This bacterium causes crown gall tumor, which can grow in culture without the plant growth hormones cytokinin and indoleacetic acid. The reason is simple: three of the seven genes on the plasmid code for enzymes that can synthesize these hormones from common precursors. Gene 4 codes for an isopentenyl transferase, which transfers an isopentenyl group to AMP to form a cytokinin (Fig. 16-1, *A*). The products of genes 1 and 2 are involved in the synthesis of indoleacetic acid. Gene 2 codes for an amidohydrolase; and although the product of gene 1 has not been positively identified, it is probably a tryptophan monooxygenase (Fig. 16-1, *B*). This two-step pathway does occur in certain bacteria; however, in plants the synthesis of indoleacetic acid normally proceeds through indole-3-pyruvic acid and indole-3-acetaldehyde. *Agrobacterium rhizogenes* uses a different strategy: its oncogenes code for glucosidases that liberate these plant mitogens from their inactive conjugates. The *rolB* and *rolC* genes produce indole β-glucosidase and cytokinin β-glucosidase, respectively.

Table 16-3
Some Oncogene Products Associated with the Endocrine System

Group	Oncogene proteins[a]	Cellular homologue
Hormones	Int-2, Hst	FGF-like
	Int-5	Aromatase
	Sis	PDGF-like
	NZ2, NZ7	VEGF
	K2[b]	IL-6
	UL146, UL147[c]	α-Chemokine
	T$_i$ plasmid[d]	Synthetic enzymes for cytokinin and indoleacetic acid and their glucosidases
	Egt[e]	Ecdysteroid UDP-glycosyltransferase
Receptors		
GPCR	vGPCR[b]	IL-8
Cytokine	Mpl	Hematopoietic factor receptor
Nuclear	Erb-A	T$_3$ receptor (dominant inhibitor)
	MC013L[f]	Hsp 40 (binds & inhibits GR and VDR)
Tyrosine kinase	Erb-B, Neu	EGF-like receptors
	Flg, Bek	FGF-like receptor
	Fms	CSF-1 receptor
	Kit	SCF-like receptor
	Met	Hepatic growth factor receptor
	Sea	Insulin-like receptor
	Trk, TrkB	Neurotropin receptors (high affinity)
Transducers		
G proteins	Ras	G-like protein (small)
	Gsp, Gip[g]	G-like protein (trimer)
	Ost[g]	RhoGEF
	Vav	RacGEF
Kinases	PKCα[g]	Ser-Thr kinase
	Raf(=Mil), Mos	Ser-Thr kinases that phosphorylate MAPKK
	Src, Lck, Fyn, etc.	STKs associated with membrane receptors
	UL13[b]	Cdk2
Miscellaneous	Crk	SH2 adaptor
	Cyclin D[b]	Cdk regulatory subunit
	V-p3k	PI3K (p110)
Transcription Factors and Modulators		
	Elk-1	TCF (part of serum response factor complex)
	Fos, Jun	PKC-regulated transcription factor (AP-1)
	Myc, Max	CK2- and MAPK-regulated transcription factor
	Ski	Smad repressor

[a] All oncogene products are from retroviruses unless otherwise noted.
[b] From a herpesvirus; cyclin D is resistant to Cdk inhibitors.
[c] From CMV; facilitate viral dissemination.
[d] From *Agrobacterium tumefaciens* and *A. rhizogenes*.
[e] From granulovirus.
[f] From DNA poxvirus.
[g] Mutated proto-oncogene; no viral counterpart known.

The granulovirus of insects uses the reverse strategy: the product of the *egt* gene is an *e*cdysteroid UDP-*g*lycosyl*t*ransferase, which conjugates and inactivates ecdysteroid. This prevents molting and pupation (see Chapter 3). As a result, the insect larvae continue feeding and growing, which leads to higher viral yields.

Oncogenes can also code for receptors. Most of these receptors are receptor tyrosine kinases (RTKs) because of the central role they play in mitogenesis. All of these receptors harbor some mutation that renders them constitutively active. The

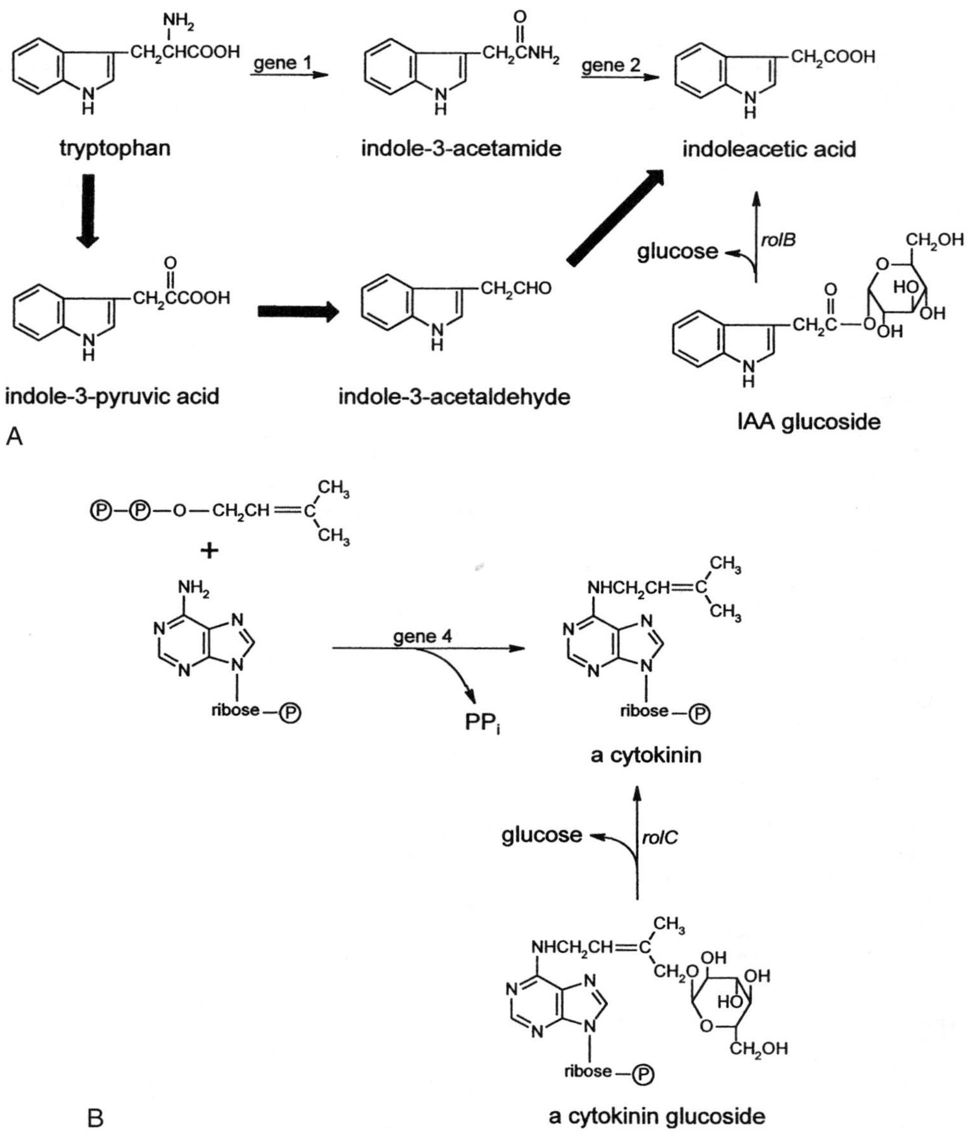

Fig. 16-1. Biosynthetic pathways for indoleacetic acid (A) and a cytokinin (B). Thin lines represent steps catalyzed by products of the T_i plasmid; heavy lines represent those steps that normally occur in plants. *IAA*, Indoleacetic acid.

only exception is Erb-A, which is a truncated triiodothyronine (T_3) receptor and a dominant inhibitor. Because T_3 often induces developmental programs, it may be considered a differentiative hormone (see Chapter 2); as such, its inhibition would favor growth. A similar situation is seen with molluscum contagiosum, a poxvirus that infects the skin. One of its gene products, MC013L, has limited homology to heat shock protein (hsp) 40, especially in the nuclear receptor (NR) box, which binds the transcription activation domain 2 (TAD2) of the glucocorticoid receptor (GR) and vitamin D receptor (VDR) and inhibits these NRs. Because cortisol inhibits keratinocyte growth and 1,25-DHCC (1,25-dihydroxycholecalciferol) promotes keratinocyte differentiation, MC013L may promote viral replication by inducing keratinocyte proliferation over differentiation.

Many components of the transduction system are also found among the oncogene products: small and trimer G proteins, STKs, Src homology domain 2 (SH2) adaptors, and some protein kinases. Finally, oncogenes can act at the level of transcription. For example, many oncogenes code for transcription factors involved with mitogenesis, only a few of which are shown in Table 16-3. Oncogenes can also code for transcription modulators, such as Ski (Sloan Kettering virus isolate), which represses Smad (Sma and mothers against dpp gene products). Smad mediates the transcriptional effects of the transforming growth factor β (TGFβ) family, which is associated with growth inhibition. Therefore the repression of Smad by Ski would favor growth.

The mitogenic pathway is frequently targeted by pathogens because both DNA and protein synthesis are very active during cell division and these processes are required by the pathogen for its own replication. However, other metabolic and cellular functions may also be useful to the pathogen and can be commandeered via hormone signaling pathways; for example, *Salmonella* and *Vibrio cholerae* produce diarrhea to disseminate their progeny. Intestinal fluid secretion is stimulated by $3'5'$-cyclic AMP (cAMP) and inhibited by IP_3 (inositol 1,4,5-trisphosphate). *Vibrio* elevates cAMP by producing cholera toxin, which inactivates the guanosine triphosphatase (GTPase) function of G_s resulting in its constitutive activation. On the other hand, *Salmonella* produces SopB, a phosphatase that not only hydrolyzes IP_3 to reduce its levels but also hydrolyzes IP_5 to IP_4, which antagonizes the action of any residual IP_3.

Hormone Regulation of Viral Expression

The relationship between hormones and oncogenes is bidirectional: not only do some oncogenes mimic the actions of hormones, but hormones can regulate the expression of oncogenes. Many viruses have HREs in the regulatory regions of their genomes. However, appearances can be deceiving; this is less an example of the host controlling the virus than of the virus sensing the endocrine status of the host to determine the best time to replicate.

The mouse mammary tumor virus (MMTV) is an excellent example. The MMTV is a retrovirus that is propagated from mother to pups. The virus is induced primarily in the mammary gland during late pregnancy and lactation, when it is secreted into the milk; the pups are infected through suckling. The virus is then carried by T lymphocytes from the intestine to the mammary gland, where its RNA is transcribed into DNA that randomly integrates into the host genome. When the female pups mature and become pregnant, the cycle repeats itself. Actually, MMTV infects many tissues in the mouse and can be naturally expressed in the immune and reproductive systems, as well as the salivary gland. However, expression in the mammary gland far exceeds that in any other tissue.

On the basis of this method of propagation, it is to the advantage of the virus to coordinate its reproductive cycle with that of its host; for example, viral production in the virgin would be futile. To this effect, the MMTV genome has a progesterone HRE in its 5' long terminal repeat (LTR); progesterone levels are elevated in pregnancy and may initiate viral transcription at a time when milk production and suckling are imminent. The human papilloma virus also has HREs for sex steroid receptors and infects reproductive tissues. In this example, the virus would be activated at the same time that the tissue is stimulated by these steroids; this stimulated state may favor viral replication. Alternatively, the presence of sex steroids could signal sexual maturity and activity, which would facilitate viral dissemination.

Reproduction in other tissues may require a different endocrine milieu. For example, in addition to mammary tumors, MMTV can cause T cell lymphomas. However, tumorigenesis in this tissue requires a deletion in the 5′ LTR; this deletion removes a tissue-specific inhibitory region and creates a new activator protein 1 (AP-1) site. The immune system is primarily dependent on the protein kinase C (PKC) transduction system, and many viruses that infect these tissues have HREs for AP-1 or NF-κB; an example of the latter would be the human immunodeficiency virus.

Antioncogenes

Mitosis is a complex process that is under both positive and negative control. In this section, the negative regulators are considered. Because they inhibit mitosis, while oncogenes stimulate it, these mitotic inhibitors are often referred to as products of *antioncogenes*. Antioncogenes are often divided into three categories based on their role in tumorigenesis. First, gatekeepers act directly to prevent tumor growth. Their inactivation is rate limiting for tumorigenesis, and restoration of their function suppresses the cancer. Second, caretakers are involved with DNA repair or stability. Their inactivation predisposes to cancer by increasing the DNA mutation rate, and restoration of their function does not suppress cancer if the secondary mutation has already occurred. Finally, landscapers modulate the microenvironment; they often give rise to polyclonal tumors. Antioncogenes can also be classified according to their intrinsic functions; for example, transcription regulators, phosphatases, cell adhesion molecules, and cell cycle regulators (Table 16-4). This latter classification is used here.

Transcription Regulators

Several antioncogene products affect transcription. The retinoblastoma (Rb) gene codes for a 105-kDa phosphoprotein, whose loss can lead to the development of retinoblastomas. In its active, hypophosphorylated state it can complex with several transcription factors and affect their activity. It binds to and inhibits Myc, TR (thyroid receptor), UBF (upstream binding factor, an RNA polymerase I cofactor), TFIID (human transcription initiation factor IID), RNA polymerase III, DRTF1 (E2F dimerization partner 1), and Elf-1, and it converts E2F (E2 transcription factor) from a transcription activator to an inhibitor. However, Rb is not an exclusive inhibitor; it is a positive cofactor for MyoD, C/EBPβ, BRG1 (Brahma-related gene 1), Jun, and Sp1. These effects are entirely consistent because Rb inhibits those factors that stimulate mitosis (Myc, E2F, RNA polymerases) and supports those factors that suppress proliferation and induce differentiation (MyoD, C/EBPβ). There is some controversy over whether Rb itself can bind DNA and activate transcription. Most data suggest that its DNA binding is nonspecific; however, it can recruit deacetylases and Abl, an STK that can phosphorylate and activate RNA polymerase II.

Rb can also inhibit mitosis by binding and inhibiting cyclin D, a regulatory subunit of the mitosis-associated kinase Cdk2 (cell cycle-dependent kinase 2) and

Table 16-4
Tumor Suppressors Related to the Endocrine System

Tumor suppressor	Function	Effects of Deficiency
Transcriptional Regulators		
Rb	Inhibits cell cycle gene transcription	Familial retinoblastoma
p53	Guardian of genome	Li-Fraumeni (breast and brain tumors, sarcomas, etc.)
WT1	Inhibits growth factor transcription	Wilms tumor (nephroblastoma)
BRCA	DNA repair	Familial breast cancer; Fanconi anemia including acute myeloid leukemia and squamous cell carcinoma (BRCA2 only)
Smad4	Mediates TGFβ effects	Juvenile polyposis
RUNX1	Mediates TGFβ effects	Leukemia
Phosphatases		
PTEN	Inositol 3′-phosphatase; protein phosphatase	Decreases PKB activity and apoptosis of tumor cells; may dephosphorylate FAK and remove contact inhibition; Cowden (hamartomas), Bannayan-Zonana (hemangiomas) and Lhermitte-Duclos (gangiocytomas) syndromes
PP2A	Dephosphorylates growth related kinases and transcription factors	Target of t oncogenes
Cell Adhesion		
APC	Inhibits β-catenin	Familial adenomatous polyposis
E-Cadherin	Cell adhesion	Familial gastric cancer
DCC	Receptor for axon morphogen	Colon cancer
Hamartin	Rho activity and cell adhesion; inhibits mTOR and β-catenin	Tuberous sclerosis (hamartomas)
NF2	Integrin-cytoskeleton interface	Neurofibromatosis, type 2
Cell Cycle		
TGFβRI	Mediates TGFβ effects	Metastatic breast cancer
TGFβRII	Mediates TGFβ effects	Head and neck squamous carcinomas
BMPR1A	Mediates BMP effects	Gastrointestinal hamartomatous polyposis
IGF-IIR	IGF clearance	Hepatocellular carcinoma
Cip/Waf	Cyclin-dependent inhibitors	Some breast cancers
INK	Cyclin-dependent inhibitors	Familial melanoma
GAP-Like Proteins		
NF1	RasGAP	Neurofibromatosis, type 1
Tuberin	RhebGAP; inhibits mTOR and β-catenin	Tuberous sclerosis (hamartomas)
Miscellaneous		
VHL	Proteolysis of O_2 sensor and RNA polymerase II	Von Hippel Lindau (cavernous hemangiomas)
Caspases	Apoptosis	Gastric and breast cancers

by recruiting H3 methylase and heterochromatin protein 1 (HP1) to the cyclin E promoter. However, this relationship is bidirectional; cyclin E binds both Rb and Cdk2, allowing the latter to phosphorylate and inactivate Rb. Raf is another kinase capable of phosphorylating Rb. Phosphorylation does not simply inactivate Rb; rather, it converts Rb from an inhibitor of mitosis to a stimulator of DNA polymerase α. Rb levels can also be regulated; TGFβ inhibits growth and elevates Rb levels. Furthermore, TGFβ, along with IFNα and IL-6, inhibits the phosphorylation of Rb. Sequestration is another mechanism of regulating Rb; p55γ, a regulatory subunit of PI3K, binds and inhibits Rb, and activation of PI3K liberates Rb.

Finally, Rb can be targeted by viral oncogene products. The active, hypophosphorylated Rb is bound by E1A from the adenovirus, the large T antigen from SV40, E7 from the papilloma virus, and EBNA-5 from the Epstein-Barr virus. These oncogene products pull Rb off of the transcription factors and stabilize the free form. E1A also binds cyclin A-Cdk2 and may recruit the kinase to phosphorylate and inactive Rb.

Another antioncogene product, p53, is a transcription factor whose loss is associated with colorectal cancers. It has an acidic, amino-terminal transcriptional activation domain (TAD), a central region that can inhibit transcription, and a carboxy-terminal DNA-binding domain (DBD) that recognizes direct repeats of the TGCCT sequence. In addition to activating its own set of genes, it blocks the DNA binding and helicase activities of the DNA polymerase α, perhaps by binding to the replication protein A. It also binds to the TATA-binding protein (TBP) and interferes with the stable binding of both the TBP and transcription factor IIA (TFIIA) to the TATA box; however, it does not affect preformed initiation complexes or elongation. Finally, p53 binds and inhibits TFIIH and several NRs, recruits histone deacetylases (HDACs), and binds and allosterically activates glycogen synthase kinase 3 (GSK3).

p53 is induced after DNA damage to arrest these cells in G1 until the DNA is repaired. For this reason p53 has been called the *guardian of the genome*. p53 accomplishes this by inhibiting the DNA replication machinery; by activating the cyclin kinase inhibitor p21$^{\text{Cip1/Waf1}}$, which will prevent Cdk2 from phosphorylating Rb; and by inducing insulin growth factor binding protein 3 (IGFBP3), which will sequester insulin growth factor I (IGF-I), a potent mitogen. In severely damaged cells, p53 will trigger apoptosis by inducing the proapoptotic Bax in the nucleus and by binding and sequestering the antiapoptotic BclXL and Bcl2 in mitochondria.

p53 can be regulated by several different modifications. Mdm2 is a ubiquitin ligase, which targets p53 for destruction. In addition, Mdm2 binds the amino-terminal TAD and blocks CBP (CREB-binding protein) binding. Upon the activation of some G protein-coupled receptors (GPCRs), β-arrestin will bind and sequester Mdm2, thereby positively affecting p53 activity. Phosphorylation is another regulatory mechanism; tyrosine phosphorylation of p53 by the STK, Abl, stabilizes p53, masks its nuclear exit signal (NES) so that p53 is retained in the nucleus, and inhibits ubiquitination by Mdm2. DNA-dependent protein kinase (DNA-PK) phosphorylation decreases Mdm2 binding; PKC increases DNA binding, although it also increases ubiquitination; casein kinase 2 (CK2) decreases the repression of some genes by p53; and Cdk2 increases the transcriptional activity of p53 at other genes. p53 is also subject to several other modifications; it is activated by acetylation and sumoylation but inhibited by tyrosine nitration and poly(ADP-ribosyl)ation. Finally, p53 may be sequestered in the cytoplasm during periods of rapid growth or metabolism; for example, p53 is excluded from the nucleus of lactating epithelial cells. This compartmentalization may be achieved by membrane receptors because it has been reported that p53 binds to the cytoplasmic tail of Trk (NGF receptor) and the IFNα-C3d receptor; this association could provide a mechanism by which hormones may directly affect p53 localization.

p53 is inactivated by several oncogene products, including the large T antigen, E1A, E1B, E6, EBNA-5, and the X antigen from the hepatitis B virus. The large T antigen and the X antigen trap p53 in inactive complexes, E1A induces p53 aggregation which inactivates its TAD, E1B and EBNA-5 repress transcription by p53, and E6 tags p53 for destruction by ubiquitination.

The WT1 (Wilms' tumor 1) gene is a zinc finger transcription factor; its loss is associated with Wilms' tumor. The amino terminus represses transcription, and its central region activates transcription. These effects are often affected by the promoter and by other factors with which WT1 associates. For example, WT1 alone activates the transcription factor, EGR-1, but the WT1-p53 complex converts EGR-1 to a repressor. However, the interaction between WT1 and p53 can be positive; WT1 enhances the transcriptional activity of p53 at the creatine kinase promoter.

BRCA1 and BRCA2 (antioncogenes 1 and 2 deleted in breast cancer) are antioncogenes whose absence predisposes women to breast cancer. Both have TADs and are associated with DNA repair. BRCA1 promotes DNA repair through homologous recombination. This type of DNA repair uses the homologous chromosome as a template to repair the damage so that the gene is faithfully reproduced. In contrast, nonhomologous repair would merely rejoin broken ends randomly and often after some degradation has occurred. As such, it is usually mutagenic. BRCA1 promotes DNA repair by first exposing the damaged site to the repair proteins by binding both the helicase BACH1 and the chromatin remodeler BRG1. Then, it recruits homologous repair proteins through Rad 51, while inhibiting nonhomologous repair through the Mre11-Rad50-Nbs1 complex. Finally, it inhibits transcription until the repair is complete; for example, it binds an E3 ligase, which ubiquinates RNA polymerase II, it recruits HDACs, it induces p53, and it can directly bind and inhibit several transcription factors like Stat5a. The only known regulation is by PKB, whose phosphorylation of BRCA1 inhibits its nuclear localization.

Less is known about BRCA2. It also suppresses nonhomologous recombination after DNA breaks, and it can repress transcription by recruiting HDACs. In addition to breast cancer, BRCA2 deletion is associated with Fanconi anemia, whose symptoms can include acute myeloid leukemia and squamous cell carcinoma.

Finally, Smad4 and RUNX1 are antioncogene transcription factors that mediate the effects of TGFβ. They will be covered later (see section on Cell Cycle).

Protein and Lipid Phosphatases

If many oncogene products are tyrosine kinases, then protein tyrosine phosphatases (PTPs) could be considered antioncogene products. Indeed, PTP genes are lost in a high percentage of some tumors, and overexpression of PTPs can suppress or even reverse transformation. Occasionally, the situation in malignancies is confused by the compensatory PTP increase, which occurs when the tumor is induced by a tyrosine kinase oncogene. In these cases, both the kinase and PTP activities are elevated; however, the greater the ratio of kinase to PTP activity, the greater the aggressiveness of the tumor because there is a relative PTP deficiency. Other factors that can influence the effects of PTPs are substrate specificity and localization.

Serine-threonine phosphatases are also antioncogene products. The A subunit of PP2A (protein phosphatase 2A) can be bound by the small t antigen of SV40 and the polyoma middle T antigen. The latter is an integral membrane protein that sequesters the normally cytosolic PP2A. These proteins may also change the substrate specificity of PP2A; for example, the PTP activity of PP2A increases 10-fold relative to the serine-threonine phosphatase activity. In addition, the antigens decrease the activity of PP2A toward phosphorylated Rb; as a result, antioncogene Rb would remain inactive. Finally, the human T-lymphotropic retrovirus Tax protein specifically binds and inhibits the IκB kinase γ (IKKγ)–associated PP2A, resulting in the constitutive activation of NF-κB.

In addition to oncogene products, several chemical cocarcinogens target serine-threonine phosphatases. Nodularin inhibits PP1, and both okadaic acid and microcystin inhibit PP1 and PP2A.

The phosphatase and tensin homologue (PTEN) is a dual protein-lipid phosphatase; it is both an S-T-Y phosphatase and an inositol 3-phosphatase. PTEN phosphorylation by either PKB or CK2 stabilizes the protein; the latter also inhibits its activity. Dephosphorylation activates PTEN and shifts it to the cell-cell junctions where it dephosphorylates Fak, resulting in the inhibition of cell migration and invasion. As an inositol 3-phosphatase, PTEN decreases the products of PI3K and inhibits the antiapoptotic PKB. Loss of PTEN function would lead to the loss of contact inhibition and the survival of abnormal cells.

Cell Adhesion

Because cell-cell and cell-substratum contacts are important in controlling cell growth, any factor promoting these contacts could also be considered a product of an antioncogene. For example, the antioncogene product in neurofibromatosis, type 2, is called *NF2, merlin*, or *schwannomin*. NF2 is a member of the ERM (ezrin-radixin-moesin) family, which is part of the integrin-cytoskeleton interface that mediates contact inhibition. Inactivation of NF2 is associated with vestibular schwannomas and multiple meningiomas.

Adenomatous polyposis of the colon (APC) was discussed in Chapter 13. Briefly, β-catenin is part of the junctional complex; upon release it migrates to the nucleus and activates the transcription of genes that diminish cell-cell contact. APC sequesters β-catenin and promotes contact inhibition; its inactivation can lead to colon cancer. The loss of DCC (deleted in colon carcinoma) is also associated with colon cancer. DCC is a transmembrane receptor for netrin, a chemoattractant for commissural axons in the spinal cord. Therefore it is involved with cell-substratum interactions during development. Finally, TSC1, also called *hamartin*, is mutated in tuberous sclerosis, a disease characterized by multiple hamartomas. These tumors consist of normal cells, but the tissue is overgrown and disorganized. TSC1 was described in Chapter 12 as an inhibitor of mTOR, but it actually has several functions. For example, it binds ERM and neurofilaments, it promotes the degradation of β-catenin by the GSK3-axin complex, and it is required for Rho activity and cell adhesion.

Cell Cycle

Obviously, mitosis is central to any proliferative disorder, and anything that checks the cell cycle would be an antioncogene product. For example, members of the TGFβ family inhibit the growth of many tissues (see Chapter 3). Mutations in TGFβRI have been associated with metastatic breast cancer; TGFβIIR, with head and neck squamous carcinoma; and BMPR1A, with gastrointestinal hamartomatous polyposis. Smad4 and RUNX1 (runt-related transcription factor 1), which are major TGFβ effectors, are mutated in some pancreatic cancers and leukemias, respectively.

The cell cycle is driven by the Cdks, and there are two families of cyclin-dependent inhibitors (CDIs): the Cip/Waf (Cdk-interacting protein/wild-type p53-activated fragment) family and the INK (inhibitor of Cdk) family. The former is inactivated in some breast cancers; the latter is another downstream target of TGFβ, and it is inactivated in some gliomas.

Finally, IGF is a potent mitogen that is cleared from circulation by the IGF-IIR. Mutations in this clearance receptor are associated with hepatocellular carcinomas.

Other Antioncogenes

Several other prominent antioncogene products do not fall into any of the previously mentioned classes. As already noted, mutated Ras, which is constitutively active, is oncogenic. However, another way to elevate Ras activity would be to inactivate GAP (GTPase activating protein), which facilitates the hydrolysis of GTP (guanosine 5′-triphosphate) bound to normal Ras and inactivates this G protein. No such mutations have been reported for RasGAP, but they have been identified in a GAP-like protein in patients with neurofibromatosis type 1. This protein, known as *NF1,* can enhance the GTPase activity of Ras; therefore it may keep Ras in check. Most genomic mutations are located in the GAP-like region, and the clinical picture reflects growth and hyperactivity: neurofibromas and an increased risk of malignancies. Somatic mutations in this gene have also been reported in colon adenocarcinomas and anaplastic astrocytomas.

TSC2 or *tuberin* is a second antioncogene product found in tuberous sclerosis and, with TSC1, it inhibits mTOR. This inhibition is a result of the RhebGAP activity of TSC2. Furthermore, most oncogenic mutations occur in the GAP region.

VHL, the antioncogene product in Von Hippel Lindau disease, was discussed in Chapter 13. Basically, it is a ubiquitin ligase for hydroxylated HIF-1α, an oxygen-sensitive transcription factor. During hypoxia, HIF-1α is not hydroxylated, escapes destruction, and induces genes that help the cell survive in reduced oxygen levels. VHL inactivation has the same effect as hypoxia because HIF-1α evades degradation. As a result, vascular endothelial growth factor (VEGF) is induced, and highly vascular tumors are generated. VHL is also the ubiquitin ligase responsible for RNA polymerase II degradation after ultraviolet (UV) irradiation.

Finally, many tumorigenic cells are eliminated by apoptosis. Therefore inactivation of the components of apoptosis would favor survival of abnormal cells. Mutated caspase 10 is found in nearly 15% of non-Hodgkin lymphomas and 3% of gastric cancers, whereas an inactivated caspase 3 occurs in the MCF-7 breast cancer cell line.

Summary

Pathogens exploit the endocrine system to facilitate their infection and replication. They can use hormone receptors to gain entry into the cell, disable the immune system by blocking cytokine actions, and hijack the mitotic apparatus to create an environment conducive to replication. The latter effect is usually accomplished by a viral homologue to a normal component of the growth factor signaling cascade; however, this homologue is mutated so that it is constitutively active. Such proteins are called *oncogenes*; their normal cellular counterparts are called *proto-oncogenes* because spontaneous mutations can render them oncogenic, as well. These oncogene products can act at any level from hormones to transcription factors, but usually two are required: a membrane-associated protein that overcomes environmental checks and a nuclear one that overcomes senescence.

In addition to stimulating the mitogenic pathway, oncogenes can inhibit regulatory checks on this process. Such checks are the products of antioncogenes, and they involve blocking the effects of proliferation-associated transcription factors, reversing the phosphorylation of kinases, mediating contact inhibition, inhibiting cell division, inactivating G proteins, and restricting the proliferation of blood vessels. The loss of these antioncogenes or the inactivation of their products is often associated with malignancies.

References

Virulence Genes and the Immune System

Alcami, A. (2003). Viral mimicry of cytokines, chemokines, and their receptors. *Nature Rev. Immunol.* 3, 36-50.

Benedict, C.A., Banks, T.A., and Ware, C.F. (2003). Death and survival: Viral regulation of TNF signaling pathways. *Curr. Opin. Immunol.* 15, 59-65.

Cuconati, A., and White, E. (2002). Viral homologues of BCL-2: Role of apoptosis in the regulation of virus infection. *Genes Dev.* 16, 2465-2478.

Favoreel, H.W., Van de Walle, G.R., Nauwynck, H.J., and Pensaert, M.B. (2003). Virus complement evasion strategies. *J. Gen. Virol.* 84, 1-15.

Gao, L.Y., and Kwaik, Y.A. (2000). Hijacking of apoptotic pathways by bacterial pathogens. *Microbes Infect.* 2, 1705-1719.

Goodbourn, S., Didcock, L., and Randall, R.E. (2000). Interferons: Cell signalling, immune modulation, antiviral responses and virus countermeasures. *J. Gen. Virol.* 81, 2341-2364.

Hay, S., and Kannourakis, G. (2002). A time to kill: Viral manipulation of the cell death program. *J. Gen. Virol.* 83, 1547-1564.

Hiscott, J., Kwon, H., and Génin, P. (2001). Hostile takeovers: Viral appropriation of the NF-κB pathway. *J. Clin. Invest.* 107, 143-151.

Kahn, R.A., Fu, H., and Roy, C.R. (2002). Cellular hijacking: A common strategy for microbial infection. *Trends Biochem. Sci.* 27, 308-314.

Lusso, P. (2000). Chemokines and viruses: The dearest enemies. *Virology* 273, 228-240.

Merrill, A.H., Sullards, M.C., Wang, E., Voss, K.A., and Riley, R.T. (2001). Sphingolipid metabolism: Roles in signal transduction and disruption by fumonisins. *Environ. Health Perspect.* 109 [Suppl 2], 283-289.

Rosenkilde, M.M., Waldhoer, M., Lüttichau, H.R., and Schwartz, T.W. (2001). Virally encoded 7TM receptors. *Oncogene* 20, 1582-1593.

Seet, B.T., Johnston, J.B., Brunetti, C.R., Barrett, J.W., Everett, H., Cameron, C., Sypula, J., Nazarian, S.H., Lucas, A., and McFadden, G. (2003). Poxviruses and immune evasion. *Annu. Rev. Immunol.* 21, 377-423.

Stebbins, C.E., and Galán, J.E. (2001). Structural mimicry in bacterial virulence. *Nature (London)* 412, 701-705.

Vincendeau, P., Gobert, A.P., Daulouède, S., Moynet, D., and Mossalayi, M.D. (2003). Arginases in parasitic diseases. *Trends Parasitol.* 19, 9-12.

Vossen, M.T.M., Westerhout, E.M., Söderberg-Nauclér, C., and Wiertz, E.J.H.J. (2002). Viral immune invasion: A masterpiece of evolution. *Immunogenetics* 54, 527-542.

Oncogenes and the Mitotic Pathway

Butel, J.S. (2000). Viral carcinogenesis: Revelation of molecular mechanisms and etiology of human disease. *Carcinogenesis (London)* 21, 405-426.

Jameson, P.E., and Clarke, S.F. (2002). Hormone-virus interactions in plants. *Crit. Rev. Plant Sci.* 21, 205-228.

Antioncogenes: General

Macleod, K. (2000). Tumor suppressor genes. *Curr. Opin. Genet. Dev.* 10, 81-93.

Antioncogenes: Transcription Regulators

Chen, Y., Lee, W.H., and Chew, H.K. (1999). Emerging roles of BRCA1 in transcriptional regulation and DNA repair. *J. Cell Physiol.* 181, 385-392.

Classon, M., and Harlow, E. (2002). The retinoblastoma tumour suppressor in development and cancer. *Nature Rev. Cancer* 2, 910-917.

Kerr, P., and Ashworth, A. (2001). New complexities for BRCA1 and BRCA2. *Curr. Biol.* 11, R668-R676.

Oren, M. (1999). Regulation of the p53 tumor suppressor protein. *J. Biol. Chem.* 274, 36031-36034.

Pugh, C.W., and Ratcliffe, P.J. (2003). The von Hippel-Landau tumor suppressor, hypoxia-inducible factor-1 (HIF-1) degradation, and cancer pathogenesis. *Semin. Cancer Biol.* 13, 83-89.

Welcsh, P.L., Owens, K.N., and King, M.C. (2000). Insights into the functions of BRCA1 and BRCA2. *Trends Genet.* 16, 69-74.

Antioncogenes: Protein Phosphatases

Maehama, T., Taylor, G.S., and Dixon, J.E. (2001). PTEN and myotubularin: Novel phosphoinositide phosphatases. *Annu. Rev. Biochem.* 70, 247-279.

Vazquez, F., and Sellers, W.R. (2000). The PTEN tumor suppressor protein: An antagonist of phosphoinositide 3-kinase signaling. *Biochim. Biophys. Acta* 1470, M21-M35.

Other Antioncogenes

Dasgupta, B., and Gutman, D.H. (2003). Neurofibromatosis 1: Closing the GAP between mice and men. *Curr. Opin. Genet. Dev.* 13, 20-27.

Kaelin, W.G. (2002). Molecular basis of the VHL hereditary cancer syndrome. *Nature Rev. Cancer* 2, 673-682.

Kwiatkowski, D.J. (2003). Tuberous sclerosis: From tubers to mTOR. *Ann. Hum. Genet.* 67, 87-96.

Molecular Bases of Endocrinopathies

Not long ago, clinical endocrinology was very simple: virtually all endocrine diseases involved too much or too little of a certain hormone. If the patient had too little, one administered the appropriate hormone or its agonist; if the patient had too much, one could either partially remove the gland or chemically suppress hormonal synthesis. Technology was not advanced enough to determine why there was too much or too little hormone, but now recombinant DNA techniques have identified the molecular bases for a large number of endocrinopathies.

Hormones

Hormone Deficiency

Peptide Hormones

Hormone Mutations. The simplest mechanism for a hypoendocrinopathy is not to have the gene for that hormone. This occurs in isolated growth hormone (GH) deficiency, type IA (IGHD IA). There are, in fact, two human growth hormone (hGH) genes, and both of them, along with three placental lactogen genes, are clustered along a 48-kb (kilobases) stretch of DNA (Fig. 17-1). hGH-N is expressed in the pituitary gland, whereas hGH-V is expressed in the placenta. Human placental lactogen (hPL)–L is inactive because an intron mutation blocks proper splicing, but both hPL-A and hPL-B are normal and contribute to serum levels during pregnancy.

In isolated GH deficiency, type IA, hGN-N is deleted. Although hGH-V remains, it is not secreted from pituitary glands, and patients are short and susceptible to hypoglycemia (see Chapter 2). Several hPL deletions have also been described. These deletions have no clinical effects in pregnant patients, suggesting that, despite the high levels of hPL found during pregnancy, it serves no essential function. Deletion of hGH-V results in severe growth retardation in the fetus but has no apparent function in the adult. These multiple deletions occur with a relatively high frequency in this region of the genome as a result of recombination among homologous, adjacent sequences.

Another possible mechanism for a hypoendocrinopathy is to have a defective hormone. Insulin deficiency results in diabetes mellitus, and several mutant insulins have been characterized (Table 17-1). V3L* in the A chain and both F24S and F25L in the B chain have been identified in patients with diabetes. All three residues occur in the receptor-binding domain of insulin. Point mutations have also been described for other members of the insulin family, the pituitary hormones, several growth factors and cytokines, members of the transforming growth factor β (TGFβ) family, and several other hormones.

Gland Defects. A hormone may also be absent if the gland that secretes it fails to form or is destroyed. Some transcription factors are highly specific for the morphogenesis of an organ, and its mutation can lead to the agenesis of that

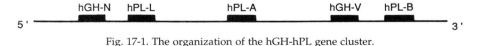

Fig. 17-1. The organization of the hGH-hPL gene cluster.

---

*Point mutations are designated by the residue number preceded by the original amino acid and followed by the new substitution. The one letter abbreviation is used for both amino acids.

Table 17-1
Clinical Syndromes Resulting in a Deficiency of Peptide Hormone Activity

Site	Mechanism	Example
Hormone	Point mutations and deletions	Insulin
Gland	Developmental failure	Anterior pituitary in Kallman syndrome
		IPF-1 in pancreatic agenesis (homozygous) and MODY4 (heterozygous)
		GCMB in parathyroid gland agenesis
	Autoimmune destruction	Antibodies to pancreas in IDDM
Hormone synthesis	Mutant transcription factor	Pit-1 in anterior pituitary
		HNF-1α, HNF-1β, HNF-4α, and NeuroD1/BETA2 in MODY3, 5, 1, and 6, respectively
	Promoter or regulatory site mutations	Insulin
	Alternate start site	ADH
Processing	Mutations in cleavage sites	Insulin and ET-3
	Mutations in protease	PC1 in proinsulin and POMC
	Mutations in precursor	Neurophysin
	Mutation at intron splice site with defective splicing	IGHD IB
Secretion	Signal sequence mutation	PTH and ADH
	Mutant releasing factors	TRHR, GHRHR, and GnRHR
	Malfunctioning sensor	Calcium receptor mutation (constitutively active) with feedback inhibition of PTH
		Activating antibodies to the calcium receptor
		Glucokinase mutation (MODY2)
Postsecretion	Sequestration by hormone antibodies	Insulin antibodies and diabetes mellitus
		EPO antibodies and pure red-cell aplasia
		PTH antibodies and hypocalcemia
		FSHβ antibodies and female infertility
		hCG antibodies and recurrent pregnancy loss

structure; defects in GCMB (glial cells missing, type B) result in parathyroid agenesis, and homozygotic defects in the insulin promoter factor 1 (IPF-1) result in pancreatic agenesis. However, one of the best-studied systems is the anterior pituitary where different transcription factors are responsible for the differentiation of each cell lineage; as a result, a mutant factor can result in the absence of cells secreting a specific hormone (Fig. 17-2).

Once formed, glands can subsequently be destroyed. There are several autoimmune diseases that target endocrine glands; for example, antibodies to glutamate decarboxylase will destroy the pancreas and cause insulin-dependent diabetes mellitus (IDDM). In addition, apoptosis is a default process in many cells and must be constantly inhibited to ensure the survival of the organ. Although this arrangement initially appears illogical and dangerous, it actually guarantees that mutant or damaged cells are automatically eliminated. Pancreatic islet-brain 1 (IB1) is a homeodomain transcription factor that normally keeps apoptosis in the pancreas in check; its mutation leads to the programmed destruction of the pancreas and causes mature onset diabetes of the young (MODY).

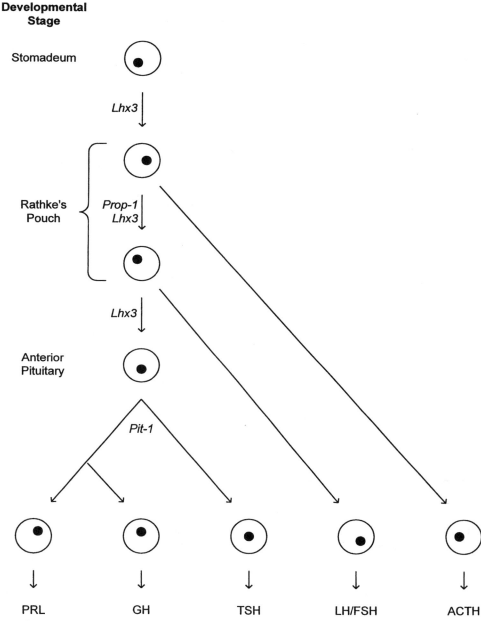

Fig. 17-2. Transcriptional regulation of the embryogenesis of the anterior pituitary gland. Transcription factors are italicized. *Lhx3*, LIM homeobox 3; *Pit-1*, pituitary transcription factor 1; *Prop-1*, prophet of Pit-1.

Hormone Synthesis.   Other transcription factors are relatively specific for genes coding for peptide hormones. The best-characterized ones regulate transcription of the insulin gene or its transcription factors; mutants in hepatic nuclear factor 1α (HNF-1α) cause MODY3; HNF-1β, MODY5; HNF-4α, MODY1; and NeuroD1 (also called BETA2), MODY6. Mutations within the hormone itself can also impair synthesis; an alternate start site in antidiuretic hormone (ADH) interferes with translation.

Hormone Processing. In addition to intrinsic mutations, processing defects have been described; insulin is synthesized as a single peptide containing a temporary C bridge between the A and B subunits (see Chapter 2). This C peptide is flanked by pairs of basic amino acids, which are substrates for a trypsin-like protease. In familial hyperproinsulinemia, the B-C junction is cleaved normally, but the A-C junction is not, and the C peptide is retained. This latter junction is demarcated by a pair of basic residues at positions 64 and 65; R65H and R65L mutants disrupt this cleavage site. The patients do not have diabetes because the semi-cleaved proinsulin has 60% of the biological activity of normal insulin and the pancreas makes up the difference by secreting more hormone to elevate the serum levels.

A similar mechanism is responsible for the failure to release endothelin 3 (ET-3) from its precursor; this defect occurs in the mouse model for Hirschsprung's disease. ET-3 is important in targeting cholinergic nerves to the colon; defects in ET-3 or its receptor will result in the inability of the colon to contract and expel feces because of a lack of neuronal stimulation. Hirschsprung's disease is treated by surgically removing the aganglionic segment of colon.

Mutations do not necessarily have to occur at the cleavage site; both ADH and its neurophysin are cleaved from a single polyprotein. In familial neurohypophyseal diabetes insipidus, the G29V mutation does not alter either ADH, which is located at the amino terminus, or its cleavage site. However, the neurophysin is required for processing and packaging and the mutation impairs this activity.

Occasionally, a processing defect can be traced to the processing enzyme rather than the hormone. In isolated, congenital adrenocorticotropin hormone (ACTH) deficiency, there is no detectable ACTH, although the precursor and some of the other products of proopiomelanocortin (POMC) are present in normal concentrations and the POMC gene has a normal sequence. In fact, the mutation is in the prohormone convertase 1 (PC1) that cleaves POMC. This enzyme also cleaves insulin, and its mutation is another cause of proinsulinemia.

Finally, mRNA processing can be impaired; a mutation at an intron splice site of hGH prevents normal splicing and produces IGHD, type IB.

Hormone Secretion. Hormone secretion can be prevented in three ways. First, there may be a mutation that blocks the cleavage of the signal sequence; this mutation occurs in ADH in the Brattleboro rat and in parathormone (PTH) in familial isolated hypoparathyroidism. Second, there may be defects in the receptors for hypothalamic releasing hormones; such mutations have been described for thyrotropin-releasing hormone receptor (TRHR), growth hormone-releasing hormone receptor (GHRHR), and gonadotropin-releasing hormone receptor (GnRHR) (see later). Finally, there may be defects in the pathway sensing other releasing stimuli. For example, PTH elevates serum calcium in response to hypocalcemia; conversely, high calcium levels will inhibit the parathyroid glands from releasing PTH. Calcium is detected by a G protein-coupled receptor (GPCR) that can be mutated to a constitutively active form; such GPCRs will misinform the parathyroid glands that calcium levels are elevated and will block PTH secretion.

Activating antibodies to the calcium receptor has the same effect. In autoimmune diseases, the immune system is unable to recognize certain endogenous antigens as "self" and generates antibodies to them. When these antigens are receptors, several interesting possibilities arise. For example, antibody binding to a receptor may prevent ligand binding; such antibodies are called blocking antibodies, although in some cases the deficiency is actually a result of increased receptor downregulation. Occasionally antibody binding can actually activate the receptor; these antibodies are called *stimulating antibodies.* Stimulating antibodies to the calcium receptor would activate the receptor in the same manner as calcium and falsely signal the parathyroids that calcium levels were high.

In the pancreatic islets, glucokinase acts as a glucose sensor; elevated blood glucose stimulates insulin secretion. Glucokinase mutations that block this function cause MODY2.

Hormone Antibodies.   Antibodies can also be directed against hormones; this commonly occurred in the past when IDDM was treated with nonhuman insulins. The titers and affinity of such antibodies were often so low as rarely to be clinically significant. However, in some autoimmune diseases, these antibodies can sequester the hormone leading to a hormone deficient-like state; antibodies against erythropoietin (EPO) produce pure red-cell aplasia; against PTH, hypocalcemia; against follicle-stimulating hormone β (FSHβ), female infertility; and against hCG, recurrent pregnancy loss.

Nonpeptide Hormones

Gland Defects.   As with peptide hormones, nonpeptide hormones can be deficient because of the absence of the appropriate endocrine gland; both developmental and autoimmune causes have been reported (Table 17-2). TTF1 is required for the survival and proliferation of thyroid follicular cell precursors; FoxE1 (formerly called *thyroid transcription factor 2* [TTF2]) is responsible for the descent of these cells from the floor of the mouth into the neck (see Chapter 2), and TSH stimulates postnatal growth of the gland. Mutations in FoxE1 result in thyroid agenesis, and severe TSH receptor (TSHR) deficiency can produce thyroid dysplasia or hypoplasia. Besides its role in the thyroid gland, TTF1 is also involved in the development of the lung and nervous system; its inactivation can lead to dyskinesia (abnormal movements) and lung defects, in addition to hypothyroidism.

Postnatally, antibodies to thyroid peroxidase (TPO) can destroy the thyroid gland in autoimmune thyroiditis, and antibodies to steroid enzymes can destroy the adrenal gland in Addison's disease.

Hormone Synthesis.   Genes do not code for nonpeptide hormones; they code for the enzymes involved in the synthesis of these hormones. Therefore nonpeptide hormone deficiencies are really enzyme deficiencies. The most thoroughly studied system is in the adrenal cortex, and impaired steroid synthesis produces a clinical and biochemical spectrum of diseases known collectively as *congenital adrenal hyperplasia.* The adrenal cortex becomes hyperplastic because no glucocorticoids are produced to inhibit the pituitary secretion of ACTH.

Almost any enzyme in the steroid biosynthetic pathway may be missing or defective (Table 17-3); however, the key to understanding the clinical picture of these patients is to determine (1) if androgen synthesis is impaired, (2) if aldosterone synthesis is impaired, and (3) what precursors are accumulating. It is also important to remember that the basic embryonic phenotype is female; the presence of androgens is required to masculinize the genitalia of males.

The 3β-hydroxysteroid dehydrogenase/5-ene-4-ene isomerase acts very early in the pathway (see Fig. 2-4); therefore its absence results in a deficiency of

Table 17-2
Clinical Syndromes Resulting in a Deficiency of Nonpeptide Hormone Activity

Site	Mechanism	Example
Absent gland	Developmental failure	TTF1, FoxE1 (TTF2), or TSHR deficiency in thyroid agenesis
	Autoimmune destruction	Antibodies to thyroid peroxidase in autoimmune thyroiditis
		Antibodies to steroid enzymes in Addison's disease
Hormone synthesis	Mutant enzymes	Steroidogenic enzymes in congenital adrenal hyperplasia
		Thyroid peroxidase and THOX2 in hypothyroidism
		1α-hydroxylase in 1,25-DHCC deficiency
		Tyrosine hydroxylase in dopamine deficiency (infantile parkinsonism)
		Choline acetyltransferase in Ach deficiency (congenital myastenic syndrome associated with episodic apnea)
		Prostaglandin synthase in some forms of essential hypertension
Miscellaneous mutations	StAR protein	Steroid synthesis
	Thyroglobulin, Na^+-I^--symporter, and apical I^- porter	Thyroid hormone synthesis
Hormone processing	5α-reductase	DHT deficiency
	Type 1 deiodinase (converts T_4 to T_3)	High T_4 in C3H mice
Hormone metabolism	Type 3 deiodinase excess (deactivating enzyme)	Consumptive hypothyroidism in certain tumors
Hormone sequestration	Antibodies to testosterone	Hypergonadotropic hypogonadism
Serum binding proteins	CBG mutations	Various fatigue syndromes

Table 17-3
Molecular Defects in Steroid Synthesis or Metabolism

Deficient activity	Genitalia	Sodium loss[a]	Hypertension[a]
StAR	Female[b]	+	0
β-Hydroxysteroid dehydrogenase/ 5-ene-4-ene isomerase	Hypomasculine[b]	+	0
P450c17:			
17α-Dehydrogenase	Female or hypomasculine[b]	0	+
17,20-Desmolase	Female or hypomasculine[b]	0	0
P450c21	Virilization[c]	+	0
P450c11:			
11β-Hydroxylase	Virilization[c]	0	+
18-Hydroxylase	Normal	+	0
18-ol-Dehydrogenase	Normal	+	0
11β-Hydroxysteroid dehydrogenase	Normal	0	+
P450aro	Pseudohermaphroditism[c]	0	0
5α-Reductase 2	Hypomasculine[b]	0	0

[a] +, Symptom is present; 0, symptom is absent.
[b] In males.
[c] In females.

aldosterone, glucocorticoids, and androgens. Without aldosterone, the patients lose sodium (salt wasting); without androgens, male infants have a hypomasculine phenotype. The reason for the latter is due to the formation of dehydroepiandrosterone; although this steroid is only a weak androgen, the hyperplastic gland produces enough of it to masculinize the fetus partially. The degree of salt wasting is variable because it is dependent on the severity of the mutation; as little as 10% of the normal enzymatic activity is sufficient to prevent salt wasting. Like many enzymes in this pathway, the dehydrogenase/isomerase has dual activities. In all of these versatile enzymes, the active sites appear to be functionally, if not structurally, distinct because clinically patients can have symptoms related to impairment of only one of the enzymatic activities. In addition, compounds can be synthesized to inhibit one activity, while leaving the other activity intact.

Theoretically, a deficiency in the 20,22-desmolase (P450scc) would exhibit a similar phenotype; however, no confirmed cases of P450scc have been reported. Earlier cases have subsequently been shown to result from mutations in the steroidogenic acute regulatory protein (StAR), which is required for cholesterol transport. In all likelihood, P450scc deficiency will never be reported because it would result in early pregnancy loss. Specifically, progesterone is required for the maintenance of pregnancy, and the corpus luteum is only active during the first trimester. Therefore the fetal placenta must be able to produce progesterone during the second and third trimesters; and this would not be possible with P450scc deficiency.

Deficiency of P450c17 also impairs androgen synthesis, resulting in female or hypomasculine phenotypes in males. Deficiency of the 17,20-desmolase component of P450c17 exclusively affects the androgens, but the absence of the 17α-dehydrogenase component also eliminates cortisol. Its precursor, 11-deoxycortisol, is a weak mineralocorticoid, but again, its accumulation compensates for its low activity. In fact, sodium is retained and hypertension develops. As already noted, mutations can selectively affect one or the other activity.

There are two P450c11 genes: CYB11B1 has 11β-and 18-hydroxylase activities and is present in the zonae fasciculata and reticularis. Deficiencies of either CYB11B1 or P450c21 (21-hydroxylase) affect both aldosterone and the glucocorticoids. Because there is no negative feedback by cortisol, the gland is hyperstimulated by ACTH; precursors accumulate and are eventually shunted into the androgen pathway. Males appear normal, but females become virilized; for example, the clitoris enlarges to resemble a penis and the labia fuse, creating the appearance of an empty scrotum. The P450c21 deficiency also leads to sodium loss, but the CYB11B1 deficiency still allows the synthesis of 11-deoxycortisol and 11-deoxycorticosterone. As noted previously, these steroids are weak mineralocorticoids whose accumulation is sufficient to overcome the loss of aldosterone; sodium is retained and hypertension results.

The second P450c11 gene is CYB11B2; it has 18-ol-dehydrogenase, as well as 11β-and 18-hydroxylase activities, and it is restricted to the zona glomerulosa. For this reason, it is often called the *aldosterone synthase*. Deficiency in CYB11B2 selectively eliminates aldosterone, so sexual development is normal and the only clinical problem is salt wasting.

However, it is the P450c21 gene that is most subject to mutations. Near the normal P450c21 gene is a second, nonfunctional copy; such copies are called *pseudogenes*. Because pseudogenes are never expressed, they are not under any selective pressure to maintain their sequences and many mutations accumulate.

Homologous, adjacent genes are subject to nonreciprocal crossing over, which can introduce some of the mutations of the pseudogene into the real gene. Such an event is called *gene conversion*, and it appears that most of the mutations in the P450c21 gene arise in this manner. It is interesting to note that some authorities have suggested that the high incidence of P450c21 mutations may also be a result of some survival advantage as has been postulated for the relationship between sickle cell trait and resistance to malaria. They note that heterozygotes for P450c21 mutations have a higher than normal cortisol response to stimuli and that such a response may allow the carrier to better cope with environmental stresses and be less susceptible to autoimmune disease.

Like steroids, thyroid hormones are enzymatically synthesized. Briefly, the thyroid accumulates iodide through a Na^+-I^--symporter and transfers it into the follicle through an iodide porter. TPO oxidizes the iodide, which is then incorporated into thyroglobulin. After coupling, the thyroid hormones are liberated by endocytosis and proteolysis of the thyroglobulin. Thyroid oxidase 2 (THOX2) generates hydrogen peroxide for TPO, and several mutations in both of these enzymes have been reported. Mutations in the iodide transporters and thyroglobulin have also been described. All known defects of the latter are truncations that remove the major iodination site. The low triiodothyronine (T_3) and thyroxine (T_4) levels resulting from both the peroxidase and thyroglobulin mutations lead to elevated TSH concentrations, which induce thyroid growth or goiter.

Other mutant enzymes involved with the synthesis of nonpeptide hormones include 1α-hydroxylase (vitamin D deficiency), tyrosine hydroxylase (dopamine deficiency in infantile parkinsonism), and choline acetyltransferase (acetylcholine deficiency in congenital myastenic syndrome associated with episodic apnea). Dopamine is the major neurotransmitter in the basal nuclei, whose degeneration results in Parkinson's disease. Acetylcholine is the neurotransmitter used by somatic efferent neurons; its deficiency results in muscular weakness. Finally, mutations have been reported for prostacyclin synthase. Prostacyclin normally inhibits platelet aggregation, smooth muscle proliferation, and vasoconstriction; the latter two effects are important in maintaining normal blood pressure. Decreased prostacyclin synthesis has been associated with essential hypertension.

Hormone Processing.   Some hormones are secreted and circulate as precursors; they are converted to their most active form in peripheral tissues. For example, the testes secrete testosterone, which is converted to dihydrotestosterone (DHT) by 5α-reductase. The type-2 enzyme is the predominant form in genital tissue and its deficiency produces male pseudohermaphroditism. Similarly, T_4 is converted to T_3 by iodothyronine deiodinase. In C3H mice, an insertion in the promoter region of the gene for the type 1 enzyme leads to a 10% decrease in this enzyme. Although T_4 levels are elevated, enough T_3 is synthesized to maintain euthyroidism.

Hormone Metabolism.   Another potential cause of a hormone deficiency is excessive metabolism. Infantile and hepatic hemangiomas may overproduce the type 3 deiodinase, which is primarily a catabolic enzyme. The elevated amounts of this enzyme will inactivate T_4 and T_3 faster than they can be synthesized; this phenomenon is known as *consumptive hypothyroidism*.

Cubilin is a membrane receptor that binds the intrinsic factor-vitamin B_{12} complex and internalizes it. It performs the same function for the vitamin D binding protein-25-hydroxyvitamin D complex in the kidney, where the sterol is recovered from the urine and converted to 1,25-dihydroxycholecalciferol (1,25-DHCC). Symptoms of mutant cubilin are dominated by vitamin B_{12} deficiency and megaloblastic anemia. Although vitamin D urine levels are elevated and blood levels are low, there are generally no symptoms related to vitamin D deficiency unless the patients consume a diet low in vitamin D.

Binding Proteins.   As with the peptide hormones, autoimmune disease can target steroids. A woman with hypergonadotropic hypogonadism was reported to have antibodies to testosterone. Although testosterone was elevated, there was no virilization because the steroid was sequestered by the antibodies. The estrogen deficiency was a result of antibody interference with the transfer of testosterone from theca interna to the granulosa cells, where it is converted to estradiol.

Although hydrophobic hormones bind serum proteins, these carriers are usually not essential for hormone function, since it is only the free hormone that is active. For example, mutations in thyronine-binding globulin (TBG) lower total serum thyroid hormone levels, even though free hormone is normal and the patient has no symptoms. However, corticosteroid-binding globulin (CBG) has an additional role in delivering cortisol to its target tissues, and mutant CBG has been implicated in the fatigue syndrome. Furthermore, such abnormal proteins can cause other problems; mutant transthyretin, also called *thyronine-binding prealbumin*, can polymerize to form long, insoluble fibrils that can precipitate in tissues and disrupt function. Such a disease is called *amyloidosis*, and the fibrils can arise from many sources. When transthyretin is the culprit, the nervous and cardiac tissues are primarily affected and the maladies are known as *familial amyloid polyneuropathy* and *familial cardiac amyloidosis*, respectively.

Hormone Excess

Historically, hormone excess was usually a result of endocrine gland hyperplasia, activating antibodies, or tumors. However, more recently defects in hormone metabolism, negative feedback, antagonists, and so on, have also been documented (Table 17-4).

Hormones
Several hormone mutations have been shown to either enhance processing or decrease proteolysis. A mutation in preproneuropeptide Y facilitates its processing and results in higher neuropeptide Y (NPY) levels following stimulation like exercise. NPY induces endothelial proliferation and elevated levels increase the risk of atherosclerosis. Fibroblast growth factor 23 (FGF23) increases renal excretion of phosphate (see Chapter 1). A mutation that increases its resistance to proteolysis leads to autosomal dominant hypophosphatemic rickets; this disease is characterized by excessive urinary loss of phosphate, low serum phosphate, and skeletal abnormalities.

Mutations that increase the intrinsic activity of hormones are more rare. There is a report of a woman with delayed puberty, menstrual irregularity, and subfertility. She had a mutant luteinizing hormone (LH) that was more potent than wild-type LH in stimulating 3'5'-cyclic AMP (cAMP) and IP_3 (inositol 4,5-trisphosphate) production *in vitro*. However, the hypogonadism was a result of the much shorter half-life of the mutant LH *in vivo*.

Table 17-4
Clinical Syndromes Resulting in an Excess of Hormone Activity

Site	Mechanism	Example
Hormone-Related		
Hormone	Increased precursor processing	Preproneuropeptide Y
	Decrease proteolysis	FGF23 in autosomal dominant hypophosphatemic rickets
	Increase potency	Possibly LH
Agonists	Stimulating antibodies to receptors	Activating antibodies to TSHR in Graves' disease
		Activating antibodies to insulin receptor in some hypoglycemia
Antagonists and inhibitors	Antagonist	Noggin mutant fails to bind and inhibit BMP-2 and BMP-4 leading to bony fusion in proximal symphalangism and related syndromes
		Sclerostin mutant fails to bind and inhibit BMP-5 and BMP-6 leading to bony overgrowth (hyperostoses)
		Fibrillin-1 mutant fails to sequester TGFβ leading to apoptosis of lung tissue and emphysema in Marfan syndrome
	Carrier protein	LAP mutant fails to bind and sequester TGFβ in Camurati-Engelmann disease (progressive diaphyseal dysplasia)
Synthesis		
Transcription factor	Elevated HIF-1α secondary to VHL (E3 ligase) mutation	Elevated EPO in Chuvash polycythemia
Promoter	Increased activity of mutated promoter	*TNFα* gene in some autoimmune diseases
	New promoter as a result of chromosomal inversion	Elevated estrogens
Repressor region	Decreased repression	Thrombopoietin gene in hereditary thrombocythemia
Splice site	Increased transcription	Excess aromatase syndrome
Iodination	Stimulating antibodies to Na^+-I^--symporter	Graves' disease
Processing	Elevated type 2 deiodinase (T_4 to T_3 conversion enzyme)	Hyperthyroidism in some thyroid cancers
$Gα_s$	Constitutively active	Mimics trophic hormones (see Table 17-12)
Secretion and Regulation		
Sensor	Ca^{2+} sensor (inactive mutants)	Neonatal severe hyperparathyroidism
	Glucokinase activating mutation	Familial hyperinsulinemia
	Glutamate dehydrogenase activating mutation	Hyperinsulinemia with hyperammonemia
	Traffic ATPase	Familial persistent hyperinsulinemic hypoglycemia of infancy
Feedback	Aromatase deficiency	Testosterone must be converted to E_2 for hypothalamic negative feedback
	Truncated GR	Somatic mutation in the pituitary (Nelson's syndrome)
	Nonhomologous recombination	Glucocorticoid remedial aldosterism
Serum-Binding Proteins		
SHGB mutants	Hyperandrogenism in females	Polycystic ovarian syndrome, idiopathic hirsutism, and ovarian failure)

continues

Table 17-4 *continued*

Site	Mechanism	Example
Metabolism		
11βHSD, type 2	Unmetabolized cortisol floods MR	Apparent mineralocorticoid excess syndrome
PAF acetylhydrolase	Increased PAF	Severe allergies
Adenosine deaminase	Adenosine is not metabolized to inosine	Form of severe combined immunodeficiency
ANFR-C	ANF is not cleared	Skeletal overgrowth
Gene Dosage		
IGFII gene	Duplication	Beckwith-Wiedemann syndrome

Many members of the TGFβ family have antagonists; defective antagonists can lead to unchecked agonist activity. For example, Noggin binds and inhibits bone morphogenetic protein 2 (BMP2) and BMP4. Inactive mutants release BMP, which overstimulates skeletal development and prevents the apoptosis necessary to form joints. This condition results in several syndromes characterized by bony fusion of the joints, such as proximal symphalangism (immobility of the proximal joints of the fingers), multiple synostoses (bony fusions) syndrome, and autosomal dominant stapes anklyosis (immobility of an ear bone). Similarly, sclerostin binds and inhibits BMP5 and BMP6. Inactive mutants produce the bony overgrowth seen in sclerosteosis, van Buchem disease, and craniotubular and generalized hyperostoses (bony overgrowth).

Defective fibrillin-1 is the cause of Marfan syndrome. Normally, fibrillin-1 sequesters TGFβ in the lung. Without this antagonist, TGFβ induces pulmonary apoptosis and emphysema, which can complicate Marfan syndrome. Finally, the latency-associated peptide (LAP) is cleaved from the TGFβ precursor and remains with TGFβ as a carrier protein. Mutations that disrupt this binding result in elevated TGFβ and progressive bone formation in Camurati-Engelmann disease, also known as progressive diaphyseal dysplasia (abnormal development of the bone shaft).

Hormone Synthesis

Increased hormone synthesis can occur at the level of transcription, synthesis, processing, and tropic stimulation. A variant tumor necrosis factor α (TNFα) promoter is more active than the wild-type and leads to elevated TNFα and to an increased risk for autoimmune disease. A mutation in the repressor region of the thrombopoietin gene results in excessive production of thrombopoietin and elevated platelets in hereditary thrombocythemia. Heterozygous inversions place the P450aro gene adjacent to constitutive active promoters that normally transcribe other genes. This chromosomal rearrangement leads to the overexpression of aromatase in many tissues. Another mutation in this gene creates a new splice site; the activity of the aromatase is unchanged but the modified protein exhibits increased expression. Males demonstrate feminization, whereas females display precocious puberty and macromastia (large breasts). Hypoxia-inducible factor 1α (HIF-1α) normally has a short half-life because of ubiquitination by Von Hippel Lindau (VHL) protein (see Chapter 13); hypoxia inhibits ubiquitination and allows HIF-1α to induce EPO, among other genes. Mutations in VHL also block HIF-1α downregulation and result in excessive EPO and red blood cell production in Chuvash polycythemia.

Graves' disease is a form of hyperthyroidism that is caused by activating antibodies. These antibodies can develop to several possible antigens, one of which is the Na^+-I^--symporter. Increased uptake of iodide can drive the synthesis of thyroid hormones. The processing of thyroid hormones can also be affected by disease. As noted previously, type 1 deiodinase regulates T_3 blood levels, and type 3 is the major catabolic enzyme. The type 2 deiodinase controls nuclear T_3 levels. This latter enzyme is frequently elevated in thyroid follicular cancer and is responsible for the increased conversion of T_4 to T_3.

Hormone regulation is frequently hierarchical (see Chapter 1); therefore excessive tropic stimulation can result in overproduction of hormones. There are two conditions that can activate this pathway. The first are activating antibodies to the receptors for tropic hormones. For example, some antibodies in Graves' disease are directed against the TSHR; they mimic TSH and stimulate the thyroid gland. Second, TSH and other tropic hormones often use cAMP to mediate their effects. cAMP, in turn, is generated by $G\alpha_s$-stimulated adenylate cyclase. Mutations in the guanosine triphosphatase (GTPase) site of $G\alpha_s$ prevent it from turning off and result in constitutive activity, which would also mimic tropic hormone stimulation (see later).

Hormone Secretion and Regulation

Sensors. Stimuli other than tropic hormones can trigger hormone secretion. PTH is secreted in response to hypocalcemia. As noted above, activating mutations of the calcium receptor mimic elevated serum calcium and inhibit PTH release. Conversely, inactivating mutations mimic hypocalcemia and homozygotic mutations cause neonatal severe hyperparathyroidism.

Several stimuli trigger insulin release. Glucokinase acts as a glucose monitor. Mutations that increase its affinity for glucose mimic hyperglycemia and lead to familial hyperinsulinemia. Amino acids also stimulate insulin secretion. Because glutamate dehydrogenase plays a central role in amino acid metabolism, it has allosteric sites for amino acids and is used as an amino acid sensor. This enzyme catalyzes the following reaction:

$$glutamate + NAD^+ \rightarrow \alpha\text{-ketoglutarate} + NH_3 + NADH + H^+$$

Gain-of-function mutations increase insulin secretion and give rise to hyperinsulinemia with hyperammonemia.

Finally, pancreatic islets are sensitive to the energy status of the body; this condition is sensed by an adenosine 5'-triphosphate (ATP)–inhibited potassium channel (K_{ATP}) and its associated adenosine triphosphatase (ATPase), the "traffic ATPase." The regulatory pathway works as follows: substrates are oxidized to generate ATP; a high ATP-to-ADP (adenosine diphosphate) ratio inhibits K_{ATP} through the associated ATPase; and the plasma membrane becomes depolarized, which stimulates insulin secretion. Insulin would then switch substrates away from ATP generation and into storage forms (see Chapter 2). ATPase or K_{ATP} defects that mimic inhibition result in familial persistent hyperinsulinemic hypoglycemia of infancy. Eventually, the islet insulin secretory capacity fails, and the hyperinsulinemic state converts to diabetes mellitus in middle age. It is worth noting that many oral hypoglycemic drugs used to treat noninsulin-dependent diabetes mellitus (NIDDM) also inhibit the traffic ATPase complex.

Feedback. The absence of negative feedback is another situation that would lead to a hypersecretory state. Sex steroids feedback to the hypothalamus, in part, through estradiol. The absence of P450aro results in a deficiency of estradiol, a lack of negative feedback, and elevated gonadotropins and androgens. Affected individuals of both sexes become virilized by the androgens, whereas the absence of estrogens induces osteoporosis and delayed bone maturation.

Nuclear receptors will be covered more completely later, but cortisol and thyroid hormones also feedback to the central nervous system to inhibit their tropic hormones. Defective nuclear receptors (NRs) would eliminate this negative feedback. For example, in Nelson's syndrome, the glucocorticoid receptor (GR) is inactivated by a somatic mutation in the corticotroph cells of the pituitary gland; there is no negative feedback and an ACTH-secreting adenoma develops. As a result, the adrenal cortex overproduces cortisol. Similarly, some TSH-secreting pituitary adenomas have been shown to be a result of an alternately spliced thyroid receptor β (TRβ), which is ineffective in mediating the T_3 negative feedback.

As noted previously, there are two P450c11 genes: CYB11B1 has 11β- and 18-hydroxylase activities and is present in the zonae fasciculata and reticularis, where it synthesizes cortisol in response to ACTH. CYB11B2 also possesses 18-ol dehydrogenase activity, which allows it to synthesize aldosterone. Because these two enzymes are 93% identical and their genes are adjacent to one another on the chromosome, the genes are subject to nonreciprocal crossing over as a result of mispairing during meoisis. This usually occurs near the 5′ end of the CYB11B2 gene and results in the CYB11B2 gene having the 5′, regulatory region of the CYB11B1 gene; that is, aldosterone synthase is now induced by ACTH. Cortisol levels fall and the absence of negative feedback then raises ACTH concentrations, which now stimulate CYP11B2. The subsequent elevation of aldosterone leads to sodium retention and hypertension; this condition can be corrected by glucocorticoids, which suppress ACTH secretion and CYB11B2 gene expression. For this reason, the disease is known as *glucocorticoid-remediable aldosterism*.

Hormone Metabolism

Hyperaldosterism can also occur with a deficiency of 11β-hydroxysteroid dehydrogenase. This enzyme occurs in aldosterone target tissues and rapidly metabolizes cortisol so that it cannot bind and activate the aldosterone receptor, which cannot distinguish between the two steroids. The absence of this enzyme allows cortisol, whose serum levels greatly exceed those of aldosterone, to flood the aldosterone receptor and activate it.

Deficiencies of several other metabolizing enzymes have also been reported. Platelet-activating factor (PAF) is a modified phospholipid involved with the inflammatory response (see Chapter 3). Patients lacking PAF acetylhydrolase activity have high levels of PAF and severe allergies. Adenosine is a parahormone that induces lymphocyte apoptosis; it is metabolized to inosine by adenosine deaminase. The deficiency of this enzyme results in a form of severe combined immunodeficiency.

Finally, GC-S is an atrial natriuretic peptide (ANP) receptor that lacks a guanylate cyclase domain and functions as a clearance receptor for ANP. The *longjohn* mouse has a mutant GC-S and ANP accumulates. Although best known for its natriuretic effect, it also plays a role in bone development and these mice show skeletal overgrowth.

Miscellaneous

There are two other conditions that can lead to excessive hormone activity. First, mutations in the sex hormone binding globulin release androgens to produce hyperandrogenism in females. Symptoms may include polycystic ovarian syndrome, idiopathic hirsutism (excessive sexual hair growth), and ovarian failure. Second, gene duplication can increase the gene dosage of a hormone. For example, Beckwith-Wiedemann syndrome is caused by a duplication of the short arm of chromosome 11, and it is characterized by organomegaly. This region contains the insulin-like growth factor II (IGF-II) gene and the organs enlarged correspond to the target tissues for IGF-II, suggesting that this aspect of the syndrome is a result of excessive IGF-II production.

Receptors

Nuclear Receptors

Genomic Mutations

Before a discussion of individual receptors and their mutations, the concept of phenotypic variation should be covered. *Phenotypic variation* refers to the phenomenon whereby patients with identical mutations in a receptor or other transduction component may have different clinical presentations. For example, males with the exact same mutation in the androgen receptor (AR) can present with intersexuality that varies from male pseudohermaphroditism to complete testicular feminization. This variation often arises from an interaction with other factors in the individual. One factor is the hormone level: males have variable 5α-reductase activity. The higher the enzyme activity, the more DHT is synthesized and the more activity that can be evoked from AR mutants having decreased ligand affinity. That is, DHT levels have a normal physiological range, and the higher concentrations can compensate for a receptor with a lower affinity. Such an individual would have less hypomasculinization.

Mosaicism is another factor. *Mosaicism* refers to the fact that one of the X chromosomes is randomly inactivated. In heterozygotes with mutations in X-linked genes, the phenotype will depend on which gene is inactivated in which cell lineages. Both the AR and $G\alpha_s$ are X-linked genes. A variation of this mechanism occurs in mutations of IKKγ, a component of the kinase complex that activates NF-κB. Such mutations are usually lethal in males but are survivable in men with Klinefelter's syndrome (XXY) because they have two X chromosomes.

Conditional mutants are also a cause for phenotypic variation. These are mutations that require a particular condition to become manifest. For example, the phenotype of some mutant vitamin D receptors (VDRs) is influenced by calcium and vitamin D intake. A temperature-sensitive $G\alpha_s$ has been described (see later). Some gain-of-function mutations in fibroblast growth factor receptor (FGFR) will only produce craniosynostosis (abnormal fusion of the cranial bones) if there is neonatal injury to the skull, as may occur in breech births. Finally, sex can also be a conditional factor; there is a TSHR mutation that allow hCG to cross-bind and activate the TSHR. hCG only exists in pregnant women, in whom it produces gestational thyrotoxicosis.

Lastly, variants can leads to phenotypic variation. The effect of any given mutation in neurofibromatosis 1 (NF1), a RasGAP (see Chapter 16), depends on the ratio of splice variants generated. In another example, limb development appears to require a certain level of total FGF rather than a specific isoform. Therefore the effect of a deficiency in one would depend on the levels of the other isoforms.

More mutations have been identified in the AR than in any other hormone receptor (Table 17-5). This is most certainly due to the fact that androgens only affect sexual differentiation and reproduction and have no major impact on the survival of the organism. Furthermore, our society often seems obsessed with sexual characteristics, so that anatomical anomalies of the sexual organs are quickly identified for possible treatment. Mutations in AR span the entire molecule, and their location frequently predicts their physiochemical effects; for example, those in the ligand-binding domain (LBD) inhibit steroid binding and others in the DNA-binding domain (DBD) reduce binding to the hormone response element (HRE). If there is complete androgen insensitivity, the fetus will develop the female phenotype, although the uterus and oviducts will be missing (see Chapter 2); this condition is known as *testicular feminization*. Less severe insensitivity will result in various degrees of male pseudohermaphroditism; females are unaffected except for their carrier status.

The most interesting mutations involve the polyglutamine stretch in the amino terminus. This track is produced by a CAG repeat that is unstable and can lengthen; normally the track is 21 glutamines long (range, 17 to 26). If the length should double, a disease called *Kennedy's disease* or *X-linked spinal and bulbar muscular atrophy* results. This syndrome is characterized by progressive muscular weakness and atrophy and a normal male phenotype, although there is frequently delayed adolescence and relative signs of androgen insensitivity, such as gynecomastia and reduced fertility. Complete inactivation of AR does not produce these muscular symptoms; rather, the toxicity is thought to arise from nuclear aggregates generated by the polyamine tracks. These aggregates do not appear to be directly responsible because heat shock protein (hsp) 70 suppresses neurodegeneration without disaggregating the nuclear inclusions. Rather, the aggregates act as sinks for other critical proteins; for example, they can sequester CBP (CREB binding protein) and target this coactivator for ubiquitination and degradation. This theory is supported by the observation that androgens are required for the disease to become manifest; clinically, females, including homozygotes, do not get symptoms. In addition, transfected male animals that are castrated at puberty do not get the disease, whereas androgen-treated females do. Because it is androgen binding that recruits coactivators to AR, unliganded AR aggregates cannot act as coactivator sinks.

Table 17-5
Clinical Syndromes Resulting from Genomic Nuclear Receptor Mutations

Receptor	Mechanism	Syndrome
AR	Inactivation	Male pseudohermaphroditism; testicular feminization
	Polyglutamate expansion	Kennedy's disease (X-linked spinal and bulbar muscular atrophy)
ER	Inactivation	Aromatase deficiency-like syndrome
VDR	Inactivation	Vitamin D resistant rickets
TR	Inactivation	Thyroid hormone resistant hypothyroidism
GR	Inactivation	Familial glucocorticoid resistance
MR	Inactivation	Pseudohypoaldosterism, type I
	Consitutively active	Early onset hypertension with exacerbation during pregnancy
PPARγ2	Inactivation	Insulin resistance with early onset of type 2 diabetes mellitus
	Consitutively active	Obesity

There are two major hypotheses on the mechanism of aggregation: (1) it arises from the interactions of β-strands, which polyglutamine tracks tend to form, or (2) it is a result of transglutaminase, which can use glutamine as a substrate in cross-linking. The former explanation is currently favored because (1) the aggregates do not appear to be covalently linked, (2) the transglutaminase does not colocalize with the aggregates, and (3) increasing transglutaminase activity does not increase aggregation. The muscular and nervous systems are most severely affected by this disease because these tissues are unable to regenerate themselves after damage.

Mutations in the estrogen receptor (ER) either are not as common as those in the AR or are more difficult to detect. After all, the default sexual phenotype is female (see Chapter 2); therefore, such defects would not produce any anatomical abnormalities. However, the girls should have evidence of hypoestrogenization at the time of puberty. Mutations in the ER resemble those of P450aro deficiency described previously.

Molecular defects in the thyroid receptor (TR) are also interesting in that those unable to bind T_3 have dominant phenotypes. TR can bind its HRE in the absence of T_3, but it still requires its ligand for transcription activation; essentially, unoccupied TR acts as an inhibitor by blocking the HRE (see Chapter 13). Mutant TRs that cannot bind T_3 would be permanent inhibitors and would compete with normal TRs for the HRE. Deletions large enough to completely inactivate the TR would be recessive.

Defects in other nuclear receptors have also been reported but they are much less common than those in the AR or TR. Several mutations in the VDR produce vitamin D-resistant rickets with alopecia (hair loss). Alopecia does not occur with vitamin D deficiency; rather, it is thought that the development of hair follicles requires that a gene or genes be inhibited by the unoccupied VDR. In addition, certain VDR genotypes have been associated with osteoporosis, osteoarthritis, and impaired intestinal calcium absorption. However, these associations have been sporadic and may depend on low dietary calcium and vitamin D to become manifest.

Defects in either GRα or GRβ give rise to familial glucocorticoid resistance. With respect to GRα, these are loss-of-function mutations; however, GRβ is subject to gain-of-function mutations. A polymorphism has been reported in the AUUUA motifs in the 3' end of the GRβ mRNA. Because these repeats signal rapid turnover (see Chapter 15), a mutation at this site would increase GRβ mRNA half-life and elevate GRβ levels. Because GRβ is a dominant inhibitor of GRα, increased GRβ concentrations would also lead to glucocorticoid resistance.

Mutations in the mineralocorticoid receptor (MR) and PPARγ2 (peroxisome proliferator activated receptor γ2) are interesting in that both gain-of-function and loss-of-function mutants have been described. Loss-of-function mutations in the MR give rise to type I pseudohypoaldosteronism, whereas gain-of-function mutants constitutively stimulate the resorption of sodium from the urine to cause early onset hypertension exacerbated during pregnancy. The latter mutation occurs in helix 5 (H5) of the LBD and increases its van der Waals interactions with H3. As such, it mimics the 21-hydroxyl group of MR ligands. This mutation has two effects: first, the MR possesses 27% of its maximal activity in the absence of ligands. Second, ligands lacking a 21-hydroxyl group, like progesterone and cortisone, are equipotent to aldosterone in further activating MR. The effectiveness of progesterone as a ligand would explain the exacerbation of hypertension during pregnancy.

PPARγ2 is a downstream mediator of many of the lipogenic actions of insulin. Because inactivating mutations in the LBD do not impair dimerization, they lead to a dominant negative phenotype characterized by insulin resistance and early onset of NIDDM. PPARγ2 is regulated, in part, by phosphorylation; in particular, in the transcriptional activation domain (TAD) there is a mitogen-activated protein kinase (MAPK) site, which inhibits the NR. Mutations in this site result in constitutive activity and obesity.

Somatic Mutations

The mutations described above occur in the germ line; however, there has been an increasing interest in somatic mutations and their possible relationship to cancer. For example, breast cancer is one of the leading causes of cancer death in women; in several animal models, it can be induced by estrogens, and in human beings antiestrogen therapy can cause regression. Unfortunately, many tumors will eventually become unresponsive to endocrine therapy. Because the ER links estradiol to its effects, several groups have sought to explain these phenomena in terms of alterations in the ER.

The MCF-7 breast cancer cell line is estrogen dependent, but a subclone is unresponsive to antiestrogens. These cells have a G400V mutation that binds the antagonist in such a way that it can activate the ER; that is, antagonists are converted to agonists. A similar phenomenon has been reported for the AR in prostate cancer; ARs with a W741C or W741L mutation can be activated by antiandrogens. Other AR mutations render the receptor susceptible to activation by estradiol, progesterone, or cortisol.

In actual breast tumors, aberrant processing that results in deleted exons is much more common than mutations; in general, there appears to be a deregulation of splicing in tumors. Of the many variant ERs that have been reported in breast cancer, three are of considerable note: deletion of either exon 3 containing the second zinc finger or exon 7 in the LBD are dominant negative mutants. These ERs are not only defective, but they would inhibit the function of normal ERs; therefore only one allele would have to be affected. In some tumors, these splice variants are reduced; the elimination of these natural ER inhibitors would make any residual estradiol much more effective. The third ER variant lacks exon 5 and is constitutively active; although this activity is only 10% to 15% of that for the normal ER, it may be enough to remove the tumor from any dependence on exogenous estrogens.

There are three other ways to increase the activity of steroid receptors (SRs) in the presence of low ligand levels. First, 30% of prostate tumors recurring after antiandrogen treatment show amplification of the AR gene; an increased receptor level can compensate for a lower hormone concentration. Second, some SRs can be activated by phosphorylation independent of ligand binding (see Chapter 6). Tumors with activation mutations in HER2, an EGFR isoreceptor, have constitutively phosphorylated ARs. Finally, there may be mutations in SR coactivators; for example, β-catenin is a coactivator for AR. An S33F mutation prolongs the β-catenin half-life such that the effect of β-catenin on AR-induced transcription doubles. This mutation has been reported in prostate cancer.

All of these examples pertained to SRs that were essential for tumor survival; the lymphomas and leukemias are endocrine-responsive tumors that represent a different paradigm. Glucocorticoids are toxic to lymphocytes and are very effective drugs in the initial treatment of these cancers. However, many of these malignant cells will eventually develop resistance to these steroids. This resistance is generally a result of defects in the glucocorticoid receptor; several such mutants have been characterized. The nuclear transfer-deficient mutants have point mutations in the DNA-binding domain, resulting in a decreased affinity for DNA. The increased nuclear transfer type has a truncated receptor lacking the amino-terminal modulatory domain. Although it actually has an increased affinity for DNA, it is incapable of triggering transcription. The r^0 mutant does not produce any receptor at all, despite the absence of any obvious deletions or rearrangements; a regulatory defect is suspected. Finally, the activation-labile receptor is extremely unstable after activation and quickly loses hormone-binding activity. The molecular basis for this latter mutant is unknown. Although these mutations occurred in human cell lines, similar defects can occur *in vivo*; for example, the deletion of all or part of GR has been documented in patients with leukemia or multiple myeloma following therapy. As with the ER, these deletions in the GR are often the result of aberrant splicing.

Another example of NR inactivation occurs in acute promyelocytic leukemia. This cancer is commonly associated with a t(15;17) chromosomal translocation, which creates a RAR/*myl* fusion gene. The function of Myl is unknown, but it replaces the amino terminus of retinoic acid receptor (RAR) and converts RAR into a dominant negative mutant. Because retinoids are often associated with differentiation, the inactivation of one of their receptors might favor growth over differentiation. For example, a deletion of the amino terminus of RAR has been reported in P19 embryonal cancer cells; this variant is also a dominant negative mutant.

Membrane Receptors

Receptor Kinases

Receptor Tyrosine Kinases. One of the most heavily studied membrane receptors is the insulin receptor, and numerous mutations have been reported for this receptor tyrosine kinase (RTK) (Table 17-6). These mutations can have many effects, depending on their locations; truncations often affect synthesis, mutations

Table 17-6
Clinical Syndromes Resulting from Receptor Tyrosine Kinase Mutations

Receptor	Mechanism	Syndrome
Insulin receptor	Inactivation	Diabetes mellitus
Kit	Inactivation	Piebaldism
TrkA	Inactivation	Congenital insensitivity to pain and anhydrosis
Ret	Inactivation	Hirschsprung's disease
VEGFR3	Inactivation	Hereditary lymphedema
Growth factor receptors	Constitutively active	Usually oncogenic
FGFR1 and FGFR2	Constitutively active (membranous ossification)	Crouzon, Jackson-Weiss, Alpert, Pfeiffer and Beare-Stevenson cutis gyrata syndromes
FGFR3	Constitutively active (endochondrial ossification)	Achondroplasia, thanatophoric syndrome, Muenke craniosynostosis, SADDAN, and Crouzon syndrome with acantosis nigricans
TIE2	Constitutively active	Venous malformations
VEGFR2 and VEGFR3	Constitutively active	Juvenile hemangioma (somatic mutations)

in the α-subunit frequently impair insulin binding, and defects in the catalytic site reduce kinase activity. In addition, some variants display processing defects. For example, the R735S mutation occurs at the α-β interface and prevents cleavage. Other mutations interfere with transport to the plasma membrane.

Most of these defects produce severe insulin resistance. The severest form is leprechaunism; the name is derived from the small size and elfin features of these infants. They also have lipoatrophy and acanthosis nigrans; the latter is a skin condition, where the epidermis is hypertrophied and darkened. It is believed to result from the cross-activation of IGF-IRs by insulin, which is quite elevated in these patients. Insulin receptors are virtually immeasurable and affected individuals die in early infancy. Growth retardation is less severe and life expectancy is longer for patients with the Rabson-Mendenhall syndrome. This syndrome is also characterized by precocious puberty and dystrophic nails and teeth. Most known mutations for both of these diseases are in the α-subunit. Type A insulin resistance occurs most commonly in adult women, who also have hirsutism and scanty menses. Mutations in either the α- or β-subunit have been found in the tissues from these patients. As noted for mutations in the nuclear receptors, the phenotype does not always correspond to the genotype; the W1193L* change has been reported for several members of a family. Although all are heterozygotes, one subject had type A insulin resistance, another had minimum symptoms, and a third had no overt disease. It is possible that individual factors may affect whether an insulin receptor mutation is dominant. Alternatively, because a tryptophan to leucine substitution would normally be considered a conservative one, the receptor may have a quasistable structure. Other compensatory factors may then determine the functionality of this protein.

In addition to mutations within the insulin receptor itself, mutations in other genes can affect insulin receptor structure and activity. Myotonic dystrophy type 1 results from a CTG expansion in the myotonic dystrophy protein kinase gene. This expansion, in turn, upregulates the CUG binding protein, which is involved with the regulation of alternate splicing and whose elevated levels produce aberrant splicing in many proteins. One of these proteins is the insulin receptor, which is shifted from the high affinity IR-B form to the low affinity IR-A form (see Chapter 8). This interchange is responsible for the insulin resistance seen in myotonic dystrophy type 1.

These syndromes are exceedingly rare, whereas NIDDM is very common. It would be far more clinically relevant if insulin receptor defects could be found in NIDDM. Although no such mutations have been definitively shown to be causative in NIDDM, two have aroused interest based on their frequency in diabetic patients: the R1152Q mutation displays normal autophosphorylation but impaired phosphorylation of a synthetic substrate; and the V985M mutation is found in both diabetic and healthy individuals, but carriers do have higher serum glucose levels following a glucose load. As such, it may be a contributing, rather than causative, factor in developing NIDDM. Another candidate is S992P; the rhesus monkey is diabetes prone and has an insulin receptor identical to that for humans except for 12 amino acids. All but the S992P substitution are conservative changes; this same alteration occurs in other diabetes-prone species.

Another interesting human disease caused by a defective RTK is piebaldism and its murine counterpart, white spotting. Stem cell factor is a tropic hormone

---

*Although the mature insulin receptor is cleaved, mutants are designated by their position in the uncleaved precursor.

for stem cells in the hematopoietic, pigmentary, and reproductive systems (see Chapter 3). In human beings, mutations in the stem cell growth factor (SCF) receptor results in depigmented patches on the skin. Because the patches are most commonly located on the ventral surface, this phenotype appears to be a result of a defect in the migration of the melanocytes from the dorsal neural crest, where they arise, to the ventral surface. This migration may depend on a gradient of the SCF receptor; if receptor activity is compromised, the entire gradient may be reduced and the lower end may drop below the threshold for melanocyte migration. In addition to depigmentation, rats also show reductions in erythrocytes and mast cells.

Alterations in the gene for at least four other human RTKs have been reported. VEGF is responsible for the formation of blood and lymphatic vessels; lymphatics fail to form properly in patients with a defective VEGFR3. As a result, hereditary lymphedema develops because lymph drainage is impaired. TrkA (NGF receptor) is required to generate pain receptors and sweat glands; its loss produces congenital insensitivity to pain with anhidrosis (absence of sweating). Glial cell line-derived neurotropic factor (GDNF) is one of several hormones like ET-3, which is important in targeting cholinergic nerves to the colon. Mutations in Ret, the RTK for GDNF, lead to Hirschsprung's disease, where the lack of innervation causes fecal retention. Finally, the colony-stimulating factor 1 (CSF-1) receptor gene is deleted in myelodysplasia. CSF-1 is a hematopoietic factor (see Chapter 3), and its loss is associated with refractory anemia and abnormal megakaryocytes.

All of the described diseases arise from loss-of-function mutations. Because many RTKs are activated by growth factors, gain-of-function mutations are often oncogenic and several were covered in Chapter 16. However, there are a few such mutations that produce other syndromes. Activating mutations of FGFRs are the most common in this group and have been the most heavily studied. Skeletal deformations dominate the phenotype, although skin abnormalities may also occur. FGFR1 and FGFR2 are primarily involved with the morphogenesis of membranous bone, such as the flat bones of the skull. Overactivity leads to premature fusion of the cranial sutures with resulting skull deformation. These diseases include Crouzon, Jackson-Weiss, Alpert, Pfeiffer, and Beare-Stevenson cutis gyrata syndromes, depending on the particular spectrum of findings. FGFR3 is primarily involved with the morphogenesis of endochondrial bones, like the long bones of the limbs. Overactivity leads to premature fusion of the epiphyses and short-limb syndromes, such as achondroplasia, thanatophoric syndrome, Muenke craniosynostosis, Crouzon syndrome with acantosis nigricans, and severe achondroplasia with developmental delay and acanthosis nigricans (SADDAN).

Most of these mutations favor FGFR dimerization; for example, mutations involving cysteine produce an uneven number of these amino acids and the free thiol group will form a disulfide bond with its counterpart on a second mutant FGFR. In the transmembrane domain, mutations to hydrophilic amino acids will induce dimerization as the hydrophobic environment forces these amino acids to cluster together. These mutations may also decrease down-regulation because this process is governed by the adjacent juxtamembrane domain. The IgIII domain is normally involved with dimerization and mutations at this site increase the number of contacts between the dimers. Mutations in the gap between IgII and IgIII have a different mechanism for activating the FGFR; they broaden the binding specificity to allow more FGF isoforms to bind and stimulate the affected FGFR.

One last topic needs to be addressed. Why can mutations in a single RTK produce so many different syndromes? The answer lies in the fact that different mutations in an RTK can have dramatically different effects. For example, FGFR3 mutations that produce thanatophoric syndrome not only activate the FGFR3 but also lead to the retention of FGFR3 in the endoplasmic reticulum, where the activated receptor has access to a different set of transducers than exist in the plasma membrane. Specifically, it will induce the phosphorylation of Stat1 but not the docking protein FGFR substrate 2α (FRS2α). FGFR3 mutants that reside in the plasma membrane would have the reverse effect.

Two other RTKs have been reported to have gain-of-function mutations. Angiopoietin-1 is involved with the morphogenesis of veins, and activating mutations of its RTK, TIE2, result in venous malformations. VEGF is also a morphogen for blood vessels, and constitutively active VEGFR2 or VEGFR3 leads to juvenile hemangiomas (benign tumors of the blood vessels).

Receptor Serine-Threonine Kinases.   The TGFβ family is very important in many developmental processes. In several instances, they serve to inhibit these processes; as such, loss-of-function mutations appear paradoxically as positive effects. For example, the fetal testes produce antimüllerian hormone (AMH) to induce apoptosis in the müllerian ducts. Inactivating mutations in either AMH or its receptor result in the persistence of these ducts, which then develop into a uterus and upper vagina in the male fetus (Table 17-7). Myostatin limits myocyte proliferation; its deficiency leads to a doubling of muscle mass in Belgian Blue and Piedmontese cattle. Similarly, loss of BMPR-II produces hyperplasia of the intima and smooth muscle in pulmonary precapillary arterioles in familial primary pulmonary hypertension; activin receptor-like kinase 1 (ALK)–1 inactivation results in vascular overgrowth in hereditary hemorrhagic telangiectasia, type II (Osler-Rendu-Weber syndrome); and mutations in BMPR-IA produce hyperplasia of the intestinal mucosa in some juvenile polyposis and related disorders with gastrointestinal polyps, such as Cowden and Bannayan-Riley-Ruvalcaba syndromes.

Finally, BMP15 inhibits FSH action in oocytes by downregulating the FSHR and decreasing FSH-stimulated cAMP and progesterone production. Mutations in either BMP15 or its receptor, BMPR-IB, increase FSH sensitivity. As a result, there is less oocyte atresia, which leads to superovulation and multiple

Table 17-7
Clinical Syndromes Resulting from Mutations in Receptor Serine-Threonine Kinases or Their Ligands

Receptor	Mechanism	Syndrome
AMH or its receptor	Failure of the müllerian ducts to regress in males	Persistent müllerian duct syndrome
Myostatin	Myocyte overproliferation	Belgian Blue and Piedmontese Cattle
ALK-1	Vascular overproliferation	Hereditary hemorrhagic telangiectasia, type II (Osler-Rendu-Weber syndrome)
BMPR-II	Hyperplasia of intima and smooth muscle	Familial primary pulmonary hypertension
ALK-3 (BMPR-IA)	Hyperplasia of intestinal mucosa	Some juvenile polyposis
ALK-6 (BMPR-IB)	Superovulation	Multiple births in Booroola sheep
BMP15 (ligand for ALK-6)	Superovulation	Multiple births in Inverdale and Hanna sheep

births in heterozyotic sheep and humans. Homozygosity results in primary ovarian failure.

Cytokine Receptors

Cytokines are most closely associated with the hematopoietic and immune systems. Granulocyte colony stimulating factor (G-CSF) and thrombopoietin are responsible for the differentiation of neutrophils and platelets, respectively; defects in their receptors cause severe congenital neutropenia (reduced numbers of neutrophils) and congenital amegakaryocytic thrombocytopenia (reduced numbers of platelets), respectively. Inactivating mutations of IL-2Rγ or IL-7R produce various forms of immunodeficiency (Table 17-8), and mutations in IFNγR or IL-12Rβ1 increase a carrier's susceptibility to tuberculosis. β_c is involved with macrophage function; its deficiency results in the inability of pulmonary macrophages to clear the surfactant protein. The accumulation of this protein leads to pulmonary alveolar proteinosis.

The other cytokines serve a variety of functions. GH is responsible for general body growth. Inactive mutants of the GHR cause Laron dwarfism. Indeed, because short stature does not affect survivability, there are more mutations documented for this receptor than any other cytokine receptor: both truncations, which lead to defects in synthesis, and point mutations, which result in processing errors, have been reported. As explained in Chapter 2, leptin is secreted by adipose tissue to signal the hypothalamus concerning the nutritional status and energy balance of the organism. A defect in the leptin receptor blocks negative feedback from adipose tissue and leads to obesity.

Fas is a member of the TNFR family and induces apoptosis of immune cells. Its deficiency results in overproliferation with subsequent lymphadenopathy (diseased lymph nodes), hepatosplenomegaly (enlarged liver and spleen), and autoimmune disease; this complex of symptoms is called the *Canale-Smith*

Table 17-8
Clinical Syndromes Resulting from Mutations in Cytokine Receptors

Receptor	Mechanism	Syndrome
G-CSFR	Deficiency	Severe congenital neutropenia
Mpl (thrombopoietin receptor)	Deficiency	Congenital amegakaryocytic thrombocytopenia
IL-2Rγ	Deficiency	X-linked severe combined immunodeficiency
IL-7R	Deficiency	Severe combined immunodeficiency
IFNγR	Deficiency	Increased susceptibility to tuberculous
IL-12Rβ1	Deficiency	Increased susceptibility to tuberculous
GHR	Deficiency	Laron dwarf
Leptin R	Deficiency	Severe early onset obesity
β_c	Deficiency	Some pulmonary alveolar proteinosis
Fas[a]	Deficiency	Canale-Smith syndrome
Ectodysplasin-A2 receptor (EDAR)[a]	Deficiency	Anhidrotic ectodermal dysplasia
Osteoprotegerin[b]	Deficiency	Juvenile Paget's disease; idiopathic hyperphosphatasia
RANK	Gain-of-function	Familial expansile osteolysis
TNFR1	Gain-of-function	TNFR-associated periodic syndromes (TRAPS)
EPOR	Gain-of-function	Primary familial polycythemia
IL-4Rα	Gain-of-function	Atopy

[a] TNFR family.
[b] Decoy receptor for RANK.

syndrome. Ectodysplasin-A2 receptor (EDAR) is another member of the TNFR family and is involved with the morphogenesis of epidermal appendages. Inactive mutants cause anhidrotic ectodermal dysplasia, which is characterized by abnormal teeth, hair, and eccrine glands. Finally, RANK is also a member of the TNFR family; its ligand, RANKL, is an osteoclastic differentiation factor. Osteoprotegerin is a decoy receptor for RANK, and its loss allows RANKL to stimulate osteoclasts, which then increase bone turnover. There are two primary phenotypes: in juvenile Paget's disease, there are fractures and progressive skeletal deformities; in idiopathic hyperphosphatasia, there are long bone deformities, kyphosis (curvature of the spine), and acetabular protrusion (pelvic deformity).

Gain-of-function mutations in RANK produce a similar syndrome, known as *familial expansile osteolysis.* This disease is characterized by osteolytic lesions, tooth loss, and deafness. A small in-frame insertion in the signal peptide blocks cleavage; retention of the signal sequence may lead to intracellular accumulation, increased self-association, and receptor activation. Three other cytokine receptors are also subject to gain-of-function mutations. Activating mutations of TNFR1 give rise to the TNFR-associated periodic syndromes (TRAPS), which are a group of diseases characterized by self-limited episodes of periodic fever and inflammation. Some cysteine missense mutants probably induce oligomerization, whereas other mutations decrease TNFR downregulation or reduce the solubilization of the extracellular domain. Because the latter normally dampens TNF activity by sequestration, its reduction would enhance TNF activity.

Deletion of the carboxy terminus of the EPO receptor (EPOR) eliminates the docking site for SHP (SH2 containing protein tyrosine phosphatase). This change would block the recruitment of SHP and prevent EPOR dephosphorylation and inactivation. Consitutively active EPOR overproduces red blood cells and is one cause of primary familial polycythemia. A point mutation in IL-4Rα also eliminates the SHP docking site. Its overactivity is associated with atopy, an allergic reaction characterized by immunoglobulin E (IgE) immediate hypersensitivity.

G Protein-Coupled Receptors

The first mutant GPCR to be described was rhodopsin, the light receptor. These defects are responsible for some forms of the disease, retinitis pigmentosa. Since then many other GPCR mutations have been described (Table 17-9). Most of these diseases are self-explanatory; for example, inactivating mutations in the receptors for tropic hormones lead to deficiencies of the down-stream hormone. However, a few topics and receptors are worth discussing in more detail.

There are several melanocortin receptor (MCR) isoforms: MCR2 is the ACTH receptor, whereas MCR1 is activated by MSH. Skin and hair color is determined, in part, by the ability of MSH to regulate epidermal melanin. Melanin has two forms: eumelanin is black-brown and phaeomelanin is red-yellow. MSH differentially stimulates the former, and loss-of-function mutations of MCR1 produce yellow mice, red cows, red-yellow chickens, chestnut horses, and red pigs. Although skin and hair color in human beings is regulated in a more complex manner, these mutations are associated with red hair, fair skin, freckles, and an increased risk of melanoma. Conversely, activating mutations generate black mice, cows, pigs, chickens, jaguars, bananaquits, and foxes (the Alaskan Silver fox).

Table 17-9
Clinical Syndromes Resulting from Mutations in G Protein-Coupled Receptors

Receptor	Mechanism	Syndrome
Releasing Hormone Receptors		
TRHR	Deficiency	Isolated central hypothyroidism
GHRHR	Deficiency	Profound GH deficiency
GnRHR	Deficiency	Hypogonadotropic hypogonadism
SST5	Deficiency	Somatostatin resistance in acromegaly
Pituitary Hormone Receptors		
TSHR	Deficiency	TSH resistance
	Gain-of-function	Familial nonimmune hyperthyroidism (genomic); hot thyroid adenomas (somatic)
LHR	Deficiency	Leydig cell hypoplasia
	Gain-of-function	Familial male precocious puberty
FSHR	Deficiency	Hypergonadotropic ovarian dysgenesis
ADH receptor (VPR)	Deficiency	Diabetes insipidus
MCR1	Deficiency	Yellow or red hair/feather color in animals/birds
	Gain-of-function	Black hair/feather color in animals/birds
MCR2 (ACTHR)	Deficiency	Hereditary isolated cortisol deficiency
	Gain-of-function	Episodic Cushing's disease
MCR4	Deficiency	Obesity through increased appetite
Other GPCRs		
PTHR1	Deficiency	Blomstrund chondroplasia
	Gain-of-function	Jansen-type metaphyseal chondroplasia; enchondromatosis (Ollier and Maffucci diseases)
ET-BR	Deficiency	Hirschsprung's disease
D1	Desensitization secondary to hyperactive GRK4 mutant	Some essential hypertension
D2	Deficiency	Hereditary autosomal dominant myoclonus dystonia
TXA$_2$R	Deficiency	Dominantly inherited bleeding disorder
P2Y$_{12}$	Deficiency	Platelet-based bleeding disorder
Hypocretin (orexin) receptor	Deficiency	Canine narcolepsy
fMLP-R	Deficiency	Localized juvenile periodonitis
CCR5	Deficiency	Resistance to HIV infections
CXCR4	Gain-of-function	WHIM syndrome
LRP5[a]	Deficiency	Osteoporosis-pseudoglioma syndrome (OPPG)
AT2	Deficiency	X-linked mental retardation
Insl3R	Deficiency	Some cryptorchidism

[a] Coreceptor for Frizzled.

There are two major mechanisms for the gain-of-function mutations in this receptor class. First, GPCRs have two main conformations: active and inactive.* The inactive state is stabilized by interhelical bonds; ligand binding disrupts these bonds and shifts the GPCR into the active state. Mutations that also disrupt these bonds can have the same effect. Second, the carboxy terminus of many GPCRs has a binding site for β-arrestin, which is required for desensitization. Mutations at this location can likewise result in constitutively active GPCRs.

---

*Some authorities have postulated various intermediate states, but these forms are not relevant to the current discussion.

Overactivity can also occur when there is a loss of an antagonist: for example, agouti is an MSH antagonist at the MCR1. Loss-of-function agouti mutants allow excessive MCR1 activity and are responsible for the black hair in the domestic cat, horses, and the Standard Silver fox. This mechanism has been reported for one other GPCR system that involves Frizzled, LRP5, and Dickkopf-1 (Dkk-1). Frizzled is a GPCR that mediates the effects of Wnt during embryogenesis; it uses LRP5 as a coreceptor (see Chapter 7). Dkk-1 binds and sequesters LRP5 to inhibit Wnt action. Inactivating mutations in Dkk-1 removes the damper on Wnt signaling to produce the high-bone-mass syndrome, which results from the Wnt-stimulated overproduction of bone.

PTH is best known for its role in calcium homeostasis in the adult; however, it plays a vital role in fetal skeletal development, where it stimulates growth while inhibiting differentiation. Inactivating mutations in PTHR1 cause Blomstrand chondroplasia, which is characterized by advanced endochondrial bone maturation and fetal death. Gain-of-function mutations in PTHR1 stimulate chrondrocytes and impair limb growth.

CCR5 is a chemokine GPCR that is used by the human immunodeficiency virus (HIV) as a coreceptor to invade immune cells. Subjects lacking this receptor do not appear to have any immune dysfunction, but they are resistant to HIV infection. Activating mutations of LH receptor (LHR) induce precocious puberty in males but not females. In males, LH is sufficient to stimulate testosterone synthesis. However, females require FSH to trigger follicular development; without ovarian follicles, there is no sex steroid synthesis and no precocious puberty.

Ion Channel Receptors

Deficiencies in several members of the Cys-loop superfamily have been reported (Table 17-10). Loss-of-function mutations in nAChR (nicotinic acetylcholine receptor) produce muscle weakness. Actually, some "gain-of-function" mutations result in a similar phenotype; such mutations may cause muscle weakness through three mechanisms. First, constitutive activity can cause nAChR desensitization; a similar phenomenon is observed with the insulin receptor, where gain-of-function mutations also cause the loss of the receptor. Second, persistently elevated calcium is toxic to myocytes. Finally, depolarization blocks voltage-gated sodium channels; the latter are important in propagating the action potential. Mutations in two other molecules can also cause congenital myastenia. First, rapsyn is a protein that recruits nAChR to the synapse; when rapsyn is inactivated, nAChR does not localize to the synapse. Second, acetylcholine (Ach) is hydrolyzed by the collagen-tailed acetylcholinesterase. Inactivating mutations actually increase ACh but, again, muscle weakness develops because of receptor

Table 17-10
Clinical Syndromes Resulting from Mutations in Ligand-Gated Ion Channels

Mechanism	Channel	Syndrome
Loss-of-function	nAChR	Congenital myastenia; slow-channel syndrome
	GlyR	Hereditary hyperekplexia
	GABA$_A$	Generalized or absence epilepsy with febrile seizures
	P2X$_1$ receptor	Severe bleeding disorder
Gain-of-function	nAchR (α4 or β2)	Nocturnal frontal lobe epilepsy
	AMPA (GluR2)	*Lurcher* mouse

downregulation. Less severe gain-of-function mutations in nAChR do lead to overstimulation and nocturnal frontal lobe epilepsy.

Glycine receptor (GlyR) and GABA$_A$ are Cys-loop channels that transmit chloride and inhibit the generation of action potentials by hyperpolarizing the cell. Loss-of-function mutations lead to hyperactivity as a result of the loss of neural inhibition. GlyR defects cause hereditary hyperekplexia, which is characterized by an exaggerated startle response, and GABA$_A$ defects produce generalized and absence epilepsy with febrile seizures.

The P2X1 receptor is a member of the epithelial sodium channel family; it is abundant on platelets, where it participates in blood clotting. Inactivating mutations lead to a severe bleeding disorder.

No mutations in the P superfamily have been reported in human beings. However, the *lurcher* mouse has an activating mutation in the AMPA (GluR2) receptor; this excessive stimulation leads to neurodegeneration.

Antibodies

Another group of endocrinopathies is related to the autoimmune diseases. These diseases are characterized by the inability of the immune system to recognize certain endogenous antigens as "self" and the subsequent development of antibodies to them. In addition to immune disorders, cancers can induce autoimmune reactions by causing tissue destruction, which exposes endogenous antigens normally sequestered from the immune system. When these antigens are hormone receptors, several interesting possibilities arise. For example, antibody binding to a receptor may prevent hormone binding and lead to a hormone deficient-like state. However, occasionally the antibody binding can actually activate the receptor and lead to a hormone excess-like state.

The latter is called a *stimulating antibody*, and the classic disease of this type is Graves' disease, a form of hyperthyroidism (Table 17-11). These patients have in their serum a "long-acting thyroid stimulator," which is in fact IgG antibodies

Table 17-11
Clinical Syndromes Resulting from Antibodies to Transduction Components

Type of antibodies	Target	Syndrome
Stimulating	TSHR, Na$^+$-I$^-$-symporter, etc.	Graves' disease
	GHR	Some acromegaly
	GluR3	Rasmussen's encephalitis
	GluR1, -4, -5, and -6	Paraneoplastic neurological syndromes
	Insulin receptor	Hypoglycemia
Blocking	nAChR	Myasthenia gravis (adult); arthrogryposis multiplex congenita (fetus)
	Ganglionic AChR	Idiopathic autonomic neuropathy
	Insulin receptor	Type B insulin resistance
	Na$^+$-I$^-$-symporter	Some autoimmune hypothyroidism
	β_1R	Idiopathic dilated cardiomyopathy
	M2R	Chagas' disease
	Ca$_V$2.1(α_1A)	Lambert-Eaton syndrome
Destructive	Pancreas	IDDM
	Thyroid peroxidase	Autoimmune thyroiditis
	Steroid synthesizing enzymes	Addison's disease
	Melanin-concentrating hormone receptor 1 (MCHR1)	Vitiligo

directed against the thyroid plasma membrane. These antibodies are heterogeneous; for example, the TSH receptor and the iodide transporter have been identified as targets of these antibodies. In some forms of acromegaly, stimulating antibodies to the GH receptor have also been identified. Acromegaly is a disease normally produced by an excess of GH in the adult. Because the epiphyses have fused, linear growth is no longer possible; instead, the bones become thicker and the facial features, coarser. Stimulating antibodies to GluR3 cause Rasmussen's encephalitis, which is characterized by severe epilepsy, dementia, and hemiplegia (unilateral paralysis). The development of stimulating antibodies to GluRs is a common occurrence with some tumors and is responsible for several paraneoplastic neurological syndromes. Stimulating antibodies to the calcium receptor have already been mentioned (see Table 17-1), and stimulating antibodies to the insulin receptor induces hypoglycemia.

Another type of antibody is called a *blocking antibody*, and the classic disease of this type is myasthenia gravis, a neuromuscular disease in adults. Clinically, patients have weak and easily fatigued skeletal muscles. Biochemically, these patients have antibodies directed primarily against the α-subunit of the nAChR, although epitopes in the γ and δ subunits have also been reported. If the patient is pregnant, these antibodies can cross the placenta and produce muscle weakness in the fetus. Lack of muscle activity causes joint contractures (arthrogryposis) in a syndrome known as *arthrogryposis multiplex congenita*. Interestingly, these antibodies do not prevent ACh binding, but they do accelerate receptor degradation and disrupt the synaptic architecture through complement-mediated attack of the post-synaptic membrane.

Blocking antibodies to the ganglionic AchR results in idiopathic autonomic neuropathy, which is characterized by orthostatic hypotension, urinary retention, constipation, anhidrosis, tonic pupils, and dry eyes and mouth. Other blocking antibodies have been documented for the insulin receptor (type B insulin resistance), Na^+-I^--symporter (some forms of autoimmune hypothyroidism), $\beta_1 R$ (idiopathic dilated cardiomyopathy), M2R (Chagas' disease), the $Ca_v2.1$ (Lambert-Eaton syndrome), and the calcium receptor (see Table 17-4).

There is a third possible consequence of autoimmune disease: the antibodies induce a severe inflammatory response that eventually leads to the destruction of the gland and hormone deficiency. Several of these situations were discussed above under hormone deficiencies arising from endocrine gland destruction: the pancreas and IDDM, the thyroid and autoimmune thyroiditis, and the adrenal cortex and Addison's disease. One other example is mentioned: vitiligo is a disease characterized by depigmented patches on the skin and results from the destruction of melanocytes. Antibodies to melanin synthetic enzymes, transcription factors, and the melanin-concentrating hormone (MCH) receptor 1 have all been described. MCH antagonizes MSH through the MCHR1; it can also regulate food intake and energy balance through other isoreceptors.

Transduction

G Proteins

The best example in this group is pseudohypoparathyroidism, type IA, also called *Albright's heredity osteodystrophy* (Table 17-12). Clinically, the patients appear to have hypoparathyroidism; that is, they have skeletal abnormalities

Table 17-12
Clinical Syndromes Resulting from Mutations in Transducers

Transducer	Mechanism	Syndrome
G Proteins		
α	Deficiency	Pseudohypoparathyroidism types IA and IB
	Gain-of-function	McCune-Albright (mosaic) and various endocrine adenomas (somatic)
Rac2	Isoform in neutrophils	Human neutrophil immunodeficiency syndrome
Rab	Rab7 deficiency	Charcot-Marie-Tooth disease, type 2B
	Rab27 deficiency	Griscelli syndrome
	Rab38 deficiency	*chocolate* mouse[a]
	Deficiency of Chm[b]	Choroideremia
	Deficiency of Rab geranyl geranyltransferase II	*gunmetal* mouse
Ras	Ras activating mutations	Various tumors
	GAP inactivating mutations	Various tumors
	Constitutively active SOS1 (RasGEF)	Hereditary gingival fibromatosis
Cyclic Nucleotides and NO		
RIα of PKA	Deficiency	Carney complex; primary pigmented nodular adrenocortical disease (ACTH-independent Cushing's disease)
PDE	Deficiency of cGMP-specific isoform	Autosomal dominant congenital stationary night blindness
NOS	Deficiency of NOS2 (iNOS)	Dahl/Rapp rat (salt-sensitive hypertension)
	NOS2 gain-of-function	Protection against malaria
	Deficiency of NOS3 (eNOS)	Association with preeclampsia and placental abruption; some essential hypertension resistant to conventional therapy
Caveolin-3	NOS inhibitor (loss-of-function)	Autosomal dominant limb-girdle muscular dystrophy 1C
PPI Pathway		
PKCα	Gain-of-function	Human pituitary adenoma (somatic)
PLCβ3	Deficiency	Multiple endocrine neoplasia
PI Phosphatases	PI(4,5)P$_2$ phosphatase deficiency	Lowe (oculocerebrorenal) syndrome
	Deficiency of MTM	X-linked myotubular myopathy
	Deficiency of MTMR	Charcot-Marie-Tooth disease, type 4B
Calpain	Deficiency of muscle-specific isoform	Limb-girdle muscular dystrophy, type 2A
Sphingomyelin Pathway		
Serine palmitoyltransferase	Activating mutations	Hereditary sensory neuropathy type I
Insulin Mediators		
HNF-4α	Transcription factor for insulin effects	Diabetes mellitus
Phosphofructokinase 1-M	Muscle and pancreatic islet isoform	Diabetes mellitus
IRS-1	Loss-of-function	NIDDM

[a] Model for oculocutaneous albinism.
[b] RabGDI and escorts Rab to geranylgeranyltransferase II.

and hypocalcemia, with resulting tetany and seizures (see Chapter 1). However, PTH levels are elevated, and hormone receptors appear normal. The adenylate cyclase is also normal because it can be stimulated by forskolin. However, fluoride and the nonhydrolyzable analogs of GTP do not activate the adenylate cyclase; unlike forskolin, which acts directly on the cyclase, fluoride and the GTP analogs act through the α subunit of G_s (see Chapter 9). There is also a decreased incorporation of labeled ADP-ribose into α_s by cholera toxin. This α_s defect was eventually shown to be due to a variety of truncations and point mutations.

A major question remains. If G_s resides in a common pool for all cAMP-dependent hormones, why is only PTH action affected? In part, the answer resides in a process called *imprinting*. A male has only one X chromosome, whereas a female has two; as a result, women potentially have a double dose of X-linked genes. In compensation for this duplication, one of the X chromosomes is inactivated. This inactivation appears to be random for many genes; however, for some genes there is a bias toward either the maternally or paternally inherited gene in certain organs. This bias is referred to as imprinting. For example, the maternal G_s is imprinted in the proximal tubules of the kidney and in endocrine tissue, but is biallelic elsewhere. Heterozygotes have pseudohypoparathyroidism, type IA, with skeletal abnormalities, multihormone resistant, and low serum calcium. However, the endocrine effects are usually subclinical; that is, they are not severe enough to produce symptoms and have only been discovered by sophisticated testing. It may be that PTH relies heavily on the cAMP pathway, whereas the other hormones may have multiple transducers; for example, in the liver glucagon activates both the adenylate cyclase and the PPI pathway. Therefore, the loss of one transducer would not be as devastating to glucagon as it would to another hormone that acted exclusively through α_s.

In contrast to the preceding text, if the mutant gene were inherited from the patient's father, the pattern of imprinting would be different. For example, there would be no hormone resistance or low serum calcium levels because the defective paternal gene would not be expressed in the kidney or endocrine glands. There would still be skeletal abnormalities because this tissue is biallelic; this clinical variant is known as *pseudopseudohypoparathyroidism*. Besides imprinting, there are at least two other factors that can contribute to the selective effects of α_s mutants. First, there are in fact some α_s mutations that selectively uncouple the G protein from the PTHR. This variant is known as pseudohypoparathyroidism, type IB. Second, there appears to be a natural variation in the binding affinities between GPCRs and G proteins; for example, CB1, a cannabinoid receptor, can sequester $\alpha_{i/o}$ from other GPCRs. As such, when competing for α_s, not all GPCRs are equal.

There are also mutations that can activate α_s. All known mutations occur at either residue 201 or 227; the former is the mono(ADP-ribosyl)ating site for cholera toxin and the latter occurs in a GAP-like sequence. These mutations impair the GTPase activity of α_s, thereby prolonging its action. It would appear that widespread activation of α_s is not compatible with life because only somatic and conditional mutations have been identified. Two types of somatic mutations have been reported. Those that occur in single cells of an adult generally produce adenomas because cAMP is often mitogenic; in secretory cells, these tumors can also overproduce hormones. For example, these mutants have been found in some GH-secreting pituitary adenomas, thyroid-secreting thyroid adenomas, steroid-secreting adrenal adenomas, and PTH-secreting parathyroid adenomas; in fact, it has been estimated that as much as 40% of acromegaly may be due to activating mutations in α_s.

The other somatic mutation occurs early in embryogenesis, before the three germ layers form. The mature individual is composed of a mixture of normal and affected tissues; such a person is said to be a genetic mosaic. The result is that certain tissues are hyperactive, particularly with respect to hormone secretion or activity; the clinical picture includes bone lesions, café au lait spots (MSH), sexual precocity (gonadotropins), hyperthyroidism (TSH), pituitary adenomas, and adrenal hyperplasia (ACTH). This disease is known as the McCune-Albright syndrome.

The one conditional α_s mutant has an alanine to serine substitution at position 366. This mutation has two effects: first, it activates α_s by increasing GNP exchange; second, it destabilizes the protein at $37°$ C. As such, the only place in the body where this mutant is active is in the testes, where the temperature is 5 to $8°$ C lower than body temperature. Therefore the only manifestations of this mutant are testosterone hypersecretion and precocious puberty.

Mutations in α_{i2} have also been identified; they inactivate α_{i2} and leave α_s unopposed. These mutations are somatic and have been found in some adrenocortical and ovarian tumors.

Mutations in small G proteins have likewise been recorded. Rac2 is an isoform found in neutrophils, where it generates ROI (reactive oxygen intermediates) for killing microorganisms. Dominant inhibitory mutants cause human neutrophil immunodeficiency syndrome. The Rab family is involved with vesicular trafficking and defects in various Rab isoforms, in a RabGDI, and in a geranylgeranyltransferase II that targets certain Rabs have all been described (Table 17–12). Many of the symptoms are due to (1) improper trafficking of melanosomes leading to partial albinism, (2) faulty immune secretion leading to immune abnormalities, and (3) defects in the formation and maintenance of dendrites leading to neurological impairment. Activating mutations of Ras and inactivating mutations of RasGAP are oncogenic and were covered in Chapter 16. The Rho family is involved with cytoskeletal rearrangement. MEGAP/ srGAP3 is a RhoGAP associated with neuron migration and axon branching; and its mutation results in severe mental retardation.

3'5'-Cyclic AMP and Nitric Oxide Pathways

3'5'-Cyclic AMP Pathway

Because transduction systems serve many hormones, defects at this level can have widespread and devastating effects. For that reason, examples in higher animals are few and usually occur in tissue-specific isoforms. As such, examples in this section are supplemented with mutations from invertebrates to provide a sampling from as many transducers as possible.

The regulatory subunit of PKA has a pseudosubstrate site that keeps the catalytic subunit inactive in the absence of cAMP. Truncations of RIα lead to an overactive catalytic subunit and a variety of benign tumors. When these tumors involve endocrine glands they can produce hyperendocrinopathies. This constellation of symptoms is called the *Carney complex*.

Several mutations in the phosphodiesterase (PDE) have also been described. The dunce$^+$ mutant in *Drosophila* has a defective cAMP-specific PDE and is characterized by impaired memory and female sterility. The cGMP-specific

PDE is restricted to the retina, and its inactivation in human beings results in autosomal dominant congenital stationary night blindness.

Nitric Oxide Pathway

Of all the functions nitric oxide (NO) has in higher organisms, the two most affected by mutations are vasodilation and toxicity. For example, inactivating mutations in NOS2 (NO synthase 2) cause salt-sensitive hypertension in the Dahl/Rapp rat, and similar defects in NOS3 in human beings are associated with preeclampsia (hypertension during pregnancy) and essential hypertension unresponsive to conventional therapy.

There are two conditions that can elevate NOS activity. NOS2[Lambarene] is a gain-of-function promoter mutation that generates NOS2 levels seven times normal. Heterozygotes are partially protected against malaria; this effect is presumably due to the toxicity of NO. In addition, caveolin-3 normally inhibits NOS along with many other transducers (see Chapter 7); two syndromes have been associated with caveolin-3 defects. In autosomal dominant limb-girdle muscular dystrophy 1C, the toxicity from high NO levels induces muscle fiber degeneration. In rippling muscle disease, the loss of caveolin-3 function produces hyperirritability that may arise from a more general lack of inhibition of several second messengers, including NO.

Lipid Pathways

Polyphosphoinositide Pathway

The polyphosphoinositide (PPI) pathway is involved with vision in *Drosophila*; because these components are specific for the eye, their loss is still compatible with a viable organism. Phospholipase Cβ (PLCβ) is essential for phototransduction, and its inactivation by an insertional mutation renders the flies blind. Protein kinase C2 (PKC2) deactivates the photoresponse in flies; its variants lead to hyperadaptation to light. Finally, retinal degeneration B in *Drosophila* is a result of a defective, membrane-bound phosphoinositide (PI) transfer protein; this protein is necessary to maintain the supply of PI for the PI cycle (see Chapter 10). Diaglycerol kinase (DGK) is also involved with recycling and its defects produce similar effects.

There are fewer examples in human beings; somatic mutations in PKCα have been reported in some pituitary adenomas, and mutations in a muscle-specific calpain cause limb-girdle muscular dystrophy, type 2A. PLCβ3 apparently acts as a tumor suppressor because its inactivation leads to a form of multiple endocrine neoplasia.

However, the most numerous mutations in the human PPI pathway occur in the PPI phosphatases. There are two inositol polyphosphate 5-phosphatases, whose mutations produce disease. ORCL is defective in oculocerebrorenal (Lowe) syndrome; the resulting accumulation of PI(4,5)P_2 affects lysosomal trafficking. The syndrome is characterized by mental retardation, hypotonia (muscle faccidity), cataracts, glaucoma, and impaired renal resorption of glucose, amino acids, phosphate, and bicarbonate. A deletion in the 3' end of the SHIP2 gene leads to overexpression. Many of the actions of insulin are mediated by PI(3)P_n (see Chapter 12), and the depletion of this second messenger by elevated levels of SHIP2 (SH2-containing inositol phosphatase) is associated with NIDDM.

There are also two inositol polyphosphate 3-phosphatases, whose deficiencies produce disease. Inactivating mutations in myotubularin causes X-linked myotubular myopathy, in which elevated IP$_3$ levels arrest muscle development prior to myofiber formation. Mutant myotubularin-related protein also increases IP$_3$ and results in Charcot-Marie-Tooth disease, type 4B, which is characterized by Schwann cell proliferation, misfolded myelin, and demyelinating motor and sensory neuropathy.

Sphingomyelin Pathway

There is only one known mutation in the sphingomyelin pathway: an activating mutation of the serine palmitoyltransferase, long chain base subunit-1. The accumulation of ceramide leads to apoptosis of the dorsal root ganglia and motor neurons in hereditary sensory neuropathy type I.

Miscellaneous Transduction

Channels and Pumps

Calcium channels and pumps play vital roles in signaling. Mutations in the ryanodine receptor produce three phenotypes (Table 17-13). Conditional mutants of RyR1 (ryanodine receptor 1) cause malignant hyperthermia. These mutant receptors are hypersensitive to several molecules, especially certain anesthetics and skeletal muscle relaxants, which trigger the release of calcium and muscle contraction. Heat is generated from ATP hydrolysis by pumps in a futile attempt to return the calcium to the sarcoplasmic reticulum, from which the calcium simply leaks back into the cytosol. RyR1 mutants that produce a slow calcium leak generate central core disease. The plasma membrane calcium pumps are able to keep up with the leakage in the periphery of the myocyte. However, excess calcium in the center of the cell is taken up by mitochondria, which are overworked and eventually self-destruct. RyR2 is the cardiac isoform and its conditional mutant generates a dangerously rapid heart rate (tachycardia) after vigorous exercise. Familial polymorphic ventricular tachycardia has a 30% mortality rate by age 30 years. Strangely, RyR2 mutations that decrease the open probability can also lead to sudden death in young adults. The arrhythmia in this latter condition, known as *arrhythmogenic right ventricular dysplasia*, results from the progressive destruction of cardiac muscle in the right ventricle.

Voltage-gated calcium channel mutations primarily produce neurological dysfunction. Deficiencies in two Trp family members have been reported: defects in polycystin-2, a calcium channel in the endoplasmic reticulum, causes polycystic kidney disease, and inactivating mutations in Trp6, a magnesium channel required for magnesium absorption in the gastrointestinal (GI) tract and resorption in the kidneys, leads to hypomagnesemia with secondary hypocalcemia.

SERCA1 (smooth endoplasmic reticulum calcium ATPase 1) is a fast twitch muscle-specific pump of the sarcoplasmic reticulum. Inactivating mutations are unable to reduce calcium levels after stimulation in a condition called *Brody's disease,* which is characterized by exercise-induced impairment of muscle relaxation with stiffness and cramps. SERCA2a is found in cardiac and slow twitch muscles and in keratinocytes. Its defects are responsible for the abnormal keratinization and psychiatric symptoms in Darier's disease. Mutations in the Golgi complex calcium pump, encoded by *ATP2C1*, produce Hailey-Hailey disease, which resembles Darier's disease except for the lack of psychiatric symptoms.

The epithelial sodium channel is induced by aldosterone. Defects in this channel cause pseudohypoaldosteronism, type 1, which is characterized by salt wasting, hyperkalemia (elevated serum potassium), and metabolic

Table 17-13
Clinical Syndromes Resulting from Mutations in Channels and Pumps

Channel or pump	Mechanism	Syndrome
Ryanodine receptor	RyR1 (skeletal muscle isoform) leakiness	Malignant hyperthermia; central core disease
	RyR2 (cardiac isoform) leakiness	Familial polymorphic ventricular tachycardia
	RyR2 (cardiac isoform) decreased open probability	Arrhythmogenic right ventricular dysplasia
Ca_V1	α_{1F} deficiency	Incomplete X-linked congenital stationary night blindness
	α_s	Hypokalemic periodic paralysis
	β_4	Human juvenile myoclonic epilepsy, generalized epilepsy, praxis-induced seizures, and episodic ataxia
$Ca_V2.1(\alpha_{1A})$	Point mutations	Familial hemiplegic migraine; episodic and progressive ataxia
	Truncations	Episodic ataxia-2
	Expanded CAG	Autosomal dominant spinocerebellar ataxia 6
Polycystin-2[a]	Deficiency	Polycystic kidney disease
Calcium pumps	SERCA1[b]	Brody's disease
	SERCA2a[c]	Darier's disease
	ATP2C1 gene product[d]	Hailey-Hailey disease
Epithelial sodium channel	Deficiency	Pseudohypoaldosteronism, type 1
	Gain-of-function	Sodium retention and hypertension
Aquaporin-2	Deficiency	Nephrogenic diabetes insipidus
CNG	cGMP-gated channel deficiency	Some autosomal recessive retinitis pigmentosa
	Human ether-a-go-go potassium channel[e] deficiency	Hereditary long QT syndrome, type 2
Glucose channels	GLUT1 deficiency	GLUT1 deficiency syndrome
	GLUT2 deficiency	Fanconi-Bickel syndrome
NHE1	Unknown	Slow-wave epilepsy in mice

[a] Trp family calcium endoplasmic reticulum channel.
[b] Fast twitch-specific pump of the sarcoplasmic reticulum.
[c] Cardiac and slow twitch-specific pump of the sarcoplasmic reticulum.
[d] Golgi complex calcium pump.
[e] cAMP-gated channel.

acidosis. Deletion or truncation of the carboxy terminus removes an inhibitory domain and generates a gain-of-function mutant that produces sodium retention and hypertension. ADH induces the migration of aquaporin-2 to the plasma membrane, where it resorbs water from the urine; its deficiency leads to nephrogenic diabetes insipidus, which is characterized by the inability to concentrate urine.

The loss of a retinal cGMP-gated channel causes some forms of autosomal recessive retinitis pigmentosa. A CNG (cyclic nucleotide-gated) channel in the heart, called the *human ether-a-go-go potassium channel*, has a mutation in the cAMP site. Its loss is responsible for the hereditary long QT syndrome, type 2, which can induce cardiac arrhythmias.

No functional mutations have been reported in GLUT4, the major glucose channel regulated by insulin. However, defects in two other isoforms are known: GLUT1 deficiency syndrome produces seizures because of low glucose in the cerebrospinal fluid. GLUT2 is specific for the liver and kidney and its loss impairs glucose export leading to an accumulation of glycogen in both organs. This disease is known as the Fanconi-Bickel syndrome.

Many mitogens elevate pH through the stimulation of NHE1 (sodium [Na^+]-hydrogen exchanger 1). No mutations have been reported in human beings; however, in mice deletion of the last transmembrane helix produces slow-wave epilepsy, which is characterized by neuronal loss in the cerebellum and brainstem.

Phosphorylation

Virtually every signaling pathway involves kinases. Indeed, their importance is such that mutations have only been found in tissue-specific isoforms or in redundant pathways. Jak3, Btk, and Lck are STKs associated with the immune system and their loss-of-function mutations cause various immune deficiency syndromes (Table 17-14). IRAK-4 (IL-1 receptor associated kinase 4) is a mediator of the actions of IL-1, another defense pathway; its loss increases the patient's susceptibility to infection by pyogenic (pus-inducing) bacteria. NIK is a MAPKKK in the NF-κB stress pathway. Its loss in mice results in the absence of lymph nodes, a histologically disorganized spleen and thymus, and immunodeficiency. Finally, DNA-PK is involved with DNA repair and its mutants produce the *scid* mouse, which exhibits severe immune deficiency.

Pak3 (p21-activated protein kinase 3) is a brain-specific Rac-activated kinase, whose deficiency causes nonsyndromic X-linked mental retardation. Rsk2, an isoform of S6KII, blocks neuronal apoptosis by phosphorylating and inhibiting Bad. Mutant Rsk2 is responsible for the Coffin-Lowry syndrome, which is characterized by neuron loss, severe X-linked psychomotor retardation, facial and digital dysmorphisms, and progressive skeletal deformations. Finally, GRK4 phosphorylates D1 to induce its desensitization. The D1 dopamine receptor

Table 17-14
Clinical Syndromes Resulting from Mutations in Kinases and Phosphatases

Mutant	Function	Syndrome
Kinases		
Jak3	Cytokine transducer	Severe combined immune deficiency
IRAK-4	IL-1 transducer	Increased susceptibility to pyogenic bacteria
NIK	NF-κB activator	Alymphoplasia in mice
DNA-PK	Transcription factor kinase	Severe immune deficiency syndrome in mice
Pak3	Brain-specific Pak	Nonsyndromic X-link mental retardation
Rsk-2	Loss-of-function	Coffin-Lowry syndrome
GRK4	Specific for D1 in kidney	Some forms of essential hypertension
Phosphatases		
PTP	Hematopoietic-specific isoform (CD45)	Severe combined immune deficiency
	Brain/liver-specific isoform	Lafora disease
	Muscle-specific isoform	X-linked recessive myotubular myopathy
SHP	Gain-of-function	Noonan syndrome
PP1	Mutation in regulatory subunit 3	Worsens insulin resistance caused by mutant PPARγ
PP2A	Excess accumulation resulting from mutant E3 ubiquitin ligase	Opitz syndrome

mediates diuresis and natriuresis in the kidney. A gain-of-function mutation in GRK4 enhances D1 deactivation and results in hypertension due to sodium and water retention.

As with kinases, many mutant PTPs are tissue specific; this restricted distribution limits the severity of their phenotypes and allows the affected individuals to survive. For example, CD45 is a hematopoietic-specific PTP, whose loss causes a type of severe combined immunodeficiency disease; the deficiency of a brain-liver-specific PTP produces Lafora disease, which is characterized by epilepsy, myoclonus, and progressive neurological deterioration, and inactivating mutations in a muscle-specific PTP are responsible for X-linked recessive myotubular myopathy, which is characterized by severe hypotonia, generalized muscle weakness, and impaired maturation of muscle fibers. In mice, SHP is primarily associated with the hematopoietic system, and loss-of-function mutations induce autoimmune disorders. In human beings, mutations in the autoinhibitory SH2 domain of SHP generate gain-of-function mutants that cause Noonan syndrome, which has a broader phenotype consisting of mental retardation, dysmorphic facies, short stature, webbed neck, heart disease, and cryptorchidism.

Defects in the S-T phosphatases have also been documented. The muscle-specific, regulatory subunit 3 localizes PP1 to the sarcoplasmic reticulum. A mutation in this subunit results in the shift of PP1 from the membrane to the cytosol and leads to a disruption of glycogen metabolism in muscle. This mutation is often found in association with a mutant PPARγ and markedly worsens the insulin resistance induced by the latter. Midline 1 (MID1) is an E3 ubiquitin ligase for PP2A. Mutant MID1 leads to a buildup of PP2A and hypophosphorylation of its substrates. MID1 is primarily involved with the morphogenesis of ventral midline structures, and the accumulation of PP2A disrupts apoptosis and cell migration required to establish these structures. The result is Opitz syndrome, which is characterized by mental retardation, corpus callosum dysplasia, hypertelorism (widely-set eyes), and cleft lip and palate.

Metabolic Pathways

Insulin has profound effects on many metabolic pathways, and defects in these pathways can mimic or aggravate diabetes mellitus. For example, HNF-4α is a transcription factor that induces the insulin gene; but it also transcribes the genes for GLUT2 and several glycolytic enzymes. Its deficiency produces MODY1. PFK1-M is the predominant phosphofructokinase isoenzyme in muscles and pancreatic islets. In muscle, mutations in PFK1-M impair glucose uptake; in islets, mutations decrease the ATP-ADP ratio, which reduces K_{ATP} activity and insulin secretion. Several of the pathways affected by insulin are located in mitochondria, and mitochondrial dysfunction can also lead to diabetes. Finally, several mutations in IRS-1 (insulin receptor substrate 1) are associated with NIDDM; mutations in the PTB (phosphotyrosine-binding) domain exhibit decreased binding to the insulin receptor and those near the SH2 domain selectively interfere with PI3K activation.

Polyamine Pathway

There is only one reported defect in the polyamine pathway; keratosis follicularis spinulosa decalvans is produced by a duplication of the spermidine-spermine N^1-acetyltransferase. The duplication leads to overexpression of the enzyme, which lowers spermidine and spermine while elevating putrescine (see

Chapter 11). The symptoms are most marked in the skin and eyes; in particular, the skin is thickened and displays hair loss.

Transcription

Transcription Factors

Like transducers, transcription factors are central to many cellular processes, and their inactivation would lead to severe cellular dysfunction. However, if the factors were specific enough, their loss would produce more limited disruption. An example would be the nuclear receptors for the sex steroids (Table 17-5); although called receptors, NRs are just ligand-activated transcription factors. Other relatively specific transcription factors would include Pit-1, which is restricted to the induction of some pituitary hormones; FoxE1, which is required for the development of the thyroid gland (see Tables 17-1 and 17-2); and Stat1, which is associated with the immune system and whose mutation results in an increased susceptibility to tuberculosis. Some transcription factors appear to mediate the effects of only one hormone. For example, several hepatic nuclear factors are involved with insulin action (Table 17-15). Stat 5b is primarily activated by GH. Inactivation of Stat 5 produces GH insensitivity with mild immunodeficiency. Microphthalmia (Mi) is activated by SCF through MAPK phosphorylation, and

Table 17-15
Clinical Syndromes Resulting from Mutations in Transcription Pathways

Transcription component	Function	Syndrome
Transcription Factors		
NRs	See Chapter 6	See Table 17-5
Pit-1, FoxE1, HNF-4α, and HNF-1α	Mediate endocrine gland development, hormone secretion, and/or hormone action	See Tables 17-1 and 17-2
Stat1	Transcription factor in immunity	Increased susceptibility to tuberculosis
Stat 5b	Transcription factor for GH	GH insensitivity with mild immunodeficiency
Microphthalmia (Mi)	Activated by SCF and Kit	Waardenburg II
Twist	Activated by FGF	Saethre-Chotzen syndrome
IGF6	Accessory transcription factor for Smad	Van der Woude and popliteal pterygium syndromes
Coactivators and Corepressors		
CBP	General coactivator	Rubinstein-Taybi syndrome
β-Catenin	AR cofactor	Somatic mutation in prostate cancer
Hairless (Hr)	TR corepressor	Alopecia universalis and papular atrichia
TGIF	Smad2 corepressor	Holoprosencephaly
Menin	JunD inhibitor	Multiple hormone-secreting tumors
Miscellaneous		
IKKγ	Phosphorylates IκB	X-linked anhidrotic ectodermal dysplasia and T cell immunodeficiency
IκBα	Ectodermal development and immunity; mutant cannot be phosphorylated by IKK	Autosomal dominant anhidrotic ectodermal dysplasia and T cell immunodeficiency

Mi mutations produce a phenotype similar to mutations in SCF or its receptor, Kit; that is, absent melanocytes and mast cells with defective development of blood and germ cells (Waardenburg II syndrome). Likewise, Twist is a bHLH transcription factor that mediates the effects of FGF; Twist mutations cause the Saethre-Chotzen syndrome, which is characterized by craniosynostosis, hypertelorism, ptosis (drooping eyelid), brachydactyly (short digits), syndactyly (fused digits), and broad great toes. Finally, IRF6 (interferon regulatory factor 6) is an accessory transcription factor for Smad. Its deficiency produces a spectrum of phenotypes from simple cleft lip and palate (Van der Woude syndrome) to cleft lip and palate with skin and genital anomalies (popliteal pterygium syndrome).

Transcription Coactivators and Corepressors

CBP is such a universal coactivator that it is surprising that mutations can be viable; however, a defect in the acetyltransferase domain causes Rubinstein-Taybi syndrome, which is characterized by mental retardation and facial and digital abnormalities. The autoimmune regulator (AIRE) mutants give rise to autoimmune polyendocrinopathy candidiasis ectodermal dystrophy. It is a coactivator for an unknown transcription factor and its deficiency is responsible for an autoimmune disease that includes Addison's disease and hypoparathyroidism. β-Catenin is a coactivator for the AR. A somatic mutation in prostate cancer prolongs the half-life of β-catenin; this mutation augments the activity of androstenedione and estradiol, which are normally poor agonists. XWBSCR11, like Smad, is a coactivator for FAST-1 and it induces an organizer for mesoderm induction. Mutations result in Williams-Beuren syndrome, which is characterized by a round face with full cheeks and lips, stellate iris, supravalvular aortic stenosis, and a friendly personality.

Mutations in several corepressors are also known. Hairless (Hr) is a TR-specific, tissue-specific corepressor that inhibits apoptosis in the skin and nervous system; its loss causes a variety of congenital hair loss disorders (alopecia universalis and papular atrichia), mental retardation, and deafness. TG-interacting factor (TGIF) is a Smad2 corepressor that acts downstream of Nodal, a TGFβ family member. Mutations result in neural midline defects, such as a narrowed skull and hypotelorism (narrow-set eyes). Finally, menin inhibits JunD, a mitogenic transcription factor; the loss of this inhibitor leaves JunD unchecked and leads to the proliferation of hormone-secreting cells in a form of multiple endocrine neoplasia.

IKKγ is neither a transcription factor nor a coactivator but is responsible for the phosphorylation and degradation of IκB. The destruction of IκB then activates NF-κB, a major transcription factor in the immune-stress response. However, there are many ways that NF-κB can be activated (see Chapter 13), so that the loss of IKK has only limited effects, which are primarily observed in the ectoderm. Total loss of IκB causes incontinentia pigmenti (also called anhidrotic ectodermal dysplasia), an X-linked dominant disease that is prenatally lethal in males. Females are genetic mosaics and exhibit inflammatory skin vesicles, verrucous patches, hyperpigmentation, and dermal scarring; the symptoms resolve when the affected cells die. Milder mutations produce viable males with hypohidrotic ectodermal dysplasia, which is characterized by dysplasia of teeth, hair, and eccrine sweat glands, as well as dysgammaglobulinemia with recurrent infections. IκBα mutants that cannot be phosphorylated by IKK produce a similar syndrome except that it is autosomal dominant.

Summary

The human body obeys Murphy's law: whatever can go wrong, will go wrong; that is, almost every possible defect in hormone action has, in fact, been observed in patients. Hormone deficiencies include gland agenesis or destruction, mutations in genes for peptide hormones or synthetic enzymes for nonpeptide hormones, improper processing, regulatory defects leading to secretion failure, sequestering antibodies, and excessive metabolism. Excess hormone activity can result in mutations that facilitate processing or impede proteolysis, or mutations in antagonists, transcription repressors, catabolic enzymes, and elements of the negative feedback pathway.

Mutations have also been documented in every class of membrane and nuclear receptor. Gain-of-function mutations mimic the active conformation of the receptor or impede downregulation; for example, mutations that favor dimerization of RTKs or eliminate docking sites for deactivating phosphatases. In addition to intrinsic receptor mutations, mutations in receptor antagonists or the development of receptor antibodies can also affect receptor activity.

Because transducers serve many systems, defects tend to occur only in genetic mosaics, somatic mutations, or isoforms whose tissue distribution is highly restricted. The muscular, nervous, and immune systems are particularly well represented in these diseases. Gain-of-function mutations can occur as a result of mutations in the GTPase domain of G proteins, in autoinhibitory domains, in the ubiquination machinery, or in regulatory sites in the promoter or mRNA.

Like transducers, mutations in the transcription machinery tend to occur in those factors with restricted distribution (Pit-1 in the pituitary gland), with multiple mechanisms of activation (IKKγ and NF-κB), or do not regulate vital functions (AR).

References

Hormone Deficiency

Betterle, C., Dal Pra, C., Mantero, F., and Zanchetta, R. (2002). Autoimmune adrenal insufficiency and autoimmune polyendocrine syndromes: Autoantibodies, autoantigens, and their applicability in diagnosis and disease prediction. *Endocr. Rev.* 23, 327-364.

Cohen, L.E., and Radovick, S. (2002). Molecular basis of combined pituitary hormone deficiencies. *Endocr. Rev.* 23, 431-442.

Dattani, M.T., and Robinson, I.C. (2000). The molecular basis for developmental disorders of the pituitary gland in man. *Clin. Genet.* 57, 337-346.

Fajans, S.S., Bell, G.I., and Polonsky, K.S. (2001). Molecular mechanisms and clinical pathophysiology of maturity-onset diabetes of the young. *N. Engl. J. Med.* 345, 971-980.

Hardelin, J.P. (2001). Kallmann syndrome: Towards molecular pathogenesis. *Mol. Cell Endocrinol.* 179, 75-81.

Hu, Y., Tanriverdi, F., MacColl, G.S., and Bouloux, P.M.G. (2003). Kallmann's syndrome: Molecular pathogenesis. *Int. J. Biochem. Cell Biol.* 35, 1157-1162.

Huopio, H., Shyng, S.L., Otonkoski, T., and Nichols, C.G. (2002). K_{ATP} channels and insulin secretion disorders. *Am. J. Physiol.* 283, E207-E216.

Imperato-McGinley, J., and Zhu, Y.S. (2002). Androgens and male physiology: The syndrome of 5α-reductase-2 deficiency. *Mol. Cell. Endocrinol.* 198, 51-59.

Kato, S., Yoshizazawa, T., Kitanaka, S., Murayama, A., and Takeyama, K. (2002). Molecular genetics of vitamin D-dependent hereditary rickets. *Horm. Res.* 57, 73-78.

Mullis, P.E. (2001). Transcription factors in pituitary gland development and their clinical impact on phenotype. *Horm. Res.* 54, 107-119.

New, M.I., and Wilson, R.C. (1999). Steroid disorders in children: Congenital adrenal hyperplasia and apparent mineralocorticoid excess. *Proc. Natl. Acad. Sci. U.S.A.* 96, 12790-12797.

Notkins, A.L., and Lernmark, Å. (2001). Autoimmune type 1 diabetes: Resolved and unresolved issues. *J. Clin. Invest.* 108, 1247-1252.

Owen, K., and Hattersley, A.T. (2001). Maturity-onset diabetes of the young: From clinical description to molecular genetic characterization. *Best Pract. Res. Clin. Endocrinol. Metab.* 15, 309-323.

Peter, M., Dubuis, J.M., and Sippell, W.G. (1999). Disorders of the aldosterone synthase and steroid 11β-hydroxylase deficiencies. *Horm. Res.* 51, 211-222.

Rapoport, B., and McLachlan, S.M. (2001). Thyroid autoimmunity. *J. Clin. Invest.* 108, 1253-1259.

Van Vliet, G. (2003). Development of the thyroid gland: Lessons from congenitally hypothyroid mice and men. *Clin. Genet.* 63, 445-455.

Winter, W.E. (2003). Newly defined genetic diabetes syndromes: Maturity onset diabetes of the young. *Rev. Endocr. Metab. Disord.* 4, 43-51.

Hormone Excess

Hussain, K., and Aynsley-Green, A. (2003). Hyperinsulinemia in infancy: Understanding the pathophysiology. *Int. J. Biochem. Cell Biochem.* 35, 1312-1317.

Karasawa, K., Harada, A., Satoh, N., Inoue, K., and Setaka, M. (2003). Plasma platelet activating factor-acetylhydrolase (PAF-AH). *Prog. Lipid Res.* 42, 93-114.

Lamberts, S.W.J. (2002). Glucocorticoid receptors and Cushing's disease. *Mol. Cell Endocrinol.* 197, 69-72.

Wilson, R.C., Nimkam, S., and New, M.I. (2001). Apparent mineralocorticoid excess. *Trends Endocrinol. Metab.* 12, 104-111.

Nuclear Receptors

Bray, P.J., and Cotton, R.G.H. (2003). Variations of the human glucocorticoid receptor gene (NR3C1): Pathological and in vitro mutations and polymorphisms. *Hum. Mutat.* 21, 557-568.

Brinkmann, A.O. (2001). Molecular basis of androgen insensitivity. *Mol. Cell Endocrinol.* 179, 105-109.

Ferrigno, P., and Silver, P.A. (2000). Polyglutamine expansions: Proteolysis, chaperones, and the dangers of promiscuity. *Neuron* 26, 9-12.

Gennari, L., Becherini, L., Falchetti, A., Masi, L., Massart, F., and Brandi, M.L. (2002). Genetics of osteoporosis: Role of steroid hormone receptor gene polymorphisms. *J. Steroid Biochem. Mol. Biol.* 81, 1-24.

Hiort, O., and Holterhus, P.M. (2003). Androgen insensitivity and male infertility. *Int. J. Androl.* 26, 16-20.

McPhaul, M.J. (2002). Androgen receptor mutations and androgen insensitivity. *Mol. Cell Endocrinol.* 198, 61-67.

Whitfield, G.K., Remus, L.S., Jurutka, P.W., Zitzer, H., Oza, A.K., Dang, H.T.L., Haussler, C.A., Galligan, M.A., Thatcher, M.L., Dominguez, C.E., and Haussler, M.R. (2001). Functionally relevant polymorphisms in the human nuclear vitamin D receptor gene. *Mol. Cell Endocrinol.* 177, 145-159.

Yong, E.L., Lim, J., Qi, W., Ong, V., and Mifsud, A. (2000). Molecular basis of androgen receptor diseases. *Ann. Med.* 32, 15-22.

Receptor Kinases

Belville, C., Josso, N., and Picard, J.Y. (1999). Persistence of mullerian derivatives in male. *Am. J. Med. Genet.* 89, 218-223.

Galloway, S.M., Gregan, S.M., Wilson, T., McNatty, K.P., Juengel, J.L., Ritvos, O., and Davis, G.H. (2002). Bmp15 mutations and ovarian function. *Mol. Cell Endocrinol.* 191, 15-18.

Ivell, R., and Hartung, S. (2003). The molecular basis of cryptorchidism. *Mol. Hum. Reprod.* 9, 175-181.

Ornitz, D.M., and Marie, P.J. (2002). FGF signaling pathways in endochondral and intra-membranous bone development and human genetic disease. *Genes Dev.* 16, 1446-1465.

Vajo, Z., Francomano, C.A., and Wilkin, D.J. (2000). The molecular and genetic basis of fibroblast growth factor receptor 3 disorders: The achondroplasia family of skeletal dysplasias, Muenke craniosynostosis, and Crouzon syndrome with acanthosis nigricans. *Endocr. Rev.* 21, 23-39.

Vikkula, M., Boon, L.M., and Mulliken, J.B. (2001). Molecular genetics of vascular malformations. *Matrix Biol.* 20, 327-335.

Cytokine Receptors

Candotti, F., Notarangelo, L., Visconti, R., and O'Shea, J. (2002). Molecular aspects of primary immunodeficiencies: Lessons from cytokine and other signaling pathways. *J. Clin. Invest.* 109, 1261-1269.

Döffinger, R., Dupuis, S., Picard, C., Fieschi, C., Feinberg, J., Barcenas-Morales, G., and Casanova, J.L. (2002). Inherited disorders of IL-12-and IFNγ-mediated immunity: A molecular genetics update. *Mol. Immunol.* 38, 903-909.

Dorman, S.E., and Holland, S.M. (2000). Interferon-γ and interleukin-12 pathway defects and human disease. *Cytokine Growth Factor Rev.* 11, 321-333.

Spivak, J.L. (2002). Polycythemia vera: Myths, mechanisms, and management. *Blood* 100, 4272-4290.

G Protein-Coupled Receptors

Achermann, J.C., Weiss, J., Lee, E.J., and Jameson, J.L. (2001). Inherited disorders of the gonadotropin hormones. *Mol. Cell. Endocrinol.* 179, 89-96.

Calvi, L.M., and Schipani, E. (2000). The PTH/PTHrP receptor in Jansen's metaphyseal chondrodysplasia. *J. Endocrinol. Invest.* 23, 545-554.

Corvilain, B., Van Sande, J., Dumont, J.E., and Vassart, G. (2001). Somatic and germline mutations of the TSH receptor and thyroid diseases. *Clin. Endocrinol. (Oxford)* 55, 143-158.

de Roux, N., and Milgrom, E. (2001). Inherited disorders of GnRH and gonadotropin receptors. *Mol. Cell Endocrinol.* 179, 83-87.

Farid, N.R., Kascur, V., and Balazs, C. (2000). The human thyrotropin receptor is highly mutable: A review of gain-of-function mutations. *Eur. J. Endocrinol.* 143, 25-30.

Kopp, P. (2001). The TSH receptor and its role in thyroid disease. *Cell Mol. Life Sci.* 58, 1301-1322.

Latronico, A.C., and Segaloff, D.L. (1999). Naturally occurring mutations of the luteinizing-hormone receptor: Lessons learned about reproductive physiology and G protein-coupled receptors. *Am. J. Hum. Genet.* 65, 949-958.

Morello, J.P., and Bichet, D.G. (2001). Nephrogenic diabetes insipidus. *Annu. Rev. Physiol.* 63, 607-630.

Themmen, A.P.N., and Huhtaniemi, I.T. (2000). Mutations of gonadotropins and gonadotropin receptors: Elucidating the physiology and pathophysiology of pituitary-gonadal function. *Endocr. Rev.* 21, 551-583.

Ion Channel Receptors

Celesia, G.G. (2001). Disorders of membrane channels or channelopathies. *Clin. Neurophysiol.* 112, 2-18.

Felix, R. (2000). Channelopathies: Ion channel defects linked to heritable clinical disorders. *J. Med. Genet.* 37, 729-740.

Gargus, J.J. (2003). Unraveling monogenic channelopathies and their implications for complex polygenic disease. *Am. J. Hum. Genet.* 72, 785-803.

Gormley, K., Dong, Y., and Sagnella, G.A. (2003). Regulation of the epithelial sodium channel by accessory proteins. *Biochem. J.* 371, 1-14.

Green, T., Heinemann, S.F., and Gusella, J.F. (1998). Molecular neurobiology and genetics: Investigation of neural function and dysfunction. *Neuron* 20, 427-444.

Hirose, S., Okada, M., Yamakawa, K., Sugawara, T., Fukuma, G., Ito, M., Kaneko, S., and Mitsudome, A. (2002). Genetic abnormalities underlying familial epilepsy syndromes. *Brain Dev.* 24, 211-222.

Hübner, C.A., and Jentsch, T.J. (2002). Ion channel diseases. *Hum. Mol. Genet.* 11, 2435-2445.

Kullmann, D.M. (2002). The neuronal channelopathies. *Brain* 125, 1177-1195.

Lester, H.A., and Karschin, A. (2000). Gain of function mutants: Ion channels and G protein-coupled receptors. *Annu. Rev. Neurosci.* 23, 89-125.

Meisler, M.H., Kearney, J., Ottman, R., and Escayg, A. (2001). Identification of epilepsy genes in human and mouse. *Annu. Rev. Genet.* 35, 567-588.

Schafer, J.A. (2002). Abnormal regulation of ENaC: Syndromes of salt retention and salt wasting by the collecting duct. *Am. J. Physiol.* 283, F221-F235.

Vincent, A., Beeson, D., and Lang, B. (2000). Molecular targets for autoimmune and genetic disorders of neuromuscular transmission. *Eur. J. Biochem.* 267, 6717-6728.

G Proteins

Lania, A., Mantovani, G., and Spada, A. (2001). G protein mutations in endocrine diseases. *Eur. J. Endocrinol.* 145, 543-559.

Radhika, V., and Dhanasekaran, N. (2001). Transforming G proteins. *Oncogene 20,* 1607-1614.

Spiegel, A.M. (2000). G protein defects in signal transduction. *Horm. Res.* 53 (Suppl. 3), 17-22.

Weinstein, L.S., Yu, S., Warner, D.R., and Liu, J. (2001). Endocrine manifestations of stimulatory G protein α-subunit mutations and the role of genomic imprinting. *Endocr. Rev.* 22, 675-705.

3'5'-Cyclic AMP and Nitric Oxide Pathways

Sandrini, F., and Stratakis, C. (2003). Clinical and molecular genetics of Carney complex. *Mol. Genet. Metab.* 78, 83-92.

Wattanapitayakul, S.K., Mihm, M.J., Young, A.P., and Bauer, J.A. (2001). Therapeutic implications of human endothelial nitric oxide synthase gene polymorphism. *Trends Pharmacol. Sci.* 22, 361-368.

Lipid Pathways

Huang, Y., and Wang, K.K.W. (2001). The calpain family and human disease. *Trends Mol. Med.* 7, 355-362.

Jones, O.T. (2002). Ca^{2+} channels and epilepsy. *Eur. J. Pharmacol.* 447, 211-225.

Lorenzon, N.M., and Beam, K.G. (2000). Calcium channelopathies. *Kidney Int.* 57, 794-802.

Miscellaneous Transduction

Barrett, T.G. (2001). Mitochondrial diabetes, DIDMOAD, and other inherited diabetes syndromes. *Best Pract. Res. Clin. Endocrinol. Metab.* 15, 325-343.

Sesti, G., Federici, M., Hribal, M.L., Lauro, D., Sbraccia, P., and Lauro, R. (2001). Defects of the insulin receptor substrate (IRS) system in human metabolic disorders. *FASEB J.* 15, 2099-2111.

Wishart, M.J., and Dixon, J.E. (2002). PTEN and myotubularin phosphatases: From 3-phosphoinositide dephosphorylation to disease. *Trends Cell Biol.* 12, 579-585.

Transcription

Razani, B., Schlegel, A., and Lisanti, M.P. (2000). Caveolin proteins in signaling, oncogenic transformation, and muscular dystrophy. *J. Cell Sci.* 113, 2103-2109.

Ryffel, G.U. (2001). Mutations in the human genes encoding the transcription factors of the hepatocyte nuclear factor (HNF)1 and HNF4 families: Functional and pathological consequences. *J. Mol. Endocrinol.* 27, 11-29.

Schussheim, D.H., Skarulis, M.C., Agarwal, S.K., Simonds, W.F., Burns, A.L., Spiegel, A.M., and Marx, S.J. (2001). Multiple endocrine neoplasia type 1: New clinical and basic findings. *Trends Endocrinol. Metab.* 12, 173-178.

Smahi, A., Courtois, G., Rabia, S.H., Döffinger, R., Bodemer, C., Munnich, A., Casanova, J.L., and Israël, A. (2002). The NF-κB signaling pathway in human diseases: From incontinentia pigmenti to ectodermal dysplasias and immune-deficiency syndromes. *Hum. Mol. Genet.* 11, 2371-2375.

CHAPTER 18

Molecular Evolution of the Endocrine System

CHAPTER OUTLINE

As the structures for more hormones and receptors are determined, interesting evolutionary relationships begin to emerge. These relationships provide valuable insights into how the endocrine system developed. In this context, it is important to realize that hormones rarely have any intrinsic biological activity; only a few are known to have enzymatic activity or be related to structural molecules. Their action is manifested only through their receptors, and if their receptors are membrane bound, the effector responsibility is further transferred to a second messenger. Therefore a change in a hormone can only be understood in terms of how it alters receptor binding, and if there is pressure for a hormone to change, its receptor may also have to change in a complementary manner. In essence, the evolution of the endocrine system requires the coevolution of many of its components.

Methodology

Because no human beings have been around for the last billion years to record evolution, evolutionary relationships must be inferred. Sequence analysis provides many clues. In addition, species surveys are helpful; by knowing the distribution of a particular component of the endocrine system, one can make reasonable estimates of when it first appeared and from what it may have been derived. However, this method assumes that the evolutionary relationships of the various classes and phyla are known. Such phylogenetic trees were originally derived from morphological data; more recently, biochemical and genetic information have augmented the initial analyses and generated considerable revisions and controversies (Fig. 18-1). For example, morphological data separated animals, plants, and fungi into separate kingdoms, but analysis of the extracellular matrix suggests that fungi and animals are monophyletic. RNA sequence analysis adds further confusion because with it some researchers argue that plants, fungi, and cnidaria (previously called coelenterata) should be grouped together. In another example, Platyhelminthes and Aschelminthes, once regarded as the most primitive members of Bilateria because of their lack of a true coelom, probably once had a coelom and subsequently lost it through degeneracy; this discovery has moved them up the evolutionary scale. Other revisions include (1) the placement of slime molds (Mycetozoa) closer to fungi-animals than to protists and (2) the supposition that land plants arose from charophytes and not chlorophytes. The implications of these changes are discussed in later sections of this chapter.

Phylogenic surveys of hormones, receptors, and transducers are further complicated by the sophisticated and highly sensitive assays available today; these assays can seemingly detect minute quantities of proteins and their mRNA almost everywhere. For example, such results have been used to support the unification theory, which states that all of the hormones present throughout the biota have always existed; they were present in the first unicellular organisms and have persisted to the modern era with relatively little change. Furthermore, although the major synthesis of these hormones may be concentrated in certain specialized glands, they are also made in many other tissues. In other words, all hormones are synthesized in all tissues of all organisms.

Such a theory is tenable for steroids, eicosanoids, and amines because they are simple compounds that can be synthesized in only a few steps from common precursors. Unfortunately, for this very reason it is difficult to evaluate the

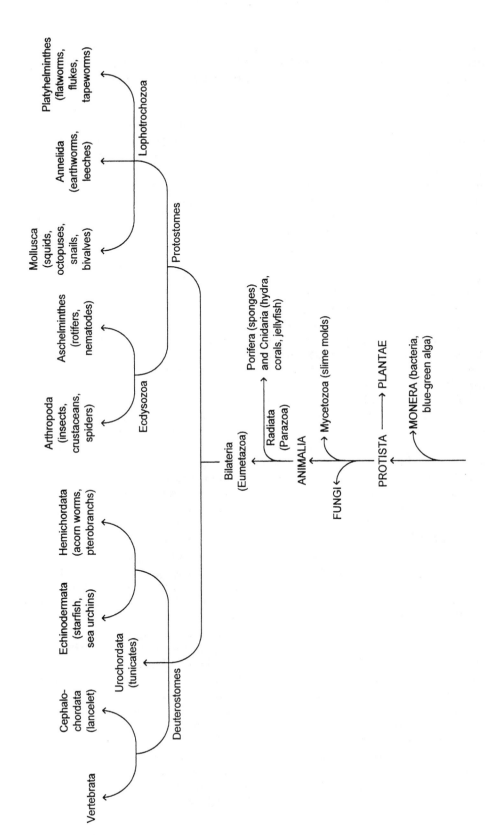

Fig. 18-1. Phylogenetic tree of the animal kingdom. Kingdoms are capitalized.

significance of their presence or absence in nonvertebrates. Data from peptide hormones should be more reliable in the determination of evolutionary relationships. However, before accepting these data, the following criteria should be met:

1. The hormone should actually be present in the invertebrate or extraglandular tissue. For this criterion to satisfied, the possibility of contamination should be eliminated. For example, it has been reported that steroid radioimmunoassays (RIAs) can be affected by oral contraceptive ingestion by the technician. Presumably, the steroids in the birth control pills were being excreted in her perspiration or other bodily fluid and contaminating the samples. RIAs are so sensitive that they do not require much contamination to be detected. In a related example, polycarbonate flasks were found to be the source of an estrogenic compound that was leached from them during autoclaving.

A second component of this criterion is that the hormone be present in reasonable amounts. Insulin is reported to be present in tetrahymena at a concentration of 75 pg/g cell; the concentration in human pancreas is 45 µg/g tissue and in serum is 1 to 5 ng/ml. The lower the concentration, the greater the chance that its presence is due to contamination.

Third, incomplete ablation of the gland must be ruled out; or if the secretion is being chemically suppressed, residual secretion must be ruled out; if cell lines are being tested, the serum, which contains many hormones, must be removed for a long enough period of time to ensure no carry over.

Fourth, the results of the RIA should be validated. Antibodies can recognize as few as two or three amino acids, resulting in the possibility of cross-reactions with unrelated material. For example, oxytocin was originally reported to be present in *Hydra* based on immunological data. However, the antibodies were actually binding to Hym-355, an entirely different hormone that coincidentally has the same last three amino acids as oxytocin, and this sequence was the basis for the cross-reaction. The use of multiple antisera directed against different parts of the protein would yield more reassuring results. Samples should also produce displacements parallel to the standards. Finally, with respect to RIAs, the decreased binding of tracer should be shown to be due to displacement and not to tracer degradation; a report about the presence of opiate peptides in *Escherichia coli* based on the fact that extracts "displaced" the tracer had to be retracted when it was discovered that the unknown substance was, in fact, a peptidase, which was destroying the tracer. Similarly, a tapeworm GH (growth hormone) detected by RIA turned out to be a cysteine protease.

Fifth, short peptides identical to or very similar to vertebrate hormones may actually be found in nonvertebrates or plants, but sequence similarities among very small peptides are difficult to evaluate. When such sequences are examined, the arguments for homology (that is, identity due to evolutionary descent) can be strengthened when (1) the sequence is longer, (2) critical residues are conserved, or (3) unusual amino acid modifications are preserved. For example, a prothoracicotropic hormone of insects (insulin-like peptide [ILP]) has two nonidentical subunits containing 20 and 28 amino acids. More important, it has a 35% identity with insulin including an identical distribution of all six cysteines. Finally, computer modeling predicts a three-dimensional structure similar to that for other members of the insulin family. The quaternary structure and cysteine placements substantially enhance the hypothesis of overall sequence homology. In another example, leucosulfakinin, a cockroach neuropeptide, is 75% identical to cholecystokinin (CCK), including the presence of a sulfated tyrosine, an unusual amino acid modification that is required for both leucosulfakinin and CCK activity. Not all examples are so clear; α-mating factor from yeast has a 60% similarity to gonadotropin-releasing hormone (GnRH), a decapeptide. However, one glaring omission is the absence of glycine in position six. This glycine is conserved in all vertebrate GnRHs, and structure-function studies suggest that it is involved in a ß-II type turn; as such, no other naturally occurring amino acid can substitute for it. This amino acid deletion results in the a-mating factor having very little activity in stimulating the release of luteinizing hormone (LH) from pituitary cells; in fact, GnRH is 10,000 times more active. However, the phylogenetic distance between yeasts and vertebrates is so great that one cannot eliminate the possibility that the two hormones are related but have different structure-function relationships. A purported human chorionic gonadotropin (hCG) from bacteria demonstrates another problem: first, the overall identity is only 25% similar; more important, numerous gaps had to be introduced into the sequence to achieve even this low level of identity. The fewer gaps there are, the more convincing the homology.

2. The hormone should actually be synthesized by the invertebrate or nonglandular tissue. Classically, this could be established by exposing the tissue to labeled amino acids and precipitating the putative hormone with specific antibodies. However, hormone synthesis is frequently too low to be detected by this technique. Methods to detect the mRNA are more sensitive, perhaps too sensitive. It appears that the genome is slightly leaky; that is, there is a very low rate of transcription of most genes in virtually all tissues. This phenomenon is called illegitimate transcription and generates less than one copy per 500 to 1000 cells. Such concentrations are probably not physiologically relevant, but are easily measured by the polymerase chain reaction technique.

Should the presence of the peptide or its mRNA be unequivocally demonstrated, four other considerations must be addressed: (1) the developmental state of the tissue, (2) the normalcy of the tissue, (3) convergent evolution, and (4) the transfer of genetic material across species. First of all, embryonic and fetal tissues have various degrees of totipotency; that is, they have the ability to differentiate into a large number of structures. When still undifferentiated, these tissues may exhibit characteristics associated with any adult structures. However, when the tissue becomes committed to a particular lineage, it only retains those characteristics associated with that adult tissue. For example, mRNA for insulin can be detected in the yolk sac and fetal liver, which, like the pancreas, are of endodermal origin. However, insulin mRNA cannot be detected in the adult liver. Therefore evidence for the expression of a hormone in early developmental stages cannot be used as proof that the tissue synthesizes the hormone in the adult. Similarly, tumors and derived cell lines are notorious for the production of ectopic hormones; this is usually a result of random gene activation and, again, cannot be reliably related to normal tissues in adults.

Another confounding factor is convergent evolution, which is the process by which unrelated proteins subjected to the same evolutionary pressures come to resemble each other. Because such a similarity is not a result of descent from a common ancestor, it is referred to as an analogy and not a homology. Plants have a gene family known as CLE (CLAVATA3/ESR-related), which determines cell fate in the shoot apical meristem. The esophageal gland of the soybean cyst nematode (*Heterodera glycines*) produces a peptide (HgCLE) that has remarkable structural and functional similarities to members of the CLE family. The nematode is a plant parasite, and HgCLE induces plant cells to become feeding cells. Because this peptide is not present in other nematodes and lateral gene transfer is unlikely (see later), HgCLE is believed to have arisen by convergent evolution. That is, the ability of the nematode to control the cellular differentiation of its host through CLE receptors resulted in the development of a ligand, and the structural requirements for binding to the CLE receptors were so strict that the nematode peptide was forced to acquire a sequence and structure similar to the original plant hormones. The similarity between cerulein, a peptide in the skin of certain amphibians, and gastrin is thought to represent another example of convergent evolution.

Finally, vertebrate hormones and their mRNA may become secondarily incorporated into invertebrates. Many pathogenic organisms can acquire host proteins; for example, *Mycoplasma* can pick up angiotensin receptors from host cells, whereas trypanosomes acquire EGFRs (epidermal growth factor receptors) from their host. This phenomenon is acute and transient; however, the transfer of genetic material can permanently alter the genome of either the pathogen or the host. Such lateral (or horizontal) gene transfer has been documented in several species; for example, the glyceraldehyde-3-phosphate dehydrogenase and glucose-6-phosphate isomerase in *E. coli* originated from eukaryotes. Bacteria are known to acquire nucleic acids from their environment, but such an exchange of genetic material among eukaryotes is less frequent. Nonetheless, there are several mechanisms that can mediate this process. First, macrophages can assimilate DNA from organisms that they phagocytize. Second, DNA can be transferred from endosymbionts to the host; for example, many prokaryotic genes have been transferred from mitochondria and chloroplasts to the nucleus. Third, parasites may provide a mechanism; there is a mite that feeds on *Drosophila* eggs in a manner that allows the mite to retain cellular inclusions, including DNA. These inclusions can then be transferred to the next egg, including those of another species. Bacterial conjugation with mammalian cells also has to be documented, and DNA viruses may perform a similar function. Finally, there is a report of DNA ingested by pregnant mice manifesting themselves as translated proteins in the fetuses.

3. Lastly, the hormone should have demonstrable and relevant effects in the nonvertebrates. This requirement would include the presence of an appropriate receptor, as well as an appropriate biological effect. A case in point would be pathogens that have evolved a sensitivity to the hormones of higher animals to adapt more effectively to their hosts; for example, the fungi responsible for vaginal infections often possess "receptors" for steroids and react to their presence, but one such receptor is now known to be an oxidoreductase that has no homology to steroid receptors (SRs), although the enzyme is modulated by estradiol. In another example, analogs of ecdysone and juvenile hormone are made by plants as a defense against insect predation and not for any internal use as hormones. The same can be said for prostaglandins in corals. In contrast, *Drosophila* has both an insulin-like hormone and an insulin-like receptor that possesses tyrosine kinase activity. The structure, activity, immunogenicity, and insulin sensitivity of this receptor resemble those of the verte-

brate insulin receptor. Furthermore, this hormone has important, defined functions in insects (see Chapter 3).

There is one final consideration when evaluating the physiological relevance of the presence of a hormone in nonglandular tissues: Is the hormone a remnant from the processing of polyproteins? Adrenocorticotropic hormone is frequently reported to be synthesized in nonpituitary tissues, but it is part of a polyprotein proopiomelanocortin (POMC), which also contains the endorphins (see Chapter 3). The latter are parahormones that have a wide distribution. Is the occurrence of adrenocorticotropin hormone (ACTH) a result of the processing of POMC for the endorphins? Is this ACTH ever secreted, or is it just degraded as a side product? For example, the ACTH from lymphocytes is synthesized in very small quantities and cannot be detected in the culture medium.

Hormones and Receptors

Origin and Evolution

Nuclear Receptors and Ligands

Hormones and receptors form such intimate complexes that they need to be considered together. Because nuclear receptors (NRs) form a homogeneous group, they are discussed first. An early hypothesis suggested that the NRs arose from steroid metabolizing enzymes that had acquired DNA-binding domains (DBDs) and transcriptional activation domains (TADs). There is precedent for an allosteric site developing from a catalytic one: the cyclic nucleotide allosteric site in several phosphodiesterase families is believed to have evolved from a duplication of the active site of the enzyme. However, there is no significant amino acid homology or three-dimensional structural similarity between the two groups. A second hypothesis, a variant of the Molecular Recognition Theory, states that hormone response elements (HREs) code for peptides that bind them; that is, HREs code for DBDs. However, such sequences are too short to establish a reliable correlation.

The most likely theory is that NRs arose from preexisting transcription factors that developed allosteric-binding sites. For example, NRs first appeared in metazoans, and SRs initially emerged in deuterostomes. However, ligand binding is not consistently found in any NR below the vertebrates; this phylogenetic distribution, along with sequences analyses, strongly suggests that ligand binding was a derived characteristic. Such factors were initially homodimers and regulated by some other mechanism such as phosphorylation; indeed, posttranslational modifications still play an important role in these receptors (see Chapter 6). The first ligands may have been intracellular, hydrophobic metabolites produced or metabolized by pathways regulated by these transcription factors. Some of these systems are still operating: for example, liver X receptor (LXR) (oxysterols), farnesoid X receptor (FXR) (bile acids), pregnenolone X receptor (PXR) (steroids), and peroxisome proliferator activated receptor (PPAR) (certain fatty acids). Eventually, the ligand-binding domain (LBD) was modified to accommodate extracellular ligands.

On the basis of the analysis of the DBD and the LBD, the nuclear receptors appear to have a common ancestor that gave rise to three major branches. As expected from their biological characteristics (see Chapter 6), the glucocorticoid receptor (GR) and the thyroid receptor (TR) families (families 3 and 1, respectively, in Fig. 18-2) comprise two of these branches. The third branch is represented by RXR (retinoic acid X receptor) and COUP (chicken ovalbumin upstream promotor) (family 2), whereas estrogen receptor (ER) is shown as separating early from the GR family. Indeed, it is believed that an ER-like receptor was the ancestor for the SR group.

It is interesting that the evolution of the HREs generally parallels that of the nuclear receptors. The original HRE is believed to have been TGACC(T), which occurred singly or in tandem repeats; this sequence corresponds to the TR response element. The introduction of variable spacing created additional HREs for family 1. Later, the original HRE was inverted with a 3-bp spacer; this sequence became the ERE (estrogen hormone response element). Finally, mutation of the HRE to TGTTCT created the GRE (glucocorticoid response element). As the nuclear receptors became more complex, they may also have become unstable; as a result, the most recent ones, ER and the GR family, required protein chaperones. At this point, the heat shock proteins were incorporated into the nonactivated complexes.

The evolution of steroids is less an evolution of hormones than it is an evolution of steroidogenic enzymes, and the development of the steroid pathway parallels the evolution of the SRs. Although steroids are nearly ubiquitous, their use as hormones is more sporadic. The simplest organism to use these compounds is a water mold whose pheromones, antheridiol and oogoniol, are steroids. In addition, plants synthesize brassinosteroids, which they use as hormones (see Chapter 3). However, the systematic used of steroids as ligands for NRs only occurs in the animal kingdom. The enzymatic machinery to synthesize the sex steroids (progesterone, estradiol, and testosterone) is present in the phyla Mollusca and Echinodermata, and estrogens stimulate egg laying in gastropods and both vitellogenesis and ovarian development in shrimp. As already noted, an ER-like receptor is believed to be the ancestor of SRs. The ER, in turn, may have developed from the RXR, which is closely linked to the ER and GR groups

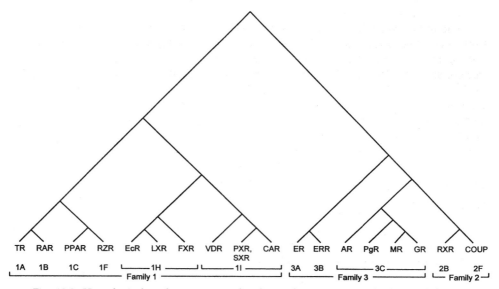

Fig. 18-2. Hypothetical evolutionary tree for the nuclear receptors. The line lengths are not proportional to the evolutionary distances between NRs. *RZR*, Retinoic acid Z receptor.

(Fig. 18-2) and whose ligand has structural similarities to steroids (Fig. 18-3). Progesterone synthesis occurs early in steroidogenesis and the progesterone receptor (PR) forms the precursor for the GR family. The additional enzymes required to make glucocorticoids appear in the elasmobranchs (sharks and rays); the glucocorticoids are also used as hormones in this class. Finally, the appearance of aldosterone in bony fishes correlates with the acquisition of 18-ol-dehydrogenase activity by P450c11. A variety of other chemicals can also act as ligands for NRs, but their heterogeneity prevents any tight correlation between the evolution of them and their NRs.

Membrane Receptors and Ligands

One of the first hypotheses on the origin of membrane receptors was based on the fact that codons and anticodons generally designate amino acids that are structurally complementary to each other. Therefore peptides synthesized off opposite DNA strands should bind one another; more specifically, the antisense strand to a peptide hormone gene could code for the receptor for that hormone. This became known as the Sense-Antisense Complementarity Theory or the Molecular Recognition Theory. However, peptides synthesized from the predicted sequence of the antisense strand only show limited specificity and low binding towards the complementary hormones. Finally, as more and more receptors are actually cloned, it has become clear that there is no homology between receptor sequences and those from the antisense strand of their ligands.

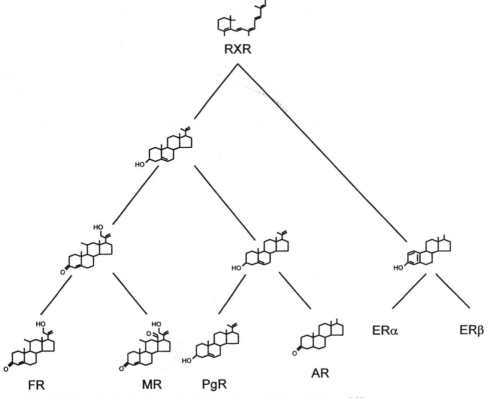

Fig. 18-3. Correlation between steroid structure and the evolution of SRs.

Another early hypothesis, the toxic defense theory, was based on the occurrence in nature of numerous toxins that target signaling components; for example, nicotine and the nAChR (nicotinic acetylcholine receptor) or cholera toxin and $G\alpha_s$. In some organisms, these interactions eventually evolved into an endocrine system, whereas in other organisms the signaling component was lost to prevent autotoxicity and only the toxin remains. This hypothesis appears to put the cart before the horse: one must first have a functioning endocrine system before a toxin can disrupt it. In truth, receptors probably had multiple, independent origins, and these are discussed later with the appropriate receptor group.

G Protein-Coupled Receptors. Certainly, the oldest and most abundant class of membrane receptors are the G protein-coupled receptors (GPCRs). These receptors exist throughout the animal kingdom; they also occur in slime molds; and a putative GPCR has been cloned from *Arabidopsis*. They may even be related to bacteriorhodopsin, a hydrogen pump in prokaryotes; although there is no obvious sequence similarity to GPCRs, the topology and function are analogous between the two groups. Sequence analysis suggests that GPCRs originated 1.2 billion years ago (bya). They probably began as food and/or sensory receptors, which allowed unicellular organisms to perceive their environment (Table 18-1). Amino acid derivatives are common ligands for GPCRs and could signal protein sources to primitive eukaryotes; other GPCRs are activated by sugars (T1R family), fatty acids (GPR40 family), and light (opsins in fungi and bacteriorhodopsin in bacteria). A 5-hydroxytryptamine-like receptor appears to have been the ancestor of the amine family; this precursor gave rise to the muscarinic and the dopamine branches with the latter diversifying into the αAR, βAR, and D groups. Most of the major GPCR groups are represented in insects, suggesting that they developed before Bilateria split.

Receptor Kinases. Another ancient group is the receptor tyrosine kinases (RTKs), which exist throughout the animal kingdom. For example, the sponge has representatives from all RTK families except for Trk (tropomyosin-related kinase) and has members from five of the eight STK (soluble tyrosine kinase) families. There are occasional reports of RTKs in fungi and protozoa, but the information

Table 18-1
Possible Origin of Receptors and Ligands

Origin	Receptor	Ligand
Molecular recognition theory	DNA complement	DNA complement
Toxin defense theory	Toxin target	Toxin
Transcription factor	Nuclear receptors	Allosteric regulators
Food receptor	GPCRs	Amines, sugars, fatty acids
Sensory receptors	GPCRs, histidine kinases	Retinal, environmental chemicals
Adhesion molecule	Cytokine, some RTKs, CX_3C chemokine receptors	Extracellular matrix fragment
Ion channels	Ion channels	Allosteric regulators
Enzymes	Enzymes	Substrates
	Variable	Enzymes
Scavenger receptors	Toll-like receptors, advanced glycation endproduct receptor	HMGB1, hsp 20, hsp 60

for these groups is too meager to draw any conclusions.* The *Arabidopsis* genome has been sequenced and does not contain any genes for either RTKs or STKs, but it does have PTPs (protein tyrosine phosphatases) and dual function kinases that operate in the MAPK (mitogen-activated protein kinase) pathways.

The LBDs for many of these receptors appear to have come from adhesion molecules via domain shuffling (see "Diversification" later). The fibronectin type III domain and the immunoglobulin loop both contain a β-structure adapted for binding and are thought to have arisen from a common ancestor; such domains are abundant in RTKs and cytokine receptors. Many of the ligands may also have been derived from matrix proteins; for example, the cysteine knot family (transforming growth factor β [TGFβ], nerve growth factor [NGF], platelet-derived growth factor [PDGF], and the glycohormones) is related to fibrillin, endostatin is generated from the carboxy-terminal globular domain of collagen XVIII, and the cysteine-rich domains in EGF are common in many matrix proteins. This relationship is logical because one of the major functions of adhesion molecules is to bind each other to secure cells to the substratum and to other cells. The use of adhesion molecules also helps to explain why RTKs are much more prevalent in animals because they have a more extensive extracellular matrix than other multicellular organisms.

The insulin receptor is probably the oldest RTK. In cnidaria, arthropods, nematodes, and mollusks, there is a carboxy-terminal extension that contains an SH2 domain and that is believed to function like a scaffolding protein. The "insulin" actually resembles relaxin more than insulin. By the lower chordates, the carboxyl-terminal extension was lost, and its function was transferred to IRS. It is also at this time that hormone diversification was occurring; for example, tunicates have a separate insulin and IGF (insulin-like growth factor).

S-T kinase receptors appear to have arisen independently several times during evolution. In animals, there is only one group, the TGFβR family, which is closely related to the RTKs. Indeed, their kinase domains differ from each other in only a few critical amino acids that determine substrate specificity. Both type I and type II receptors have been cloned from sponges. Bone morphogenetic protein (BMP) is the oldest ligand and occurs in planaria and coral, whereas TGFβ is not found in any organism below the nematodes. All members of this family are involved with pattern formation in metazoans.

Kinases can be divided into two branches (Fig. 18-4): the S-T kinases and the tyrosine-like kinases. In animals, the Raf ancestor split to give rise to Raf and the RTKs, which became the dominant receptor form. In plants, Raf never split and the receptor-like kinases (RLKs) became the dominant form. The only member of the RLK that has persisted in vertebrates is IRAK (IL-1 receptor associated kinase), which is associated with the IL-1R; Pelle is the insect homologue of IRAK. All plant RLKs are monophyletic and the different groups were generated by fusion with various extracellular domains, like the LRR (leucine-rich repeat) module.

S-T kinase receptors in bacteria have not been well characterized. They have been found in at least three species; the membrane topology is the same as for eukaryotic receptor kinases, and dimerization and autophosphorylation have been shown for some of these proteins. Most bacteria receptor kinases are involved with spore, fruiting body, and biofilm formation.

---

* Although tyrosine kinases are present in slime molds, some authorities believe that these organisms are facultative metazoans that belong at the base of the animal kingdom rather than among the protists (see Fig. 18-1).

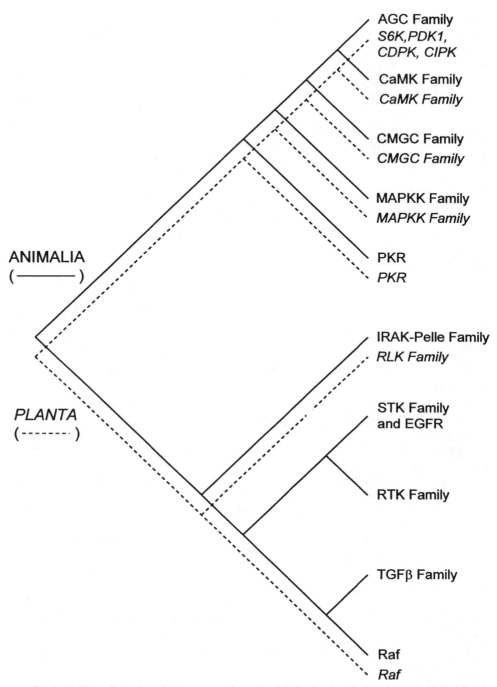

Fig. 18-4. Hypothetical evolutionary tree for animal *(solid lines)* and plant kinases *(dashed lines)*. The latter are shown in italics.

Cytokine Receptors. Although the target of cytokines, the Stat transcription factors, are quite old, having been cloned from *Dictyostelium*, the cytokine receptors themselves are the youngest receptor group and have only been found in Bilateralia. Tracing the lineage of these receptors is difficult because much exon shuffling has happened during evolution, and the duplication of various domains seems to have occurred independently in several lines. The long chain cytokines (GH, PRL [prolactin], leptin, G-CSF [granulocyte colony stimulating factor], EPO [erythropoietin], and TPO [thrombopoietin]) probably appeared first because their homodimeric receptors are simple and nonredundant. The interleukin-6 (IL-6) family was derived from G-CSF, whereas the type II receptors (interferon (IFN)α/β and IL-10) are the most recent addition. The short chain cytokines initially used RTKs, and M-CSF (macrophage colony stimulating factor), SCF (stem cell growth factor), and VEGF (vascular endothelial growth factor) still do. The β_c family was the first cytokine receptor group to bind these cytokines; the β_c family was then followed by the γ_c family, whereas the type II receptors (IFN$_\gamma$) were again the last to appear.

Ligand-gated Ion Channels.   Ion channels are ancient; for example, voltage-gated and rectifying potassium channels occur in bacteria. Like NRs, these channels were converted to receptors by the addition of an allosteric site for hormones and neurotransmitters. Such a modification could arise from exon shuffling (Fig. 18-5). The primitive ion channel may have been a simple H5 domain flanked by a transmembrane helix on either side. This basic structure is retained in the rectifying channels, whereas the voltage-gated channels acquired four additional helices. In a third branch, the transmembrane helices expanded to three, and a LBD was added through exon shuffling; this branch became the glutamate receptor family. The LBD originated from the bacterial periplasmic protein; this is a protein that binds small molecules between the inner and outer membranes of gram-negative bacteria and transfers the molecules to transporters in the inner membrane. It is believed that this domain was subsequently passed on to several GPCRs, where it became the new amino terminus of the mGluR (metabotropic glutamate receptor) and GABA$_B$(γ-aminobutyric acid).

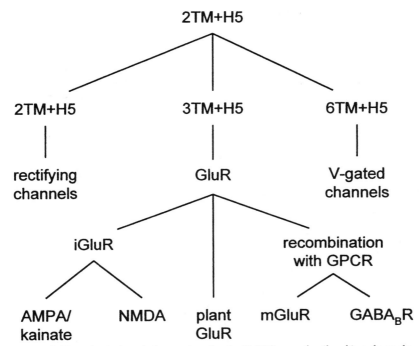

Fig. 18-5. Hypothetical evolutionary tree for the P (H5) superfamily of ion channels.

The acquisition of ligand gating was thought to be restricted to animals until a GluR was cloned from *Arabidopsis*. However, despite the presence of a periplasmic protein homologue, there is no evidence for endogenous ligand regulation. Some authorities have postulated that it simply unloads calcium from the xylem and does not participate in plant hormone signaling.

The other ion channel group used as a receptor for extracellular ligands is the Cys-loop superfamily. This channel had already split into an anionic (GlyR-GABA$_A$) and a cationic channel (AChR—5-HT) 2.5 bya. The latter further differentiated into a distinct AChR and a 5-HT channel by the time the Radiata and Bilateralia separated; the GlyR and GABA$_A$ did not split until 0.5 to 1 bya in the invertebrates.

Miscellaneous Groups.   When one thinks of high affinity, high specificity interactions, enzyme-substrate binding often comes to mind, and indeed, several hormones either are enzymes or were derived from enzymes (Table 18-2). Examples of enzyme-related receptors are fewer: (1) neurotactin is homologous to cholinesterase and acts as a receptor for amalgam in adhesion, and (2) σ_1 is a sterol C_8-C_7 isomerase and is proposed to be a membrane-bound receptor for progesterone. In plants, the extracellular domain of the Nod-factor receptor kinase (NFR) binds the Nod factor, a polysaccharide, and is homologous to the binding domains of enzymes that digest proteoglycans. However, with the exceptions of the thrombin-protease-activated receptor and the progesterone-σ_1

Table 18-2
Enzyme-Related Hormones

Hormone	Enzyme	Receptor
Homology Only		
Endothelial monocyte activating protein	Tyr-tRNA synthetase	IL-8R$_A$
Imaginal disc growth factors (IDGFs)	Chitinase	
Human cartilage glycoprotein 39 (HC-gp39)	Chitinase	
Active Enzyme but Catalytic Activity Not Required		
Differentiation inhibitory factor	Nucleoside diphosphokinase	
PLA$_2$	PLA$_2$	Mannose receptor homologue
Angiostatin	Plasminogen (first four kringle domains)	
Autocrine motility factor	Phosphoglucose isomerase	
Active Enzyme and Catalytic Activity Required		
Autotaxin[a]	LysoPLD	GPCR
Platelet-derived endothelial cell growth factor[a]	Thymidine phosphorylase	
Thrombin	Protease	GPCR
Angiogenin	Pancreatic RNase A	
Macrophage migration inhibitory factor (MIF)	Thioredoxin	
Insect-derived growth factor (IDGF)	Adenosine deaminase	
Relationship Between Hormone and Enzyme Activity Unknown		
Hsp 60 and hsp 70	Chaperone-ATPase	TLR2 and TLR4

[a] It has been subsequently discovered that these enzymes actually synthesize the hormones rather than become hormones themselves. Autotaxin generates lysoPA, and platelet-derived endothelial cell growth factor generates 2-deoxy-D-ribose.

complexes, these examples do not represent true enzyme-substrate interactions. Rather, many of these enzyme-related hormones are simply markers for specific cellular events, such as inflammation (thrombin and phospholipase A2 [PLA$_2$]) or stress and cell lysis (heat shock proteins). As a result, they have become appropriated as signaling molecules for these events and their enzyme activity is simply fortuitous. The last potential origin of receptors and ligands listed in Table 18-1 are the scavenger receptors. RAGE (receptor for advanced glycation endproducts) triggers inflammatory reactions in response to the presence of abnormal proteins, such as proteins glyoxidized during hyperglycemia in diabetes. However, it can also bind HMGB1, which is secreted by macrophages and monocytes and released by necrotic cells. TLRs (toll-like receptors) were originally part of the innate immunity system and initiate defense responses upon binding various constituents of bacteria. However, some of them can bind heat shock proteins, as well. Again, HMG (high mobility group protein) and heat shock proteins are being used as convenient markers for inflammation and stress.

Diversification

Mechanisms

The simplest way for a hormone or receptor gene to diversify is by duplication. Evidence for this mechanism comes from the many examples of homologous hormone receptors occurring either in tandem or clustered on a restricted region of a chromosome: PDGFR (platelet-derived growth factor receptor), CSF-1R (colony-stimulating factor 1 receptor), and Kit on chromosome 5q; PRLR (prolactin receptor) and GHR on 5p; and several of the GPCRs on 5q. One of the best examples of gene duplication is found in the FGF family, where a series of duplications closely corresponds to major changes in body form. The split between FGF1 and FGF2 occurred in vertebrates; the FGF10-FGF11 separation is associated with the appearance of limbs; and the FGF4-FGF6 split is linked with the divergence between mammals and birds.

A more controversial mechanism of gene duplication is reverse transcription, whereby a processed or partially processed mRNA from one gene is somehow transcribed into DNA and reintegrated into the genome to form a second gene, called a *retroposon*. Such genes can often be recognized by a deficiency of introns, a poly(A) tail, and a chromosomal location widely separated from the original gene; highly homologous genes arising from gene duplication should be located very close to each other. Most retroposons are nonfunctional because they lack a promoter. However, expression from a retroposon could occur if the original mRNA had an aberrant start site that was upstream from the promoter or if it had been inserted near another promoter.

One of the best examples of reverse transcription in endocrinology is the murine insulin. Unlike all other mammals, rats and mice have two insulin genes; they are both expressed and are highly homologous. However, insulin gene I lacks one of the two introns in gene II, has a poly(A) tail, and is located either on a separate chromosome from that containing gene II (mice) or on the same chromosome but more than 9 kb from gene II (rats). The conclusion is that gene I is a functional retroposon of gene II.

Another mechanism of diversification is exon shuffling. Eukaryotic genes are split into short coding regions, called *exons*, widely separated by long introns. It is thought that each exon encodes a functional unit: for example, in the insulin receptor gene, the ligand-binding domain, tyrosine kinase, and transmembrane region are all encoded on separate groups of exons. Because these exons are usually separated by very long introns, it is easy to imagine how they may undergo aberrant recombination resulting in a particular exon, with its intrinsic activity being transferred to the gene of another protein. This process is called *exon shuffling*, and it allows a protein to accrue many functions by the accumulation of selected exons. The extracellular domains of many receptors are replete with examples of modules that have been shared among receptor groups (see Table 7-1). Other modules are common to the intracellular domains and transducers (see Table 7-2).

Gene conversion is a mechanism of diversification where nonreciprocal crossing over between adjacent homologous genes can transfer amino acid variations from one gene into another. Point mutations are random and more likely to be detrimental than advantageous, whereas gene conversion between functional genes will disseminate amino acid variants that have already proven themselves. This mechanism is most closely associated with rapidly evolving genes still clustered on the chromosome. An example are the chemokine receptors, which are under selective pressure to evolve quickly because of their defense functions.

A final mechanism of diversification is branch jumping, which is best illustrated by the GH-PRL family. This family has two major branches, GH and PRL, which arose by gene duplication in the earliest vertebrates. The receptors appear to have duplicated at about the same time. With one exception, there is no cross-binding of either GH or PRL to the receptor of the other; the exception is the primate GH, which binds and activates the PRLR, as well as the GHR. Because the binding sites overlap, a common core of residues must have been conserved in both GHs and PRLs, necessitating the change of only a few amino acids for GH to reacquire PRL-binding activity. At this point, primate GH is straddling the branches. Recently, primate GH has duplicated to form a new hormone expressed in the placenta; this placental lactogen is 85% identical to GH but has absolutely no GH activity. Additional mutations made the placental lactogen incapable of binding the GH receptor, although binding to the PRL receptor is unaffected. The placental lactogen has now completely switched over to a receptor on another branch.

The insulin family represents a more dramatic leap. Although insulin and IGF-I bind RTKs, members of the relaxin subfamily bind GPCRs This switch is not a result of any common binding motif. Indeed, the two receptor groups have completely different binding domains: the RTKs use two β-helices separated by a Cys-rich region, whereas the GPCRs use LRRs. The crossover of the short chain cytokines between RTKs and the cytokine receptors and the switch of GDNF, a TGFβ family member, from a S-T receptor kinase to a RTK (Ret) are equally striking.

The GPCRs represent an example of branch jumping at both the ligand and the G protein levels (Fig. 18-6). Although there are many ways to construct a phylogenetic tree for these receptors, most of them show the βARs (β-adrenergic receptors) and MRs (muscarinic acetylcholine receptors) clustered together. However, other isoforms are inexplicably split: for example, the α-agonists and dopamine bind to receptor isoforms scattered between several distinctly separate subfamilies. Because all of these hormones are amines and make similar contacts with the receptor (see Fig. 7-13), a slight change in the ligand-binding pocket of one receptor may favor the binding of an amine that normally binds to a totally

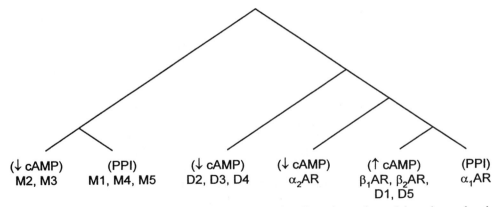

(↓ cAMP)	(PPI)	(↓ cAMP)	(↓ cAMP)	(↑ cAMP)	(PPI)
M2, M3	M1, M4, M5	D2, D3, D4	α_2AR	β_1AR, β_2AR, D1, D5	α_1AR

Fig. 18-6. Hypothetical evolutionary tree for the GPCRs. Transducers for each branch are placed in parentheses. *αAR*, α-Adrenergic receptor; *βAR*, β-adrenergic receptors; *D*, dopamine receptors; *M*, muscarinic receptors; *PPI*, polyphosphoinositide pathway.

different receptor subfamily. Figure 18-6 also demonstrates the fluidity of contacts with G proteins. Again, there appears to be only a few residues critical in determining G protein selectivity, and changes in these amino acids can alter the GPCR output. This branch jumping between transducers and receptors can also be seen in the long chain cytokine family: the earliest homodimeric receptors coupled to Stat5; the IL-6 family switched to Stat3; and the late appearing type II receptors use Stat1. This progression presumably arose via changes in the phosphotyrosine (pY) flanking regions, allowing them to be recognized by different classes of SH2 domains. Branch jumping, at whatever level, allows a hormone to acquire new functions and/or tissue distributions.

Timing

Although gene duplication appears to be a simple and easy way to diversify hormones and receptors, there is one major problem to overcome: conversion to a pseudogene. Specifically, there is no known mechanism to duplicate both a hormone and its receptor simultaneously, and if there are two hormones for the same receptor or vice versa, there is no selective pressure to keep the duplicate, which would then rapidly accumulate deleterious mutations and become a pseudogene. This section discusses several conditions that might select for a duplicate gene pending the duplication of its binding partner.

There are several factors that could act to retain a duplicated hormone when there is only one receptor. First, if gene duplication is combined with fusion to the original gene, then the new hormone would be kept as parasitic DNA until the receptor gene duplicates. Examples of such fusion genes include the glucagon-GLP1,2 and the PTH-PTH-like genes. Second, one of the hormones may become defective through deleterious mutations; at that point, it may act as an antagonist and a duplicate receptor would never be needed. The IL-1 antagonist developed in this manner. Third, the new hormone may be regulated differently; for example, epigenetic silencing of the duplicate gene may occur in a tissue-or developmentally specific manner to allow distinct selection and retention. An excellent example of this mechanism is LH and hCG, which both bind the same GPCR; however, LH only occurs in the anterior pituitary gland and regulates the mentrual cycle, whereas hCG is restricted to the placenta where it is produced during pregnancy.

Fourth, the new hormone may bind a heterodimer of existing, related receptors; that is, a new receptor is created from existing receptors. As noted previously, hPL (human placental lactogen) recently split off from hGH (human growth hormone) and primarily uses the PRL receptor. However, it also has some unique biological activities, and there is evidence that this activity is mediated by a heterodimer between the GH and PRL receptors.

Fifth, the two hormones may have different kinetics or half-lives; these differences may impart sufficiently distinct bioactivity to each hormone to cause both hormones to be retained. EGF and TGFα bind the same EGFR, but EGF is more resistant than TGFα to dissociation from the EGFR in the acidic endosome. As a result, TGFα has a lower receptor occupancy and a shorter duration of activity. Some second messengers are sensitive to these parameters; for example, TGFα but not EGF can activate MAPK. In another example, lymphotoxin α is more stable than TNF on the TNFR; this difference results in the former being mitogenic, whereas the latter is cytotoxic. Finally, insulin is degraded more quickly than IGF-I in the endosome; again, the longer receptor residency time favors mitogenesis, whereas the shorter activation time favors metabolic effects.

Lastly, the two hormones may bind the receptor differently so as to elicit a different spectrum of activities. The mineralocorticoid receptor can double as a second GR because it cannot distinguish between aldosterone and cortisol. However, aldosterone binding induces TAD1a to recruit RNA helicase A and CBP, while cortisol does not. Both normal LH and a variant LH bind to a single GPCR; the variant LH-bound LHR couples more efficiently to G_q, whereas the normal LH-bound LHR couples more efficiently to G_s. Finally, both betacellulin and neureglin bind HER4 (human EGF receptor homologue 4), but each triggers a different pattern of autophosphorylation.

Conversely, there are several factors that can select for duplicate receptors when there is only one hormone. First, isoreceptors allow hormones to have divergent effects on different tissues (see Table 7-9). Second, the isoreceptors may have different internalization kinetics, which could produce disparate effects; for example, the δ opiate receptor internalizes within 10 minutes, but the κ receptor requires at least an hour with the same ligand. Finally, a recently duplicated receptor may use an existing ligand; for example, there are numerous intermediates in steroid metabolism, and a newly duplicated SR could quickly adapt to one of these metabolites. This process has been called *ligand exploitation*.

Once the binding partner duplicates, the new hormone-receptor pair can start coevolving. Often it does not require major structural changes to achieve binding specificity; ectodysplasin A1 and A2 differ by only two amino acids, yet the former strictly binds EDAR (ectodysplasin-A2 receptor), a member of the TNFR family, and the latter only binds XEDAR (X-linked EDAR).

Transducers

Cyclic Nucleotides, Nitric Oxide, and G Proteins

The small G proteins that are involved with translation and organelle trafficking are ubiquitous; this is no surprise considering the vital nature of these processes. However, this is not true of the signal transducing Ras. Ras is present throughout animals, fungi, and protists. Indeed, the Grb2 adaptor system used by the RTKs to activate Ras is not only present in mammals, *Drosophila,* and the nematode *Caenorhabditis elegans,* but their components are interchangeable. There is no Ras gene in the *Arabidopsis* genome. Although a sequence for a purported Ras protein in bacteria has been published, its low identity with other Ras proteins (19%) makes its designation as Ras unlikely.

The trimeric G proteins are also widely distributed; sequences from mollusk, *C. elegans,* fungi, plants, and protists have been reported. In fact, all four groups (α_s, α_i, α_q, and $\alpha_{12/13}$) are present in Porifera (sponges). Although identities vary between 35% and 45%, functional interactions are highly conserved; for example, the mammalian α-subunit can bind the yeast $\beta\gamma$-dimer, and the *Drosophila* α-subunit can replace the mammalian α-subunit in stimulating adenylate cyclase. In addition, the mammalian βAR can activate insect adenylate cyclase by means of the insect G protein. Compared with animals, plants have very few trimer G isoforms and only one GPCR, which has no known ligand and does not appear to be coupled to G proteins. However, G protein null mutants are unresponsive to abscisic acid (ABA) and brassinosteroids, suggesting that G protein trimers may have some limited signaling functions in plants.

One of the targets of the trimeric G proteins is the adenylate cyclase. This enzyme resembles the voltage-gated ion channels in its topology; in fact, in paramecium it functions as a potassium channel, from which it may have been derived (Fig. 18-7). The active site in the enzymes from animals, fungi, and protists are all homologous; for this reason, the catalytic site is known as the *universal adenylate cyclase.* In metazoans, there are two catalytic sites and each is coupled to six transmembrane helices. Protozoans have a receptor-like adenylate cyclase with a single transmembrane helix; however, dimerization is still required. Fungi and cyanobacteria have a large adenylate cyclase that is membrane associated, but eubacteria and the testis have a soluble enzyme. There is no adenylate cyclase in archaebacteria or plants; and the one in enterobacteria is unrelated to the universal cyclase.

The catalytic site of the guanylate cyclase is homologous to that from the universal adenylate cyclase. Indeed, because only a few amino acid substitutions are required to convert one into the other, it is believed that the guanylate

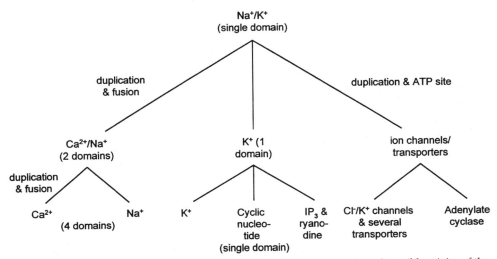

Fig. 18-7. Hypothetical evolutionary tree of the voltage-gated channels and possible origins of the CNG channels, the IP$_3$ and ryanodine receptors, and the adenylate cyclase.

cyclase arose multiple times during evolution. However, the classic soluble and membrane guanylate cyclases occur only in metazoans.

The presence of cyclic nucleotides in plants is still controversial. The evidence for cGMP is good; *Arabidopsis* has a soluble guanylate cyclase although it is highly divergent and insensitive to NO, and cGMP functions in light transduction and regulates a CNG (cyclic nucleotide-gated) potassium channel. A cAMP system is clearly present and operational in *Chlamydomonas*, but some authorities place these organisms within the protista kingdom. Furthermore, these green algae are members of the chlorophytes, whereas higher plants are believed to have arisen from the charophytes. The concentrations of cAMP in higher plants are very low and may originate from the guanylate cyclase, which has some residual adenylate cyclase activity. No protein kinase A (PKA) has been identified. If functionally activity, cAMP appears to play a minor role in higher plant signal transduction.

In contrast to plants, PKA has been unequivocally demonstrated in protists, fungi, and animals. In protists, the regulatory (R) and catalytic (C) components are part of the same protein; in these organisms the kinase is still a monomer. After the protein kinase G (PKG) split off, R and C became separated into individual genes. By the time the fungi appeared, an amino-terminal extension had been added to allow for R dimerization. In animals, NO and cGMP have been closely associated. Unfortunately, the determination of the phylogenetic distribution of the nitric oxide synthase (NOS) system is complicated by the multiple ways of synthesizing nitric oxide (NO). The classic NOS only exists in higher animals; it is found in vertebrates, insects, and mollusks, but not in nematodes, fungi, protists, or plants. However, virtually all organisms are capable of producing NO; for example, plants synthesize NO through the reduction of nitrate to nitrite. In healthy plants the latter would be rapidly reduced, but in the absence of photosynthesis or under anaerobic conditions, the nitrite can be converted to NO. Another pathway found in plants uses the pyridoxal phosphate-containing (P-like) subunit of the glycine decarboxylase complex. Normally, the full complex converts two glycines to one serine, carbon dioxide, and ammonia; however, the P-like subunit by itself has NOS activity. There is little information on the role of NO in plants. Indoleacetic acid (IAA) elevates NO; NO donors mimic the effects of ABA on stomatal closure and IAA on root organogenesis; and an NO scavenger blocks these two effects.

The coupling between NO and cGMP is highly variable. For example, an NO donor increases cGMP in spruce needles, but the soluble guanylate cyclase in *Arabidopsis* is unresponsive to NO. In myxomycete, the effect of NO on starvation-induced sporulation is mediated by cGMP, and NO is coupled to cGMP production in mollusks. However, NO does not activate soluble guanylate cyclase in *Hydra* or lobster. Interestingly, in many invertebrates it is arachidonic acid rather than NO that stimulates this cyclase. In summary, the consistent coupling of NO and cGMP appears to be a late acquisition in higher animals, although sporadic associations may occur elsewhere in eukaryotes.

Calcium and Phospholipids

The original function of the PPI (polyphosphoinositide) pathway was probably in the regulation of vesicular trafficking; both the charge and reversibility of PPI accumulation made it and ideal candidate for this process. In animals, the role of the PPI pathway was expanded to include signal transduction. In animals, it can be traced as far down as the sponges, where an aggregation factor stimulates PI (phosphoinositide) turnover, IP_3 (inositol 1,4,5-trisphosphate) release, and cytosolic calcium elevation. In addition, the effects of this factor are mimicked by a calcium ionophore. Finally, $PLC\beta$, $PLC\gamma$, and $PLC\delta$ are all found in cnidaria. In *Dictyostelium*, IP_3 is used as the second messenger of extracellular cAMP, an ectohormone in this species. Its production is coupled to a PI-specific PLC whose sequence resembles that of mammalian $PLC\delta$. Yeast and plants also have only $PLC\delta$, which appears to be the PLC ancestor. However, neither the yeast nor plant genomes have genes for IP_3R or RyR (ryanodine receptor), and electro-physiological data do not support the existence of CICR (calcium-induced calcium release) channels in plants. Indeed, nematodes are the lowest organisms from which IP_3Rs have been cloned. Data from protozoans are incomplete: IP_3 and cADPR (cyclic ADP-ribose) elevate calcium from separate pools in *Euglena*, but no IP_3R or RyR have been isolated.

The existence of this pathway in plants is controversial. There is no question that calcium is an important cellular regulator in plants, but it is not clear if IP_3 is responsible for elevating this cation. First, the major sources of calcium in plant cells are the plasma membrane and the vacuole. Second, although IP_3 can elevate calcium, most reports are based on indirect or conflicting data. Furthermore, PPI levels are far below those seen in mammalian systems, usually accounting for only 0.5% of PI; there is no IP_3R; and hormones do not elevate IP_3. Finally, ABA does elevate IP_6, which both mimics the action of this hormone and is also more potent than IP_3. As such, the effects of IP_3 may actually be mediated by IP_6. Some authorities have suggested that the major functions of the PPI pathway in plants are to regulate vesicular trafficking and to generate both IP_3 for IP_6 synthesis and DG (diacylglycerol) for phosphatidic acid production.

Other calcium elevating pathways are less well-studied. cADPR has been detected in sponges and plants and elevates calcium in both groups. In plants, cADPR is reported to mediate the effects of ABA on transcription. NAADP (nicotinic acid ADP) is synthesized by beet and cauliflower and releases calcium from the endoplasmic reticulum rather than the vacuole, which is the target of IP_3 and cADPR action. However, there are so little data on these second messengers that their roles in lower animals and in other kingdoms are still undefined.

Calcium has several targets: PKC, camodulin (CaM), calpain, and PP2B. PKC is present in animals and fungi; only the cPKC and nPKC groups are found in sponges, whereas nematodes also have aPKC. There is no PKC in protists or plants; however, plants do have several groups of protein kinases regulated by calcium through separate CaM subunits or intrinsic CaM-like domains (calcium-dependent protein kinase [CDPK]). In addition, plants have a unique kinase group that is activated by a calcineurin-like B subunit; the calcineurin B-like interacting protein kinases (CIPKs). Interestingly, calcineurin (PP2B) itself has not been cloned from plants, although its activity has been detected in plants and it is present in animals, fungi, and protists. CaM exists in all eukaryotes, but the presence of CaM in prokaryotes is more controversial. Bacteria clearly have peptides and proteins with calcium-binding EF-hand domains; however, these motifs are highly variable in number and structure. In addition, these domains either stabilize the overall structure of the protein or buffer or transport calcium. It is unlikely that any of these peptides are homologous to CaM. Finally, calpain is also present in all eukaryotes.

Likewise, the PI3K (phosphatidylinositol-3 kinase) pathway is present in all eukaryotes, but its role in signaling appears to be restricted to animals. For example, plants do not have class I or II PI3K; they only have class III, which is involved with vesicular trafficking. Furthermore, *Trypanosoma*, a protist, has protein kinase B (PKB), but the kinase lacks a pleckstrin homology (PH) domain and is not regulated by PI(3)P$_n$ (polyphosphoinositide phosphorylated at the 3′ position). Finally, plants have both PLD and PLA$_2$. Indeed, plants have at least five PLD isoforms that differ in the calcium and phospholipid requirements and in their substrate preferences. However, they appear to be primarily involved in vesicular trafficking and phospholipid metabolism rather than signaling. Plants only have the secreted form of PLA$_2$.

Miscellaneous Protein Kinases and Phosphatases

Plants and animals have representatives from every soluble S-T protein kinase group (see Fig. 18-4). Although plants lack all three prototypic members of the AGC group (PKA, PKG, and PKC), they do have other members, including S6K, PDK1, CDPK, and CIPK. They also have protein kinases that can use CaM as an accessory subunit, although they are less common than the CDPK and CIPK groups.

The CMGC (Cdk-MAPK-GSK-Cdk-like) family is also well-represented. The casein kinase 2 (CK2) is ubiquitous and highly conserved. Most kinases from mammals, yeasts, and plants have between 40% and 50% identity in their cata-lytic domains; however, for CK2 the residues are 75% to 80% identical. In fungi and peas, the kinase is a tetramer and is stimulated by polyamines. Polyamines themselves are present in all life forms. In plants, they are important in growth, the stabilization of cell membranes, and the retardation of senescence. They appear to operate as hormone mediators for some of the actions of indoleacetic acid, the cytokinins, and the gibberellins. Unfortunately, there are not enough data from any single system to satisfy all of the requirements for designating them a second messenger; however, data from several different studies indicate that these hormones do elevate polyamine levels and ODC (ornithine decarbox-ylase) activity. Furthermore, their effects are inhibited by MGBG (methylgloxal bis[guanylydrazone]), and this inhibition is relieved by spermidine. Finally, polyamines alone can at least cause growth. In plants, polyamines are also negatively controlled. Both ABA and ethylene promote senescence, and they tend to have effects on polyamine metabolism that are opposite to those just described.

The proline-directed kinases are widely distributed. The MAPK and its activation cascade have been documented in fungi and plants; however, plants only have the ERK (extracellular-signal regulated kinase) pathway, whereas fungi have both the ERK and an ancestral JNK (Jun amino[N]-terminal kinase)–p38 pathway. The kinases are so conserved that the plant MAPKK can phosphorylate and activate *Xenopus* MAPK. In addition, mammalian Raf can phosphorylate yeast MAPKK and bind yeast Ras. GSK3 and Cdk2 are other proline-directed kinases present in plants and fungi. Finally, PKR is found in both plants and animals.

The other major branch in the protein kinase tree includes the tyrosine kinases, S-T kinase receptors, and related enzymes. In animals, Raf split to produce the tyrosine kinases, which diversified early; as noted previously, all RTK classes except for Trk, and five of the eight STK classes, are present in sponges. Tyrosine kinases also exist in slime molds, but not in plants. Plants do have dual specificity kinases that are restricted to the MAPK pathway. There are reports of pY occurring in bacteria, but no bacterial genome contains genes homologous to animal tyrosine kinases.

This branch of the protein kinase tree also gave rise to two S-T kinase receptor groups: the TGFβ and the RLK families. The former occurs only in animals, and although the latter has members in both plants and animals, they have flourished in plants, whereas only a single group, the IRAK-Pelle family, persisted in animals.

The S-T protein phosphatases are nearly ubiquitous. There is even a PP1-like phosphatase in bacteria, but there are no regulatory subunits. PP-1, PP-2A, and PP-2C have been identified in plants, although the PP2C isoforms are more numerous and diverse than either PP1 or PP2A. PP-1, PP-2A, and PP-2B are also present in fungi and protists, and the *Neurospora* PP-2B can form functional heterodimers with the mammalian enzyme. The fungal phosphatases are between 50% and 86% identical to their mammalian counterparts. Soluble protein tyrosine phosphatases are similarly widespread in animals and fungi. Receptor PTPs first appear in the sponge; the extracellular domains primarily consist of FNIII (fibronectin type III) and Ig-like domains. PTP sequences from plants may be dual specificity phosphatases for the MAPK pathway.

Summary

When examining the distribution of transducers among the various kingdoms (Fig. 18-8), one is struck by the antiquity of many of them: calcium (as a second messenger), CaM, all the major S-T kinase and phosphatase groups (with the possible exception of PP2B), trimeric G proteins, and cGMP are present in all eukaryotes. If one assumes that the chlorophyta belong in the plant kingdom, then one can add cAMP to this list and make the assumption that the more advanced plants have secondarily lost the cAMP pathway. There are several other conclusions that can be drawn regarding the phylogenetic relationship of the various groups. First of all, the transduction apparatus in sponges is very similar to that in animals; for example, the presence of PKC, RTKs, STKs, TGFβRs, and the PPI pathway. This finding strongly supports the original placement of sponges in the animal kingdom. Second, the shared transducers between the animal and fungal kingdoms would tend to support the view that these groups had a monophyletic origin. This same evidence also supports the contention that mycetozoa are closer to animals and fungi than to protists. Finally, virtually all transducers are in place by the time the Radiata and Bilateria split; this observation supports the placement of the various phyla in the latter group on the same level rather than in a hierarchical relationship.

By examining the degree of conservation among these various systems, one can also speculate about how the system was originally established and then coevolved. The peptide hormones are the least conserved; with rare exception, they have no intrinsic biological activity and only interact with a single component, the receptor complex. With few demands, there are few structural requirements to keep constant. On the other hand, membrane receptors are more conserved because they interact with more components. This phenomenon is well-illustrated by the cytokine receptors, where the intracellular domain is evolving slower than the extracellular domain and where heterooligomers are evolving more slowly than homooligomers. The extracellular domain only has to bind the hormone and possibly participate in dimerization, whereas the intracellular domain must engage numerous transducers, interact with the endocytotic

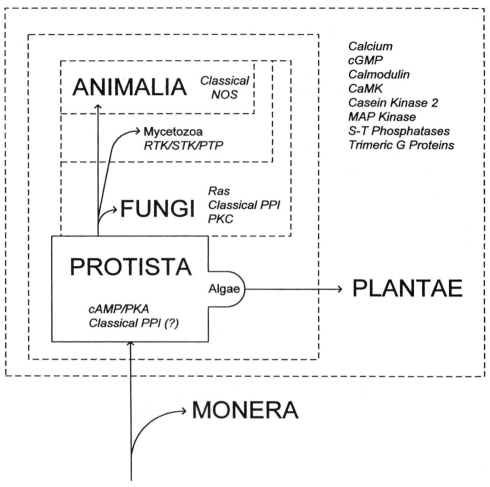

Fig. 18-8. Distribution of components of transduction pathways in the biological kingdoms. The transducers present within the boxed kingdoms are shown in italics.

machinery, and conserve phosphorylation sites; heteroligomers must further interface with different receptor subunits. Basically, the greater the number of interactions, the greater the structural constraints on a protein, and the slower it evolves. For this reason, many transducers are the most highly conserved components of the entire pathway. For example, CaM interacts with kinases, metabolic enzymes, transcription factors, cytoskeletal elements, channels, receptors, etc.; it is not surprising that plant and vertebrate CaM are 90% identical. There are two basic hypotheses concerning how signaling pathways originated. In the first scenario, the pathway is built up gradually: second messengers appeared first as intracellular regulators. As organisms became multicellular, hormones were developed to coordinate cells and tissues; the easiest mechanism by which hormones could act was for the hormone to hook up to the preexisting machinery. Receptors were developed at the same time as an interface between the two components. The second hypothesis postulates that all the components of the signaling pathway were present from the beginning as relatively independent entities. For example, G proteins functioned in translation and vesicular trafficking; early receptors were originally adhesion molecules and pumps (bacteriorhopsin); calcium, calcium channels, and calcium pumps existed to maintain low intracellular calcium levels to avoid forming precipitates with phosphate; the adenylate cyclase may have evolved as an ion or voltage regulator, as it appears to have done in paramecium; and NO was generated for defensive purposes because of its toxicity, and so on. It is then envisioned that chance encounters connected them. These two hypotheses are not mutually exclusive; the GPCRs are very old and their pathways may have been slowly built up over time, whereas cytokine receptors appeared recently, when many of the components of their pathways already existed.

Once established, the system could then be expanded by gene duplication or, more rarely, by reverse transcription. For example, multiple receptors could develop to increase the spectrum of effects of a hormone; hormone duplication and specialization of function would then follow. Alternatively, the hormone could duplicate first. Although the new hormone would have to act on preexisting receptors, it may have sufficiently different metabolism or binding kinetics to maintain itself until its own receptor evolved. In addition, new functions could be acquired by receptor jumping or by exon shuffling among the receptors.

Transcription Factors

Conservation of Transcriptional Machinery

Although yeasts do not have any detectable nuclear receptors for mammalian steroids, T_3 (triiodothyronine), 1,25-DHCC (1,25,-dihydroxycholecalciferol), or retinoids, these receptors can be introduced into yeasts and activate transcription. This activity suggests that the factors necessary for nuclear receptor function are present in yeast and can interact with the mammalian receptors. In fact, the glucocorticoid receptor will even function in plant cells. This phenomenon is not restricted to nuclear receptors; mammalian Jun-Fos can stimulate the transcription of wheat glutenin, which has two AP-1 sites. Factors involved with chromatin modifications are similarly conserved; plants have all major HAT (histone acetyltransferase) and HDAC (histone deacetylase) families and a large number of SET (su[var]/E[z]/trithorax) methyltransferases.

Transcription Factors

The reason that mammalian transcription factors can function so effectively in different kingdoms is that many of these factors are themselves present and highly conserved in eukaryotes. The bZIP factors are ubiquitous: proteins homologous to CREB (cAMP response element binding protein) in the DBD and dimerization domain have been isolated from tobacco and wheat and recognize CRE (cAMP response element) sequences. However, because of the low concentration of cAMP and complete absence of PKA in plants, these factors must be regulated in some alternate fashion. The ABA response element binding protein is also a bZIP factor, although it is not known how ABA regulates its function. The control of another plant bZIP factor is known; the G-box binding factor 1 is activated by phosphorylation by CK2.

bHLH-ZIP (basic helix-loop-helix zipper) factors are somewhat more restricted. For example, Myc is present in many coelemates, where the CK2 sites are conserved; however, they are not found in yeasts. The Ah receptor, a bHLH factor, has only been identified in vertebrates. However, it is difficult to draw any conclusions about the evolution of these bZIP variants, as sequence analysis suggests that some of them arose independently; that is, bHLH-ZIP factors do not form a monophyletic group.

Like bZIP transcription factors, members of the MADS-box family are widely distributed. In yeast, this family includes MCM1 and ARG80, which are involved with the pheromone response and arginine metabolism, respectively. Furthermore, MCM1 will both heterodimerize with mammalian SCF and bind TCF (ternary complex factor) on a modified HRE. In plants, the homologues include DEF A and AG, which are both involved with flowering; however, these factors will not dimerize with SRF or recruit TCF to an HRE. The latter finding may be due to the fact that TCF, or Elk, cannot be detected in plants or fungi, although they are present in sponges.

Homeodomains are another ancient group and predate the divergence of plants, animals, and fungi; they may have arisen from the fusion between H1 and a bacterial HTH motif. The POU factors have been found at least as far back as nematodes and planaria, and the POU_S domain probably arose by exon shuffling between a homeodomain and a Cro-like repressor. The Brn subfamily appears to have been generated by reverse transcription because its members have no introns and have remnants of a poly(A) tail.

Immune-related and morphology-related transcription factors are more recent additions to the genome. NF-κB has been identified in insects and can be activated by PKC; it is conserved enough to bind the HRE from mammalian immunoresponsive genes. However, it is not present in either nematodes or yeast. The original genes for NF-κB and IκB arose from a common fused ancestor. Later, the NF-κB—IκB gene split into separate genes, although the original arrangement still exists for some members, like p50. Stat and Jak do not exist in protists but have been cloned from the slime mold, *Dictyostelium*. Smad can be traced as far back as the cnidaria, and it is regulated by TGFβ in nematodes.

Evolution of Function

The examples just given demonstrate how hormones, receptors, transducers, and transcription factors can evolve and alter their functions, but because hormones are removed from their effects, it is also possible for a hormone to remain structurally stable and still have its function change dramatically during evolution. In essence, as old functions become less important, the hormone can be recruited to regulate new functions. Such a process is called *exaptation*. Prolactin is an excellent example of a hormone that has undergone this process.

PRL is best known for its function in lactation; indeed, it is this activity which has given the hormone its name. However, only mammals synthesize milk, and PRL is an ancient hormone that is present in most vertebrates, including bony fish. There are usually five functions attributed to PRL, and several of them show a definite phylogenetic trend.

1. PRL is important in osmoregulation, especially in those species whose life cycles bring them into dramatically different environments. For example, certain fish migrate between fresh and salt water, and amphibians undergo early development in water before moving onto land; PRL is involved in both of these transitions.

2. PRL has several reproductive functions in higher vertebrates: it is gonadotropic in rodents, is mammotropic in all mammals, and induces brooding behavior in birds.

3. PRL has been reported to have epidermal functions, but most of these can be reclassified into one of the first two categories. For example, the effects of PRL on ion and water permeability in skin can be related to osmoregulation.

4. A fourth function is growth, which may be either specific, such as mammary hyperplasia or immune cell proliferation, or general, such as larval growth. Juvenile and adult growth are governed by GH.

5. PRL has several metabolic functions, but most of them simply support the other activities listed and show no phylogenetic specificity.

Figure 18-9 shows a gradual shift from one function to another as different vertebrate classes appeared. The first function PRL had was osmoregulation. This activity continued in amphibians because their life cycle required them to return to the water for reproduction and their larvae developed in water; in this way, osmoregulation became linked to reproduction through the water drive and larval growth. Little is known about the actions of PRL in reptiles; but in mammals and birds, the osmoregulatory functions have nearly disappeared because both classes are primarily terrestrial. However, lactation can be considered an extension of both osmoregulation and larval growth; the mammary gland originated as an apocrine gland associated with hair, and its secretion kept permeable eggs moist and supplied them with nutrients. Mammary patch secretions were then coopted to feed the hatchlings. Eventually, these glands evolved to produce milk for live-born young. The gonadotropic function, which is required for the maintenance of pregnancy in rodents (that is, intrauterine growth), can be considered an extension of larval growth. Finally, it has been suggested that the newest member of this family, placental lactogen, may play some direct role in fetal growth. PRL, therefore, is a hormone that has gradually assumed new functions and shed old ones during its history.

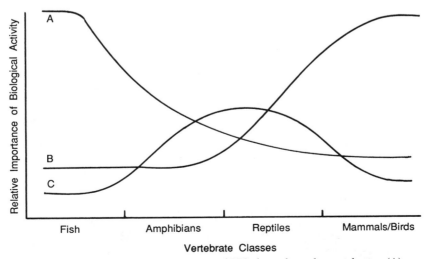

Fig. 18-9. Distribution of the different actions of PRL throughout the vertebrates: (A) osmoregulation, (B) reproduction, and (C) larval growth.

Summary

Evolution implies a preexisting structure or system. Kinases, G proteins, and transcription factors are essential for any type of internal regulation and therefore were present in the earliest organisms. In the unicellular organisms, there was also a need to interact with the environment and to locate food sources. As such, membrane proteins developed to bind the substratum and to detect nutrients, especially amines. When unicellular organisms became multicellular, there was a further need to coordinate activities, and messenger molecules were developed. Because of their endless variability and ease of synthesis (all organisms have translational machinery), peptides became one of the earliest and most versatile groups of hormones; amines and secondary metabolites were other sources. These external messengers recruited membrane proteins to plug into a preexisting regulatory system, consisting of second messengers and transcription factors.

Because peptide hormones only need to interact with their receptors, they had the least structural constraints and evolved rapidly. Receptors had to interact with more components and changed more slowly. The transducers were critically involved with so many regulatory processes that their evolution was the slowest of any component in the signaling pathway. The mechanisms of diversification include gene duplication, exon shuffling, and branch jumping. As organisms became more complex, additional second messengers were developed; Ras, PKC, and the classical PPI pathway appeared in the animal-fungi group, while the dedicated tyrosine phosphorylation system and the classical NOS pathway came late and are most closely associated with animals. Finally, few hormones have any intrinsic biological activity; rather, their activity actually resides in the target cells. That is, the biological activity of a hormone is really nothing more than how a cell responds to the presence of that hormone. As a result, the effects of a hormone can change dramatically over time as cells reprogram how they respond. This functional evolution can occur independent of any structural changes in the hormone.

References

Methodology

Adoutte, A., Balavoine, G., Lartillot, N., Lespinet, O., Prud'homme, B., and de Rosa, R. (2000). The new animal phylogeny: Reliability and implications. *Proc. Natl. Acad. Sci. U.S.A.* 97, 4453-4456.

Koonin, E.V., Makarova, K.S., and Aravind, L. (2001). Horizontal gene transfer in prokaryotes: Quantification and classification. *Annu. Rev. Microbiol.* 55, 709-742.

Ponting, C.P., Schultz, J., Copley, R.R., Andrade, M.A., and Bork, P. (2000). Evolution of domain families. *Adv. Protein Chem.* 54, 185-244.

Nuclear Receptors and Ligands

Anderson, P.A.V., and Greenberg, R.M. (2001). Phylogeny of ion channels: Clues to structure and function. *Comp. Biochem. Physiol.* 129B, 17-28.

Enmark, E., and Gustafsson, J.Å. (2001). Comparing nuclear receptors in worms, flies and humans. *Trends Pharmacol. Sci.* 22, 611-615.

Harte, R., and Ouzounis, C.A. (2002). Genome-wide detection and family clustering of ion channels. *FEBS Lett.* 514, 129-134.

Kumánovics, A., Levin, G., and Blount, P. (2002). Family ties of gated pores: Evolution of the sensor module. *FASEB J.* 16, 1623-1629.

Nelson, R.D., Kuan, G., Saier, M.H., and Montal, M. (1999). Modular assembly of voltage-gated channel proteins: A sequence analysis and phylogenetic study. *J. Mol. Microbiol. Biotechnol.* 1, 281-287.

Owen, G.I., and Zelent, A. (2000). Origins and evolutionary diversification of the nuclear receptor superfamily. *Cell Mol. Life Sci.* 57, 809-827.

Sluder, A.E., and Maina, C.V. (2001). Nuclear receptors in nematodes: Themes and variations. *Trends Genet.* 17, 206-213.

Thornton, J.W. (2001). Evolution of vertebrate steroid receptors from an ancestral estrogen receptor by ligand exploitation and serial genome expansions. *Proc. Natl. Acad. Sci. U.S.A.* 98, 5671-5676.

Membrane Receptors and Ligands

Claeys, I., Simonet, G., Poels, J., Van Loy, T., Vercammen, L., De Loof, A., and Vanden Broeck, J. (2002). Insulin-related peptides and their conserved signal transduction pathway. *Peptides (N.Y.)* 23, 807-816.

Cock, J.M., Vanoosthuyse, V., and Gaude, T. (2002). Receptor kinase signalling in plants and animals: Distinct molecular systems with mechanistic similarities. *Curr. Opin. Cell Biol.* 14, 230-236.

Davenport, R. (2002). Glutamate receptors in plants. *Ann. Bot.* 90, 549-557.

Garofalo, R.S. (2002). Genetic analysis of insulin signaling in *Drosophila*. *Trends Endocrinol. Metab.* 13, 156-162.

Shiu, S.H., and Bleecker, A.B. (2001). Receptor-like kinases from *Arabidopsis* form a monophyletic gene family related to animal receptor kinases. *Proc. Natl. Acad. Sci. U.S.A.* 98, 10763-10768.

Tichtinsky, G., Vanoosthuyse, V., Cock, J.M., and Gaude, T. (2003). Making inroads into plant receptor kinase signalling pathways. *Trends Plant Sci.* 8, 231-237.

Cyclic Nucleotides, Nitric Oxide, and G Proteins

Assmann, S.M. (2002). Heterotrimeric and unconventional GTP binding proteins in plant cell signaling. *Plant Cell* 14, S355-S373.

Davies, S.A. (2000). Nitric oxide signalling in insects. *Insect Biochem. Mol. Biol.* 30, 1123-1138.

Fujisawa, Y., Kato, H., and Iwasaki, Y. (2001). Structure and function of heterotrimeric G proteins in plants. *Plant Cell Physiol.* 42, 789-794.

Jones, A.M. (2002). G-protein-coupled signaling in *Arabidopsis*. *Curr. Opin. Plant Biol.* 5, 402-407.

Lamattina, L., García-Mata, C., Graziano, M., and Pagnussat, G. (2003). Nitric oxide: The versatility of an extensive signal molecule. *Annu. Rev. Plant Biol.* 54, 109-136.

Millner, P.A. (2001). Heterodimeric G-proteins in plant signaling. *New Phytol.* 151, 165-174.

Moroz, L.L. (2001). Gaseous transmission across time and species. *Am. Zool.* 41, 304-320.

Morton, D.B., and Hudson, M.L. (2002). Cyclic GMP regulation and function in insects. *Adv. Insect Physiol.* 29, 1-54.

Rockel, P., and Kaiser, W.M. (2002). NO production in plants: Nitrate reductase versus nitric oxide synthase. *Prog. Botany* 63, 246-257.

Valster, A.H., Hepler, P.K., and Chernoff, J. (2000). Plant GTPases: The Rhos in bloom. *Trends Cell Biol.* 10, 141-146.

Vernoud, V., Horton, A.C., Yang, Z., and Nielsen, E. (2003). Analysis of the small GTPase gene superfamily of *Arabidopsis*. *Plant Physiol.* 131, 1191-1208.

Wendehenne, D., Pugin, A., Klessig, D.F., and Durner, J. (2001). Nitric oxide: Comparative synthesis and signaling in animal and plant cells. *Trends Plant Sci.* 6, 177-183.

Calcium and Phospholipids

Harmon, A.C., Gribskov, M., Gubrium, E., and Harper, J.F. (2001). The CDPK superfamily of protein kinases. *New Phytol.* 151, 175-183.

Meijer, H.J.G., and Munnik, T. (2003). Phospholipid-based signaling in plants. *Annu. Rev. Plant Biol.* 54, 265-306.

Michiels, J., Xi, C., Verhaert, J., and Vanderleyden, J. (2002). The functions of Ca^{2+} in bacteria: A role for EF-hand proteins? *Trends Microbiol.* 10, 87-93.

Mueller-Roeber, B., and Pical, C. (2002). Inositol phospholipid metabolism in Arabidopsis. Characterized and putative isoforms of inositol phospholipid kinase and phosphoinositide-specific phospholipase C. *Plant Physiol.* 130, 22-46.

Reddy, A.S.N. (2001). Calcium: Silver bullet in signaling. *Plant Sci.* 160, 381-404.

Sanders, D., Pelloux, J., Brownlee, C., and Harper, J.F. (2002). Calcium at the crossroads of signaling. *Plant Cell* 14, S401-S417.

Snedden, W.A., and Fromm, H. (2001). Calmodulin as a versatile calcium signal transducer in plants. *New Phytol.* 151, 35-66.

Stevenson, J.M., Perera, I.Y., Heilmann, I., Persson, S., and Boss, W.F. (2000). Inositol signaling and plant growth. *Trends Plant Sci.* 5, 252-258.

Wang, X. (2000). Multiple forms of phospholipase D in plants: The gene family, catalytic and regulatory properties, and cellular functions. *Prog. Lipid Res.* 39, 109-149.

White, P.G. (2000). Calcium channels in higher plants. *Biochim. Biophys. Acta* 1465, 171-189.

Protein Kinases and Phosphatases

Hardie, D.G. (2000). Plant protein-serine/threonine kinases: Classification into subfamilies and overview of function. *Adv. Bot. Res.* 32, 1-44.

Jonak, C., and Hirt, H. (2002). Glycogen synthase kinase 3/SHAGGY-like kinases in plants: An emerging family with novel functions. *Trends Plant Sci.* 7, 457-461.

Kennelly, P.J. (2003). Archaeal protein kinases and protein phosphatases: Insights from genomics and biochemistry. *Biochem. J.* 370, 373-389.

Luan, S. (2003). Protein phosphatases in plants. *Annu. Rev. Plant Biol.* 54, 63-92.

Manning, G., Plowman, G.D., Hunter, T., and Sudarsanam, S. (2002). Evolution of protein kinase signaling from yeast to man. *Trends Biochem. Sci.* 27, 514-520.

Meskiene, I., and Hirt, H. (2000). MAP kinase pathways: Molecular plug-and-play chips for the cell. *Plant Mol. Biol.* 42, 791-806.

Morris, P.C. (2001). MAP kinase signal transduction pathways in plants. *New Phytol.* 151, 67-89.

Plowman, G.D., Sudarsanam, S., Bingham, J., Whyte, D., and Hunter, T. (1999). The protein kinases of *Caenorhabditis elegans*: A model for signal transduction in multicellular organisms. *Proc. Natl. Acad. Sci. U.S.A.* 96, 13603-13610.

Tena, G., Asai, T., Chiu, W.L., and Sheen, J. (2001). Plant mitogen-activated protein kinase signaling cascades. *Curr. Opin. Plant Biol.* 4, 392-400.

Zhang, L., and Lu, Y.T. (2003). Calmodulin-binding protein kinases in plants. *Trends Plant Sci.* 8, 123-127.

Zhang, S., and Klessig, D.F. (2001). MAPK cascades in plant defense signaling. *Trends Plant Sci.* 6, 520-527.

Transcription Factors

Sibéril, Y., Doireau, P., and Gantet, P. (2001). Plant bZIP G-box binding factors: Modular structure and activation mechanisms. *Eur. J. Biochem.* 268, 5655-5666.

Evolution of Function

Forsyth, I.A., and Wallis, M. (2002). Growth hormone and prolactin—Molecular and functional evolution. *J. Mam. Gland Biol. Neoplasia* 7, 291-312.

Oftedal, O.T. (2002). The mammary gland and its origin during synapsid evolution. *J. Mam. Gland Biol. Neoplasia* 7, 225-252.

List of Abbreviations

A	(1) *A*denosine receptor; (2) *a*denine; (3) *a*lanine
AA	(1) *A*rachidonic *a*cid; (2) *a*mino *a*cid
αAR	α-*A*drenergic *r*eceptor
ABA	*Ab*scisic *a*cid
AC	*A*denylate *c*yclase
ACh(R)	*A*cetyl*ch*oline (*r*eceptor)
ACTH	*A*dreno*c*ortico*t*ropic *h*ormone
ActR	*Act*ivin *r*eceptor
ADAM	*A* *d*isintegrin *a*nd *m*etalloprotease domain
ADF	*A*nti*d*iuretic *f*actor
ADH	*A*nti*d*iuretic *h*ormone; also called vasopressin (VP)
ADM	*A*dreno*m*edullin
AH	*A*ndrogenic *h*ormone
Ah	*A*romatic *h*ydrocarbon receptor
AHP	*A*rabidopsis *h*istidine *p*hosphotransfer protein
AIRE	*A*uto*i*mmune *re*gulator (coactivator)
AKAP	*A* *k*inase (PKA) *a*nchoring *p*rotein
AKH	*A*dipo*k*inetic *h*ormone
ALK	(1) *A*ctivin receptor-*l*ike *k*inase (TGFβ family type I receptor); (2) *a*naplastic *l*ymphoma *k*inase (pleiotrophin receptor)
ALS	*A*cid-*l*abile *s*ubunit of IGFBP-3
AMH	*A*nti*m*üllerian *h*ormone; also called müllerian inhibiting substance (MIS)
AMPA	α-*A*mino-3-hydroxy-5-*m*ethyl-4-isoxazole *p*ropionic *a*cid
ANP(R)	*A*trial *n*atriuretic *p*eptide (*r*eceptor)
ANS	*A*utonomic *n*ervous *s*ystem
ANT-1	*A*R amino(*N*)-terminal domain *t*ransactivating protein *1* (splicing factor)
AP-1	(1) *A*ctivator *p*rotein *1* (transcription factor); also called Jun-Fos; (2) *a*daptor *p*rotein *1* (associated with clathrin)
APC	*A*denomatous *p*olyposis of the *c*olon (antioncogene)
APOBEC-1	*A*po*l*ipoprotein *B* *e*diting *c*atalytic subunit *1*
APS	*A*daptor *p*rotein containing a *PH* and an *SH2* domain
AR	(1) *A*ndrogen *r*eceptor; (2) *a*mphi*r*egulin
Arf	*A*DP-*r*ibosylation *f*actor (small G protein)
ARIA	*A*cetylcholine *r*eceptor *i*nducing *a*ctivity; also called Schwannoma-derived growth factor (SDGF)
Arnt	*A*romatic hydrocarbon (Ah) *r*eceptor *n*uclear *t*ranslocator
ARR	*A*rabidopsis *r*esponse *r*egulator
ARTN	*Art*emi*n*; also called enovin
ASK	*A*poptosis *s*timulating *k*inase

A-Smase	Acid sphingomyelinase
Asn	Asparagine (N)
Asp	Aspartic acid (D)
ASP	Acylation stimulating protein
AT	Angiotensin II; when followed by a numerical subscript, it refers to an angiotensin isoreceptor
ATF	Activating transcription factor
ATP	Adenosine 5′-triphosphate
Axl	Anexelekto (Greek, uncontrolled); a RTK
B	Bradykinin receptor
βAR	β-Adrenergic receptor
βARK	β-Adrenergic receptor kinase; now called G protein-coupled receptor kinase (GRK)
BDNF	Brain-derived neurotrophic factor
BH	Breakpoint cluster homology domain
bHLH	Basic helix-loop-helix transcription factor
bHLH-ZIP	Basic helix-loop-helix zipper transcription factor
BK	Large conductance, calcium-activated potassium (big K^+) channel
BMP	Bone morphogenetic protein
bp	Base pairs of nucleotides
BPCH	Black pigment-concentration hormone
BRCA	Antioncogene deleted in breast cancer
BRG	Brahma-related gene (chromatin remodeler)
BTC	Betacellulin
BXR	Benzoic acid X receptor
bya	Billions of years ago
bZIP	Basic zipper transcription factor
C	(1) Catalytic subunit of PKA; (2) cytosine; (3) cysteine
C1, C2	PKC conserved region 1 and 2 (calcium-binding domains)
cADPR	Cyclic ADP-ribose
CAH	Carbonic anhydrase-like domain
Cain	Calcineurin inhibitor; also called Cabin
Calpain	Calcium-activated, papain-like protease
CaM	Calmodulin
CAM	Cell adhesion molecule
CaMK	CaM-dependent protein kinase
CaMKK	CaM-dependent protein kinase kinase
cAMP	3′,5′-Cyclic AMP
CAP	(1) Cardioactive peptide; (2) Cbl-associated protein
CaR	Calcium receptor
CAR	Constitutive androstane receptor
CARD	Caspase activity recruiting domain
CARM	Coactivator-associated arginine methyltransferase
Cas	Crk-associated substrate (a docking protein); also called p130^{Cas}
Caspase	Cysteine aspartase; formerly called ICE
CBC	RNA cap binding complex
CBG	Corticosteroid-binding globulin
CBF	Core binding factor (a component of CSL)
Cbl	Casitas B-lineage lymphoma (a scaffold-E3 ligase)
CBP	CREB binding protein
CCAP	Crustacean cardioactive peptide

CCK	Cholecystokinin
CCR	β Chemokine receptor
cCMP	3′,5′-Cyclic CMP
CDI	Cyclin-dependent inhibitor
Cdk	Cell cycle-dependent kinase
cDNA	DNA synthesized from (and therefore complementary to) mRNA
CDP/cut	CCAAT displacement protein/cut homologue (repressor)
CDPK	Calcium-dependent protein kinase
C/EBP	CCAAT/enhancer-binding protein
cGMP	3′,5′-Cyclic GMP
CGRP	Calcitonin gene-related peptide
CH	Cysteine-histidine-rich domain
CHD	Chromo-helicase-DNA-binding
CHH	Crustacean hyperglycemic hormone
CHO	Chinese hamster ovarian cells
CI	Cubitus interruptus (a transcription factor)
CICR	Calcium-induced calcium release
Cip/Waf	Cdk-interacting protein/wild-type p53-activated fragment
CIPK	Calcineurin B-like interacting protein kinase
CIS	Cytokine inducible SH2-containing protein
CISK	Cytokine-independent survival kinase; also called SGK
CK2	Casein kinase 2
CKIP-1	CK2 interacting protein (an anchor)
CNG	Cyclic nucleotide-gated channel
CNTF(R)	Ciliary neurotropic factor (receptor)
CO	Carbon monoxide
Cos-2	Costal-2 (kinesin-like molecule)
COUP	Chicken ovalbumin upstream promotor (nuclear receptor)
CRABP	Cellular retinoic acid binding protein
CRBP	Cellular retinal binding protein
CRE	cAMP response element
CREB	cAMP response element binding protein
CREM	cAMP response element modulator
CRF	Corticotropin (ACTH) releasing factor
CRK	CT10 (avian sarcoma virus) regulator of kinase
CRLR	Calcitonin receptor-like receptor
CSF-1(R)	Colony-stimulating factor-1 (receptor): also called macrophage colony stimulating factor (M-CSF)
CSL	CBF1/Suppressor of hairless/LAG-1 (a Runt transcription factor)
CT	Calcitonin
CT-1	Cardiotrophin-1
CtBP	Carboxy-terminal binding protein (a transcriptional repressor)
CTD	Carboxy-terminal domain of RNA polymerase II
CTP	Cytidine 5′-triphosphate
CURL	Compartment for uncoupling receptor and ligand
CXCR	α Chemokine receptor
Cys	Cysteine (C)
D	(1) Dopamine receptor; (2) aspartic acid
DAPK	Death-associated protein kinase
DARPP	Dopamine-and cAMP-regulated phosphoprotein

DBD	*DNA-binding domain*
dbpB	*DNA binding protein B*
DCC	*Deleted in colon carcinoma (antioncogene)*
DD	*Death domain*
DDR	*Discoidin domain receptor*
DED	*Death effector domain*
DEP	*Dishevelled/EGL-10/pleckstrin-related domain*
DFMO	*α-Difluoromethylornithine*
DG	*Diacylglycerol; also abbreviated DAG*
DGK	*Diacylglycerol kinase*
DH	*(1) Diuretic hormone; (2) Diapause hormone*
DHCC	*Dihydroxycholecalciferol*
DHEA	*Dehydroepiandrosterone*
DHEAS	*Dihydroepiandosterone sulfate*
DHP	*Dihydropyridine*
DHT	*Dihydrotestosterone*
DIG	*Detergent-insoluble, ganglioside-enriched complex*
DIT	*Diiodothyronine*
Dkk-1	*Dickkopf-1 (a receptor antagonist)*
DNA-PK	*DNA-dependent protein kinase*
DNMT	*DNA methyltransferase*
DOT	*Disruptors of telomeric silencing (lysine methyltransferase)*
DOPA	*Dihydroxyphenylalanine*
DR	*Direct repeat*
DRE	*DREAM response element*
DREAM	*Downstream regulator element antagonist modulator (a transcription repressor)*
DRIP	*Vitamin D receptor interacting protein; also called TR associated protein (TRAP)*
DSL	*Delta/serrate/LAG-2 (ligands for Notch)*
dsRNA	*Double-stranded RNA*
E	*(1) Epinephrine; (2) glutamic acid*
E_2	*Estradiol (estrogen with 2 hydroxyl groups)*
4E-BP	*eIF-4E binding protein*
EC_{50}	*Effective concentration necessary to give 50% of the maximal effect*
Eck	*Epithelial cell kinase*
EcR	*Ecdysone receptor*
EDAR	*Ectodysplasin-A2 receptor*
EDHF	*Endothelium-derived hyperpolarizing factor; also called 11,12-epoxyeicosatrienoic acid (11,12-EET)*
EDNH	*Egg development neurosecretory hormone; also called ovarian ecdysteroidogenic hormone (OEH)*
eEF	*Eukaryotic (translation) elongation factor*
EET	*Epoxyeicosatrienoic acid; also called endothelium derived hyperpolarizing factor (EDHF)*
EGF(R)	*Epidermal growth factor (receptor)*
EGTA	*Ethylene glycol-bis(β-aminoethyl ether)-N,N,N′, N′-tetraacetic acid*
EH	*Eclosion hormone*
eIF	*Eukaryotic (translation) initiation factor*
EKLF	*Erythroid Krüppel-like factor (transcription factor)*
Elk	*Eph-like kinase*

ENF	A group of insect defense peptides beginning with the sequence glutamic acid (*E*)-asparagine (*N*)-phenylalanine (*F*)
ENTH	*E*psin amino(*N*)-*t*erminal *h*omology (PPI binding motif)
Eph	RTK cloned from an *E*PO *p*roducing *h*epatocellular carcinoma
EPO(R)	*E*rythro*po*ietin (*r*eceptor)
ER	(1) *E*strogen *r*eceptor; (2) *E*ndoplasmic *r*eticulum
ERE	*E*strogen hormone *r*esponse *e*lement
ERK	*E*xtracellular-signal *r*egulated *k*inase; a MAP kinase (MAPK)
ERM	*E*zrin-*r*adixin-*m*oesin family of actin binding proteins
ERR	*E*strogen *r*eceptor-*r*elated *r*eceptor
ET	*E*ndo*t*helin; when followed by a third letter, it refers to an endothelin isoreceptor
ETH	*E*cdysis-*t*riggering *h*ormone
Ets	*E*26 (*twenty-six*) avian erythroblastosis virus oncogene
ETYA	5,8,11,14-*E*icosa*t*etra*y*noic *a*cid
F	Phenylalanine
FADD	*F*as-*a*ssociated *d*eath *d*omain protein: also called mediator of receptor-induced toxicity (MORT)
Fak	*F*ocal *a*dhesion *k*inase
Fan	*F*actor-*a*ssociated with *N*-SMase (an adaptor)
FAT	*F*actor *a*cetyl*t*ransferase
FERM	4.1 (*F*our point one)-*e*zrin-*r*adixin-*m*oesin (PPI binding motif)
FGF(R)	*F*ibroblast *g*rowth *f*actor (*r*eceptor)
FHA	*F*ork*h*ead *a*ssociated domain (pT binding motif)
FHL-2	*F*our and a *h*alf *L*IM-only protein *2* (adaptor)
FKBP	*F*K506 *b*inding *p*rotein
fMLP	*F*ormyl*m*ethionyl*l*eucyl*p*henylalanine
FMRFamide	Phenylalanylmethionylarginylphenylalaninamide (a neuroactive tetrapeptide)
FNIII	*F*ibro*n*ectin type *III* domain
Fox	*F*orkhead b*ox* transcription factor
FRAP	*F*K506 binding protein-*r*apamycin *a*ssociated *p*rotein (a kinase); also called RAFT1 or TOR
FRGY2	*F*rog *Y*-box protein *2*
FRS	*F*GF *r*eceptor *s*ubstrate (docking protein)
FSH(R)	*F*ollicle-*s*timulating *h*ormone (receptor)
Fu	*Fu*sed (S-T kinase)
FXR	*F*arnesoid *X* *r*eceptor
FYVE	*F*ab1p-*Y*OTP-*V*ac1p-*EE*A1 (PPI binding motif)
G	(1) *G*lycine; (2) *G*TP-binding protein (usually followed by a subscript designating the transduction system to which it is associated); (3) *g*uanine
Gab1	*G*rb2-*a*ssociated *b*inder *1* (docking protein)
GABA	γ (*G*amma)-*a*mino*b*utryic *a*cid
GAF	Interferon-γ (*g*amma) *a*ctivated *f*actor (Stat1 homodimer)
GAP	*G*TPase *a*ctivating *p*rotein
GAS	Interferon-γ (*g*amma) *a*ctivated *s*ite (HRE for the Stat1 homodimer)
GBP	*G*rowth-*b*locking *p*eptide
GC	*G*uanylate *c*yclase
GCMB	*G*lial *c*ells *m*issing, type *B* (transcription factor)
GCN2	*G*eneral *c*ontrol *n*onderepressible *2* (eIF2α kinase)

G-CSF(R)	Granulocyte colony stimulating factor (receptor)
GDI	GDP dissociation inhibitor
GDNF	Glial cell line-derived neurotropic factor
GDS	GDP dissociation stimulator; also called GNP exchange facilitator (GEF)
GEF	GNP exchange facilitator; also called GDP dissociation stimulator (GDS)
Gem	Immediate early gene expressed in mitogen-stimulated T cells
GFRα	GDNF family receptor α
GGF-II	Glial-derived growth factor II
GGL	Gγ (gamma)-like domain
GH(R)	Growth hormone (receptor)
GHRF	Growth hormone-releasing factor
GI	Gastrointestinal
GIH	Gonad-inhibiting hormone
GIP	(1) Gastric inhibitory peptide; (2) GTPase inhibiting protein
GIRK	G protein regulated inward rectifying potassium (K^+) channel
Gln	Glutamine
GLP	Glucagon-like peptide
Glu(R)	Glutamate (receptor)
Gly	Glycine (G)
GM-CSF(R)	Granulocyte-macrophage colony-stimulating factor (receptor)
GNAT	GCN5-related N-acetyltransferase
GnRH	Gonadotropin-releasing hormone; also called LH-releasing hormone (LHRH)
GPCR	G protein-coupled receptor
GR	Glucocorticoid receptor
GRAB	Guanine nucleotide exchange factor for Rab3A
Grb2	Growth factor receptor-bound protein 2
GRE	Glucocorticoid response element
GRIF	GABA$_A$ receptor interacting factor (anchoring protein)
GRK	G protein-coupled receptor kinase; formerly called β-adrenergic receptor kinase (βARK)
GRP	Gastrin-releasing peptide
GSK3	Glycogen synthase kinase 3
GTP	Guanosine 5'-triphosphate
H	(1) Histamine receptor; (2) histidine
Hπ	Hydrophilic amino acid
HΦ	Hydrophobic amino acid
HAT	Histone acetyltransferase
HAVA	α-Hydrazino-δ-aminovaleric acid
HB-EGF	Heparin-binding EGF-like growth factor
HBNF	Heparin-binding neurite-promoting factor; also called pleiotrophin or heparin-binding growth-associated molecule (HB-GAM)
HB-GAM	Heparin-binding growth-associated molecule; also called pleiotrophin or heparin-binding neurite-promoting factor (HBNF)
hCG	Human chorionic gonadotropin
HC-gp39	Human cartilage glycoprotein 39
HDAC	Histone deacetylase
HEAT	Huntingtin-EF3-A subunit of PP2A-TOR1 (protein-protein interaction domain)

HER	*Human EGF receptor homologue*
HETE	*Hydroxy-6,8,11,14-eicosa-tetraenoic acid*
HgCLE	*Heterodera glycines CLAVATA3/ESR-related peptide*
HGF(R)	*Hepatocyte growth factor (receptor)*
hGH	*Human growth hormone*
HIF-1α	*Hypoxia-inducible factor 1α*
Hip	*Hsp 70 interacting protein; also called p48*
His	*Histidine*
HIV	*Human immunodeficiency virus*
HMG	*High mobility group protein; may be followed by a fourth letter designating the class*
hn	*Heterologous nuclear; a prefix used with such abbreviations as RNA and RNP*
HNF	*Hepatic nuclear factor*
Hop	*Hsp 70/hsp 90 organizing protein; also called p60*
HP1	*Heterochromatin protein 1*
HPETE	*5-Hydroxyperoxy-6,8,11,14-eicosatetraenoic acid*
HPK	*Hematopoietic progenitor kinase*
hPL	*Human placental lactogen (human chorionic somatomammotropin)*
Hr	*Hairless (corepressor)*
HRE	*Hormone response element*
HRI	*Heme regulated inhibitor kinase*
hsp	*Heat shock protein (usually followed by a number designating its molecular weight)*
5-HT	*(1) 5-Hydroxytryptamine; also called serotonin; (2) 5-HT receptor (when followed by a subscript)*
HTH	*(1) Hypertrehalosemic hormone; also called hyperglycemic hormone; (2) helix-turn-helix transcription factor*
I	*Isoleucine*
IAA	*Indoleacetic acid*
IAP	*Inhibitor of apoptosis protein*
IAPP	*Islet amyloid polypeptide; also called amylin*
IB1	*Pancreatic islet-brain 1 (transcription factor)*
IC	*Intracellular (loop)*
ICE	*IL-1β converting enzyme; now called caspase*
ICER	*Inducible cAMP early repressor*
IDDM	*Insulin-dependent diabetes mellitus*
IDGF	*(1) Imaginal disc growth factor; (2) insect-derived growth factor*
IFN	*Interferon*
Ig	*Immunoglobulin*
IGF(R)	*Insulin-like growth factor (receptor)*
IGFBP	*Insulin-like growth factor binding protein*
IGHD	*Isolated growth hormone deficiency*
IκB	*Inhibitor of NF-κB*
IKK	*IκB kinase*
IL	*Interleukin*
ILK	*Integrin-linked kinase*
ILP	*Insulin-like peptide; also called insulin-related peptide (IRP)*
INK	*Inhibitor of cdk*

INSL3	*Ins*ulin-*l*ike *3* protein; also called the relaxin-like factor (RLF) or the Leydig insulin-like protein (LEYI-L)
IP-1	*I*nhibitor of A*P-1*
IP$_2$	*I*nositol 1,4-bis*p*hosphate
IP$_3$(R)	*I*nositol 1,4,5-tris*p*hosphate (receptor)
IP$_4$	*I*nositol 1,3,4,5-tetrakis*p*hosphate
IPF-1	*I*nsulin *p*romoter *f*actor *1*
IQGAP	*IQ* (CaM-binding motif) and Ras*GAP* domain containing protein
IRAK	*I*L-1 *r*eceptor *a*ssociated *k*inase
IRAP	*I*nsulin-*r*egulated *a*mino*p*eptidase
IRE	*I*ron *r*esponse *e*lement
IRF	(1) *I*ron *r*esponse *f*actor; (2) *i*nterferon *r*egulatory *f*actor
IRP	*I*nsulin-*r*elated *p*eptide; also called insulin-like peptide (ILP)
IRS	*I*nsulin *r*eceptor *s*ubstrate
ISG15	*I*FN-*s*timulated *g*ene *15* (a ubiquitin homologue)
ISGF3	*I*FN-*s*timulated *g*ene *f*actor *3* (Stat1-Stat2-IRF9 complex)
ISRE	*I*FN-*s*timulated *r*esponse *e*lement (HRE for the Stat1-Stat2-IRF9 complex)
ISWI	Chromatin remodeler that is an *i*mitation of *sw*itch
ITP	*I*on-*t*ransport *p*eptide
JAB	*Ja*nus *b*inding protein
JAB1	*J*un *a*ctivation domain *b*inding protein-*1*
Jak	*Ja*nus *k*inase
JG	*J*uxta*g*lomerular
JH	*J*uvenile *h*ormone
JNK	*J*un amino(*N*)-terminal *k*inase; also called stress-activated protein kinase (SAPK)
JNKK	*J*un amino(*N*)-terminal *k*inase *k*inase; also called stress-ERK kinase (SEK)
kb	*K*ilo*b*ases of nucleotides
kDa	*K*ilo*da*lton
KGF	*K*eratinocyte *g*rowth *f*actor
KID	*K*inase *i*nteracting *d*omain in CREB
Kir	Tyrosine *k*inase-*i*nducible *R*as-like small G protein
KIX	*K*ID *i*nteracting domain in CBP
KL	*K*it *l*igand; also called MGF, SCF, and SLF
KO	*K*nock*o*ut (the process in which a gene is deleted from the genome of an animal)
KSR	*K*inase *s*uppressor of *R*as (scaffold)
L	*L*eucine
LAP	*L*atency-*a*ssociated *p*eptide
LBD	*L*igand-*b*inding *d*omain
LBS	*L*igand *b*inding *s*ubunit of the Ah receptor
LERK	*L*igand for *e*phrin-*r*elated *k*inase
Leu	*Leu*cine
LEYI-L	*Ley*dig *i*nsulin-*l*ike protein; also called insulin-like 3 protein (INSL3) or the relaxin-like factor (RLF)
LH(R)	*L*uteinizing *h*ormone (receptor)
LHRH	*L*uteinizing *h*ormone-*r*eleasing *h*ormone; also called gonadotropin-releasing hormone (GnRH)

Lhx3	*LIM homeobox 3* (transcription factor)
LIF(R)	*Leukemia inhibitory factor* (receptor)
LIMK	*LIM* motif-containing protein *kinase*
LMW-PTP	*Low molecular weight PTP*; also called the cysteine phosphatase
LNB-TM7	*Long-amino(N)-terminus* in GPCR family *B* with seven *transmembrane* helices
LNGFR	*Low-affinity* (or low-molecular-weight) *NGF* receptor
LRP	*LDL* (low density lipoprotein) *receptor-related protein*
LRR	*Leucine-rich repeat*
LT	(1) *Leukotriene* (when followed by an English letter); (2) *lymphotoxin* (when followed by a Greek letter)
LTR	*Long terminal repeat*
LXR	*Liver X receptor*
M	(1) *Methionine*; (2) *muscarinic* acetylcholine receptor; also abbreviated mAChR; (3) *mammalian* (lower case), when used as a prefix
mAChR	*Muscarinic acetylcholine receptor*; also abbreviated M
Mad	*Max dimerization* partner
MADS	*MCM1-AG-DEFA-SRF* transcription factor family
MAPK	*Mitogen-activated protein kinase*
MAPKAP	*MAPK activated* protein kinase
MARCKS	*Myristoylated alanine-rich C kinase substrate*
MB	*Myc box*
MBD	*Methylcytosine binding domain* protein
MCH(R)	*Melanin-concentrating hormone* (receptor)
MCIP1	*Modulatory calcineurin* (PP2B) *interacting protein 1*
MCP	*Monocyte chemoattractant protein*
MCR	*Melanocortin* (MSH-ACTH) *receptor*
M-CSF	*Macrophage colony stimulating factor*; also called CSF-1
MDBP-2-H1	Methylated *DNA binding protein* 2-histone *H1*
MeCP2	*Methylcytosine binding protein 2*
MEF2	*Myocyte enhancer factor 2*
MEK	*MAPK-ERK-activating kinase*; also called MAPK kinase (MAPKK)
Met	(1) *Methionine*; (2) HGF receptor
MF	*Methylfarnesoate*
MGBG	*Methylgloxal bis*(guanylhydrazone)
MGF	*Mast cell growth factor*; also called KL, SCF, and SLF
mGluR	*Metabotropic glutamate receptor*
MGSA	*Melanocyte growth stimulatory activity*; also called GRO
MH2	*(S)mad homology domain 2* (pS binding motif)
MHC	*Major histocompatibility complex*
MID1	*Midline 1* (an E3 ligase)
MIF	*Macrophage migration inhibitory factor*
MIH	*Molt inhibiting hormone*
MIP	*Macrophage inflammatory protein*
MIS	*Müllerian inhibiting substance*; also called antimüllerian hormone (AMH)
MIT	*Monoiodothyronine*
MISS	*Membrane-initiated steroid signaling*
MK	*Midkine*
MKP	*MAPK phosphatase*

MLCK	Myosin light chain kinase; also called myotonic dystrophy kinase related Cdc42-binding kinase (MRCK)
MLK	Mixed lineage kinase
MLTK	MLK-like mitogen-activated protein triple kinase (MLK-like MAPKKK)
MMTV	Mouse mammary tumor virus
Mnk	MAPK interacting kinase
MODY	Maturity onsent diabetes of the young
MOIH	Mandibular organ-inhibiting hormone
MORT	Mediator of receptor-induced toxicity; also called Fas-associated death domain protein (FADD)
Mpl	Cellular homologue of the myeloproliferative leukemia viral oncogene (thrombopoietin receptor); also called TPOR
MR	Mineralocorticoid receptor
MRCH	Melanization and reddish coloration hormone
MRCK	Myotonic dystrophy kinase related Cdc42-binding kinase; also called myosin light chain kinase (MLCK)
mRNA	Messenger RNA
MRP	Multidrug resistance protein
MSH	Melanocyte-stimulating hormone
MSP	Macrophage stimulating protein
MTM	Myotubularin (an inositol phosphatase mutated in X-linked myotubular myopathy)
MTMR	MTM-related protein (an inositol phosphatase)
Multi-CSF	Multi-colony stimulating factor; also called IL-3
MuSK	Muscle specific RTK
Mxi1	Max interactor 1
MYST	MOZ/Ybf-Sas3/Sas2/Tip60 (HAT)
N	Asparagine
NAADP	Nicotinic acid ADP
nAChR	Nicotinic acetycholine receptor
NAD	Nicotinamide adenine dinucleotide
NAG	N-acetylglucosamine
NCoA	Nuclear receptor coactivator; also abbreviated N-CoA
NCoR	Nuclear receptor corepressor; also abbreviated N-CoR
NDF	Neu differentiation factor
NDGA	Nordihydroguaiaretic acid
NE	Norepinephrine
Nedd	Neural precursor cell-expressed developmentally downregulated (a ubiquitin-like peptide)
NES	Nuclear exit signal
NF	(Antioncogene deleted in) neurofibromatosis
NFAT	Nuclear factor of activated T cells
NF-κB	Nuclear factor that stimulates the κ chain gene in B cells
NGF	Nerve growth factor
NHE	Sodium (Na^+)-hydrogen exchanger
NHERF	Sodium (Na^+)-hydrogen exchanger regulatory factor
nHRE	Negative hormone response element
NID	Nuclear receptor interaction domain
NIDDM	Noninsulin-dependent diabetes mellitus
NIK	(1) Nck interacting kinase; (2) NF-κB inducing kinase

NISS	Nuclear-initiated steroid signaling
NKSF	Natural killer cell stimulatory factor; also called IL-12
NLK	Nemo-like kinase
NLS	Nuclear location signal
NMDA	N-Methyl-D-aspartate
NO	Nitric oxide
NOS	Nitric oxide synthase
NPY	Neuropeptide Y
NR	Nuclear receptor
NRG	Neuregulin
NRTN	Neurturin
NSC-CC	Neurosecretory cells-corpus cardiacum
N-Smase	Neutral sphingomyelinase
NT	(1) Neurotropin; (2) Neurotensin
NuRD	Nuclear remodeling and histone deacetylation (a repressor)
OCIF	Osteoclastogenesis inhibitory factor; also called osteoprotegerin (OPG)
OCRL	Oculocerebrorenal (Lowe) syndrome protein
ODC	Ornithine decarboxylase
ODF	Osteoclast differentiation factor; also called osteoprotegerin ligand (OPG-L) or RANK ligand (RANKL)
OEH	Ovarian ecdysteroidogenic hormone; also called egg development neurosecretory hormone (EDNH)
OGT	O-NAG transferase
OIP	OGT interacting protein (anchoring protein)
OPG	Osteoprotegerin; also called osteoclastogenesis inhibitory factor (OCIF)
OPG-L	Osteoprotegerin ligand; also called osteoclast differentiation factor (ODF) or RANK ligand (RANKL)
OPPG	Osteoporosis-pseudoglioma syndrome
OSM(R)	Oncostatin M (receptor)
OT	Oxytocin
P	Proline
p58^{IPK}	Phosphoprotein (58-kDa) inhibitor of protein kinase R
p62^{Dok}	Phosphoprotein (62-kDa) downstream of tyrosine kinase (a docking protein); also called Dok-1
p85	Phosphoprotein of 85 kDa (regulatory subunit of PI3K)
p110	Phosphoprotein of 110 kDa (catalytic subunit of PI3K)
p130^{Cas}	Phosphoprotein (130-kDa) that is a Crk-associated substrate (a docking protein); also called Cas
PA	Phosphatidic acid
PABP	Poly(A) binding protein
PACAP	Pituitary adenylate cyclase-activating polypeptide
PACT	PKR activating protein
PAF	Platelet-activating factor
Pak	p21-activated protein kinase
PAR	Partitioning defective protein (a scaffold)
PARP	Poly(ADP-ribose) polymerase
PAS	Per-Arnt-Sim homology domain
PBAN	Pheromone biosynthesis activating neuropeptide
PC	(1) Phosphatidylcholine; (2) prohormone convertase

PCAF	*p300/CBP associated factor;* also abbreviated P/CAF and pCAF
PCNA	*Proliferating cell nuclear antigen*
PC-PLA$_2$	*PC-specific phospholipase A$_2$*
PC-PLC	*PC-specific phospholipase C*
PC-PLD	*PC-specific phospholipase D*
PDE	*Phosphodiesterase*
PD-ECGF	*Platelet-derived endothelial cell growth factor*
PDGF(R)	*Platelet-derived growth factor (receptor)*
PDH	*Pigment dispersing hormone*
PDK	*PIP$_2$-dependent kinase*
PDZ	*PSD-95/DLG/ZO-1* (protein-protein interaction domain)
PE	*Phosphatidylethanolamine*
PEK	*Pancreatic eIF2α kinase;* also called PERK
PEP	*Phosphoenolpyruvate*
PERK	*PKR-like endoplasmic reticulum kinase;* also called PEK
PEST	Domain rich in the amino acids proline (*P*), glutamic acid (*E*), serine (*S*), and threonine (*T*)
PETH	*Pre-ecdysis-triggering hormone*
PG	*Prostaglandin* (usually followed by a third letter designating the head group)
PGC-1	*PPARγ (gamma) coactivator 1* (splicing factor)
pGlu	*Pyroglutamic acid*
PH	*Pleckstrin homology domain* (PPI binding motif)
Phe	*Phenylalanine* (F)
phox	*NADPH oxidase subunit;* usually shown as a superscript to the molecular weight given in kDa
PI	*Phosphoinositide (phosphatidylinositol)*
PIAS	*Protein inhibitor of activated Stat*
PICK	*Protein interacting with C kinase* (PKC)
PIKE	*PI(3)kinase enhancer* (a small G protein)
PI-glycan	*Phosphatidylinositol-glycan*
PI3K	*Phosphatidylinositol-3 kinase*
PIMT	*PRIP interacting protein with methyltransferase domain*
PIP$_2$	*Phosphatidylinositol 4,5-bisphosphate*
PI(3)P$_n$	*Polyphosphoinositide phosphorylated at the 3′ position among other possible sites*
Pit-1	*Pituitary transcription factor 1*
PITP	*PI transfer protein*
PKA	*Protein kinase A* (cAMP-dependent protein kinase)
PKB	*Protein kinase B*
PKC	*Protein kinase C* (calcium-activated, phospholipid-dependent protein kinase)
PKG	*Protein kinase G* (cGMP-dependent protein kinase)
PKI	*Protein kinase inhibitor*
PKN	Novel PKC-like protein kinase
PKR	dsRNA-activated protein kinase
PLA	*Phospholipase A*
PLC	*Phospholipase C*
PLD	*Phospholipase D*
PLP	*Pyridoxal phosphate*
PMCA	*Plasma membrane calcium ATPase* (calcium pump)

PML	*Promyelocytic leukemia oncogene*
PNUTS	*Protein phosphatase 1 nuclear targeting subunit*
POH	*Permanent ovarian hormone*
POMC	*Proopiomelanocortin*
POU	*Pit-1/Oct/Unc-86 family of transcription factors*
PP	*(1) Pancreatic polypeptide; (2) protein phosphatase (when followed by a number)*
PPAR	*Peroxisome proliferator activated receptor*
PPI	*Polyphosphoinositide*
PR	*Progesterone receptor*
PRE	*Progesterone response element*
PRIP	*PPAR interacting protein*
PRL(R)	*Prolactin (receptor)*
PRMT	*Protein arginine methyltransferase*
Pro	*Proline*
Prop-1	*Prophet of Pit-1 (transcription factor)*
PrRP	*Prolactin releasing peptide*
pS	*Phosphoserine*
PS	*Phosphatidylserine*
PSF	*Polypyrimidine tract binding protein-associated splicing factor*
PSP	*Plasmatocyte spreading peptide*
PSPN	*Persephin*
pT	*Phosphothreonine*
PTB	*Phosphotyrosine binding domain*
Ptc	*Patched (receptor for Sonic hedgehog)*
PTEN	*Phosphatase and tensin homologue*
PTH	*Parathormone*
PTN	*Pleiotrophin*
PTP	*(1) Protein tyrosine phosphatase; (2) permeability transition pore*
PTPA	*Protein tyrosine phosphatase activator*
PTTH	*Prothoracicotropic hormone*
PUFA	*Polyunsaturated fatty acids*
PX	*Phox homology (PPI binding motif)*
PXR	*Pregnenolone X receptor*
pY	*Phosphotyrosine*
Pyk2	*Proline-rich tyrosine (Y) kinase 2*
PYY	*Peptide YY*
Q	*Glutamine*
R	*(1) Arginine; (2) regulatory subunit of PKA*
Rab	*Ras-related rat brain small G protein*
Rac	*Ras-related C$_3$-botulinum toxin substrate (small G protein)*
RACK	*Receptor for activated C kinase*
Rad	*Ras associated with diabetes (small G protein)*
RAFT1	*Rapamycin and FK506 target 1 (a protein kinase); also called FRAP or TOR*
RAGE	*Receptor for advanced glycation endproducts*
RAIDD	*RIP-associated ICE-like death domain protein*
Ral	*Ras related small G protein*
RANK	*Receptor activator of NF-κB*
RANKL	*RANK ligand; also called osteoclast differentiation factor (ODF)*
RANTES	*Regulated upon activation, normal T cell expressed and secreted*

Rap	*Ras p*roximate small G protein
RAR	*R*etinoic *a*cid *r*eceptor
Rb	(Antioncogene deleted in) *retinob*lastoma
RCP	*R*eceptor *c*omponent *p*rotein
Ret	*Re*arranged in *t*ransformation (GDNF receptor)
RGS	Negative *r*egulators of *G* protein *s*ignaling
RHA	*R*NA *h*elicase *A*
Rheb	*R*as *h*omologue highly *e*nriched in *b*rain
Rho	*R*as *h*omologue small G protein
RIA	*R*adio*i*mmuno*a*ssay
RIP140	140-kDa *r*eceptor *i*nteracting *p*rotein (a transcriptonal repressor)
RING	*R*eally *i*nteresting *n*ew *g*ene (protein binding motif)
RIP	*R*eceptor *i*nteracting *p*rotein
RKIP	*R*af *k*inase *i*nhibitory *p*rotein
RLF	*R*elaxin-*l*ike *f*actor; also called insulin-like 3 protein (INSL3) or the Leydig insulin-like protein (LEYI-L)
RLK	*R*eceptor-*l*ike *k*inase
Rnd	*R*ou*nd* up secondary to the inhibition of actin stress fibers (small G protein)
RNP	*R*ibo*n*ucleo*p*rotein
ROCC	*R*eceptor-*o*perated *c*alcium *c*hannel
ROCK	*R*ho-associated *c*oiled coil protein *k*inase; also abbreviated as ROK
ROI	*R*eactive *o*xygen *i*ntermediates; also called reactive oxygen species (ROS)
Ron	Macrophage stimulating protein receptor; also called the MSP receptor
ROR	*R*etinoic acid-related *o*rphan *r*eceptor
ROS	*R*eactive *o*xygen *s*pecies; also called reactive oxygen intermediates (ROI)
RPCH	*R*ed *p*igment-*c*oncentration *h*ormone
rRNA	*r*ibosomal *R*NA
RSK	*R*ibosomal *S*6 *k*inase; also called S6 kinase II (S6KII) or p90^{s6k}
RTK	*R*eceptor *t*yrosine *k*inase
RUNX	*Run*t-related transcription factor
RUSH	*R*ING protein that binds the *u*teroglobin gene and is a *S*WI/SNF *h*elicase
RXR	*R*etinoic acid *X* *r*eceptor
RyR	*Ry*anodine *r*eceptor
S	*S*erine
S6KI	*S6* kinase *I*; also called p70^{s6k}
S6KII	*S6* kinase *II*; also called ribosomal S6 kinase (RSK) or p90^{s6k}
SADDAN	*S*evere *a*chondroplasia with *d*evelopmental *d*elay and *a*canthosis *n*igricans
SAM	(1) *S*-*A*denosyl*m*ethionine; (2) *s*terile *a*lpha *m*otif
SAPK	*S*tress-*a*ctivated *p*rotein *k*inase; also called Jun amino-terminal kinase (JNK)
Sara	*S*mad *a*nchor for *r*eceptor *a*ctivation
SCaMPER	*S*phingolipid *ca*lcium release-*m*ediating *p*rotein of the *e*ndoplasmic *r*eticulum
SCAP	*S*REBP *c*leavage *a*ctivating *p*rotein
SCF	*S*tem *c*ell growth *f*actor; also called KL, MGF, and SLF

SCP-2	Sterol carrier protein-2
SDGF	Schwannoma-derived growth factor; also called acetylcholine receptor inducing activity (ARIA)
SEK	Stress-ERK kinase; also called Jun amino-terminal kinase kinase (JNKK)
Ser	Serine (S)
SERCA	Smooth endoplasmic reticulum calcium ATPase
Serpin	Serine protease inhibitor
SET	Su(var)/E(z)/trithorax (lysine methyltransferase)
SF	Splicing factor
sGC	Soluble guanylate cyclase
SGK	Serum and glucocorticoid-regulated kinase; also called CISK
SH	Src homology domain
SHBG	Sex hormone binding globulin
Shc	SH2-containing protein
SHH	Sonic hedgehog (a ligand for Patched)
SHIP	SH2-containing inositol phosphatase
SHP	SH2-containing protein tyrosine phosphatase
SIE	sis-Inducible element (HRE for the Stat1-Stat3 heterodimer)
SIF	sis-Inducible factor (Stat1-Stat3 heterodimer)
SIR	Silencing information regulator
SK	Small conductance, calcium-activated potassium (K^+) channel
Ski	Sloan Kettering virus isolate
SKIP	Ski interacting protein (transcription coactivator)
SLF	Steel factor; also called KL, MGF, and SCF
Smad	Sma and mothers against dpp gene products (transcription factors for the TGFβ family)
SMase	Sphingomyelinase
SMDF	Sensory and motor neuron derived growth factor
SMN	Survival motor neuron gene product (spliceosome assembly factor)
Smo	Smoothened (a GPCR)
SMRT	Silencing mediator of RAR and TR; also called TR associated cofactor (TRAC)
SOC	Store-operated channel
SOCS	Suppressor of cytokine signaling
SODD	Suppressor of the death domain
SR	(1) Sarcoplasmic reticulum; (2) serine (S)-arginine (R)-rich protein (a splicing factor)
SRC	Steroid receptor coactivator
SRCR	Scavenger receptor cysteine-rich domain
SRE	Serum response element
SREBP	Sterol response element binding protein
SRF	Serum response factor
SRIF	Somatotropin (GH) release-inhibiting factor (somatostatin)
SSI	Stat-induced Stat inhibitor
SSTR	Somatostatin receptor
StAR	Steriodogenic acute regulatory protein (cholesterol shuttle)
Stat	Signal transducers and activators of transcription
STK	Soluble tyrosine kinase
Su(fu)	Suppressor of Fused
Sumo	Small ubiquitin-like modifier

SVS IV	Seminal vesicle secretion protein IV
SWI/SNF	Chromatin remodeler required for the expression of mating type switch and sucrose nonfermentors genes
SXR	Steroid X receptor
τ	Transcriptional activation domain in the GR
T	(1) Testosterone; (2) threonine; (3) thymine
T$_3$	Triiodothyronine (thyroid hormone with 3 iodines)
T$_4$	Thyroxine (thyroid hormone with 4 iodines)
TAB	TAK binding protein
TACE	TNFα (alpha) converting enzyme
TAD	Transcriptional activation domain
TAF	(1) Transcriptional activation function (TAD in ER); (2) TBP-associated factor
TAK	TGFβ activated kinase
TAL1/SCL	T cell acute leukemia 1/stem cell leukemia (transcription factor)
TANK	TRAF-associated NF-κB aktivator (sic)
TBG	Thyronine-binding globulin
TBP	TATA box binding protein
TBPA	Thyronine-binding prealbumin
TCF	Ternary complex factor; also called Elk1
TCF/LEF	T cell factor/lymphoid enhancer factor
TEF1A	Transcription enhancer factor 1A
TF	Transcription factor; followed by a Roman numeral and letter to designate the particular subunit
TGF(R)	Transforming growth factor (receptor)
TGIF	Thymine-guanine interacting factor (corepressor)
Thr	Threonine (T)
TIE	Tyrosine kinase receptor with immunoglobulin and EGF domains
TLR	Toll-like receptor
TMOF	Trypsin-modulating oostatic factor
7-TMS	Seven (7)-transmembrane-segment receptor
TNF(R)	Tumor necrosis factor (receptor)
TOP	Terminal oligopyrimidine tract
TOR	Target of rapamycin (a protein kinase); also called FRAP or RAFT1
TPA	12-O-tetradecanoylphorbol-13-acetate
Tpl2	Tumor progression locus 2 (a MAPKKK)
TPO	(1) Thyroid peroxidase; (2) thrombopoietin
TPO(R)	Thrombopoietin (receptor); TPOR is also called Mpl
TPR	Tetratricopeptide repeats
TR	Thyroid receptor
TRAC	TR associated cofactor; also called silencing mediator of RAR and TR (SMRT)
TRADD	TNFR-associated death domain protein
TRAF	TNFR-associated factor
TRAK	TNFR-associated kinase
TRAP	TR associated protein; also called vitamin D receptor interacting protein (DRIP)
TRAPS	TNFR-associated periodic syndromes
TRBP	TR binding protein
TRH	Thyrotropin (TSH)-releasing hormone
Trk	Tropomyosin-related kinase (RTK for NGF)

tRNA	Transfer RNA
Trp	(1) Tryptophan (W); (2) transient receptor potential channel
TRRAP	Transformation-transcription domain associated protein
TSC1	Tuberous sclerosis 1 (also called hamartin)
TSC2	Tuberous sclerosis 2 (also called tuberin)
TSH	Thyroid-stimulating hormone (thyrotropin)
TSLP(R)	Thymic stromal lymphopoietin (receptor)
TTF	Thyroid transcription factor
TX	Thromboxane
Tyr	Tyrosine (Y)
USP	Ultraspiracle (the insect equivalent of RXR)
V	(1) Valine; (2) vasopressin receptor (when followed by a number)
VDR	Vitamin D receptor
VEGF	Vascular endothelial growth factor; also called vascular permeability factor (VPF)
VHL	(Antioncogene deleted in) Von Hippel Lindau disease
VIP	Vasoactive intestinal peptide
VOCC	Voltage-operated calcium channel
VP	Vasopressin; also called antiduretic hormone (ADH)
VPF	Vascular permeability factor; also called vascular endothelial growth factor (VEGF)
VPR	Vasopressin (ADH) receptor
VSOH	Vitellogenin-stimulating ovarian hormone
W	Tryptophan
wHTH	Winged helix-turn-helix transcription factor
(N-)WASP	(Neural) Wiskott-Aldrich syndrome protein (a cytoskeletal scaffold)
Wnt	Wingless (Drosophila mutant) + int (integration site for MMTV); ligand for Frizzled
WT1	(Antioncogene deleted in) Wilms' tumor
WW	Proline binding motif named for two conserved Trp
WD-40	Protein-protein interaction domain named for a conserved Trp (W) and Asp (D) and the number of amino acids
X	Any amino acid
XIAP	X chromosome-linked inhibitor of apoptosis protein (an adaptor)
Y	Tyrosine
YY1	Yin Yang 1 (repressor)
ZONAB	ZO-1 associated nucleic acid binding protein

Index

1,25-Dihydroxycholecaliferol, 11, 432, 479
1α-hydroxylase, in vitamin D deficiency, 521
3′5′-cyclic AMP pathway mutations, 543–544
5–Hydroxytryptamine, 13

A

A/B region, of nuclear receptors, 126–128
α-helix, as most common secondary structure in DNA
    binding, 388
A kinase anchorage proteins, 259
α-subunit, of G proteins, 259–260
A/U binding protein, mechanisms and examples of, 475
Acantosis nigricans, 533
Accessory proteins, for membrane receptors, 152
Accessory transcription factors, 412, 427
Accuracy, of kinetic assays, 115
Acetylation
    criteria for studies in, 454–455
    functions of selected transcription factors in, 455
    polyamines inactivated by, 322–323
    as regulatory mechanism of transcription factors, 425
    role in histone and HMG protein modifications,
        451–456
Acetylcholine (ACh), 13, 39
    receptors for, 190
Achondroplasia, 531
Acromegaly, 539
Actin accessory proteins, 335
Actin stress fibers, stimulated by Rho family, 248
Activating transcription factor family, 405–407
Activation loops, of protein kinases, 351
Activin, 47, 166
Activin-like receptors, 167
Activin receptor-like kinase 1, inactivation leading to
        hereditary hemorrhagic telangiectasia type II, 534
Acute responses, PKA I in cAMP pathway, 258
Adaptor proteins, 106, 152
    kinases acting as, 351
    SHP acting as, 367
    tyrosine phosphorylated cytoskeletal components as,
        342
Addison's disease, 518, 519, 539
Adenohypophysis, 26
Adenomatous polyposis of colon
    β-catenin and DCC in, 508
    receptors for, 189
Adenosine 5′-triphosphate (ATP), 81–82
    binding domains in tyrosine kinase receptors, 158
    hormonal mediation of, 10
Adenosine receptors, 18
Adenylate cyclase (AC)
    in cAMP pathway, 251
    origin and evolution of, 575
Adenylate cyclases
    activation and inhibition of, 255
    in cAMP metabolism, 253
    lacking GAP activity, 256

regulating CaM in phosphoinositide cycle, 295
    stimulated by large G proteins, 249
Adhesion molecules, 338, 566
Adipocytes, 216, 370
Adipokines, 57
Adiponectin, 57
Adipose tissue, as source of hormones, 57
ADP-ribosylation factors (Arfs), 248
Adrenal cortex, androgen production in females in, 44
Adrenal glands, 32
    medulla of, 38–40
    zona fasciculata and reticularis of, 32–36
    zona glomerulosa of, 36–38
Adrenal hyperplasia, 519
Adrenal medulla, 38–39, 81
Adrenal precursors, of sex steroids, 7
Adrenal steroidogenesis, 30, 31
Adrenal tumor cells, effects of intermediate filaments
    on, 339
Adrenergic receptors, 188
Adrenocortical and ovarian tumors, association with α_{i2}
    mutations, 543
Adrenocorticotropic hormone (ACTH), 11, 31
    congenital deficiencies in, 517
    lack of glucocorticoids to inhibit production of,
        518
    opiate peptide relationships to, 81
    receptors for, 188
    regulating glucocorticoids, 32–36
    role of lipoxygenase pathway in, 314
    VOCC calcium channels and, 280
Adrenomedullin (ADM), 18
Aequorin, 275
Affinity
    affected by dimerization, 187
    determining, 113–114
    for DNA among thyroid receptors, 136–137
    effects of phosphorylation on, 230
    factors affecting, 221–223
    increased by alkalinization, 280
    of nonreceptor proteins, 119
    of receptors, 112
AGC family, 352
Aggression, induced by androgens, 45
Ah receptor, HREs for, 432
α_{i2} mutations, ovarian and adrenocortical tumors
    associated with, 543
Alarmones, 4
Albright's hereditary osteodystrophy, 40
Aldosterone, 12, 36
    bound by mineralocorticoid receptor, 134
    deficiency conditions, 518, 520
    as example of temporal relationships in regulation
        of PPI pathway, 301
    roles of lipoxygenase and angiotensin in synthesis
        of, 314
Aldosterone receptors, 134